DISEASES
OF TEMPERATE ZONE
TREE FRUIT
AND
NUT CROPS

DISEASES
OF TEMPERATE ZONE
TREE FRUIT
AND
NUT CROPS

BY JOSEPH M. OGAWA
AND HARLEY ENGLISH

UNIVERSITY OF CALIFORNIA

**DIVISION OF AGRICULTURE
AND NATURAL RESOURCES**

1991

Ordering

For information about ordering this publication, write to:

Publications
Division of Agriculture and Natural Resources
University of California
6701 San Pablo Avenue
Oakland, California 94608-1239

or telephone (510) 642-2431

Publication 3345

ISBN 0-931876-97-4

Library of Congress Catalog Card No. 91-65409

© 1991 by The Regents of the University of California, Division of Agriculture and Natural Resources

All rights reserved.

No part of this publication may be reproduced, stored in a retrieval system, or transmitted, in any form or by any means, electronic, mechanical, photocopying, recording, or otherwise, without the written permission of the publisher and the authors.

Printed in the United States of America.

This book is an extensive revision and expansion of Wilson and Ogawa's 1979 *Fungal, Bacterial, and Certain Nonparasitic Diseases of Fruit and Nut Crops in California*.

Please use this standard form when citing this book:
Ogawa, Joseph M., and Harley English. 1991. *Diseases of Temperate Zone Tree Fruit and Nut Crops*. University of California, Division of Agriculture and Natural Resources, Oakland, CA. Publication 3345. 461 pp.

General Warning on the Use of Chemicals

Pesticides are poisonous. Always read and carefully follow all precautions and safety recommendations given on the container label. Store all chemicals in their original labeled containers in a locked cabinet or shed, away from foods or feeds, and out of the reach of children, unauthorized persons, pets, and livestock.

Confine chemicals to the property being treated. Avoid drift onto neighboring properties, especially gardens containing fruits and/or vegetables ready to be picked.

Mix and apply only the amount of pesticide you will need to complete the application. Spray all the material according to label directions. Do not dispose of unused material by pouring down the drain or the toilet. Do not pour on ground: soil or underground water supplies may be contaminated. Follow label directions for disposing of container. **Never burn pesticide containers.**

PHYTOTOXICITY: Certain chemicals may cause plant injury if used at the wrong stage of plant development or when temperatures are too high. Injury may also result from excessive amounts or the wrong formulation or from mixing incompatible materials. Inert ingredients, such as wetters, spreaders, emulsifiers, diluents, and solvents, can cause plant injury. Since formulations are often changed by manufacturers, it is possible that plant injury may occur, even though no injury was noted in previous seasons.

To simplify information, trade names of products have been used. No endorsement of named products is intended, nor is criticism implied of similar products that are not mentioned.

Production: Jim Coats, Senior Editor, and Franz Baumhackl, Senior Artist

The dustjacket artwork is made up of original mixed-media illustrations (ca. 1860 to 1906) from the Corti Collection, which are reproduced courtesy of the Department of Special Collections, University of California Library, Davis, California.

The University of California, in compliance with Titles VI and VII of the Civil Rights Act of 1964, Title IX of the Education Amendments of 1972, Sections 503 and 504 of the Rehabilitation Act of 1973, and the Age Discrimination Act of 1975, does not discriminate on the basis of race, religion, color, national origin, sex, mental or physical handicap, or age in any of its programs or activities, or with respect to any of its employment policies, practices, or procedures. Nor does the University of California discriminate on the basis of ancestry, sexual orientation, marital status, citizenship, medical condition (as defined in section 12926 of the California Government Code) or because individuals are special disabled veterans or Vietnam era veterans (as defined by the Vietnam Era Veterans Readjustment Act of 1974 and Section 12940 of the California Government Code). Inquiries regarding this policy may be addressed to the Affirmative Action Director, University of California, Agriculture and Natural Resources, 300 Lakeside Drive, 6th Floor, Oakland, CA 94612-3560. (510) 987-0097.

Dedication

The authors recognize the very important role of the late Professor E. E. Wilson in the authorship of both the first edition—published as *Fungal, Bacterial, and Certain Nonparasitic Diseases of Fruit and Nut Crops in California*—and the present, revised and enlarged edition of this book. The original edition was an outgrowth of a graduate course on the diseases of fruit and nut crops first taught by Dr. Wilson, and later by Joseph M. Ogawa.

Dr. Wilson came to the Department of Plant Pathology at the University of California at Davis in 1929 after obtaining his doctorate at the University of Wisconsin. He retired as Professor Emeritus in 1968 after some 39 years of productive research and distinguished teaching. For many years he was the only pathologist in California whose attention was devoted exclusively to the diseases of fruit and nut crops. Dr. Wilson published extensively on his studies of the bacterial, fungal, and abiotic diseases of these crops. Of note are his investigations of diseases such as scab and European canker of apple; olive knot; phloem cankers and branch wilt of walnut; bacterial canker, brown rot, and shothole of stone fruit; and both infectious and noninfectious bud failure of almond. His research on the many facets of plant diseases caused by various biotic and abiotic agents led to his selection in 1951 as the ninth faculty research lecturer on the Davis Campus of the University of California. He was able to strike a happy balance between basic and applied research, and thus his contributions are well remembered by both the scientific community and California's agriculture industry.

Dr. Wilson was a kind and considerate man who took a keen interest in the development and welfare of his colleagues and graduate students. We miss this talented and productive scientist and friend, and so we dedicate this book in memory of Edward E. Wilson, Professor Emeritus, who spent nearly half a century in dedicated research and teaching on the diseases of California's fruit and nut crops.

Preface

This book has special significance for me, since it represents a collaborative effort with my former graduate advisor, Professor Harley English. I am grateful to him for his encouragement, guidance, and friendship through my professional career. I would like to thank him for these wonderful years and memories.

Joseph M. Ogawa

Acknowledgments

The authors express their thanks to the University of California, and especially to the Department of Plant Pathology at Davis, for providing the broad teaching and research experience that made possible the writing of this text. The authors' combined experiences in fruit and nut crop pathology of around 80 years and their opportunities to work and exchange ideas with other pathologists here and around the world are reflected in this publication. Thus, the book has had inputs far beyond the authors' capabilities.

The authors are grateful to the following people who played a special role in the preparation of this text: Dr. J. K. Uyemoto, University of California, Davis, who authored the sections on apple graft-incompatibility, almond union disorders, cherry mottle leaf, cherry necrotic rusty mottle, cherry rasp leaf, cherry rusty mottle, *Prunus* necrotic ringspot, cherry stem pitting, and *Prunus* stem pitting; Dr. J. E. Adaskaveg, a postdoctorate at the University of California, Davis, who wrote the section on wood decay; and Drs. G. Nyland, S. M. Mircetich, and B. C. Kirkpatrick of the University of California, Davis, and Drs. R. P. Covey and G. I. Mink of Washington State University, who critically reviewed portions of the text. We also appreciate the cooperation of other investigators in this and other countries who contributed many of the color photographs used to illustrate this book. Others in the Department of Plant Pathology here who helped in the preparation of the manuscript are research associate B. T. Manji, postdoctorates T. J. Michailides and A. J. Feliciano, and graduate student J. M. Osorio.

Special thanks are extended to Mrs. Rhonda O'Brien, who did an excellent job typing and compiling the manuscript, to Mr. M. M. Hatamiya, who edited a portion of the manuscript, to free-lance editor Andrew Alden, and to Jim Coats and Franz Baumhackl of the Division of Agriculture and Natural Resources, University of California, for editorial and design help.

Contents

		Text	Color plates
1	Introduction	1	
2	Pome Fruit and Their Diseases	7	383
3	Stone Fruit and Their Diseases	125	403
4	Diseases Attacking Several Genera of Fruit and Nut Trees	243	423
5	The English Walnut and Its Diseases	301	431
6	The Pistachio and Its Diseases	325	434
7	The Olive and Its Diseases	337	435
8	The Fig and Its Diseases	351	436
9	Minor Crops and Their Diseases	367	438
10	Pesticides	375	
	Index	439	

Chapter 1

INTRODUCTION

1

This book is a complete revision and enlargement of Edward E. Wilson and Joseph M. Ogawa's 1979 text, *Fungal, Bacterial, and Certain Nonparasitic Diseases of Fruit and Nut Crops in California,* issued as Publication 4090 by the Division of Agricultural Sciences, University of California. The new edition presents a thorough coverage of the biotic and abiotic diseases of temperate-zone tree fruit and nut crops in California. In addition, a few diseases of these crops that occur in other areas and that pose a potential threat to California fruit and nut culture are briefly discussed. The crops include pome fruit (apple, pear, quince, and loquat), stone fruit (apricot, peach, nectarine, plum, prune, and sweet cherry), nut crops (almond, chestnut, filbert, pecan, pistachio, and walnut), and miscellaneous fruit crops (feijoa, fig, olive, persimmon, and pomegranate). Most of the diseases of these fruit and nut crops in California also occur in other parts of the United States and in other regions of the world where these crops are grown. Furthermore, much of the presented material on the symptomatology, etiology, epidemiology, and control of these diseases is pertinent worldwide. The text does not cover the diseases of subtropical fruits (avocado, citrus, date, etc.), nor does it include the diseases of small fruits (grapes, berries, kiwifruit, etc.).

This book is designed to serve a broad audience of agriculturists and plant scientists who have an interest in the diseases of fruit and nut crops. With its descriptions and color photographs of symptoms and its thorough discussion of control measures, the text should appeal to people in agricultural extension and in industry whose principal concern is disease diagnosis and control. Teachers and students studying the diseases of fruit and nut crops should find the book invaluable as a text or reference work. With its thorough treatment of the etiology, epidemiology, and control of disease and its extensive, up-to-date bibliographies, the book should have appeal to researchers involved in all aspects of fruit and nut crop pathology. It should also be of value to agriculturists whose primary interests are the development of programs in sustainable agriculture and integrated pest management.

The major addition in this revision is the inclusion of diseases caused by mycoplasmalike organisms, viruses, and viruslike pathogens. Also included are several newly diagnosed diseases caused by fungi and bacteria. In all, some 40 additional biotic diseases and several additional abiotic disorders are discussed. Also included is a section on wood decay as it relates to orchard crops, and a chapter on the role of pesticides in the control of biotic diseases of these crops. The authors have thoroughly reviewed world literature on the diseases discussed, and cited the important references. As an aid for disease diagnosis, one or more color photographs illustrate the symptoms of almost every disease described.

The format of the book has been changed so that the diseases of a specific crop or group of crops are arranged in chapters, with color illustrations of the included diseases also organized by chapter toward the end of the book. Page edges are marked for easy comparison of photographs and text descriptions. A general index is provided at the end of the book.

In order that readers may better understand the extent and diversity of the tree fruit and nut crops grown in California and the environmental parameters of the various production districts, a brief picture of the state's fruit and nut culture is presented. The challenge to California plant pathologists to develop effective control measures for the diseases of these crops is most rewarding under an umbrella where private industry (including growers) and the state and federal governments provide support for basic and applied research.

California's Tree Fruit and Nut Culture

The mission fathers introduced systematic agriculture to California by planting seeds brought on their earliest expeditions. Farming began at the 21 missions established from San Diego to Sonoma between 1769 and 1823. Workers at the missions planted several of the fruit and nut cultivars (from *cultivated varieties*) that are grown in California today.

California, the most southwesterly of the Pacific Coast states, is bounded on the north by Oregon, on the east by Nevada and Arizona, on the south by Mexico's Baja California, and on the west by the Pacific Ocean. It extends from 32° 30′ to 42° North latitude, and its longitudinal limits are approximately 114° and 124° 29′ West. The length of its north-south medial line is 770 miles (1,232 km), and its breadth varies from 150 to a maximum of 250 miles (241 to 402 km) in an east-west direction. The total area is 158,693 square miles (410,856 km^2), including 2,120 square miles (5,488 km^2) of inland water. California's physiography provides a splendid central valley 20 to 50 miles (32 to 80 km) wide between the Coast Ranges and the magnificent snow-covered Sierra Nevada on the eastern side. This valley is about 450 miles (720 km) long, and has about 18,000 square miles (46,602 km^2) of irrigated agricultural land. Through this interior basin flow the Sacramento and San Joaquin rivers, for which the northern and southern portions of the valley are named.

The climate during most of the growing season is warm and dry, and irrigation is required for production of most crops. Most of California's agricultural areas have only two seasons—a mild, relatively wet winter alternating with a dry summer. Average precipitation varies from around 50 to 70 inches (127 to 178 cm) in the northwestern part of the state to less than 2 inches (5 cm) in Death Valley in the southeast. Water from winter rain and snow in the Sierra Nevada provides for irrigation as it flows down rivers or is dammed for later release; additional irrigation comes from groundwater. In southern California, much of the irrigation water comes from the Colorado River.

The map at left shows six main agricultural districts for production of fruit and nut crops in California. The districts are described below.

California's major fruit and nut producing districts: (1) San Joaquin Valley; (2) Sacramento Valley; (3) central coast; (4) north coast; (5) Sierra Nevada foothills; and (6) southern California.

San Joaquin Valley. Alluvial deposits from the Stanislaus, Tuolumne, Merced, Fresno, San Joaquin, Kings, Kern, and Kawaeah rivers make the surface of most of the San Joaquin Valley relatively flat. The maximum temperature during the hottest part of the year (June through August) may exceed 40°C (104°F). In the northern part of the valley (Stockton, San Joaquin County), the average temperature in July is 24°C (75°F), and in January, 7°C (45°F). In the southern part (Bakersfield, Kern County), the average temperature in July is 29°C (84°F) and in January 8°C (47°F). Over 80 percent of the precipitation occurs in winter, with an average 5.7 inches (14.5 cm) per year at Bakersfield and 14.3 inches (36.3 cm) per year at Stockton. Dense fogs lasting a week or more are common in winter.

4 • Introduction

Tree fruit and nut crops grown in the San Joaquin Valley include almond, apple, avocado, cherry, chestnut, citrus, fig, jojoba, loquat, nectarine, olive, peach, pear, pecan, persimmon, pistachio, plum, pomegranate, prune, quince, and English and black walnut.

Sacramento Valley. The Sacramento Valley, like the San Joaquin, is relatively flat. It is drained by the south-flowing Sacramento River, which is joined by rivers from the Sierra Nevada—the Feather, Yuba, American, Cosumnes, and Bear. Winters are cool and moist, with fogs that may persist for a week or more; summers are clear, hot, and dry. The average January temperature in Sacramento is 7.2°C (45°F); in July it is 23.9°C (75°F). Annual precipitation is 17.2 inches (43.7 cm). Redding, in the north of the valley, averages 40.9 inches (103.9 cm) of rain per year.

The main fruit and nut crops of the Sacramento Valley are almond, apple, apricot, olive, orange, peach, pear, pecan, pistachio, plum, prune, and English and black walnut.

Central Coast. The Central Coast area extends from the San Francisco Bay area, just north of Walnut Creek (Contra Costa County), south to the mountain ranges of Santa Barbara County. The climate is mild, with

California's principal tree fruit and nut crops

Crop	Total acreage	Ranking in U.S.A.[a]	Harvest season	Years from planting to bearing
Almond	423,375	1	Aug 1–Oct 31	4
Apple	29,396	5	June 15–Oct 30	3
Apricot	23,100	1	June 1–Aug 15	4–5
Avocado	75,300	1	Continuous	3–5
Cherry, sweet	11,512	4	May 20–June 25	4–6
Citrus fruit (all types)	266,237	—	—	5–6
Grapefruit	21,481	2	Continuous	5–6
Lemon	49,365	1	Continuous	5–6
Lime	1,022	2	Continuous	5–6
Orange (all types)	185,528	2	Continuous	5–6
Tangerine, tangelo, and tangor	8,841	1	Nov 1–Apr 30	5–6
Date	5,695	1	Oct 1–Dec 15	6
Feijoa	900	1	Oct 15–Dec 15	3–4
Fig	18,184	1	June 10–Sept 15	5
Jojoba	8,894	1	July 10–Aug 30	4–5
Nectarine	25,586	1	June 10–Sept 15	4
Olive	33,175	1	Sept 25–Mar 15	7
Peach (all types)	63,284	1	May 10–Sept 15	2–3
Clingstone	35,781	1	July 15–Sept 15	2–3
Freestone	27,503	1	May 10–Sept 15	2–3
Pear	24,837	1	July 5–Oct 5	5–6
Pecan	2,708	—	Nov 15–Dec 15	8
Persimmon	1,434	1	Sept 25–Dec 10	5
Pistachio	54,120	1	Sept 1–Dec 10	6
Plum	42,752	1	May 25–Sept 20	4
Pomegranate	3,549	1	Sept 15–Nov 10	4
Prune	82,032	1	Aug 10–Oct 10	6
Walnut, English	196,439	1	Sept 5–Nov 10	6–9
Walnut, black	2,659	—	Sept 15–Oct 30	8

[a]Based on quantity produced.

cool summers on the coast and warm summers in the interior. Fog is common along the coast in summer. Salinas (Monterey County), about 10 miles (16 Km) from the ocean, has an average January temperature of 10°C (50.0°F) and a September average of 23.3°C (73.9°F). Annual precipitation in Salinas is 13.7 inches (34.8 cm).

The tree fruit and nut crops grown in the Central Coast area are almond, apple, apricot, cherry, pear, plum, prune, and English and black walnut.

North Coast. The North Coast is a mountainous region with a few small valleys and coastal plains. The alluvial lowland soils are generally acid and the interior valleys have calcareous soils. This is the wettest region of the state, receiving from 25 to 40 inches (63.5 to 101.6 cm) of rain per year, and irrigation is limited.

Tree fruit and nut crops grown on the North Coast are apple, pear, prune, and English and black walnut.

Sierra Nevada Foothills. The narrow, hilly zone on the east side of the Central Valley makes up the Sierra Nevada Foothills. Fruit and nut crops are grown in the better soils of this region at elevations ranging from approximately 500 to 3,000 feet (153 to 915 m). The soils tend to be shallow and rocky, although a few narrow valleys have alluvial, valley bottom, and terrace soils. Tree fruit and nut production in this region has decreased during the past 30 years, largely because of competition from higher-yielding orchards in the Central Valley. Annual precipitation ranges from about 20 to 40 inches (51 to 102 cm), and spring frosts, especially at the higher altitudes, frequently cause problems.

Apple, cherry, peach, pear, pistachio, quince, persimmon, plum, prune, and English walnut are the principal fruit and nut crops grown.

Southern California. The Southern California region has a moderate climate, with some subtropical areas. Except for desert regions, temperatures over the region are uniform, with a July average of 20.5°C (68.9°F) in the Los Angeles area. Coastal fog is common. Precipitation along the coast is as high as 16.2 inches (41.1 cm) per year at Santa Barbara and as low as 9.3 inches (23.6 cm) in San Diego. Mountain ranges border this region, acting as a barrier between the coastal areas and the deserts. The region sometimes suffers droughts, especially in the interior areas (Goldhamer 1989); essentially, all crops are irrigated (Snyder, Hanson, and Coppock 1986).

In addition to citrus, the main tree fruit and nut crops grown in southern California are apple, avocado, date, jojoba, macadamia, olive, peach, persimmon, and English walnut.

REFERENCES

Anon. 1982. *Agricultural resources of California counties.* University of California, Division of Agricultural Sciences, Special Publ. 3275. 134 pp.

———. 1983. *Average daily air temperatures and precipitation in California.* University of California, Division of Agricultural Sciences, Special Publ. 3285. 125 pp.

———. 1987. *California agriculture and statistical review, 1986.* California Agricultural Statistical Service, Sacramento. 10 pp.

Fox, J. A., and J. L. Hatfield. 1983. *Soil temperatures in California.* Univ. Calif. Agric. Exp. Stn. Bull. 1908, 164 pp.

Goldhamer, D. A. 1989. *Drought irrigation strategies for deciduous orchards.* University of California, Division of Agriculture and Natural Resources, Publ. 21453. 15 pp.

Gubler, W. D., A. H. McCain, H. D. Ohr, A. O. Paulus, and B. Teviotdale. 1986. *California plant disease handbook and study guide for agricultural pest control advisors.* University of California, Division of Agriculture and Natural Resources, Publ. 4046. 157 pp.

Salitore, E. D., ed. 1971. *California, past, present, and future.* E. V. Salitore, Lakewood, California. 670 pp.

Snyder, R. L., B. R. Hanson, and R. Coppock. 1986. *How farmers irrigate in California.* University of California, Division of Agriculture and Natural Resources, Leaflet 21414. 3 pp.

Tippett, J., D. Kleweno, E. N. Severson, and C. Claypoole. 1988. *1987 California Fruit and Nut Acreage.* California Agricultural Statistical Service, Sacramento. 28 pp.

Wildman, W. E., and K. D. Gowans. 1978. *Soil physical environment and how it affects plant growth.* University of California, Division of Agricultural Sciences, Leaflet 2280. 10 pp.

Wilson, E. E., and J. M. Ogawa. 1979. *Fungal, bacterial, and certain nonparasitic diseases of fruit and nut crops in California.* University of California, Division of Agricultural Sciences, Publ. 4090. 190 pp.

Chapter 2

POME FRUIT AND THEIR DISEASES

Chapter 2 Contents

Page numbers in bold indicate photographs.

INTRODUCTION	10	
The Apple *Malus domestica* Borkh.	10	
The Pear *Pyrus communis* L. and *P. Pyrifolia* (Burm.) Nakai	11	
The Quince *Cydonia oblonga* Mill.	12	
The Loquat *Eriobotrya japonica* (Thundb.) Lindl.	12	
REFERENCES	13	

DISEASES OF POME FRUIT CAUSED BY BACTERIA AND MYCOPLASMALIKE ORGANISMS	13	
Bacterial Blast and Canker of Pome Fruit Pathovars of *Pseudomonas syringae* van Hall	13	383
REFERENCES	15	
Crown Gall of Pome Fruit *Agrobacterium tumefaciens* (Smith and Townsend) Conn	15	
Fireblight of Pome Fruit *Erwinia amylovora* (Burrill) Winslow et al.	16	383
REFERENCES	21	
Hairy Root of Apple and Quince *Agrobacterium rhizogenes* (Riker et al.) Conn	25	384
REFERENCES	26	
Pear Decline Mycoplasmalike Organism (MLO)	27	384
REFERENCES	31	

FUNGAL DISEASES OF POME FRUIT	32	
Anthracnose and Perennial Canker of Pome Fruit *Pezicula malicorticis* (Jacks.) Nannf.	32	385
REFERENCES	38	
Dematophora Root Rot of Pome Fruit *Rosellinia necatrix* Prill.	39	385
REFERENCES	41	
European (Nectria) Canker of Apple and Pear *Nectria galligena* Bres.	41	385
REFERENCES	45	
Fabraea Leaf Spot of Pome Fruit *Diplocarpon mespili* (Sorauer) Sutton	46	386
REFERENCES	47	
Phytophthora Root and Crown Rot of Pome Fruit *Phytophthora* spp.	48	386
REFERENCES	49	
Powdery Mildew of Pome Fruit *Podosphaera leucotricha* (E. & E.) Salmon	50	387
REFERENCES	52	
Rusts of Pome Fruit *Gymnosporangium* spp.	54	387
REFERENCES	55	
Sappy Bark of Apple *Trametes versicolor* (L.:Fr.) Pil.	56	388
REFERENCES	57	
Scab of Apple *Venturia inaequalis* (Cke.) Wint.	58	388
REFERENCES	63	
Scab of Loquat *Spilocaea pyracanthae* (Otth.) v. Arx	66	
REFERENCES	67	
Scab of Pear *Venturia pirina* Aderh.	67	389
REFERENCES	69	
Shoestring (Armillaria) Root Rot of Apple and Pear *Armillaria mellea* (Vahl.:Fr.) Kummer	70	
Southern Blight of Apple *Sclerotium* (*Athelia*) *rolfsii* Sacc.	70	389
REFERENCES	72	

VIRUS AND VIRUSLIKE DISEASES OF POME FRUIT	74	
Apple Flat Limb Graft-Transmissible Pathogen	75	389
Apple Graft-Incompatibility Cause Unknown	75	389
Apple Green Crinkle Unidentified Graft-Transmissible Pathogen	76	389
Apple Mosaic Apple Mosaic Virus (AMV)	76	390
Apple Rubbery Wood Unidentified Graft-Transmissible Pathogen	78	
Apple Star Crack Unidentified Graft-Transmissible Pathogen	79	390
Apple Stem Pitting Apple Stem Pitting Virus	79	390
Apple Union Necrosis and Decline Tomato Ringspot Virus (TmRSV)	80	390
Pear Bark Measles Unidentified Graft-Transmissible Pathogen	81	

Pear Blister Canker	82	391	ABIOTIC DISORDERS OF APPLE IN STORAGE	111	
Unidentified Graft-Transmissible Pathogen			Carbon Dioxide Injury	112	397
Pear Stony Pit	82	391	Chemical Injury	112	397
Unidentified Graft-Transmissible Pathogen			Friction Marking	113	398
Pear Vein Yellows and Red Mottle	83	391	Heat Injury	113	399
Apple Stem Pitting Virus			Jonathan Spot	113	399
Quince Sooty Ring Spot	83		Low Oxygen (Alcohol) Injury	113	399
Unidentified Graft-Transmissible Pathogen			Scald	114	400
REFERENCES	84		Soft Scald	114	400
			Sunscald	114	400
ABIOTIC ORCHARD DISORDERS OF POME FRUIT	87		Core Browning	115	400
			Internal Browning	115	400
Bitter Pit of Apple and Pear	87	393	Internal Carbon Dioxide Injury	116	
REFERENCES	89		Internal Breakdown	116	401
Black-End of Pear	91	393	Water Core	117	
REFERENCES	92		Bitter Pit	117	
Internal Bark Necrosis (Measles) of Apple	92	393	Freezing Injury	117	402
REFERENCES	94		ABIOTIC DISORDERS OF PEAR IN STORAGE	118	
Water Core of Apple	95	394	Chemical Injury	118	
REFERENCES	96		Friction Marking	118	
			Senescent Scald	118	402
POSTHARVEST DISEASES OF POME FRUIT	97		Superficial Scald	118	
			Carbon Dioxide Injury	119	
BIOTIC DISEASES OF POME FRUIT IN STORAGE	97		Core Breakdown	119	
Alternaria Rot of Apple and Pear	97	394	Freezing Injury	119	402
Alternaria alternata (Fries) Keissler			Bitter Pit (Anjou Pit, Cork Spot)	119	393
REFERENCES	98		Premature Ripening (Pink End)	119	402
Blue-Mold Rot or Soft Rot of Pome Fruit	99	394	REFERENCES	119	
Penicillium expansum Lk.					
REFERENCES	101		**MISCELLANEOUS DISEASES OF POME FRUIT**	121	
Bull's-Eye Rot of Apple and Pear	103	395	REFERENCES	123	
Pezicula malicorticis (Jacks.) Nannf.					
REFERENCES	104				
Gray Mold Rot of Apple and Pear	104	395			
Botrytis cinerea Pers.:Fr.					
REFERENCES	105				
Phomopsis Rot of Apple	107	395			
Phomopsis mali Roberts					
REFERENCES	107				
Miscellaneous Fungal Rots of Pome Fruit	108	396			
Black rot, brown rot, Cladosporium-Hormodendrum rots, Coprinus rot of pear and apple, Mucor rot, side rot, Stemphylium and Pleospora rots					
REFERENCES	110				

2

INTRODUCTION

The Apple

Malus domestica Borkh.

The cultivated apple probably originated in southwest Asia where forests of wild *Malus* species were known to grow. It was apparently derived from *Malus domestica* (*Pyrus malus* L., *M. sylvestris* Mill.) or from hybridization between *M. domestica* and other wild *Malus* species. Some early apple trees grown in the United States probably were hybrids between *M. domestica* and wild North American species such as *M. ioensis* Britt. Certain horticultural cultivars probably are derived from *M. baccata* Borkh., the wild Siberian crab (Beach 1905, Chandler 1951).[1]

The apple has been cultivated in Europe and Asia for many hundreds of years, but wild apples were eaten long before attempts were made to domesticate them. Reputedly, several horticultural cultivars were grown in Greece as early as 325 B.C. (Smock and Neubert 1950). From such early centers of civilization, the apple was taken to other parts of Europe, and gradual selection both natural and human resulted in types adapted to the climatic conditions of that continent.

Vegetative propagation was practiced in Europe long before America was colonized. Nevertheless, the first domestic apple trees of the New World grew from seeds. Consequently, most if not all early American cultivars were developed on this continent. Some Russian cultivars, later to become popular in the United States, were introduced in the nineteenth century (Butterfield 1937–38). Commercial apple orchards as known today were practically nonexistent in North America until the beginning of the twentieth century. Western New York and Virginia were the first places in which apple growing became a successful business. The extensive apple orchards of the Pacific Northwest came into existence during the early years of this century. The Hood River Valley of Oregon had developed apple culture by 1910, the Wenatchee and Yakima valleys of Washington shortly thereafter.

Important apple production areas in the United States and Canada are (1) New York, New England, Nova Scotia; (2) New Jersey, Delaware, Maryland, West Virginia, Virginia, and the Carolinas (commercial apple-growing is not successful south of northern Georgia); (3) the Ohio River Valley, northern Ohio, Michigan, Wisconsin, southwestern Missouri, northwestern Arkansas, and eastern Oklahoma; (4) California, Washington, Oregon, and British Columbia. The leading state in apple production is Washington; California ranks fifth.

Because of diverse climatic conditions, cultivars grown in one of these areas may not be grown in another. Delicious (and its red mutants), Winesap, Rome Beauty, and Golden Delicious are grown in most of the areas with different degrees of success. Jonathan and Grimes Golden are grown extensively in the mid-Atlantic states, midwestern states, and to some extent in the Pacific Northwest, but are of little importance in California and the North Atlantic region. Other important cultivars in one area or another are or have been McIntosh, Gravenstein, Yellow Newtown, Spitzenberg, Stayman Winesap, Golden Russett, Fameuse, Wealthy, Courtland, Baldwin, and Northern Spy.

California's 29,396 acres of apple orchards, which constitute about 7 percent of the nation's total, are planted primarily with Granny Smith, Red Delicious, Yellow Newtown (Pippin), Golden Delicious, Gravenstein, Rome Beauty, and Jonathan, with new cultivars of interest to growers being Gala, Fuji, Mutsu, Akane, and Jonagold. Currently, Granny Smith and Fuji are being planted extensively in the San Joaquin Valley. The Gravenstein, a German cultivar, is mentioned in the American literature of 1824 and could have been first planted at Fort Ross in 1820. Apples were growing in 1792 at Mission Santa Clara and Mission San Buena Ventura, brought here by J. F. Eschscholtz (who also gave the botanical name to the California poppy in 1824). Named cultivars such as Rhode Island Greening and Winesap were planted in Napa Valley near Calistoga in 1850, and in 1851 sample

---

[1] Names in parentheses identify "References," which follow each section.

fruit brought to Sacramento from Oregon sold for a dollar each. Smith's catalog of 1856 listed 56 cultivars, and by 1858 apples were selling for 25 to 50 cents a pound. Santa Clara was then the leading county in apple production, thanks to Captain Joseph Aram, who later started a nursery close to the old Southern Pacific Railroad depot in San Jose.

In 1925 the leading apple-producing counties were Santa Cruz, Sonoma, and Kern, and in 1987 they continue to lead, with new plantings in El Dorado County and several Central Valley counties. The first report of tree decline, possibly due to oak root fungus, was in 1858 in San Jose. The University of California established an apple test orchard in 1874; in 1894 it listed 200 cultivars under investigation. Later, the California Standard Apple Act ensured consumers of a pack free of defective and diseased fruit. Control of the codling moth without use of lead was one of the priority programs (Butterfield 1937–38).

Many of the oldest orchards in California were planted on so-called French roots (Chandler 1951), which were nothing more than the roots of trees produced from seed imported from France and Austria. In recent years, however, nursery operators have employed roots of seedlings of the common American cultivars. The East Malling Experiment Station in England has developed the so-called Malling (e.g., M26, M7A) and Malling-Merton rootstocks (e.g., MM111, MM26). These are selections from European types which can be readily propagated from root cuttings. Some of these rootstocks induce various amounts of dwarfing in the scion tree. Their use in California and other parts of the United States has markedly increased (Chandler 1951; Tyler et al. 1983).

The Pear

Pyrus communis L. and *P. pyrifolia* (Burm.) Nakai

The ancient Greeks knew the pear and had distinct cultivars of it. The Chinese developed edible cultivars many hundreds of years ago.

The pear has been grown in California since the establishment of the first Spanish mission in San Diego in 1769. Mission Monterey (established in 1770) was also one of the centers of distribution. During the 1848 gold rush a young man settled 2 miles above Sacramento on 50 acres of land and established the first important nursery in California. In 1858 pears were selling for 25 to 50 cents a pound. In 1859 California had about 212,650 pear trees; in 1930, Sacramento County alone had over 728,000 bearing trees. Current estimates indicate that California has approximately 26,000 acres of pears, with only 1,700 acres of nonbearing trees.

Asian pears or apple pears (*P. pyrifolia*) in 1987 comprised over 3,500 acres, with cultivars such as 20th Century, Hosui, Kikisui, Shinseki, Ishiiwase, Yali, Chojuro, and Niitaka. Fowler Nurseries, Inc., located in Newcastle, California, is one of the prime movers of Asian pears, and is credited with selecting the *P. betulaefolia* rootstock for its vigor as well as its moderate resistance to the fireblight disease. Major plantings of Asian pears are now located in the central and southern San Joaquin Valley where fireblight decimated the pear industry in the early 1900s.

Modern domestic pear cultivars were derived from at least two species, *Pyrus communis* L. and *P. pyrifolia* (*P. serotina* Rehd.). *Pyrus communis* is a native of southern Europe and southern Asia. French and Belgian orchardists of the eighteenth and nineteenth centuries played leading roles in developing improved cultivars from this species (Hedrick 1921). Many of these were introduced into the United States and retained, in many cases, their French names. As early as 1830 a nursery in the eastern United States was offering 30 or 40 cultivars for sale. *Pyrus pyrifolia*, the sand pear of horticulturists, is a native of China and also grows wild in Japan; the Orient's domesticated pears have largely been developed from this species. A few of the cultivars (Kieffer and Clapp's Favorite) grown in America were developed from this species, and some of the newer selections, as indicated above, are becoming increasingly popular here.

Although the pear is almost as widely grown as the apple, commercial production is much more restricted in the United States and is largely confined to the western states, particularly Washington, Oregon, and California. Pears are important as food in western Asia, South Africa, and Europe.

The most widely grown cultivar in the United States is Bartlett, which is known as Williams or Williams Bon Cretien outside of the United States. More than 80 percent of the pears in California are Bartletts. Other cultivars of importance in California are Beurre Hardy, Winter Nelis, Comice, Beurre Bosc, Morettini, and d'Anjou. Those grown to some extent in different parts of the United States are Clapp's Favorite, Conference, Flemish Beauty, Lawson, Kieffer, and Seckel. Most of the cultivars just mentioned are diploid (34 chromosomes). Mutants with a triploid number of chromosomes are Beurre Diel and Vicar of Winkfield, the latter an English cultivar. A tetraploid mutant of Bartlett, the Max-Red Bartlett, discovered in Washington state some years ago, is gaining favor in the Pacific Northwest. In California, the red pears are also gaining popularity, with cultivars such as Super Red Bartlett, Canal Red, Reimer, Red Sensation, Red Anjou, Rosi-Red Bartlett, and Autumn Red being increasingly planted.

Research at the agricultural experiment stations of Oregon and California has provided much information on rootstocks for pear. In his search for fireblight-resistant understocks, Reimer (1925) in Oregon assembled and tested most known wild species of *Pyrus* and all available horticultural cultivars of domestic pear from Europe, Africa, and Asia. Day (1947) and others have studied the performance of the more widely used understocks under California conditions.

The following understocks have been used: seedlings of the oriental species, *Pyrus betulaefolia* Bge., *P. calleryana* Decne., *P. pyrifolia* (Burm.) Nakai, and *P. ussuriensis* Maxim; seedlings of domestic cultivars grown in America; seedlings from seed imported from France (these are predominantly *P. communis* but possibly are also other European species such as *P. nivalies* Jacq.); and vegetatively propagated plants of quince (*Cydonia oblonga* Mill.).

The fruit tend to develop the black-end disease on scion cultivars when oriental species (except *P. calleryana*) are used as rootstocks. For this reason they have been all but abandoned in recent years. Furthermore, the importation of seeds from France was not resumed after World War II. Seedlings of Bartlett and Winter Nelis are extensively used in California. Occasionally, *P. calleryana* seedlings are budded or grafted onto Old Home stock. In a few localities the quince is in demand as a rootstock. Because several pear cultivars do not do well when grafted on quince, the Beurre Hardy, which makes a good union with quince, is often used as an intermediate stock when quince rootstock is desired.

The Quince

Cydonia oblonga Mill.

The domestic quince, *Cydonia oblonga* Mill., also has a long history. It is used only for preserves, jams, and jellies, so it is of slight commercial importance in most fruit-growing countries. Cultivars grown in the United States are Smyrna, Pineapple, Champion, and Orange. Trees are propagated by grafting onto quince rootstocks that have been produced by layering or from cuttings.

Quince has been grown in California since the founding of the first mission in San Diego in 1769. In 1848, four cultivars were named in the Nurseryman's Catalog in Sacramento. By 1859 there were 45,821 quince trees in the state. The Smyrna was imported by George Roeding of the Fancher Creek Nurseries of Fresno in 1899. The 1930 census showed a reduced trend in quince production, with only 32,633 bearing trees and 10,988 nonbearing trees (Butterfield 1937–38). In 1987, there were 345 acres of bearing and nonbearing trees in California.

The Loquat

Eriobotrya japonica (Thundb.) Lindl.

The loquat is a small, evergreen fruit tree native to eastern China. It reaches a maximum height of 25–30 feet and has been grown both as an ornamental and for its fruit. It has been cultivated extensively in China, Japan, and India, to some extent in Chile and the Mediterranean basin, and in a more limited fashion in California and the southern United States. The tree grows best in a subtropical to warm-temperate climate; it does well on a variety of soils but appears to do best on clay loam.

The fruit, which are borne in large loose clusters, ripen in spring. They are pear-shaped to oval or round and in the best cultivars may reach a length of 3 inches. They vary in color from pale yellow to deep orange and have a tough plumlike skin. The flesh is white to orange, firm, juicy, and flavorful. The fruit are used either fresh or in jams, jellies, and pies. For the fresh market they should not be picked before full maturity; otherwise they are entirely too acid. If properly handled they can be shipped to distant markets.

The tree requires a minimum of pruning, but some thinning of fruit for optimum size may be required. It is propagated mainly by budding or grafting onto seedling rootstocks; quince root can be used if a dwarf tree is desired. The tree comes into bearing in three to four years. Several improved cultivars (Advance, Champagne, Premier, Victor, and Early Red) have been developed in California. Two diseases that sometimes create problems in California are fireblight (*Erwinia amylovora* [Burrill] Winslow et al.) and loquat scab (*Spilocaea pyracanthae* [Otth.] v. Arx).

Although the production of loquats in California is currently limited to a few backyard "orchards" (less than a total of 50 acres), there appears to be some interest in expansion of the industry, especially in the southern San Joaquin Valley.

REFERENCES

Beach, S. A. 1905. *Apples of New York; Report of the Agricultural Experiment Station for the year 1903*. New York Department of Agriculture, vol. 1, 409 pp., vol. 2, 360 pp.

Bethel, R. S., tech. ed. 1978. *Pear pest management*. University of California, Division of Agricultural Sciences, Publ. 408b, 234 pp.

Butterfield, H. M. 1937–38. History of deciduous fruits in California. *Blue Anchor*, vols. 14 and 15, 38 pp.

Chandler, W. H. 1951. *Deciduous orchards*. Lea & Febiger, Philadelphia, 436 pp.

Day, L. H. 1947. *Apple, quince and pear rootstocks in California*. Univ. Calif. Agric. Exp. Stn. Bull. 700, 44 pp.

Griggs, W. H., and B. C. Iwakiri. 1977. *Asian pear varieties in California*. University of California, Division of Agricultural Sciences, Publ. 4068. 60 pp.

Hedrick, U. P. 1921. *The pears of New York*. New York Dept. Agric. 29th Ann. Rept. vol. 2, part 2, 636 pp.

Reimer, F. C. 1925. *Blight resistance in pears and characteristics of pear species and stocks*. Oregon State Coll. Agric. Exp. Stn. Bull. 214, 99 pp.

Smock, R. M., and A. M. Neubert. 1950. *Apples and apple products*. Interscience, New York, 486 pp.

Taylor, H. V. 1948. *The apples of England*, 3d ed. Crosby Lockwood & Son, London, 214 pp.

Tyler, R. H., W. C. Micke, D. S. Brown, and F. G. Mitchell. 1983. *Commercial apple growing in California*. Univ. Calif. Coop. Ext. Leaflet 2456, 20 pp.

Westwood, M. N. 1978. *Temperate-zone pomology*. Freeman & Co., New York, 428 pp.

DISEASES OF POME FRUIT CAUSED BY BACTERIA AND MYCOPLASMALIKE ORGANISMS

Bacterial Blast and Canker of Pome Fruit

Pathovars of *Pseudomonas syringae* van Hall

Two pathovars of *P. syringae* are occasionally involved in a blossom blast and canker disease of pome fruit in California. The disease is rare on apple, and is of some importance on loquat (Lai, Morin, and Weigle 1972), but causes great concern as a blossom blast of pear (Bethell et al. 1977). It is not known on quince either in California or in other parts of the world.

The blast and canker disease on pear is generally distributed throughout the world, but its occurrence on apple appears to be more restricted and less severe. The disease occasionally causes serious damage in Chile (Cancino, Latorre, and Larach 1974; Latorre 1988) and has recently attracted considerable attention on both pear and apple in South Africa (Hattingh, Roos, and Mansvelt 1989; Mansvelt and Hattingh 1986a, 1987). It causes sporadic damage to pear orchards in the three Pacific Coast states and British Columbia, and occasionally is destructive to both pear and apple in states east of the Rocky Mountains (Coyier 1963; McKeen 1955; Sands and McIntyre 1977; Sprague and Covey 1969). A blister spot of apple fruit has been recognized in eastern states for many years; it recently has been prevalent on the Mutsu cultivar (*Malus pumila*) in New York and Ontario (Burr and Hurwitz 1979; Burr and Katz 1984; Dhanvantari 1969). The pathovar causing this disorder is also reported to cause a scurfy-bark canker of apple (Rose 1917; Lacey and Dowson 1931). A third pathovar causes canker of loquat in California (Lai, Morin, and Weigle 1972) and Australia (Wimalajeewa, Pascoe, and Jones 1978), and both canker and bud blight in Japan (Morita 1978).

Pear blossom blast in California has three common causes: boron deficiency, lack of winter chilling, and infection by *P. syringae* pv. *syringae*. The latter, generally, is the most damaging and in some years has a marked effect on production. Bacterial blast is most common in California's north coastal counties and in the Sierra Nevada foothills. Affected blossoms turn dark brown but necrosis seldom extends far into the supporting twig (**fig. 1**).

Causal organism. Three pathovars of *P. syringae* are involved in the blast, canker, and fruit-spot syndrome of pome fruit. *Pseudomonas syringae* pv. *syringae* van Hall causes the blossom blast and canker disease of pears and apples, and sometimes, at least in pears, induces a killing of buds and a spotting of fruit and leaves. *Pseudomonas syringae* pv. *papulans* (Burr et al. 1988; Dhanvantari 1969; Rose 1917) causes a blister spot of apple fruit and, according to some workers (Lacey and Dowson 1931, Rose 1917), is implicated in a scurfy-bark canker of apple branches. *Pseudomonas syringae* pv. *eriobotryae* (Dhanvantari 1977; Young et al. 1978) causes branch cankers, dead buds, and leaf spots on loquat (Lai, Morin, and Weigle 1972). Only the pathovars *syringae* and *eriobotryae* are known to occur in California.

Disease development. Little is known about the epidemiology of the diseases caused by the pathovars *eriobotryae* and *papulans* or about the blast and canker of apple caused by pv. *syringae*. This discussion, therefore, will be restricted to pear blast and canker induced by the last-mentioned pathovar. Cold, wet weather in early spring favors infection of buds, blossoms, leaves, and fruit (**fig. 1**). Snowfall during the bloom or early postbloom period is often associated with serious blossom blast and fruit infection. Experimental evidence in England (Panagopoulos and Crosse 1964b) and circumstantial evidence in California indicate that frost injury is an important predisposing factor in blossom blast. Whether the ice-nucleating capacity of the pathogen plays a role in pathogenesis has not been established. Research in Australia (Jenkins, Wauchope, and Freeman 1961) indicates that inadequate pollination also is a predisposing factor in blossom blast. Under favorable conditions, the calyx cup, sepals, receptacles, pedicels, peduncles, leaves, and young fruitlets are susceptible to infection (Panagopoulos and Crosse 1964a). The inoculum comes from high populations of the pathogen that build up epiphytically on pear trees (Luisetti and Paulin 1972; Panagopoulos and Crosse 1964a; Waissbluth and Latorre 1978). A recent study in South Africa (Mansvelt and Hattingh 1987) indicates that the causal organism probably overwinters in dormant buds of both pear and apple. In Italy, Ercolani (1969) found that the bacteria could invade both leaf and fruit scars and migrate via vascular traces to axillary buds, resulting in severe blast symptoms the following spring. Meteoric water, wind, and bees are thought to be the factors largely responsible for dissemination of the bacteria (Crosse 1966; Waissbluth and Latorre 1978).

Control. Control of this disease is difficult because favorable conditions for infection are periodic and unpredictable. Once symptoms appear, control is not possible. Attempts to control pear blast by means of chemical sprays have produced somewhat erratic results. In New Zealand (Dye 1956), blossom sprays of streptomycin provided good control whereas Bordeaux mixture was ineffective. However, both materials caused excessive fruit drop. In Australia (Jenkins, Wauchope, and Freeman 1961), neither Bordeaux mixture nor streptomycin (four bloom applications) was effective. In this country, recent tests in Connecticut and California have shown that blossom sprays of streptomycin provide effective control of blast (Bethell et al. 1977; Sands and McIntyre 1977). Recommendations in Chile call for the application of copper oxychloride at bud swell and one or more blossom sprays of either the same material or streptomycin (Latorre 1988). In California, the application of fixed copper at the green-tip stage followed by streptomycin at early bloom provided the best results (Bethell et al. 1977).

Since frost injury often is a predisposing factor for blast, protecting blossoms from this type of injury is a means of reducing this disorder. Overhead sprinklers are being successfully used in California to protect pears from frost damage and subsequent blast. In Australia, the severity of blast in the cultivar Packham's Triumph has been reduced by providing adequate pollination or by spraying the blossoms with a solution of gibberellic acid (Jenkins, Wauchope, and Freeman 1961).

More information on the symptoms of blast and canker and on its causal organisms is provided in Chapter 4.

REFERENCES

Bethell, R. S., J. M. Ogawa, W. H. English, R. R. Hansen, B. T. Manji, and F. J. Schick. 1977. Copper-streptomycin sprays control pear blossom blast. *Calif. Agric.* 31 (6):7–9.

Burr, T. J., and B. Hurwitz. 1979. The etiology of blister spot of "Mutsu" apple in New York State. *Plant Dis. Rep.* 63:157–160.

Burr, T. J., and B. H. Katz. 1984. Overwintering and distribution pattern of *Pseudomonas syringae* pv. *papulans* and pv. *syringae* in apple buds. *Plant Dis.* 68:383–85.

Burr, T. J., J. L. Norelli, B. Katz, W. F. Wilcox, and S. A. Hoying. 1988. Streptomycin resistance of *Pseudomonas syringae* pv. *papulans* in apple orchards and its association with a conjugative plasmid. *Phytopathology* 78:410–13.

Cancino, L., B. A. Latorre, and W. Larach. 1974. Pear blast in Chile. *Plant Dis. Rep.* 58:568–70.

Coyier, D. L. 1963. Pseudomonas blight and its control on pome fruits in Oregon. *Ore. State Hort. Soc. Ann. Rept.* 55:42.

Crosse, J. E. 1966. Epidemiological relations of the pseudomonad pathogens of deciduous fruit trees. *Ann. Rev. Phytopathol.* 4:291–310.

Dhanvantari, B. N. 1969. Bacterial blister spot of apple in Ontario. *Can. Plant Dis. Survey* 49(2):36–37.

———. 1977. A taxonomic study of *Pseudomonas papulans* Rose 1917. *N. Z. J. Agric. Res.* 20:557–61.

Dye, D. W. 1956. Blast of pear. *Orch. N. Z.* 29(7):5.

Ercolani, G. L. 1969. Sopravivenza di *Pseudomonas syringae* van Hall sul pero in rapporto all'epoca della contaminazione in Emilia. *Phytopathol. Mediterranea* 8(3):207–16.

Hattingh, M. J., I. M. M. Roos, and E. L. Mansvelt. 1989. Infection and systemic invasion of deciduous fruit trees by *Pseudomonas syringae* in South Africa. *Plant Dis.* 73:784–89.

Jenkins, P. T., D. G. Wauchope, and H. Freeman. 1961. Blossom blast of pears. *Proc. Plant Pathol. Conf.*, Adelaide, Australia, vol. 1, Paper 35.

Lacey, M. S., and W. J. Dowson. 1931. XVIII. A bacterial canker of apple trees. *Ann. Appl. Biol.* 18:30–36.

Lai, Mingtan, C. W. Morin, and C. G. Weigle. 1972. Two strains of *Pseudomonas eriobotryae* isolated from loquat cankers in California. *Phytopathology* 62:310–13.

Latorre, G. B. 1988. *Enfermedades de las plantas cultivadas*. Ediciones Universidad Católica de Chile, Santiago, Chile. 307 pp.

Luisetti, J., and J. P. Paulin. 1972. Études sur les bactérioses des arbres fruitiers. III. Recherche du *Pseudomonas syringae* (van Hall) à la surface des organes aëriens du poirier et étude de ses variations quantitatives. *Ann. Phytopathol.* 4:215–27. (*Rev. Plant Pathol.* 52:2974. 1973).

Mansvelt, E. L., and M. J. Hattingh. 1986a. Pear blossom blast in South Africa caused by *Pseudomonas syringae* pv. *syringae*. *Plant Pathol.* 35:337–43.

———. 1986b. Bacterial blister bark and blight of fruit spurs of apple in South Africa caused by *Pseudomonas syringae* pv. *syringae*. *Plant Dis.* 70:403–05.

———. 1987. *Pseudomonas syringae* pv. *syringae* associated with apple and pear buds in South Africa. *Plant Dis.* 71:789–92.

McKeen, W. E. 1955. Pear blast on Vancouver Island. *Phytopathology* 45:629–32.

Morita, A. 1978. [Studies on the loquat canker caused by *Pseudomonas eriobotryae* (Takimoto) Dowson. II. Grouping of the bacterial isolates on the basis of their pigment producibility and pathogenicity.] *Ann. Phytopathol. Soc. Japan* 44:6–13.

Panagopoulos, C. G., and J. E. Crosse. 1964a. Pseudomonas blight and related symptoms caused by *Pseudomonas syringae* van Hall on pear trees. *Rept. E. Malling Res. Stn.* 1963:119–22.

———. 1964b. Frost injury as a predisposing factor in blossom blight of pear caused by *Pseudomonas syringae* van Hall. *Nature* 202:1352.

Rose, D. H. 1917. Blister spot of apples and its relation to a disease of apple bark. *Phytopathology* 7:198–208.

Sands, David C., and J. L. McIntyre. 1977. Possible methods to control pear blast caused by *Pseudomonas syringae*. *Plant Dis. Rep.* 61:311–12.

Sprague, R., and R. P. Covey. 1969. Fungous and bacterial pear diseases in eastern Washington. *Wash. Agric. Exp. Stn. Circ.* 498, 14 pp.

Waissbluth, J. E., and B. A. Latorre. 1978. Source and seasonal development of inoculum for pear blast in Chile. *Plant Dis. Rep.* 62:651–55.

Wimalajeewa, D. L. S., I. G. Pascoe, and D. Jones. 1978. Bacterial stem canker of loquat. *Aust. Plant Pathol.* 7:33 (*Rev. Plant Pathol.* 58:3374. 1979).

Young, J. M., D. W. Dye, J. F. Bradbury, C. G. Pangopoulos, and D. F. Robbs. 1978. A proposed nomenclature and classification for plant pathogenic bacteria. *N. Z. J. Agric. Res.* 21:153–77.

Crown Gall of Pome Fruit

Agrobacterium tumefaciens
(Smith and Townsend) Conn

Crown gall caused by *Agrobacterium tumefaciens* (Smith and Townsend) Conn may cause losses of nursery trees. The disease has rarely been observed on mature apple trees, but on mature pear trees it is common.

Control measures in the nursery have involved seed treatment with sodium hypochlorite and soil treatment with methyl bromide.[1] Direct seeding in nursery rows has become a common practice to avoid injury to roots during transplanting. Details on crown gall and its control are provided in Chapter 4.

Fireblight of Pome Fruit

Erwinia amylovora (Burrill) Winslow et al.

"Fireblight" is the name most often applied to this pome fruit disease. In California, however, it is best known on the pear and often is called "pear blight." On apples, except on cultivars Jonathan, Rhode Island Greening, and Fuji, it is a minor problem in California, while on quince it can be severe. Other names formerly used are "spur blight," "fruit blight," "twig blight," and "blossom blight."

Fireblight is believed to be of American origin—it was known in this country many years before it was discovered abroad. In 1794, William Denning wrote that for some years he had known the disease on apple, pear, and quince in the Hudson Valley of New York. Before the middle of the next century it became established in orchards in the midwestern states. On the Pacific Coast the disease was apparently first identified in 1887 near Chico, California. By 1900, severe epidemics of fireblight had destroyed one-third of the pear trees in Fresno, Kern, and Tulare counties, and the pear industry has never regained its importance in those counties. Pear growing was next established in the Sacramento Valley. Fireblight struck that region in the 1920s and 1930s and periodically thereafter, the most recent outbreak being in 1970 (Baker 1971).

Fireblight now occurs in many of the important pome fruit producing countries (van der Zwet and Oitto 1972). It is present in North America from Canada to Guatemala, in the British Isles, and in most of the countries of northern and central Europe. It is not known to occur in the Soviet Union or Asia, including Japan and Taiwan, and its presence in the southern hemisphere appears limited to New Zealand. Although the disease has been reported in Chile (van der Zwet and Oitto 1972), extensive surveys over a period of years by Chilean and American pathologists have failed to confirm its presence there. Most pome fruit producing countries free of fireblight have adopted strict quarantine measures in an effort to exclude the disease.

Symptoms. The leaves, green shoots, fruit, mature branches, and roots are attacked (**fig. 2A**). Symptoms first appear on the blossoms, which wither and die (**fig. 2B**). Under humid conditions the affected blossoms and fruit exude creamy yellow drops of ooze (**fig. 2C**). From the blighted blossom the bacteria move into the spur and thence into the branch where they kill the bark down to the cambium. During the period of activity, affected bark appears water-soaked and reddish. After canker extension ceases, the diseased bark becomes brown and dry as it shrinks, and a fissure commonly develops along the margin. This is followed by the formation of callus which may entirely surround the canker. Girdled branches die in the summer, and these "flags," together with blighted blossom clusters, are the most noticeable symptoms of the disease.

Twig blight, produced by infection of the succulent leafy shoots, often develops into major outbreaks of fireblight. In pears, blighted twigs and leaves turn black; in apple, quince, and loquat, they tend to be brown. After sucker infection, the bacteria sometimes extend into the roots. Similarly, infection of suckers which develop on the body of the tree may lead to the invasion of the trunk and death of the tree.

Fruit infection sometimes occurs. Infected green fruit quickly develop a dark-green, water-soaked appearance, and under some conditions are covered by droplets of ooze or tendrils of bacteria. Sometimes the pear fruit are infected after they are harvested and packed for shipment; on such mature fruit, infection results in circular, black, sunken lesions which seldom are more than an inch in diameter and more than one-fourth inch deep.

In early stages of shoot infection the bacteria occur intercellularly, forming what Nixon (1927) described as "zoogloeal" masses which migrate through the tissue by means of pseudopod-like extension. The host cells adjacent to the bacterial masses collapse and separate, producing schizogenous cavities. The bacteria then enter the cells, which undergo dissolution. The resulting lysigenous cavities gradually enlarge. As bacterial activity wanes, the bacterial masses go into a cyst stage. Haber (1928) described a similar situation in invaded apple leaves. Miller (1929), however, found no evidence of zoogloea; he said that the bacteria, in early stages of infection at least, were free-swimming

---
[1]Permission from County Agricultural Commissioner required for purchase, possession, or use.

and migrated individually from one cell to another through apertures. The cells first exhibited plasmolytic effects and later separated along the middle lamella, presumably as a result of enzymatic activity. Miller, however, failed to demonstrate the formation of pectinase by the bacteria. A recent electron micrograph study of infection in apple shoots by Suhayda and Goodman (1981) indicates that bacterial proliferation and systemic migration is in mature xylen vessels rather than in the cortex and pith.

Causal organism. In 1878, Burrill published the first results of studies which proved that fireblight is caused by a bacterial organism, and established the concept that bacteria were able to produce plant diseases. Burrill (1883) later described the bacterium under the name *Micrococcus amylovorus*, but this was changed to *Bacillus amylovorus* by Trevisan. Under the system of nomenclature later adopted by the Society of American Bacteriologists in 1920, the accepted name is *Erwinia amylovora* (Burrill) Winslow et al.

Erwinia amylovora is a short, gram-negative, capsulated, nonspore-forming rod 0.7 to 1.0 μm long. It is motile by peritrichous flagella, and occurs singly, in pairs, or sometimes in short chains. On agar media, colonies are circular, grayish white, moist, and glistening, with irregular margins. The bacteria produce acid without gas from the common sugars. Minimum temperature for growth is between 3° and 8°C, the maximum 35° to 37°C, and the optimum about 23.3°C. Strains differ in growth rate with generation times of 45 to 100 minutes at 30°C. Growth occurs between pH 4.0 and 8.8, the optimum being pH 6.8.

Strains vary in virulence, morphology, serology (Schroth et al. 1974), and phage typing. The bacteria are unstable when cultured continuously, and single-celled isolates are more stable than are single-colony isolates (Hildebrand 1937a). Of the many tests used to identify populations of virulent strains of *E. amylovora*, the differential medium developed by Miller and Schroth (1972) has enabled positive identification and enumeration within 60 hours. Another efficient, differential medium that permits colony identification after 36 hours has been developed by Ritchie and Klos (1978). For even more rapid identification, but not for population counts, the use of antiserum in agglutination tests has been suggested (Rosen 1935). Biological tests for pathogenicity are still used to confirm resistance of rootstocks (Coyier 1969; Cummins and Aldwinckle 1973; Thompson 1971). Pathogenicity tests have involved the inoculation of green pear fruits (Billing, Crosse, and Garrett 1960), shoots of pear and apple (Parker, Luepschen, and Jones 1974), pyracantha (Ritchie and Klos 1974), and apple seedlings (Powell 1965).

Pugashetti and Starr (1972) used conjugation to transfer virulence in *E. amylovora* from virulent donor strains to avirulent recipient strains during a three-hour mating period.

Initially, ammonia was suspected as being the necrotoxin produced by *E. amylovora* (Goodman, Plurad, and Lovrekovich 1970; Lovrekovich, Lovrekovich, and Goodman 1970a, b). This agent was shown by Hildebrand (1939) to be thermostable and able to cause cell plasmolysis. The toxin was later shown by Eden-Green (1972) to be a high-molecular-weight heteropolysaccharide with galactose, glucose, mannose, and glucuronic acid residues. Goodman and Huang (1974) isolated a polysaccharide from diseased apple tissue infected with *E. amylovora* and named it amylovorin, which is a polymer of galactose (98 percent) with 0.375 percent protein and traces of undetermined substances. This host-specific, wilt-inducing toxin was shown by Beer and Woods (1978) to be coextensive in inoculated apple shoots with high populations of the pathogen. It could not be detected in stem sections from which *E. amylovora* could not be isolated. Amylovorin was shown by Suhayda and Goodman (1981) to be involved in occlusion of the xylem in inoculated apple shoots. Bennett (1978) found that capsulation and extracellular polysaccharide production by *E. amylovora* were correlated with virulence, but also determined that at least one other factor must be involved in pathogenesis.

Hosts. Inoculation tests have shown that *Erwinia amylovora* will invade the succulent parts of a large variety of plants, and in some it produces typical fireblight symptoms. Virtually all of the plants that demonstrate susceptibility to the bacterium belong to the family Rosaceae. Moreover, only certain genera of the subfamily Pomoideae exhibit susceptibility comparable to that of *Pyrus* and *Cydonia*. For example, species of *Spiraea, Aruncus, Kageneckia* and *Holodiscus* of the subfamily Spiraeoides develop mild symptoms on inoculation, but none of these species is seriously blighted and none is known to develop the disease under natural conditions. Among genera of the subfamily Pomoideae, however, certain species of *Cotoneaster, Crataegus, Aronia, Eriobotrya, Chaenomeles, Docynia,* and *Pyracantha* are readily infected by inoculation, and species in five of the genera are probably reservoirs of the pathogen under California conditions. These are *Cotoneaster pannosa, Crataegus oxyacantha* (English hawthorn), *Eriobotrya japonica* (loquat), *Heteromeles arbutifolia* (toyon), and the firethorns: *Pyrancantha angustifolia, P. crenulata, P. formosiana* (= *P. koidzumii*), and *P. gibbsii yunnanensis* (Thomas and Ark 1934a).

Susceptibility of the host and host parts has been related to the arbutinhydroquinone complex (Goodman and Burkowicz 1970; Hildebrand, Powell, and Schroth 1969; Powell and Hildebrand 1970). Beta-glucosidase activity was found to be low in tissues most susceptible to fireblight, tissues such as the nectary and inner parts of blossoms, the midrib and petioles of leaves, and the bark of stems. Arbutin was found in greater amounts in fireblight-resistant pear cultivars such as Old Home, Keiffer, and *P. serotina* than in the susceptible Bartlett and Forelle (Hildebrand, Powell, and Schroth 1969). The relationship of ammonia production to symptoms has been suggested (Lovrekovich, Lovrekovich, and Goodman 1970b). For a more comprehensive discussion of the biochemistry of fireblight, the reader is referred to the excellent monograph on this disease by van der Zwet and Keil (1979).

Disease development. The fireblight pathogen overwinters in the infected bark tissues of the apple and pear hosts. Other hosts such as loquat, quince, toyon, and some species of firethorn also are reservoirs of the pathogen. Twigs as small as 6 mm in diameter, as well as large branches, are "holdover" sites. Keil and van der Zwet (1972a) were able to isolate highly virulent *E. amylovora* from internal tissues of symptomless side shoots on artificially inoculated apple and pear. Isolations from apparently healthy suckers in blighted Bartlett trees showed *E. amylovora* to occur in 60 percent of the shoots. Whether they overwintered in healthy tissues, as suggested, is questionable, since no field data are presented (Schroth et al. 1974). The primary source of inoculum in the spring appears to be both extending and nonextending cankers plus latent infections (Schroth et al. 1974). The pathogen will survive for weeks in detached, infected branches and remain alive in the soil for a month or more, but probably does not overwinter there. There also is limited evidence for overwintering survival of the bacteria in infected pear and apple fruit (Schroth et al. 1974). At one time, beehives were suspected of harboring the bacteria, but it was shown that the pathogen remains alive for only a short time in honey (Ark 1932; Burrill 1878; Erskine 1973; Gossard and Walton 1922; Miller 1929; Whetzel 1906). Several investigators have reported isolating fireblight bacteria from dormant buds of pear or apple in early spring (Baldwin and Goodman 1963; Dueck and Morand 1975; Morrill 1969). However, similar isolations from pear buds in California yielded negative results (Schroth et al. 1974). Although van der Zwet and Keil (1979) believe that epiphytic bacteria associated with dormant buds play an important role in fireblight epidemiology, only further research can establish the validity of this hypothesis.

When humidity is high in the spring, the bacteria ooze to the surface of holdover cankers. Though it is agreed that the ooze is the principal source of inoculum for primary infection, considerable disagreement exists as to the manner in which the bacteria are transmitted from the cankers to the blossoms, the first parts to become infected in the spring. Some (Arthur 1885; Gossard and Walton 1922; Hildebrand 1939) have cited the development of cone-shaped areas of infected blossoms beneath holdover twigs as evidence that rainwater performs this function. Another view is that although rainwater undoubtedly transmits the bacteria downward in the tree, it is not the principal agent responsible for the extensive lateral transmission that has been noted (Goodman 1954). Moreover, primary infection can occur during rainless weather in the arid regions of the western United States. Thus it is contended that insects are the major agents of transmission from canker to blossom and from blossom to blossom. Waite (1898a) and many others have shown that the honeybee is capable of transmitting the pathogen from blossom to blossom, but this insect is not known to visit the cankers. Ants, beetles, and flies, however, are known to feed on the bacterial exudate. As these insects also visit blossoms in search of nectar, it is believed they are the principal agents in transmission of the bacterium from holdover cankers to the blossom (Goodman 1954). Many additional insects are reported to be involved with the pathogen either as vectors or in the infection process (Stahl and Luepschen 1977; van der Zwet and Keil 1979). However, the value of insect control in reducing the severity of blight has not been demonstrated (Aldwinkle and Beer 1979). Wind also is thought to be an important disseminating agent by carrying the bacterial strands or droplets of dew or rain containing the pathogen (van der Zwet and Keil 1979). There is circumstantial evidence that birds play a role in at least the long-distance dispersal of the pathogen (Schroth et al. 1974; van der Zwet and Keil 1979).

Once a small percentage of blossoms are infected, honeybees, other insects, and rain disseminate the bacteria from blossom to blossom (Bauske 1971; Keitt and Ivanoff 1941; Schroth et al. 1974). According to Miller (1929), bacteria enter the flowers through the stomata on the outside of the receptacle cup. However, Hildebrand noted that invasion is primarily through nectaries, stigmas being next in importance (1937a). Avenues of infection are the stomata and hydathodes (Tullis 1929) of leaves, and stomata on green shoots and young fruit. Fresh wounds and lenticels provide the avenues for branch infection (Crosse, Goodman, and Shafer 1972; Ritchie and Klos 1974).

Many investigators have noted that extensive blossom infection by the fireblight organism occurs during rainy or humid weather, and investigation of this revealed that fireblight infection is favored by a low amount of sugars (dextrose, levulose, and sucrose) in the nectar of blossoms and hindered by a high amount. The sugar concentration of nectar varies greatly; in humid weather it may be as low as 2 percent, and as high as 30 to 40 percent in dry weather. Earlier work indicated that the bacteria grow well at sugar concentrations of 3 to 4 percent but little if any at 30 percent, so it was proposed that the increase in volume of nectar and the corresponding decrease in sugar concentration during periods of high humidity are important factors in the initiation of blossom infection (Goodman, Plurad, and Lovrekovich 1970; Ivanoff and Keitt 1941; Keitt and Ivanoff 1941; Thomas and Ark 1934a). It was also supposed that a decrease in the volume of nectar and the resulting increase in sugar concentration inhibits growth of the bacteria in the nectaries. Other studies indicate, however, that the bacteria can grow in 30 percent glucose, 35 percent artificial nectar, or 58 percent sucrose, and that blossom infections can readily occur at the sugar concentration at which nectar is secreted (Hildebrand 1939; Hilderand and Phillips 1936; Schroth et al. 1974).

Temperature plays an important role in the rates at which the bacteria move through the host tissue. At 15°C or below, the progress of the bacteria through succulent host tissue is relatively slow, while at 24°C to 27°C it is rapid. Temperature may be one factor responsible for increased canker activity in spring and early summer, and for waning canker activity in the hot months. Canker increase may occur again with the advent of lower temperatures in early autumn, although it should be pointed out that an increase in atmospheric humidity—a condition which would favor progress of blight in the tissue—also occurs as temperature decreases in autumn.

In some regions, winter temperature affects blight development indirectly by influencing the length of the blossoming period the following spring. Following abnormally warm winters in California, the pear comes into bloom slowly and the blooming period is unusually prolonged. This extends the period during which blossoms may be inoculated. Thomas and Ark (1934a) cited a season in which the blossoming period of Bartlett was 23 days, as compared to 13, 13, and 12 in the three succeeding years.

Cultural practices influence blight development by increasing or decreasing tree growth. Vigorously growing shoots are more severely damaged by this disease than are shoots of low vigor. Therefore, abundant soil moisture and high soil fertility increase the susceptibility of trees by favoring the development of vigorous shoots.

Control. Though none of the common types of high-quality apples and pears is highly resistant to fireblight, some are so susceptible that they require extensive control measures during years when blight is a serious problem. Among the most susceptible cultivars of apples are Willow Twig, Jonathan, Rome Beauty, and Yellow Transparent; among highly susceptible cultivars of pears are Bartlett, Forelle, Beurre Hardy, and Buerre Bosc (Aldwinckle 1974; Aldwinckle and Presczewski 1976; Jones and Sutton 1984; van der Zwet and Keil 1979). Because of their high susceptibility, it is thought that the disease could also become a serious problem in the extensive new plantings of Asian pears in the San Joaquin Valley.

The use of a resistant understock for pear has met with some success. Seedlings of Old Home, a pear of relatively low quality, are sometimes used as a framework on which commercial cultivars are grafted. Oriental species such as *Pyrus calleryana*, *P. betulaefolia*, and *P. ussuriensis* have been planted to some extent in the western United States, but the tendency is to avoid most of these species owing to unfortunate experiences with both *P. ussuriensis* and *P. pyrifolia*. Although these species are more resistant than the French rootstock (*P. communis*), considerable variation in susceptibility occurs among their seedlings (Cameron, Westwood, and Lombard 1969). Also, the fruit disorder known as "hard end" or "black end" often is serious when cultivars are grown on some of these stocks, especially *P. pyrifolia* and *P. ussuriensis* (Wilson and Ogawa 1979). An additional factor is the severity of the pear decline disease when these two species are used as rootstocks. Because of their vigor, moderate resistance to fireblight, and moderate to high tolerance of pear decline, both *P. calleryana* and *P. betulaefolia* are recommended rootstocks for some pear-producing areas of California (Bethell 1978). For a more comprehensive treatment of cultivar and rootstock susceptibility to fireblight and a discussion of breeding programs for control of this disease, the reader is referred to the monograph by van der Zwet and Keil (1979).

Surgical treatment, such as the excision of blighted branches and the scarification of cankers on limbs and trunks, is practiced only where the returns for the crop are high. In the pear orchards of California, for example, growers may employ a crew of "blight cutters" whose duty is to patrol the orchard in late spring and summer and remove all visible blight infections. It is essential that cuts be made 8 to 12 inches below visible bark symptoms and that this work be done only in dry weather. Cutting tools should be dipped in disinfestant; one can be made by diluting 1 cup of household bleach (5.25 percent sodium hypochlorite) with 9 cups of water. Since this liquid corrodes metals, treated tools should be rinsed with tap water, dried, and oiled frequently (Bethell 1978). An equally effective, noncorrosive disinfestant is 70 percent ethanol (Aldwinckle and Beer 1979).

Two kinds of chemical treatments were used in conjunction with surgical treatment in the early 1900s. One was the scarification of established cankers and subsequent treatment with a solution containing mercuric chloride and mercuric cyanide, each at 1 part of chemical to 500 parts of water. To increase the wetting properties of the solution and to retard its drying, 10 percent glycerine was added. The second type of treatment involved brushing the diseased branches with a solution containing zinc chloride, hydrochloric acid, denatured alcohol, and hot water (Day 1928, 1930). For several years after its introduction in 1928, the zinc chloride treatment was used extensively in certain pear orchards of California. When applied by experienced persons it proved valuable for arresting the extension of established cankers and for reducing the survival level of bacteria in the orchard. It has been employed much less frequently in recent years, however, mainly because when improperly done it often led to excessive branch injury. Recent work in New York has shown that a prebloom spray of concentrated Bordeaux mixture and oil reduces the primary inoculum from holdover cankers (Aldwinckle and Beer 1979). This practice is now recommended in New York.

In 1929, Monroe McCown of Purdue University reported the successful control of fireblight in apple blossoms with weak Bordeaux mixture. A preparation containing 2 pounds of copper sulfate and 6 pounds of lime in 100 gallons of water was sprayed on the open blossoms. Later, two treatments of weak Bordeaux mixture were used with some success in commercial orchards. This led to the introduction of copper lime dust, which has been employed successfully in some California pear orchards.

Because copper materials induce damage to fruit lenticels in certain cultivars, streptomycin applied as a spray or as a dust was quickly accepted. Recommendations for control of the disease were as follows: Bordeaux mixture 0.5–0.5–100, proprietary copper sprays as manufacturer directs, or copper lime dust (10–90 or 20–80 at 13 to 25 pounds per acre) applied at five-day intervals starting when 10 percent of the blossoms are open and continuing until all late blossoms have shed their petals; or wettable streptomycin spray at 60 ppm or streptomycin dust at 1,000 pm and 40 pounds per acre applied according to the above schedule. Recent isolations in orchards where streptomycin has been ineffective have revealed the presence of bacteria that are resistant to that chemical.

In the initial laboratory evaluation of streptomycin, English and Van Helsema (1954) reported the rapid development of streptomycin-tolerant *Erwinia amylovora* and showed that a combination of streptomycin and oxytetracycline delayed the emergence of resistant strains. Early formulations of streptomycin for field testing often contained 1.5 percent oxytetracycline (Terramycin). Studies by Ark (1953), Goodman (1954), Leupschen (1960), and others showed that effective control can be obtained without damaging fruit by using low dosages of streptomycin applied at three- to five-day intervals. Commercial applications of streptomycin have been made since 1968 (Bailey and Morehead 1970). In 1972 Schroth et al. reported severe blight development in streptomycin-sprayed orchards; the predominant strains of *Erwina* isolated were resistant to streptomycin levels over 300 ppm. Schroth et al. (1974) later reported that resistance was not caused by mutation from a single strain. In orchards with resistant bacteria, growers have returned to the use of copper sprays (Moller et al. 1973). As of 1981, streptomycin-resistant strains of *E. amylovora* had been detected only in California, Oregon, and Washington (Moller, Schroth, and Thomson 1981; Ogawa et al. 1983).

On pears, using the principle established by Brooks (1926), Mills (1955), and Powell (1963, 1965), Miller and Schroth (1972) developed a monitoring system for the bacterial population on the blossoms and correlated this information with climatological data. The relation of weather to populations of the bacteria on healthy pear blossoms and subsequent disease development was reported by Thomson et al. (1975, 1976, 1982; Zoller and Sisevich 1976, 1979). Data accumulated thus far indicate that infection is closely correlated with temperatures over 16.5°C and the presence of bacteria on the blossoms. Accurate disease forecasting would enable growers to apply chemicals only when required, and treatments could be timed for greatest efficacy. A thorough discussion of predictive systems as related to fire blight control is presented by Van der Zwet, Zoller, and Thomson (1988).

On apples, Steiner (1989) developed a predictive system for blossom, canker, shoot, and trauma (hail, frost) blight using the comprehensive MARYBLYT model. The empirically derived model can anticipate and identify infection periods and predict symptom appearance for the four distinct phases of fire blight of apples. As an example, for blossom infection the following conditions must be met: (1) flowers must be open with petals intact; (2) accumulation of at least 110 degree hours >18.3°C from full pink; (3) a wetting event such as rain ≥2.5 mm rain the previous day; (4) an average daily temperature of ≥15.6°C; and (5) all of the above in the sequence shown. Symptoms usually appear 50 degree days >12.7°C after infection. Systemic canker activity begins routinely with the accumulation of at least 72.2 degree days >12.7°C after green tip and only rarely occurs before petal fall. Canker blight symptoms follow regularly at 122.2 degree days >12.7°C after green tip. If suitable insect vectors are present when blossom or canker blight appear, then shoot blight symptoms can be expected with an additional 50 degree days >12.7°C. Following hail or late frost incidents that occur after at least 300 degree hours >18.3°C have accummualted, a similar 50-degree-day interval is observed before symptom appearance. Forecasts based on these thresholds and intervals have been generally reliable to within one day.

Timing of the first spray during bloom is critical to the control of blight in California pears. When blooming commences, applications begin when daily mean temperatures during March approach 16.7°C; during April, 15.5°C; and during May, 14.4°C (Bethell 1978; Wilson and Ogawa 1979). Before foliation, alternate rows are sprayed preferably at night or early morning every five days until at least 30 days after full bloom. If rain or hail should occur, orchards are resprayed immediately. If humid weather occurs in June, treatment of "rat-tail" (late) bloom is needed (**fig. 2B**). Chemicals used are Bordeaux mixture, copper hydroxide, copper oxychloride sulfate, streptomycin, and oxytetracycline. In orchards known or suspected to contain streptomycin-resistant fireblight strains, copper or oxytetracycline sprays are recommended. In California apples, fireblight generally is much less severe than in pears, and the number of sprays can be markedly reduced. Current recommendations call for applications at full bloom, petal fall, and, if required, at late secondary bloom. The bactericides used are the same as those for pears.

Other methods of disease control, such as the use of phage (Erskine 1973), antagonistic bacteria (Thomson et al. 1976), and nonvirulent *Erwinia* (Thomas and Henderson 1952) are being tested. As yet, however, no biocontrols have proved commercially effective. An effective treatment for the eradication of *E. amylovora* from the surface of mature apple fruit has been recently reported (Janisiewicz and Van der Zwet 1988). This treatment involves dipping the fruit in a solution of benzalkonium chloride. It could possibly be a way to meet quarantine requirements for shipment of pome fruit to countries free of fireblight. In central Washington, much of the infection of pears occurs in the secondary bloom (Covey and Fischer 1988), and a recommended control practice is the removal of these blossoms.

REFERENCES

Aldwinckle, H. S. 1974. Field susceptibility of 46 apple cultivars to fire blight. *Plant Dis. Rep.* 58:819–21.

Aldwinckle, H. S., and S. V. Beer. 1976. Nutrient status of apple blossoms and their susceptibility to fire blight. *Ann. Appl. Biol.* 82:159–63.

———. 1979. Fireblight and its control. *Hort. Rev.* 1:423–74.

Aldwinckle, H. S., and J. L. Preczewski. 1976. Reaction of terminal shoots of apple cultivars to invasion by *Erwinia amylovora*. *Phytopathology* 66:1439–44.

Ark, P. A. 1932. The behavior of *Bacillus amylovorus* in the soil. *Phytopathology* 22:657–60.

———. 1937. Variability in the fire-blight organism, *Erwinia amylovora*. *Phytopathology* 27:1–28.

——— 1953. Use of streptomycin dust to control fire blight. *Plant Dis. Rep.* 37:404–06.

Arthur, J. C. 1885. Proof that the disease of trees known as pear-blight is directly due to bacteria. *N.Y. (Geneva) Agric. Exp. Stn. Bull.* 2:1–4.

Bailey, J. B., and G. W. Morehead. 1970. Streptomycin control of pear fireblight in California. *Calif. Agric.* 24(9):14–15.

Baker, K. F. 1971. Fire blight of pome fruits: The genesis of the concept that bacteria can be pathogenic to plants. *Hilgardia* 40:603–33.

Baldwin, C. H., Jr., and R. N. Goodman. 1963. Prevalence of *Erwinia amylovora* in apple buds as detected by phage typing. Phytopathology 53:1299–1303.

Bauske, R. J. 1971. Wind dissemination of waterborne *Erwinia amylovora* from *Pyrus* to *Pyracantha* and *Cotoneaster*. Phytopathology 61:741–42.

Bayot, R. G., and S. M. Ries. 1986. Role of motility in apple blossom infection by *Erwinia amylovora* and studies of fireblight control with attractant and repellent compounds. Phytopathology 76:441–45.

Beer, S. F., and D. C. Opgenorth. 1976. *Erwinia amylovora* on fire blight canker surfaces and blossoms in relation to disease occurrence. Phytopathology 66:317–22.

Beer, S. F., and A. C. Woods. 1978. Distribution of *Erwinia amylovora* in apple (*Malus pumila*) shoots inoculated with the fireblight pathogen. *Proc. 4th Internat. Conf. Plant Pathogen. Bact.* 2:471-78.

Bennett, R. A. 1978. Characteristics of *Erwinia amylovora* in relation to virulence. *Proc. 4th Internat. Conf. Plant Pathogen Bact.* 2:479–81.

Bethell, R. S., ed. 1978. *Pear pest management*. Division of Agricultural Sciences, University of California, 243 pp.

Billing, E., J. E. Crosse, and C. M. E. Garrett. 1960. Laboratory diagnosis of fire blight and bacterial blossom blight of pear. *Plant. Pathol.* 9:19–25.

Brooks, F. N. 1926. Studies of the epidemiology and control of fire blight of apple. Phytopathology 16:665–96.

Burrill, T. J. 1878. Pear blight. *Trans. Ill. St. Hort. Soc.* 11:114–16.

———. 1879. Fireblight. *Trans. Ill. St. Hort. Soc.* 12:77–80.

———1881. Anthrax of fruit trees. *Proc. Am. Assoc. Adv. Sci.* 29:583–97.

———. 1883. New species of micrococcus (bacteria). *Am. Naturalist* 17:319.

Cameron, H. R., M. N. Westwood, and P. B. Lombard. 1969. Resistance of *Pyrus* species and cultivars to *Erwinia amylovora*. Phytopathology 59:1813–15.

Carpenter, T. R., and J. R. Shay. 1953. The differentiation of fireblight resistant seedlings within progenies of interspecific crosses of pear. Phytopathology 43:156–62.

Covey, R. P., and W. R. Fischer. 1973. Short-term population dynamics of *Erwinia amylovora* in succulent pear tissue. Phytopathology 63:844–46.

———. 1988. The significance of secondary bloom to fireblight development on Bartlett pears in eastern Washington. *Plant Dis.* 72:911.

Coyier, D. L. 1969. Inoculation of apple and pear roots with *Erwinia amylovora*. *Workshop Fire Blight Res. Proc. 1st*, 57–58.

Crosse, J. E., R. N. Goodman, and W. H. Shaffer. 1972. Leaf damage as a predisposing factor in the infection of apple shoots by *Erwinia amylovora*. Phytopathology 62:39–42.

Cummins, J. N., and H. S. Aldwinkle. 1973. Fire blight susceptibility of fruiting trees of some apple rootstock clones. *HortScience* 8:176–78.

Day, Leonard H. 1928. Pear blight control in California. *Calif. Agric. Exp. Stn. Circ.* 20, 50 pp.

———. 1930. Zinc chloride treatment of pear blight cankers. *Calif. Agric. Exp. Stn. .Circ.* 45, 13 pp.

Delp, C. J., ed. 1988. Fungicide resistance in North America. APS Press. The American Phytopathological Society, St. Paul, Minnesota. 133 pp.

Denning, W. M. 1794. On decay of apple trees. *Proc. Soc. Prom. Agric. Sci.* 1:219–22.

Dueck, J., and J. B. Morand. 1975. Seasonal changes in the epithytic population of *Erwinia amylovora* on apple and pear. *Can. J. Plant Sci.* 55:1007–12.

Dunegan, J. C., H. H. Moon, and R. A. Wilson. 1953. Further results of testing susceptibility of pear seedlings to *Erwinia amylovora* (abs.). Phytopathology 43:405.

Eden-Green, S. J. 1972. Studies in fire blight disease of apple, pear and hawthorn (*Erwinia amylovora* [Burrill] Winslow et al.). Ph.D. thesis, University of London, 202 pp.

Eden-Green, S. J., and E. Billing. 1972. Fireblight: Occurrence of bacterial strands on various hosts under glasshouse conditions. *Plant Pathol.* 21:121–23.

English, A. R., and G. Van Halsema. 1954. A note on the delay in the emergence of resistant *Xanthomonas* and *Erwinia* strains by the use of Streptomycin plus Terramycin combinations. *Plant Dis. Rep.* 38:429–31.

Erskine, J. M. 1973. Characteristics of *Erwinia amylovora* bacteriophage and its possible role in the epidemiology of fire blight. *Can. J. Microbiol.* 19:837–45.

Fulton, H. R. 1911. The persistence of *Bacillus amylovorus* in pruned apple twigs (abs.). *Phytopathology* 1:68.

Goodman, R. N. 1954. Development of methods for use of antibiotics to control fireblight (*Erwinia amylovora*). *Missouri Agric. Exp. Stn.. Res. Bull.* 540, 16 pp.

Goodman, R. N. and A. Burkowicz. 1970. Ultrastructural changes in apple leaves inoculated with virulent or an avirulent strain of *Erwinia amylovora*. *Phytopathol. Z.* 68:258–68.

Goodman, R. N., and J. S. Huang. 1974. Host specific phytotoxic polysaccharide from apple tissue infected by *Erwinia amylovora*. *Science* 183:1081–82.

Goodman, R. N., S. B. Plurad, and L. Lovrekovich. 1970. Similarity of ultrastructural modifications in tobacco leaf tissue caused either by pathogenic bacteria or exogenous ammonia (abs.). *Phytopathology* 60:1293.

Gossard, H. A., and R. C. Walton. 1922. Dissemination of fire blight. *Ohio Agric. Exp. Stn. Bull.* 357:81–126.

Gowda, S. S., and R. N. Goodman. 1970. Movement and persistence of *Erwinia amylovora* in shoot, stem and root of apple. *Plant Dis. Rep.* 54:576–80.

Haber, Julia. 1928. The relation between *Bacillus amylovorus* and leaf tissues of the apple. *Penn. Agric. Exp. Stn. Bull.* 228:3–13.

Hildebrand, D. C. 1970. Fire blight resistance in *Pyrus*: Hydroquinone formation as related to antibiotic activity. *Can J. Bot.* 48:177–81.

Hildebrand, D. C., C. C. Powell, and M. N. Schroth. 1969. Fire blight resistance in *Pyrus*: Location of arbutin and beta-glucosidase. *Phytopathology* 59:1534–39.

Hildebrand, E. M. 1937a. The blossom-blight phase of fire blight, and methods of control. *N.Y. (Cornell) Agric. Exp. Stn. Mem.* 207, 40 pp.

———. 1937b. Infectivity of the fire-blight organism. *Phytopathology* 27:850–52.

———. 1939. Studies on fire blight ooze. *Phytopathology* 29:142–56.

———. 1954. Relative stability of fire blight bacteria. *Phytopathology* 44:192–97.

Hildebrand, E. M., and E. F. Phillips. 1936. The honeybee and the beehive in relation to fire blight. *J. Agric. Res.* 52:789–810.

Ivanoff, S. S., and G. W. Keitt. 1941. Relation of nectar concentration to growth of *Erwinia amylovora* and fire blight infection of apple and pear blossoms. *J. Agric. Res.* 62:733–43.

Janisiewicz, W. J., and T. Van der Zwet. 1988. Bacterial treatment for the eradication of *Erwinia amylovora* from the surface of mature apple fruit. *Plant Dis.* 72:715–18.

Jones, A. L., and T. B. Sutton. 1984. *Diseases of tree fruits*. N. Central Reg. Ext. Publ. 45, Coop Ext. Serv. Mich. State Univ. 59 pp.

Jones, D. H. 1909. Bacterial blight of apple, pear and quince. *Ontario Agric. Coll. Bull.* 176, 63 pp.

Keill, H. L., and T. van der Zwet. 1972a. Recovery of *Erwinia amylovora* from symptomless stems and shoots of Jonathan apple and Bartlett pear trees. *Phytopathology* 62:39–42.

———. 1972b. Aerial strands of *Erwinia amylovora*; structure and enhanced production by pesticide oil. *Phytopathology* 62:355–61.

Keitt, G. W., and S. S. Ivanoff. 1941. Transmission of fire blight by bees and its relation to nectar concentration of apple and pear. *J. Agric. Res.* 62:745–53.

Luepschen, N. S. 1960. Fire blight control with streptomycin, as influenced by temperature and other environmental factors and by spray adjuvants added to sprays. *N. Y. Agric. Exp. Stn. Mem.* 375, 39 pp.

Lovrekovich, L., H. Lovrekovich, and R. N. Goodman. 1970a. Ammonia as a necrotoxin in the hypersensitive reaction caused by bacteria in tobacco leaves. *Can. J. Bot.* 48:167–71.

———. 1970b. The relationship of ammonia to symptom expression in apple shoots inoculated with *Erwinia amylovora*. *Can. J. Bot.* 48:999–1000.

McClintock, J. A., and H. L. Facker. 1933. Canker treatment for fire blight control. *Tenn. Agric. Exp. Stn. Circ.* 36, 4 pp.

McCown, Monroe. 1929. Bordeaux spray in the control of fire blight of apple. *Phytopathology* 19:285–93.

Miller, P. W. 1929. Studies of fire blight of apple in Wisconsin. *J. Agric. Res.* 39:579–621.

Miller, T. D., and M. N. Schroth. 1972. Monitoring the epiphytic population of *Erwinia amylovora* on pear with a selective medium. *Phytopathology* 62:1175–82.

Mills, W. D. 1955. Fire blight development on apple in western New York. *Plant Dis. Rep.* 39:206–07.

Moller, W. J., J. A. Beutel, W. O. Reil, F. J. Perry, and M. N. Schroth. 1973. Fire blight: Streptomycin-resistant control studies, 1972. *Calif. Agric.* 27(6):4–5.

Moller, W. J., M. N. Schroth, and S. V. Thomson. 1981. The scenario of fireblight and streptomycin resistance. *Plant Dis.* 65:563–68.

Morrill, G. D. 1969. Overwintering of *Erwinia amylovora* inside living host tissue in Cache Valley, Utah. M.S. thesis, Utah State University, Logan, 40 pp.

Nixon, E. L. 1927. The migration of *Bacillus amylovorus* in apple tissue and its effects on the host cell. *Penn. Agric. Exp. Stn. Tech. Bull.* 212:3–16.

Ogawa, J. M., B. T. Manji, C. R. Heaton, J. Petrie, and R. M. Sonoda. 1983. Methods for detecting and monitoring the resistance of plant pathogens to chemicals. In *Pest resistance to pesticides*, 117–162. Plenum Publishing Corp., New York, NY. 809 pp.

Parker, K. G., N. S. Luepschen, and A. L. Jones. 1974. Inoculation trials with *Erwinia amylovora* to apple rootstocks. *Plant Dis. Rep.* 58:243–47.

Powell, C. C., Jr., and D. C. Hildebrand. 1970. Fireblight resistance in *Pyrus*; involvement of arbutin oxidation. *Phytopathology* 60:337–40.

Powell, D. 1963. Prebloom freezing as a factor in the occurrence of the blossom blight phase of fire blight of apples. *Trans. Ill. State Hort. Soc.* 97:144–48.

———. 1965. Factors influencing the severity of fire blight infections on apple and pear. *Mich. State Hort. Soc.* 94:1–7.

Pugashetti, B. K., and M. P. Starr. 1972. Conjugational transfer of genes determining plant virulence in *Erwinia amylovora*. *J. Bacteriol.* 122:485–91.

Ritchie, D. F., and E. J. Klos. 1974. A laboratory method of testing pathogenicity of suspected *Erwinia amylovora* isolates. *Plant Dis. Rep.* 58:181–83.

———. 1978. Differential medium for isolation of *Erwinia amylovora*. *Plant Dis. Rep.* 62:167–69.

Roberts, R. G., S. T. Reymond, and R. J. McLaughlin. 1989. Evaluation of mature apple fruit from Washington State for the presence of *Erwinia amylovora*. *Plant Dis.* 73:917–21.

Rosen, H. R. 1929. The life history of the fire blight pathogen, *Bacillus amylovorus*, as related to the means of overwintering and dissemination. *Ark. Agric. Exp. Stn. Bull.* 244, 96 pp.

———. 1933. Further studies on the overwintering and dissemination of the fire blight pathogen. *Ark. Agric. Exp. Stn. Bull.* 283, 102 pp.

———. 1935. The mode of penetration of pear and apple blossoms by the fire blight pathogen. *Science* 81:26.

Schroth, M. N., J. A. Beutel, W. J. Moller, and W. O. Reil. 1972. Fire blight of pears in California, current status and research progress, I. Streptomycin resistance. *Calif. Plant Pathol. Publ.* 7, 10 pp.

Schroth, M. N., W. J. Moller, S. V. Thomson, and D. C. Hildebrand. 1974. Epidemiology and control of fire blight. *Ann. Rev. Phytopathol.* 12:389–412.

Shaffer, W. H., and R. N. Goodman. 1970. Control of twig blight (fireblight) on "Jonathan" apple trees with a combination spray of streptomycin, sulfur and glyodin. *Plant Dis. Rep.* 54:203–05.

Scholberg, P. L., A. P. Gaunce, and G. R. Owen. 1988. Occurrence of *Erwinia amylovora* of pome fruit in British Columbia in 1985 and its elimination from the apple surface. *Can. J. Plant Pathol.* 10:178–82.

Smith, Clayton O. 1931. Pathogenicity of *Bacillus amylovorus* on species of *Juglans*. *Phytopathology* 21:219–23.

Stahl, F. J., and N. S. Luepschen. 1977. Transmission of *Erwinia amylovora* to pear fruit by *Lygus* spp. *Plant Dis. Rep.* 61:936–39.

Steiner, P. W. 1989. Predicting apple blossom infections by *Erwinia amylovora* using the MARYBLYT model (9 pp.) and Predicting canker, shoot and trauma blight phases of apple fire blight epidemics using the MARYBLYT model (8 pp.). Vth I.S.H.S. International fire blight workshop. 19–22 June 1989. Diepenbeek, Belgium.

Suhayda, C. G., and R. N. Goodman. 1981. Early proliferation and migration and subsequent xylem occlusion by *Erwinia amylovora* and the fate of its extracellular polysaccharide (EPS) in apple shoots. *Phytopathology* 71:697–707.

Thomas, H. E., and P. A. Ark. 1934a. Fire blight of pears and related plants. *Calif. Agric. Exp. Stn. Bull.* 586, 43 pp.

———. 1934b. Nectar and rain in relation to fire blight. *Phytopathology* 24:682–85.

Thomas, W. D., Jr., and W. J. Henderson. 1952. Spray experiments for the control of fire blight on apples and pears, 1947–50. *Plant Dis. Rep.* 36:273–75.

Thompson, J. M. 1971. Effect of rootstock on fireblight in several apple cultivars. *Hort. Science* 6:167.

Thomson, S. V., M. N. Schroth, W. J. Moller, and W. O. Reil. 1975. Relation of water and epiphytic populations of *Erwinia amylovora* to the occurrence of fire blight. *Phytopathology* 65:353–58.

———. 1976. Effects of bactericides and saprophytic bacteria in reducing colonization and infection of pear flowers by *Erwinia amylovora*. *Phytopathology* 66:1457–59.

———. 1982. A forecasting model for fireblight of pear. *Plant Dis.* 66:576–79.

Thornberry, H. H. 1966. Thomas Jonathan Burrill's contribution to the history of microbiology and plant pathology. *Trans. Ill. St. Hort. Soc.* 100:1–42.

Tullis, E. C. 1929. Studies on the overwintering and modes of infection of the fire blight organism. *Mich. Agric. Exp. Stn. Tech. Bull.* 97, 32 pp.

Van der Zwet, T., and H. L. Keil. 1968a. Recent spread and present distribution of fire blight in the world. *Plant Dis. Rep.* 52:698–702.

———. 1968b. Review of fire blight control measures in the United States. *Trans. Ill. St. Hort. Soc.* 101:63–71.

———. 1972. Incidence of fire blight in trunks of "Magness" pear trees. *Plant Dis. Rep.* 56:844–47.

———. 1979. Fireblight, a bacterial disease of rosaceous plants. USDA Agric. Handbook 510, 200 pp.

Van der Zwet, T., and W. A. Oitto. 1972. Further evaluation of the reaction of "resistant" pear cultivars to fire blight. *HortScience* 7:395–97.

Van der Zwet, T., B. G. Zoller, and S. V. Thomson. 1988. Controlling fire blight of pear and apple by accurate prediction of the blossom blight phase. *Plant Dis.* 72:464–72.

Veldeman, I. R. 1972. [Discovery of *Erwinia amylovora* (Burrill) Winslow et al. in Belgium.] *Rev. Agric.* (Brussels) 25:1587–94.

Waite, W. B. 1898a. Pear blight and its treatment. *East N. Y. Hort. Soc. Proc.* 2:779–90.

———. 1898b. The life history and characteristics of the pear blight germ. (abs.) *Proc. Am. Assoc. Adv. Sci.* 47:427–28.

Whetzel, H. H. 1906. The blight canker of apple trees. *N. Y. (Cornell) Agric. Exp. Stn. Bull.* 236:103–38.

Wilson, E. E., and J. M. Ogawa. 1979. *Fungal, bacterial, and certain nonparasitic diseases of fruit and nut crops in Califor-*

nia. University of California, Division of Agricultural Sciences, Berkeley, 190 pp.

Wrather, H. A., J. Kuc, and E. B. Williams. 1973. Protection of apple and pear fruit tissue against fireblight with nonpathogenic bacteria. *Phytopathology* 63:1075–76.

Zoller, B. G., and J. Sisevich. 1976. Effect of temperature on blossom populations of *Erwinia amylovora* in Bartlett pear orchards in California during 1972–1976. *Proc. Am. Phytopathol. Soc.* 3:322.

———. 1979. Blossom populations of *Erwinia amylovora* in pear orchards vs. accumulated degree hours over 18.3°C (65°F), 1972–1976. (abs.) *Phytopathology* 69:1050.

Hairy Root of Apple and Quince

Agrobacterium rhizogenes (Riker et al.) Conn

Early in the studies of crown gall of apple, investigators had noted the development of excessive numbers of fibrous roots from an area at the base of the plant. Hedgecock reported in 1910 that this "hairy root" condition had some definite relation to the malformation known as crown gall because variations existed that ranged from formation of superabundant fibrous roots to definite enlargements of the crown-gall type. In 1911, Smith, Brown, and Townsend called attention to the different symptoms produced and the cultural characteristics of various isolates. They questioned whether they were dealing with variations in a species or with different species.

Brown (1929), Muncie and Suit (1930), and Siegler (1929) later studied the hairy-root disease and concluded that it was a manifestation of crown gall. In 1930, however, Riker et al. (1930) and Wright, Hendrickson, and Riker (1929) reported that the hairy-root disease differed fundamentally from crown gall, and gave the pathogen a name other than that of the crown gall organism.

Although hairy-root disease of apple is widespread in the eastern United States (Suit 1933), it is relatively uncommon on that host in California. However, it has been reported on roses in California (Anon. 1954) and on quince in the Pacific Coast states and some eastern states (U.S. Department of Agriculture, 1960). Artificial inoculations have induced the disease in *Prunus persica*, *Rubus* spp., *Juglans* sp., and more than 30 other plant species (DeCleene and DeLey 1981; Keane, Kerr, and New 1970).

Symptoms. Hairy-root disease occurs most commonly on one- to three-year-old grafted apple trees (**fig. 3**). At the union between scion and root piece, an enlargement somewhat resembling a newly formed crown gall appears. From this arise numerous roots, fleshy or fibrous in texture, with many containing numerous branches. In addition to the enlargements with excessive rootlet development, there may be other malformations of similar nature but with few or no rootlets. The surface of these enlargements bears numerous convolutions with fissures extending deep into the interior of the enlargements.

Causal organism. Because of differences in bacteriological characteristics and host responses, Riker et al. (1930) and Wright, Hendrickson, and Riker (1929) considered the hairy-root organism to be a new species and named it *Phytomonas rhizogenes* in accordance with the generic classification system of plant-pathogenic bacteria then in use. According to the system now in use, the name should be *Agrobacterium rhizogenes* (Riker et al.) Conn. See the later studies of Huisingh and Durbin (1967) and Holmes and Roberts (1981) for further elaboration in this area.

The pathogen is a nonspore-forming, gram-negative rod, 0.55 to 2.54 μm by 0.15 to 0.75 μm, motile by one to several peritrichous flagella, and containing capsules. *A. tumefaciens* produces some nitrite from nitrates, but *A. rhizogenes* does not. Acid is formed from arabinose, cellobiose, galactose, glucose, fructose, lactose, maltose, mannnose, raffinose, rhamnose, sucrose, trehalose, xylose, dulcitol, inositol, erithritol, mannitol, sorbitol, and salicin. No acid is produced from melezitose, dextrin, inulin, and aesculin. Starch is not hydrolyzed. The optimum growth temperature is 20° to 28°C.

Agrobacterium rhizogenes harbors three large plasmids, one of which, the Ri (root-inducing) plasmid, contains genes for virulence. Genes for sensitivity to agrocin 84 reside on a second plasmid (White and Nester 1980). Both types of plasmids have been transferred from a virulent to an avirulent strain of the pathogen (White and Nester 1980). In other research (Colak and Berderbeck 1979), the genes for tumor induction have been transferred from *A. tumefaciens* to *A. rhizogenes*. All of the pathogenic strains of *A. rhizogenes* that have been studied appear to be members of the group of *Agrobacterium* strains designated by Keane, Kerr, and New (1970) as biotype (biovar) 2. Valuable techniques for the isolation and identification of *Agrobacterium* species are discussed by Kerr and Brisbane (1983). Further information on this and the other pathogenic species of *Agrobacterium* can be found in the discussion of crown gall in Chapter 4.

Disease development. Inoculations have shown that at favorable temperatures symptoms in the form of enlargements at the inoculation point develop in two or three weeks; this is followed within a month by the development of numerous roots on the enlargements. The bacterium enters the plant only through relatively fresh wounds (Hildebrand and Riker 1932). During the infection process the T-DNA of the pathogen is transferred to the host cell and becomes an integral part of its genome. These new bacterial genes induce root proliferation and synthesis of novel amino acid compounds called opines. Two of these opines, agropine and mannopine, have been identified in hairy-root infections; they can serve as the sole source of carbon and nitrogen for the invading pathogen (Kerr and Brisbane 1983). The difference between root proliferation with *A. rhizogenes* and gall formation with *A. tumefaciens* is thought to be related to the balance of auxin and cytokinin production (Kerr and Brisbane 1983).

The pathogen apparently is capable of living free of the host in soil for a considerable period (Hildebrand and Riker 1932). It occurs in large numbers in the tissues of the basal enlargements and readily escapes to the soil, where it is doubtless spread about by water. Other important agencies of dissemination are insects, such as white grubs, which not only feed upon the basal enlargements, thereby picking up the pathogen, but subsequently feed on roots of uninfected plants and inoculate them (Hildebrand and Riker 1932).

Control. In apple, infection by the hairy root pathogen most commonly occurs during or following the grafting process. During the grafting process, a root piece from a seedling is fitted onto a scion and the two are secured by wrapping with a suitable material. Formerly, a waxed string was used, but this left spaces through which bacteria in soil water could enter the cut surfaces of scion and root piece. By using a special adhesive tape which better protects cut surfaces, and by dipping roots and scion in a disinfectant, infection at this stage is greatly reduced. Grafts then should be planted in soil known to be free of the pathogen.

The incidence of hairy root in roots has been reduced by planting clean stock in soil fumigated with either chloropicrin or a dichloropropane-dichloropropene mixture (Munnecke, Chandler, and Starr 1963).

When the trees are dug up from the nursery row and planted in the orchard, care should be taken to avoid wounding them. Trees should not be planted in soils where the disease has already occurred.

Although some, if not most, pathogenic strains of *A. rhizogenes* are sensitive to agrocin 84 (Moore, Warren, and Strobel 1978; White and Nester 1980), the control of hairy root through the use of strain K84 has not been established.

REFERENCES

Anon. 1954. Thirty-fourth annual report for the period ending December 31, 1953. *Calif. Dept. Agric. Bull.* 42:234–72.

Brown, Nellie A. 1929. The tendency of the crown-gall organism to produce roots in conjunction with tumors. *J. Agric. Res.* 39:747–66.

Colak, O., and R. Berderbeck. 1979. Die übertragung der Fähigkeit zur tumorinduktion von *Agrobacterium tumefaciens* auf *A. rhizogenes. Phytopath. Z.* 96:268–72.

DeCleene, M., and J. DeLey. 1981. The host range of infectious hairy-root. *Bot. Rev.* 47:147–94.

Hedgecock, C. G. 1910. Field studies in the crown-gall and hairy-root of the apple tree. *USDA Plant Indus. Bull.* 186, 108 pp.

Hildebrand, E. M. 1934. Life history of the hairy-root organism in relation to its pathogenicity on apple trees. *J. Agric. Res.* 48:857–85.

Hildebrand, E. M., and A. J. Riker. 1932. *Meeting agriculture's old and new problems with the aid of science.* Wisconsin Agr. Exp. Stn. Ann. Rept. for 1931, 152 pp.

Holmes, B., and P. Roberts. 1981. The classification, identification, and nomenclature of *Agrobacteria*, incorporating revised descriptions for each of *Agrobacterium tumefaciens* (Smith and Townsend) Conn 1942, *A. rhizogenes* (Riker et al.) Conn 1942, and *A. rubi* (Hildebrand) Starr and Weiss 1943. *J. Appl. Bacteriol.* 50:443–67.

Huisingh, D., and R. D. Durbin. 1967. Physical and physiological differentiation among *Agrobacterium rhizogenes, A. tumefaciens,* and *A. radiobacter. Phytopathology* 57:922–23.

Keane, P. J., A. Kerr, and P. B. New. 1970. Crown gall of stone fruit, II. Identification and nomenclature of *Agrobacterium* isolates. *Austral. J. Biol. Sci.* 23:585–95.

Kerr, A., and P. G. Brisbane. 1983. *Agrobacterium*. In *Plant bacterial diseases: A diagnostic guide*, edited by P. C. Fahy and G. J. Persley, 27–43. Academic Press, Sydney.

Moore, L. W., G. Warren, and G. Strobel. 1978. Plasmid involvement in the hairy root infection caused by *Agro-*

bacterium rhizogenes. *Proc. 4th Internat. Conf. Plant Pathogen. Bact.* 127–31.

Muncie, J. H., and R. F. Suit. 1930. Studies on crown gall, overgrowths, and hairy root of apple nursery stock. *Iowa State Coll. J. Sci.* 4:263–300.

Munnecke, D. E., P. A. Chandler, and M. P. Starr. 1963. Hairy root (*Agrobacterium rhizogenes*) of field roses. *Phytopathology* 53:788–99.

Riker, A. J., W. M. Banfield, W. H. Wright, G. W. Keitt, and H. E. Sagan. 1930. Studies on the infectious hairy root of nursery apple trees. *J. Agric. Res.* 41:507–40.

Riker, A. J., and E. M. Hildebrand. 1934. Seasonal development of hairy root, crown gall, and wound overgrowth in apple trees in the nursery. *J. Agric. Res.* 48:887–912.

Siegler, E. A. 1929. The woolly knot type of crown gall. *J. Agric. Res.* 39:427–50.

Smith, E. F., N. A. Brown, and C. O. Townsend. 1911. Crown-gall of plants: Its cause and remedy. *USDA Bur. Plant Indus. Bull.* 312, 215 pp.

Suit, F. R. 1933. *Pseudomonas rhizogenes* R. B. W. K. & S.: Its host relations and characteristics. *Iowa State Coll. J. Sci.* 8:131–73.

U.S. Department of Agriculture. 1960. *Index of plant diseases in the United States*. USDA Handbook 165, 531 pp.

White, F. F., and E. W. Nester. 1980. Hairy root: Plasmid encodes virulence traits in *Agrobacterium* rhizogenes. *J. Bacteriol.* 141:1134–41.

Wright, W. H., A. A. Hendrickson, and A. J. Riker. 1929. Studies on the progeny of single-cell isolations from hairy-root and crown-gall organisms. *J. Agric. Res.* 41:541–47.

Pear Decline

Mycoplasmalike Organism (MLO)

During the summer of 1959, a sudden and widespread collapse of pear trees in California resulted in an estimated loss of 10,000 mature bearing trees (Nichols et al. 1960). This was followed by losses of 100,000 and 700,000 trees, respectively, during the following two years (Nichols and Shalla 1960; Rackham et al. 1964). This disease was first noted in British Columbia in 1948 by McLarty (1948) and during the same year was found in the state of Washington (Blodgett et al. 1955; Woodbridge, Blodgett, and Diener 1957). A few years later the disease was reported in the Medford area of southern Oregon (Blodgett et al. 1955). In North America the disease has also been reported from Connecticut (McIntyre et al. 1978). By 1963 about one million trees in the Pacific Coast region of North America had been destroyed (Stout 1963). Since California was the leading pear-producing state, the potential importance of this disease to California agriculture was quite evident. Now, over 25 years later, much of the pear-growing land in Placer and El Dorado counties has been replaced by housing developments. The remaining orchardists are having difficulty applying pesticides for fear of drift onto nearby residences. In 1984, California still ranked first in pear production (288,390 tons), compared to Washington and Oregon with 99,101 and 41,034 tons, respectively.

This sudden pear decline epidemic convinced Klinkowski (1970) that it should be included, along with late blight of potato, Dutch elm disease, and St. Anthony's fire, in his list of catastrophic plant diseases.

With regard to the worldwide distribution of pear decline, "moria del pero," now considered to be identical with pear decline, has been known to occur in northern Italy in the Trentino–Alto Adige area since 1934 (Shalla and Chiarappa 1961; Shalla et al. 1961). According to Mader (1908) and Conti (1934), there is a possibility that the disease occurred in Italy as early as 1908. According to Refatti (1948) and Shalla and Chiarappa (1961), during the period 1945–47 over 50,000 trees were destroyed in the Po Valley, the main fruit-production region of Italy. A disease thought to be identical to pear decline has been reported from Greece (Agrios 1972), France (Lemoine 1975), Switzerland (Schmid 1974), West Germany (Benke, Schaper, and Seemuller 1980), Czechoslovakia (Blattny and Vana 1974), Yugoslavia (Grbic 1974), Israel (Gur 1957), and Argentina (Sarasola 1960). Although originally thought to be a disease affecting only pear, recent research by Parish (1988) in Washington indicates that apple also suffers from a disorder induced by a mycoplasmalike organism (**fig. 4E**). Whether the pathogen attacking apple is the same as the one in pear has not been determined.

Symptoms. Pear decline symptoms (Bethell 1978; Nichols et al. 1960) are influenced by environmental conditions, inherent vigor of trees, and rootstock tolerance. Reactions to the pathogen are expressed as quick decline or collapse of the trees, slow decline, or leaf curl.

The most serious effect is quick decline, which is seen most commonly in older vigorous trees following seasons with heavy infestation of the psylla insect. There is sudden wilting of leaves during the heat of the summer, and within a few days the entire tree may collapse. This symptom is most common on pear trees on oriental rootstocks (*Pyrus serotina* and *P. ussuriensis*). On diseased trees the sequence of symptoms is as follows: The fruit stop sizing and both fruit and leaves suddenly wilt within a two-week period; this is followed by some scorching and death of leaves; the top symptoms resemble those of trees severely girdled by rodents; in the early fall the foliage shows premature and abnormal reddening; the following spring the new growth is sparse, with little or no terminal growth (diseased trees show less than 100 mm growth whereas healthy trees usually show over 600 mm) and the leaves are small and light green; the graft union shows a brown line between the oriental pear understock and the domestic pear overstock; the feeder roots are sparse or lacking, as if infected by water-mold fungi; starch accumulation occurs above the bud union and is deficient in the understock, indicating interference in carbohydrate translocation. Trees on domestic pear seedlings are rarely affected.

The characteristics of slow decline are small fruit, reduced shoot growth, and premature reddening of fall foliage in contrast to the normal yellowing of healthy trees (**fig. 4A, 4B**). Trees on sensitive oriental rootstocks and sometimes those on quince show a distinct brown line at the graft union (**fig. 4C**). Trees on other rootstocks seldom show the brown line but may exhibit distinct fluting or ropelike ridging on the wood and bark at the bud union. In the spring following infection, trees with slow decline express yellow leaves, low vigor, and erratic bloom. With improved psylla control and good cultural practices, slow-decline trees will produce good shoot growth.

The leaf curl phase of pear decline occurs most commonly in young vigorous trees on decline-tolerant rootstock, such as seedlings of *Pyrus communis*, *P. calleryana*, and *P. betulaefolia*. Symptoms are expressed in late summer or fall following colder nights. The leaves show downward curling and thickening, their color is a purplish-red (Millecan, Gotan, and Nichols 1963), and they drop prematurely. Curl symptoms are striking in cultivars Bartlett, Old Home, and Hardy, while Comice shows leaf bronzing but only slight curl; oriental species and cultivars rarely show curl symptoms (Griggs et al. 1967).

Decline symptoms expressed as reduced growth can occur after exposure to heavy psylla populations during the previous year. Also, conditions which reduce feeder roots or food distribution between the rootstock and the scion will show decline symptoms on both young and old trees.

Today, quick decline symptoms are not common because most of the trees on oriental rootstocks have been eliminated. Symptoms of slow decline and leaf curl act as a guide to growers to provide better psylla control as well as better orchard culture.

Cause. Pear decline was first reported (Lindner, Burts, and Benson 1962) to be caused by a toxin produced by pear psylla, *Capopsylla* (*Psylla*) *pyricola* (**fig. 4D**), as an outgrowth of the generally accepted relationship of psylla to decline (Gonzales et al. 1963; Jensen and Erwin 1963). Later, however, graft transmission was shown by Blodgett, Aichele, and Parsons (1963), Schneider (1970), and Shalla, Chiarappa, and Carroll (1963a, b). This suggested the cause to be a viruslike agent (Jensen et al. 1964; Kaloostian, Oldfield, and Jones 1968; Shalla, Chiarappa, and Carroll 1963a), but studies by Hibino and Schneider (1970) and Hibino, Kaloostian, and Schneider (1971) implicated a mycoplasmalike organism (MLO) as the probable pathogen.

The cause of pear decline, initially called "summer wilt" in the state of Washington, was thought by Degman (1954) to be impaired conduction between the tops and roots that possibly resulted from winter injury or faulty irrigation. Blodgett et al. (1955) in Washington observed no correlation between soil type and decline but indicated that trees on Japanese (oriental) rootstocks were more prone to slow decline than were those on French rootstocks. Later, however, Blodgett and Aichele (1959) showed that pears on oriental rootstocks could also develop "quick decline" symptoms. Sprague (1955, 1957) observed a scarcity of small feeder roots on weaker trees, and McIntosh (1960) isolated the fungus *Phytophthora cactorum* from orchards with decline in British Columbia. Yet Nichols et al. (1964) could not specifically relate either Pythiaceous fungi or parasitic nematodes to decline. Kienholz (1958) questioned whether viruses were implicated in pear decline. Meanwhile, Batjer and Schneider (1961) presented data associating necrosis of the sieve-tubes below the bud union and the accumulation of starch above the union with pear decline. Furthermore, they confirmed that trees on oriental rootstocks (*P. ussuriensis* Maxim, *P. serotina* Rehd., and *P. calleryana* Decne) were severely affected by decline, whereas less than half of those on French rootstock were affected, and then only mildly. Trees on Bartlett seedling rootstock were rarely affected, and those on *P. betulaefolia* as well as scion-rooted Bartlett trees were free of the disease. Jensen et al. (1964) demonstrated that *Psylla pyricola* is a vector of the pear-decline "virus," while Kaloostian, Oldfield, and Jones (1968) concluded that the decline symptoms were induced by a toxin injected into the tree by the insect. Other workers, however, obtained evidence that the disease is produced by an entity which can be transmitted by grafting (Shalla, Chiarappa, and Carroll 1963b; Shalla, Carroll, and Chiarappa 1964). At first the causal agent was considered to be a virus (Jensen et al. 1964; Shalla, Chiarappa, and Carroll 1963a), but later studies indicated that it probably is a mycoplasmalike organism (Hibino and Schneider 1970; Hibino, Kaloostian, and Schneider 1971; Kaloostian, Hibino, and Schneider 1971). More recently, Raju, Nyland, and Purcell (1983) indicated that a spiroplasma, similar if not identical to *Spiroplasma citri*, is present in some pear trees with decline but is not the cause of the disease.

A mycoplasmalike organism is now generally accepted as the cause of pear decline. This organism has been studied by electron microscopy in pear (Hibino and Schneider 1970), in its psyllid vector (Hibino, Kaloostian, and Schneider 1971), and in inoculated periwinkle plants (Kaloostian, Hibino, and Schneider 1971). It is a small prokaryotic, polymorphic organism with a unit membrane but without a cell wall. The mycoplasmalike bodies show considerable variation in morphology—spherical, oblong, filamentous, or irregular. These bodies vary from 50 to 1,500 nm in diameter, reaching their largest size in the vector. They were found only in the sieve tubes of pear, in sieve tubes and rarely in parenchyma cells of periwinkle, and in the foregut and salivary glands of the pear psylla. The pear MLO has not been cultured in vitro.

Disease development. The pear decline mycoplasmalike organism (MLO) overwinters in infected pear trees, including those of the more commonly used rootstocks (Batjer and Schneider 1961). It does not survive in oriental rootstock trees. Although the pear decline MLO is not thought to be related to the peach yellow leaf roll MLO, recent studies (Raju, Nyland, and Purcell 1983) have shown some relationship between pear orchards with pear decline and peach trees with yellow leaf roll disease. The pear decline MLO resides in the sieve tube elements of the phloem tissue (Batjer and Schneider 1961; Tsao, Schneider, and Kaloostian 1966; Tsao and Schneider 1967) as spherical to oblong bodies that cause these cells to become necrotic and to collapse. The population of the MLO is high during the summer months and low during the early spring growth period. Thus, during May and June normal shoot growth occurs. In severely affected trees the new sieve cells are killed almost as fast as they are produced through being crushed by the enlargement of adjacent ray cells. The presence of this abnormal phloem in the annual phloem rings is indicative of the number of years the trees have been affected. The first pathological change seen in affected pear tissue is the abnormal accumulation of a substance called "callose" on the sieve areas of the sieve tubes in the phloem of the rootstock just below the bud or graft union. Important is the damage to sieve cells in the leaves which reduces the normal movement of carbohydrates and nitrogen from the leaves to the woody parts. Also extremely damaging is the injury to the phloem at the graft union which reduces the food supply to the roots. As a result of the overall phloem injury, the first decline symptoms are expressed in early July, with red or purple foliage occurring in August, and premature defoliation in October or November.

The relationship of the vector to pear decline was established by Jensen et al. (1964), who obtained pathogen-free psylla and let them feed on caged pear trees. Their initial conclusion was that slow decline was related to toxins produced by the psylla and quick decline was caused by a virus transmitted by the psylla. In later studies, Nichols et al. (1966) compared the spread of the psylla with incidences of decline and established a correlation between the presence of psylla and pear decline.

The relationship of rootstock to pear decline was first established by researchers at Wenatchee, Washington in the late 1950s (Batjer and Schneider 1961, Blodgett et al. 1955; Blodgett and Aichele 1959). They implicated both *Pyrus serotina* (*pyrifolia*) and *P. ussuriensis*. These Japanese rootstocks were introduced by Reimer (1925) from 1915 to the 1930s because they were resistant to fireblight (*Erwinia amylovora*). However, because trees on these stocks developed a black-end condition of the fruit, later plantings were made on pear seedlings from France. Before and during the introduction of the Japanese rootstocks, California pear orchards were established on seedlings of *P. communis* imported from France. The use of Winter Nelis (*P.*

communis) seedlings as rootstocks for pears began in the 1930s. The fruit of this cultivar often are used for baby food, and the trees serve as pollinators for the Bartlett cultivar. Today, Winter Nelis is one of the more popular rootstocks used for the control of pear decline. The variability in the *P. communis* rootstocks used today is attributed to the use of seedlings instead of cuttings, which are difficult to root.

Control. The guidelines to successfully control pear decline and to maintain high yields involve pear psylla control, rootstock selection, chemical treatment of infected trees, and good orchard culture.

The severity of pear decline is related to the number of pear psylla in the orchard. If pear psylla are excluded, very little decline is observed in trees grown under good culture. Even with slowly declining trees on susceptible rootstocks, the control of psylla will result in the trees regaining almost full productivity. Orchards on more tolerant rootstocks may start to show decline and reduced yield with increases in psylla populations. Pear psylla control is becoming more and more difficult because the psylla are becoming resistant to chemicals and because the presence of new housing developments adjacent to pear orchards restrict the use of certain pesticides.

Of the rootstocks used, low decline susceptibility ratings are given for Winter Nelis seedlings, Bartlett rooted cuttings, *P. betulaefolia* seedlings and cuttings, and quince (*Cydonia oblonga*) seedlings. Both inherent susceptibility of rootstocks to the MLO and natural vigor of the trees play a role in expression of decline symptoms. Thus highly tolerant stocks with moderate or low vigor (domestic French or rooted Bartlett) may show more decline than moderately susceptible stocks with high vigor (*P. calleryana*), which may outgrow the effects of the disease. Highly tolerant and vigorous stock (*P. betulaefolia*) can resist decline most successfully and does well in marginal foothill soils but may be too vigorous if used on highly fertile valley soils. Therefore it is important to use rootstocks that are adapted to specific sites.

Another important point to consider is that orchards established on *P. communis* rootstocks are carriers of the decline MLO (Doyle et al. 1970). Attempts to inarch declining trees on susceptible rootstocks with more tolerant stocks have provided erratic results.

Interplanting with tolerant rootstocks in declining orchards is possible, but the replants require special cultural attention in order to compete with older trees. The scion cultivar affects the expression of decline symptoms, with Winter Nelis and Bartlett being most susceptible and Flemish Beauty and Comice somewhat more tolerant (Bethell 1978).

Therapeutic chemical treatment of diseased trees was introduced by Nyland and Moller (1973) with the registration of oxytetracycline hydrochloride (Terramycin) for this purpose in 1973. This treatment involves postharvest injection of trees not later than about two weeks before anticipated leaf fall with a Terramycin tree formulation. Injections can be made with a presssure injector (Reil 1979); Reil and Beutel 1976) or by a gravity-flow procedure (Moller et al. 1978). The liquid is introduced into the tree trunk through horizontal holes, 1½ to 2 inches deep, made with a ¼-inch bit. These holes are spaced 3 to 4 inches apart, for a total of six to eight holes per mature tree; proportionately fewer holes are used for smaller trees. Annual injections markedly improve the condition of affected trees, but complete recovery has not yet been achieved.

Good management practices that include the establishment of trees on the proper rootstock, effective psylla control, and chemical injection of affected trees ensure crop yield gains greater than the added costs. Some of the cultural programs requiring special attention to provide vigorous growth are increased nitrogen fertilization, more frequent irrigation, reduced crop load, heavier pruning of declining trees, and good mite control. Good orchard management requires a system for monitoring tree health and the population of pear psylla (Bethell 1978).

Useful guidelines for monitoring the vigor of slow-decline trees have been designed. A system developed in the Pacific Northwest measures shoot growth on a scale of 1 to 4 as follows: Normal vigorous trees are graded 1; trees with 20 to 40 percent growth reduction are graded 2; those reduced 50 to 80 percent (shoots only 6 to 8 inches long) are graded 3; and those with little or no growth are graded 4. Psylla numbers are monitored by means of spur samples, beating-tray samples, and leaf samples taken at different times of the year. On the basis of these data appropriate psylla control measures are taken (Bethell 1978). An increase in vigor rating of a full grade during a one-year period after establishment of a disease management program is considered excellent.

REFERENCES

Agrios, G. N. 1972. A decline disease of pear in Greece—Pear decline or graft incompatibility. *Phytopathol. Mediterr.* 11:87–90.

Batjer, L. P., and H. Schneider. 1961. Relation of pear decline to rootstocks and sieve-tube necrosis. *Proc. Am. Soc. Hort. Sci.* 76:85–97.

Benke, H. K., V. Schaper, and E. Seemuller. 1980. Association of mycoplasmalike organisms with pear decline symptoms in the Federal Republic of Germany. *Phytopath. Z.* 97:89–93.

Bethell, R. S., ed. 1978. *Pear pest management*. Division of Agricultural Sciences, University of California, Berkeley, 234 pp.

Blattny, C., and V. Vana. 1974. Pear decline accompanied with mycoplasma-like organisms in Czechoslovakia. *Biol. Plant (Prague)* 16:474–75.

Blodgett, E. C., and M. D. Aichele. 1959. Progress report on pear decline research. *The Goodfruit Grower* 9(21):2.

Blodgett, E. C., M. D. Aichele, and J. L. Parsons. 1963. Evidence of a transmissible factor in pear decline. *Plant Dis. Rep.* 47:89–93.

Blodgett, E. C., C. G. Woodbridge, N. Benson, and W. J. O'Neill. 1955. Report of pear decline committee. *Proc. Wash. State Hort. Assoc.* 51:231–33.

Conti, G. 1934. Departmenti e mortalità dei fruttetti nella venezia tridentina. *Alm. Agr. pel 1934. Cons. Prov. Ec. Corp.-Trento* 52:24–40.

Degman, E. S. 1954. Is pear decline caused by over-irrigation? *Proc. Wash. State Hort. Assoc.* 50:66–67.

Doyle, F. J., H. J. O'Reilly, G. Nyland, B. Bearden, R. Bethell, C. Hemstreet, G. Morehead, and S. Sibbett. 1970. Distribution of pear decline virus in California orchards. *Calif. Agric.* 24(6):14–15.

Gonzales, C. G., W. H. Griggs, D. D. Jensen, and S. M. Gotan. 1963. Orchard tests substantiate role of pear psylla in pear decline. *Calif. Agric.* 17(1):4–5.

Grbic, V. 1974. Some injurious species of the family Psyllidae in pear orchards in Vojvodina. *Zastita Bilja* 25:121–31.

Griggs, W. H., D. D. Jensen, B. T. Iwakiri, and J. A. Beutel. 1967. Effects of different rootstocks, and degree of psylla infestation on leaf curl in young pear trees. *Calif. Agric.* 21(10):16–20.

Gur, A. 1957. *The compatibility of the pear with quince rootstocks*. Israel Mn. Agric. Res. Stn., Div. Hort. Fac. Agric. Hebrew Univ. Spec. Bull. 10.

Hibino, H., G. H. Kaloostian, and H. Schneider. 1971. Mycoplasma-like bodies in the pear psylla, vector of pear decline. *Virology* 43:34–40.

Hibino, H., and H. Schneider. 1970. Mycoplasma-like bodies in sieve tubes of pear trees affected with pear decline. *Phytopathology* 60:499–501.

Jensen, D. D., and W. R. Erwin. 1963. The relation of pear psylla to pear decline. *Calif. Agric.* 17(1):2–3.

Jensen, D. D., W. H. Griggs, C. Q. Gonzales, and H. Schneider. 1964. Pear decline virus transmission by pear psylla. *Phytopathology* 54:1346–51.

Kaloostian, G. H., H. Hibino, and H. Schneider. 1971. Mycoplasma-like bodies in periwinkle: Their cytology and transmission by pear psylla from pear trees affected with pear decline. *Phytopathology* 61:1177–79.

Kaloostian, G. H., G. N. Oldfield, and L. S. Jones. 1968. Effect of pear decline virus and psylla (*Psyllla pyricola*) toxin on pear trees. *Phytopathology* 58:1236–38.

Kienholz, J. R. 1958. Are viruses concerned in pear decline? *Plant Dis. Rep.* 42:1194.

Klinkowski, M. 1970. Catastrophic plant diseases. *Ann. Rev. Phytopathol.* 8:37–59.

Lemoine, J. 1975. A dieback of pears observed in France and resembling pear decline or moria. *Acta hort.* 44:131–38.

Lindner, R. C., E. C. Burts, and N. R. Benson. 1962. A decline condition in pears induced by pear psylla. *Plant Dis. Rep.* 46:59–60.

Mader, C. 1908. La mortalità dei peri nella plaga di Bolzano-Gries. *Alm. Agr. pel 1908. Sez. Trento Cons. Prov. Agr. Tirolo* 26:347–50.

McIntosh, D. L. 1960. The infection of pear rootlets by *Phytophthora cactorum*. *Plant Dis. Rep.* 44:262–64.

McIntyre, J., A. Dodds, G. S. Walton, and G. H. Lacy. 1978. Declining pear trees in Connecticut: Symptoms, distribution, symptom remission by oxytetracycline, and associated mycoplasma-like organisms. *Plant Dis. Rep.* 62:503–07.

McLarty, H. R. 1948. Killing of pear trees. *Ann. Rept. Can. Plant Dis. Survey* 28:77.

Millecan, A. A., S. M. Gotan, and C. W. Nichols. 1963. Red-leaf disorders of pear in California. *Calif. Dept. Agric. Bull.* 3:186–92.

Moller, W. J., J. A. Beutel, R. S. Bethell, and W. O. Reil. 1978. Pear decline. In *Pear pest management* edited by R. S. Bethell, 150–59. Division of Agricultural Science, University of California, Berkeley.

Nichols, C. W., F. L. Blanc, A. A. Millecan, and G. D. Barbe. 1966. Spread of pear psylla and pear decline virus in California (abs.). *Phytopathology* 56:150.

Nichols, C. W., S. M. Garnsey, R. L. Rackham, S. M. Gotan, and C. N. Mahannab. 1964. Pythiaceous fungi and plant-parasitic nematodes in California pear orchards. *Hilgardia* 35:577–602.

Nichols, C. W., H. Schneider, H. J. O'Reilly, T. A. Shalla, and W. H. Griggs. 1960. Pear decline in California. *Bull. Calif. Dept. Agric.* 49:186–92.

Nichols, C. W., and T. A. Shalla. 1960. Pear decline disease and its apparent southward spread along the Pacific coast of North America. *FAO Plant Protec. Bull.* 9:39–42.

Nyland, G., and W. J. Moller. 1973. Control of pear decline with a tetracycline. *Plant Dis. Rep.* 57:634–37.

Parish, C. L. 1988. Apple decline: Characterization, cause, and cure. In *XIV international symposium on fruit tree virus* diseases and V international symposium on small fruit virus diseases. p. 34. Min. Agric. and Hellenic Phytopath. Soc. 12–18 June 1988. Hellexpo, Thessaloniki, Greece.

Rackham, R. L., D. C. Alderman, H. H. O'Reilly, and C. W. Nichols. 1964. Pear decline incidence and severity in California in 1960, 1961, and 1962. *Plant Dis. Rep.* 48:204–05.

Raju, B. C., G. Nyland, and A. H. Purcell. 1983. Current status of the etiology of pear decline. *Phytopathology* 73:350–53.

Refatti, E. 1948. Su di una grave malattia dei peri nella provincia di Trento e Bolzano. *Cicerca Scient.* 18:856–60.

Reil, W. O. 1979. Pressure-injecting chemicals into trees. *Calif. Agric.* 33(6):16–19.

Reil, W. O., and J. A. Beutel. 1976. A pressure machine for injecting trees. *Calif. Agric.* 30(12):4–5.

Reimer, G. C. 1925. Blight resistance in pears and characteristics of pear species and stocks. Or. State Coll. Agric. Exp. Stn. Bull. 214, 99 pp.

Sarasola, A. A. 1960. El decaimiento del peral en el valle del Rio Negro. Inst. Nac. Technol. Agro. Inst. Patol. Vegetal (Argent.) Publ. Tec. 60.

Schmid, G. 1974. Pear decline. *Schweiz. Z. Obst. Weinbau* 1210:297–301.

Schneider, H. 1970. Graft transmission and host range of the pear decline causal agent. *Phytopathology* 60:204–07.

Shalla, T. A., T. W. Carroll, and L. Chiarappa. 1964. Transmission of pear decline by grafting. *Calif. Agric.* 18(3):4–5.

Shalla, T. A., and L. Chiarappa. 1961. Pear decline in Italy. *Calif. Dept. Agric. Bull.* 50:213–16.

Shalla, T. A., L. Chiarappa, E. C. Blodgett, E. Refatti, and E. Baldacci. 1961. The probable coidentity of the moria disease of pear trees in Italy and pear decline in North America. *Plant Dis. Rep.* 45:912–15.

Shalla, T. A., L. Chiarappa, and T. W. Carroll. 1963a. Evidence for virus as a cause of pear decline. *Calif. Agric.* 17(1):8.

———. 1963b. A graft transmissible factor associated with pear decline. *Phytopathology* 53:366–67.

Sprague, R. 1955. Fungi isolated from roots of pear trees at Wenatchee. *Proc. Wash. State Hort. Assoc.* 51:93.

———. 1957. Fungi isolated from roots and crowns of pear trees. *Plant Dis. Rep.* 41:74–76.

Stout, G. L. 1963. Bureau of plant pathology. *Calif. Dept. Agric. Bull.* 52:109–17.

Tsao, P. W., and H. Schneider. 1967. Pathological anatomy of pear tissues sensitive to pear decline virus (abs.). *Phytopathology* 57:103.

Tsao, P. W., H. Schneider, and G. H. Kaloostian. 1966. A brown leaf-vein symptom associated with greenhouse-grown pear plants infected with pear decline virus. *Plant Dis. Rep.* 50:270–74.

Woodbridge, C. G., E. C. Blodgett, and T. O. Diener. 1957. Pear decline in the Pacific Northwest. *Plant Dis. Rep.* 41:569–72.

Fungal Diseases of Pome Fruit

Anthracnose and Perennial Canker of Pome Fruit

Pezicula malicorticis (Jacks.) Nannf.

Apple-tree anthracnose, or "Northwestern anthracnose," has been known in the Pacific Northwest since about 1890 but did not receive critical attention until the early 1900s (Cordley 1900a, b, c). The disease usually is manifested as a cankerous condition of the branches, but under some conditions it occurs on fruit before harvest or after fruit is picked and stored (Cordley 1900 a, b, c; Lawrence 1904). As a result of his pathogenicity studies, Cordley (1900c) gave the first comprehensive description of symptoms in the 1900s as well as a description of the conidial stage of the causal fungus, which he named *Gloeosporium malicorticis*. Between 1911 and 1913, Jackson (1911, 1912, 1913) contributed further information on the life history of the fungus and described the ascigerous stage, which he named *Neofabraea malicorticis*.

In 1925, Zeller and Childs (1925a, b) reported another anthracnose disease that was widely distributed in Oregon, Washington, and British Columbia. This disease, originally known to growers as "false anthracnose," was named "perennial canker" by Zeller and Childs. The fungus attacks the bark of the host and produces a ripe-fruit rot. The conidial stage of the causal fungus was named *Gloeosporium perennans* Zeller and Childs, and the ascigerous stage *Neofabraea perennans* (Zeller and Childs) Kienholz (Kienholz 1939).

Although anthracnose and perennial canker were originally thought to be caused by two different fungus species, G. *malicorticis* and G. *perennans*, Kienholz (1939) was unable to distinguish the organisms by means of "morphological, physiological, or mycological studies." He found that both the asexual and sexual stages of the organisms were indistinguishable, but because of apparent pathological differences he recommended retaining them as distinct species. Most of the world's mycologists and pathologists familiar with this problem favor recognition of a single species, now generally classified as *Pezicula malicorticis* (Jacks.) Nannf. (anamorph: *Cryptosporiopsis curvispora* (Pk.) Gremmen), as the cause of both anthracnose and perennial canker (Borecka, Borecki, and Millikan 1971; Corke and Sneh 1979; Guthrie 1959; Hawksworth, Sutton, and Ainsworth 1983; Pierson, Ceponis, and McColloch 1971; Sharples 1962; Sutton 1980). Irrespective of the view one might take regarding the separation of species on the basis of pathogenic differences, there remains the question of whether symptomatic dissimilarities between anthracnose and perennial canker are due to the parasite itself or to the environment. Since no one has demonstrated the existence of two distinct pathotypes of the pathogen, it is likely that anthracnose and perennial canker represent different symptom expressions of a single disease with the environment being the determinant.

In the Pacific Northwest, these reputedly distinct diseases tend to be confined to different climatic zones. On the whole, anthracnose is most prevalent west of the Cascade Range, a region characterized by high annual rainfall and moderate winter temperatures. In the relatively dry climate east of these mountains, perennial canker predominates. Both diseases occur in the White Salmon–Hood River Valley apple districts where the Columbia River breaks through the Cascade Range, and where climatic conditions are intermediate between those to the east and west.

The anthracnose–perennial canker complex is sometimes found in other parts of the west, including Idaho, California (Barnett 1944), and Montana (U. S. Department of Agriculture 1960). This disease has been reported also from Nebraska (Heald 1920, 1926), Massachusetts (Boyd 1939), Illinois (Anderson 1940), Michigan (Pepper 1959), Maine (Hilborn 1938), Virginia, Pennsylvania (McColloch and Watson 1966), and Oklahoma (U.S. Department of Agriculture 1960). It also occurs in western Canada (Lopatecki and Burdon 1966; McLarty 1933), eastern Canada, the British Isles (Corke and Sneh 1979; Edney, Tan, and Burchill 1977; Wilkinson 1943, 1945), central and northern Europe (Borecka, Borecki, and Millikan 1971; Borecki and Puchala 1976; Gram, Weber, and Schnicker 1932; van Poeteren 1934–35; Wilson and Ogawa 1979), New Zealand (Brien 1932; Curtis 1953; Wollenweber 1939), Australia, and Zimbabwe.

Most of the research on this disease has been concerned with apples, but in the Pacific Northwest the pathogen causes a serious canker and fruit-rot disorder of pears and occasionally is found on quince. Therefore, most of our discussion of this disease is based on apple studies; however, the evidence suggests that the symptoms and disease cycle on apple and pear are similar.

Symptoms: Anthracnose. Elliptical sunken necrotic cankers develop in the bark of the host (**fig. 5A**). Smaller branches are most commonly infected, but at times the scaffold limbs or the trunk may also be involved. The cankers enlarge and then become separated from surrounding healthy tissue by a crack which may extend completely around the affected area. The periderm over the necrotic area cracks and becomes separated into numerous small pieces which curl up at the edges exposing the acervuli. At first, the acervuli appear as small, cream-colored pustules projecting through the periderm. Later, however, they turn dark. On old cankers, the bark, except for the bast fibers, breaks away. The remnants of the bast fibers running lengthwise on the canker have been referred to as "fiddle strings."

Symptoms: Perennial canker. Elliptical, roughened cankers (**fig. 5B**) develop on large or small branches, commonly occurring where lateral branches had been removed in pruning. Their distinguishing characteristic is a series of concentric callus ridges that form each year after the canker has ceased to expand. An old canker may have numerous ridges, usually one for each year, and so its age may be approximately reckoned. Because of pressure exerted by the developing ridges, the periderm breaks and falls away from the face of the canker. Acervuli form in the outer bark as in anthracnose, but the fiddle-string appearance of older cankers generally is not evident. The recurrent annual extension of the cankered area suggested the name "perennial canker."

Another type of canker sometimes forms. This is a circular necrotic area, often surrounding a lenticel, usually shallow, and rarely active for more than one season. These lesions, called "button cankers" by McLarty (1933), although apparently caused by the perennial-canker fungus, resemble small anthracnose lesions in location and general appearance. These cankers enlarge the following year only if the surrounding callus has become infested with the woolly apple aphid.

Symptoms: Fruit rot. The fruit rot phase of this disease has been termed "bull's-eye rot" (van Poeteren 1934–35; Wollenweber 1939). Although most infection occurs in the orchard, the rot usually does not become evident until after the fruit has been stored for some time or reaches the market. Infection occurs through the lenticels, breaks in the skin, and, occasionally, at the calyx. The lesions begin as small brown specks that slowly enlarge to spots from 1 to 3 cm in diameter. They may be a pale, yellowish cream or uniformly brown, but are most often brown with a pale center that forms a bull's-eye. The spots are flat to slightly sunken. The rotted tissue is relatively firm, and the skin over the surface does not rupture easily. The rot may be shallow or nearly as deep as it is wide. The decayed tissue is somewhat mealy and does not separate readily from the healthy tissue. Acervuli sometimes develop on the surface of the rot; they are small, moist, cream-colored, and protrude through the skin. Bull's-eye rot seldom involves the entire fruit, but it frequently opens the way for secondary rot fungi to complete the decay process.

Causal organism. In view of the confusion regarding the classification and nomenclature of the fungus causing the anthracnose–perennial canker disease it is necessary to provide a brief history of this problem. The imperfect stage of the anthracnose fungus was incorrectly described as *Macrophoma curvispora* by Peck (1900). Later, Cordley (1900c) gave it the name *Gloeosporium malicorticis* and subsequently the name was changed to *Cryptosporiopsis curvispora* (Pk.) Gremmen. Jackson (1911, 1912, 1913), having discovered the ascigerous stage in 1913, named it *Neofabraea malicorticis* (Cord.) Jack., erecting a new genus. The acervulus stage of the perennial canker pathogen was named *Gloeosporium perennans* by Zeller and Childs in 1925 (1925 a, b). According to Kienholz (1939), the ascigerous stage of this fungus was discovered in 1928 by Cooley, who noted its similarity to *N. malicorticis*, but preferred to retain the specific name *perennans*. Accordingly, in 1939 Kienholz formally described and named it *N. perennans* (Zeller and Childs) Kienholz. In doing so, he rejected Nannfeldt's (1932) transfer of these fungi to *Pezicula*, a genus erected by the Tulasnes in 1865. Nannfeldt based this transfer on a study of *Neofabraea corticola* (Edg.) Jorg. Thompson (1939) also objected to the extinction of *Neofabraea*, which he maintained is related to *Pezicula* but is not identical with it. Wollenweber (1939), however, recommended the transfer.

Pezicula malicorticis and *Neofabraea perennans* are so similar as to be indistinguishable. The chief differences are found in the shape of the conidia and in the size of asci and ascospores. *Neofabraea malicorticis* usually produces curved or hooked conidia, *N. perennans* straight or slightly curved conidia. The former produces slightly larger asci and ascospores than the latter (Kienholz 1939; Zeller and Childs 1925a, b). Differences in physiological characters are a somewhat greater conversion of starch into dextrin and maltose, and a greater sensitivity to certain chemicals (tannic acid and malachite green) on the part of *N. malicorticis* (Kienholz 1939; Miller 1932).

With both the morphological and physiological characters, however, greater variations may exist between different isolates of either species than between the species themselves (Kienholz 1939; Miller 1932). Moreover, in the White Salmon–Hood River Valley district, where the climate is intermediate between the coastal and interior regions, the fungi themselves tend to be intermediate in type. To quote Kienholz (1939, p. 662), "A separation of these forms as species based wholly on morphological or physiological grounds appears unwarranted." He also said, "Because of the practical aspects of disease control, however, it seems desirable to consider them as distinct species, for the diseases caused by them must be handled as distinct diseases in the orchard."

In the Pacific Northwest, perennial canker symptoms are more prevalent in regions of light rainfall and relatively cold winters. Low temperature plays an important part in infection by the perennial canker fungus, but whether or not temperature has a similar effect on infection by the anthracnose fungus has not been fully explored. Kienholz (1939) remarked on the tendency for symptoms induced by the anthracnose fungus in the Hood River Valley to resemble those of perennial canker, but he was unable to obtain conclusive experimental evidence on this point. He also stated, "A distinction between apple rots produced from inoculations with the two fungi could not be made."

Though Kienholz (1939) did not feel justified in combining the two species, he pointed out the intergradations in morphological characters displayed by various isolates of the two species. He and others (Fisher and Reeves 1928, 1929) noted, for example, that certain isolates of the perennial canker fungus produced conidia indistinguishable from the anthracnose fungus. Keinholz (1939) observed that saltation was common with both fungi, arising either in colonies from conidia or colonies from ascospores. The saltant from the colony of one species often produced conidia similar to those produced by the other species, so it is suggested that these fungi might be strains arising from mutation of a common parent.

The acervuli of both fungi are uniformly distributed over the surface of the canker. They are at first subperidermal then erumpent. The acervulus stroma is a closely packed structure from the surface of which arise numerous simple or branched conidiophores. Conidia of *P. malicorticis* usually are curved, 4 to 6 μm broad and 15 to 24 μm long; those of *N. perennans* typically are straight, 4 to 6 μm broad and 12 to 20 μm long. The shape and size of the conidia, however, are greatly influenced by environment, and vary greatly among different isolates (Heald 1926; Kienholz 1939, Owens 1928). Conidia produced by either fungus on apple fruit are of the perennial-canker type (Kienholz 1939).

With the abandonment of the genus *Neofabraea* and with the general acceptance of the belief that anthracnose and perennial canker are caused by a single fungus, the name of the pathogen becomes *Pezicula malicorticis* (Jacks.) Nannf. (Borecki, Czynczyk, and Millikan 1978; Chen, Spotts, and Mellenthin 1981; Hawksworth, Sutton, and Ainsworth 1983; Nannfeldt 1932; Pierson, Ceponis, and McColloch 1971). Also, with the rejection of the genus *Gloeosporium* the correct binomial for the imperfect stage is *Cryptosporiopsis curvispora* (Pk.) Gremmen.

The apothecia, which are 0.5 to 1 mm in diameter, develop within stromata of old acervuli. Conditions of temperature and moisture influence the size of these structures. They are sessile, waxy in texture, and gray to flesh-colored. In wet weather they often persist for a considerable time, but quickly disintegrate during dry periods. The asci are indehiscent, clavate, and slightly pedicellate. They are on average 13.1 μm broad and 100.4 μm long. The ascospores are one- or two-seriate in the ascus. They are unicellular until germination when they may develop one to four septa; their shape is ellipsoidal but sometimes flattened on one side. The contents are coarsely granular or finely guttular, and colorless or slightly amber in old spores. The ascospores average 6.2 μm wide by 18.7 μm long (Heald 1926; Kienholz 1939; Owens 1928).

Small, rod-shaped, single-celled microconidia are produced in culture. Their development in positions near hyphal coils, which appear to be the beginnings of apothecia, suggested to Kienholz (1939) that they perform the same function that microconidia (spermatia) perform in certain other Discomycetes.

Host relations and cultivar susceptibility. The following rosaceous plants have proved susceptible to *P. malicorticis*: commercial apple, Siberian crab, Oregon crab, pear, quince, peach, plum, apricot, cherry, flowering quince, serviceberry, hawthorn, and mountain ash. Only on Oregon crab and the *Pyrus* species, however, does the fungus produce cankers similar to those on cultivated apple in nature. Pear and quince are infected naturally but have not proved as susceptible as apple. It is thought that the Oregon crab, a widely distributed native in the Pacific Northwest, may be a natural host of the fungus (Kienholz 1939).

Few data are available on the comparative susceptibility of apple cultivars to anthracnose and perennial canker. Baldwin is especially susceptible, but Northern Spy and Ben Davis are rarely affected. Such cultivars as Esopus Spitzenberg, Yellow Newtown, Rome Beauty, and Jonathan are severely affected in the Pacific Northwest; Gravenstein, McIntosh, Winter Banana, Delicious, Arkansas Black, Winesap, and Wealthy are much less affected (Childs 1929; Owens 1928, Wilson and Ogawa 1979). Susceptibility to the pathogen is believed to be conditioned, at least in part, by the susceptibility of the tree to freezing injury and to attack by the woolly apple aphid (Childs 1929; Cooley and Shear 1931; Crenshaw and Cooley 1931; Kienholz 1939; McLarty 1933; Wilson and Ogawa 1979). In England, Worcester Pearmain, Bramley's Seedling, Allington Pippin, Laxton's Superb, and Cox's Orange Pippin are some of the most susceptible cultivars (Corke and Sneh 1979; Edney, Tan, and Churchill 1977; Wilkinson 1943, 1945). Based on artificial inoculations in Poland, Golden Delicious, Starkrimson, and Idared showed considerably more resistance than McIntosh and the three scab-resistant cultivars Priam, Prima, and Priscilla (Borecki, Czynczyk, and Millikan 1978). In New Zealand, Sturmer is considered to be one of the most susceptible cultivars (Brook 1957). No information has been found on cultivar susceptibility in pears.

Disease development. Although the perfect stage, through the production of air-borne ascospores, may play an important role in long-distance dissemination, the greater abundance of acervuli on the cankers suggests that the imperfect stage is probably more important in year-to-year infection in the orchard. The fungus survives from one season to the next in branch cankers. It may exist in the cankered bark for several years, producing conidia during rainy periods at almost any time of the year. The conidia are disseminated by rain, and they infect branches mainly in autumn and winter (Cooley and Shear 1931; Shear and Cooley 1933; Wilson and Ogawa 1979). The cankers develop during the winter and early spring, but generally cease activity in late spring. Acervuli may appear on new cankers as early as the first summer following infection. In the Pacific Northwest, apothecia apparently are not produced until approximately two years after infection (Kienholz 1939; Owens 1928), but in England Sharples (1962) found mature apothecia in November, one year after infection. He reported that ascospores were discharged only on rainy days when the temperature exceeded 5°C. Apothecia continued to form the following spring and summer.

The anthracnose-type cankers develop commonly in uninjured bark, with most infection apparently occurring through lenticels (Kienholz 1939). This type of canker is typically annual and, in the Pacific Northwest, is found mainly in areas of high rainfall and mild winter temperatures (Kienholz 1939). The so-called perennial cankers usually develop at pruning wounds and are found mainly in the drier and colder regions of the Pacific Northwest (Kienholz 1939). With both types of cankers, a ridge of callus is formed around the diseased area in the spring. The enlargement of perennial cankers was originally thought to be caused by the annual extension of fungus mycelium into healthy tissue around the periphery of the canker, but this now is said to be caused by the annual infection of injured callus tissues by spores produced on the canker itself. Sloughing of the bark eventually removes most of the fungus from both types of cankers.

Although most perennial canker infections originate at pruning wounds, fruit-stem scars and injuries caused by sunburn, freezing temperatures, toxic chemicals, and insects may also be significant (Brook 1957; Kienholz 1939; Reeves, Yothers, and Murray 1939; Wilson and Ogawa 1979). The year-to-year expansion of these cankers is possible only if open infection courts are present in the uninfected tissue around the margin of the canker. The agencies which produce these wounds are the woolly apple aphid, *Eriosoma lanigerum* (Haus.), and freezing temperatures (Childs 1929; Kienholz 1939; Wilson and Ogawa 1979). The insect is attracted to the succulent tissue produced by the healing of the bark around the margin of the inactive, original canker. In feeding on this tissue it produces galls that are more prone than normal bark tissue to injury by low temperature (Crenshaw and Cooley 1931). If temperatures of -17.8°C or lower prevail in winter, the galls rupture and provide avenues of entry for the fungus. When temperatures are extremely low, even the normal callus around wounds or cankers becomes ruptured and infected. This, however, is rare compared with the infection of aphid-infested callus tissue. According to Childs (1929), Northern Spy, which is almost immune to aphid attack, rarely develops perennial canker.

The fruit of apple and pear become infected when prolonged rains carry spores from branch cankers onto the fruit. They are susceptible from petal fall until harvest, although there is some evidence for a decrease in susceptibility in mid-summer (Kienholz 1956; Pierson, Ceponis, and McColloch 1971; Washington State University 1984). The rots begin mostly at open lenticels and skin injuries, but the infection is usually latent until after harvest (Edney 1956; Pierson, Ceponis, and McColloch 1971). The decay develops slowly at cold-storage temperatures, appearing late in the storage season, during transit, and on the market. Infection of lenticels is most rapid at about 20°C, but rotting can occur at as low as 0°C (Edney 1956). Edney, Tan, and Burchill (1977) have shown that subjecting apples to simulated rain before inoculation increased their susceptibiblity to lenticel infection. Bull's-eye rot does not spread by contact from one fruit to another (Pierson, Ceponis, and McColloch 1971).

Control: Anthracnose. The anthracnose phase of this disease complex can be effectively controlled through the autumn application of one to three fungicidal sprays (Oregon State University 1984; Washington State University 1984; Wilson and Ogawa 1979). The objective of these applications is to maintain a protective layer of fungicide on the branches during their susceptible period in fall and winter. Since rain is essential for spore production, dissemination, and infection, it is essential that the first application be made ahead of the rainy period. In the Pacific Northwest, where anthracnose is most significant, this usually involves both a preharvest and a postharvest application (Heald 1926; Owens 1928; Wilson and Ogawa 1979). The preharvest treatment has the additional advantage of providing at least partial protection of the fruit against bull's-eye rot (Oregon State University 1984; Washington State University 1984; Wilson and Ogawa 1979). Materials used in Oregon and Washington for the preharvest application on apples are captan, ziram, and mancozeb; on pears the preferred fungicides are ziram and mancozeb (Oregon State University 1984; Washington State University 1984). Recommended postharvest fungicides are Bordeaux mixture, fixed copper compounds, and dichlone (Oregon State University 1984; Washington State University 1984; Wilson and Ogawa 1979). Other measures sometimes employed are canker excision and the removal of badly affected branches (Owens 1928; Wilson and Ogawa 1979).

Control: Perennial canker. Control measures for perennial canker are aimed largely at removing the inoculum sources and preventing infestation by the woolly apple aphid. Such measures are much more likely to produce beneficial results if started before the orchard becomes severely infected, but in badly infected orchards they are likely to prove expensive and ineffectual. In extreme cases, removal of old trees may be the only solution. In replanting the orchard, attention should be given to cultivars most resistant to the disease; for example, the Delicious apple cultivar has been recommended in Oregon because of its resistance to the woolly apple aphid and subsequent freezing injury (Childs 1929).

Complete removal of affected smaller branches where feasible and excision of cankers occurring on larger limbs and on trunks should be done in July or August (Childs 1929). Surgical work consists of scraping dead bark from the surface of the canker and cutting out the remaining dead tissue with a sharp knife. The cuts should be made in the noninfected tissue no further from the margin of the canker than is necessary to excise all diseased bark. The edge of the excised area should be smooth, because ragged cuts leave places under which the aphids can establish themselves. For the exclusion of woolly aphids from these wounds and ordinary pruning cuts, Cooley (1942) recommended application of a preparation consisting of 8 parts rosin and 3 parts sardine oil.

Since the woolly aphid plays such an important role in the development of perennial canker, it is essential that its numbers be kept to the absolute minimum. In many orchards of the Pacific Northwest, the parasite *Alphelinas mali* and several predators have reduced populations of these aphids to low levels; however, it is necessary in some cases to use insecticides selected for their effectiveness against this insect (Washington State University 1984).

Unlike anthracnose, perennial canker cannot be controlled by fall applications of copper fungicides (Kienholz 1951, 1956; Wilson and Ogawa 1979). In Poland, however, there is some evidence that benzimidazole fungicides are effective (Borecka, Borecki, and Millikan 1971; Borecki and Puchala 1976). It has not been determined whether similar results could be obtained in this country. Another approach to control is the use of antisporulants. Corke and Sneh (1979) found significantly less sporulation on cankers treated with preparations containing benomyl, 8-hydroxyquinoline, or miscible tar-petroleum oil than on untreated cankers. How this reduction in sporulation might relate to canker or fruit-rot control was not determined. Other recommended control measures are late winter pruning and maintenance of tree vigor (Kienholz 1951; McLarty 1933).

Control: Bull's-eye rot. The most effective control of fruit rot is through the use of fungicidal sprays in the orchard (Pierson, Ceponis, and McColloch 1971). In the Pacific Northwest, apple and pear fruit are susceptible to infection for approximately a month after petal fall and again from about mid-August until harvest (Kienholz 1956; Washington State University 1984). In this area, adequate rot control can be achieved with the application of a fungicide immediately after petal fall and with one or two applications prior to harvest. On apples, recommended fungicides for the petal-fall spray are maneb, ziram, captan, and tricarbamix; on pears only ziram is recommended (Washington State University 1984). Preharvest materials used on apples are mancozeb, ziram, and captan; on pears only ziram is recommended (Oregon State University 1984; Washington State University 1984). Reports from England (Edney 1956, 1970) and some European countries (Katschinski and Ramson 1975) indicate that benzimidazole compounds used as preharvest sprays and postharvest dips on apples reduce losses from bull's-eye rot. This decay is also reduced by postharvest dip treatments with the scald-control agents diphenylamine and ethoxyquin (Lopatecki and Burdon 1966). A recent report from Oregon (Chen, Spotts, and Mellenthin 1981) indicated that pears inoculated with *P. malicorticis* and stored in low oxygen developed markedly less stem-end decay than those stored in air. In general, the most effective control measure for bull's-eye rot is the preharvest application of fungicidal sprays.

REFERENCES

Anderson, H. W. 1940. Apple tree anthracnose in Illinois. *Plant Dis. Rep.* 24:475.

Barnett, H. L. 1944. Anthracnose and other diseases in California apple-growing regions. *Plant Dis. Rep.* 28:717–18.

Boyd, O. C. 1939. Northwestern anthracnose of apple reported from Massachusetts. *Plant Dis. Rep.* 28:125–26.

Borecka, H., Z. Borecki, and D. .F. Millikan. 1971. Control of apple scab, bitter rot and sawfly in Poland with the use of some newer fungicides. *Plant Dis. Rep.* 55:828–31.

Borecki, Z., A. Czynczyk, and D. F. Millikan. 1978. Susceptibility of several cultivars of apple to bark canker fungi. *Plant Dis. Rep.* 62:8171–19.

Borecki, Z., and Z. Puchala. 1976. [Anthracnose of apple bark caused by *Pezicula* fungi.] *Roczniki Nauk Rolniczych* 5(2):55–72. (*Rev. Plant Pathol.* 56:763. 1977.)

Boyd, O. C. 1939. Northwestern anthracnose of apple reported from Massachusetts. *Plant Dis. Rep.* 28:125–26.

Brien, R. M. 1932. "Delicious spot" on apples due to *Gloeosporium perennans*. *J. N. Z. Dept. Agric.* 45:215–18.

Brook, P. J. 1957. Ripe spot of apples in New Zealand. *N. Z. J. Sci. Tech.* 38:735–41.

Chen, P. M., R. A. Spotts, and W. M. Mellenthin. 1981. Stem-end decay and quality of low oxygen stored d'Anjou pears. *J. Am. Soc. Hort. Sci.* 106:695–98.

Childs, Leroy. 1929. The relation of wooly apple aphid to perennial canker infection with other notes on the disease. *Ore. Agric. Exp. Stn. Bull.* 243, 31 pp.

Cooley, J. S. 1942. Wound dressings on apple trees. *U.S. Dept. Agric. Circ.* 656, 18 pp.

Cooley, J. S., and E. V. Shear. 1931. Relation of perennial canker to its environment (abs.). *Phytopathology* 21:1000.

Cordley, A. B. 1900a. Apple-tree anthracnose. *Ore. State Bd. Hort. Bien. Rept.* 6:405–09.

———. 1900b. Some observations on apple-tree anthracnose. *Botan. Gaz.* 30:48–58.

———. 1900c. Apple-tree anthracnose: A new fungus disease. *Ore. Agric. Exp. Stn. Bull.* 60, 8 pp.

Corke, A. T. K., and B. Sneh. 1979. Antisporulant activity of chemicals towards fungi causing cankers on apple branches. *Ann. Appl. Biol.* 91:325–30.

Crenshaw, J. H., and J. S. Cooley. 1931. Experimental freezing of apple trees (abs.). *Phytopathology* 21:997–98.

Curtis, K. M. 1953. Cankers in apple and pear trees caused by the ripe-spot fungus, *Neofabraea malicorticis*. *Orchard. N. Z.* 26:2.

Edney, K. L. 1956. The rotting of apples by *Gloeosporium perennans* Zeller and Childs. *Ann. Appl. Biol.* 44:113–28.

———. 1970. Some experiments with thiabendazole and benomyl as postharvest treatments for the control of storage rots of apples. *Plant Pathol.* 19:189–93.

Edney, K. L., A. M. Tan, and R. T. Burchill. 1977. Susceptibility of apples to infection by *Gloeosporium album*. *Ann. Appl. Biol.* 86:129–32.

Fisher, D. F., and E. L. Reeves. 1928. Perennial canker. *Proc. Wash. State Hort. Assoc.* 24:55–61.

———. 1929. Recent studies on perennial canker. *Proc. Wash. State Hort. Assoc.* 25:155–66.

Gram, E., A. Weber, and J. L. Schnicker. 1932. Platesygdomme in Denmark. 1931. Oversigt samlet ved statens plantpatologiske forsog. *Tidsskr. Plantev.* 38:349–90.

Guthrie, E. J. 1959. The occurrence of *Pezicula alba* sp. nov. and *P. malicorticis*, the perfect states of *Gloeosporium album* and *G. perennans*, in England. *Brit. Mycol. Soc. Trans.* 42:502–06.

Hawksworth, D. L., B. C. Sutton, and G. C. Ainsworth. 1983. *Ainsworth and Bisby's dictionary of the fungi* 7th ed. Commonwealth Mycological Institute, Kew, England, 445 pp.

Heald, F. D. 1920. Apple anthracnose or blackspot canker. *Wash. Agric. Exp. Stn. Ser. Bull.* 64, 4 pp.

———. 1926. Apple-tree anthracnose. In *Manual of plant pathology*, 1st ed, 500–11. McGraw-Hill, New York.

Hilborn, M. J. 1938. Northwestern apple tree anthracnose found in Maine. *Plant Dis. Rep.* 22:354.

Jackson, H. S. 1911. Apple-tree anthracnose. *Ore. Agric. Exp. Stn. Circ.* 17, 4 pp.

———. 1912. The development of *Gloeosporium malicorticis* Cordley (abs.). *Phytopathology* 2:95.

———. 1913. Apple-tree anthracnose: A preliminary report. *Ore. Agric. Exp. Sta. Bien. Crop Pest Hort. Rept.* 1911–12:178–97.

Katschinski, K. H., and A. Ramson. 1975. [Contribution to the biology and control of *Gloeosporium* rots of apple.] *Nachricht. Pflanz DDR* 29:168–72. (*Rev. Plant Pathol.* 55:1851. 1976.)

Kienholz, J. R. 1932. Perennial canker and anthracnose fungi: Host relations and cultural difference (abs.). *Phytopathology* 22:995–96.

———. 1939. Comparative study of the apple anthracnose and perennial canker fungi. *J. Agric. Res.* 59:635–65.

———. 1951. The bull's-eye-rot problem of apples and pears. *Ore. State Hort. Soc. Ann. Rept.* 43:975–77.

———. 1956. Control of bull's-eye rot on apple and pear fruits. *Plant Dis. Rep.* 40:872–77.

Lawrence, W. H. 1904. Black-spot canker. *Wash. Agric. Exp. Stn. Bull.* 66, 35 pp.

Lopatecki, L. E., and H. Burdon. 1966. Suppression of bull's-eye of apples and growth of *Neofabraea perennans* by scale control agents diphenylamine and ethoxyquin. *Can. J. Plant Sci.* 46:633–38.

McColloch, L. P., and A. J. Watson. 1966. Perennial canker rot of apples in West Virginia and Pennsylvania. *Plant Dis. Rep.* 50:348–49.

McLarty, H. R. 1933. Perennial canker of apple trees. *Can. J. Res.* 8:492–507.

McLarty, H. R., and J. C. Roger. 1931. Perennial canker survey in the Okanogan Valley. *Can. Dept. Agric. Rept.* 1930:108–11.

Miller, Erston V. 1932. Some physiological studies of *Gloeosporium perennans* and *Neofabraea malicorticis*. *J. Agric. Res.* 45:65–77.

Nannfeldt, J. A. 1932. Studien über die Morphologie und Systematik der Nichtlichenisierten Inoperculaten Discomyceten. *Nova Acta Reg. Soc. Sci. Upsala Sec.* 48(2):1–386.

Oregon State University Extension Service. 1984. 1984 pest management guide for tree fruits in the mid-Columbia area. Hood River, Ore., 23 pp.

Owens, C. E. 1928. Apple-tree anthracnose. In *Principles of plant pathology*, 288–99. John Wiley & Sons, New York.

Peck, C. H. 1900. New species of fungi. *Bul. Torrey Botan. Club* 27:14–21.

Pepper, E. H. 1959. Northwestern apple tree anthracnose reported from Michigan. *Plant Dis. Rep.* 43:920–21.

Pierson, C. F., M. J. Ceponis, and L. P. McColloch. 1971. *Market diseases of apples, pears, and quinces.* USDA Agric. Handbook 376, 112 pp.

Reeves, E. L., M. A. Yothers, and C. W. Murray. 1939. Unusual development of apple perennial canker following application of toxic wound dressings. *Phytopathology* 29:739–42.

Sharples, R. O. 1962. The perfect stage of *Gloeosporium perennans* on apple trees. *Plant Pathol.* 11:180–82.

Shear, E. V., and J. S. Cooley. 1933. Relation of growth cycle and nutrition to perennial apple-canker infection (abs.). *Phytopathology* 23:33.

Sutton, B. C. 1980. *The coelomycetes: Fungi imperfecti with pycnidia, acervuli, and stromata.* Commonwealth Mycological Institute, 696 pp.

Thompson, G. E. 1939. A canker disease of poplars caused by a new species of *Neofabraea*. *Mycologia* 31:455–64.

U.S. Department of Agriculture. 1960. *Index of plant diseases in the United States.* USDA Agric. Handbook 165, 531 pp.

van Poeteren, N. 1934–35. *Verslag over de werkzaamheden van den Plantenziektenkundigen dienst in het Jaar 1933, 1934,* 76 and 80. Plantenziektenkundigen Dienst, Verslag. Mededelingen, Wagenigen.

Washington State University Cooperative Extension. 1984. *1984 Spray guide for tree fruits in eastern Washington.* Pullman, Wash., 86 pp.

White, E. W. 1922. Apple tree anthracnose or blackspot canker control. *Sci. Agric.* 6:186–91.

Wilkinson, E. H. 1943. Perennial canker of apple trees in England. *Gardner's Chronical Sec.* 3114:159.

———. 1945. Perennial canker of apple trees in England. *J. Pomol.* 21:180–85.

Wilson, E. E., and J. M. Ogawa. 1979. *Fungal, bacterial, and certain nonparasitic diseases of fruit and nut crops in California.* University of California, Division of Agricultural Science, Berkeley, 190 pp.

Wollenweber, H. W. 1939. Diskomyzetenstudien (*Pezicula* Tul. und *Occlaria* Tul.). *Arb. Biol. Anst. (Reichsanst) Berlin* 22:521–70.

Woodhead, C. E., and H. Jacks. 1952. Nature and control of ripe-spot of apples. *N. Z. J. Agric.* 84:239–40.

Zeller, S. M., and Leroy Childs. 1925a. Another apple-tree anthracnose in the Pacific Northwest and a comparison with the well-known apple-tree anthracnose (abs.). *Phytopathology* 15:728.

———. 1925b. Perennial canker of apple trees: A preliminary report. *Ore. Agric. Exp. Stn. Bull.* 217, 17 pp.

Dematophora Root Rot of Pome Fruit

Rosellinia necatrix Prill.

Dematophora root rot (white root rot) affects a wide range of hosts including 170 species and varieties in 63 genera and 30 families (Khan 1955). Among tree fruits and nuts are apple, apricot, almond, avocado, peach, pear, plum, quince, sweet and sour cherry, citrus, feijoa, olive, pistachio, and walnut. Other plants affected are boysenberry, loganberry, strawberry, loquat, fig, mango, and grapes.

The disease occurs on all continents but appears to be most serious in the warm temperate and subtropical regions. It has attracted considerable attention in Japan (Matuo and Sakurai 1959; Tanaka, Yamamoto, and Yamada 1966), India (Gupta 1977; Gupta and Verma 1978), Israel (Sztejnberg et al. 1987), Italy, and France (Tourvieille de Labrouhe 1982; Viala 1891), but is of somewhat less importance in England (Nattrass 1926). A similar disease occurs in New Zealand (Wight 1890), but here there is some question as to whether the same species or a different species of *Rosellinia* is involved (Cunningham 1925).

Dematophora root rot was reported in the eastern United States more than 50 years ago, but its occurrence there has not been confirmed in recent years. In California the disease has been most prevalent in Santa Cruz and Santa Clara counties, but it has been found in at least 18 counties from Orange and Riverside in the south to Napa and Butte in the north (Thomas, Wilhelm, and MacLean 1953).

Symptoms. Trees show symptoms of the disease first by yellowing of the foliage and poor growth of the branches. One or all limbs may show this effect. This is followed by defoliation and tree death (Gupta 1977). Below ground, the fibrous roots are rotted away, and a white wefty mycelium (**fig. 6A**) may remain in the soil around the rotted roots. Under the cortex of the affected roots is found a whitish wefty to floccose layer of mycelium (**fig. 6B**),—this is distinct from the more or less leathery, slightly yellow mycelial fans produced by *Armillaria mellea*. Another difference between *Dematophora* and *Armillaria* root rots is in the occurrence of rhizomorphs. Though some writers mention *Dematophora* as producing rhizomorphs, Thomas, Hansen, and Thomas (1934) stated that well-defined rhizomorphs are absent on roots affected by this fungus. Nattrass (1926) mentioned a structure which he said is a whitish wefty strand or ribbon bearing no resemblance to the dense rhizomorphs of *A. mellea*. The most definite diagnostic feature of *Dematophora* is the production of the coremia on diseased roots that have been kept in a moist chamber for several days.

Causal organism. Root rots of this type in England, Europe, and California are produced by the same fungus. Hartig (1883) first named the conidial stage *Rhizomorpha (Dematophora) necatrix*, and suggested its relationship to *Rosellinia*, an ascomycete. Viala (1891) and Prilleaux (1902) described a *Rosellinia* associated with the imperfect stage, but did not produce evidence of their genetic connection. Hansen, Thomas, and Thomas (1937) provided such evidence and also confirmed that the fungus produces an ostiolate perithecium. Its currently recognized binomial is *R. necatrix* Prill. As previously mentioned, the specific identity of the white root rot fungus in New Zealand remains somewhat in doubt (Cunningham 1925).

Coremia arising from sclerotia produce small, single-celled, colorless conidia on multiple-branched conidiophores. Perithecia are spheroidal, black, erumpent, and situated on a mycelial crust. Asci are cylindrical, and ascospores are one-celled, with a hyaline epispore. Characteristic swellings (geniculations) occur at the septa of the vegetative mycelium.

Disease development. Mycelium of the fungus in affected roots remains viable for several years even under the most unfavorable conditions. Ascospores are produced so rarely as to be of little importance in development of the disease. Conidia produced under some conditions are of low viability (Hansen, Thomas, and Thomas 1937). Infection is believed to occur almost entirely by penetration of the mycelium into susceptible roots. The fungus is spread by growth of the mycelium throughout the soil. Spread of the fungus over long distances occurs through the transportation of affected nursery stock.

Young roots are penetrated directly by the mycelium. Nattrass (1926) and Gupta (1977) believed that old roots were not subject to direct invasion but became infected as the fungus grew into them from attached infected young roots. Tourvieille de Labrouhe (1982) showed that the pathogen forms an infection cushion on apple roots from which hyphae grow into the cortex and develop an extensive subcortical prosenchymatous thallus.

Although the incubation period of the disease is unknown, Hansen, Thomas, and Thomas (1937) found that young apple trees were killed within six weeks after inoculation. The fungus develops most rapidly in comparatively cool, moist soils.

Susceptibility. Among apples, Golden Delicious, McIntosh, most clones of Malling and Malling Merton rootstocks, and many apple seedlings are susceptible (Gupta and Verma 1978; van der Merwe and Mathee 1974). Some resistance is shown by seedlings of *Malus floribunda* and *M. toringoides*. Among cherries, Mahaleb, mazzard, and sour cherry (*Prunus cerasus*) rootstocks are all susceptible. Most species of pear (*Pyrus*) are susceptible, though certain seedlings of *P. communis* appear to be tolerant of the fungus. Marianna and myrobalan plums are probably the most resistant of the stone fruits used for rootstocks. Persian (English) walnut (*Juglans regia*), and the northern California black walnut (*J. hindsii*) are susceptible, as are the butternut (*J. cinerea*) and the eastern black walnut (*J. nigra*). The Himalaya and Young blackberries show fairly high degrees of resistance, while the loganberry and the boysenberry are highly susceptible. Rough lemon, sweet and sour orange, trifoliate orange, and Windsor grapefruit are all susceptible (Khan 1955). Pecan and persimmon trees appear to be resistant, since neither showed symptoms after growing for more than four years in infested soil (Sztejnberg and Madar 1980).

Control. No entirely satisfactory control methods have been devised. Godfrey (1934) reported tests in which a soilborne *Dematophora* species was killed with chloropicrin, but Khan (1955) found that neither chloropicrin, carbon disulfide, ethylene bromide, allyl bromide, nor formaldehyde were very effective against the fungus. Matuo and Sakurai (1959) found that chloropicrin killed the hyphae of *R. necatrix* in the soil but not in infected mulberry roots. Efficacy of the treatment was enhanced by covering the soil with a vinyl sheet. In Israel, the fungus in naturally infected avocado roots was killed by deep injection with methyl bromide (Sztejnberg, Omary, and Pinkus 1983). Gupta (1977) reported that applications of carbendazim to infected apple trees resulted in their recovery and development of new roots. There is some evidence that legumes inhibit spread of the pathogen in the soil (Tanaka, Yamamoto, and Yamada 1966). Recent research in Israel (Sztejnberg et al. 1987) indicated that soil solarization is effective in the control of this disease. Biological control through the use of *Trichoderma harzianum* was ineffective, but there was some evidence that control could be enhanced by combining solarization with an application of this fungus to the soil.

REFERENCES

Cunningham, G. H. 1925. *Fungus diseases of fruit-trees in New Zealand.* Brett Printing Co., Auckland, 382 pp.

Godfrey, G. H. 1934. Control of soil fungi by fumigation with chloropicrin (abs.). *Phytopathology* 24:1145.

Gupta, V. K. 1977. Root rot of apple and its control by carbendazim. *Pesticides* 11(9):49–52.

Gupta, V. K., and K. D. Verma. 1978. Comparative susceptibility of apple root-stocks to *Dematophora necatrix*. *Indian Phytopathol.* 31:377–78.

Hansen, H. N., Harold E. Thomas, and H. Earl Thomas. 1937. The connection between *Dematophora necatrix* and *Rosellinia necatrix*. *Hilgardia* 10:561–64.

Hartig, Robert. 1883. *Rhizomorpha (Dematophora) necatrix* n. sp. *Untersuch. Forstbotan. Ins. Munchen* 3:95–153.

Khan, Abdul Hamid. 1955. Dematophora root rot. *Calif. Dept. Agric. Bull.* 44(4):167–70.

Massee, George. 1896. Root diseases caused by fungi. *Kew Bull.* 109:1–5.

Matsuo, T., and Y. Sakurai. 1959. On the fungicidal effect of chloropicrin and other few drugs upon *Rosellinia necatrix* and *Corticium centrifugum* in the soil. *J. Seric. Sci. Japan* 28:395–401.

Nattrass, R. M. 1926. The white rot of fruit trees caused by *Rosellinia necatrix* (Hart.) Berl. *Long Ashton Agric. Hort. Res. Stn. Ann. Rept.* 1926:66–72.

Pierce, Newton B. 1892. *The California vine disease.* U.S.D.A. Div. Veg. Pathol. Phys. Bull. 2, 222 pp.

Prilleaux, E. 1902. Les périthèces de *Rosellinia necatrix*. C. R. Acad. Sci. (Paris) 135:275–78.

Sztejnberg, A., S. Freeman, I. Chet, and J. Katan. 1987. Control of *Rosellinia necatrix* in soil and in apple orchard by solarization and *Trichoderma harzianum*. *Plant Dis.* 71:365–69.

Sztejnberg, A., and Z. Madar. 1980. Host range of *Dematophora necatrix*, the cause of white root rot disease in fruit trees. *Plant Dis.* 64:662–64.

Sztejnberg, A., N. Omary, and Y. Pinkas. 1983. Control of *Rosellinia necatrix* by deep placement and hot treatment with methylbromide. *OEPP/EPPO Bull.* 13:483–85.

Tanaka, H., S. Yamamoto, and S. Yamada. 1966. Influence of various soil conditions on the growth of white root rot fungus, *Rosellinia necatrix*, and of Satsuma Mandarin trees inoculated with the fungus. *Bull. Hort. Res. Sta. Japan.* Ser. B 5:119–39.

Thomas, Harold E., H. N. Hansen, and H. Earl Thomas. 1934. Dematophora root rot (abs.). *Phytopathology* 24:1145.

Thomas, H. E., S. Wilhelm, and N. A. MacLean. 1953. Two root rots of fruit trees. In *Plant diseases, the yearbook of agriculture 1953*, 702–05. U. S. Department of Agriculture.

Tourvieille de Labrouhe, D. 1982. Pénétration de *Rosellinia necatrix* (Hart.) Berl. dans le racines du pommier en conditions décontamination artificielle. *Agronomie* 2:553–60.

Van der Merwe, J. J. H., and F. N. Mathee. 1974. *Rosellinia* root rot of apple and pear trees in South Africa. *Phytophylactica* 6:119–20.

Viala, P. 1891. *Monographie du pourridie.* Paris, Georges Masson, 95 pp.

Wight, R. A. 1890. Root fungi of New Zealand. *J. Mycol.* 5:199.

European (Nectria) Canker of Apple and Pear

Nectria galligena Bres.

European canker, also known as "apple canker" and "Nectria canker," affects apple and pear trees. The disease occurs in almost all countries in which pome fruit are grown (Swinburne 1975).

European canker is one of the major apple diseases in the British Isles, where it has received more attention than in any other area (Cayley 1921; Corke and Hunter 1979; Crowdy 1949, 1952; Marsh 1939; Munson 1939; Swinburne 1975, 1978; Wiltshire 1921, 1922). It also is an important canker and fruit-rot disease in northern Europe (Bulit 1957; Swinburne 1975; Zagaja et al. 1971), New Zealand (Brook and Clarke 1975), and Chile. It is a relatively minor disease in the eastern and midwestern apple-growing districts of the United States, but sometimes causes serious damage in the more humid regions of the Pacific Coast states (Dubin and English 1975a; English, Dubin, and Moller 1972; Wilson and Nichols 1964; Zeller 1926; Zeller and Owens 1921). In countries with frequent autumn rains, fruit infection can cause crop losses of 10–60 percent (Swinburne 1975).

The disease was reported from California by Smith in 1909 and by Fawcett in 1912. It is found chiefly in the high-rainfall apple-growing districts north of San Francisco Bay and to a limited extent in the Santa Cruz Mountains south of San Francisco (Dubin and English 1975b; Wilson and Nichols 1964). In most instances earlier outbreaks were relatively unimportant, but since 1955 the disease has caused serious damage in Sonoma County orchards, sometimes killing young trees and the shoots and branches of older trees (Dubin and English 1975a; Nichols and Wilson 1956; Wilson and Nichols 1964). It has not been found in apple-growing districts in the Sierra Nevada foothills, the Central Valley, or in Southern California (Dubin and English 1975a; English, Dubin, and Moller 1972). The disease has been found on California pears only in Sonoma County (Dubin and English 1975a).

Symptoms. Though occasionally fruit and rarely leaves are affected, the most destructive phase is twig and branch infection (**fig. 7A,B**). New cankers, which commonly develop at the nodes of twigs, are elliptical, sunken, dead areas of bark over which the periderm loosens and breaks away. As the causal agent spreads, it encircles and kills the twigs. In England, the annual peripheral expansion of the diseased area produces an elongated canker with series of roughly concentric ridges. In western Oregon, where winters are comparatively mild, expansion is said to be a more or less continuous process, resulting in a very long nonzonate necrotic area (Zeller 1926).

In cool, moist, autumn weather, white to cream-colored, waxy conidial fructifications (sporodochia) develop over the cankers (**fig. 7E**). In winter, clusters of small red perithecia appear on cankers that are one or more years old (**fig. 7F**).

In moist regions the fruit is frequently infected (English, Dubin, and Schick 1979; McCartney 1967; Swinburne 1975). This infection, called "eye-rot" if at the calyx end of the fruit, results in the rotting and collapse of the fruit tissue in this region (McCartney 1967) (**fig. 7C**). Lenticel infection (English, Dubin, and Moller 1972; Swinburne 1975) and core rot are also known to occur. The symptoms of lenticel infections are a brown, slightly depressed, dry decay, and the center of the circular lesions often has a light tan color (**fig. 7D**). No evidence of the fungus is present at first, but in well-developed rots whitish conidial pustules (sporodochia) may occur.

Causal organism. Although Petch (1938, p. 265) suggested that the fungus be transferred to the genus *Dialonectria*, this has not been done by mycologists. The presently accepted named for the perithecial stage is *Nectria galligena* Bres. Earlier workers had called it *N. coccinea* Fr. or *N. ditissiman* Tul. Zeller (1926) and others, however, showed it to be distinct from these species. In addition to attacking pome fruit, *N. galligena* is known to cause cankers in several other species of broad-leafed trees (Flack and Swinburne 1977). Some, but not all, of the biotypes attacking these trees can also infect apple and pear (Dubin and English 1975a; Flack and Swinburne 1977). Flack and Swinburne (1977) found that the biotypes in apple and ash, although morphologically the same, were pathogenically different. They proposed that *N. galligena* from apple be designated f. sp. *mali* and that from ash f. sp. *fraxini*. Although the asexual stage has been known for many years as *Cylindrocarpon mali* (Allesch.) Wollenw. (Wollenweber 1913), there is evidence (Swinburne 1975) that the correct binomial is *C. heteronema* (Berk. & Br.) Wollenw.

The macroconidia of *N. galligena* are produced in white-to-cream-colored masses (sporodochia) scattered over the surface of the canker. Macroconidia are hyaline, cylindrical, straight or slightly curved, 3- to 7-septate structures with rounded ends. They vary greatly in size but average about 52 to 62 µm long and 4.5 to 5.5 µm broad. Microconidia may be produced in abundance by abstriction from hyphal branches. They measure 4 to 7 μm long 1 to 2 μm broad and are hyaline and one-celled (Zeller 1926).

The bright red perithecia of *N. galligena* (Wilson and Nichols 1964) are produced in clusters or scattered over the surface of the older cankers, but seldom on newly developed cankers. At times the clusters of perithecia surround sporodochia. Mature perithecia are generally ovoid-pyriform, 300 to 400 μm in diameter and about 300 to 450 μm in height. A stroma may or may not be present. The ostiole is raised (Munson 1939). The asci are stalked, cylindrical to clavate in shape, and bear eight ascospores. These spores are hyaline, two-celled, 14 to 20 μm long by 5 to 7 μm broad, and elliptical in shape.

The fungus is heterothallic and bisexual. Light is essential in vitro to the formation of both conidia and perithecia. The optimum temperature for germination of conidia and ascospores is 20°C (Swinburne 1975), and there is little growth in vitro at 4° to 10° and 30°C.

Susceptibility. In California, Delicious and its red mutants are the most susceptible apple cultivars. Gravenstein, Jonathan, McIntosh, and Rome Beauty are infected to some extent, but Golden Delicious is rarely attacked. Zeller (1926) reported that Bismark, Delicious, Winter Bellflower, Spitzenberg, and Newtown apples and d'Anjou, Howell, and Beurre Bosc pears are susceptible in Oregon. In Europe, Alkmene, Cox's Orange Pippin, Idared, Maigold, McIntosh, Orangenburg, Priam, Prima, Priscilla, White Transparent, and Winter Gold Pearmain apple cultivars are said to be highly susceptible. The literature on cultivar susceptibility tends to be confusing, and artificial inoculations do not necessarily predict field performance (Swinburne 1975).

Disease development. The fungus survives from one infection season to the next as mycelium in twig and branch cankers where it produces both conidia and ascospores. Where rain occurs throughout the year conidia are present at all times (Bulit 1957; Swinburne 1975), but are most abundant in September and October and least abundant in the summer (Munson 1939). In the drier climate of California, however, they are produced largely during the rainy fall and winter months (Dubin and English 1975a), and are mostly disseminated by spattering rain.

Perithecia are initiated in autumn and, in California, from late fall until spring. In rainy periods, the ascospores may be either forcibly ejected from the perithecia and disseminated by air currents or extruded in a sticky mass and rain disseminated (Dubin and English 1975a). Bulit (1957) in France, Munson (1939) in England, and Zeller and Owens (1921) in Oregon recorded the discharge of ascospores for all months of the year. The heaviest discharges occur in winter.

Although Swinburne (1971) postulated that ascospores play a dominant role in infection in Northern Ireland, they appear to be a minor factor in disease epidemiology in California (Dubin and English 1975a).

Infection most commonly occurs through leaf scars, the major period for infection in most countries being in autumn during leaf-fall (Crowdy 1952; Dubin and English 1975a; Wiltshire 1921). In Northern Ireland, however, most infection is reported to occur in spring and summer through scars left by the abscission of bud scales and primordial leaves (Swinburne 1975). According to Crowdy (1949), the leaf scar is highly susceptible to infection during the first hour after the leaf falls, and becomes much less susceptible in the next hour or so. Tests in California (Dubin and English 1974b; Wilson 1966) indicated that some leaf scars remained susceptible to infection for 10 to 28 days after leaf fall. According to Wiltshire (1921), another period when leaf scars become susceptible to infection occurs in spring, at which time the fungus can enter through small cracks that form in the leaf scar as the result of growth of the twig. Zeller and Owens (1921) found that in Oregon the greater part of infection occurs in fall. Evidence in California (Dubin and English 1975a) also indicated essentially no spring infection. Temperature within the range generally occurring in California winters does not limit infection. In the winter of 1958–59, for example, if leaf scars were wounded by cutting with a knife they became infected at various times in winter and early spring. The periods at which little or no infection occurred were those during which little rain fell. Experiments under controlled moisture conditions showed that at temperatures of 14° to 15.5°C, the trees had to be wet for at least six hours before appreciable infection of leaf scars occurred. The percentage infection of leaf scars increased rapidly with increases in the length of the moisture period (Dubin and English 1974a, 1975a).

Other infection courts, although generally of minor importance in California, include pruning wounds and other mechanical injuries, freezing damage, insect injuries, fruit-stem scars, and lesions caused by *Venturia inaequalis* and *Cryptosporiopsis corticola*. As with leaf scars, both fruit-stem scars and pruning wounds become progressively resistant to infection with time (Swinburne 1975). Less infection (artificial inoculation with conidia) developed in pruning wounds made in February than in those made later in winter or in early spring (Sealey and Swinburne 1976).

The incubation period of the fungus in the twig may range from a few days in early fall to several weeks or possibly months during the coldest part of winter (Zeller 1926). In years of early autumn rains in California, twig lesions may become visible in late fall, but in most years they appear in late winter and early spring.

After infection, the fungus colonizes the cortex and invades the xylem by means of the cell-wall pits. The host cells die in advance of the hyphae, suggesting the in vivo production of a toxin. The development of phelloderm around the canker appears associated with an increase in indole acetic acid in the host tissues (Swinburne 1975).

Several cultural factors appear to influence the development of European canker (Swinburne 1975). Disease incidence and severity are greater in trees that are overly vigorous and succulent and in those located at high water-table sites or in heavy, clay, acid soils. Disease severity can be reduced by judicious use of fertilizers.

Although most of the inoculum for the infection of apple trees undoubtedly comes from within the orchard, Flack and Swinburne (1977) showed that in Northern Ireland, poplar, beech, and hawthorn harbor the apple pathotype and thus could serve as external sources of inoculum. Mountain ash is still another host of the apple pathogen. In California, isolates of *N. galligena* from aspen and birch did not infect apple trees when inoculated at leaf scars, but apple isolates were pathogenic to pear (Dubin and English 1975a). There is no evidence that external sources of inoculum are important in California.

The fruit rot phase is rare in California during the normal harvest season, but where the harvest is delayed fall rains can induce both calyx infection (eye-rot) and lenticel rot (English, Dubin, and Moller 1972; English, Dubin, and Schick 1979; McCartney 1967). Infection can also occur through wounds and scab lesions (Swinburne 1975). In some cultivars infection is latent, and rotting does not become evident until after three or more months of storage (Swinburne 1975). Research in Northern Ireland has shown that the resistance of unripe fruit is due to the production of benzoic acid at the infection site (Swinburne 1975). This compound is fungitoxic to the pathogen under acid conditions, but with the decrease in acidity and increase in sugar during ripening, its toxicity decreases and the fungus resumes growth. A nonconstitutive protease secreted by *N. galligena* was demonstrated to be the elicitor of benzoic acid production in immature apples (Swinburne 1975).

Control. Although probably the most certain way of avoiding canker epidemics is to cut out and destroy all infected wood, this procedure, in commercial orchards, is quite impractical (Byrde, Crowdy, and Roach 1952; Swinburne 1975). Therefore, the main reliance for control has been placed on the application of properly timed protective or eradicative fungicidal sprays (Brook and Clarke 1975; Byrde, Crowdy, and Roach 1952; Dubin and English 1974a; English, Dubin, and Moller 1972; English, Dubin, and Schick 1979; Marsh 1939; Swinburne 1975; Wilson 1968). Because of climatic differences, an effective spray schedule in one area is not necessarily effective in another region (Swinburne 1975). In California, where autumn-exposed leaf scars are the principal infection courts, excellent control is obtained by spraying at the onset of leaf-fall and again at mid- to late leaf-fall (Dubin and English 1974a; English, Dubin, and Schick 1979; Wilson 1968). The first application is far more important than the second, and in some instances this spray alone has provided adequate canker control (Corbin 1971; English, Dubin, and Moller 1972; English, Dubin, and Schick 1979; Wilson 1968). In the British Isles and in some other regions, an additional application at bud burst is necessary to control infection at scars left by the abscission of bud scales and primordial leaves, and cracks formed in autumnal leaf scars (Swinburne 1975).

Research in California (Dubin and English 1974a; English, Dubin, and Moller 1972; English, Dubin, and Schick 1979; Wilson 1968) has shown that copper compounds and captafol are highly effective when used as fall sprays. Benomyl was effective in two of three years it was tested (English, Dubin, and Schick 1979). Dithianon and, especially, carbendazim gave promising results in Northern Ireland (Swinburne 1975). In addition to their protective action, some of the fungicides, namely, benomyl, captafol, carbendazim, and pyridinitril, show antisporulant activity (Dubin and English 1974a; Swinburne 1975). Other materials with the ability to suppress sporulation are 2,4,5,6-tetrachlorophenol and miscible tar-petroleum oil (Corke and Sneh 1979). The copper materials are the only fungicides currently registered for use against European canker in California.

The control of fruit rot, although not a serious problem in California, is best achieved by controlling the canker phase of the disease (Swinburne 1975). Additional fruit protection has been obtained by spray applications in summer or preharvest with such fungicides as benomyl, captan, carbendazim, captafol, and thiophanate-methyl (Swinburne 1975). Postharvest dips containing benomyl or 2-aminobutane also are effective (Swinburne 1975). In controlled-atmosphere storage, the optimal concentration of carbon dioxide for decay control and fruit quality is 5–8 percent (Swinburne 1975).

Several other measures are sometimes employed to reduce the severity of European canker. In the British Isles, infection was less when pruning was done in midwinter than at other times (Swinburne 1975). Also, fresh pruning wounds are reported to be adequately protected if treated with an aqueous suspension or paste of thiophanate-methyl. Fungicide pastes containing such chemicals as copper, benomyl, or thiophanate-methyl are used to some extent on cankers. The treatment is made more efficient by scarifying the surface of the canker (Swinburne 1975). Recent experiments with *Bacillus subtilis* and other microorganisms indicate that biological control of both leaf-scar and pruning-wound infection is at least feasible (Corke and Hunter 1979; Swinburne 1975, 1978).

REFERENCES

Brook, P. J., and A. D. Clark. 1975. Comparison of benomyl and captafol with Bordeaux mixture for control of European canker of apple. *N. Z. J. Exp. Agric.* 3:271–72.

Bulit, J. 1957. Contribution à l'étude biologique du *Nectria galligena* Bres. agent du chancre du pommier. *Ann. Epiphytol.* 8:67–89.

Byrde, R. J. W., S. H. Crowdy, and F. A. Roach. 1952. Observations on apple canker. V. Eradicant spraying and canker control. *Ann. Appl. Biol.* 39:581–87.

Cayley, D. M. 1921. Some observations on the life history of *Nectria galligena* Bres. *Ann. Bot. London* 35:79.

Corbin, J. B. 1971. Benomyl beneficial for European canker control. *Orchard. N. Z.* 44:55–57.

Corke, A. T. K., and T. Hunter. 1979. Biocontrol of *Nectria galligena* infection of pruning wounds on apple shoots. *J. Hort. Sci.* 54:47–55.

Corke, A. T. K., and B. Sneh. 1979. Antisporulant activity of chemicals towards fungi causing cankers on apple branches. *Ann. Appl. Biol.* 91:325–30.

Crowdy, S. H. 1949. Observations on apple canker. III. The anatomy of the stem canker. *Ann. Appl. Biol.* 36:483–95.

———. 1952. Observations on apple canker. IV. The infection of leaf scars. *Ann. Appl. Biol.* 39:569–80.

Dubin, H. J., and H. English. 1974a. Factors affecting control of European apple canker by Difolatan and basic copper sulfate. *Phytopathology* 64:300–06.

———. 1974b. Factors affecting apple leaf scar infection by *Nectria galligena* conidia. *Phytopathology* 64:1201–03.

———. 1975a. Epidemiology of European apple canker in California. *Phytopathology* 65:42–50.

———. 1975b. Effects of temperature, relative humidity, and desiccation on germination of *Nectria galligena* conidia. *Mycologia* 47:83–88.

English, W. H., H. J. Dubin, and W. J. Moller. 1972. European canker of apple in California. *Univ. Calif. Agric. Ext.* AXT-n85, 4 pp.

English, H., H. J. Dubin, and F. J. Schick. 1979. Chemical control of European canker of apple. *Plant Dis. Rep.* 63:998–1002.

Fawcett, H. S. 1912. Two apple cankers due to the fungi *Nectria ditissima* and *Nectria cinnabarina*. *Calif. State Comm. Hort. Monthly Bull.* 1:247–49.

Flack, N. J., and T. R. Swinburne. 1977. Host range of *Nectria galligena* Bres. and the pathogenicity of some Northern Ireland isolates. *Trans. Brit. Mycol. Soc.* 68:185–92.

Marsh, R. W. 1939. Observations on apple canker. I. Experiments on the incidence and control of shoot infection. *Ann. Appl. Biol.* 26:458–69.

McCartney, W. O. 1967. An unusual occurrence of eye rot of apple in California due to *Nectria galligena*. *Plant Dis. Rep.* 51:279–81.

Munson, R. G. 1939. Observations on apple canker. I. The discharge and germination of spores of *Nectria galligena* Bres. *Ann. Appl. Biol.* 26:440–69.

Nichols, C. W., and E. E. Wilson. 1956. An outbreak of European canker in California. *Plant Dis. Rep.* 40:952–53.

Petch, J. 1938. British Hypocreales. *Trans. Brit. Mycol. Soc.* 21:243–301.

Sealey, D. A., and T. R. Swinburne. 1976. Protection of pruning wounds on apple trees from *Nectria galligena* Bres. using modified pruning shears. *Plant Pathol.* 25:50–54.

Smith, R. E. 1909. Report of the plant pathologist. *Calif. Univ. Agric. Exp. Stn. Bull.* 203:56.

Swinburne, T. R. 1971. The seasonal release of spores of *Nectria galligena* from apple cankers in Northern Ireland. *Ann. Appl. Biol.* 69:97–104.

———. 1975. European canker of apple (*Nectria galligena*). *Rev. Plant Pathol.* 54:787–99.

———. 1978. The potential value of bacterial antagonists for the control of apple canker. *Ann. Appl. Biol.* 89:94–96.

Wilson, E. E. 1966. Development of European canker in a California apple district. *Plant Dis. Rep.* 50:182–86.

———. 1968. Control of European canker of apple by eradicative and protective fungicides. *Plant Dis. Rep.* 52:227–31.

Wilson, E. E., and C. W. Nichols. 1964. European canker of apple. *Calif. Dept. Agric. Bull.* 53:151–55.

Wiltshire, S. P. 1921. Studies on the apple canker fungus. I. Leaf scar infection. *Ann. Appl. Biol.* 8:182–92.

———. 1922. Studies on the apple canker fungus. II. Canker infection of apple trees through scab wounds. *Ann. Appl. Biol.* 9:275–81.

Wollenweber, H. W. 1913. *Ramularia, Mycosphaerella, Nectria, Calonectria*—Eine morphologisch-pathologisch Studie zur Abgrenzung von Pilzgruppen mit cylindrischen und sichelformigen Konidienformen. *Phytopathology* 3:197–242.

Zagaja, S. W., W. F. Millikan, W. Kaminski, and J. Myszka. 1971. Field resistance to *Nectria* canker in apple. *Plant Dis. Rep.* 55:445–47.

Zeller, S. M. 1926. European canker of pomaceous fruit trees. Oregon State Agric. Exp. Stn. Bull. 222, 52 pp.

Zeller, S. M., and C. E. Owens. 1921. European canker on the Pacific slope. *Phytopathology* 11:464–68.

Fabraea Leaf Spot of Pome Fruit

Diplocarpon mespili (Sorauer) Sutton

Fabraea leaf and fruit spot, reported in the United States as early as 1854, is a common and at times serious disease on pear trees in the eastern United States (Anderson 1956; Sorauer 1908) and has been reported in apple seedling nurseries adjacent to severely affected pear trees (Goldsworthy and Smith 1938b). The disease has also been reported from Canada, New Zealand, Australia, South America, South Africa, Asia, and Europe (Goldsworthy and Smith 1938a). In California, the disease is not of economic importance in commercial pome fruit orchards but occasionally develops in pear nurseries after an unusually wet spring, such as in 1981. Since then the disease has developed in pear nurseries but is not severe enough to require control measures. Other hosts are quince (*Cydonia oblonga*), India-hawthorn (*Raphiolepis indica*), firethorn (*Pyracantha* spp.), hawthorn (*Crataegus* spp.), Contoneaster spp., *Aronia* spp., *Amelanchier* spp., loquat (*Eriobotrya* spp.), medlar (*Mespilus* spp.), *Photinia* spp., and mountain-ash (*Sorbus* spp.) (Anderson 1956; Inman 1962, Pierce and McCain 1983; Strong 1960).

Symptoms. Disease symptoms are found on leaves, stems, and fruit (**fig. 8**). The leaf spot first appears as purplish specks that expand to reddish or brown spots, 1 to 3 mm in diameter, and develop acervuli at their centers. Under optimum environmental conditions, leafspots become visible about four to seven days after infection. Once acervuli are formed (two to four weeks), the cuticle ruptures, exposing the glistening, cream-white, conidial mass. Severely infected leaves become necrotic, and early defoliation occurs. Twig lesions are mostly restricted to current year's growth and rarely persist as cankers into two-year-old wood (Goldsworthy and Smith 1938a). Severely infected fruit may crack.

Causal organism. The perfect stage of the fungus is *Diplocarpon mespili* (Sorauer) Sutton (*Fabraea maculata* (Lév.) Atk. = *D. sorauri* (Kleb.) Nannf.; anamorph = *Entomosporium mespili* (DC.) Sacc.). The perithecia are formed within the overwintering leaves on the soil during late winter to early summer and produce clavate, uniseptate, hyaline, footprint-shaped ascospores. The conidia are composed of four round cells, with two smaller cells connected laterally between two larger ones. Setae, formed on each of three upper cells, give the conidia the insectlike appearance for which the imperfect stage was named. Within an hour's exposure to free moisture, the four-celled conidia dissociate. Each cell can germinate and produce one or more germ tubes (Anderson 1956).

Cultures on potato-dextrose agar grow slowly as erumpent colonies that may extend 3 to 4 mm above the agar surface and produce conidia after 30 to 45 days as creamy-white to yellow masses on the surface of the colony. For growth in culture, the optimum temperature is 22° to 26°C, with a pH of 4.0 to 7.0 (Horie and Kobayashi 1979). Cultures kept on potato-dextrose agar show reduced pathogenicity within three months to a year and lose the ability to produce spores in two years (Rosenberger 1981). The addition of thiamine to potato-dextrose agar increases growth as well as sporulation. Light intensity of 100 lux stimulates sporulation while 500 lux inhibits both growth and sporulation (Van der Zwet and Stroo 1985).

Disease development. In the winter, perithecia with mature ascospores, as well as acervuli and conidia (Klebahn 1918), have been observed on fallen leaves (Atkinson 1909; Ishchenko, Yakovlev, and Dzhigadlo 1983; Piehl and Hildebrand 1936; Plakidas 1941; Sorauer 1908). Goldsworthy and Smith (1938a) and Plakidas (1941) provided evidence that the initial inoculum source was not ascospores but conidia produced in twig cankers. Perithecia were not found on overwintering leaves and only a few conidia were present; disease incidence on trees was not significantly different between trees where the leaves were removed or not removed. In New York, however, Piehl and Hildebrand (1936) considered that the primary inoculum was the ascospores formed in perithecia in overwintering leaves. Overwintered diseased fruit did not form conidia the following spring (Goldsworthy and Smith 1938a). Studies on the source of the primary inoculum have not been made in California, but cankers sometimes can be observed on young diseased nursery stock. This fungus often is present on container-grown nursery stock in garden shops where the plants receive overhead irrigation.

In studies on the effect of temperature and wetting period on infection of Bosc pear leaves by conidia, Rosenberger (1981) showed that minimum wetting periods required for infection were 12, 10, 8, and 8 hours at 10°, 15°, 20°, and 25°C, respectively, and that about seven days' incubation were required for symptom expression; affected leaves began to abcise in 15 days.

Control. Effective fungicides are Bordeaux mixture (Duggar 1898), thiram, ferbam, and captan (Palmiter 1961; Rosenberger 1981), dodine (Bauske 1963), Zineb, captafol, and chlorothalonil (Chandler 1967), and benomyl (Rosenberger 1981; Vonica et al. 1978). Spray applications (timing and number) were related to summer weather only, since host parts, as they age, do not decrease in susceptibility to the leaf-spot organism (Chandler 1967). The first spray treatment is applied immediately after ascospores discharge (Vonica et al. 1978). Cultural control could involve reducing the number of infected leaves on the orchard floor and avoiding overhead watering. California growers would not expect disease problems in the major fruit production areas because of the dry, hot climate during the growing season. In nurseries with overhead sprinklers, both removal of leaves from the field during the winter and fungicide treatments with benomyl or copper after initial infection should assure prevention of epidemics. Resistant cultivars have not been developed for pears, but the inheritance of resistance was shown in seedlings of *Pyrus ussuriensis* and Red Williams cultivar (*P. communis*) (Piehl and Hildebrand 1936; van der Zwet and Keil 1977).

REFERENCES

Anderson, H. W. 1956. Leaf blight and fruit spot of pear and quince. In *Diseases of fruit crops*, 149–53. McGraw-Hill, New York.

Atkinson, G. F. 1909. The perfect stage of leafspot of pear and quince. *Science* 30:770.

Bauske, R. J. 1963. Quince leafspot (*Fabraea maculata*). Fungicide and nematicide tests: Results of 1963. *Am. Phytopathol. Soc.* 19:47.

Chandler, W. A. 1967. Fungicidal control of Fabraea leafspot of pear. *Plant Dis. Rep.* 51:257–61.

Duggar, B. M. 1898. Some important pear diseases. *Cornell Univ. Agric. Exp. Stn. (Ithaca) Bull.* 145:267–99.

Goldsworthy, M. C., and M. A. Smith. 1938a. The comparative importance of leaves and twigs as overwintering infection sources of pear leaf blight pathogen, *Fabraea maculata*. *Phytopathology* 28:574–82.

———. 1938b. An apple leafspot associated with *Fabraea maculata*. *Phytopathology* 28:938.

Horie, H., and T. Kobayashi. 1979. Entomosporium leaf spot of Pomoideae (Rosaceae) in Japan. I. Distribution of the disease; morphology and physiology of the fungus. *Eur. J. For. Pathol.* 9:366–79.

Inman, R. E. 1962. Cycloheximide and the control of hawthorn leaf blight under nursery conditions. *Plant Dis. Rep.* 46:827–30.

Ishchenko, L. A., S. P. Yakovlev, and E. N. Dzhigadlo. 1983. The inheritance of resistance in pear to leaf blight under conditions of natural and artificial infection. *Mikolog. Fitopatolog.* 17(3):218–22.

Klebahn, H. 1918. *Entomopeziza soraueri*. In *Haupt. und Hebenfruchtiformen der Askomyzeten*, 317–44. Ersterteil, Leipzig.

Palmiter, D. H. 1961. Pear and leaf fruit spot. Fungicide and nematicide tests: Results of 1961. *Am. Phytopathol. Soc.* 17:38.

Piehl, A. E., and E. M. Hildebrand. 1936. Growth relations and stages in the life history of *Fabraea maculata* in pure culture. *Am. J. Bot.* 23:663–68.

Pierce, L., and A. H. McCain. 1983. Entomosporium leaf spot. *Calif. Plant Pathol.* 64:3–4.

Plakidas, A. B. 1941. The mode of overwintering of *Entomosporium maculatum* in Louisiana. *Phytopathology* 31:18.

Rosenberger, D. A. 1981. Fabraea leaf spot blight of pear. In *Bad apple* (Proc. Apple and Pear Dis. Workshop) 2(1):81–86.

Sorauer, P. 1908. *Handbuch der Pflanzenkrankheiten*, vol. 2. Paul Parey, Berlin, 237 pp.

Southworth, E. A. 1889. Leaf blight and cracking of pear. U.S. Agric. Dept. Rept. 1888:357–64.

Strong, F. C. 1960. Control of hawthorn leaf blight. Plant Dis. Rep. 44:396–98.

van der Zwet, T., and H. L. Keil. 1977. Resistance to Fabraea leaf spot in *Pyrus* species and *P. calleryana* seedling progenies (abs.). Proc. Am. Phytopathol. Soc. 4:220.

van der Zwet, T., and H. F. Stroo. 1985. Effects of cultural conditions on sporulation, germination and pathogenicity of *Entomosporium maculatum*. Phytopathology 75:94–97.

Vonica, I., M. Olangiu, V. Sevcenco, and N. Minoiu. 1978. [Forecast of preventative treatment and control of quince and pear leaf brown spots caused by *Entomosporium maculatum* Lev.] Prod. Veg. Hort. 27(9):33–36. (Rev. Plant Pathol. 58:3894. 1979.)

Phytophthora Root and Crown Rot of Pome Fruit

Phytophthora spp.

This disease, which is particularly severe in apple orchards located in British Columbia, has also been reported from Europe and New Zealand. In California, Mircetich (1975) and Mircetich, Matheron, and Tyler (1976) reported that the disease is as serious on apples as it is on stone fruit, but on pears the occurrence is sporadic. It is known to occur on loquat in California, but not on quince. The disease may occur at any age of the tree, but rootstocks are most susceptible during the first years of fruit bearing.

In Switzerland as early as 1912, death of young grafted apple trees was attributed to *Phytophthora omnivora*, now considered synonymous with *P. cactorum*. In the United States between 1900 and 1922, collar rot of several apple cultivars was reported in New York, Ohio, Indiana, and Illinois, and the symptoms described were similar to those now known to be induced by *P. cactorum*. The 1933–34 experimental results in Indiana were the first to implicate *P. cactorum* as the causal agent of collar rot and decline or death of apple trees in the United States (Mircetich and Browne 1985). Root and crown rot is a serious disease of apple in New York, where at least five species of *Phytophthora* reportedly are involved (Jeffers and Aldwinkle 1986).

Symptoms. Apple trees with initial infections of *Phytophthora* species are not outwardly distinguishable from healthy trees. However, as the disease progresses, the trees show a lack of terminal growth and have small leaves that are chlorotic and usually drooping; this is followed by defoliation and dieback of terminal shoots. Infected trees may undergo a slow decline or may collapse during the same growing season in which foliage symptoms occur. The aboveground disease symptoms may vary to some extent with geographic region, tree age, or cultivar (Mircetich and Browne 1985).

Although the feeding rootlets of affected pear and apple trees are sometimes black and dead (**fig. 9B**), the most common symptom is an extensive discoloration of the bark tissues at the crown of the tree (**fig. 9C**) (Mircetich 1960; McIntosh and O'Reilly 1963). The newly affected bark shows various shades of light and dark brown and has a marbled appearance. There is usually a well-defined margin between diseased and healthy bark. Diseased trees occur at random throughout the planting, evidence that the causal organism is not transmitted from one affected tree to another.

When young, vigorously growing trees become infected, the leaf midrib develops a purplish-red color that gradually spreads into the entire leaf by midsummer. Mature trees partially girdled with crown rot may show delayed growth the following season on the side affected, with branches having small leaves, small, highly colored fruit, and poor terminal growth.

Death of trees after infection (**fig. 9A**) depends largely on the species of *Phytophthora* involved, the type of rootstock, and the soil, water, and climatic conditions. Generally, trees with crown rot die sooner than trees affected with root rot. Likewise, tree decline and death are more common on young than older trees.

Causal organisms. *Phytophthora cactorum* (Leb. & Cohn) Schroet is generally regarded (Braun and Nienhaus 1959; Jones and Sutton 1984; McIntosh 1964) as the usual causal organism associated with crown rot of apple. McIntosh (1964) found it to be widely distributed in the irrigated soils of the apple-growing areas of British Columbia. It was not found in virgin soils nor in nonirrigated cultivated soils.

Phytophthora cactorum, along with *P. cryptogea* Pethybridge & Lafferty and *P. cambivora* (Petri) Buisman, was found to infect the rootlets of pear, cherry, apricot, and peach (McIntosh 1960, 1964; McIntosh and O'Reilly 1963). In California, Mircetich and coworkers (Mircetich 1975; Mircetich, Matheron, and Tyler 1976) found in their isolations from apples that the most common *Phytophthora* species were *P. cacto*-

rum, *P. cambivora*, *P. megasperma* Dreschler, and *P. dreschleri* Tucker; also present were six unidentified *Phytophthora* species. Planting one-year-old apple seedlings in soil infested by *Phytophthora* species showed *P. cambivora* to be highly pathogenic (fig. 9B).

Both *P. cactorum* and *P. cinnamomi* Rands attack pear in Oregon (Cameron 1962); species attacking pear in California are *P. cambivora* and *P. megasperma*. In experiments in Oregon, Bartlett and Winter Nelis pears proved susceptible, whereas Old Home, Old Home × Farmingdale and *Pyrus calleryana* were not seriously damaged (Cameron 1962). *Phytophthora cactorum* has been found on loquat in California.

Disease development. Factors influencing the occurrence, incidence, and severity of the disease include the *Phytophthora* species present in the soil, the level of soil moisture, the temperature, the relative resistance of the rootstock (Sewell and Wilson 1959), and contamination of the planting stock (Jeffers and Aldwinkle 1988).

On apple seedlings, apple isolates of *P. cambivora* and *P. cryptogea* were more pathogenic than *P. cactorum*, *P. megasperma*, *P. drechsleri*, and four unidentified *Phytophthora* species (Mircetich 1975; Mircetich, Matheron, and Tyler 1976). These *Phytophthora* species are soilborne and can survive for a long period in moist soil, but not in dry soil. A temperature range of 27° to 32°C suppresses the activities of these pathogens, and temperatures over 32°C shorten their survival in the soil. The spread of *Phytophthora* species takes place largely by surface irrigation and runoff water. Flood irrigation with prolonged saturation of the soil favors fungal survival and increases production of sporangia and zoospores, which are very important in local dissemination of the pathogen from diseased to healthy trees within the orchard. In flooded soil, *P. cryptogea* is more likely than *P. cactorum* to cause disease in apple seedlings, whereas in nonflooded soil *P. cambivora* causes more disease than *P. cryptogea* or *P. cactorum* (Browne and Mircetich 1988; Mircetich and Browne 1985).

Control. It is important, when developing a new orchard where the site is free of the *Phytophthora* species pathogenic to apple trees, to avoid introducing the pathogen in infested soil on farm machinery, in infected nursery stock, or in contaminated irrigation and runoff water. Once the pathogen is established eradication is impossible, but soil fumigation with methyl bromide before replanting could temporarily reduce the number of propagules.

The use of rootstocks resistant to *Phytophthora* species is the most economical method of disease control. There are reports of marked differences in resistance of various rootstocks to certain *Phytophthora* species, yet rootstocks resistant in some regions are susceptible in other areas. There is some agreement among researchers that rootstocks MM104 and MM106 are more susceptible to *P. cactorum* than MM111 and M7, and M2, M4, M9, M26, and Antonovka seedling rootstocks are more resistant than MM111 and M7. None of the rootstocks available is immune to infection by *Phytophthora* species. Field resistance of other rootstocks to species of *Phytophthora* is not known.

It is important in the planting operation to ensure that the graft union will not be in contact with the soil, regardless of the rootstock used. For that reason trees in California are commonly planted on mounds above the grade of the orchard.

Most important to good culture after planting is soil-water management, avoiding prolonged and periodic soil saturation or standing water around the base of the trees. Selecting orchard sites with well-drained soil or improving internal and surface water drainage will minimize losses from root and crown rot.

Fungicide applications at the base of the tree have resulted in few successes. Bordeaux mixture and fixed copper fungicides have been used with limited success. Recent studies using metalaxyl (n-(2,6,-dimethylphenyl)-n-(methoxyacetyl) alanine methyl ester) and fosetyl-A1 (aluminum tris-(o-ethyl phosphonate)) in greenhouse tests and limited field trials appear encouraging, but their potential is still under study (Mircetich and Browne 1985; Pinto de T., Carreño, and Alvarez 1988). Biological control using drench applications of the bacterium *Enterobacter aerogenes* significantly increased shoot growth in one plot but not in another orchard site (Utkhede, Li, and Owen 1987).

Details of *Phytophthora* root and crown rots are discussed in Chapter 4.

REFERENCES

Braun, H., and F. Nienhaus. 1959. Fortgeführt Untersuchungen über die Krangenfaule des Apfels (*Phytophthora cactorum*). *Phytopathol. Z.* 36:169–208.

Browne, G. T., and S. M. Mircetich. 1988. Effects of flood duration on the development of Phytophthora root and crown rots of apple. *Phytopathology* 78:846–51.

Cameron, H. R. 1962. Susceptibility of pear roots to *Phytophthora*. *Phytopathology* 52:1295–97.

Fitzpatrick, R. E., F. C. Mellor, and M. F. Welsh. 1944. Crown rot of apple trees in British Columbia—Rootstock and scion resistance trials. *Sci. Agric.* 24:533–41.

Goldhamer, D. A., and R. L. Snyder. 1989. *Irrigation scheduling—A guide for effecient on-farm water management.* University of California, Division of Agriculture and Natural Resources, Publ. 21454. 67 pp.

Houten, J. G. ten. 1958. Resistance trials against collar rot of apples caused by *Phytophthora cactorum*. *Tijdschr. Plantenz.* 65:422–31.

Jeffers, S. N., and H. S. Aldwinkle. 1986. Seasonal variation in extent of colonization of two apple rootstocks by five species of *Phytophthora*. *Plant Dis.* 70:941–45.

———. 1988. Phytophthora crown rot of apple trees: Sources of *Phytophthora cactorum* and *P. cambivora* as primary inoculum. *Phytopathology* 78:328–35.

Jones, A. L., and T. B. Sutton. 1984. Diseases of tree fruits. *N. Central Reg. Ext. Publ.* 45, Coop. Ext. Serv., Mich. State Univ. 59 pp.

McIntosh, D. L. 1960. The infection of pear rootlets by *Phytophthora cactorum*. *Plant Dis. Rep.* 44:262–64.

———. 1964. *Phytophthora* spp. in soils of the Okanagan and Similkameen Valleys of British Columbia. *Can. J. Bot.* 42:1411–15.

McIntosh, D. L., and F. C. Mellor. 1953. Crown rot of apple trees in British Columbia. II. Rootstock and scion resistance trials of apple, pear and stone fruits. *Can. J. Agric. Sci.* 33:615–19.

McIntosh, D. L., and H. J. O'Reilly. 1963. Inducing infection of pear rootlets by *Phytophthora cactorum*. *Phytopathology* 53:1447.

Mircetich, J. M. 1975. *Phytophthora* root and crown rot of orchard trees. *Calif. Plant Pathol.* 24:1–2.

Mircetich, S. M., and G. T. Browne. 1985. *Phytophthora* root and crown rot of apple trees. *Proc. Oregon Hort. Soc. 76th Ann. Rept.* (99th Ann. Mtg., January 29–31, 1985):44–54.

Mircetich, S. M., M. E. Matheron, and R. H. Tyler. 1976. Apple root rot and crown rot caused by *Phytophthora cambivora* and *Phytophthora* sp. (abs.). *Proc. Am. Soc. Phytopathol.* 3:229.

Pinto de T., A., I. Carreño I., and M. Alvarez A. 1988. Effecto de pulverizaciones foliares con Aliette en manzanas con pudricion del tronco y raices, causada por *Phytophthora* spp. *Agric. Tech.* (Chile) 48(4):331–33.

Sewell, G. W. F., and J. F. Wilson. 1959. Resistance trials of some apple rootstock varieties to *Phytophthora cactorum* (L. & C.) Schroet. *J. Hort. Sci.* 34:51–58.

Smith, H. C. 1955. Collar-rot and crown rot of apple trees. *Orchard. N. Z.* 28:16.

Utkhede, R., T. Li, and G. Owen. 1987. Chemical and biological control 1. Replant problem: Apple. *Res. Repts. 61st Ann. Western Orchard Pest Dis. Mgmt. Conf.* (Portland, Ore., Jan. 14–16, 1987):5.

Waterhouse, G. M. 1963. Key to species of *Phytophthora* de Bary. *Commonwealth Mycol. Inst. Mycol. Paper* 92.

Powdery Mildew of Pome Fruit

Podosphaera leucotricha (E. & E.) Salmon

While powdery mildew of apple occurs to a greater or lesser extent wherever apples are grown, it is a major disease of orchard trees in semiarid regions. In California it is serious on Jonathan and Gravenstein, and it is destructive in central Washington where summers are almost rainless. It is quite serious on nursery trees in California. The disease was first reported in the United States about 1877, and the fungus was named in 1888.

Considerable differences in susceptibility exist among apple cultivars; Jonathan, Rome Beauty, Gravenstein, and Granny Smith are among the most susceptible. Golden Delicious, Winesap, and Stayman Winesap are moderately susceptible, while Delicious and its red mutants are moderately resistant and suffer damage only under conditions of very high inoculum pressure. Pear (**fig. 10D**) (Spotts 1984) and quince are also attacked by the mildew. In the d'Anjou pear cultivar the foliage is only moderately susceptible to infection, but the fruit is highly susceptible. The discussion of powdery mildew here relates primarily to its occurrence on apple.

Symptoms. Leaves, flowers, green shoots, and fruit are attacked. On leaves, lesions first appear as whitish feltlike patches along the margin, commonly on the underside. The fungus later spreads over the entire leaf blade, down the leaf petiole, and over the new shoot (**fig. 10A, B**). Infected leaves are narrower than normal, folded longitudinally, or curled, crinkled, and stiff. When infection is severe, many leaves fall during the summer. With the more resistant cultivars such as Winesap and Delicious, secondary infection results in lesions that are at first ill-defined pale spots with reddish or lavender borders.

On infected blossoms the white mycelial mat may cover the petals, sepals, receptacles, and peduncles. Blossom infection is less common than infection of leaves and green shoots, but is important because infected blossoms fail to set fruit. The fruit are often attacked when young but seldom after they have reached some size. Infected fruit may be stunted and their surface severely russeted (**fig. 10C**).

Terminal or leafy shoots are stunted by the fungus and often killed outright. Those that survive the summer may be killed during fall or winter. Infected shoots that survive the winter and infected buds act as a source of inoculum for the coming season.

Causal organism. The most common powdery mildew fungus on apple is *Podosphaera leucotricha* (E. & E.) Salm. *Podosphaera oxycanthae* also occurs occasionally on apple. *Podosphaera leucotricha* was described in 1888 by Ellis and Everhart as *Sphaerotheca leucotricha*. It had previously been named *Podosphaera kunzei* and later *S. mali*, but in 1900 Salmon gave it its present name.

Like other powdery mildew fungi, *P. leucotricha* produces a white, wefty, loose layer of mycelium on the surface of the host. The fungus obtains nutriment by means of haustoria, which are sent into the epidermal cells of the host. Conidiophores arising from the mycelium produce chains of ellipsoid-to-barrel-shaped conidia 28 to 30 μm long and 12 μm broad.

According to Coyier (1968), conidia are very sensitive to temperature even at 97 percent relative humidity. They failed to germinate in 6 hours at 4.3° or 10° C but germinated well in 24 hours. Exposure to a temperature of 32.2°C for 6 hours resulted in very limited germination, and in 24 hours there was only 10 percent germination. At a temperature of 21°C, 70 percent germination occurred in 6 hours and 88 percent in 24 hours. Diurnal conidial dispersal pattern peaked in early afternoon with a distinct subsidiary peak after dark, and correlated positively with wind velocity, temperature, and solar radiation, and negatively with relative humidity and leaf wetness (Sutton and Jones 1979). Mycelial growth also was best at 21°C but was still high at 15.6° and 26.7°C.

Cleistothecia produced on twigs and those that occasionally form on fruit and peduncles are gregarious, 75 to 96 μm in diameter, and subglobose. Appendages are of two kinds: 3 to 11 erect, apical appendages with undivided ends, or sometimes dicotomously branched one or two times; basal appendates which are short, tortuous, pale brown, and simple or irregularly branched.

The asci contain eight spores which are 22 to 26 μm long and 12 to 14 μm wide. Coyier (1974) showed heterothallism in mildew cultures collected in Oregon. The shortest interval for cleistothecial production was 24 days and the longest 109 days. Cleistothecia are not common in California, but in 1974 and 1975 they were detected in trees at Davis, and since then they have occurred yearly on diseased flowering apple and Rome Beauty trees.

Disease development. The fungus overwinters in leaf and flower buds infected during the summer. It is located on the apices of the inner bud scales and attacks leaves as they emerge in the spring. The older leaves are less susceptible. Susceptibility of leaves was suggested as being dependent on the amount of kinetin in the tissue (Kirkham, Hignett, and Ormerod 1974). Mildew infection significantly reduces photosynthesis and transpiration of apple leaves (Ellis, Ferree, and Spring 1981). In colder climates infected leaf buds are sometimes killed by low winter temperatures (Spotts and Chen 1984).

Even though Woodward (1927) claimed that secondary infection does not occur, most subsequent workers (Burchill 1960; Petherbridge and Dillon-Weston 1928–29; Seem and Gilpatrick 1980) considered it to be an important factor in disease development and spread. In any event, fruit infection occurs during the bloom period (Daines et al. 1984) and thus is secondary. Coyier (1968) showed a reduction in conidial germination with increases in temperature, and showed that a relationship exists between seasonal increase in temperature and a reduction in the severity of mildew (Coyier 1971). Working in a milder climate, Burchill (1960) related the severity of leaf infection only to subsequent infection of the growing shoot apex. Lalancette and Hickey (1985) reported that leaf aging has a deleterious effect on established mildew colonies.

The principal inoculum for infection is the conidia. Most workers believe that ascospores play little or no role in the disease cycle.

Control. Early work on control of apple mildew was done by Ballard and Volk (1914) in the Pajaro Valley of California, and by Fisher at Wenatchee, Washington. These workers developed spray programs based on sulfur. Sprague (1953, 1955a, b) developed spray schedules for central Washington that took into account the sulfur tolerance of various cultivars. He also pointed out the need for an effective wetting agent when using dinocap.

Byrde and coworkers (1963, 1965) in England made extensive tests of organic fungicides, including dinocap, the benzoates, and dimethyl acrylates of eight phenols, against apple powdery mildew. Butt (1971), also in England, found that high-volume sprays of dinocap (0.019 percent) controlled powdery mildew on Cox's Orange Pippin during the current year, and reduced blossom infection the following year. Ward (1965) in Australia reported that a new fungicide, binapacryl (dinitro-alkyl-phenyl-acrylate), at 1 and 2 lb per 100 gallons of water gave better control than dinocap.

Delp and Klopping (1968) reported that benomyl acts as a preventive, is resistant to rainfall, and eradicates the fungus in leaves up to four days after infection. Systemic uptake of benomyl sufficient to control mildew on one-year-old apple trees in clay pots occurred when the soil was drenched with benomyl (Cimanowski, Masternak, and Millikan 1970). On older trees, however, control was not possible from soil applications.

On susceptible cultivars a typical spray recommendation could be as follows: A prepink application of benomyl or dinocap followed by one at the pink (prebloom) stage, then one of the same materials as a calyx application, followed by a first cover spray 15 days after petal fall and a second cover spray 15 or more days later. To protect buds from infection, Butt (1972) recommended repeated sprays at 10-day intervals from postbloom until new leaf production ceases. In central Washington, effective control of fruit russet is obtained with just pink and petal fall sprays of dinocap (R. P. Covey, personal communication). Daines et al. (1984) also stressed the importance of pink-bud sprays for controlling fruit infection.

For mildew control, sprays should be applied with reference to new leaf development. Usually, however, the sprays are timed for control of both mildew and scab. In California, the first spray is applied at the pink-bud stage, the second at petal-fall, and cover sprays are then done as needed. The chemicals used are benomyl, sulfurs, and dinocap.

The most promising new fungicides for the control of pome powdery mildew are the broad-spectrum, sterol-inhibiting compounds (bitertanol, etaconazole, fenarimol, triadimefon, etc.) (Yoder and Hickey 1983). These materials have been tested extensively in both eastern (Berkett, Hickey, and Cole 1988; Cimanowski and Szkolnik 1985; Hickey and Yoder 1981) and western states (Daines et al. 1984; Spotts and Cervantes 1986; Spotts, Covey, and MacSwan 1981), and some have been superior in mildew control to currently employed fungicides. Because some of these compounds are also effective against several other apple diseases, their increased use in pome-fruit disease-management programs appears highly probable unless pathogen resistance problems emerge (Delp 1988).

Eradication of mildew in the buds was shown to occur with the use of 4,6-dinitro-o-cresol (DNOC) in 3 percent petroleum oil at bud burst (Burchill 1960). Eradication also has been achieved with dormant sprays containing methyl esters of fatty acids or other surface-active compounds (Burchill et al. 1979; Frick and Burchill 1972; Hislop and Clifford 1976). In part because of phytotoxicity these treatments have not come into commercial use; they have not been tested in California. Another control strategy that is used to some extent is the removal of infected terminal shoots during the winter pruning operation (Burchill 1960; Sprague 1953; Yoder and Hickey 1983). By reducing inoculum pressure, this practice enhances the efficacy of chemical control measures.

REFERENCES

Ballard, W. S., and W. H. Volk. 1914. Apple powdery mildew and its control in the Pajaro Valley. *U.S. Dept. Agric. Bull.* 120, 26 pp.

Bennett, M., and M. H. Moore. 1963. The relative merits of DNOC/petroleum spray in the field control of apple mildew. *Rept. E. Malling Res. Stn.* 1962: 105–08.

Berkett, L. P., K. D. Hickey, and H. Cole, Jr. 1988. Relation of application timing to efficacy of triadimefon in controlling apple powdery mildew. *Plant Dis.* 72:310–13.

Berwith, C. E. 1936. Apple powdery mildew. *Phytopathology* 26:1071–73.

Burchill, R. T. 1960. The role of secondary infections in spread of apple powdery mildew, *Podosphaera leucotricha* (Ell. & Ev.) Salm. *J. Hort. Sci.* 35:66–72.

Burchill, R. T., E. L. Frick, M. E. Cook, and A. A. J. Swait. 1979. Fungitoxic and phytotoxic effect of some surface-active agents applied for control of apple powdery mildew. *Ann. Appl. Biol.* 91:41–49.

Butt, D. J. 1971. The role of the apple spray programme in the protection of fruit buds from powdery mildew. *Ann. Appl. Biol.* 68(2):149–57.

———. 1972. The timing of sprays for the protection of terminal buds on apple shoots from powdery mildew. *Ann. Appl. Biol.* 72:239–48.

Byrde, R. J. W., G. M. Clark, and C. W. Harper. 1963. Spraying experiments against apple mildew and apple scab at Long Ashton, 1963. *Rept. Agric. Hort. Res. Stn. Bristol* 1962:98–102.

Byrde, R. J. W., D. R. Clifford, E. D. Evans, and D. Woodcock. 1965. Potential fungicides for apple mildew. Progress report 1964. *Rept. Agric. Hort. Res. Stn. Bristol* 1964:135–42.

Cimanowski, J., A. Masternak, and D. F. Millikan. 1970. Effectiveness of benomyl for controlling apple powdery mildew and cherry leaf spot in Poland. *Plant Dis. Rep.* 54:81–83.

Cimanowski, J., and M. Szkolnik. 1985. Postsymptom activity of ergosterol inhibitors against apple powdery mildew. *Plant Dis.* 69:562–63.

Covey, R. P. 1971. The effect of methionine in the development of apple powdery mildew. *Phytopathology* 61:346.

Coyier, D. L. 1968. Effect of temperature on germination of *Podosphaera leucotricha* conidia (abs.). *Phytopathology* 58:1047–48.

———. 1971. Control of powdery mildew on apples with various fungicides as influenced by seasonal temperature. *Plant Dis. Rep.* 55:263–66.

———. 1974. Heterothallism in the apple powdery mildew fungus, *Podosphaera leucotricha*. *Phytopathology* 64:246–48.

Daines, R., D. J. Webster, E. D. Bunderson, and T. Roper. 1984. Effect of early sprays on control of powdery mildew fruit russet on apples. *Plant Dis.* 68:326–28.

Delp, C. J., ed. 1988. *Fungicide resistance in North America*. APS Press, The American Phytopathological Society, St. Paul, Minnesota. 133 pp.

Delp, C. J., and H. L. Klopping. 1968. Performance attributes of a new fungicide and mite ovicide candidate. *Plant Dis. Rep.* 52:95–99.

Ellis, J. B., and B. M. Everhart. 1888. New species of fungi from various localities. *J. Mycol.* 4:49–59.

Ellis, M. A., D. C. Ferree, and D. E. Spring. 1981. Photosynthesis, transpiration, and carbohydrate content of apple leaves infected by *Podosphaera leucotricha*. *Phytopathology* 71:392–95.

Fisher, D. F. 1928. Control of apple powdery mildew. *USDA Farmer's Bull.* 1120, 14 pp.

Frick, E. L., and B. T. Burchill. 1972. Eradication of apple powdery mildew from infected buds. *Plant Dis. Rep.* 56:770–72.

Hickey, K. D., and K. S. Yoder. 1981. Field performance of sterol-inhibiting fungicides against apple powdery mildew in the mid-Atlantic apple growing region. *Plant Dis.* 65:1002–06.

Hislop, E. C., and D. R. Clifford. 1976. Eradication of apple powdery mildew (*Podosphaera leucotricha*) with dormant sprays of surface-active ingredients. *Ann. Appl. Biol.* 82:557–68.

Kirkham, D. S., R. C. Hignett, and P. J. Ormerod. 1974. Effects of interrupted light on plant disease. *Nature* 247:158–60.

Lalancette, N., Jr., and K. D. Hickey. 1985. Apple powdery mildew disease progress on sections of shoot growth: An analysis of leaf maturation and fungicide effects. *Phytopathology* 75:130–34.

Lawrence, W. H. 1905. The powdery mildews of Washington. *Coll. Agric. Exp. Stn. Bull.* 70, 16 pp.

Petherbridge, F. R., and W. A. R. Dillon-Weston. 1928–29. Observations on the spread of the apple mildew fungus *Podosphaera leucotricha* (Ell. & Ev.) Salm. *Trans. Brit. Mycol. Soc.* 14:109–11.

Salmon, E. S. 1900. *A monograph of the Erysiphaceae*. Torrey Botan. Club Memoir 9, 292 pp.

Seem, R. C., and J. D. Gilpatrick. 1980. Incidence and severity relationships of secondary infection of powdery mildew on apple. *Phytopathology* 70:851–54.

Spotts, R. A. 1984. Infection of Anjou pear fruit by *Podosphaera leucotricha*. *Plant Dis.* 68:857–59.

Spotts, R. A., and L. A. Cervantes. 1986. Effects of fungicides that inhibit ergosterol biosynthesis on apple powdery mildew control, yield, and fruit growth factors. *Plant Dis.* 70:305–06.

Spotts, R. A., and P. M. Chen. 1984. Cold hardiness and temperature responses of healthy and mildew-infected terminal buds of apple during dormancy. *Phytopathology* 74:542–44.

Spotts, R. A., R. P. Covey, and I. C. MacSwan. 1981. Apple powdery mildew and scab control studies in the Pacific Northwest. *Plant Dis.* 65:1006–09.

Sprague, R. 1953. Powdery mildew of apples. In *Plant diseases*, 667–70. USDA Yearbook 1953.

———. 1955a. Study of apple powdery mildew. *Western Fruit Grower* 9(3):17.

———. 1955b. Fungus disease control. *Western Fruit Grower* 9(4):34–36.

Sutton, T. B., and A. L. Jones. 1979. Analysis of factors affecting dispersal of *Podosphaera leucotricha* conidia. *Phytopathology* 69:380–83.

Ward, J. R. 1965. Binapacryl: A new fungicide for the control of apple powdery mildew (*Podosphaera leucotricha*). *Austral. J. Exp. Agric. Anim. Husb.* 4(12):52–54.

Wieneke, J., R. P. Covey, and N. Benson. 1971. Influence of powdery mildew infection on ^{35}S and ^{45}Ca accumulations in leaves of appled seedlings. *Phytopathology* 61:1099–1103.

Woodward, R. C. 1927. Studies on *Podosphaera leucotricha* (Ell. & Ev.) Salm. I. The mode of perennation. *Trans. Brit. Mycol. Soc.* 12:173–204.

Yarwood, C. E., and L. Jacobsen. 1955. Accumulation of chemicals in diseased areas of leaves. *Phytopathology* 45:43–48.

Yoder, K. S., and K. D. Hickey. 1983. Control of apple powdery mildew in the mid-Atlantic region. *Plant Dis.* 67:245–48.

Rusts of Pome Fruit

Gymnosporangium spp.

Some ten species of rust fungi attack pome fruit, but only the four that occur in the western states will be discussed here. Two of these, *Gymnosporangium fuscum* and *G. libocedri*, are present in California, and *G. kernianum* and *G. nelsoni* occur in several other western states. *Gymnosporangium juniperi-virginianae*, important on apple in the eastern and central states (Aldwinckle, Pearson, and Seem 1980, Pearson, Seem, and Meyer 1980) is not known to occur in the western United States.

Pear trellis rust. This disease, caused by *G. fuscum*, is considered to be serious in Europe (Metzler 1982; Mijuskovic 1979; Vukovits 1979), Asia Minor (Dinc and Karaca 1975), and North Africa. It was first reported in North America in Victoria, British Columbia, in 1961 (Ziller) and in Lafayette, California, in the same year (McCain 1961; McCain and Rosenberg 1961). Although it was thought at one time to have been eradicated from the British Columbia mainland (Creelman 1966), it evidently was not eradicated and continues to cause trouble in that region (Hunt and O'Reilly 1978; Ormrod et al. 1984; Ziller 1974). In California, extensive control programs have limited spread of the disease, and today it is found only occasionally on juniper in the San Francisco Bay Area.

Damage is confined to leaves, fruit, twigs, and branches of pear. No appreciable damage is shown on the alternate host, *Juniperus sabina* var. *tamariscifolia*, known to the nursery industry as "Spanish," "Tam," and "Tamarix" junipers. The host for the pycnial and aecial (roestelia) stages are the common commercial cultivars of pears as well as the commonly used rootstocks. Telial hosts are the *Juniperus* species commonly found as landscaping plants. The telial stage of the fungus has sometimes been identified on juniper nursery stock entering the United States and Canada (McCain 1961; Ziller 1961).

The causal fungus, *Gymnosporangium fuscum* DC., is described by Ziller as follows: Pycnia and aecia develop on yellow to bright-orange tuberculate leaf spots (**fig. 11**), on swollen twigs, and on pear fruit. Pycnia are 170 to 190 μm wide by 150 to 170 μm high. Aecia are acorn-shaped (balanoid), rupturing along the sides, the apex remaining conic, 0.5 to 1 mm in diameter and up to 6 mm high; sidewalls of peridial cells verrucose; aeciospores 20 to 28 μm wide and 25 to 35 μm long and light chestnut brown. Telia are conic or laterally compressed, dark brown, up to 5 mm wide and 5 mm high, on spindle-shaped enlargements of the twigs and branches of juniper (**fig. 11**).

In spring, basidiospores are windborne and may infect young pear leaves of trees within a radius of 1,000 feet from the juniper host. Within 13 to 17 days after infection, pycnia appear as yellow blotches on the upper leaf surface. Later in the summer aecial horns develop on the lower leaf surface, and aeciospores are disseminated from these in late summer and fall. Aeciospores infect the juniper but not the pear. Unlike basidiospores, aeciospores are relatively resistant to temperature and moisture changes and may be carried for miles by wind currents. Infected areas on juniper remain inconspicuous and may escape detection for many months until the first telial horns develop in the second spring. Infected areas may subsequently produce a new crop of telial horns every spring without apparent damage to the juniper host. Thus the fungus requires a two-year period to complete its life cycle. In some areas the fungus can overwinter in hypertrophied pear branches and produce a crop of aeciospores the second year after infection (Hunt and O'Reilly 1978). However, it generally does not survive into the third year.

Eradication of the junipers is considered the only practical method of control in a recently infected area. Copper and carbamate fungicides such as ferbam and maneb (Ziller 1961) can act as protectants. Other effective fungicides are mancozeb and triforine (Mijuskovic 1979; Ormrod et al. 1984; Vukovits 1979).

Pacific Coast pear rust. This rust has been reported in Oregon and northern California (Dietel 1895; Jackson 1914; O'Gara 1914). Although pear, quince, and apple are attacked, the disease has been most serious on pear. Pear fruit are malformed while still young and drop from the tree. Yellowish spots with numerous cup-shaped aecia at their center develop over the surface of affected fruit. Green shoots and leaves are also attacked but not so frequently as fruit.

The aecial stage also occurs on *Pyrus fusca* Doug. (*Malus rivularius* Roem.), a flowering crab known to nursery operators as *Malus floribundus* Sieb., and on *Sorbus sambucifolia* Roem., *S. spuria* Pero., *Amelanchier alnifolia* Nutt., *A. pallida* Greene, *Chaenomeles japonica* (Thunb.) Lindl., and *Crataegus douglasii* Lindl.

The telial host of the fungus is *Libocedrus decurrens* Torr., on the leaves of which the fungus produces small dark pustules but no marked distortion. A witches'-broom condition, however, commonly develops on affected *Libocedrus* trees.

The causal fungus is now known as *Gymnosporangium libocedri* (P. Henn.) Kern, and the following binomials are considered to be synonymous: *Phragmidium libocedri* P. Henn., *Aecidium blasdaleanum* Diet. & Holw., *Gymnosporangium blasdaleanum* (Diet & Holw.) Kern. and *Gynotelium blasdaleanum* (Diet. & Holw.) Arth.

Uredia are wanting. Reddish brown pulvinate telia develop chiefly on leaves and green stems of *L. decurrens*. Teliospores are brown, linear-oblong, two- to five-celled, 19 to 30 μm broad by 35 to 87 μm long, and blunt at both ends. Cup-shaped aecia are produced on the fruit, leaves, and green stems of the host. Aeciospores are globoid 12 to 20 μm wide and 14 to 32 μm long.

Both Oriental and European cultivars of pears are affected (O'Gara 1914). Winter Nelis is severely attacked, but Bartlett is said to be little affected (Jackson 1915).

The fungus attacks fruit and occasionally the leaves. Jackson (1914) believed that infection occurs at blossoming time. Symptoms develop two to four weeks after infection and aecia mature about a week later. By July, after affected fruit have fallen, it is difficult to find the disease in the orchard.

Little has been published on control of this disease. Jackson (1915) suggested the application of Bordeaux mixture as pear flowers emerge from winter buds. Destruction of the alternate host should reduce the disease, although such a procedure would not be practical in some areas.

Kern's pear rust. The aecial stage of *Gymnosporangium kernianum* Bethel (Bethel 1911) occurs on pears in certain western states. Another aecial host is the serviceberry, *Amelanchier alnifolia*. The geographic range of this fungus is from Idaho and Oregon to New Mexico and Arizona. The telial stage of the fungus occurs on *Juniperus utahensis* (Englem.) Lemm., *J. occidentalis* Hook., and *J. pachyphlea* Torr. On *J. utahensis* it produces a compact, globose witches'-broom, 5 to 60 cm. in diameter. The dark-brown telial sori are hemispherical, scattered, and solitary. Teliospores are two-celled, narrowly ellipsoid, 21 to 26 μm broad by 55 to 74 μm long, slightly constricted at the septum, yellowish, and smooth. The pedicels are long, hyaline, and cylindrical.

The incospicuous aecia are cylindrical and 2 to 2.5 mm high. The cinnamon-brown aeciospores are globose and 21 to 32 μm in diameter. As with other species of *Gymnosporangium*, the uredial stage has not been found.

Rocky Mountain pear rust. The rust fungus *Gymnosporangium nelsoni* Arth. (*G. durum* Kern.) (Kern 1907) occurs in the Rocky Mountain states, producing its telial stage on several species of juniper, including *Juniperus utahensis* and *J. scopulorum*, and its aecial stage on the native serviceberries, *Amelanchier* spp.; mountain ashes, *Sorbus* spp.; hawthorn, *Crataegus* sp.; and wild crabapple, *Malus* sp. (Arthur 1934; Weier and Hubert 1917). Uredia are wanting.

On pear leaves and fruit the aecia arise from the surfaces of yellowish lessions. The peridial walls are cylindrical, 2 to 4 mm high, splitting at the apex and more or less down the sides. The aeciospores are globoid, 19 to 26 μm broad by 21 to 29 μm long, with chestnut-brown, finely verrucose walls pierced by six to eight pores. On juniper, the irregularly flattened telial horns arise from woody globose galls that range from 6 to 12.5 mm in diameter. The teliospores are ellipsoid, more or less pointed at each end, 18 to 22 μm broad by 50 to 65 μm long, and have cinnamon-brown walls. The pedicels are long and cylindrical.

REFERENCES

Aldwinckle, H. S., R. C. Pearson, and R. C. Seem. 1980. Infection periods of *Gymnosporangium juniperi-virginianae* on apple. *Phytopathology* 70:1070–73.

Arthur, J. C. 1934. *Manual of rusts in the United States and Canada*. Science Press, Lancaster, Pa., 376 pp.

Bethel, Ellsworth. 1911. Notes on some species of *Gymnosporangium* in Colorado. *Mycologia* 3:156–60.

Creelman, D. W. 1966. Summary of the prevalence of plant diseases in Canada in 1965. *Can. Plant Dis. Survey* 46(2):37–79.

Dietel, P. 1895. New North American *Uredineae*. *Erythea* 3:77.

Dinc, N., and I. Karaca. 1975. Researches on pear rust disease (*Gymnosporangium fuscum* DC.) in Elazig and Malatya. *J. Turk. Phytopathol.* 4(2):63–73.

Hunt, R. S., and H. J. O'Reilly. 1978. Overwintering of pear trellis rust in pears. *Plant Dis. Rep.* 62:659–60.

Jackson, H. S. 1914. A new pomaceous rust of economic importance, *Gymnosporangium blasdaleanum*. *Phytopathology* 4:261–70.

———. 1915. A Pacific Coast rust attacking pears, quince, etc. *Oregon Agric. Exp. Stn. Second Bien. Crop. Pest Hort. Rept.* 1913–14:202–12.

Kern, F. D. 1907. New Western species of *Gymnosporangium* and *Roestelia*. *Bull. Torrey Bot. Club* 34:459–63.

———. 1908. Studies on the genus *Gymnosporangium* and *Roestelia*. *Bull. Torrey Bot. Club* 35:499–571.

McCain, A. H. 1961. *Gymnosporangium fuscum* found on pears in California. *Plant Dis. Rep.* 45:151.

McCain, A. H., and D. Y. Rosenberg. 1961. Pear-juniper rust, a disease new to California and the United States. *Calif. Dept. Agric. Bull.* 50(1):13–19.

Metzler, B. 1982. Untersuchungen an Heterobasidiomyceten (23): Basidiosporenkeimung und Infektionsvorgang beim Birnengitterrost. *Phytopathol. Z.* 103:126–38.

Mijuskovic, M. 1979. [The importance of *Gymnosporangium sabinae* (Dicks.) Wint. as the parasite of pear in southern parts of Montenegro and the possibility of its control.] *Zbor. Radova Kupari* 1:353–63. (*Rev. Plant Path.* 60:4555. 1981.)

O'Gara, P. J. 1914. A rust new on apples, pears and other pome fruits. *Science* 39:620–21.

Ormrod, D. J., H. J. O'Reilly, B. J. VanderKamp, and C. Borno. 1984. Epidemiology, cultivar susceptibility, and chemical control of *Gymnosporangium fuscum* in British Columbia. *Can. J. Plant Pathol.* 6:63–70.

Pearson, R. C., R. C. Seem, and F. W. Meyer. 1980. Environmental factors influencing the discharge of basidiospores of *Gymnosporangium juniperi-virginianae*. *Phytopathology* 70:262–66.

Smith, R. E. 1941. Diseases of fruits and nuts. *California Agr. Ext. Serv. Cir.* 120, 168 pp.

Vukovits, G. 1979. Birnengitterrost—Eine Gefahr für den Obstbau. *Pflanzenarzt* 32(5):47–48.

Weier, J. R., and E. E. Hubert. 1917. Recent cultures of forest tree rusts. *Phytopathology* 7:106–09.

Ziller, W. G. 1961. Pear rust (*Gymnosporangium fuscum*) in North America. *Plant Dis. Rep.* 45:90–94.

———. 1974. *The tree rusts of Western Canada*. Can. Forest Serv. Publ. 1329, 272 pp.

Sappy Bark of Apple

Trametes versicolor (L.:Fr.) Pil.

The disease known as "papery bark" and "sappy bark" is occasionally troublesome in apple trees growing in the Pajaro Valley, the San Joaquin Valley, and Sonoma County of California. According to Wormald (1955), the disease was prevalent early in this century in Australia and Tasmania, but is unknown in Great Britain. The causal fungus, however, is found in England on dead stumps of orchard trees. Hotson (1920) reported it to be associated with a collar rot disease of apple in central Washington. During the last 15 years, the disease has again attracted attention in Tasmania, where it is regarded as the most serious disorder affecting the structural part of apple trees (Darbyshire, Wade, and Marshall 1969; Kile 1976; Kile and Wade 1974, 1975; Wade 1968). Reports in central Washington in 1981 indicated that the causal fungus can be troublesome in both young and mature apple trees (Covey et al. 1981; Dilley and Covey 1981). There also is evidence of its occurrence in Oregon (Dilley and Covey 1981).

Symptoms. The causal fungus usually infects the larger branches, causing areas of the bark to become discolored, spongy, and moist and the underlying wood to become discolored. The affected areas, which commonly occur at pruning cuts (**fig. 12A**), sometimes exude a dark sap. The papery bark (phellem) peels away, exposing a disorganized spongy cortex and phloem. Papery bark (**fig. 12B**) is a characteristic but not specific symptom of the disease (Kile and Wade 1974). In apple, the fungus appears to be confined to the sapwood (Kile and Wade 1974) where it causes a soft, white, spongy rot. A narrow black zone is always present at the margin of the decayed wood (Dilley and Covey 1981; Kile and Wade 1974, 1975). The fungus causes a sapwood rot of many hardwood species and is also reported in several pome- and stone-fruit species. The small, leathery brackets (sporophores) of the pathogen develop on the dead bark, especially in fall and winter (Kile 1976). Severely affected branches develop dieback. Young trees infected at the pruning cut made above the scion bud are sometimes killed (Covey et al. 1981; Kile and Wade 1974).

Causal organism. The basidiomycetous fungus causing the paper bark-sapwood-rot syndrome is *Trametes versicolor* (L.:Fr.) Pil. Names previously given to this organism are *Polystictus versicolor* Fr., *Polyporus versicolor* L. ex Fr., and *Coriolus versicolor* (L. ex Fr.) Quél. On killed areas of the tree, the fungus produces thin, tough, leathery, densely imbricate annual conks (sporophores) up to 5 cm. wide, and sometimes wider (**fig. 12C**). They have a hairy or velvety upper surface varying in color from white through brown to grayish black, with more or less conspicuous multicolored zones.

The under surface of the sporophores consists of a light-colored layer of minute pores or tubes within which the basidia and basidiospores are formed. The basidiospores are allantoid, smooth, hyaline, and measure 4 to 6 by 1 to 2 μm. They are forcibly abjected from the basidia, fall out through the pores, and are disseminated by the wind.

Disease development. Airborne spores from sites within or external to the orchard or nursery initiate tree infection most commonly through pruning wounds. The fungus invades living sapwood, but apparently not dead sapwood or heartwood (Kile and Wade 1974). According to Smith (1924) the fungus attacks the cellulose and hemicellulose of the xylem elements. Sapwood invasion occurs throughout the year, but is more rapid in summer than in fall or winter (Kile 1976). The papery-bark symptom develops in fall and winter (Kile and Wade 1974). The dark zone formed at the margin of the decayed wood contains oxidized polyphenols and polysaccharide gum (Kile and Wade 1975). Unspecified bacteria, whose function, if any, in the disease syndrome is unknown, are present in this discolored zone. Although disease susceptibility appears to increase with tree age (Kile 1976; Kile and Wade 1975), under some conditions very young trees can become infected and killed (Covey et al. 1981). Fresh pruning wounds are highly susceptible to infection, but after 30 days they become highly resistant (Kile 1976). It is not known whether old wounds are resistant because of physiological changes or colonization by saprophytic organisms.

Wade (1968) reported that trees in sand culture that was deficient in phosphorus were highly susceptible to infection, whereas those with deficient magnesium, potassium, or calcium were less susceptible. Trees deficient in nitrogen or receiving complete nutrients developed no visible symptoms following inoculation. However, Darbyshire, Wade, and Marshall (1969), in an orchard survey in Tasmania, failed to establish any correlation between the phosphorus, potassium, and calcium levels of new wood and the incidence of dieback. They did find a correlation between polyphenol oxidase production and wood decay, and found that the formation of this enzyme was suppressed at high sugar levels. They concluded that "dieback of apple trees due to *T. (Coriolus) versicolor* is a low sugar disease." Kile and Wade (1974) reported that some pruning-wound infections resulted in only limited decay of the sapwood, but the mechanism of resistance was not investigated.

Control. Little information is available on the control of this sappybark and dieback disorder. Where the disease is troublesome the following measures are suggested: (1) Good site selection and orchard management should be followed to maintain tree vigor. (2) Pruning cuts on framework branches should be made in late summer or early fall and treated with wound protectants (Gendle, Clifford, Mercer, and Kirk 1983) when no inoculum is present. (3) Dead branches should be removed by cutting well below any evidence of infection; they should be destroyed to prevent the formation of new inoculum.

Trichoderma viride has protected wood blocks from decay by *T. versicolor* (Hulme 1970), but no one has determined whether biocontrol would be effective in the orchard. Observations in California indicate that topworked trees under overhead irrigation develop considerably more infection that those irrigated by other means.

Further information on the wood rots of fruit trees is given in Chapter 4.

REFERENCES

Covey, R. P., H. J. Larsen, T. J. Fitzgerald, and M. A. Dilley. 1981. *Coriolus versicolor* infection of young apple trees in Washington state. *Plant Dis.* 65:280.

Darbyshire, B., G. C. Wade, and K. C. Marshall. 1969. In vitro studies of the role of nitrogen and sugars on the susceptibility of apple wood to decay by *Trametes versicolor*. *Phytopathology* 59:98–102.

Dilley, M. A., and R. P. Covey. 1981. Association of *Coriolus versicolor* with a dieback disease of apple trees in Washington State. *Plant Dis.* 65:77–78.

Gendle, P., D. R. Clifford, P. C. Mercer, and S. A. Kirk. 1983. Movement, persistence and performance of fungitoxicants applied as pruning wound treatments on apple trees. *Ann. Appl. Biol.* 102(2):281–92.

Hotson, J. W. 1920. Collar-rot of apple trees in the Yakima Valley. *Phytopathology* 10:465–86.

Hulme, M. A., and J. K. Shields. 1970. Biological control of decay fungi in wood by competition for nonstructural carbohydrates. *Nature* 227(5255):300–01.

Kile, G. A. 1976. The effect of season of pruning and of time since pruning upon changes in apple sapwood and its susceptibility to invasion by *Trametes versicolor*. *Phytopathol. Z.* 87:231–40.

Kile, G. A., and G. C. Wade. 1974. *Trametes versicolor* on apple. I. Host-pathogen relationship. *Phytopathol. Z.* 81:328–38.

———. 1975. *Trametes versicolor* on apple. II. Host reaction to wounding and fungal infection and its influence on susceptibility to *T. versicolor*. Phytopathol. Z. 82:1–24.

Smith, R. E. 1941. *Diseases of fruits and nuts*. Calif. Agric. Ext. Serv. Circ. 120, 168 pp.

Smith, R. G. 1924. A chemical and pathological study of decay of the xylem of the apple caused by *Polystictus versicolor*. Phytopathology 14:114–18.

University of California. 1924. Plant pathology. *Univ. Calif. Agric. Exp. Stn. Ann. Rept.* 1922–23:179–87.

Wade, G. C. 1968. The influence of mineral nutrition on the susceptibility of apple trees to infection by *Trametes (Polystictus) versicolor. Austral. J. Exp. Agric. Anim. Husb.* 8:436–39.

Wormald, H. 1955. *Diseases of fruit and hops*. C. Lockwood & Son, London, 325 pp.

Scab of Apple

Venturia inaequalis (Cke.) Wint.

Apple scab, also called "black spot" in some countries, occurs wherever apples are grown, but it is much more severe where springs and summers are moist and temperate. For example, in the northeastern United States, including New England and states bordering the Great Lakes, apple scab is more damaging than in the southeastern part of the country. Similarly, outbreaks are more common in the coastal areas of California than in the interior and in the apple-growing areas of Oregon, but less severe in eastern Washington. In the Wenatchee area of Washington, scab emerged as a problem for the first time in 1985. Apple scab probably was known in Europe soon after the commercial apple was introduced, and the causal fungus was first called *Spilocaea pomi* in Sweden by Fries in 1819. Botanists noted its presence in the eastern United States as early as 1834. It was undoubtedly introduced into the Pacific Coast states soon after the region was settled.

Economic importance. When not controlled, scab can cause almost total destruction of an apple crop. Loss comes about principally through fruit infection. Some of the infected young fruit drop; those remaining on the tree become so malformed and unsightly as to be unmarketable. On highly susceptible cultivars, the tree may be damaged by defoliation following heavy leaf infection. In California, crop losses are minimal in orchards yearly sprayed with fungicides, but if sprays are not applied at the proper time high percentages of fruit scab can occur, even in the central San Joaquin Valley where rainfall is limited during the spring and summer months, but where trees commonly receive overhead sprinkler irrigation.

Symptoms. The first symptoms of apple scab usually appear on the emerging leaves (green bud) as dark green velvety spots. Later, similar lesions may appear on sepals of the young developing fruit. As the fruit develop, additional lesions appear on the sides of the fruit adjacent to the infected sepals (**fig. 13A**). As the lesions expand and coalesce they produce large, dirty green, scabby patches beneath which the host tissue stops growing, resulting in misshapen fruit (**fig. 13B**).

The leaf lesions are of two types: (1) well-defined, more or less circular, greenish scabby areas, which generally occur on the upper side of the leaf (**fig. 13C**), and (2) wefty lesions with indefinite margins which may cover most of the lower surface of the leaf.

Spotts and Ferree (1979) evaluated the effects of scab on leaves and found a significant reduction of photosynthesis 14 days after inoculation of Delicious leaves and 28 days after inoculation of McIntosh leaves. Reduced carbon dioxide assimilation was detected in diseased leaves only after visible symptoms had appeared. Transpiration was not affected, but water potential was decreased in McIntosh leaves.

Twig infection occurs occasionally on some cultivars of apple (Cook 1974; Hill 1975), but this phase of the disease is usually of minor importance. The scabby patches are composed largely of fungal tissue in the form of stromatic cushions from which arise short, unbranched conidiophores bearing conidia shaped like elongated pears. On leaves, the stromatic cushion extends between the cuticle and the outer walls of the epidermis. Rarely do hyphae from the stroma penetrate into the interior of the living leaf.

Causal organism. Apple scab is caused by the ascomycetous fungus *Venturia inaequalis* (Cke.) Wint. In the initial report, the imperfect stage was identified as *Spilocea pomi* Fries but later changed to *Fusicladium dendriticum* and recently back to *Spilocea pomi*, based on the manner in which the conidia are borne on the conidiophore.

In vitro colonies of the fungus are effuse, dark olivaceous brown, and velvety. The conidiophores are cylindrical, pale to mid-brown or olivaceous brown, up to 90 μm long, 5 to 6 μm thick, and sometimes swollen to 10 μm at the base. The conidia are obpyriform or obclavate, pale to mid-olivaceous brown, smooth, 0 to 1 septate, 16 to 24 (ave. 20.5) μm long, 7 to 10 (ave. 8.5) μm thick in the broadest part, and with a truncate base 4 to 5 μm wide (Ellis 1971).

Ascospores are produced in flask-shaped pseudothecia in dead leaves (fig. 13D). An ostiolum from the pseudothecium projects through the leaf surface. Eight ascospores (11 to 15 μm long, 4 to 8 μm broad) are produced in cylindrical asci. The ascospores have two cells, one cell being broader and shorter than the other; this cell is uppermost in the ascus. This feature distinguishes *V. inaequalis* from *V. pirina*, in which the reverse arrangement occurs.

Several races or pathotypes of *V. inaequalis* are known to occur in the United States and Japan (Sawamura 1988; Williams and Brown 1968). Their presence drastically complicates the breeding of resistant apple cultivars.

Venturia asperata Samuels & Sivanesan is a second species of *Venturia* detected on apples. This species was originally reported in 1975 from New Zealand (Samuels and Sivanesan 1975) and later from Canada (Corlett 1984). It is not known to occur in California. *V. asperata* has morphological characteristics similar to those of the peach scab organism (*V. carpophila*).

Disease development. Primary infection in spring is caused almost entirely by ascospores developed in pseudothecia in dead leaves on the ground. Twig infections—that could produce primary conidial inoculum—have not been observed on apple trees in California. Fruit and leaf lesions produce conidia that cause secondary infection both before and after the ascosporic inoculum is exhausted.

Keitt and Jones (1926) found that the optimum temperature for infection is 20°C; some infection, however, will occur at a temperature as low as 5°C. The rate of germination and appressorium formation of both ascospores and conidia is directly proportional to temperatures from 5° to 20°C, but conidia germinate and form appressoria more quickly than ascospores (Turner, MacHardy, and Gadoury 1986). The fungus spore germinates only if surrounded by a film of water. The moisture must be present for at least 6 hours at 20°C or 18 hours at 6°C to produce germination and penetration of the host tissue. Temperatures above the optimum are unfavorable to infection. At 27.8°C, for example, little infection was found even under optimum moisture conditions (Keitt, Clayton, and Langford 1941). A table (see table 1) for calculation of infection periods was developed by Mills and La Plant (1954), and has been used extensively as a guide in the application timing of scab-control chemicals. Recent research (MacHardy and Gadoury 1985, 1986, 1989; Gadoury and MacHardy 1986; Gadoury, MacHardy, and Rosenberger 1989) indicates the possible need for some modification of the Mill's predictive system. Incorporation of high humidity periods after rain, time of ascospore discharge, ascospore longevity and dose, and cultivar susceptibility is being considered in the revision of this system.

Once the germ tube has established itself in the tissue of the fruit or leaf, the fungus is no longer dependent on an external water supply. Its further development is, however, influenced by temperature. Keitt and Jones (1926) reported that whereas at 20° to 24°C, 8 to 12 days elapsed between infection and appearance of lesions, at 7.2°C the incubation process required 17 days.

Temperature and moisture conditions, therefore, restrict fungus infection to the cooler, moister periods of the growing season. Consequently, little disease will develop during the warmer, drier part of the summer. Leaf and fruit age also influence infection: young leaves are subject to infection on both surfaces, but as they become older the cuticle on the upper surface becomes resistant to penetration by the fungus, while the lower surface remains susceptible throughout most of the life of the leaf. Thus, abundant infection on the underside of leaves commonly follows rainy periods when the temperature has moderated in early autumn. This type of infection greatly increases pseudothecial development.

Although fruit are most susceptible to infection when young, even mature fruit develop small multiple infections when prolonged rains occur in autumn. If moisture is present, this type of infection may even develop after the fruit is picked and stored.

Mummified fruit with scab lesions from the previous season do not sporulate the following spring.

Rain is an important vehicle of spore dissemination. Even strong air currents do not always detach conidia from the conidiophores, but water does, and consequently spattering and windblown raindrops are very important in the spread of conidia. Striking evidence of rain dissemination is the development of numerous secondary lesions below infected sepals. For a time after the fruit are set, the calyx end of the fruit remains

Table 1. Approximate wetting period required for primary apple scab infection at different air temperatures and time required for development of conidia[a]

Average temperature		Wetting period (hr)[b]			Incubation period[c] (days)
(°F)	(°C)	Light infection	Moderate infection	Heavy infection	
78	25.6	13	17	26	—
76	24.4	9.5	12	19	—
63–75	17.2–23.9	9	12	18	9
62	16.7	9	12	19	10
60	15.6	9.5	13	20	11
58	14.4	10	14	21	12
56	13.3	11	15	22	13
54	12.2	11.5	16	24	14
52	11.1	12	18	26	15
50	10.0	14	19	29	16
48	8.9	15	20	30	17
46	7.8	16	24	37	—
44	6.6	19	28	43	—
42	5.5	23	33	50	—
41	5.0	26	37	53	—
40	4.4	29	41	56	—
38	3.3	37	50	64	—
33–36	0.5–2.2	48	72	96	—

[a]Adapted from Mills, 1944, as modified by A. L. Jones and prepared by A. R. Biggs for publication by Jones, A. L. and H. S. Aldwinckle, ed. 1990. *Compendium of Apple and Pear Diseases*. APS Press. 100 pp.
[b]The infection period is considered to start at the beginning of the rain.
[c]Approximate number of days required for conidial development after the start of the infection period.

uppermost, and rainwater running down over the infected sepals can carry conidia to the cheek of the fruit.

As long as the leaf is alive, the scab fungus is restricted to the subcuticular position. Soon after the leaf falls in autumn, fungus hyphae from the stromatic layer penetrate between the epidermal cells and grow into the leaf's interior. If the leaf remains moist, the initiation of pseudothecia soon follows in the interior of the leaf (Wilson 1928).

Because hyphae from a single lesion are relatively limited, both extension of hyphae through the leaf interior and the number of lesions per leaf are important in determining the number of pseudothecia that develop in the leaf. The well-defined circular lesions resulting from early infection of the upper surface of the leaf may produce pseudothecia only near the edges of the lesion. In contrast, the diffuse lesions on the lower surface of the leaf, where the subcuticular stroma of the fungus grows out extensively from the point of infection, produce pseudothecia scattered over larger areas (Wilson 1928). Thus, when autumnal infection of the lower surfaces of leaves is abundant, pseudothecial development may likewise be abundant.

The optimum temperature for pseudothecial initiation is about 12.8°C, as compared to around 25°C for spore germination and mycelial growth. Pseudothecia consequently are initiated during the autumn and early winter. They develop more or less rapidly depending on temperature, and the ascospores mature in spring at about the time the blossom buds begin to open (Wilson 1928).

Gadoury and MacHardy (1982) showed that the pseudothecia increase in diameter most rapidly at 10°C; for ascospore maturation, the optimum temperature is 20°C. The number of pseudothecia formed per unit of leaf was found to be inversely proportional to temperature. Most pseudothecia form within 28 days after leaf fall. The number of asci that develop is determined by the temperature between the appearance of the ascus initials and the appearance of ascospores, with 6°C producing more asci than 12°C (Gadoury, MacHardy, and Hu 1984). James and Sutton (1982) showed that the development of asci and ascospores is initiated in the spring only after a dormant or rest period during which no development is observed in the lumina of the pseudothecia, regardless of temperature or moisture conditions. The optimum temperature range for ascogonial development is from 8° to 12°C, while for ascospore maturation it is 16° to 18°C. Neither cultivar nor date of leaf fall affect the date of ascospore maturation in the spring. Ascospore maturity can be determined by microscopic observation of the contents of the crushed pseudothecia extracted from leaves. Szkolnik (1969) examined about 20 asci from each of at least 25 crushed pseudothecia taken from several leaves and categorized them according to the maturation of the contents: 1 = no spore formation evident, 2 = spores forming, 3 = spores formed but not colored, 4 = spores formed and colored, and 5 = spores discharged. Palmiter (1962) observed 100 pseudothecia to obtain the percent maturity categories. Both methods have routinely provided satisfactory information on ascospore maturity for approximately 25 years in conjunction with grower spray programs in New York (Gilpatrick and Szkolnik 1978).

After maturation, the ascospores are forcibly discharged into the air through the pseudothecial neck (ostiolum), which extends to the surface of the leaf. Discharge occurs when the leaves are wetted, and is repeated at each rain as long as ascospores are produced in the pseudothecium. In Wisconsin, Keitt and Jones (1926) found that the period of ascospore discharge lasted seven to nine weeks in April and May, while in New York, Heye and Andrews (1983) found that the discharge period began in April and continued through the first week of July. Ascospores were therefore present in the orchard air at the time of green tip stage of growth; thereafter, the number of spores increased with time, reaching a peak at the late pink to early bloom stage; then spore output decreased until 95 percent or more of the ascospore supply in the old leaves was exhausted, about six weeks after petal fall (Gilpatrick and Szkolnik 1978). In California, in 1979, after a scab epidemic in 1978, ascospore emission started in late dormancy and was virtually complete before any cultivars reached full bloom, a period of only four or five weeks (Moller 1980).

The number of ascospores in the air can be determined by using spore traps. Critical studies on spore release can be measured by use of the Burkard seven-day trap or the Hirst 24-hour trap (Hirst and Stedman 1961, 1962a, 1962b). Ascospore release may occur within a few seconds after leaf wetting, but spore-dose peaks appear in orchards about two hours after the start of wetting (Hirst and Stedman 1961). Several investigations (Brook 1969; MacHardy and Gadoury 1989) have noted that much greater numbers of spores are released during daylight than during the night; also, more are released during the afternoon than in the morning.

The conidial stage (Sutton 1978) is largely responsible for the increase of scab in the orchard in late spring, summer, and fall following initial infection in the spring by ascospores. The lesions produce conidia before they become macroscopic. Sporulation occurs at temperatures ranging from 16° to 20°C. Conidia are produced at 60 to 100 percent relative humidity, with an optimum of 90 percent. Day length does not affect sporulation; however, in continuous darkness sporulation is reduced by about 32 percent. Spore production generally stops in 30 to 36 days, when lesions become necrotic or bronzed and reddish brown. The viability of conidia that remain attached to the conidiophores decreases with high temperature and high relative humidity. Rainfall is the most important agent for conidial dispersal, and neither temperature nor wind velocity has any effect on the pattern of dissemination by rainfall. However, conidia can be released in the absence of free moisture. Sutton, Jones, and Nelson (1976) found that dissemination of airborne conidia is generally associated with increasing temperature, sunshine, increasing wind speeds, low relative humidity, and dry foliage. Hirst and Stedman (1961) report that conidia have been caught over the English Channel at an altitude of 2,000 ft and suggest that they may play an important role in long-distance dissemination of the pathogen. The survival period of nongerminated dry conidia is increased by low temperature and low relative humidity. Once spores germinate, they are highly sensitive to drying, and infection is sharply reduced from exposures of 10 to 24 hours between wetness periods.

Susceptibility of apple cultivars. Schneiderhan and Fromme (1924) reported that under Virginia conditions, Winesap, Rome Beauty, Stayman, and Delicious cultivars were highly susceptible to scab infection, and Jonathan, York Imperial, and Grimes somewhat so. In California, in 1978 after heavy rains during and after bloom, scab was found throughout a 1,700-acre (688-ha) solid planting of apples in which a test plot with 39 cultivars was located. Paulared, Gravenstein, Spigold, Jon Grimes, Jonathan, Rhode Island Greening, and Golden Delicious show tolerance of scab on leaves and fruit, while less tolerance is observed in Monroe, Blushing Gold, Yellow Newtown, Mutsu, and Idared (Szkolnik, Ogawa, and Nagaoka, unpublished).

According to Williams et al. (1972) and Williams (1978), a cooperative program for the development of apple cultivars resistant to scab was initiated in 1945 by L. F. Hough of the University of Illinois and J. R. Shay (Shay, Dayton, and Hough 1953) of Purdue University as the PRI (Purdue-Rutgers-Illinois) group. The group inoculated young seedlings with the scab organism, then observed fruit quality and susceptibility to scab in orchard plantings for four to seven years. Selections introduced with promise of scab resistance are Priscilla, Sir Prize, Co-op 12, and Co-op 13 in the United States, Prima in France, and Easygro and Macfree in Canada. Aldwinckle, Gustafson, and Lamb (1976) reported two major genes for resistance to *V. inaequalis* derived from *Malus floribunda* and *M. pumila*. By forcing the flowering of seedlings in 16 to 20 months, they developed numbered selections, such as New York 55140–19, that are immune to scab and resistant to mildew (Aldwinckle and Lamb 1977). Valsangiacomo and Gessler (1989) found that resistance to scab was not related to penetration of the cuticular membrane of apple leaves. Resistance in the Liberty cultivar was expressed as an inhibition of stroma growth only after two or three days. The mechanism of resistance remains undetermined.

Control. Preventing the infection of leaves, which are first exposed at the green-tip stage (**fig. 13E**) of spring growth, and the later infection of flower sepals is an important step toward successful control of scab on later-formed leaves and developing fruit. As dormant buds open in the spring, the first susceptible host parts exposed are the leaves (green bud). Once these leaves are infected and the fungus is allowed to sporulate, it becomes difficult to prevent sepal infection. These first exposed host parts are infected by ascospores, which normally mature before or during the green-tip stage and are present throughout the blossoming period. These host parts require preventive treatment because once they become infected control becomes more difficult, even though fairly effective postinfection fungicides are available. In California, two or more sprays are applied during the critical April–May period when ascospores are being discharged and susceptible host parts emerge. The state's warm, dry summer weather is not conducive to the development of secondary infections, but exceptional conditions, such as the use of overhead sprinklers or the occurrence of heavy dews or unseasonable rains, can sometimes cause scab to become serious. Fungicides that provide temporary protection against infection by the scab fungus are lime sulfur (5 days), dodine (10 days), captan (10 days), benomyl (10 days), and triforine (7 days). Eradicative or curative action of fungicides was first shown with mercury-containing fungicides such as phenylmercury acetate and penylmercury chloride. Later, such curative action from applications made a few hours or days after infection was demonstrated for such fungicides as maneb (9 hours), captan (26 hours), benomyl (36 to 48 hours), and dodine (36 to 72 hours), as well as for some of the new ergosterol biosynthesis-inhibiting (EBI) fungicides such as triforine (about 4 to 5 days). In a Michigan study, mixtures of two of the newer EBI fungicides (flusilazole and pyrifenox) with protective fungicides provided effective control of scab when applied on a 14-day schedule (O'Leary, Jones, and Ehret 1987). Some of the EBI fungicides (e.g., triforine) not only have protective and curative action against the scab organism, they also effectively control powdery mildew caused by *Podosphaera leucotricha* and cedar-apple rust caused by *Gymnosporangium juniperi-virginianae*.

Resistance of the scab organism to fungicides such as dodine (Gilpatrick and Blowers 1974; Szkolnik and Gilpatrick 1969) and benomyl (Sholberg, Yorston, and Warnock 1989; Wicks 1974) has resulted in loss of their use in many instances (Delp 1988). There are no methods known to prevent or delay the selection of resistant strains of the pathogen that results from multiple application of fungicides (Gilpatrick 1983; Ogawa et al. 1983). Mixtures of two fungicides effective in scab control, such as the combination of benomyl and captan, have had no success (Delp 1988). Resistance problems have been avoided only where conditions limit the use of a particular fungicide and thus prevent the selection pressure that occurs with more numerous applications. This procedure was suggested by Jones (1981) in his proposal to introduce EBI fungicides for the control of apple scab. His apprehension was later substantiated when field isolates of *V. inaequalis* with reduced sensitivity to sterol-inhibiting fungicides were found (Stanis and Jones 1985; Brent and Hollomon 1988). Once benomyl-resistant populations have been established, sporulation cycles do not change the ratio of sensitive and resistant isolates in an environment free of benomyl (McGee and Zuck 1981).

Studies on chemical control of apple scab have probably received greater attention than those on any other disease of fruit crops. From 32 fungicide applications per season in scab-prone areas, the number of applications in recent years has been reduced by more than half. This reduction is due to improved knowledge of the maturation and discharge of ascospores and the phenology of the host with reference to the early periods of infection. The environmental conditions of duration of wetness and average temperature play a major role in the timing of fungicide treatments. More recent studies have revealed that, like the mercury fungicides and liquid lime-sulfur, some of the newer fungicides have curative effects after the establishment of infection (Szkolnik 1981; Zuck and MacHardy 1981). Thus it is possible, for the first time, to reduce the inoculum in the orchard and thereby reduce secondary infection. Along with this technique, another suggested concept is the single application treatment, where a high dosage of captafol is applied at the green-tip stage and followed by fewer postbloom sprays than usual. Today, integrated pest management (IPM) places emphasis on developing automated data collection and prediction devices for monitoring infection periods in each orchard (Ellis, Madden, and Wilson 1984). The prediction methods include more critical data on environmental conditions (Jones et al. 1980; Tomerlin and Jones 1983b), biology of the organism, including inoculum levels (Sutton, James, and Nardacci 1981; MacHardy and Gadoury 1989), and phenology of crop susceptibility (Schwabe, Jones, and Jonker 1984; Tomerlin and Jones 1983a). Researchers in Chile (Pinto and Carreño 1986; Pinto, Carreño, and Moller 1984) obtained effective scab control when fungicides were applied on either a "calendar" or "climatic conditions" basis. The latter was the more efficient system. Emphasis has also gone toward developing fungicide applicators that use sprays of ultralow volume (14 L/ha), as compared to low-volume (190 L/ha) or dilute sprays (1,900 L/ha) (Barrat et al. 1981).

Keitt and coworkers (Keitt and Jones 1937; Keitt et al. 1941) in Wisconsin earlier studied the so-called eradicant method for combating apple scab. This consisted of treating apple leaves with a chemical that prevented development of the pseudothecial stage. In early tests the spray was applied to the trees in the autumn after harvest but before leaf fall. Later the spray was applied in early spring to overwintered leaves on the ground. The first successful chemicals were monocalcium arsenite and later a proprietary dinitro-o-cresol compound (Elgetol Extra). Workers in England (Burchill and Hutton 1965; Hutton 1954) reported success with phenylmercuric chloride applied to the trees before leaf fall. Researchers in the United States and Chile (Keitt and Palmiter 1937; Latorre 1972) found that phenylmercuric chloride, like dinitro cresol, suppresses the development of ascospores. Benomyl and thiophanate methyl were effective in controlling apple scab because, in part, they also reduce ascospore development (Ross 1973). New apple cultivars with greater resistance to the scab organism are important in reducing the use of fungicides for apple scab control (Jones and Sutton 1984). However, new introductions have had difficulty competing with established cultivars in commercial marketing.

Efforts toward biological control of apple scab have indicated antagonism to *V. inaequalis* by such fungi as *Trichoderma viride*, *Aureobasidium pullulans*, *Athelia bombacina*, and *Chaetomium globosum* (Andrews, Berbee, and Nordheim 1983; Heye and Andrews 1983). The reduction of primary inoculum by microbial action has been enhanced by the autumnal application of foliar sprays of urea (Burchill et al. 1965; Carreño and Pinto 1982; Moller 1981).

REFERENCES

Aderhold, R. 1900. Die Fusicladien unserer Obstbaume II. *Tiel. Landw. Jahrb.* 29:541–88.

Aldwinckle, H. S., H. L. Gustafson, and R. C. Lamb. 1976. Early determination of genotypes of apple scab resistance by forced flowering of test cross progenies. *Euphytica* 25:185–91.

Aldwinckle, H. S., and R. C. Lamb. 1977. Controlling apple diseases without chemicals. *New York's Food & Life Sci.* 10(2):12–14.

Andrews, J. H., F. M. Berbee, and E. V. Nordheim. 1983. Microbial antagonism to the imperfect stage of the scab pathogen, *Venturia inaequalis*. *Phytopathology* 73:228–34.

Barrat, R. E., J. L. Maas, H. J. Retzer, and R. E. Adams. 1981. Comparison of spray droplet size, pesticide deposition, and drift with ultralow-volume, low-volume, and dilute pesticide application on apple. *Plant Dis.* 65:872–75.

Brent, K. J., and D. W. Hollomon. 1988. Risk of resistance against sterol biosynthesis inhibitors in plant protection. Chap. 13:332–46. In *Sterol biosynthesis inhibitors*, eds. D. Berg and M. Plempel. Ellis Horwood Ltd., Chichester, England.

Brook, P. J. 1969. Effects of light, temperature, and moisture on release of ascospores by *Venturia inaequalis* (Cke.) Wint. *N. Z. J. Agric. Res.* 12:214–27.

Burchill, R. T., and K. E. Hutton. 1965. The suppression of ascospore production to facilitate the control of apple scab (*Venturia inaequalis* [Cke.] Wint.). *Ann. Appl. Biol.* 56:285–92.

Burchill, R. T., K. E. Hutton, J. E. Crosse, and C. M. E. Garrett. 1965. Inhibition of the perfect stage of *Venturia inaequalis* (Cke.) Wint. by urea. *Nature (London)* 205:520–21.

Carreño, I., and A. Pinto. 1982. Efecto de las pulverizaciones otoñales de urea en la reducción del inóculo primario de *Venturia inaequalis* (Cke.) Wint., en manzanos de la zone de Curico, Chile. *Agric. Tech. (Chile)* 42(3):235–38.

Clinton, G. P. 1901. Apple scab. *Illinois Agric. Exp. Stn. Bull.* 67:109–56.

Cook, R. T. A. 1974. Pustules on wood as sources of inoculum in apple scab and their response to chemical treatments. *Ann. Appl. Biol.* 77:1–9.

Corlett, M. 1984. *Venturia asperata*, a second species of *Venturia* occurring on apple in Canada. *Can. J. Plant Pathol.* 6(3):261.

Crosse, J. E., C. M. E. Garrett, and R. T. Burchill. 1968. Changes in microbial population of apple leaves associated with the inhibition of the perfect stage of *Venturia inaequalis* after urea treatment. *Ann. Appl. Biol.* 61:203–16.

Delp, C. J., ed. 1988. *Fungicide resistance in North America*. APS Press, The American Phytopathological Society, St. Paul, Minnesota. 133 pp.

Ellis, M. B. 1971. *Dematiaceous Hyphomycetes*. Commonwealth Mycological Institute, Kew, England, 608 pp.

Ellis, M. A., L. V. Madden, and L. L. Wilson. 1984. Evaluation of an electronic apple scab predictor for scheduling fungicides with curative activity. *Plant Dis.* 68:1055–57.

Fries, E. 1819. *Spilocaea pomi* Fr. *Novit. fl. Svec.* 5:79.

Gadoury, D. M., and W. E. MacHardy. 1982. Effects of temperature on the development of pseudothecia of *Venturia inaequalis*. *Plant Dis.* 66:464–68.

———. 1985. Negative geotropism in *Venturia inaequalis*. *Phytopathology* 75:856–59.

———. 1986. Forecasting ascospore dose of *Venturia inaequalis* in commercial apple orchards. *Phytopathology* 76:112–18.

Gadoury, D. M., W. E. MacHardy, and C. Hu. 1984. Effects of temperature during ascus formation and frequency of ascospore discharge on pseudothecial development of *Venturia inaequalis*. *Plant Dis.* 68:223–25.

Gadoury, D. M., W. E. MacHardy, and D. A. Rosenberger. 1989. Integration of pesticide application schedules for disease and insect control in apple orchards of the northeastern United States. *Plant Dis.* 73:98–105.

Gilpatrick, J. D. 1983. Management of resistance in plant pathogens. pp. 735–67. In *Pest resistance to pesticides*, eds. G. P. Georghiou and T. Saito. Plenum Publishing Corp., New York. 809 pp.

Gilpatrick, J. D., and D. R. Blowers. 1974. Ascospore tolerance to dodine in relation to orchard control of apple scab. *Phytopathology* 64:649–52.

Gilpatrick, J. D., and M. Szkolnik. 1978. Maturation and discharge of ascospores of the apple scab fungus. Proc. Apple Scab and Pear Scab Workshop. Kansas City, Missouri, July 11, 1976. *N. Y. Agric. Exp. Stn. Spec. Rept.* 28:1–6.

Heye, C. C., and J. H. Andrews. 1983. Antagonism of *Athelia bombacina* and *Chaetomium globosum* to the apple scab pathogen, *Venturia inaequalis*. *Phytopathology* 73:650–54.

Hill, S. A. 1975. The importance of wood scab caused by *Venturia inaequalis* (Cke.) Wint. as a source of infection for apple leaves in the spring. *Phytopathol. Z.* 82:216–23.

Hirst, J. M., and O. J. Stedman. 1961. The epidemiology of apple scab (*Venturia inaequalis* [Cke.] Wint.) I. Frequency of airborne spores in orchards. *Ann. Appl. Biol.* 49:290–305.

———. 1962a. The epidemiology of apple scab (*Venturia inaequalis* [Cke.] Wint.) II. Observations on the liberation of ascospores. *Ann. Appl. Biol.* 50:525–50.

———. 1962b. The epidemiology of apple scab (*Venturia inaequalis* [Cke.] Wint.) III. The supply of ascospores. *Ann. Appl. Biol.* 50:551–67.

Hutton, K. E. 1954. Scab of apples. In *USDA Yearbook of Agriculture*, 646–52.

James, J. R., and T. B. Sutton. 1982. Environmental factors influencing pseudothecial development and ascospore maturation of *Venturia inaequalis*. *Phytopathology* 72:1073–80.

Jones, A. L. 1981. Fungicide resistance: Past experience with benomyl and dodine and future concerns with sterol inhibitors. *Plant Dis.* 65:990–92.

Jones, A. L., S. L. Lillevik, P. D. Fisher, and T. C. Stebbins. 1980. A microcomputer-based instrument to predict apple scab infection periods. *Plant Dis.* 64:69–72.

Jones, A. L., and T. B. Sutton. 1984. Diseases of tree fruits. *N. Central Reg. Ext. Publ.* 45. Coop Ext. Serv., Mich. State Univ. 59 pp.

Keil, H. L., H. J. Retzer, R. E. Barrat, and J. L. Maas. 1980. Ultralow-volume spray control of three apple diseases. *Plant Dis.* 64:681–84.

Keitt, G. W. 1953. Scab of apples. In *USDA Yearbook of Agriculture*, 646–52.

Keitt, G. W., C. N. Clayton, and M. H. Langford. 1941. Experiments with eradicant fungicides for combating apple scab. *Phytopathology* 31:296–322.

Keitt, G. W., and L. K. Jones. 1926. *Studies of the epidemiology and control of apple scab.* Wisc. Agric. Exp. Stn. Res. Bull. 73, 104 pp.

Keitt, G. W., and D. H. Palmiter. 1937. Potentialities of eradicant fungicides for combating apple scab and some other plant diseases. *J. Agric. Res.* 55:397–438.

Latorre, B. A. 1972. Late season fungicide applications to control apple scab. *Plant Dis. Rep.* 56:1079–82.

MacHardy, W. E., and D. M. Gadoury. 1985. Forecasting the seasonal maturation of ascospores of *Venturia inaequalis*. *Phytopathology* 75:381–85.

———. 1986. Patterns of ascospore discharge by *Venturia inaequalis*. *Phytopathology* 76:985–90.

———. 1989. A revision of Mill's criteria for predicting apple scab infection periods. *Phytopathology* 79:304–10.

McGee, D. C., and M. G. Zuck. 1981. Competition between benomyl-resistant and sensitive strains of *Venturia inaequalis* on apple seedlings. *Phytopathology* 71:529–32.

Mills, W. D., and A. A. La Plante. 1951. Diseases and insects in the orchard. *Cornell Ext. Bull.* 711:20–22.

Moller, W. J. 1980. Effects of apple cultivar on *Venturia inaequalis* ascospore emission in California. *Plant Dis.* 64:930–31.

———. 1981. Efficacy of autumn urea in reducing spring inoculum of apple scab. *Calif. Plant Path.* 52, 2 pp.

Ogawa, J. M., B. T. Manji, C. R. Heaton, J. Petrie, and R. M. Sonoda. 1983. Methods for detecting and monitoring the resistance of plant pathogens to chemicals. pp. 117–62. In *Pest resistance to pesticides.* Plenum Publishing Corp., New York. 809 pp.

O'Leary, A. L., A. L. Jones, and G. R. Ehret. 1987. Application rates and spray intervals for apple scab control with fluzilazol and pyrifenox. *Plant Dis.* 71:623–26.

Palmiter, D. H. 1962. Eastern New York disease notes: Apple scab fungus development. In *Weekly report on insects, diseases, and crop development to County Agricultural Agents,* 27. New York State College of Agriculture.

Pickel, C., and R. S. Bethell. 1985. Apple scab management. Univ. Calif. Div. Agric. Nat. Resour. Leaf. 21412.

Pinto, A., and I. Carreño. 1986. Eficiencia de nuevos fungicidas en el control de Venturia. *Agric. Tech.* (Chile) 46(4):423–28.

Pinto, A., I. Carreño, and W. Moller. 1984. Control químico de Venturia en manzanos. Aplicaciones a calendario fijo o cuando el tiempo favorece la infeccion. Niveles de inoculo primario. *Agric. Tech.* (Chile) 44(2):123–30.

Ross, R. G. 1973. Suppression of perithecium formation in *Venturia inaequalis* by seasonal sprays of benomyl and thiophanate-methyl. *Can. J. Sci.* 53:601–02.

Samuels, G. J., and A. Sivanesan. 1975. *Venturia asperata* sp. nov. and its Fusicladium state on apple leaves. *N. Z. J. Bot.* 13:645–52.

Sawamura, K. 1988. Apple scab in Japan. pp. 32–38. In *Deciduous tree fruit (apple) disease workshop proc.* Faculty Agric., Hirosaki Univ. Hirosaki, Aomori, Japan.

Schneiderhan, F. J., and F. D. Fromme. 1924. Apple scab and its control in Virginia. *Virg. Agric. Exp. Stn. Bull.* 236, 29 pp.

Schwabe, W. F. S., A. L. Jones, and J. P. Jonker. 1984. Changes in the susceptibility of developing apple fruit to *Venturia inaequalis*. *Phytopathology* 74:118–21.

Shay, J. R., D. F. Dayton, and L. F. Hough. 1953. Apple scab resistance from a number of *Malus* species. *Proc. Am. Soc. Hort. Sci.* 62:348–56.

Sholberg, P. L., J. M. Yorston, and D. Warnock. 1989. Resistance of *Venturia inaequalis* to benomyl and dodine in British Columbia, Canada. *Plant Dis.* 73:667–69.

Spotts, R. A., and D. C. Ferree. 1979. Photosynthesis, transpiration, and water potential of apple leaves infected by *Venturia inaequalis*. *Phytopathology* 69:717–19.

Stanis, V. F., and A. L. Jones. 1985. Reduced sensitivity to sterol-inhibiting fungicides in field isolates of *Venturia inaequalis*. *Phytopathology* 75:1098–1101.

Sutton, T. B. 1978. Role of conidia of *Venturia inaequalis* in the epidemiology of apple scab. *N. Y. Agric. Exp. Stn. Spec. Rept.* 28:6–9.

Sutton, T. B., J. R. James, and J. F. Nardacci. 1981. Evaluation of a New York ascospore maturity model for *Venturia inaequalis* in North Carolina. *Phytopathology* 71:1030–32.

Sutton, T. B., A. L. Jones, and L. A. Nelson. 1976. Factors affecting dispersal of conidia of the apple scab fungus. *Phytopathology* 66:1313–17.

Szkolnik, M. 1969. Maturation and discharge of ascopores of *Venturia inaequalis*. *Plant Dis. Rep.* 53:534–37.

———. 1978. Techniques involved in greenhouse evaluation of deciduous tree fruit fungicides. *Ann. Rev. Phytopathol.* 16:103–29.

———. 1981. Physical modes of action of sterol-inhibiting fungicides against apple diseases. *Plant Dis.* 65:981–85.

Szkolnik, M., and J. D. Gilpatrick. 1969. Apparent resistance of *Venturia inaequalis* to dodine in New York apple orchards. *Plant Dis. Rep.* 53:861–64.

Tomerlin, J. R., and A. L. Jones. 1983. Development of apple scab on fruit in the orchard and during cold storage. *Plant Dis.* 67:147–50.

———. 1983b. Effect of temperature and relative humidity on the latent period of *Venturia inaequalis* in apple leaves. *Phytopathology* 73:51–54.

Turner, M. L., W. E. MacHardy, and D. M. Gadoury. 1986. Germination and appressorium formation by *Venturia inaequalis* during infection of apple seedling leaves. *Plant Dis.* 70:658–61.

Valsangiacomo, C., and C. Gessler. 1988. Role of the cuticular membrane in ontogenic and Vf-resistance of apple leaves against *Venturia inaequalis*. *Phytopathology* 78:1066–69.

Wicks, T. 1974. Tolerance of the apple scab fungus to benzimidazole fungicides. *Plant Dis. Rep.* 58:886–89.

Williams, E. B. 1978. Current status of apple scab resistant varieties. *N. Y. Agric. Exp. Stn. Spec. Rept.* 28:18–19.

Williams, E. B., and A. G. Brown. 1968. A new physiologic race of *Venturia inaequalis*, incitant of apple scab. *Plant Dis. Rep.* 52:799–801.

Williams, E. B., J. Janick, F. H. Emerson, L. F. Hough, C. Bailey, D. F. Dayton, and J. B. Mowry. 1972. Priscilla, a fall red apple with resistance to apple scab. *Fruit Variety Hort. Dig.* 26:34–35.

Wilson, E. E. 1928. Studies of the ascigerous stage of *Venturia inaequalis* (Cke.) Wint. in relation to certain factors of the environment. *Phytopathology* 18:375–418.

Zuck, M. G., and W. E. MacHardy. 1981. Recent experience in timing sprays for control of apple scab: Equipment and test results. *Plant Dis.* 65:995–98.

Scab of Loquat

Spilocaea pyracanthae (Otth.) v. Arx

Scab of loquat occurs to some extent in almost all countries in which this fruit and ornamental tree is grown. Its occurrence is widespread in the United States, and in California it is common in the San Francisco Bay area (Raabe and Gardner 1972). In some countries the disease is serious enough to require the use of fungicidal control measures (Salerno, Somma, and Rosciglione 1971b; Somma and Rosciglione 1977). In California, the causal fungus also attacks pyracantha, toyon, and *Kageneckia oblonga*; it is not pathogenic to apple and pear. It is thought that the fungus was brought into this country from Europe on either pyracantha or loquat. It was found on toyon in San Francisco in 1881 and was first reported on loquat in California in 1901 (Gardner and Raabe 1966; Raabe and Gardner 1972).

Symptoms. The disease is characterized by the formation of dark, velvety, circular spots on the young leaves, fruit, and shoots. The spots may coalesce to form large darkened areas and some deformity of infected fruit. Heavy leaf infection can cause premature defoliation. The symptoms are similar to those of apple scab, which is caused by a closely related fungus.

Causal organism. There has been considerable confusion regarding the nomenclature of the scab fungus on loquat and pyracantha (Raabe and Gardner 1972).

The fungus on loquat was named *Fusicladium eriobotryae* (Cav.) Br. & Cav. in 1891. In 1953, Hughes correctly transferred it to the genus *Spilocaea* as *S. eriobotryae* (Cav.) Hughes. However, since the scab fungi on loquat and pyracantha are now considered to represent a single species and since the fungus on pyracantha was apparently described first, the name should be *S. pyracanthae* (Otth.) v. Arx (Raabe and Gardner 1972).

The fungus forms a subcuticular, dark, septate mycelium that becomes locally thickened to form a pustule on which the characteristic annellate conidiophores are produced. The conidiophores are hyaline, one-celled, erect to undulating, stocky, and 50 to 60 by 6 to 8 μm. The conidia are hyaline, solitary, terminal, obclavate, one-celled, and 12 to 22 by 6 to 9 μm. Although reported once (D'Oliviera and D'Oliviera 1946), there is no conclusive evidence that the fungus has a sexual stage. The teleomorph has not been found in California (Raabe and Gardner 1972).

Disease development. Although relatively few epidemiological studies of loquat scab have been conducted, some of the salient points pertinent to the disease cycle have been elucidated. Since the pathogen apparently has no sexual stage, its life cycle is much simpler than that of the species that cause apple and pear scab. Furthermore, inasmuch as the loquat is an evergreen tree, there is no need for a sexual stage to provide primary inoculum each year. In Sicily, leaf infection was reported to occur in autumn and spring, and fruit infection in winter and spring (Salerno, Somma, and Rosciglione 1971a). In California, conidial inoculations during humid periods in winter and spring resulted in symptom development within three to five weeks, with infection occurring in flowers, leaves, and fruit (Raabe and Gardner 1972). In greenhouse inoculations, only young leaves and shoots are susceptible. At a temperature of 15.5°C, symptoms first appear in 15 days and are well developed in 21 days. In Sicily, leaf infection is favored by temperatures below 20°C, and the incubation period in late March is 11 to 13 days (Salerno, Somma and Rosciglione 1971a). In southern Russia, the incubation period at 21° to 25°C was reported to be 16 days. The disease, like apple and pear scab, appears to develop best under cool, humid conditions, with rain the principal disseminating agent and free moisture necessary for spore germination. The fungus survives from year to year in lesions on the susceptible aerial parts of its host.

Control. There appears to be no published research on control of this disease in California. In Sicily, two or more fungicidal sprays applied in winter and spring provided satisfactory control (Salerno, Somma, and Rosciglione 1971b). Effective materials were ziram, dodine, benomyl, thiophanate, and TBZ. Two treatments with 0.2 percent benomyl in December and March gave excellent control and were more effective than similar applications of ziram (Somma and Rosciglione 1977).

REFERENCES

Andrade, S. N., C. Aruta M., and J. Montealegre A. 1984. La sarna del nispero: *Spilocaea pyracanthae* (Otth.) von Arx. *Rev. Frut.* 5(2):60–61.

D'Oliviera, B., and M. de L. D'Oliviera. 1946. Nota sobre os corpos M do tricoginio e do anteridio nas *Venturia inaequalis, V. pirina,* e *V. eriobotryae. Agron. Lusit.* 8:291–301.

Gardner, M. W., and R. D. Raabe. 1966. Scab of loquat, toyon, and *Kageneckia* (abs.). *Phytopathology* 56:147.

Hughes, S. J. 1953. Some foliicolous Hyphomycetes. *Can. J. Bot.* 31:560–76.

Raabe, R. D., and M. W. Gardner. 1972. Scab of pyracantha, loquat, toyon, and *Kageneckia. Phytopathology* 62:914–16.

Salerno, M., V. Somma, and B. Rosciglione. 1971a. Richerche sull' epidemiologia della ticchiolatura del nespolo del Giappone. *Tec. Agric. Catania* 23(1):3–15. (*Rev. Plant Pathol.* 52:802. 1973).

———. 1971b. Confronto di fungicidi nella lotta contro la ticchiolatura del nespolo de Giappone. *Tec. Agric. Catania* 23:947–56. (*Rev. Plant Pathol.* 55:1377. 1976).

Smith, R. E. 1941. *Diseases of fruits and nuts.* Calif. Agric. Ext. Circ. 120, 168 pp.

Somma, V., and B. Rosciglione. 1977. [Further field trials against loquat scab.] *Rev. Plant Pathol.* 56:5378.

Scab of Pear

Venturia pirina Aderh.

Pear scab, also known as "black spot" in Australia and New Zealand (Cunningham 1925), occurs to some extent wherever pears are grown, but it is most prevalent in regions of high summer rainfall. Thus, when not controlled, it is occasionally the source of much economic loss in some areas of the Pacific Northwest. It occurs sporadically in the coastal areas of California, but seldom in the drier interior valleys (Beardon, Moller, and Reil 1976; Smith 1905). Losses from fruit infection occurred in Lake County during 1975 and in Sacramento County during 1984 and 1985.

Symptoms. Blossoms, fruit, leaves, and occasionally twigs are infected (**fig. 14**). In spring and summer superficial, more or less circular, wefty, olivaceous lesions develop on the leaf blade, leaf petiole, flower peduncle, and cheek of the fruit. Lesions on the fruit frequently occur in greater numbers at the stylar end, and the coalescence of several lesions creates large irregular, dull green, feltlike patches. Growth of fruit tissue under such areas is checked, and the fruit become malformed and unsalable. The dark, feltlike surface of the lesion is composed of numerous conidiophores and conidia. On living leaves and fruit, the fungus occupies the area between the cuticle and epidermis, where it develops a stroma from which conidiophores arise.

Twig lesions appear first as small (1 to 2 mm) blisterlike pustules. Later the corky periderm breaks away, exposing the fungus stroma (Kienholz and Childs 1937, 1951).

Cultivar susceptibility. Most if not all of the commonly grown cultivars are susceptible to scab. Fruit of d'Anjou, Bartlett, Comice, Winter Nelis, Easter Beurre, Forelle, Flemish Beauty, and Seckel are highly susceptible. Very young fruit of Beurre Bosc are highly susceptible, but older fruit are comparatively resistant to infection. Stanton (1953a, 1953b) analyzed the field resistance of selected pear cultivars and found that Beurre Gifard was immune, Conference resistant, Doyenne du Comice moderately resistant, and Winter Nelis and Williams (Bartlett) susceptible. By making crosses between cultivars he concluded that resistance to individual *V. pirina* clones appeared to be dominant to susceptibility, but he indicated that further data are required to verify this interpretation. The recently developed Bulgarian cultivar Trapezitsa is reported to be highly resistant to the pear scab pathogen (Komitov 1979).

The Asian pear industry in California is rapidly expanding with plantings in both the Sacramento and San Joaquin valleys. Yet no evidence of scab on these oriental cultivars has been reported, and we have found no published data indicating that these cultivars are susceptible to *V. pirina*. Asian pears (*Pyrus serotina* var. *culta* and *P. usuriensis* var. *sinensis*), however, are susceptible to scab caused by *V. nashicola* Tanaka & Yamamoto, a species not known to occur in the United States. The European pear (*P. communis*) is not mentioned as a host of this scab species (Kitajima 1989).

Causal organism. *Venturia pirina* Aderh., the causal fungus of pear scab, produces the conidial stage (*Fusicladium*) on fruit, leaves, and twigs in summer and the pseudothecial stage in dead leaves on the ground in winter. Conidia of *V. pirina* are light yellowish green, pyriform, at first nonseptate, becoming septate with age, and 28 to 30 µm long by 7 to 9 µm wide; they are produced on short, wavy conidiophores. After each conidium is liberated, the end of the conidiophore to one side of the conidial scar elongates slightly, leaving the conidial scar as knee-shaped protrusion (Aderhold 1896).

Pseudothecia are almost completely immersed in the leaf, with only the neck projecting through the cuticle. They are globose and 120 to 160 µm in diameter. The cylindrical asci produce eight yellowish green, two-celled ascospores, and are 14 to 20 by 5 to 8 µm in dimension. The cells of the ascospores are unequal in length, the longer cell being oriented toward the apex of the ascus. This feature distinguishes *V. pirina* from *V. inaequalis*, which produces ascospores with the short cell oriented upward in the ascus.

Langford and Keitt (1942) found the pathogen to be a species of many strains that differ in cultural characteristics and pathogenicity. For example, none of their isolates infected Bartlett and Keiffer, although neither of these cultivars is immune or even highly resistant to scab.

According to Shabi, Rotem, and Loebenstein (1973), physiological races of the fungus occurring in Israel are specific to differential hosts of *Pyrus communis* and *P. syriaca* origin. Environmental conditions did not change the specific pathogenic properties of the races tested.

Disease development. The fungus survives from one season to the next on twig lesions or in diseased leaves on the ground. Some investigators (Cheal and Dillon Weston 1938; Kienholz and Childs 1937, 1951) asserted that twig lesions are often the more important overwintering sites. Conidia are produced on twig lesions, and these are largely disseminated by rain. After leaves fall, the fungus penetrates from its subcuticular position into the interior and initiates pseudothecia. The ascospores mature in early spring; they are forcibly ejected into the air during wet weather and are blown to susceptible parts of the tree. Thus, the inoculum for primary infection in spring may be both conidia and ascospores. Recent Chilean research showed that ascospores are first ejected at the green-tip stage of bud development; they reach maximum numbers at the white cluster to full bloom stage and progressively decrease until December (Latorre, Yañez, and Rauld 1985).

As with apple scab, among the first parts of the pear to be infected in spring are the leaves covering the emerging flowers. The fruit and new shoots are then in turn infected. Kienholz and Childs (1951) state that while twigs may be attacked at any time during the growing season, infection is most frequent in spring.

Temperature and moisture play important roles in development of the disease. Rain disseminates the conidia about the tree, and provides the moisture necessary for ascospore ejection and spore germination. The fungus can attack the host over a relatively wide range of temperatures, but variations in this factor greatly influence both the amount of infection and the length of the incubation period. For example, infection will occur at 23.9°C when the host part remains continually wet for only 5 hours, whereas at 4.4°C continually moist period of 48 hours is necessary for infection. The incubation period varies from 10 to 25 days, also depending on temperature (Kienholz and Childs 1951).

In contrast to the apple scab fungus, which releases few ascospores at night, *V. pirina* frequently ejects ascospores during nighttime but in smaller numbers than during the day (Beardon, Moller, and Reil 1976; Latorre, Yañez, and Rauld 1985).

Bearden, Moller, and Reil (1976) monitored ascospore emissions of the pear scab fungus in Mendocino County (California), and showed that the Mills's Chart appears to give a reasonably accurate guideline for predicting infection. Their calculations for quick assessment—(mean temperature [°F]–30) × hours of wetness—were 300 for light infection, 400 for moderate infection, and 600 for heavy infection.

Sugar and Lombard (1981) monitored scab incidence in a high-density orchard of Bartlett and d'Anjou pear trees. Infections on fruit nearly doubled between June 1 and late August in plots irrigated from sprinklers above the trees, but they did not increase significantly in plots irrigated at ground level.

Control. The most effective control of pear scab is attained with fungicidal sprays. Kienholz and Childs

(1951) showed that liquid lime-sulfur is highly effective in destroying the conidia in twig lesions. They therefore recommended an application of liquid lime-sulfur in water at the delayed dormant stage of flower bud development, and to further suppress infection, a second spray (ferbam or ziram) at the pre-pink stage. This application might be delayed until the pink stage if the weather is dry. If the weather is unusually wet, both the pre-pink and pink treatment were advised. The next treatment, using ferbam or ziram, is applied at the calyx stage (after most of the petals have fallen). Additional treatments at 10 to 20-day intervals were recommended for regions where rains extend into late spring and summer.

Kienholz and Childs (1951) and later Coyier and Mellenthin (1969) reported that lime-sulfur applied after the buds open is injurious to the d'Anjou cultivar. Sprays of lime-sulfur plus wettable sulfur should be restricted to the period spanning the green-tip and cluster-bud stages of bloom.

In most of California's pear districts, one or possibly two protective sprays are applied, the first at green tip and the second at early bloom. But in the wetter coastal counties an additional cover spray is needed to prevent heavy calyx infection of young fruit (Beardon, Moller, and Reil 1976). Treatment with benomyl sprays applied up to 48 hours after the three predicted infection periods gives effective scab control. The three benomyl sprays are equivalent in scab control to eight protective applications that include one liquid lime-sulfur followed by one wettable sulfur and subsequent dodine cover sprays. Both benomyl and dodine, in addition to protective action, have eradicative properties if used within 24 to 48 hours after infection. Trunk injections with the benzimidazole fungicide carbendazim, although reducing late leaf infection, do not provide adequate control of fruit scab (Shabi, Solel, and Pinkas 1979).

Studies in Israel (Shabi and Katan 1979; Shabi, Koenraadt, and Katan 1986) demonstrated the occurrence of benomyl- and carbendazim-resistant isolates of *Venturia pirina* in orchards in which benomyl failed to control scab. These isolates retained their resistance in the absence of the fungicide, produced typical scab lesions, and transmitted the resistance to new conidia. Four levels of resistance were recognized. Crosses between resistant and sensitive types showed that the various levels of resistance are conferred by four allelic mutations that constitute a polymorphic series at a single gene.

Recent research in Israel (Shabi, Elisha, and Zelig 1981) and New Zealand (Gaelan 1983) showed that some of the new ergosterol biosynthesis-inhibiting fungicides are effective in the control of scab. Sprays of bitertanol, prochloraz, triforine, and a captan-triforine mixture provided good disease conotrol, but, at least in New Zealand, bitertanol at 0.0024 percent caused some fruit deformity.

Williamson and Burchill (1974) reported that, as with the apple scab fungus, relatively low amounts of phenylmercuric chloride, benomyl, thiabendazole, and urea applied to scab-infected pear leaves in autumn prevent ascospore formation the following spring. Applied to scab-infected twigs in autumn or spring, however, these materials failed to prevent conidial development. In a recent Chilean study (Latorre and Marin 1982), fall applications of bitertanol, fenarimol, and urea also were found to significantly reduce ascospore production. Greatest reduction was obtained with fenarimol.

REFERENCES

Aderhold, R. 1896. Die Fusicladien unsere Obstbaum. *Landw. Jahrb.* 25:875–914.

Beardon, B. E., W. J. Moller, and W. D. Reil. 1976. Monitoring pear scab in Mendocino County. *Calif. Agric.* 30(4):16–19.

Cheal, W. F., and W. A. R. Dillon Weston. 1938. Observations on pear scab (*Venturia pirina* Aderh.) *Ann. Appl. Biol.* 25:206–08.

Coyier, D. L., and W. M. Mellenthin. 1969. Effect of lime-sulfur oil sprays on d'Anjou pear. *Hortscience* 4(2):91.

Cunningham, G. H. 1925. *The fungous diseases of fruit trees.* Brett Printing Co., Auckland, 382 pp.

Geelan, J. A. 1983. New fungicides for the control of black spot in process pears. In *Proc. 36th N. Z. Weed and Pest Control Conf.*, 273–76. [*Rev. Plant Pathol.* 63:729].

Kienholz, J. R. 1953. Scab on the pear. In *Plant diseases,* 674–77. USDA Yearbook of Agriculture.

Kienholz, J. R., and L. Childs. 1937. Twig lesions as a source of early spring infection by the pear scab fungus. *J. Agric. Res.* 55:667–81.

———. 1951. Pear scab in Oregon. *Oregon State Coll. Agric. Exp. Stn. Tech. Bull.* 21, 31 pp.

Kirk, J. W. 1913. Pear scab. *N. Z. J. Agric.* 7:1–59.

Kitajima, H. 1989. *Diseases of orchard crops.* Yokendo Company, Tokyo. 581 pp.

Komitov, R. 1979. [Biological and economic features of the new early-ripening pear variety Trapezitsa]. *Rev. Plant Pathol.* 58:2852.

Langford, M. H., and G. W. Keitt. 1942. Heterothallism and variability in *Venturia pirina. Phytopathology* 32:357–69.

Latorre, B. A., and G. Marin. 1982. Effect of bitertanol, fenarimol, and urea as fall treatments on *Venturia pirina* ascospore production. *Plant Dis.* 66:585–86.

Latoree, B. A., P. Yañez, and E. Rauld. 1985. Factors affecting release of ascospores by the pear fungus (*Venturia pirina*). *Plant Dis.* 69:213–16.

Perlberger, J. 1944. The occurrence of apple and pear scab in Palestine in relation to weather conditions. *Palestine J. Bot. Rehovot Ser.* 4:157–61.

Shabi, E., and Y. Ben-Yephet. 1976. Tolerance of *Venturia pirina* to benzimidazole fungicides. *Plant Dis. Rep.* 60:451–54.

Shabi, E., S. Elisha, and Y. Zelig. 1981. Control of pear and apple diseases in Israel with sterol-inhibiting fungicides. *Plant Dis.* 65:992–94.

Shabi, E., and T. Katan. 1979. Genetics, pathogenicity, and stability of carbendazim-resistant isolates of *Venturia pirina. Phythopathology* 69:267–69.

———. 1980. Fitness of *Venturia pirina* isolates resistant to benzimidazole fungicides. *Phytopathology* 70:1124–28.

Shabi, E., H. Koenraadt, and T. Katan. 1986. Further studies on the inheritance of benomyl resistance in *Venturia pirina* isolated from pear orchards in Israel. *Plant Pathol.* 35:310–13.

Shabi, E., J. Rotem, and G. Loebenstein. 1973. Physiological races of *Venturia pirina* on pear. *Phytopathology* 63:41–43.

Shabi, E., Z. Solel, and Y. Pinkas. 1979. Advantages and disadvantages of carbendazim injected into pear trees for scab control. *Plant Dis. Rep.* 63:376–78.

Smith, R. E. 1905. Pear scab. Calif. Univ. Agric. Exp. Stn. Bull. 1963.

Stanton, W. R. 1953a. Breeding pears for resistance to the pear scab fungus *Venturia pirina* Aderh. I. Variation in the pathogenicity of *Venturia pirina. Ann. Appl. Biol.* 40:184–191.

———. 1953b. Breeding pears for resistance to the pear scab fungus *Venturia pirina* Aderh. II. The study of field resistance of selected pear seedlings and the inheritance of resistance in seedling pear families under controlled conditions. *Ann. Appl. Biol.* 40:192–96.

Sugar, D., and P. B. Lombard. 1981. Pear scab influenced by sprinkler irrigation above the tree or at ground level. *Plant Dis.* 65:980.

Tanaka, S., and S. Yamamoto. 1963. Studies on the pear scab (*Venturia* spp.). III. Infection on the leaves by conidia. *Bull. Hort. Res. Sta., Japan. Ser. B.* 2:181–92.

University of California Statewide IPM Project. 1989. *Pear pest management guidelines.* UCPMG16. University of California, Davis, 43 pp.

Wiesmann, R. 1931. Untersuchungen über Apfel und Birnschorfpilz *Fusicladium dendriticum* (Wallr.) Fckl. und *Fusicladium pirinum* (Lib.) Fckl. sowie die Schorfanfälligkeit einzelner Apfel und Birnsorten. *Landw. Jahr. Schweiz* 45:109–56.

Williamson, C. J., and R. J. Burchill. 1974. The prevention and control of pear scab (*Venturia pirina* Aderh.). *Plant Pathol.* 23:67–73.

Shoestring (Armillaria) Root Rot of Apple and Pear

Armillaria mellea (Vahl.:Fr.) Kummer

Shoestring root rot is not considered an important disease on apples or pears in California. Apple rootstocks are moderately resistant to the disease. Pear rootstocks developed from Bartlett and Winter Nelis (both of French origin) seeds are quite resistant, but Oriental pear rootstocks are as a group more susceptible than the French types. Pear roots weakened by the *Prionus* beetle were found to be infected with shoestring root rot, but this was an exception. Thus, pears and apples could be possible replacement crops in soils infested with *Armillaria*.

Details of shoestring root rot disease are discussed in Chapter 4.

Southern Blight of Apple

Sclerotium (Athelia) rolfsii Sacc.

Southern blight is a serious disease of many dicotyledonous plants (Aycock 1966; Punja 1985a), including apples, *Malus domestica* Borkh. The first report of *Sclerotium* disease of Northern Spy stocks was in South Africa in 1922 (Anonymous). In the United States, pathogenicity of *Sclerotium rolfsii* on young apple trees was reported by Turner (1936) in North Carolina, and in the same year Cooley (1936) reported this disease in Maryland. Cooley experienced negative results in many of his isolation attempts and suggested that this was due to the early death of the *Sclerotium* fungus in killed host tissue. He believed that the disease was of more common occurrence than his isolations indicated. In one nursery where the *Sclerotium* disease was identified, Cooley reported a loss of 5 percent of the trees over an eight-year period. Southern blight has also been reported on apples in Indiana (Shay 1953), Georgia (Taylor and McGlohon 1975; Brown and Hendrix 1980), Oklahoma (Tomasino and Conway 1987), and some other southern states.

Smith (1941), in his publication on the diseases of California fruit and nut crops, did not include *S. rolfsii* as a pathogen of either stone or pome fruit. Southern blight of apple was first identified by L. D. Leach of UC Davis in 1980, in a young orchard in Santa Cruz County. Since then, W. D. Gubler (personal communication, 1987) has identified the disease in Stanislaus County (1984), Kern and Yuba counties (1985), and Tulare County (1986). Southern blight has also been reported on black walnut (*Juglans nigra*) seedings in the southern San Joaquin Valley (French 1987). The recent increase in apple plantings in California coupled with the fact that *S. rolfsii* has been here for many years as a pathogen of field and vegetable crops is probably largely responsible for the rather sudden and widespread incidence of southern blight in apple. A further source of the disease may be infected or contaminated nursery stock from California or other states.

Symptoms. On young nursery trees and in orchard plantings, the first sign of the disease is a web of white mycelium at the soil surface around the tree trunk (**fig. 15**). A few days later, the white mycelium disappears and is replaced by masses of brown sclerotia of the fungus. The size and color of the sclerotia are those of a mustard seed. By the time sclerotia of the fungus are observed, the root bark is rotted, resulting in complete girdling and death of the tree (Cooley 1936). The fungus may advance a few centimeters up the trunk but usually is confined to the root tissue, including the epidermis, cortex, and stele (Mordue 1974). Older trees appear to be resistant to infection.

Causal organism. The sclerotial state of the fungus is *Sclerotium rolfsii* Sacc. Its teleomorph was originally called *Corticium rolfsii* Curzi but its name is now *Athelia rolfsii* (Curzi) Tu & Kimbrough. According to Mordue (1974), colonies on potato dextrose agar are white and usually have many narrow mycelial strands in the aerial mycelium. Cells of the primary hyphae at the advancing edge of the colony usually are 4.5–9 μm wide and up to 350 μm long, with one or more clamp connections at the septa; the secondary hyphae arise immediately below the distal cell septum and often grow appressed to the primary hyphae; and the tertiary and subsequent branches are narrow (1.5–2 μm wide), with comparatively short cells and a wide angle of branching not closely associated with septation, and are usually without clamps. On host plants, appresoria develop as swollen tips of short hyphal branches behind the advancing margin of the mycelium.

On potato dextrose agar, the nearly spherical sclerotia (1–2 mm in diameter) develop on the colony surface; they are slightly flattened below with a smooth or shallow-pitted, shiny surface. The sclerotia show a sharply differentiated rind, with evenly thickened, strongly pigmented walls, a cortex with faintly pigmented walls, and a medulla with colorless, unevenly thickened walls. The cortex and medulla both contain vesicles of reserve food materials, the rind cells do not (Mordue 1974). On the host, sclerotia often do not develop until after the death of the host.

The basidospores are smooth, hyaline, globose-pyriform, 4.5–6.75 by 3.5–4.5 μm and borne on a white to ochraceous hymenium (Mordue 1974). The role of the basidiospores in the disease cycle has not been established (Punja and Grogan 1983).

Disease development. The fungus overwinters in naturally infested fields as clustered sclerotia (Punja et al. 1985; Rodriguez-Kabana, Backman, and Wiggins 1974). The inoculum density can be assessed by enumeration of the sclerotia by wet-sieving (Leach and Davey 1938; Punja et al. 1985), flotation-sieving (Rodriguez-Kabana, Backman, and Wiggins 1974), or elutriation (Shaw and Beute 1984). Sclerotium viability can be determined on water agar (Punja 1986) or on a selective medium containing potassium oxalate and gallic acid (Bateman and Beer 1965; Backman and Rodriguez-Kabana 1976; Rodriguez-Kabana, Backman, and Wiggins 1974). Disease incidence can be measured by using a trap crop such as carrot (Punja 1986). Disease incidence and severity are increased by fluctuating periods of high temperature and soil moisture (Punja and Grogan 1981b; Smith 1972a; Sztejnberg and Golan 1980) as well as by the presence of organic substrates which favor mycelial growth (Beckman and Finch 1980; Beute and Rodriguez-Kabana 1979). Yet, according to Punja (1985), precise temperature and moisture levels have not been correlated with disease incidence or severity. Initial infections by *S. rolfsii* occur at the soil surface (Punja and Grogan 1981b), and free moisture is not required (Punja 1986). Plant-to-plant spread is extensive in closely cropped plants such as carrots and sugar beets. In apple plantings, however, root-to-root spread is unlikely, and the infection foci appear to be determined by the foci of the disease in the previous crop. Survival of the sclerotia is reduced by high temperature (Mathur and Sarbhoy 1976; Yuen and Raabe 1984) and moist soils (Beute and Rodriguez-Kabana 1981). The mycelium of the fungus survives better in sandy soils than in heavy soils (Chatopadhyay and Mustafee 1977). Factors that increase nutrient leakage from the sclerotia predispose them to microbial degradation; such factors are drying (Coley-Smith, Ghaffar, and Javed 1974; Henis and Papavizas 1983; Punja and Grogan 1981b; Smith 1972a; Smith 1972b), heating (Elad, Katan and Chet 1980; Lifshitz et al. 1983; Yuen and Raabe 1984), deep burial (Punja and Jenkins 1984), and exposure to chemicals (Henis and Papavizas 1983; Linderman and Gilbert 1973; Punja and Grogan 1981b).

Control. Chemical control methods have utilized fungicides (Abeygunawardena and Wood 1957; Davey and Leach 1941; Harrison 1961; Punja and Grogan 1982); soil fumigants such as methyl bromide, chloropicrin, or metham-sodium; nematicides (Rodriguez-Kabana, Backman, and King 1976); insecticides (Backman and Hammond 1981); herbicides (Grinstein, Elad, Katan and Chet 1979); and the general biocide potassium azide (Rodriguez-Kabana et al. 1972). The application of chemical control in the field, however, requires improvement. The addition of fertilizers to increase the nitrogen level of the soil (Leach and Davey 1942), to provide ammonia which inhibits sclerotial germination and retards mycelial growth, and to increase calcium levels in the host to partially offset the action of oxalic acid and cell wall-degrading enzymes produced by the pathogen (Punja, Huang, and Jenkins 1985) has received much attention, but is not a solution to the disease. Soil solarization, using polyethylene tarps, reduces sclerotial numbers (Elad, Katan, and Chet 1980; Grinstein, Katan, et al. 1979; Mihail and Alcorn 1984) and can eliminate sclerotia from soil to a depth of 6–20 cm (Elad, Katan, and Chet 1980; Mihail and Alcorn 1984). The application of *Trichoderma harzianum* (Wells, Bell, and Jarworski 1972) in combination with solar heating has been more effective than either method alone (Elad, Katan, and Chet 1980). Deep plowing of fields, to essentially bury the sclerotia and the infected crop debris, has been effective on crops such as peanuts (Gurkin and Jenkins 1985; Punja et al. 1985). Rotations with such crops as corn, wheat, or oats can significantly reduce the amount of inoculum in the soil (Gautum and Kolte 1979; Mehrotra and Caludius 1972).

In the planting of new apple orchards, suggested approaches to the control of southern blight are to avoid establishing orchards in infested soils through the monitoring of fields for the population of sclerotia (Punja et al. 1985; Rodriguez-Kabana, Beute, and Backman 1980); eliminate the sclerotia at the planting site by removing infested soil and replacing it with noninfested soil; reduce the likelihood of infection, if planting in infested sites, by applying biocontrol agents (Conway 1986), solar heating, or pesticides at the planting sites; and plant nonsusceptible weeds along with the trees at the planting sites (Israeli researchers, personal communication). Information regarding the levels of apple rootstock susceptibility to *S. rolfsii* has not been found.

REFERENCES

Abeygunawardena, D. V. W., and R. K. S. Wood. 1957. Effect of certain fungicides on *Sclerotium rolfsii* in the soil. *Phytopathology* 47:607–09.

Anonymous. 1922. The *Sclerotium* disease of Northern Spy stocks. *S. Afr. J. Agric.* 4:405.

Aycock, R. 1966. *Stem rot and other diseases caused by Sclerotium rolfsii*. N. Car. State Univ. Tech. Bull. 174, 202 pp.

Backman, P. A., and J. M. Hammond. 1981. Suppression of peanut stem rot with the insecticide chlorpyrifos. *Peanut Sci.* 8:129–30.

Backman, P. A., and R. Rodriguez-Kabana. 1976. Development of a medium for the selective isolation of *Sclerotium rolfsii*. *Phytopathology* 66:234–36.

Bateman, D. F., and S. V. Beer. 1965. Simultaneous production and synergistic action of oxalic acid and polygalacturonase during pathogenesis by *Sclerotium rolfsii*. *Phytopathology* 55:204–11.

Beckman, P. M., and H. C. Finch. 1980. Seed rot and damping-off of *Chenopodium quinoa* caused by *Sclerotium rolfsii*. *Plant Dis.* 64:497–98.

Beute, M. K., and R. Rodriguez-Kabana. 1979. Effect of volatile compounds from remoistened plant tissues on growth and germination of sclerotia of *Sclerotium rolfsii*. *Phytopathology* 69:802–05.

———. 1981. Effects of soil moisture, temperature, and field environment on survival of *Sclerotium rolfsii* in Alabama and North Carolina. *Phytopathology* 71:1293–96.

Brown, E. A., and F. F. Hendrix 1980. Distribution and control of *Sclerotium rolfsii* on apple. *Plant Dis.* 64:205–06.

Chatopadhyay, S. B., and T. P. Mustafee. 1977. Behavior of *Macrophomina phaseoli* and *Sclerotium rolfsii* with relation to soil texture and soil pH. *Curr. Sci.* 46:226–28.

Clayton, C. N., A. J. Julis, and T. B. Sutton. 1976. Root rot diseases of apple in North Carolina. *N. Car. State Univ. Bull.* 455, 11 pp.

Coley-Smith, J. R., A. Ghaffar, and Z. U. R. Javed. 1974. The effect of dry conditions on subsequent leakage and rotting of fungal sclerotia. *Soil Biol. Biochem.* 6:307–12.

Conway, K. E. 1986. Use of fluid-drilling gels to deliver biological control agents to soil. *Plant Dis.* 70:835–39.

Cooley, J. S. 1936. *Sclerotium rolfsii* as a disease of nursery apple trees. *Phytopathology* 26:1081–83.

Davey, A. E., and L. D. Leach. 1941. Experiments with fungicides for use against *Sclerotium rolfsii* in soils. *Hilgardia* 13:523–47.

Elad, Y., J. Katan, and I. Chet. 1980. Physical, biological and chemical control integrated for soilborne diseases in potatoes. *Phytopathology* 70:418–22.

French, A. M. 1987. *Califonria plant disease host index Part 1: Fruits and nuts.* California Department of Food and Agriculture, Division of Plant Industry, Sacramento. 39 pp.

Gautum, M., and S. J. Kolte. 1979. Control of *Sclerotium* of sunflower through organic amendments of soil. *Plant Soil* 53:233–38.

Grinstein, A., Y. Elad, J. Katan, and I. Chet. 1979. Control of *Sclerotium rolfsii* by means of herbicide and *Trichoderma harzianum*. *Plant Dis. Rep.* 63:823–26.

Grinstein, A., J. Katan, A. A. Razik, O. Zeydan, and Y. Elad. 1979. Control of *Sclerotium rolfsii* and weeds in peanuts by solar heating of the soil. *Plant Dis. Rep.* 63:1056–59.

Gurkin, R. S., and S. F. Jenkins. 1985. Influence of cultural practices, fungicides and inoculum placement on Southern blight and *Rhizoctonia* crown rot of carrot. *Plant Dis.* 69:477–81.

Harrison, A. L. 1961. Control of *Sclerotium rolfsii* with chemicals. *Phytopathology* 51:124–28.

Henis, Y., and G. C. Papavizas. 1983. Factors affecting germinability and susceptibility to attack of sclerotia of *Sclerotium rolfsii* by *Trichoderma harzianum* in field soil. *Phytopathology* 73:1469–74.

Leach, L. D., and A. E. Davey. 1938. Determining the sclerotial population of *Sclerotium rolfsii* by soil analysis and predicting losses of sugar beets on the basis of these analyses. *J. Agric. Res.* 56:619–32.

———. 1942. Reducing southern *Sclerotium* rot of sugarbeets with nitrogenous fertilizers. *J. Agric. Res.* 64:1–18.

Lifshitz, R., M. Tabachnik, J. Katan, and I. Chet. 1983. The effect of sublethal heating on sclerotia of *Sclerotium rolfsii*. *Can. J. Microbiol.* 29:1607–10.

Linderman, R. G., and R. G. Gilbert. 1973. Behavior of sclerotia of *Sclerotium rolfsii* produced in soil or in culture regarding germination stimulation by volatiles, fungistasis, and sodium hypochlorite treatment. *Phytopathology* 63:500–04.

Mathur, S. B., and A. K. Sarbhoy. 1976. Physiological studies on *Sclerotium rolfsii* causing root rot of sugarbeets. *Indian Phytopathol.* 29:454–55.

Mehrotra, R. S., and G. R. Caludius. 1972. Biological control of the root rot and wilt diseases of *Lens culinaris*. Medic. *Plant Soil.* 36:657–64.

Mihail, J. D., and S. M. Alcorn. 1984. Effects of soil solarization on *Macrophomina phaseolina* and *Sclerotium rolfsii*. *Plant Dis.* 68:156–59.

Mordue, J. E. M. 1974. *Corticum rolfsii*. CMI Descrip. Pathogen. Fungi Bact. 410, 2 pp.

Punja, Z. K. 1985. The biology, ecology, and control of *Sclerotium rolfsii*. *Ann. Rev. Phytopathol.* 23:97–127.

———. 1986. Relationships among soil depth, soil texture, and inoculum placement in infection of carrot roots by eruptively germinating sclerotia of *Sclerotium rolfsii*. *Phytopathology* 76:976–80.

Punja, Z. K., J. D. Carter, G. M. Campbell, and E. L. Rossell. 1985. Effects of calcium and nitrogen fertilizers, fungicides, and tillage practices on incidence of *Sclerotium rolfsii* on processing carrots. *Plant Dis.* 70:819–24.

Punja, Z. K., and R. G. Grogan. 1981a. Eruptive germiantion of sclerotia of *Sclerotium rolfsii*. *Phytopathology* 71:1092–99.

———. 1981b. Mycelial growth and infection without a food base by eruptive germinating sclerotia of *Sclerotium rolfsii*. *Phytopathology* 71:1099–1103.

———. 1982. Chemical control of *Sclerotium rolfsii* on golf greens in northern California. *Plant Dis.* 66:108–11.

———. 1983. Germination and infection by basidiospores of *Athelia (Sclerotium) rolfsii*. *Plant Dis.* 67:875–78.

Punja, Z. K., R. G. Grogan, and G. C. Adams, Jr. 1982. Influence of nutrition, environment, and the isolate on basidiocarp formation, development, and structure in *Athelia (Sclerotium) rolfsii*. *Mycologia* 74:917–26.

Punja, Z. K., J. S. Huang, and S. F. Jenkins. 1985. Relationship of mycelial growth and production of oxalic acid and cell wall degrading enzymes to virulence in *Sclerotium rolfsii*. *Can. J. Plant Pathol.* 7:109–17.

Punja, Z. K., and S. F. Jenkins. 1984. Influence of temperature, moisture, modified gaseous atmosphere, and depth in soil on eruptive sclerotial germination of *Sclerotium rolfsii*. *Phytopathology* 74:749–54.

Punja, Z. K., V. L. Smith, C. L. Campbell, and S. F. Jenkins. 1985. Sampling and extraction procedures to estimate numbers, spatial pattern, and temporal distribution of sclerotia of *Sclerotium rolfsii* in soil. *Plant Dis.* 69:469–74.

Rodriguez-Kabana, R., P. A. Backman, H. Ivey, and L. L. Farrar. 1972. Effect of post-emergence application of potassium azide on nematode populations and development of *Sclerotium rolfsii* in a peanut field. *Plant Dis. Rep.* 56:362–67.

Rodriguez-Kabana, R., P. A. Backman, and P. S. King. 1976. Antifungal activity of the nematicide ethoprop. *Plant Dis. Rep.* 60:255–59.

Rodriguez-Kabana, R., P. A. Backman, and E. A. Wiggins. 1974. Determination of sclerotial populations of *Sclerotium rolfsii* in soil by a rapid flotation-sieving technique. *Phytopathology* 64:610–15.

Rodriguez-Kabana, R., M. K. Beute, and P. A. Backman. 1980. A method for estimating numbers of viable sclerotia of *Sclerotium rolfsii* in soil. *Phytopathology* 70:917–19.

Shay, J. R. 1953. Southern blight on apple nursery stock in Indiana. *Plant Dis. Rep.* 37:121.

Shaw, B. B., and M. K. Beute. 1984. Effects of crop management on the epidemiology of southern stem rot of peanut. *Phytopathology* 74:530–35.

Smith, A. M. 1972a. Drying and wetting sclerotia promotes biological control of *Sclerotium rolfsii* Sacc. *Soil Biol. Biochem.* 4:119–23.

———. 1972b. Nutrient leakage promotes biological control of dried sclerotia of *Sclerotium rolfsii* Sacc. *Soil Biol. Biochem.* 4:125–29.

Smith, R. E. 1941. *Diseases of fruits and nuts.* Calif. Agric. Exp. Stn. Circ. 120, 168 pp.

Sztejnberg, A., and J. Golan. 1980. Changes in the occurrence of fruit crop diseases as a result of changes in irrigation measures. In *Proceedings of the Congress of Mediterranean Phytopathological Union, 5th.* 127–128.

Taylor, J., and N. E. McGlohon. 1975. Severity of some common apple diseases in middle Georgia orchards. *Ga. Agric. Res.* 16:8–11, 14.

Tomasino, S. F., and K. E. Conway. 1987. Spatial pattern, inoculum density-disease relationship, and population dynamics of *Sclerotium rolfsii* on apple rootstock. *Plant Dis.* 71:719–24.

Turner, T. W. 1936. Pathogenicity of *Sclerotium rolfsii* for young apple trees (abs.) *Phytopathology* 26:111.

Wells, H. D., D. K. Bell, and C. A. Jarworski. 1972. Efficacy of *Trichoderma harzianum* as a biocontrol for *Sclerotium rolfsii*. *Phytopathology* 62:442–47.

Yuen, G. Y., and R. D. Raabe. 1984. Effects of small-scale aerobic composting on survival of some fungal plant pathogens. *Plant Dis.* 68:134–36.

Virus and Viruslike Diseases of Pome Fruit

During the past 50 years a large number of graft-transmissible diseases have been described on pome fruit trees. Since no living organism could be associated with these disorders, it was generally assumed that they were caused by viruses. However, for only a few of these diseases is there definitive evidence that the cause is a virus. A few are now known to be incited by mycoplasmalike organisms, but for many the identity of the causal agent still remains uncertain. Since most of these disorders of unknown etiology have many of the characteristics of virus diseases, they often are referred to as "viruslike" diseases. At present, it seems best to indicate merely that these diseases are caused by one or more unidentified, graft-transmissible pathogens.

According to Mayhew (1981) in his "Index of plant virus diseases in California," the recorded virus diseases of *Malus sylvestris (M. domestica)* in California are apple stem pitting, apple green crinkle, apple mosaic, apple star cracking, and apple flat limb. On *Pyrus communis*, he reported pear blister canker, pear stony pit, pear vein yellows, red mottle, and pear decline (now known to be caused by a mycoplasmalike organism). On *Cydonia oblonga*, only quince sooty ringspot was listed. Mayhew did not differentiate known virus diseases from those caused by either mycoplasmalike organisms or unidentified, graft-transmissible pathogens. Since most of the virus and viruslike diseases of pome fruit are not considered to be of major importance in California and since most of them have not been extensively studied here, only a rather limited description of most of them is presented in this publication. Virus and viruslike diseases of apple that occur in the Pacific Northwest but not, apparently, in California (**fig. 26**), include blister bark, false sting, flat apple, rough skin, and russet ring (C. L. Parish, personal communication). Information on virus and viruslike diseases of pome fruit has been obtained primarily from publications by Fridlund (1989), Mayhew (1981), Németh (1986), Posnette (1963a), Smith (1972), and Wood (1979).

Apple Flat Limb

Graft-Transmissible Pathogen

Flat limb, or *twist* as it is known in Australia (Broadfoot 1956), is a disease caused by a graft-transmissible pathogen that occurs in most apple-growing countries. According to Hockey (1943), the disease was described in Nova Scotia in 1887 and in the United States in 1906 (Clinton). Symptoms of flat limb are not common because in many cultivars the disease is latent. While the Gravenstein cultivar is probably the most susceptible, the disease has been noted in others, including Wagener, Golden Pearmain, and Ontario. When the new cultivar Idared was introduced and topworked onto old Gravenstein trees without symptoms, some of the Idared scions developed severe flat-limb symptoms (Stouffer 1981). Symptoms resembling flat limb have been observed in pear and quince (McCrum et al. 1960), mountain ash, *Pyracantha*, cherry, and walnut (Fridlund and Waterworth 1989). The Bradford pear (*Pyrus calleryana*), however, is reported to be a symptomless carrier of the disease (Waterworth 1976). Thomas (1942) considered the disease to be caused by a virus when he showed that rough-bark symptoms developed on *Pyracantha* sp. after inoculation with apple flat-limb material.

The disease on the Gravenstein cultivar and its sports can seriously reduce the production of fruit, but its economic effects on symptomless cultivars are not known (Wood 1979, p. 26).

Symptoms. Unequal enlargement of one or more sides of the trunk or a branch causes a flattening or furrowing that may result in severe twisting (**fig. 16**) (Wood 1979). The tissue in the concave areas of the limb may later develop secondary decay. A lack of wood (xylem) development is responsible for the concavity. Apparently, either cambial activity is arrested or new cells developing from the cambium differentiate only into bark (phloem and accompanying tissue) (McCrum et al. 1960). McAlpine (1912) published an excellent photograph of Gravenstein apple limbs showing symptoms typical of those of the flat-limb disease. No noticeable symptoms are expressed in the leaves or fruit (Fridlund and Waterworth 1989).

Cause. No definitive research has been done to establish the cause of this graft-transmissible disorder. Because of its infectious nature and because no organism could be isolated from infected trees, the disease, for many years, was assumed to be caused by a virus (Posnette 1963a). In 1976, Schmidt reported the association of a mycoplasmalike organism with flat limb, and Németh (1979) concluded that the disease is caused by the apple rubbery wood mycoplasma. A later report with photographs suggests that the rubbery wood disease is caused by xylem-limited bacteria that resemble Rickettsia (Minoiu and Cracium 1982). Until further research substantiates the pathogen of flat limb, it appears prudent to consider the cause to be an unidentified graft-transmissible pathogen.

Disease development. Symptoms of the disease develop slowly. A period of four to eight years may elapse between inoculation and definite flat-limb symptoms (Hockey 1943). Certain apple cultivars may carry the infectious agent but remain symptomless. Sensitive cultivars such as Gravenstein when grafted onto diseased root pieces show less severe disease symptoms than when grafted onto aboveground parts (Hockey 1943, 1957). Kegler (1964), by means of extensive transmission tests, showed that the pathogen only moves upward.

Control. The only certain way to avoid the disease is to utilize cultivars that do not develop flat limb or to propagate disease-free cultivars on disease-free stock. In the event the understock has a latent infection, it is important to know which cultivars are highly susceptible to flat limb (Atkinson 1955). The Gravenstein cultivar is a satisfactory indicator plant, with symptoms often evident the first year following inoculation but sometimes not until after several years. If the pathogen is a mycoplasmalike or Rickettsia-like organism, then healthy propagation material should be easily obtained through thermotherapy (Fridlund and Waterworth 1989).

Apple Graft-Incompatibility

Cause Unknown

A form of graft-incompatibility was observed recently on second-leaf trees of Granny Smith apple cultivars grafted onto M26 rootstock (J. K. Uyemoto, unpublished). Affected trees showed a swollen union that broke easily. Close examination of bench grafts (cut joints made with the aid of a grafting machine), by means of a longitudinal cut through the union, revealed a necrotic area confined to and outlining the original cut surfaces of the xylem tissues of both the Granny Smith scion and the M26 rootstock (**fig. 17**). Correspondingly, removal of bark tissues from the union area revealed a woody cylinder with an irregular growth pattern consisting of two opposite sides that were flattened or depressed, alternating with two sides that had large overgrowths.

In the literature, graft-incompatibility among dwarfing apple rootstocks was ascribed, in the main, to a genetic cause (Mosse 1962). When M26 were grafted to different scions (including Granny Smith) (Simons 1982; Simons and Chu 1979, 1985), some tree combinations exhibited swollen unions, necrotic xylem tissues, and poorly developed phloem- and xylem-ray cells at or near contact points between the scion and rootstock. However, despite such graft union anomalies, trees of Granny Smith/M26 have had commercial success in Washington state (Auvil 1981).

An alternative explanation for apple graft-incompatibility is the possible involvement of a latent virus. The cause of this disorder is under investigation.

Apple Green Crinkle

Unidentified Graft-Transmissible Pathogen

Apple green crinkle is a disease of worldwide occurrence. It is especially important in New Zealand because of its adverse effect on the Granny Smith cultivar (Wood 1979), but it is not common on other apple cultivars (Thomsen 1989a). In California, the disease has been reported from Humboldt, Santa Clara, and Santa Cruz counties, and with the increase in acreage of Granny Smith in the Central Valley the disease could become more widespread. Green crinkle has been transmitted only to *Malus* species; transmission is by grafting and budding. Experiments in Canada and New Zealand indicate that the incubation period is three to eight years.

Symptoms. The symptoms are largely restricted to the fruit, but on some cultivars growth retardation can also occur. First symptoms appear a few weeks after flowering as small creases in the fruit surface. Fruit distortion occurs as the creases deepen, and in some cultivars is expressed as severe cracking or wartlike swellings (**fig. 18**). The vascular tissue below the swellings or creases is green and distorted. In affected trees, symptomatic fruit may be present on some branches and absent on others. Severe fruit symptoms have been noted on Golden Delicious, Granny Smith, Gravenstein, McIntosh, and Northern Spy. In Granny Smith, the disease causes no apparent retardation of tree growth but markedly reduces the quality of the fruit (Fridlund and Drake 1987). A similar disorder of Red Delicious in the Pacific Northwest is called *false sting* (**fig. 26B**); it is thought to be caused by a different strain of the green crinkle pathogen.

Cause. Green crinkle is attributed to an unidentified graft-transmissible agent. The Golden Delicious and Granny Smith cultivars are useful indicator hosts (Anonymous 1983).

Control. Control is through the use of healthy propagating material. Diseased trees should be eliminated to prevent possible root grafting and to eliminate a possible source of inoculum (Thomson 1989a).

Apple Mosaic

Apple Mosaic Virus (AMV)

Apple mosaic, or infectious variagation, is one of the most commonly occurring virus diseases of apple (Bradford and Joley 1933; McCrum et al. 1960; Mink 1989; Smith 1972). The disease was described early in the nineteenth century (McCrum et al. 1960) and is known in England, New Zealand, North America, and many other apple-growing areas. In the eastern United States, Stewart (1910) observed variegated foliage on apple trees in Long Island, New York, in 1896, and Morse (1916) made similar observations in Maine orchards in 1915. In the Pacific Coast states, Blodgett (1923) found that mosaic could be transmitted by means of grafts; this was apparently the first record of transmission of a virus by grafting and served as an important supplement to Beijerinck's classic work on transmission of tobacco mosaic in 1896 (McCrum et al. 1960). In California, Thomas (1937) noted that symptoms of genetic variagation are similar in some respects to those of apple mosaic. Yarwod (1955) described another sap-transmissible mosaic-type disease of apples that was later classified as Tulare apple mosaic by Gilmer (1958) to set it apart from other known types of apple mosaic.

Symptoms. Some of the leaves on infected apple trees have cream-colored areas. At times the chlorosis is confined to the midvein and veinlets. All leaves on a shoot will not necessarily exhibit symptoms, and some may appear normal or contain only a few light green areas. The chlorotic areas may later become necrotic and the leaves may fall. Leaves that exhibit symptoms early in the season may develop large necrotic areas. Severe defoliation may cause the fruit to sunburn. Chlorotic leaves are also sensitive to spray materials (McCrum et al. 1960; Yarwood 1955). In general, no fruit symptoms have been observed (Atkinson and Chamberlain 1948), but according to Posnette and Cropley (1956) severely infected trees of the red-fruited cultivar Lord Lambourne may develop conspicuous cream-colored patches on the fruit and a reddish brown color of the young shoots. Palmiter (1963) noted fruit deformity on apple mosaic–diseased trees. On inoculated trees, Christow (1934) found phloem necrosis in the root tip as well as in the taproot and stems.

Trees infected with apple mosaic have smaller leaves and less foliage (Corte and Scaramuzzi 1959), reduced girth of trunks (Mallach 1956), and yield reductions that vary from slight to as much as 55 percent (Harris 1949; Mallach 1956; Posnette and Cropley 1956, 1959).

Temperature plays an important role in symptom expression, with temperatures above 80°F tending to mask the mosaic pattern on the leaves (Gilmer 1958). Thus, symptoms may be severe one year and not severe the next. Latent infections have also been reported.

Different types of symptoms (**fig. 19**) induced by strains of the virus were described by Posnette and Cropley (1952) as follows: (1) white or pale yellow flecks often on a veinlet, but frequently bearing no regular relationship to the veins; (2) large white or pale yellow patches extending over two or more of the interveinal areas separated by the main veins; (3) vein-banding, where the primary and secondary veins are banded by white or pale yellow strips; and (4) necrotic areas, usually developing on the chlorotic patches described under type 2.

Cause. The apple mosaic virus is synonymous with the common apple mosaic virus, apple infectious variegation virus, rose infectious chlorosis virus, rose mosaic virus, plum line pattern virus, and European plum line pattern virus (Gilmer 1956; McCrum et al. 1960). The virus particle is about 26 nm in diameter, isometric, and very labile in extracts. As with other Ilar viruses, some anisometric particles are present. The thermal inactivation point of the virus in stabilized extract is 54°C (Mink 1989). Strains ranging from mild to virulent in apple have been described (Kristensen and Thomson 1963; Posnette and Cropley 1959), and strains differ in their ability to infect particular herbaceous species. In apple, mild strains of the virus protect against more virulent strains. The three virus strains—severe, vein-banding, and mild apple mosaic—were categorized based on the symptoms they induced on Bramley's Seedling, Cox's Orange Pippin, Hoomead Pearmain, and Lord Lambourne (Smith 1972). Apple mosaic virus also cross-reacts weakly with the antiserum to *Prunus* necrotic ringspot virus (DeSequeira 1967; Fulton 1968).

Diagnostic host species are cucumber (*Cucumus sativus*), which develops prominent chlorotic primary lesions on cotyledons, with systemic invasion that results in extreme stunting; cowpea (*Vigna sinensis*), which produces chlorotic lines and rings on trifoliate leaves when inoculated with rose isolates (but some isolates from apple do not infect this host); and Lord Lambourne and Golden Delicious cultivars of apple (Anonymous 1983), in which a prominent mosaic develops. Unrelated viruses that cause similar symptoms in some of their woody hosts are the Tulare apple mosaic and American plum line pattern viruses (Fulton 1972; Mink 1989).

Transmission. The apple mosaic virus can be transmitted by grafting (Hunter, Chamberlain, and Atkinson 1958). When young, vigorously growing trees are graft-inoculated, the spring symptoms develop that year or the following year. Cool springtime temperatures and stimulation of new growth by pruning favor symptom development. The minimum incubation period from the time of inoculation by budding or grafting to the development of symptoms is 34 days (Atkinson and Chamberlain 1948; Stewart 1910; Yarwood 1955). There is also a suggestion of seed transmission (Posnette 1963a).

Although the spread of apple mosaic in the field has been reported, the method by which the virus is transmitted from tree to tree has not been established.

One worker suggested that pruning spreads the virus; experiments, however, have not supported this hypothesis. The virus is spread during nursery propagation through the use of infected scions or clonal rootstocks. Transmission tests with sucking insects (McCrum et al. 1960) such as *Myzus persicae, Macrosiphon eriosoma,* and *Apis pomi* have been negative. It is most likely that the virus passes from one tree to another by means of natural root grafts.

The common apple mosaic virus has been transmitted by grafting to a wide range of hosts, including *Malus, Pyrus, Sorbus, Cydonia, Crataegus, Chaenomeles, Prunus,* and *Fragaria* (Fulton 1968; Gilmer 1958; Gotlieb and Berbee 1973; Kilpatrick 1955; McCrum 1968; McCrum and Hilborn 1962; McCrum and Rich 1966; McMorran and Cameron 1983; Posnette 1963a). Serological tests have shown that the apple and rose mosaic viruses are the same (Casper 1973).

Control. The best means of disease control is to obtain nursery stock certified as free of apple mosaic. The benefits of removing trees with symptoms and eradicating infected wild hosts are not clear. Heat therapy has proved largely ineffective in destroying the virus in apple tissue. Apparently, however, the virus is inactivated when young infected plants are held at 37°C for a two- or three-week period.

Treatments with chemicals such as elgetol, hydroquinone, sodium salicylate, and urea have likewise proved ineffective in inactivating the virus (Cheplick and Agrios 1983). Orchardists should closely monitor their production, and when the disease appreciably reduces the yield, severely affected trees should be removed and replanted with healthy stock. Cross-protection with a mild strain of the apple mosaic virus has also been suggested, but Posnette and Cropley (1959) found that the mild strains can seriously reduce crop yields.

Apple Rubbery Wood

Unidentified Graft-Transmissible Pathogen

Rubbery wood disease was first observed in the cultivar Lord Lambourne in England in 1935, but was not described until 1944 by Wallace, Swarbrick, and Ogilvie (1944). In the United States the disease was first detected by Brase and Gilmer (1959) in EM I rootstocks that had been imported from East Malling, England some 31 years earlier. While the apple cultivar Lord Lambourne is most sensitive to this disease, symptoms have been observed on other cultivars such as Gala, Golden Delicious, Jonathan, Red Delicious, and Dartmouth Crab (Cropley 1963). In addition, the pathogen has been detected in many other apple and pear cultivars and rootstocks (Waterworth and Fridlund 1989). Most apple cultivars and clonal rootstocks have latent infections (Németh 1986). Its occurrence in California is suspected but not proven.

Symptoms. The most characteristic symptom of the disease is the conspicuous flexibility of apple stems and branches resulting from the lack of lignification of xylem vessels and fibers (Minoui et al. 1980). The decreased lignification can be demonstrated by staining a transverse section of the wood with phloroglucinol and hydrochloric acid (Németh 1986). The natural weight of the branches alone causes them to droop, and branches 2 to 3 cm in diameter may be bent easily using finger pressure (Atkinson, Chamberlain, and Hunter 1959). Even in young trees, vigorous side shoots of normal rigidity emerge a few centimeters above the ground and grow taller than the main shoot. Symptoms are lacking in the fruit and leaves.

Initial symptoms in young trees may occur one to three years after infection, but in later years the trees may again produce normal wood. Infected trees are very susceptible to winter injury (terminal dieback), which results in proliferation of secondary shoots (Wood and Cassidy 1978).

Cause. The rubbery wood pathogen was first suspected to be a virus and its transmissibility was demonstrated in 1950. In 1971, Beakbane, Mishra, Posnette, and Slater observed mycoplasmalike organisms in the phloem cells of petioles, pedicels, stems, and roots from diseased Lord Lambourne trees. However, Minoui and Cracium (1983), using electron microscopy, showed the presence of xylem limited bacteria in affected trees. Because temperature affects both symptom expression and indexing for the pathogen, the causal agent could be a mycoplasmalike organism as suggested by Németh (1986). Interestingly, a source of rubbery wood when transmitted to mazzard cherry seedling has caused flat limb symptoms. However, the relationship, if any, between the pathogens that cause rubbery wood and apple flat limb remains obscure

(Waterworth and Fridlund 1989). It appears prudent at this time to consider the cause to be an unidentified graft-transmissible pathogen.

Disease development. The pathogen is spread in propagating material selected from diseased trees and in symptomless young nursery stock. Symptom expression on shoots of inoculated, well-fertilized, and rapidly growing indicator trees varies with the strain of the pathogen, and ranges from six months for severe to two years for mild strains. Thus, the disease has probably spread throughout the world through exchange of diseased stocks. In stoolbeds, nurseries, and orchards, transmission can occur by natural root grafting.

Control. As rubbery wood is not frequently transmitted in nature, control is by use of disease-free wood for propagation and topworking. The pathogen can be eliminated by heat therapy or by culturing sphaeroblasts (Németh 1986). It can be partially eliminated from infected scions by a cobalt irradiation technique using a dose of 1 kilorad during the month of March (Minoui and Cracium 1983).

Apple Star Crack

Unidentified Graft-Transmissible Pathogen

Apple star crack occurs in all of the important fruit-growing areas of England and in many other countries, including the United States (Hamdorf 1989; Németh 1986; Posnette 1963a). In California, it has been reported only from Santa Cruz County (Mayhew 1981). In England, 28 of 45 inoculated apple cultivars developed fruit symptoms, with Golden Delicious and Cox's Orange Pippin being especialy susceptible (Németh 1986). Several cultivars, including Granny Smith and Lord Lambourne, are apparently not affected by the disease (Posnette 1963a). Star crack is transmissible by grafting, budding, and chip budding, and has an incubation period of two to four years. Apples are the only known hosts (Hamdorf 1989).

Symptoms. The symptoms on Cox's Orange Pippin include the development of star-shaped cracks on the side and calyx end of the fruit (**fig. 20**) (Posnette 1963a). Cankers that develop around the buds of one-year-old shoots during January often kill the shoot tips. Affected shoots of Cox's Orange Pippin, Queen Cox, and James Grave are delayed one to three weeks in leafing and blossoming, and often show a reduced set of fruit. Some cultivars, such as Idared and Merton Beauty, leaf out four or more days earlier than healthy trees (Hamdorf 1989). Several strains of the pathogen are suspected, the milder ones producing symptoms only on the fruit and the severe ones affecting the fruit, leaves, and shoots (Posnette 1963a; Smith 1972).

Cause. Star crack is attributed to an unidentified graft-transmissible agent. The Golden Delicious and Cox's Orange Pippin cultivars are useful indicator hosts (Anonymous 1983; Hamdorf 1989).

Control. Control is through the use of healthy propagating material. The star crack disease can be eliminated by thermo-therapy (Hamdorf 1989).

Apple Stem Pitting

Apple Stem Pitting Virus

Apple stem pitting has been recorded on Red Delicious in Monterey County (Mayhew 1981). The disease occurs in New Zealand (Wood 1979) and many other parts of the world, but little is known of its economic importance. Of 29 cultivars indexed in New Zealand, all were found to be infected. Of the rootstocks indexed, those found to be infected were M IX, M XII, M XVI, and Northern Spy, and those uninfected were M793, MM104, MM106, and MM115. Transmission tests on apples are difficult to interpret because of the high incidence of this disease in apples. The stem-pitting pathogen has been transmitted by grafting, budding, and chip budding, with symptoms developing one year after inoculation (Posnette 1963a). Natural hosts include one or more species of *Malus*, *Pyrus*, *Crataegus*, and *Sorbus* (Németh 1986).

Symptoms. The wood-pitting symptom (**fig. 21**) is of common occurrence in stems of apple trees, but some cultivars and clonal rootstocks are symptomless carriers of the pathogen. The disease is best identified from symptoms on indicator hosts such as Virginia Crab (**fig. 21**) and Spy 227. On Virginia Crab, elongated or ovate depressions in the wood with corresponding raised areas on the inner surface of the bark are expressed. Where pitting is severe, the overlying bark is thick, dark, and depressed, with longitudinal fissures. Infected trees produce small fruit with grooves that extend from the calyx end to the stem ("flute fruit"), and with a partially russeted skin (Posnette 1963a; Wood 1979). On Spy 227, the symptoms are severe, including epinasty of the leaves, splitting of the bark, and a decline in growth (Wood 1979). Welsh and Uyemoto (1980) reported mild stem pitting and severe stem pitting syndromes for the disease.

Cause. The causal agent has been isolated and identified as a flexuous virus 12 to 15 nm wide and 800 nm long (Stouffer 1989). According to Koganezawa and Yanase (1990), the virus particles readily form end-to-end aggregates, have four prominent peaks in particle length distribution, contain a single species of RNA (M_r 3.1×10^6) and a major coat protein of M_r 48,000; this virus does not appear to fall into any recognized group of plant viruses. Virginia Crab and Spy 227 are useful indicator hosts (Anonymous 1983; Stouffer 1989). In a more rapid and sensitive indexing procedure under greenhouse/laboratory conditions, Radiant or Sparkler Crab apples are used (Fridlund 1980).

Control. Control is through the use of healthy propagating material. The pathogen can be eliminated from the tips of plants kept at 38°C from 14 to 77 days, depending on the cultivar involved (Németh 1986).

Apple Union Necrosis and Decline

Tomato Ringspot Virus (TmRSV)

Union necrosis and decline is a recognized disease of apple orchards in the northeastern United States (Cummins, Uyemoto, and Stouffer 1978; Rosenberger, Harrison, and Gonsalves 1983; Stouffer, Hickey, and Welsh 1977; Stouffer and Powell 1989), and in the state of Washington (Parish and Converse 1981). It was first observed in California (El Dorado County) in 1986 on Red Delicious on MM106 rootstock. The disease could become a serious problem here if there is extensive propagation on this rootstock of highly sensitive cultivars such as Red Delicious, McIntosh, and Tydeman's Early. The TmRSV has been detected in nonbudded MM106 rooted layers from Oregon (Stouffer and Uyemoto 1976).

Symptoms. Affected trees appear unthrifty and frequently show symptoms of decline similar to those resulting from trunk girdling (**fig. 22A**). The leaves are sparse and small and have a pale, dull green color. The terminal growth is reduced and the twigs express some rosetting. Diseased trees tend to flower heavily and commonly set a large number of small fruit with a brightly colored blush. The lateral leaf and flower buds may be killed, and the internodal length of the shoots is reduced. On severely affected trees, the portion of the trunk above the graft union is frequently swollen and may show partial or complete cleavage at the union (Stouffer, Hickey, and Welsh 1977). Removal of bark from the swollen area reveals abnormally thick, spongy bark and a distinct line at the scion-stock interface. At the wood interface there may be small, deep pits or a smooth, narrow suture. The configuration of the union aberration and presence or absence of necrotic tissue are dependent on the scion-rootstock combination. The weakening of the graft union is related to an increase of ray and axial parenchyma cells and a decrease of vessel and fiber cells (Tuttle and Gotlieb 1982). In severely affected trees, there is a proliferation of rootstock sprouts, and breakage at the union is not uncommon (**fig. 22B**).

Cause. There is strong evidence that the causal agent is tomato ringspot virus (TmRSV) (Rosenberger, Cummins, and Gonsalves 1989). This virus, when sap-inoculated on indicator plants such as *Nicotiana tabacum* cv. Turkish, *N. clevelandii*, *Chenopodium quinoa*, and *C. amaranticolor*, incites necrotic lesions on inoculated leaves within 48 to 72 hours. Systemically infected leaves of *Gomphrena globosa* and *Vinca rosea* were found to be good sources of virus inoculum, and serological tests with antiserum specific for TmRSV served to properly identify it. The TmRSV isolates from the state of Washington are serologically distinct from the Pennsylvania isolate (Parish and Converse 1981); likewise, the Oregon isolates are distinct from the New York and Pennsylvania isolates (Stouffer and Uyemoto 1976).

Detection of TmRSV in woody plants has been erratic even with those known to be infected. Bitterlin, Gonsalves, and Cummins (1984) used the ELISA technique to record the irregular distribution of the

TmRSV in apple trees. In unbudded MM106 trees, they detected the virus most consistently in leaves, slightly less consistently in bark, and only erratically in roots. Also, the titer of the virus declines toward the end of the growing season but remains highest in leaves and lowest in roots. In the case of symptomatic trees, TmRSV has been detected in the MM106 rootstock but not in the scions. Rosenberger, Harrison, and Gonsalves (1983), in their study of the incidence of apple union necrosis and decline in New York orchards, isolated the virus from the inner bark (just below the graft union) of a high percentage of symptomatic trees on this rootstock.

Epidemiology. The apple industry in California is rapidly expanding, especially in the Sacramento and San Joaquin valleys where stone fruit historically have been grown. Furthermore, because of favorable horticultural qualities, the highly susceptible MM106 rootstock is being increasingly used. Other TmRSV-susceptible rootstocks include M26, MAC-30, MAC-39, and P-2 (Cummins and Gonsalves 1982). Thus, there is a strong possibility that apple union necrosis will become a serious problem in California orchards.

Other strains of TmRSV cause diseases in California such as yellow bud mosaic of peaches, sweet cherries, and almonds, and brown line of prunes. The nematode vector, *Xiphinema californicum*, is of widespread occurrence in California and is directly involved in the transmission of the virus from tree to tree. Common weed hosts that can serve as virus reservoirs include dandelion (*Taraxacum officinale*), chickweed (*Stellaria media*), henbit (*Laminum amplexicaule*), creeping woodsorrel (*Oxalis corniculata*), common plantain (*Plantago major*), strawberry (*Fragaria virginiana*), sorrel (*Rumex acetosella*), and red clover (*Trifolium pratense*).

Control. The control of apple union necrosis, as discussed by Cummins, Uyemoto, and Stouffer (1978), begins with obtaining disease-free rootstocks produced under a certification program. This may be accomplished by first choosing a nursery site without a history of TmRSV. Next, the soil should be fallowed and fumigated for *Xiphinema* control. This site should be planted with virus-free liners on which TmRSV-free scionwood should be grafted. Early in the second nursery year, rootstock shoots should be indexed on suitable indicator plants or tested by the ELISA serological procedure (Lister et al. 1980) to establish the presence or absence of TmRSV.

Orchard sites with a history of TmRSV should be avoided but, as additional insurance, cultural programs aimed at reducing the population of the nematode vectors and eliminating carryover of the TmRSV in plant roots should be considered. In established orchards, the floor should be maintained free of broadleaved weeds by establishing a permanent grass sod.

Once TmRSV is identified in an orchard but before the onset of disease symptoms, inarching with a nonsensitive rootstock such as M7A or MM111 may prolong the productive life of affected trees. Trees with definite disease symptoms should be killed with chemicals and removed. To avoid spread of the nematode vector, mechanical cultivation should be avoided. Sensitive scion-rootstock combinations, such as Red Delicious/MM106, should be avoided when replanting in sites with a history of union necrosis. However, if the scionwood on MM106 is Golden Delicious or York Imperial, few or no disease symptoms develop even in infected trees. Also, McIntosh and Stayman Winesap trees may be productive for several years after initial infection. For replanting in a diseased orchard, the semidwarfing rootstocks M7A and MM111 would be preferable to MM106 because of their apparent resistance to TmRSV (Cummins, Uyemoto, and Stouffer 1978).

Pear Bark Measles

Unidentified Graft-Transmissible Pathogen

Pear bark measles was observed in California in early 1927 and now occurs also in Oregon and Colorado (Hansen and Waterworth 1989). This disease could be the same as the *blister canker* reported in Europe (Cropley 1960) and in California by Mayhew (1981).

Symptoms. On two- and three-year-old shoots, water-soaked areas in the bark form small blisters that become necrotic and develop cracks. During succeeding years the affected areas expand in a series of concentric rings and older branches develop rough bark with marked ridges and flutes. These symptoms have been observed in nearly all commercial cultivars grown in California except Old Home, but there is some evidence that the disease was spread through interstocks of Old Home used in attempts to control the fireblight disease (Millecan, Nichols, and Brown, Jr. 1962).

Cause. The causal entity is an unidentified graft-transmissible pathogen that may be the same as the one causing pear blister canker (Németh 1986). Seed transmission has been reported by Cordy and Mac-Swan (1961), but such transmission appears to be a minor source of spread. No vectors have been reported for this pathogen.

Control. Propagating material from a source with no history of disease symptoms should be used. If the pathogen is the same as that of blister canker, indicator plants are available.

Pear Blister Canker

Unidentified Graft-Transmissible Pathogen

Pear blister canker is of worldwide distribution (Németh 1986; Thomsen 1989b), but in California it has been found only in El Dorado County (Mayhew 1981). Pear is the only known natural host of blister canker (Thomson 1989b). The disease may be related to pear rough bark in Denmark (Thomsen 1961) and to bark measles in California (Griggs et al. 1961) and Oregon (Cordy and MacSwan 1961). The pathogen is transmitted by budding and grafting, with an incubation period of two to three years (Németh 1986). Preferred indexing hosts are cultivars Bartlett and Winter Nelis (Thomsen 1989b).

Symptoms. On Bartlett and Winter Nelis, the symptoms appear as small bark blisters that enlarge and split (**fig. 23**). The outer part of the affected bark becomes light brown, in contrast to the darker brown of the healthy bark. Necrosis of the phloem rays and sieve tube elements occurs in May and extends into the cortex by late summer (Wolfswinkel 1961). Lesion enlargement may cause girdling and death of affected shoots.

Cause. The cause is an unidentified graft-transmissible agent. Williams (Bartlett) and A20 (both *Pyrus communis*) are indicator hosts (Anonymous 1983).

Control. Control is through the use of healthy propagating material.

Pear Stony Pit

Unidentified Graft-Transmissible Pathogen

Pear stony pit probably occurs wherever pears are grown. (Thomsen 1989c). It has been recorded in four California counties and is considered a serious disease because it can ruin the trees as well as the fruit. Some of the most susceptible cultivars are Beurre Bosc, Beurre d'Anjou, Comice, Conference, Hardy, Old Home, and Winter Nelis. Pear and quince are the only known natural hosts (Németh 1986). Although quince is considered to be susceptible to the disease in Europe, Kienholz (1953) was unable to infect this species with the strain of the pathogen occurring in Oregon. The Williams' Bon Chretien (Bartlett) cultivar is a symptomless carrier of most strains of the pathogen, but expresses symptoms of one U.S. and several European strains (Wood 1979). Slow natural spread of the disease has been observed (Kienholz 1939). Pear trees inoculated by budding developed fruit symptoms in one to two years (Kienholz 1939).

Symptoms. Fruit of sensitive cultivars may show dark green areas under the epidermis within 10 to 20 days after petal fall. These dark green areas make less growth than the surrounding healthy tissue and become pitted (**fig. 24A**) with borders of dark green rings or haloes. Necrotic tissue develops at the base of the pits. A characteristic of stony pit is the formation of sclerenchymatous cells in the tissue adjacent to the pits. All or only a few of the fruit on an affected tree may develop symptoms, and the number of pits per fruit may vary from one to many. According to Kegler et al. (1961), diseased fruit have higher dry matter, ash, potassium, calcium, ascorbic acid, and malic acid than healthy fruit. They also show increased activity of peroxidase and phenoloxidase.

Although fruit symptoms are the most common, Kienholz (1939) described both bark and leaf symptoms on the Bosc cultivar. The bark symptoms are first evident as small pimples on one- and two-year-old twigs. Later, the epidermis in the affected areas cracks, and the underlying tissue collapses to form a pattern resembling oak bark (**fig. 24B**). Leaf symptoms include either narrow chlorotic areas along the veins or a faint mottling; they are most evident on young leaves. Posnette (1963a), however, believes that the leaf symptoms could be attributable to the pear vein yellows pathogen.

Cause. Stony pit for many years was considered to be caused by a virus (Posnette 1963a), and more recently Kegler et al. (1976, 1977) provided additional evidence in support of this view. They characterized a virus isolated from pear trees with stony pit and found it to consist of isometric particles 32 nm in diameter. The virus was sap-transmissible to several herbaceous plants and was vectored by the nematode *Longidorus macrosoma*. Kegler et al. concluded that pear stony pit, pear vein yellows and red mottle, apple stem pitting, and apple spy epinasty and decline are caused by strains of this virus. Although this virus is present in some pear trees with stony pit, conclusive evidence that it is the cause of this disease appears to be lacking. In view of this fact, it seems prudent merely to list the cause as an unidentified graft-transmissible pathogen.

Beurre Bosc is a satisfactory indicator host (Anonymous 1983; Thomsen 1989c).

Control. Control is through the use of healthy propagating material. Top grafting with a tolerant cultivar such as Williams's Bon Chretien (Bartlett) has been recommended (Kienholz 1953).

Pear Vein Yellows and Red Mottle

Apple Stem Pitting Virus

Pear vein yellows and red mottle affect tree growth and crop yield, and probably occurs wherever pears are grown. Although Mayhew (1981) reported vein yellows in five California counties and red mottle in one county, other researchers believe that these disorders represent different expressions of the same disease. Pear and quince are the only natural hosts of the pathogen; quince is a symptomless carrier. According to Wood (1979), the Winter Cole cultivar shows the most conspicuous symptoms of both vein yellows and red mottle. Other susceptible cultivars include d'Anjou, Bartlett, Hardy, Comice, and Conference. Double grafting of healthy and infected buds on the same rootstock results in symptoms on the indicator cultivar the first season.

Symptoms. The symptoms are most pronounced on scion shoots developed during their first year's growth; infected mature trees are frequently symptomless. Small, pale yellow flecks, spots, or blotches, usually associated with the finer veins, develop in spring and become most conspicuous by midsummer (**fig. 25**). The red mottle symptoms appear as scattered, dark red spots or flecks associated with the finer veins; they are most pronounced in late summer and autumn.

Cause. Viruslike, flexuous rods (800 × 20 nm) were detected by Hibino and Schneider (1971) in the leaves of pear trees with symptoms of vein yellows. On the other hand, Kegler et al. (1976, 1977) reported that a virus with isometric particles was associated with this disease. Recently, pear vein yellows, apple stem pitting, and pear necrotic leaf spot were shown to be caused by strains of the same filamentous virus (Yanase, Koganezawa, and Fridlund 1988). *Pyronia veitchii* and *Pyrus communis* (Nouveau Poiteau and A20) are indicator hosts (Anonymous 1983).

Control. Control is through the use of healthy propagating material. By thermotherapy (three weeks at 36°C), shoot tips free of the pathogen can be obtained (Posnette 1963a).

Quince Sooty Ring Spot

Unidentified Graft-Transmissible Pathogen

Quince sooty ring spot is of worldwide distribution (Waterworth 1989), but in California it has been reported only from Placer County (Mayhew 1981). It occurs on quince and pear cultivars as latent infections. Symptoms of epinasty develop on the Quince E and quince indicator clone C7/1 in early summer of the season following inoculation. Later, small black spots or rings, accompanied by vein clearing and yellowing, develop on the upper leaves, which usually fall prematurely. The sooty mold appearance comes from a black pigment that develops in the cuticle of the leaf (Posnette 1957; Posnette and Cropley 1958). The pathogen is readily transmitted by grafting and budding. When buds of a sensitive cultivar (Quince C7/

1, a seedling of Quince E) are inserted into infected rootstocks, they produce weak shoots that grow only a few centimeters long before becoming necrotic at the base, wilting, and dying (Posnette 1963a). Seedlings of *Chaenomeles japonica* are tolerant, but some produce inconspicuous chlorotic leaf patterns (Posnette 1963a).

Cause. Some investigators believe that quince sooty ring spot is a virus disease and that it may be caused by the same virus that is reported to cause pear vein yellows (Németh 1986), and apple stem pitting (Waterworth 1989). Since the evidence for this causal relationship appears inconclusive, it seems best to merely list the cause as an unidentified graft-transmissble pathogen.

Indicator hosts are *Cydonia oblonga* (C 7/1) and *Pyronia veitchii* (Németh 1986).

Control. Control is through the use of healthy propagating material.

REFERENCES

Anonymous. 1983. *Indexing of virus and virus-like diseases of fruit trees.* Working Group Fruit Tree Virus Diseases, Internat. Soc. Hort. Sci., 22 pp.

Atkinson, J. D. 1955. The performance of Malling apple stocks in New Zealand. *Internat. Hort. Cong. 14th Reptg. Sect.* 2D:856–58.

Atkinson, J. D., and E. E. Chamberlain. 1948. Apple-mosaic in New Zealand. *N.Z. J. Sci. Technol. Sect. A* 30:1–4.

Atkinson, J. D., E. E. Chamberlain, and J. A. Hunter. 1959. Apple rubbery wood in New Zealand. *Orch. N. Z.* 32:2–3.

Auvil, G. 1981. What are we doing with Granny Smith on M.26 using Red Delicious as pollenizers. *Compact Fruit Tree* 14:19–22.

Beakbane, A. B., M. D. Mishra, A. F. Posnette, and C. H. Slater. 1971. Mycoplasma-like organisms associated with chat fruit and rubbery wood diseases of apple, *Malus domestica* Borkh., compared with those in strawberry with green petal disease. *J. Gen. Micro.* 66:55–62.

Bitterlin, M. W., D. Gonsalves, and J. N. Cummins. 1984. Irregular distribution of tomato ringspot virus in apple trees. *Plant Dis.* 68:567–71.

Blodgett, F. M. 1923. A new host for mosaic. *Plant Dis. Rep.* 7:11.

Bradford, F. C., and L. Joley. 1933. Infectious variegation in the apple. *J. Agric. Res.* 46:901–08.

Brase, K. D., and R. M. Gilmer. 1959. The occurrence of rubbery wood virus of apple in New York. *Plant Dis. Rep.* 43:157–58.

Broadfoot, H. 1956. Control of "twist" in the Gravenstein apple. *Agr. Gaz. New South Wales* 67:180–88.

Cameron, H. R. 1989. Pear vein yellows. pp. 175–81. In *Virus and viruslike diseases of pome fruits and simulating noninfectious disorders.* ed. P. R. Fridlund. Washington State University, Pullman, WA. 330 pp.

Casper, R. 1973. Serological properties of Prunus necrotic ringspot and apple mosaic isolates from rose. *Phytopathology* 63:238–40.

Cheplick, S. M., and G. N. Agrios. 1983. Effect of injected antiviral compounds on apple mosaic, scar skin, and dapple apple diseases of apple trees. *Plant Dis. Rep.* 67:1130–33.

Christow, A. 1934. Mosaikkrankheit oder Virus-Chlorose bei Apfeln: Eine neue Virus-Krankheit. *Phytopathol. Z.* 7:521–36.

Cordy, C. B., and I. C. MacSwan. 1961. Some evidence that pear bark measles is seed-borne. *Plant Dis. Rep.* 45:891.

Corte, A., and G. Scaramuzzi. 1959. Ricerche sui "mosaico" del melo. *Not. Mal. Piante* 47–48 (N.S. 26–7):122–39.

Cropley, R. 1960. Pear blister canker: A virus disease. East Malling Res. Stn. Rept. 1959, A43:104.

———. 1963. Apple rubbery wood. In *Virus diseases of apples and pears.* Tech. Commun. Bur. Hort. E. Malling. 30:69–72.

Cummins, J. N., and D. Gonsalves. 1982. Recovery of tomato ringspot virus from inoculated apple trees. *J. Am. Soc. Hort. Sci.* 107:798–800.

Cummins, J. N., J. K. Uyemoto, and R. F. Stouffer. 1978. "Union necrosis and decline"—A newly recognized disease of apple trees, associated with tomato ringspot virus. *New York State Hort. Soc. 1978 Proc.* 123:125–28.

DeSequeira, O. A. 1967. Purification and serology of an apple mosaic. *Virology* 30:314–22.

Fridlund, P. R. 1980. The IR-2 program for obtaining virus-free trees. *Plant Dis.* 64:826–30.

Fridlund, P. R., and S. R. Drake. 1987. Influence of apple green crinkle disease on the quality of Granny Smith apples. *Plant Dis.* 71:585–87.

Fridlund, P. R., and H. E. Waterworth. 1989. Apple Flat Limb. pp. 89–93. In *Virus and viruslike diseases of pome fruits and Simulating noninfectious disorders.* ed. P. R. Fridlund. Washington State University, Pullman WA 330 pp.

Fulton, R. W. 1968. Serology of viruses causing cherry necrotic ringspot, plum line pattern, rose mosaic, and apple mosaic. *Phytopathology* 58:635–38.

———. 1972. Apple mosaic virus. *Common. Mycol. Inst. Assoc. Appl. Biol. Descrip. Plant Viruses* 83, 4 pp.

Gilmer, R. M. 1956. Probable co-identity of Shiro line pattern virus and apple mosaic virus. *Phytopathology* 46:127–28.

———. 1958. Two viruses that induce mosaic of apple. *Phytopathology* 48:432–34.

Gotlieb, A. R., and J. G. Berbee. 1973. Line pattern of birch caused by apple mosaic virus. *Phytopathology* 63:1470–77.

Griggs, W. H., H. T. Hartmann, A. A. Millecan, and C. J. Hansen. 1961. Old Home pear is proving valuable as a rootstock in combating pear decline and fire blight. *Calif. Agric.* 15(10):11–13.

Hamdorf, G. 1989. Apple star crack. pp. 69–76. In *Virus and viruslike diseases of pome fruits and simulating noninfectious disorders.* ed. P. R. Fridlund. Washington State University, Pullman, WA. 330 pp.

Hansen, A. J., and H. E. Waterworth. 1989. Other minor infectious diseases of pear. pp. 206–12. In *Virus and viruslike diseases of pome fruits and simulating noninfectious disorders.* ed. P. R. Fridlund. Washington State University, Pullman, WA. 330 pp.

Harris, R. V. 1949. V. Plant pathology. C. Virus diseases. *East Malling Res. Stn. Rept.* 1948:40.

Hibino, H., and H. Schneider. 1971. Virus-like flexuous rods associated with pear vein yellows. *Arch. Ges. Virusforsch.* 33:347–55.

Hockey, J. F. 1943. Mosaic, false sting, and flat limb of apple. *Sci. Agric.* 23:633–46.

———. 1957. Further observations on flat-limb of Gravenstine. *Can. J. Plant Sci.* 37:259–61.

Hunter, J. A., E. E. Chamberlain, and J. D. Atkinson. 1958. Note on transmission of apple mosaic by natural root grafting. *N. Z. J. Agric. Res.* 1:80–82.

Kegler, H. 1964. Untersuchungen über Virosen des Kernobstes. IV Die Flachästigkeit des Apfels. *Phytopathol. Z.* 50:297–312.

Kegler, H., H. Kleinhempel, and T. Verderevskaja. 1976. Investigations on pear stony pit virus. *Acta Hort.* 67:209–18.

Kegler, H., H. Opel, and H. Herzmann. 1961. Untersuchungen über Virosen des Kernobstes. III. Zur Histologie und Pysiologie steinfruchtiger Birnen. *Phytopathol. Z.* 41:42–54.

Kegler, H., T. Verderevskaja, E. Proll, R. Fritzsche, H. B. Schmidt, J. A. Kalasjan, O. I. Kosakovskaja, H. Kleinhempel, and E. Hermann. 1977. Isolierung und Identifizierung eines Virus von Birnen mit Steinfruchtigkeit (pear stony pit). *Arch. Phytopathol. Pflanzenschutz.* 13:297–310.

Kienholz, J. R. 1939. Stony pit, a transmissible disease of pears. *Phytopathology* 29:260–67.

———. 1953. Stony pit of pears. In *Yearbook of agriculture 1953: Plant diseases,* 670–73. U. S. Department of Agriculture.

Kirkpatrick, H. C. 1955. Infection of peach with apple mosaic virus. *Phytopathology* 45:292–93.

Koganezawa, H., and H. Yanase. 1990. A new type of elongated virus isolated from apple trees containing the stem pitting agent. *Plant Dis.* 74:610–14.

Kristensen, H. R., and A. Thomson. 1963. Apple mosaic virus—host plants and strains. *Phytopathol. Mediterr.* 2:97–102.

Lister, R. M., W. R. Allen, D. Gonsalves, A. R. Gottlieb, C. A. Powell, and R. F. Stouffer. 1980. Detection of tomato ringspot virus in apple and peach by ELISA. *Acta Phytopathol. Acad. Sci. Hung.* 15:47–55.

Mallach, N. 1956. Die wirtschlaftliche Bedeutung des Apfelmosaiks. *Prakt. Bl. Pflanzenbau-Pflanzenschutz* 51:225–29.

Mayhew, D. 1981. *Index of plant virus diseases in California.* California State Department of Food and Agriculture, 49 pp.

McAlpine, D. 1912. *Bitter pit investigations: First progress report.* (Australia) Government Printer, Melbourne, 97 pp.

McCrum, R. C. 1968. *Prunus virginiana* a natural host for a mosaic virus of apple. *Plant Dis. Rep.* 52:794.

McCrum, R. C., J. G. Barrat. M. T. Hilborn, and A. E. Rich. 1960. An illustrated review of apple virus diseases. Published jointly as *Maine Agric. Exp. Stn. Bull.* 595 and *New Hampshire Agric. Exp. Stn. Tech. Bull.* 101, 63 pp.

McCrum, R. C., and M. T. Hilborn. 1962. A comparison of apple virus inoculum from stem pitted, dapple apple, flat limb, and mosaic infected trees. *Plant Dis. Rep.* 46:40–44.

McCrum, R. C., and A. E. Rich. 1966. Transmission studies with selective apple virus indicators. *Plant Dis. Rep.* 50:40–44.

McMorran, J. P., and H. R. Cameron. 1983. Detection of 41 isolates of necrotic ringspot, apple mosaic, and prune dwarf virus in *Prunus* and *Malus* by enzyme-linked immunosorbent assay. *Plant Dis.* 67:536–38.

Millecan, A. A., C. W. Nichols, and W. M. Brown, Jr. 1962. Pear bark measles and its association in California with Old Home interstocks. (Abs.) *Phytopathology* 52:363.

Mink, G. I. 1989a. Apple mosaic virus. pp. 34–39. In *Virus and viruslike diseases of pome fruits and simulating noninfectious disorders.* ed. P. R. Fridlund. Washington State University, Pullman, WA. 330 pp.

———. 1989b. Tulare apple mosaic virus. pp. 40–42. In *Virus and viruslike diseases of pome fruits and simulating noninfectious disorders.* ed. P. R. Fridlund. Washington State University, Pullman, WA. 330 pp.

Minoui, N., and C. Cracium. 1989. Electron microscopy detection of apple rubbery wood pathogens and cobalt-therapy of infected fruit trees. *Acta Hort.* 130:313–15.

Minoui, N., M. Isac, K. Pattantyus, M. Stirban, C. Cracium, and M. Straulea. 1980. Experimental results concerning the apple rubbery wood mycoplasma in Romania. *Acta Phytopath. Acad. Sci. Hung.* 15:267–71.

Morse, W. J. 1916. Spraying experiments and apple diseases in 1915. *Maine Agric. Exp. Stn. Bull.* 252:186–87.

Mosse, B. 1962. Graft-incompatibility in fruit trees. *Commonw. Agric. Bur. England Tech. Commun.* 28, 36 pp.

Németh, M. 1979. (*Virus, mycoplasma, and rickettsia diseases of fruit trees*). Hungarian Academy of Science, Budapest, 628 pp.

———. 1986. Virus, mycoplasma and rickettsia diseases of fruit trees. Martinus Nijhoff, Dordrecht, 841 pp.

Palmiter, D. H. 1963. A fruit deformity associated with apple mosaic. *Plant Dis. Rep.* 47:477–78.

Parish, C. L., and R. H. Converse. 1981. Tomato ringspot virus associated with apple union necrosis and decline in western United States. *Plant Dis.* 65:261–63.

Posnette, A. F. 1957. Virus diseases of pears in England. *J. Hort. Sci.* 32:53–61.

———. 1963a. *Virus diseases of apples and pears.* Commonw. Bur. Hort. Plant Crops Tech. Commun. 30, 141 pp.

———. 1963b. Apple mosaic. In *Virus diseases of apples and pears*, 19–21. Commonw. Bur. Hort. Plant Crops Tech. Commun. 30.

Posnette, A. F., and R. Cropley. 1952. A preliminary report on strains of the apple mosaic virus. *East Malling Res. Stn. Rept.* 1951:128–30.

———. 1956. Apple mosaic viruses: Host reactions and strain interactions. *J. Hortic. Sci.* 31:119–33.

———. 1958. Quince indicators for pear viruses. *J. Hortic. Sci.* 33:289–91.

———. 1959. The reduction in yield caused by apple mosaic. *East Malling Res. Stn. Rept.* 1958:89–90.

Powell, C. A., L. B. Forer, and R. F. Stouffer. 1982. Reservoirs of tomato ringspot virus in fruit orchards. *Plant Dis.* 66:583–84.

Rosenberger, D. A., J. N. Cummins, and D. Gonsalves. 1989. Evidence that tomato ringspot virus causes apple union necrosis and decline: Symptom development in inoculated apple trees. *Plant Dis.* 73:262–65.

Rosenberger, D. A., M. B. Harrison, and D. Gonsalves. 1983. Incidence of apple union necrosis and decline, tomato ringspot virus, and *Xiphinema* vector species in Hudson Valley orchards. *Plant Dis.* 67:356–60.

Schmidt, G. 1976. Indexing for flat limb and line pattern. *Mitteil. Biol. Bundesanst. f. Land-und Forstwirtschaft.* H. 170:73.

Simons, R. K. 1982. Scion/rootstock incompatibility in young trees. *Compact Fruit Tree* 16:30–32.

Simons, R. K., and M. C. Chu. 1979. Graft union characteristics: 2. Scarlet Red Stayman/M. 7-A; Double Red Staymared/M.9 and Winesap/M.26. *Compact Fruit Tree* 12:141–44.

———. 1985. Graft union characteristics of M.26 apple rootstock combined with Red Delicious strains—Morphological and anatomical development. *Sci. Hortic.* 25:49–59.

Smith, K. M. 1972. *A textbook of plant virus diseases*, 3d ed. Academic Press, New York, 684 pp.

Stewart, F. C. 1910. Notes on New York plant diseases. *N. Y. Agric. Exp. Stn. Geneva Tech. Bull.* 328:318.

Stouffer, R. F. 1981. Virus diseases of economic importance. *The Bad Apple* 2(1):27.

———. 1989. Apple stem pitting. pp. 138–144. In *Virus and Viruslike Diseases of Pome Fruits and Simulating Noninfectious Disorders.* ed. P. R. Fridlund. Washington State University, Pullman, WA. 330 pp.

Stouffer, R. F., K. D. Hickey, and M. F. Welsh. 1977. Apple union necrosis and decline. *Plant Dis. Rep.* 61:20–24.

Stouffer, R. F., and C. A. Powell. 1989. Apple union necrosis. pp. 148–56. In *Virus and viruslike diseases of pome fruits and simulating noninfectious disorders.* ed. P. R. Fridlung. Washington State University, Pullman, WA. 330 pp.

Stouffer, R. F., and J. K. Uyemoto. 1976. Association of TmRSV with apple union necrosis and decline. *Acta Hortic.* 67:203–08.

Thomas, H. E. 1937. Apple mosaic. *Hilgardia* 10:581–87.

———. 1942. Transmissible rough-bark diseases of fruit trees. *Phytopathology* 32:435–36.

Thomsen, A. 1961. Split bark of pears. In *Proceedings of the 4th Symposium on Virus Diseases of Fruit Trees in Europe.* (Tidsskr. Planteavl 65:69–72).

———. 1989a. Green crinkle. pp. 55–57. In *Virus and viruslike diseases of pome fruits and simulating noninfectious disorders.* ed. P. R. Fridlund. Washington State University, Pullman, WA. 330 pp.

———. 1989b. Pear rough bark and blister canker. pp. 202–05. In *Virus and viruslike diseases of pome fruits and simulating noninfectious disorders.* ed. P. R. Fridlund. Washington State University, Pullman, WA. 330 pp.

———. 1989c. Pear stony pit. pp. 182–87. In *Virus and viruslike diseases of pome fruits and simulating noninfectious disorders.* ed. P. R. Fridlund. Washington State University, Pullman, WA. 330 pp.

Tremaine, J. H., R. S. Willison, and M. F. Welsh. 1963. Viruses mechanically transmitted to herbaceous hosts from apple trees with mosaic symptoms. *Phytopathol. Med.* 2:144–46.

Tuttle, M. A., and A. R. Gotlieb. 1982. Histological study of graft union necrosis in apple associated with TmRSV infection (abs.). *Phytopathology* 72:267.

Wallace, T., T. Swarbrick, and L. Ogilvie. 1944. Some new troubles in apples with special reference to the variety Lord Lambourne. *Fruitgrower* 98:427.

Waterworth, H. E. 1976. Effect of some pome fruit viruses, mycoplasmalike, and other pests on Bradford ornamental pear trees. *Plant Dis. Rep.* 60:104–05.

———. 1989. Viruslike diseases of quince. pp. 213–17. In *Virus and viruslike diseases of pome fruits and simulating noninfectious diseases.* ed. P. R. Fridlund. Washington State University, Pullman, WA. 330 pp.

Waterworth, H. E., and P. R. Fridlund. 1989. Apple rubbery wood. In *Virus and viruslike diseases of pome fruits and simulating noninfectious disorders.* ed. P. R. Fridlund. Washington State University, Pullman, WA. 330 pp.

Welsh, M. F., and J. K. Uyemoto. 1980. Differentiation of syndromes in apple by graft-transmissible, xylem-affecting agents. *Phytopathology* 70:349–53.

Wolfswinkel, L. D. 1961. Studies in viruses infecting the pear, *Pyrus communis* L. Thesis, University of London.

Wood, G. A. 1979. Virus and viruslike diseases of pome fruits and stone fruits in New Zealand. N. Z. Dept. Sci. Ind. Res. DSIR Bull. 226, 87 pp.

Wood, G. A., and D. C. Cassidy. 1978. Effect of rubbery wood disease on growth and yield of Golden Delicious apple trees. *Orch. N. Z.* 51:66–7.

Yanase, H., H. Koganezawa, and P. R. Fridlund. 1988. Correlation of pear necrotic spot with pear vein yellows and apple stem pitting, and a flexuous filamentous virus associated with them. (abs.) *Acta Hort.* 235:157.

Yarwood, C. E. 1955. Mechanical transmission of an apple mosaic virus. *Hilgardia* 23:613–28.

ABIOTIC ORCHARD DISORDERS OF POME FRUIT

Bitter Pit of Apple and Pear

First described on apple in Germany under the name "stippen," bitter pit appears under this name in some earlier American literature. Other names are "fruit spot" and "Baldwin spot." Cobb, who reported the disease in Australia, probably was the first to call it "bitter pit" (Cobb 1892, 1895, 1898; Jaeger 1869; Jones 1891; Jones and Orton 1899; Mix 1916). A similar disorder, sometimes called "cork spot" or "Anjou pit," occurs to some extent in pears, especially in the Anjou and Packham's Triumph cultivars.

Bitter pit was once confused with the internal cork manifestion of boron deficiency, Jonathan spot, and Phoma fruit spot (Brooks and Fisher 1914, 1918). The disorder has received much attention from investigators throughout the world. Early in this century it was extensively studied by an Australian government commission, which published five reports totaling more than 900 pages (McAlpine 1911–16).

The disease received little attention in the United States until 1914, the year Brooks and Fisher (1918) began investigations in the Wenatchee Valley of central Washington. Their studies showed a close relationship between orchard practices, especially irrigation methods, and the incidence of the disorder.

British studies of bitter pit were published under the auspices of the Imperial Bureau of Fruit Production and the Food Investigation Board of the Department of Scientific and Industrial Research (Barker 1928, 1932, 1934; Kidd and West 1923; Smith 1926; Wallace and Jones 1939). These were mainly concerned with storage aspects of the disorder in apples from overseas. Contemporary studies on field aspects of bitter pit were conducted in Australia, New Zealand, and South Africa (Adam 1924; Atkinson 1935; Carne 1927, 1928, 1948; Carne and Martin 1934; Carne, Pittman, and Elliott 1929, 1930; Evans 1909, 1911). Other studies on bitter pit were made at Cornell University (Smock 1936, 1941; Smock and Van Doren 1937).

Symptoms. Bitter pit shows as brown spots in the fruit flesh. In apple, many of the spots occur in the outer cortex and are visible as dark pits in the skin (**fig. 27A**). These pits are 2 to 6 mm in diameter, more or less circular in shape, and sharply delimited in outline. The internal affected areas are globose in shape, brown, and spongy. Frequently, they are more numerous in the flesh around the calyx half of the fruit. Some have a distinctly bitter taste. In pear, the lesions are less visible externally than in apple.

Affected spots on the fruit are located around vascular bundles, commonly near the end of a bundle. A slight plasmolysis of the cells first develops, then the walls of some of these cells collapse. By this time the discoloration is visible to the unaided eye as a small brown spongy area. There is little evidence of suberization, cutinization, or other chemical changes in the cell walls, though some changes in pectin content occur in the middle lamella. A striking microscopic feature of the cells is the presence of starch grains, which have disappeared from the surrounding unaffected cells but are retained in the cells affected by bitter pit. In the early stages of bitter pit development, the skin is not visibly affected, but as the underlying flesh cells collapse, the skin becomes somewhat distorted although seldom broken. Some workers have thought that histological differences exist between bitter pit that develops while the fruit is on the tree and bitter pit that develops after the fruit is stored. Others, however, have disagreed (Carne 1928; Carne, Pittman and Elliott 1929, 1930; MacArthur 1940; Smock and Van Doren 1937).

Cultivar susceptibility. Notable for their susceptibility to bitter pit in most apple regions of North America are Baldwin, Northern Spy, Gravenstein, York Imperial, and Stayman Winesap. The cultivars Fameuse, Grimes Golden, Yellow Bellflower, Rhode Island Greening, Delicious, McIntosh, Jonathan, and Rome Beauty exhibit the disorder more commonly in some apple-growing regions than in others. Other cultivars known to be susceptible include Cox's Orange Pippin, Fuji, Golden Delicious, Granny Smith, Jonagold, Maigold, Prima, Starking, and Starkrimson. In certain years, few if any of the better cultivars have remained free of the disorder. With respect to the newer apple cultivars, Haralson, Lawfam, Lobo, Red Gravenstein, Spartan, and Stonetosh are said to be less susceptible than some of the older cultivars (Brooks and Fisher 1918; Cummings and Dunning 1940; Jones 1891; Jones and Orton 1899; Porritt, Meheriuk, and Lidster 1982; Smock 1941). Highly susceptible pear cultivars are Anjou and Packham's Triumph (Porritt, Meheriuk, and Lidster 1982).

Development of the disorder. Before discussing etiology, it is well to consider how this puzzling disorder is affected by certain factors.

Bitter pit develops either before or after the fruit is picked. The preharvest outbreaks occur when the fruit is ripening; the postharvest outbreaks occur in the first six or eight weeks after the fruit is stored. Fruit with no visible symptoms when picked may become severely affected in storage. Symptoms that develop while the fruit are on the tree become more pronounced in storage.

The disorder varies greatly from year to year. In the same year, it may be more severe in one orchard than in another, more prevalent on some trees than on others, and may even vary greatly between fruit in the same cluster. Fruit borne at the ends of fruit clusters may develop less bitter pit than those borne laterally (Heinicke 1921).

Judging from these variations from one season to the next, one would expect to find certain environmental factors to be influential in the development of bitter pit. One such factor is soil moisture. Both high soil moisture and moisture stress are reported to enhance the development of bitter pit, and so is high soil moisture accompanied by a large fluctuation in temperature. Evidence concerning the effects of fluctuation in temperature alone is conflicting: some have linked bitter pit to a diurnal temperature that varies greatly, whereas others have failed to find such a relationship (Brooks and Fisher 1914, 1918; Smock 1941; Wallace and Jones 1939; Wilton 1975).

The fruit from trees high in vegetative vigor develop bitter pit more severely than do fruit from low-vigor trees. High levels of nitrogen or potassium in the tree are said to predispose fruit to the disorder, but phosphorous and sulfur apparently counteract the effects of these elements (Burrell 1938; Butler and Dunn 1941; Hill and Davis 1936–37; Mix 1916; Mulder 1951; Smock 1941).

At one time bitter pit was thought to be associated with a boron deficiency in the soil, but injections of boron, iron, and certain other minor elements do not affect the development of the disorder. "Internal cork," as the boron deficiency disorder is called, differs significantly from bitter pit (Atkinson 1935; Wallace and Jones 1939).

Bitter pit is more serious when the crop is light than when it is heavy; also, it is more severe on large than on small fruit. Fruit from the more shaded parts of the tree are more prone to bitter pit than fruit from less shaded parts of the tree. Apparently this is not a temperature effect, as the cheek of the apple exposed to direct sunlight is no less liable to develop bitter pit than is the opposite cheek (McAlpine 1912–13; Palmer 1943; Smock 1936, 1941).

Severely pruning the tree in late summer increases the susceptibility of the fruit to bitter pit (Barker 1928, 1932; Scott, Hardisty, and Stafford 1980).

Fruit that are picked before maturity often become more severely affected than fruit allowed to ripen on the tree. The first fruit to mature on a tree develop more bitter pit in storage than fruit that ripen later (Adam 1924; Allen 1932; Brooks and Fisher 1914; Carne 1928; Carne, Pittman, and Elliott 1929; Smith 1926).

Storage conditions influence development of the disease. Fruit held for long periods are usually stored at temperatures of 0° to 1.1°C, but before being sold they are ripened at about 21°C. Bitter pit develops more slowly during storage than at ripening temperature; however, cold-stored fruit that are then ripened at 21°C may develop more of the disorder than fruit held constantly at 21°C until ripe (Adam 1924; Allen 1932; Carne, Pittman, and Elliott 1929; Scott, Hardisty, and Stafford 1980; Smith 1926). An initial period of precooling, in which the temperature of the fruit is quickly lowered to 1.1°C prior to storage at that temperature, suppresses bitter pit development more than storage at 1.1°C (Smith 1926; Smock 1941). In the Granny Smith cultivar, less bitter pit developed in fruit stored at 0°C than at 3°C (Scott and Wills 1979).

A high relative humidity surrounding the fruit or a high carbon dioxide atmosphere inhibits development of bitter pit. The removal of these conditions is followed by development of the disorder (Hewett 1984; Smith 1926).

Coating fruit with certain types of waxes delays the appearance of bitter pit, the delay being somewhat greater at ripening temperatures than at storage temperatures. Waxing retards ripening and reduces water loss, so it is not known whether one or both of these effects are responsible for the delay (Smock 1941).

Cause. Numerous views as to the cause of bitter pit have been entertained. Those featuring microorganisms or insects have long since been abandoned. Fungi isolated from bitter pit tissue proved to be secondary invaders (Farmer 1907; Jones 1891; Jones and Orton 1899). Stigmonose, a malformation of fruit caused by the apple aphid that at one time was considered identical with bitter pit, differs from it in important respects (Brooks and Fisher 1914, 1918).

In 1934, bitter pit was reported to be a virus disease transmissible by budding (Atanasoff 1933–34, 1934–35). In 1962, Jackson reported that, of numerous chemical elements tested, only calcium, as calcium nitrate, significantly altered the incidence of bitter pit when applied as a preharvest spray to the trees or postharvest spray to the fruit. Hilkenbaumer (1970) later reported that bitter pit is closely related to low calcium levels in the fruit: areas of calcium deficiency in the fruit are affected by bitter pit symptoms. Calcium applied to the fruit reduces the incidence of the disorder.

Control. That a shortage of calcium can cause bitter pit has been confirmed by several investigations (Jackson 1962; Martin, Wade, and Stockhouse 1962; Melville, Hardisty, and Shorter 1964). Therefore, the control recommended is to spray trees with calcium nitrate or calcium chloride; calcium nitrate is recommended, applied shortly after the trees are through blooming and again one to two months later (Jackson 1967; Melville, Hardisty, and Shorter 1964; Raphael and Richards 1962; Stiles 1964).

Recent studies (Scott, Hardisty, and Stafford 1980; Scott and Wills 1979; Watkins et al. 1982; Webster and Forsyth 1979) have shown that the incidence of bitter pit can be markedly reduced by the use of postharvest dips containing calcium chloride and a suitable wetting agent. With some cultivars, adequate absorption of calcium was obtained only through the use of vacuum or pressure infiltration. Some injury to cultivars with an open calyx was reported (Scott and Wills 1979). Minimizing water stress in trees has reduced bitter pit in irrigated orchards. Fruit harvest at prime maturity is important in that maturity and rate of ripening affect disease development. Controlled atmosphere storage also reduces incidence of bitter pit (Hewett 1984).

REFERENCES

Adam, D. B. 1924. Experiments in the storage of fruit bitterpit. *J. Dept. Agric. Victoria, Austral.* 22:577–90.

Allen, F. W. 1932. Maturity and rate of ripening of Gravenstein apples in relation to bitter pit development. *Proc. Am. Soc. Hort. Sci.* 28:639–45.

Atanasoff, D. 1933–34. Bitter pit of apples: A virus disease? *Yearbook Univ. Sofia Fac. Agric.* 12:31–67.

———. 1934–35. Bitter pit of pome fruit is a virus disease. *Yearbook Univ. Sofia Fac. Agric.* 13:1–8. (Also *Phytopathol. Z.* 7:145–68, 1934.)

Atkinson, J. D. 1935. Progress report on the investigations of corky pit of apples. *N. Z. J. Sci. Tech.* 16:316–19.

Barker, J. 1928. Wastage in Australian fruit exported to England. *J. Counc. Sci. Indus. Res. Austral.* 1:261–67.

———. 1932. The prevention of wastage in New Zealand fruit. Dept. Sci. Indus. Res. Food Invest. Bd. (Eng.) Spec. Rept. 39.

———. 1934. Annotated bibliography on bitter pit. Imp. Bur. Fruit Prod. Occ. Paper 3, 28 pp.

Bowles, E. A. 1912. Bitter pit in apples. *J. Roy. Hort. Soc.* 37:part 3.

Breidahl, H. C., and A. C. H. Rothera. 1914–15. Bitter pit and sensitivity of apples to poisons. *Proc. Roy. Soc. Victoria Austral.* N. S. 27:191–97.

Brooks, C., and D. F. Fisher. 1914. Jonathan spot, bitter pit and stigmonose. *Phytopathology* 4:402–03.

———. 1918. Irrigation experiments on apple spot diseases. *J. Agric. Res.* 12:109–37.

Burrell, A. B. 1938. Control of internal cork of apple with boron (abs.). *Phytopathology* 28:4.

Butler, O. R., and S. Dunn. 1941. Studies on the bitter pit disease of apples. *New Hamp. Agric. Exp. Stn. Tech. Bull.* 78, 10 pp.

Carne, W. M. 1927. A preliminary note on a theory as to the origin of bitter pit in apples. *J. Dept. Agric. West. Austral.* 2d ser. 4:382–85.

———. 1928. Bitter pit in apples: Some recent investigations. *J. Counc. Sci. Indus. Res. Austral.* 1:358–65.

———. 1948. The non-parasitic disorders of apple fruit in Australia. *Austral. Counc. Sci. Indus. Res. Bull.* 238, 83 pp.

Carne, W. M., and D. Martin. 1934. Apple investigations in Tasmania: Miscellaneous notes. I. The virus theory of bitter pit. *J. Counc. Sci. Indus. Res. Austral.* 7:203–14.

Carne, W. M., H. A. Pittman, and H. G. Elliott. 1929. Studies concerning the so-called bitter pit of apples in Australia; with special reference to the variety Cleopatra. *J. Counc. Sci. Indus. Res. Austral. Bull.* 41, 88 pp.

———. 1930. Notes on wastage of non-parasitic origin in sorted apples. *J. Counc. Sci. Indus. Res. Austral.* 3:167–203.

Cobb, N. A. 1892. Another obscure disease of the apple. *Agric. Gaz. New South Wales* 3:1004.

———. 1895. Bitter pit of apples. *Agric. Gaz. New South Wales* 6:859.

———. 1898. Bitter pit. *Agric. Gaz. New South Wales* 9:663.

Crabill, C. H., and H. E. Thomas. 1916. Stippen and spray injury. *Phytopathology* 6:51–54.

Cummings, M. B., and R. G. Dunning. 1940. Bitter pit of apples. I. In orchard and in storage. *Vermont Agric. Exp. Stn. Bull.* 467:3–30.

Evans, I. B. P. 1909. Bitter pit of apple. *Transvaal Dept. Agric. Pretoria Tech. Bull.* 1, 18 pp.

———. 1911. Bitter pit of the apple. *South Afr. Dept. Agric. Tech. Bull.* 2.

Ewart, A. J. 1911–15. On bitter pit and the sensitivity of apples to poisons. *Proc. Roy. Soc. Victoria Austral.* N.S. 24:367–419, 1911–12; 26:12–44, 1913–14; 26:228–42, 1913–14; 27:342–49, 1914–15.

———. 1917. Cause of bitter pit. *Proc. Roy. Soc. Victoria Austral.* N.S. 30:15–20.

Farmer, J. B. 1907. Bitter pit in cape apples. *Kew. Bull. Misc. Inform.* 6:250.

Greene, G. M., and C. B. Smith. 1988. Physiological disorders and calcium nutrition in apples. *Compact Fruit Tree.* 21:121–29.

Greig-Smith, R. 1911. Note on bitter pit of apples. *Proc. Linn. Soc. New South Wales* 36:158.

Heinicke, A. J. 1921. The seed content and the position of the fruit as factors influencing stippen in apples. *Proc. Am. Soc. Hort. Sci.* 17:225–32.

Herbert, D. A. 1922. Bitter pit in apples: The crushed cell theory. *Phytopathology* 12:489–91.

Hewett, E. W. 1984. Bitter pit reduction in "Cox's Orange Pippin" apples by controlled and modified atmosphere storage. *Sci. Hort.* 23:59–66.

Hilkenbaumer, F. 1970. The present position of bitter pit investigations. *Qual. Pl. Mater. Veg.* 19(1–3):267–74.

Hill, H., and M. B. Davis. 1936–37. Physiological disorders of apples. *Sci. Agric.* 17:199–208.

Hutton, K. E. 1947. Bitter pit of pome fruit. *Agric. Gaz. New South Wales* 58:205–08.

Jackson, D. I. 1962. The effect of calcium and other minerals on incidence of bitter pit on Cox's Orange apples. *N. Z. J. Agric. Res.* 5:302–09.

———. 1967. Bitter pit and calcium nitrate. *Orchard. N. Z.* 40:395.

Jaeger, G. 1869. Über das Pelsig- oder Stippigwerden der Kernobstfrucht. *Illus. Monatsch. Obst- und Weinbau:* 318–19.

Jones, L. R. 1891. A spot disease of the Baldwin apple. *Vermont Agric. Exp. Stn. Ann. Rept.* 5:133–34.

Jones, L. R., and W. A. Orton. 1899. The brown spot of the apple. *Vermont Agric. Exp. Stn. Ann. Rept.* 12:159–64.

Kaiser, P. 1923. Die Stippfleckenkrankheit der Äpfel. *Gartenwelt* 57:204–05.

Kidd, F., and C. West. 1923. Brown heart, a functional disease of apples and pears. *Dept. Sci. Indus. Res. (Eng.) Food Invest. Bd. Spec. Rept.* 12:34–36.

MacArthur, Mary. 1940. Histology of some physiological disorders of the apple fruit. *Can. J. Agric.* 18(sec. C):26–34.

Martin D., G. C. Wade, and K. Stockhouse. 1962. Bitter pit in the apple variety Sturmer in a pot experiment using low levels of major elements. *Austral. J. Exp. Agric. Anim. Husb.* 2(5):92–96.

McAlpine, D. 1911–12. *The past history and present position of the bitter pit question.* (Bitter pit investigation first progress report.) Melbourne, 197 pp.

———. 1912–13. *The cause of bitter pit: Its contributing factors together with an investigation of susceptibility and immunity in apple varieties.* (Bitter pit investigation second progress report.) Melbourne, 124 pp.

———. 1913–14. *The control of bitter pit in the growing fruit.* (Bitter pit investigation third progress report.) Melbourne, 176 pp.

———. 1914–15. *The experimental results in their relation to bitter pit and a general summary of the investigation.* (Bitter pit investigation fourth progress report.) Melbourne, 178 pp.

———. 1915–16. *The cause and control of bitter pit with the results of the investigation.* (Bitter pit investigation.) Melbourne, 144 pp.

———. 1921. Bitter pit in apples and pears. The latest results in preventive measures. *Phytopathology* 11:366–70.

Melville, F., S. E. Hardisty, and N. S. Shorter. 1964. The control of bitter pit in apples. *J. Agric. West Austral.* ser. 4, 5:938–40.

Mix, A. J. 1916. Cork, drought spot and related diseases of the apple. *New York (Geneva) Agric. Exp. Stn. Bull.* 426:473–522.

Moller, W. J., and W. C. Micke. 1975. Bitter pit of apples. *Univ. Calif. Div. Agric. Sci. Leaf.* 2712. 3 pp.

Mulder, D. 1951. [Bitter pit in apples as cultural phenomenon.] *Meded. Direct. Tuinb.* (Wageningen) 14:26–27.

O'Gara, P. J. 1911. Absorption of arsenic by apples from spray. *Better Fruit* (February): p. 28.

Palmer, R. C. 1943. The influence of amount of crop and harvesting maturity on bitter pit on Okanagan-grown Newtown apples. *Proc. Am. Soc. Hort. Sci.* 43:63–68.

Porritt, S. W., M. Meheriuk, and P. D. Lidster. 1982. *Postharvest disorders of apples and pears.* Publ. 1737E. Communications Branch, Agriculture Canada, Ottawa. 66 pp.

Raphael, J. D., and R. R. Richards. 1962. Bitter pit-control with calcium nitrate spray. *Tasman. J. Agric.* 33:60–63.

Scott, K. J., S. E. Hardisty, and I. A. Staford. 1980. Control of bitter pit in early picked Granny Smith apples from western Australia. *CSIRO Food Res. Quart.* 40(2):29–32.

Scott, K. J., and R. B. H. Wills. 1979. Effects of vacuum and pressure infiltration of calcium chloride and storage temperature on the incidence of bitter pit and low temperature breakdown of apples. *Austral. J. Agric. Res.* 30:917–28.

Smith, A. J. M. 1926. Bitter pit in apples: A review of the problem. *Dept. Scientif. Indus. Res. (Eng.) Food Invest. Bd. Spec. Rept.* 28, 24 pp.

Smock, R. M. 1936. Bitter pit of Gravenstein apples. I. The effect of environmental temperature during the growing season. *Proc. Am. Soc. Hort. Sci.* 34:179–86.

———. 1941. Studies of bitter pit of the apple. New York (Cornell) Agric. Exp. Stn. Mem. 234, 45 pp.

Smock, R. M., and A. Van Doren. 1937. Histology of bitter pit in apples. *Proc. Am. Soc. Hort. Sci.* 35:179.

Stiles, W. C. 1964. Influence of calcium and boron tree sprays on York spot and bitter pit of York Imperial apples. *Proc. Am. Soc. Hort. Sci.* 84:39–43.

Wallace, T., and J. O. Jones. 1939. Pot experiments on bitter pit of apples. *Univ. Bristol Agric. Hort. Res. Stn. Ann. Rept.* 1939:79–84.

Watkins, C. B., J. E. Harman, I. B. Ferguson, and M. S. Reid. 1982. The action of lecithin and calcium dips in the control of bitter pit in apple fruit. *J. Am. Soc. Hort. Sci.* 107:262–65.

Webster, D. H., and F. R. Forsyth. 1979. Partial control of bitter pit in Northern Spy apples with a post-harvest dip in calcium chloride solution. *Can. J. Plant Sci.* 59:717–23.

White, J. 1910. Bitter pit and the enzymes of the apple. *J. Dept. Agric. Victoria Austral.* 8:805.

Wilton, W. J. W. 1975. Bitter pit of apples—Understanding the orchard factors a key to control. *Orchard. N. Z.* 48:372–75.

Wortmann, J. 1892. Über die sogenannte "Stippen" der Äpfel. *Landw. Jahrb.* 21:663–75.

Black-End of Pear

"Black-end" and "hard-end" are names given to a pear fruit disorder in which the tissue of the calyx end of the ripe fruit is hard and dry and the pH of the tissue is high (Davis and Moore 1938). In advanced stages of the disorder the tissue becomes dark. This disorder has been reported from California (Davis and Moore 1938), Oregon (Barss 1921), South Africa (Davis and Tufts 1931), and Japan (Ryugo and Davis 1968).

Black-end occurs chiefly in fruit (**fig. 28**) of most European cultivars when they are grown on rootstocks such as *Pyrus serotina, P. ussuriensis, P. betulaefolia,* and Keiffer, but when cultivars are grown on French (*P. communis*) rootstock, the disease is of little importance. The disorder is most severe on trees with a light crop.

Causal factors. Aside from influence of the rootstock, no other causal factor is known. Tests have indicated (Ryugo and Davis 1968) that the disorder is not caused by a virus, nor does it spread from tree to tree in the orchard (Davis and Tufts 1931). Soil and tree injections have failed to reveal any chemical or combination of chemicals that affect the development of the disorder (Davis and Tufts 1931). It is postulated (Ryugo 1988) that black-end is caused by certain metabolites produced in Asian pear rootstocks and translocated to the fruit of French pear scions. These toxic metabolites, which most cultivars apparently cannot catabolize, accumulate in the calyx end and induce the necrosis typical of the disorder.

Control. The inarching of affected trees on rootstocks that are not known to cause the disorder has been tried with varying success. Only when the original rootstock is severed and the tree is allowed to grow on the inarches is the disease controlled (Davis 1948). The only practical way of avoiding the disease is to propagate the trees on French rootstock. Beurre Hardy cultivar interstock was ineffective in counteracting the rootstock influence (Davis and Ryugo 1968).

REFERENCES

Barss, H. P. 1921. Physiological disorders of developing fruit. *Oregon Agric. Exp. Stn. Third Crop Pest Hort. Rept.* 159–66.

Davis, L. D. 1948. Black-end of pear. *Calif. Agric.* 2(6):10,15.

Davis, L. D., and N. P. Moore. 1948. Black-end of pear: Seasonal changes in pH of the fruit. *Proc. Am. Soc. Hort. Sci.* 35:131–38.

Davis, L. D., and K. Ryugo. 1968. The influence of Buerre Hardy interstock on the occurrence of black-end of "Bartlett" pears. *Proc. Am. Soc. Hort. Sci.* 92:167–69.

Davis, L. D., and W. P. Tufts. 1931. Black-end and its occurrence in selected pear orchards. *Proc. Am. Soc. Hort. Sci.* 29:634–38.

———. 1936. Black-end pears II. *Proc. Am. Soc. Hort. Sci.* 33:304–15.

Heppner, M. L. 1927. Pear black-end and its relation to different rootstocks. *Proc. Am. Soc. Hort. Sci.* 24:139.

Ryugo, K. 1988. *Fruit culture: Its science and art.* John Wiley and Sons, New York. 344 pp.

Ryugo, K., and L. D. Davis. 1968. Yuzuhada, a physiological disorder of Oriental pear, and its possible relation to black-end and hard-end of Bartlett. *Hort. Sci.* 3(1):15–17.

Internal Bark Necrosis (Measles) of Apple

The name "measles" was first applied to a disorder of apple tree in northwestern Arkansas described by Hewitt and Truax (1912), who first observed the disease in 1908. Two types of branch symptoms were associated with this disease. The more prevalent symptoms were scattered red papular excrescences; the less prevalent were scurfy cankers. In 1914, Rose in Missouri discovered a disorder of apple bark which he called "pimple canker." His descriptions and illustrations show this disorder to have many characteristics of the papular type of measles mentioned by Hewitt and Truax.

By 1922, according to Haskell and Wood (1922), a similar apple disorder was found in Arkansas, Missouri, Pennsylvania (Adams 1919), New Mexico (Anonymous 1918–28), Maryland, Virginia, Alabama, Michigan, Illinois, and Kansas. According to Ark and Thomas (1940), California has been added to the list since the disease was found in both young and old orchards near Sebastopol. The disorder is also known to occur in Australia (Holbreche 1946) and Italy (Refatti and Ciferri 1952).

Although the disease has been found in several cultivars—notably Jonathan, Gano, and Stayman Winesap—Delicious and its mutants are most often affected. Grimes Golden, Rome Beauty, and Golden Delicious are seldom affected.

Symptoms. The description of measles (**fig. 29**) furnished by Hewitt and Truax (1912) follows:

> Scattered red pycnidia-like pustules occur on otherwise smooth bark of young twigs. The pustules are 0.3 to 1 mm in diameter and 0.3 to 0.5 mm high, dark red at their centers and light red at the periphery. The latter feature suggested the name "measles." Pustules may become so numerous as to cause the bark to be very rough. The necrotic centers of these pustules are located either in the cortex or, at times, deep in the phloem.

> The less prevalent scurfy-type of canker involves a considerable area of bark and is produced by an irregular thickening of the bark periderm, which is dark red to nearly black in color. The thickened, scurfy condition is relatively shallow, whereas the papular condition extends into the phloem regions of the bark.

Rhoads (1924), who gave an extensive account of the symptoms of measles in Missouri, said there are three more or less distinct types. For convenience, he designated these as the isolated pustular type, the aggregate pustular or scurfy type, and the canker type. Various gradations exist between the first and second and, less frequently, between the second and third types, depending upon the stage of their development and the variety of apple affected. Judging from his description, the macroscopic aspects of the isolated pustular type are similar to those of the pustular type described by Hewitt and Truax, except that the pustules are not so deep-seated.

Descriptions of measles from other parts of the United States conform to those from Arkansas and Missouri. From his studies in West Virginia, Berg (1934) distinguished two types of measleslike disorders that appeared to be nonparasitic; a third disease, black pox, is caused by a fungus, he stated. The two other diseases he tentatively named "internal bark necrosis" and "measles"; the latter he believed to represent true measles as described by Hewitt and Truax.

Berg (1934) stated that, in time, the number of dead areas in the bark increases, radial development of the bark is retarded, and numerous short cracks appear in affected areas. The cracking of the bark and sloughing of the periderm may continue for several years, thereby producing a characteristically scaly, cracked, and roughened condition of the surface.

Berg (1934) distinguished internal bark necrosis from measles as it occurred in West Virginia on the basis of size, color, and distribution of the papules. The papules of measles were, he said, smaller and redder in color than those of internal bark necrosis and occurred in closely aggregated groups, while those of internal bark necrosis were more or less uniformly distributed over the surface of the affected branch. However, Hewitt and Truax (1912) did not mention aggregate patches of pustules as a feature of the disease they described, although they mentioned a scurfy bark canker as a type of measles. Rhoads (1924) mentioned both the scattered and aggregate type of pustule distribution in his description of measles.

In view of the many similarities, internal bark necrosis should not be separated from measles on the basis of symptoms alone. Although Berg was inclined to believe that several distinct and widely distributed papular diseases have been called measles, he realized that until the cause or causes of all measleslike symptoms were known such a separation as he proposed was tentative.

Cause. None of the numerous attempts to isolate a pathogenic organism from measles-affected branches proved successful, though some of the organisms obtained were, for a time, considered the possible cause of the disorder. Rose (1914), for example, described *Pseudomonas papulans* as the probable cause of a papular canker condition of apple bark in Missouri. Although he did not specifically associate this condition with measles, which in an earlier report he had called "pimple canker," others began to do so. This bacterium is no longer considered to have any causal connection with measles. Berg (1934) found a condition of apple twigs characterized by well-defined, scattered, papillate lesions to be caused by the fungus *Helminthosporium papulosum* Berg, but failed to connect this organism or any other fungus or bacterium with internal bark necrosis or measles.

Experiments conducted in New Mexico (Anonymous 1918–28), Ohio (Thomas, Mack, and Fagan 1947), and Illinois (Anderson 1933) indicate that the condition is not transmissible by budding or grafting diseased scions onto healthy trees. Berg (1934) reported that internal bark necrosis was not contracted by young trees planted around badly affected orchard trees. In New Mexico studies (Anonymous 1918–28), young apple trees planted in soil taken from affected orchards developed the disorder, but not those planted in soil taken from healthy orchards. The soils in which trees developed measles were abnormally high in soluble salts.

Young and Winter (1937) of Ohio and Hildebrand (1939) of New York reported that symptoms similar to those described by Berg for internal bark necrosis were produced in Delicious grown in sand cultures from which boron was omitted. Hildebrand also reported that orchard trees were cured of the disorder by applying borax to the soil. However, Berg and Clulo (1946) said that they were "unable to produce any necrotic lesions or other symptoms typical of internal bark necrosis on the bark of Red Delicious trees growing in boron-free sand cultures. Other symptoms of boron deficiency, such as stunting of the trees, procumbent growth of the twigs, and rosetting were in evidence. The application of boron to the soil of diseased trees in orchards did not correct the disease."

Berg and Clulo (1946) obtained evidence that the disease they called internal bark necrosis was induced by an excess of manganese in the soil, but in this connection they did not mention the disorder Berg had called measles. Later, Clulo (1949) published additional evidence on internal bark necrosis; she reported that this disorder was induced by adding 64 ppm of manganous sulfate to sand and soil cultures, but failed to mention measles. Furthermore, the addition of manganous sulfate to soil around Delicious trees which had grown normally for 20 years was followed by a marked development of internal bark necrosis.

Hildebrand (1939) observed an orchard in which internal bark necrosis of Delicious was not benefited by boron treatment. The soil in this orchard was of a different type than the soil of orchards that did respond to boron treatment.

The bark of trees having internal bark necrosis is abnormally high in manganese and iron, indicating a possible causal relationship between an excess of one or both of these elements and the disorder. However, leaf analyses reported by Thomas, Mack, and Fagan (1947) afforded no evidence that an excess of either of these elements is a causal factor; nor do they support the viewpoint that boron deficiency is involved. Later, Eggert and Hayden (1970), Rogers, Thompson, and Scott (1965), and Shelton and Zeiger (1970) presented strong evidence that an excess of manganese causes symptoms of internal bark necrosis to develop, and Shelton and Zeiger (1970) found that manganese accumulated in the necrotic bark areas. Further evidence supporting the causal relationship of manganese to this disorder has been presented by Domoto and Thompson (1976), Miller and Schubert (1977), and Scibisz and Sadowski (1979).

Little is known about the relationship of measles to other soil conditions. Although the incidence of measles is not constantly associated with any particular topography, drainage condition, or soil structure, it often varies from one area of the orchard to another. There appears to be a strong correlation between soil acidity and severity of the disorder, as measles-producing soil in New York had a pH reaction of 4.4 or below, and low soil pH in West Virginia was definitely correlated with the disorder (Miller and Schubert 1977). In the latter study, leaves, wood, and bark from severely diseased trees had a higher manganese concentration than tissue from healthy trees. The application of ammonium sulfate or other acidifying fertilizers to measles-conducive soils is said to have increased the incidence of the disorder, whereas the application of alkalizing materials such as calcium hydroxide or magnesium carbonate is said to have suppressed the disorder.

In New York, a measles-producing soil was found to be low in calcium and magnesium, which probably accounted for its low pH. The potassium, nitrogen, and phosphorus contents apparently were neither excessive nor deficient. The manganese, iron, and aluminum contents, however, were very high. The disorder in that orchard was not alleviated by boron treatment.

According to Hildebrand (1939), soils on which trees responded to boron treatments had a pH well over 5.0. In a soil nutrient regime of low calcium and high manganese, internal bark necrosis increased with increasing potassium (Domoto and Thompson 1976). Studies reported by Shannon in 1954 are interpreted as meaning that both a deficiency of boron and an excess of manganese result in the development of histologically indistinguishable symptoms in young apple trees.

Except for one report (Ark and Thomas 1940), bark measles in apple trees has not been described as a prominent feature of symptoms known to be caused by boron deficiency. The symptoms most commonly mentioned in connection with the boron deficiency disorder are corky areas in the flesh or the skin of the fruit, dwarfing, yellowing, and rosetting of the foliage, and necrosis of the smaller branches. In severely affected trees, the larger branches or even the entire tree may die. All these symptoms may be concurrent in a given orchard, but not often are they concurrent on the same trees. Such variations in symptoms on trees have come to be regarded as the identifying features of boron deficiency. However, they have not been identified as a part of the syndrome of internal bark necrosis,

Control. According to Clulo (1949), the application of calcium hydroxide or other alkaline materials to the soil suppressed internal bark necrosis for five years. Scibisz and Sadowski (1979) explained that lime applications are effective in reducing the disorder by reducing the absorption of manganese by trees.

REFERENCES

Adams, J. F. 1919. Notes on plant diseases in Pennsylvania in 1916. *Penn. Agric. Exp. Stn. Rept.* 1916–17:329–36.

Anderson, H. W. 1931. Apple measles. *Ill. Agric. Exp. Stn. Rept.* 1931:68–70.

———. 1933. Apple measles. *Ill. Agric. Exp. Stn. Ann. Rept.* 1933:44.

Anonymous. 1918–28. Apple measles. *New Mex. Agric. Exp. Stn. Ann. Repts.* 30:18, 1918–19; 36:18, 1924–25; 37:17, 1925–26; 38:20, 1926–27; 39:18, 1927–28.

Ark, P. A., and H. E. Thomas. 1940. Apple die-back in California. *Phytopathology* 30:148–54.

Berg, A. 1934. Black pox and other apple-bark diseases commonly known as measles. *West Va. Agric. Exp. Stn. Bull.* 260, 24 pp.

Berg, A., and G. Clulo. 1946. The relation of manganese to internal bark necrosis of apples. *Science* 104:265–66.

Clulo, G. 1949. The production of internal bark necrosis of apple in sand and soil cultures (abs.). *Phytopathology* 39:502.

Domoto, P. A., and A. H. Thompson. 1976. Effect of interactions of calcium, potassium and manganese supply on "Delicious" apple trees as related to internal bark necrosis. *J. Am. Hort. Soc.* 101:44–47.

Eggert, D. A., and R. A. Hayden. 1970. Histo-chemical relationship of manganese to internal bark necrosis of apple. *J. Am. Hort. Soc.* 95:715–19.

Haskell, R. J., and Jessie I. Wood. 1922. Diseases of fruit and nut crops in the United States in 1921. *U.S. Dept. Agr. Bur. Pl. Indus. Plant Dis. Bull. Supp.* 20:51–52.

Hewitt, J. L., and H. E. Truax. 1912. An unknown apple disease. *Ark. Agric. Exp. Stn. Bull.* 112:481–91.

Hildebrand, E. M. 1939. Internal bark necrosis of Delicious apples, a physiogenic "boron-deficiency" disease (abs.). *Phytopathology* 29:10.

———. 1947. Internal bark necrosis (measles) of Delicious apples in New York in relation to pH, minor element toxicity, and nutrient balance of soil. *Plant Dis. Rep.* 31:99–106.

Holbreche, J. A. 1946. Boron deficiency in apples: Observations at New England Experiment Farm. *New South Wales Agric. Gaz.* 57:17–21.

Miller, S. S., and O. E. Schubert. 1977. Plant manganese and soil pH associated with internal bark necrosis in apple. *West Va. Acad. Sci.* 49(2/3/4):97–102.

Refatti, E., and R. Ciferri. 1952. [Defoliation, measles, stunting, bark necrosis of apple trees due to boron deficiency in the region of Trentino-Alto Adige.] *Not. Malat. Piante* 18:1–20.

Rhoads, A. H. 1924. Apple measles with special reference to the comparative susceptibility and resistance of apple varieties to the disease in Missouri. *Phytopathology* 14:289–314.

Rogers, B. L., A. H. Thompson, and L. E. Scott. 1965. Internal bark necrosis (measles) on Delicious apple trees under field conditions. *Proc. Am. Soc. Hort. Sci.* 86:46–54.

Rose, D. H. 1914. Report of the pathologist: Pimple canker. *Missouri State Fruit Exp. Stn. Bienn. Rept.* 1913–14 (Bull. 24):29–30.

———. 1917. Blister spot of apples and its relation to a disease of apple bark. *Phytopathology* 7:198–208.

Scibisz, K., and A. Sadowski. 1979. Internal bark necrosis in Starkrimson Delicious apple trees in relation to mineral nutrition. *Gartenbauwissenschaft* 44:177–81.

Shannon, Leland M. 1954. Internal bark necrosis of the Delicious apple. *Proc. Am. Soc. Hort. Sci.* 64:165–74.

Shelton, J. E., and D. C. Zeiger. 1970. Distribution of manganese[54] in "Delicioius" apple trees in relation to the occurrence of internal bark necrosis (IBN). *J. Am. Hort. Sci.* 95(6):758–62.

Thomas, W., W. B. Mack, and F. N. Fagan. 1947. Foliar diagnosis: Internal bark necrosis in young apple trees. *Proc. Am. Soc. Hort. Sci.* 50:1–9.

Young, H. C., and H. F. Winter. 1937. The effect of boron, manganese, and zinc on the control of apple measles. *Ohio Agric. Exp. Stn. Bi-Mon. Bull.* 22:147–52.

Water Core of Apple

This disorder, known as "water core" or "glassiness" ("vitrescenza" in Italy), occurs worldwide but is particularly prevalent in arid and semiarid regions (Campbell 1905; Clinton 1914; Cobb 1891; Norton 1911; Petri, 1926). Water core is a misnomer because, except for a few cultivars, the disorder is usually manifested only in the fruit cortex (Brown 1943). Some have called it "glassiness" but this name has not been widely accepted. Water core occurs to some extent in pears, especially in Asian pear selections. Because fruit with water core at harvest are sweeter with no reduction in quality, they are preferred for immediate retailing in some foreign markets, but not for extended storage.

Symptoms. The fruit cortex can become water-soaked in distinct areas or more or less uniformly throughout the flesh (**fig. 30A**). This condition is due to the presence of water and soluble materials in the intercellular spaces; it first appears around the vascular bundles. The influx of water into the affected areas may increase the moisture content as much as 20 percent and produce sufficient pressure to cause excretion of liquid through the lenticels.

In the initial stages, the disorder causes few if any histological changes in the apple flesh (MacArthur 1940)—there is little collapse, discoloration, or abnormal growth of affected cells. The water-soaked condition is rarely discernible from the exterior. Affected fruit transmit light more readily than nonaffected ones, the water-soaked areas being visible as bright luminous spots, particularly where such areas occur near the surface of the fruit (Kemp 1939).

Susceptibility of cultivars. Summer and fall cultivars are more susceptible to water core than are winter cultivars. Tompkins King, Fall Pippin, Yellow Transparent, Early Harvest, Rambo, Winesap, and Stayman Winesap are notable for their susceptibility to this disorder. Cultivars ordinarily not affected by the disorder may develop it if exposed to a prolonged period of high temperature (Fisher, Harley, and Brooks 1931; Harley 1939).

Cause. Some researchers believe that the influx of water into intercellular spaces is a consequence of an abnormally high osmotic concentration produced by a rapid conversion of starch to sugar (starch conversion is a normal process in ripening fruit). Reports vary regarding the comparative rate at which starch is converted into sugar in affected apples (Brown 1943; Fisher, Harley, and Brooks 1931; Harley 1939; MacArthur 1940). According to the proponents of this theory, starch conversion occurs prematurely in affected tissue and is accelerated if the fruit is exposed to high temperature during ripening. After the water-soaked condition develops, the osmotic concentration remains high, but the soluble sugars decrease and ethyl alcohol appears (Fisher, Harley, and Brooks 1931; Harley 1939).

Brown (1943), however, did not regard the rapid conversion of starch to sugar as the basic cause of water core. He believed that something happening elsewhere in the tree (probably the spur or in the fruit cluster base) causes an influx of water and solutes into the fruit with enough pressure to fill the intercellular spaces of the cortex or sometimes the core tissue. He found the juice from affected tissue to be abnormally high in soluble solids and low in acidity. According to Ryugo (1988), the diurnal alteration in temperature in different parts of the fruit results in a gradient in water activity in one direction during the day and in the opposite direction at night. This process causes cells to lose their ability to retain solutes within the vacuoles. Hence, the sorbitol-rich solutes diffuse into the core zone, giving it a glazed or water-soaked appearance.

Development of the disorder. Water core begins to develop as the fruit approach maturity. It increases as the fruit ripen, and reaches its maximum stage of development in fruit that are left on the tree until fairly ripe. It seldom if ever develops after the fruit are picked, and the disorder, if not too severe, often will disappear gradually from fruit that showed symptoms at harvest. The rate of disappearance is faster at moderate than at low temperatures. Severely affected fruit held too long at storage temperatures (0° to 1.1°C) will undergo secondary breakdown (**fig. 30B**) (Brown 1943; Kemp 1939).

Environmental conditions in the orchard exert a marked influence on water core. Its incidence is increased if the fruit are exposed to high temperatures while growing. Fruit exposed to direct sunlight and thus subjected to high temperatures develop the disorder more often than do fruit shaded by leaves. Low soil moisture during the latter part of the growing season tends to increase the number of fruit affected by the disorder (Fisher, Harley, and Brooks 1931; Kemp 1939).

Conditions that increase vegetative growth tend to intensify the disorder. Harley (1939) reported a direct correlation between water core intensity and the ratio of leaf area to the number of fruit. The fruit produced on young trees, which normally develop few fruit but abundant foliage, are likewise more prone to water core than are fruit produced on older trees where the leaf-fruit ratio is lower. Consequently, the forcing of leaf growth by heavy nitrogen applications may result in an increase of the disorder. This increase is associated with an increase in the ability of the leaves to synthesize and transport carbohydrates to the fruit (Harley 1939). Recent Chinese research (Tong et al. 1980) showed that water core is inversely correlated with calcium levels in the fruit.

Control. No highly effective method of controlling water core is known, but overfertilization with nitrogenous fertilizers should be avoided. Where irrigation is practiced, trees should be supplied with adequate soil moisture throughout the growing season. Since water core often increases rapidly as apples become overmature, "stop-drop" sprays should be used judiciously and the fruit picked at optimum maturity (Pierson, Ceponis, and McColloch 1971).

REFERENCES

Birmingham, W. A. 1925. An uncommon water core condition in apples. *Agric. Gaz. New South Wales* 36:59–62.

Brooks, C., and D. F. Fisher. 1926. Water-core of apples. *J. Agric. Res.* 32:223–60.

Brown, D. S. 1943. Notes and observations on a study of water core in Illinois apple during the 1942 season. *Proc. Am. Soc. Hort. Sci.* 42:267–69.

Campbell, A. G. 1905. Constitutional diseases of fruit trees. *J. Victoria Agric. Dept. Austral.* 3:463–65.

Carne, W. M., H. A. Pittman, and H. G. Elliott. 1930. Notes on wastage of non-parasitic origin in stored apples. *J. Austral. Counc. Sci. Indus. Res.* 3(3):167–82. (*Rev. Appl. Mycol.* 10:114–15. 1931.)

Clinton, G. P. 1914. Report of the Botanist: Water core. *Conn. Agric. Exp. Stn. Ann. Rept.* 1913:8–9.

Cobb, N. A. 1981. Water core in apples. *New South Wales Agric. Gaz.* 2:286–87.

Fisher, D. F., C. P. Harley, and C. Brooks. 1931. The influence of temperature on the development of watercore. *Proc. Am. Soc. Hort. Sci.* 28:276–80.

Harley, C. P. 1939. Some associated factors in the development watercore. *Proc. Am. Soc. Hort. Sci.* 36:435–39.

Kemp, H. K. 1939. Detection of water core in apples. *J. Austral. Inst. Agric. Sci.* 5:227–29.

Kemp, H. K., and J. A. Beare. 1939. The effect of water core on the keeping quality of apples. *J. Dept. Agric. South Austral.* 43:22–28.

MacArthur, Mary. 1940. Histology of some physiological disorders of the apple fruit. *Can. J. Res. Sect. C.* 18:26–34.

Norton, J. B. S. 1911. Water core of apples. *Phytopathology* 1:126–28.

O'Gara, P. J. 1913. Studies on the water core of apples. *Phytopathology* 3:121–28.

Petri, L. 1926. La "vitrescenza" dell mele. *Boll. Ris. Staz. Pat. Veg.* 6(3):253–60.

Pierson, C. F., M. J. Ceponis, and L. P. McColloch. 1971. Market diseases of apples, pears and quinces. USDA-ARS Agric. Handbook 376, 112 pp.

Ryugo, K. 1988. *Fruit culture: Its science and art.* John Wiley and Sons, New York. 344 pp.

Tong, Y.-A., H.-J. Zhou, R.-L. Yang, and G.-F. Zhang. 1980. [The inorganic constituents and nitrogen contents of Delicious apple fruits in relation to occurrence of watercore on calcareous soils.] *Sci. Agric. Sinica* 2:67–71. (*Rev. Plant Path.* 61:5834. 1982).

POSTHARVEST DISEASES OF POME FRUIT

BIOTIC DISEASES OF POME FRUIT IN STORAGE

Nonparasitic disorders of stored apple and pear fruit are discussed later in this chapter. The diseases to be discussed here are all caused by fungi. Almost no research has been done in California on the postharvest parasitic diseases of pome fruit. Much of our information on this group of disorders has resulted from research in Washington and Oregon, but because of the omnipresence of most of the storage-rot fungi, much of this information is pertinent to apples and pears in California.

Alternaria Rot of Apple and Pear

Alternaria alternata (Fries) Keissler

Infection of apples and pears by *Alternaria* is most frequently found on fruit held in ordinary storage, but it seldom causes extensive loss of fruit in cold storage. This type of rot (**fig. 31A**) was described by Cook and Martin (1913).

Alternaria mali Roberts is a common cause of apple fruit rot in the northeastern United States, and in Illinois (Tweedy and Powell 1962) it causes about 1 percent loss of Jonathan and Delicious apple crops each year. The disease caused by *A. mali* was named "cork rot" to distinguish it from that caused by *A. tenuis* (*alternata*) (Tweedy and Powell 1962). Another phase of the disease is a carpel discoloration and decay (**fig. 31B**) described by Miller (1959) on Wagener and Red Delicious apples and by Ceponis, Kaufman, and Butterfield (1969) during surveys of apples in New York City. The open calyx tube in the fruit allows *Alternaria* and other fungi to enter the carpel area; as much as 5 percent of the Pacific Northwest shipments sampled had moldy carpels and 2.9 percent developed *Alternaria* core rot. Susceptible cultivars of apples have deep calyx-sinuses or pits (Miller 1959). Recent studies emphasize the continuing problem of *Alternaria* moldy-core and core rot in certain apple cultivars (Brown and Hendrix 1978; Ellis and Barrat 1983; Ogawa, Feliciano, Manji, and Adaskaveg 1990).

Alternaria has been recognized for years as a minor postharvest pathogen of pears (English 1940; Lockhart and Forsyth 1974) and it recently attracted attention as the cause of a serious stem decay of pears in Washington (Sitton and Pierson 1983).

Causal organism. The causal organism is a fungus classified by Neergaard (1945) as *Alternaria tenuis* Auct., and more recently by Simmons (1967) as *A. alternata* (Fries) Keissler (=*A. tenuis* Nees). The species *A. mali* was established by Roberts (1924) because its morphology appeared to be somewhat different from that of *A. tenuis*, an earlier-described species on apple. However, most current researchers recognize *A. alternata* (*A. tenuis*) as the valid binomial for the *Alternaria* that causes decay of pome fruit (Brown and Hendrix 1978; Lockhart and Forsyth 1974; Prusky and Ben-Arie 1981; Sitton and Pearson 1983). *Alternaria mali* is the name preferred by some Oriental researchers for the biotype that causes a leaf-spot disease of apple in Japan and Korea (Lee and Kim 1980; Saito, et al. 1983; Sawamura 1962).

Alternaria alternata is described by Simmons (1967) as follows:

> Conidophores and conidia dilute yellow brown to medium golden brown in color. Conidiophores

simple, straight or curved, smooth, 1 to 3 septate, 20 to 46 μm long by 4 to 6 μm wide, apically uniperforate, sometimes with basal cell slightly swollen. Conidia ovoid, obclavate, obpyriform, or rarely simple ellipsoidal in shape, usually with an easily visible basal pore; beakless when ellipsoidal or with a short conical, narrowly tapered, or cylindrical beak 2 to 3 μm in diameter, the apex of the beak may be narrow and rounded without a terminal pore, or abruptly blunt with a well-defined pore; beak length 25 μm, never equalling the length of the conidium body, but commonly representing one-fourth to one-third of the total conidium length; the beak is usually lighter in color than the body. Conidium body (10–) 18 to 47 × (5–) 7 to 18 μm, av. 30.9 × 12.6 μm; l/w = 1.7 to 3.4, av. 2.4; with (1–) 3 to 8 transverse septa, one or two longitudinal septa in each of the 1 to 6 transverse divisions, and commonly, a strongly oblique septum occurs in the basal division, distinctly but not deeply constricted at major transverse septa. The conidium wall is smooth or very minutely roughened.

Alternaria alternata produces several identified toxins (Maeno et al. 1984; Woodhead et al. 1975), but it is not known if these compounds play an active role in fruit decay.

Symptoms. Several types of Alternaria rot symptoms are found on apples grown in the Pacific Northwest. In cold storage, small, firm, slightly sunken areas, brown around the edge but covered by a black crust, will sometimes occur. Another type that develops on apples after removal from cold storage is characterized by firm, slightly sunken, rotted areas that commonly are dark brown to black (but sometimes light yellowish brown to gray) and may show mold growth on the surface if stored in high relative humidity. The third type is characterized by the development of a black rot on areas of the fruit affected by scald. A *Cladosporium* may also cause a black rot indistinguishable from that produced by *Alternaria* (Rose, Fisher, and Bratley 1933). Rick (1970) characterized *Alternaria* as a weak pathogen that develops slowly on immature apples and produces a brown dry rot. The lesions commonly occur at the calyx and in the depression around the stem, and occasionally can be found within the fruit core.

On pear, decay lesions may also occur at the calyx but are more commonly centered at injuries that occur during and after harvest. They are firm, medium to dark brown, sometimes slightly zonate, and slow-growing under cold-storage conditions (English 1940). Spores of the fungus commonly occur as contaminants on the fruit and leaf surfaces (Kuss and Harnish 1972). This fungus grows saprophytically on dead or weakened plant tissues and sporulates abundantly.

Control. Except for core rot, effective control is generally achieved by careful handling during harvest and postharvest and by prompt storage at −0.6° to 0.6°C. Benzimidazoles, effective against decay caused by *Botrytis* and *Penicillium*, fail to control Alternaria rot (Szkolnik 1968), but washes in chlorine solution reduce the number of contaminating spores (Segall 1968). Conflicting results have been obtained in the use of orchard sprays for the control of moldy-core and core rot. Brown and Hendrix (1978) reported that application of various fungicides during bloom significantly reduced this type of infection. On the other hand, Ellis and Barrat (1983) using similar applications failed to reduce the incidence of moldy-core at harvest. In pears, Sitton and Pierson (1983) obtained effective control of both Alternaria stem decay and blue mold (*Penicillium expansum*) by using a postharvest drench containing prochloraz and benomyl, and Ogawa, Feliciano, Manjii, and Adaskaveg (1990) found that an iprodione spray or drench provided control of *Alternaria*, *Penicillium*, and *Botrytis* on Fuji apple. Prusky and Ben-Arie (1981) found that a postharvest drench containing imazalil inhibited development of *Alternaria* rot in inoculated apples and naturaly infected pears.

REFERENCES

Adams, R. E., and S. E. Tamburo. 1957. The West Virginia spot-rot complex of apple in 1956. *Plant Dis. Rept.* 41:760–65.

Brooks, C., J. S. Cooley, and D. F. Fisher. 1935. Diseases of apples in storage. *USDA Farmer's Bull.* 1160, 20 pp.

Brown, E. A., and F. F. Hendrix. 1978. Effect of certain fungicides sprayed during apple bloom on fruit set and fruit rot. *Plant Dis. Rep.* 62:739–41.

Ceponis, M. J., J. Kaufman, and J. E. Butterfield. 1969. Moldy carpels in Delicious apples on the greater New York markets. *Plant Dis. Rep.* 53:136–38.

Cook, M. T., and G. W. Martin. 1913. Alternaria rot of apples. *Phytopathology* 3:72.

Ellis, ,M. A., and J. G. Barrat. 1983. Colonization of Delicious apple fruits by *Alternaria* spp. and effect of fungicide sprays on moldy-core. *Plant Dis.* 67:150–52.

English, W. H. 1940. *Taxonomic and pathogenicity studies of the fungi which cause decay of pears in Washington.* Ph.D. Thesis, Washington State University, Pullman, 270 pp.

———. 1944. Notes on apple rots in Washington. *Plant Dis. Rep.* 26:610–22.

Heald, F. D., and G. D. Ruehle. 1931. The rots of Washington apples in cold storage. *Wash. Agric. Exp. Stn. Bull.* 253. 48 pp.

Kuss, F. R., and W. N. Harnish. 1972. *Alternaria* species associated with apple leaf spot. *Plant Dis. Rep.* 56:721–22.

Lee, C. U., and M. H. Kim. 1980. Effects of fungicides on sporulation of apple leafspot. *Alternaria mali* Roberts. *Korean J. Plant Protect.* 19:169–74.

Lockhart, C. L., and F. R. Forsyth. 1974. *Alternaria alternata* storage decay of pears. *Can. Plant Dis. Survey* 54:101–02.

Maeno, S., K. Kohmoto, H. Otani, and S. Nishimura. 1984. Different sensitivities among apple and pear cultivars to AM-toxin produced by *Alternaria alternata* apple pathotype. *J. Fac. Agric. Tottori Univ.* 19:8–19.

Miller, P. M. 1959. Open calyx tubes as a factor in contributing to carpel discoloration and decay of apples. *Phytopathology* 49:520–23.

Neergaard, P. 1945. *Danish species of Alternaria and Stemphylium.* Oxford University Press, London, 560 pp.

Ogawa, J. M., A. J. Feliciano, B. T. Manji, and J. E. Adaskaveg. 1990. Postharvest decay organisms of Fuji apples and disease control with iprodione (abs.). *Phytopathology* 80:892.

Pierson, C. F., M. J. Ceponis, and L. P. McColloch. 1971. Market diseases of apples, pears, and quinces. *USDA-ARS Agric. Handb.* 376, 112 pp.

Prusky, D., and R. Ben-Arie. 1981. Control by imazalil of fruit storage rots caused by *Alternaria alternata*. *Ann. Appl. Biol.* 98:87–92.

Rick, A. E. 1970. Calyx-end rot of apples. *Phytopathology* 60:1152.

Roberts, J. W. 1914. .Experiments with apple leaf-spot fungi. *J. Agric. Res.* 2:57–66.

———. 1924. Morphological characters of *Alternaria mali* Roberts. *J. Agric. Res.* 27:699–708.

Rose, D. H., D. F. Fisher, and C. O. Bratley. 1933. Market diseases of fruit and vegetables: Apples, pears, quinces. USDA Misc. Publ. 168, 70 pp.

Saito, K., M. Niizeki, M. Wake, and Y. Hidano. 1983. [Fundamental studies on breeding of the apple. IX. Variation of pathogenicity and chemical tolerance to polyoxin in *Alternaria mali* Roberts.] *Bull. Fac. Agric. Hirosaki Univ.* 40:23–30. (*Rev. Plant Pathol.* 63:3444. 1984.)

Sawamura, K. 1962. Studies on spotted disease of apples. 1. Causal agent of Alternaria blotch. (In Japanese; English summary). *Tohoku National Exp. Stn. Bull.* 23:163–75.

Segall, R. H. 1968. Fungicide effectiveness of chlorine as infuenced by concentration, temperature, pH and spore exposure time. *Phytopathology* 58:1412–14.

Simmons, E. G. 1967. Typification of *Alternaria, Stemphyllium,* and *Ulocladium.* *Mycologia* 59:67–92.

Sitton, J. W., and C. F. Pierson. 1983. Interaction and control of Alternaria stem decay and blue mold in d'Anjou pears. *Plant Dis.* 67:904–07.

Stakman, E. C., and R. C. Rose. 1914. A fruit spot of the Wealthy apple. *Phytopathology* 4:333–35.

Szkolnik, M. 1968. Apple fruit rots (*Botrytis* sp. and *Penicillium* sp.) Am. Phytopathol. Soc. *Fungicide-Nematicide Tests, Results of 1968,* 24:26.

Tweedy, B. G., and D. Powell. 1962. Cork rot of apples and its causal organism, a pathogenic strain of *Alternaria mali*. *Phytopathology* 52:1073–79.

Wilshire, S. P. 1933. The foundation species of *Alternaria* and *Macrosporium*. *Trans. Brit. Mycol. Soc.* 18:135–60.

Woodhead, S. H., G. E. Templeton, W. L. Meyer, and R. B. Lewis. 1975. Procedures for crystallization and further purification of tentoxin. *Phytopathology* 65:495–96.

Blue-Mold Rot or Soft Rot of Pome Fruit

Penicillium expansum Lk.

Blue-mold rot is the most common and important storage rot of apples, quinces, and pears in the United States. In Washington state alone, the loss from blue-mold decay is said to have been well over a million dollars per year. The disease is also the cause of serious loss of apples in England (Brooks, Cooley, and Fisher 1935; Kidd and Beaumont 1924; Wiant and Bratley 1948). Recent studies showing mycotoxin production by certain strains of *Penicillium* make decay control highly desirable (Harwig, Chen, and Kennedy 1973; Pytel and Borecka 1985).

Jonathan and Delicious are the most susceptible apple varieties grown on the Pacific Coast. Baker and Heald (1936) stated that the storage life of such cultivars as Jonathan, Delicious, and Winesap is related to their susceptibility to blue-mold. Winesap, which is less susceptible than the other two cultivars, has the longest storage life. In the eastern United States such cultivars as McIntosh, Fameuse, York Imperial, Stayman Winesap, Winesap, Rome Beauty, and Baldwin are frequently affected by blue-mold (Baker and Heald 1936).

Among pear cultivars, the Winter Nelis, d'Anjou, and Comice are highly susceptible, while Columbia,

Doyenne, d'Alencon, and Seckel are said to be relatively immune (Baker and Heald 1936).

Symptoms. Infection is first visible as pale-brown to brown, sunken, more or less circular lesions. The maceration of the tissue is caused by a pectolytic enzyme produced by the fungi (Cole and Wood 1961). Lesions commonly are located at lenticels or at breaks in the skin. Lesions expand rapidly; the affected tissue, which at first is yellowish brown or dark brown is soft and watery, with a strong musty odor and an unpleasant taste. The apple skin becomes wrinkled as the underlying tissue collapses. If fruit is held under humid conditions, the fungus produces numerous bluish to greenish blue tufts of spores on the skin (fig. 32).

Causal organism. Although blue-mold decay can be caused by several species of *Penicillium*, *P. expansum* is the most common (Borecka 1977; English 1940; Heald and Ruehle 1931). Synonyms of the fungus are *P. glaucum* Link, *Coremium glaucum* Link, *Floccaria glauca* Grenville, and *Coremium vulgare* Corda. The fungus grows rapidly on Czapek's medium and attains a diameter of 40 to 50 mm in 8 to 10 days at room temperature. Sporulation is heavy throughout, with conidiophores regularly arising from the substrate. Small colonies are white but turn quickly to yellow-green shades with maturation of conidia. Conidiophores are mostly 150 to 700 μm long and 3.0 to 3.5 μm in diameter, with walls smooth or finely roughened and terminating in large penicilli commonly measuring up to 75 to 100 μm in length. These bear tangled chains of spores 150 to 200 μm long. Conidia are elliptical when first formed and continue to show some ellipticity, with measurements of 3.0 to 3.5 μm in diameter (Raper and Thom 1949).

Disease development. Although this fungus cannot penetrate uninjured fruit skin, it gains entrance through lenticels and through breaks sustained during harvesting and packing (Baker and Heald 1932, 1934a; English, Ryall, and Smith 1946). Lenticel infection is common in such cultivars as Jonathan and Delicious. The susceptibility of these openings is influenced by a variety of factors and increased if the fruit skin is subjected to pressure bruising during harvesting and packing. Pressure bruising ruptures the layer of cutinized cells in the lenticel basins (Baker and Heald 1932, 1934a; English, Ryall, and Smith 1946).

Washing fruit with alkaline or acid solutions, such as were employed for the removal of arsenical residues, increases the infectibility of lenticels. The lenticels on fruit that have been washed are often found to be "open"—that is, lacking a compact cutinized and suberized layer of cells in the lenticel basin. The open lenticels develop a distinct stained halo when fruit is immersed in a methylene blue solution; the closed lenticels remain uncolored. Exposure of fruit to warm dry air is effective in closing lenticels. Storing fruit at 0°C also tends to close the lenticels and reduce lenticel infection (English, Ryall, and Smith 1946).

Susceptibility of lenticels increased with fruit maturity, probably because they become more easily injured by pressure and because changes occur in the composition of cells in the lenticel basin.

Mechanical injuries sustained when one fruit strikes the stem of another fruit or the sharp edge of a box during grading and packing, or from small breaks made by sand particles, are also common avenues for entry of the fungus. Susceptibility to bruising is greatest with ripe fruit (Wright and Smith 1954). Newly made bruises are more susceptible to infection than are old bruises. With pears, the fruit stem is one of the most important sites of infection.

Because the fungus is a ubiquitous saprophyte, the fruit can become contaminated either in the orchard or in the packing shed. Favorable conditions for contamination and infection occur when the fruit is handled after it is wetted by rain or stored in boxes that had contained rotting fruit.

Control. Better handling of fruit to reduce mechanical injuries has markedly reduced blue mold decay. Some growers periodically steam-sterilize the fruit boxes used in the orchard. Others employ the same boxes for harvesting in the orchard and ultimate packing; because these are new boxes, they are rarely contaminated with the spores of rot fungi (Wellman and Heald 1938). Harvest bins now are usually padded, and dump tanks commonly are disinfested of mold organisms. Spotts and Cervantes (1985) showed that dump tanks could be effectively disinfested by increasing the water temperature instead of using chlorine treatments, which corrode metals.

In the pome fruit-growing regions of both the eastern and northwestern United States, there has been a great increase in the treatment of harvested fruit to destroy unwanted mold spores. One of the first chemical washes used was sodium hypochlorite solution. Sodium chlororthophenylphenate was also found to be effective (English 1948), but because of fruit injury it was replaced by a drench of sodium orthophenylphenate followed by a water rinse (Kienholz, Robinson, and Degman 1949; Pierson, Ceponis, and McColloch 1971). Bertrand and Saulie-Carter (1979), however, obtained unsatisfactory control of blue mold rot of punctured, inoculated pears with dip treatments containing either sodium orthophenylphenate or chlorine. Other Oregon investigators (Spotts and Cervantes 1987; Spotts and Peters 1982) found that certain surfactants added to a chlorine solution improved control of blue mold rot and two other common rots. However, surfactant added to solutions of sodium orthophenylphanate, captan, or iprodione did not improve control of blue mold rot (Spotts 1984).

Benzimidazole compounds used as drenches have offered decay control generally superior to that of other treatments (Beattle and Orthred 1970; Ben-Arie and Guelfat-Reich 1973; Blanpied and Purnasiri 1968; Cargo and Dewey 1970; Daines and Snee 1969; Maas and MacSwan 1970; Sitton and Pierson 1983; Tepper and Yoder 1982). However, the development of resistance to these and several other fungicides has complicated the control of blue mold rot. (Bertrand and Saulie-Carter 1979; Burton and Dewey 1981; Rosenberger and Meyer 1981). Fortunately, at least some of the resistant isolates are less virulent than the wild-type isolates (Rosenberger and Meyer 1981). In Washington and Oregon, benomyl is restricted to postharvest use in order to delay the development of resistant strains of the blue-mold organism. Other fungicides that have shown promise in controlling this disease are iprodione, prochloraz, vinclozolin, and CGA-64251 (Burton and Dewey 1981; Little, Taylor, and Peggie 1980; Rosenberger and Meyer 1981; Tepper and Yoder 1982). Rosenberger and Meyer (1985) recently reported that diphenylamine and benzimidazole fungicides provided better control of resistant *P. expansum* in inoculated apples when used combined than when used alone. Other effective dips involving several fungicides in combination with diphenylamine or calcium chloride have been reported (Burton and Dewey 1981; Little, Taylor, and Peggie 1980).

Calcium chloride dips are thought to aid in control by forming cell wall materials resistant to degradation by fruit-decay fungi (Conway and Sams 1984; Conway et al. 1988). Conway, Gross, and Sams (1987) found that as the inoculum concentration decreases, the relative effectiveness of increased cell wall-bound calcium in reducing blue-mold rot increases. In other research, Sams and Conway (1987) demonstrated the additive effects of calcium chloride and controlled-atmosphere storage in reducing decay and improving the keeping quality of apples.

Prompt cooling to storage temperatures near 0°C and maintaining such temperatures will arrest some mold infections (Pierson and Couey 1970), but spores not chemically destroyed will germinate and may produce lesions within 30 days; these may enlarge up to 1 inch in diameter in 60 days (U. S. Department of Agriculture 1965).

Other treatments used with some effectiveness are applications of 5,000 ppm 2-aminobutane (Eckert, Kolbezen, and Sleesher 1962; Pierson 1966) to the fruit, and gamma radiation of 100,000 rep (roentgen equivalent physical), which arrests progress of decay (Beraha et al. 1961; Maxie et al. 1966). Also of interest is the recently demonstrated biological control of blue mold by means of an antagonistic bacterium and a yeast (Janisiewicz 1987), and the control of both blue mold and gray mold with *Pseudomonas cepacia* (Janisiewicz and Roitman 1988). According to R. G. Roberts (personal communication), the yeast *Cryptococcus laurentii* showed promise in controlling Penicillium decay by competitive action and not by production of an antibiotic.

REFERENCES

Baker, K. F., and F. D. Heald. 1932. Some problems concerning blue mold in relation to cleaning and packing of apples. *Phytopathology* 22:879–98.

———. 1934a. An investigation of factors affecting the incidence of lenticel infection of apples by *Penicillium expansum*. *Wash. Agric. Exp. Stn. Bull.* 298, 48 pp.

———. 1934b. Investigations on methods of control of blue-mold decay of apples. *Wash. Agric. Exp. Stn. Bull.* 304, 32 pp.

———. 1936. The effect of certain cultural and handling practices in the resistance of apples to *Penicillium expansum*. *Phytopathology* 26:932–48.

Beattle, B. B., and W. L. Outhred. 1970. Benzimidazole derivatives as postharvest fungicides to control rotting of pears, cherries, and apricots. *Austral. J. Exp. Agric. Anim. Husb.* 10:651–56.

Ben-Arie, R., and S. Guelfat-Reich. 1969. Postharvest heat treatment to control storage rots of Spadona pears. *Plant Dis. Rep.* 53:363–67.

———. 1973. Preharvest and postharvest applications of benzimmidazoles for control of storage decay of pears. *Hort. Sci.* 8:181–83.

Beraha, L., G. B. Ramsey, M. A. Smith, et al. 1961. Gamma radiation in control of decay in strawberries, grapes, and apples. *Food Tech.* 15:94–98.

Bertrand, P., and J. L. Saulie-Carter. 1979. Postharvest decay control of apples and pears after immersion dumping. *Oregon State Univ. Agric. Exp. Stn. Spec. Rep.* 545, 9 pp.

Blanpied, G. D., and A. Purnasiri. 1968. Thiabendazole control of Penicillium rot of McIntosh apples. *Plant Dis. Rep.* 52:867–71.

Borecka, H. 1977. Fungi of the genus *Penicillium* on apples and pears during the storage period. *Acta Agrobot.* 30:213–27.

Brooks, C., J. S. Cooley, and D. F. Fisher. 1935. Diseases of apples in storage. *USDA Farmer's Bull.* 1160, 20 pp.

Brown, G. E. 1989. Management of disease resistance in harvested fruits and vegetables. Host defenses at the wound site on harvested crops. *Phytopathology* 79:1381–84.

Burton, C. L., and D. H. Dewey. 1981. New fungicides to control benomyl-resistant *Penicillium expansum* in apples. *Plant Dis.* 65:881–83.

Cargo, C. A., and D. H. Dewey. 1970. Thiabendazole and benomyl for the control of postharvest decay of apples. *Hortscience* 5:259–60.

Cole, M., and R. K. S. Wood. 1961. Types of rot, rate of rotting, and analysis of pectin substances in apples rotted by fungi. *Ann. Bot.* 25:417–34.

Conway, W. S., K. C. Gross, C. D. Boyer, and C. E. Sams. 1988. Inhibition of *Penicillium expansum* polygalacturonase activity by increased apple cell wall calcium,. *Phytopathology* 78:1052–55.

Conway, W. S., K. C. Gross, and C. E. Sams. 1987. Relationship of bound calcium and inoculum concentration to the effect of postharvest calcium treatment on decay of apples by *Penicillium expansum*. *Plant Dis.* 71:78–80.

Conway, W. S., and C. E. Sams. 1984. Possible mechanisms by which postharvest calcium treatment reduces decay in apples. *Phytopathology* 74:208–10.

Daines, R. H., and R. D. Snee. 1969. Control of blue mold of apples in storage. *Phytopathology* 59:792–94.

Eckert, J. W., M. J. Kolbezen, and R. L. Sleesher. 1962. Control of postharvest fruit decay with 2-aminobutane. *Phytopathology* 52:730.

English, H. 1948. Disinfectant washes for the control of decay in apples and pears (abs.). *Phytopathology* 38:914.

English, W. H. 1940. *Taxonomic and pathogenicity studies of the fungi which cause decay of pears in Washington*. Ph.D. Thesis, Washington State University, Pullman, 270 pp.

English, H., A. L. Ryall, and E. Smith. 1946. Blue mold decay of Delicious apples in relation to handling practices. USDA Circ. 751, 20 pp.

Harwig, J., Y. K. Chen, and B. P. C. Kennedy. 1973. Occurrence of patulin and patulin-producing strains of *Penicillium expansum* in natural rots of apple in Canada. *Can. Inst. Food Technol. J.* 6:22–25.

Heald, F. D., and G. D. Ruehle. 1931. The rots of Washington apples in cold storage *Wash. Agric. Exp. Stn. Bull.* 253, 48 pp.

Janisiewicz, W. J. 1987. Postharvest biological control of blue mold on apples. *Phytopathology* 77:481–85.

Janisiewicz, W. J., and J. Roitman. 1988. Biological control of blue mold and gray mold on apple and pear with *Pseudomonas cepacia*. *Phytopathology* 78:1697–1700.

Kidd, M. N., and A. Beaumont. 1924. Apple rot fungi in storage. *Trans. Brit. Mycol. Soc.* 10:98–118.

Kienholz, J. R., R. H. Robinson, and E. S. Degman. 1949. Reduction of pear rots in Oregon by the use of chemical washes. *Oregon Agric. Exp. Stn. Inf. Circ.* 160, 7 pp.

Lakshminarayana, S., N. F. Sommer, V. Polito, and R. J. Fortlage. 1987. Development of resistance to infection by *Botrytis cinerea* and *Penicillium expansum* in wounds of mature apple fruit. *Phytopathology* 77:1674–78.

Little, C. R., H. J. Taylor, and I. D. Peggie. 1980. Multiformulation dips for controlling storage disorders of apples and pears. I. Assessing fungicides. *Sci. Hort.* 13:213–19.

Maas, J. L., and I. C. MacSwan. 1970. Postharvest fungicide treatments for reduction of Penicillium decay of Anjou pears. *Plant Dis. Rep.* 54:887–90.

Maxie, E. C., N. F. Sommer, C. J. Muller, and H. L. Rae. 1966. Effect of gamma radiation on the ripening of Bartlett pears. *Plant Physiol.* 41:437–42.

Pierson, C. F. 1966. Fungicides for the control of blue-mold rot of apples. *Plant Dis. Rep.* 12:913–15.

Pierson, C. F., M. M. Ceponis, and L. P. McColloch. 1971. Market diseases of apples, pears, and quinces. *USDA-ARS Agric. Handb.* 376, 112 pp.

Pierson, C. F., and H. M. Couey. 1970. Control of decay in Anjou pears with hot water and with heated and unheated suspensions of benomyl or thiabendazole (abs.). *Phytopathology* 60:1308.

Pytel, M., and H. Borecka. 1985. [Production of patulin by isolates of *Penicillium expansum* (Link) Thom.] *Acta Agrobot.* 35:235–42. (*Rev. Plant Pathol.* 64:4998. 1985.)

Raper, K. B., and C. Thom. 1949. *A manual of the Penicillia*. Williams and Wilkins, Baltimore, 875 pp.

Rosenberger, D. A., and F. W. Meyer. 1981. Postharvest fungicides for apples: Development of resistance to benomyl, vinclozolin, and iprodione. *Plant Dis.* 65:1010–13.

———. 1985. Negatively correlated cross-resistance to diphenylamine in benomyl-resistant *Penicillium expansum*. *Phytopathology* 75:74–79.

Sams, C. E., and W. S. Conway. 1987. Additive effects of controlled-atmosphere storage and calcium chloride on decay, firmness retention, and ethylene production in apples. *Plant Dis.* 71:1003–05.

Sitton, J. W., and C. F. Pierson. 1983. Interaction and control of Alternaria stem decay and blue mold in d'Anjou pears. *Plant Dis.* 67:904–07.

Sommer, N. F. 1989. Management of disease resistance in harvested fruits and vegetables. Manipulating the postharvest environment to enhance or maintain resistance. *Phytopathology* 79:1377–80.

Spalding, D. H., and R. E. Hardenburg. 1971. Postharvest chemical treatments for control of blue mold of apples in storage. *Phytopathology* 61:1308–09.

Spalding, D. H., H. C. Vaught, R. H. Day and G. A. Brown. 1969. Control of blue mold rot development in apple treated with heated and unheated fungicides. *Plant Dis. Rep.* 53:738–42.

Spotts, R. A. 1984. Effect of a surfactant on control of decay of Anjou pear with several fungicides. *Plant Dis.* 68:860–62.

Spotts, R. A., and L. A. Cervantes. 1985. Effects of heat treatments on populations of four fruit decay fungi in sodium orthophenylphate solutions. *Plant Dis.* 69:574–76.

———. 1987. Effects of the nonionic surfactant Ag-98 on three decay fungi of Anjou pear. *Plant Dis.* 71:240–42.

Spotts, R. A., and B. B. Peters. 1982. Use of surfactants with chlorine to improve pear decay control. *Plant Dis.* 66:725–27.

Tepper, B. L., and K. S. Yoder. 1982. Postharvest chemical control of Penicillium blue mold of apple. *Plant Dis.* 66:829–31.

U. S. Department of Agriculture, Agricultural Research Service. 1965. *A review of literature on harvesting, handling, storage and transportation of apples.* ARS 51-4, 215 pp.

Walker, J. R. L. 1970. Phenolase inhibitor from cultures of *Penicillium expansum* which may play a part in fruit rotting. *Nature* 227:298–99.

Wellman, R. H., and F. D. Heald. 1938. Steam sterilization of apple boxes for blue mold. Wash. Agric. Exp. Stn. Bull. 357, 16 pp.

Wiant, J. S., and C. O. Bratley. 1948. Spoilage of fresh fruits and vegetables in rail shipments unloaded at New York City, 1935–1942. *USDA Circ.* 773, 62 pp.

Wilson, C. L. 1989. Management of disease resistance in harvested fruit and vegetables. Managing the microflora of harvested fruits and vegetables to enhance resistance. *Phytopathology* 79:1387–90.

Wilson, D. M., G. J. Nuovo, and W. B. Darby. 1973. Activity of o-diphenol oxidase in post-harvest apple decay by *Penicillium expansum* and *Physalospora obtusa*. *Phytopathology* 63:1115–18.

Wright, J. R., and E. Smith. 1954. Relation of bruising and other factors to blue mold decay of Delicious apples. *USDA Circ.* 935, 15 pp.

Bull's-Eye Rot of Apple and Pear

Pezicula malicorticis (Jacks.) Nannf.

Bull's-eye rot can be serious on apple and pear grown in Washington, Oregon, and British Columbia (Spotts 1985a). In California it occurs occasionally on apple in the central and northern coast districts (Smith 1941); total decay in experimental lots was a fraction of a percent in the Watsonville area in 1975 (Combrink et al. 1976). The causal organism is *Pezicula malicorticis* (Jacks.) Nannf., the imperfect stage being *Cryptosporiopsis curvispora* (Pk.) Gremmen. The fungus was previously reported as *Neofabraea malicorticis* (Cord.) Jack., and the imperfect stage *Gloeosporium malicorticis*.

Infection of fruit can occur from as early as petal-fall stage of bloom until harvest. The rot usually begins at open lenticels and develops very slowly at cold-storage temperatures of near 0°C. The most susceptible apple cultivars are Yellow Newtown, Winesap, Delicious, and Golden Delicious, although all cultivars can be infected (Pierson, Ceponis, and McColloch 1971).

Symptoms. Symptoms on fruit appear as pale yellowish cream or uniformly brown spots, but are often brown with a pale center resembling a bull's-eye. (**fig. 33**). The spots are flat to sunken, and the rotted tissues are relatively firm. The skin does not break easily under pressure. The asexual, spore-bearing tufts of the fungus may or may not develop in storage (Pierson, Ceponis, and McColloch 1971). Orchard infections are latent and usually do not become evident as decay lesions until after fruit storage for several months. Studies suggest that the primary mechanism responsible for initiating and maintaining this latency is the accumulation of inhibitory concentrations of benzoic acid (Noble and Drysdale 1983). Conidia of the pathogen on fruit in the orchard often survive less than two weeks (Spotts 1985b).

Control. Partial control is accomplished by rapidly cooling fruit after harvest and maintaining it at temperatures of −0.6° to 0°C. A forecast method for determining the presence of the disease is to hold samples of fruit at 18.3° to 21.1°C and a high relative humidity to hasten disease development. Fruit from lots developing the disease are then marketed early in the storage season (Kienholz 1951, 1956, Pierson 1958). Sprays containing either dithiocarbamate fungicides or benomyl applied during the last month before harvest are used for control in some regions (Oregon State University 1984; Svedelius 1982; Washington State University 1984), but in California such applications are unnecessary because disease incidence is low. Further information on the canker and fruit-rotting diseases caused by *P. malicorticis* is discussed under "Anthracnose and Perennial Canker of Pome Fruit."

REFERENCES

Combrink, J. C., N. F. Sommer, R. H. Tyler, and R. J. Fortlage. 1976. Postharvest Phomopsis rot of apple fruits. *Plant Dis. Rep.* 60:1060–64.

Kienholz, J. R. 1951. The Bull's-eye rot (*Neofabraea*) problem of apples and pears. *Oregon State Hort. Soc. Proc.* 66:75–77.

———. 1956. Control of Bull's-eye rot on apple and pear fruits. *Plant Dis. Rep.* 40:872–77.

Noble, J. P., and R. B. Drysdale. 1983. The role of benzoic acid and phenolic compounds in latency in fruits of two apple cultivars infected with *Pezicula malicorticis* and *Nectria galligena*. *Phys. Plant Pathol.* 23:207–16.

Oregon State University Extension Service. 1984. Pest management guide for tree fruits in the mid-Columbia area. Hood River, 23 pp.

Pierson, C. F. 1958. Forecasting bull's-eye rot in northwest-grown apples in storage. *Plant Dis. Rep.* 42:1394–96.

Pierson, C. F., M. J. Ceponis, and L. P. McColloch. 1971. Market diseases of apples, pears, and quinces. *USDA-ARS Agric. Handb.* 376:21–23.

Smith, R. E. 1941. Diseases of fruits and nuts. *Calif. Agric. Ext. Serv. Circ.* 120.

Spotts, R. A. 1985a. Effect of preharvest pear fruit maturity on decay resistance. *Plant Dis.* 69:388–90.

———. 1985b. Environmental factors affecting conidial survival of five pear decay fungi. *Plant Dis.* 69:391–92.

Svedelius, G. 1982. [Gloesporium storage rot of apples, trials with minimized fungicide treatment.] *Vaxtskyddnotizer* 46(3):62–66 (*Rev. Plant Pathol.* 62:269. 1983.)

Washington State University Cooperative Extension. 1984. *1984 Spray guide for tree fruits in eastern Washington.* Washington State University, Pulllman, 86 pp.

Gray Mold Rot of Apple and Pear

Botrytis cinerea Pers.:Fr.

Gray mold rot occurs wherever pears and apples are grown and is especially serious in pears held for extended periods in cold storage. On pears it is considered second in importance to blue mold rot. It is one of the more important rots of apples and pears in the Pacific Northwest and in the eastern United States (English 1940, 1944; Heald and Ruehle 1931; Meheriuk and McPhee 1984; Pearson, Rosenberger, and Smith 1980; Pierson et al. 1971, Spotts 1985a).

Symptoms. The decay caused by *Botrytis cinerea* originates at the stems or calyces, or at punctures in the skin (**fig. 34A**). A water-soaked spot on the fruit turns grayish and then becomes light brown with darker brown areas. The decaying tissue is at first firm, but later becomes slightly soft and readily crumbles when rubbed between the fingers (Cooley 1932). Sporulation of the fungus occurs under high-moisture conditions but may not be evident when fruit are stored under low-moisture conditions. Decaying tissue does not separate easily from the healthy as with blue mold decay. The fungus moves from one fruit to the next causing a "nesting effect" (**fig. 34B**) (Daines 1968), and at this stage of decay small black sclerotia, the resting bodies of the fungus, may be detected. The decaying tissue usually has a pleasant fermented odor rather than the musty odor associated with blue mold rot. Symptoms vary according to cultivars and stage of fruit maturity. On red apple cultivars a reddish brown ring (2 to 3 mm in diameter) around the lenticels amid the light brown decaying areas has been described as "spot rot" (Adams and Tamburo 1957; Heald and Sprague 1926). Only certain strains of *Botrytis* cause spot rot, and this symptom is found only when apples are stored at low temperatures. In 1932 Cooley reported that in pears a high percentage of stem infection started as blackened areas with a definite line of demarcation. The infected areas later spread into the fruit and to adjacent fruit. *Botrytis* infection on pears is common in orchards in which abundant spores of the fungus occur on infected cover crops such as vetch and clover. Severe sepal infection on Rome Beauty apples was reported from the Hudson River Valley, New York (Palmiter 1951) in 1951 when rains fell for 28 to 34 hours and the temperature was 7.8° to 8.9°C during the end of bloom. Blanpied and Purnasiri (1968) found that with an increase in spore concentration in the wash water there was an increase in the number of punctured fruit infected.

Causal organism. The causal organism of gray mold rot is *Botrytis cinerea* Pers.: Fr. The perfect stage is *Botryotinia fuckeliana* (de Bary) Whetzel (= *Sclerotinia fuckeliana* (de Bary) Fuckel [Whetzel 1945]). The only other species occurring on apple was reported from Washington by Ruehle (1931), who named it *Botrytis mali* Ruehle in 1931. This species produces exceptionally large sclerotia in culture but appears to be relatively uncommon (Pierson et al. 1971).

Colonies of B. cinerea on potato-dextrose agar consist at first only of mycelium. Later, erect fasciculate conidiophores develop; these are 2 mm or more in length, mostly 16 to 30 μm thick, and olive-colored. They turn darker with exposure to light. Conidiophores form open clusters of short branches which are smooth, clear, brown below, and paler near the apex, with ends often quite colorless. These branches, which are usually hyaline and terminally swollen, bear dense clusters of conidia on sterigmata. Conidia are smooth, unicellular, hyaline to light brown, ovate to subglobose or subpyriform. They are usually most profuse when the atmosphere is dry. They are 6 to 18 by 4 to 11 μm (mostly 8 to 14 by 6 to 9 μm) in dimension. Older cultures form a stroma at the edges of the petri dish—these are black sclerotia and are characteristically flattened, loaf-shaped, or hemispherical.

Spermatia are formed on branching spermatiophores with the entire structure enveloped in a mucilaginous matrix. Apothecia, rare in California, are cupulate, stalked, and brown-colored. Ascospores are hyaline, unicellular, and ellipsoidal (Ellis and Waller 1974; Whetzel 1945). The conidia of B. cinerea survive quite well in the orchard, their half-life on pear fruit being approximately 14 days (Spotts 1985b).

Disease development. The fungus develops in stored apples and pears at injuries and from incipient infections commonly located at the stem or calyx end. Of particular significance is the fact that gray mold infections can continue to develop in storage temperatures of $-0.6°$ to $0°C$ (Heald and Ruehle 1931; Rose, McColloch, and Fisher 1951; Schneider-Orelli 1911). As an example, healthy fruit stored at the above temperature in autumn may be completely decayed by the following February or March. The fungus forms a nest rot by moving from infected to adjacent healthy fruit. It infects by forming masses of appressoria on the hyphal tips. Spotts (1985a) found that pear fruit inoculated three or four months before harvest were resistant to decay, but this resistance decreased as the fruit approached harvest maturity.

Control. Control of fruit rot is related to the reduction of inoculum or infection in the orchard and the prevention of fruit injuries during postharvest handling. Prompt cooling of fruit below 0°C and maintenance of low temperature throughout storage are essential (Pierson et al. 1971). Partial control of Botrytis in the orchard can be achieved with benzimidazole sprays. Scab sprays with these materials are certainly a benefit if resistant strains of Botrytis are not presnt. Postharvest drenches of either chlorine or sodium orthophenylphenate have been used with some success in control of a gray mold rot in the Pacific Northwest (English 1948; Pierson 1960; Pierson et al. 1971; Spotts 1984; Spotts and Peters 1980, 1982b). A surfactant was found to increase the effectiveness of chlorine but to reduce the performance of sodium orthophenylphenate, captan, or iprodione (Spotts 1984, Spotts and Peters 1982b). Postharvest treatments with benzimidazoles provide fair to good control, but the development of resistant strains of the pathogen has complicated control through use of these fungicides (Bertrand and Saulie-Carter 1978; Meheriuk and McPhee 1984; Pearson, Rosenberger, and Smith 1980). However, some recently synthesized homologues of benomyl have shown toxicity to both resistant and sensitive strains (Chiba and Northover 1988). Thiabendazole is a protectant but not an eradicant; however, it is compatible with storage scald inhibitors such as diphenylamine and ethoxyquin (Thomas and Blanpied 1971). Little, Taylor, and Peggie (1980) reported that multiformulation dips control storage disorders of apples and pears. They successfully combined calcium chloride and diphenylamine with benomyl, iprodione, imazalil, or mancozeb in the control of certain physiogenic and parasitic diseases, including gray mold rot. This rot has also been reduced by cold-storing pears in low oxygen (1 percent oxygen, 0.05 percent carbon dioxide, $-1.1°C$) (Chen, Spotts, and Mellenthin 1981). Some control of "dry eye rot" of apple caused by B. cinerea has been achieved by spraying the flowers with a conidial suspension of the antagonistic fungus Trichoderma harzianum (Tronsmo and Ystaas 1980). Other research (Janisiewicz 1988) indicates that the antagonistic fungus Acremonium breve could possibly be used postharvest to reduce the incidence of gray-mold rot. Spread of gray mold rot from one fruit to another in the packed box can be largely prevented by the use of copper-impregnated paper wrappers (Pierson et al. 1971).

REFERENCES

Adams, R. E., and S. E. Tamburo. 1957. The West Virginia spot-rot complex of apple in 1956. *Plant Dis. Rep.* 41:760–65.

Bertrand, P., and J. L. Saulie-Carter. 1978. Occurrence of benomyl-tolerant strains of *Penicillium expansum* and *Botrytis cinerea* in the mid-Columbia region of Oregon and Washington. *Plant Dis. Rep.* 62:302–05.

Blanpied, G. D., and A. Purnasiri. 1968. Penicillium and Botrytis rot of McIntosh apples handled in water. *Plant Dis. Rep.* 52:865–67.

Brooks, C., and J. S. Cooley. 1917. Temperature relations of apple-rot fungi. *J. Agric. Res.* 8:139–64.

Brooks, C., J. S. Cooley, and D. F. Fisher. 1935. Diseases of apples in storage. *USDA Farmer's Bull.* 1160, 20 pp.

Brown, G. E. 1989. Management of disease resistance in harvested fruits and vegetables. Host defenses at the wound site on harvested crops. *Phytopathology* 79:1381–84.

Chen, P. M., R. A. Spotts, and W. M. Mellenthin. 1981. Stem-end decay and quality of low oxygen stored d'Anjou pears. *J. Am. Soc. Hort. Sci.* 106:695–98.

Chiba, M., and J. Northover. 1988. Efficacy of new benzimidazole fungicides against sensitive and benomyl-resistant *Botrytis cinerea*. *Phytopathology* 78:613–18.

Cooley, J. S. 1932. Botrytis stem infection of pears. *Phytopathology* 22:269–70.

Cooley, J. S., and J. H. Crenshaw. 1931. Control of Botrytis rot of pears with chemically treated wrapper. *USDA Circ.* 177, 10 pp.

Conway, W. S. 1989. Management of disease resistance in harvested fruits and vegetables. Altering nutritional factors after harvest to enhance resistance to postharvest disease. *Phytopathology* 79:1384–87.

Daines, R. H. 1968. Soft rot (blue mold) and nest rot (*Botrytis* sp.) of apples and their control in storage. *Proc. Mass. Fruit Growers' Assoc.* 74:85–89.

Ellis, M. B., and J. M. Waller. 1974. *CMI descriptions of pathogenic fungi and bacteria. No. 431, Sclerotinia fuckeliana*. Commonwealth Mycological Institute, Kew, England.

English, H. 1940. *Taxonomic and pathogenicity studies of the fungi which cause decay of pears in Washington*. Ph.D. Thesis, Washington State University, Pullman, 270 pp.

———. 1944. Notes on apple rots in Washington. *Plant Dis. Rep.* 28:610–22.

———. 1948. Disinfectant washes for the control of decay in apples and pears (abs.). *Phytopathology* 38:914.

Heald, F. D., and G. D. Ruehle. 1931. The rots of Washington apples in cold storage. *Wash. Agr. Exp. Stn. Bull.* 253, 48 pp.

Heald, F. D., and R. Sprague. 1926. A spot-rot of apples in storage caused by Botrytis. *Phytopathology* 16:485–88.

Janisiewicz, W. J. 1988. Biocontrol of postharvest diseases of apples with antagonist mixtures. *Phytopathology* 78:194–98.

Little, C. R., H. J. Taylor, and I. D. Peggie. 1980. Multi-formulation dips for controlling storage disorders of apples and pears. I. Assessing fungicides. *Sci. Hort.* 13:213–19.

Meheriuk, M., and W. J. McPhee. 1984. Postharvest handling of pome fruits, soft fruits and grapes. *Comm. Branch Agric. Can. Pub.* 1768E, 50 pp.

Palmiter, D. H. 1951. A blossom-end rot of apples in New York caused by Botrytis. *Plant Dis. Rep.* 35:435–36.

Pearson, R. C., D. A. Rosenberger, and C. A. Smith. 1980. Benomyl-resistant strains of *Botrytis cinerea* on apples, beans, and grapes. *Plant Dis.* 64:316–18.

Pierson, C. F. 1960. Postharvest fungicide treatments for reduction of decay in Anjou perars. *Plant Dis. Rep.* 44:64–65.

Pierson, C. F., M. M. Ceponis, and L. P. McColloch. 1971. Market diseases of apples, pears, and quinces. *USDA-ARS Agric. Handb.* 376, 112 pp.

Rose, D. H. 1924. Diseases of apples on the market. *USDA Bull.* 1253, 24 pp.

Rose, D. H., L. P. McColloch, and D. F. Fisher. 1951. Market diseases of fruits and vegetables: Apples, pears, quinces. *USDA Misc. Pub.* 168, 72 pp.

Ruehle, G. D. 1931. New apple-rot fungi from Washington. *Phytopathology* 21:1141–52.

Schneider-Orelli, O. 1911. Versuche über die Wachstumsbedingungen und Verbreitung der Fäulnispilze des Lagerobstes. *Landw. Jahrb. Schweiz.* 25(3):225–46.

Sinclair, W. A. 1972. Prevention of Botrytis and Penicillium rots and scald of apples in storage with postharvest dip treatments (abs.). *Phytopathology* 62:500.

Spotts, R. A. 1984. Effect of a surfactant on control of decay of Anjou pear with several fungicides. *Plant Dis.* 68:860–62.

———. 1985a. Effect of preharvest pear fruit maturity on decay resistance. *Plant Dis.* 69:388–90.

———. 1985b. Environmental factors affecting conidial survival of five pear decay fungi. *Plant Dis.* 69:391–92.

Spotts, R. A., and B. B. Peters. 1980. Chlorine and chlorine dioxide for control of d'Anjou pear decay. *Plant Dis.* 64:1095–97.

———. 1982a. Effect of relative humidity on spore germination of pear decay fungi and d'Anjou pear decay. *Acta Hort.* 124:75–78.

———. 1982b. Use of surfactants with chlorine to improve pear decay control. *Plant Dis.* 66:725–27.

Thomas, O. C. N., and G. D. Blanpied. 1971. Further evaluation of thiabendazole as a post-harvest fungicide for apples. *Plant Dis. Rep.* 55:791–94.

Tronsmo, A., and J. Ystaas. 1980. Biological control of *Botrytis cinerea* on apple. *Plant Dis.* 64:1009.

Whetzel, H. H. 1945. A synopsis of the genera and species of the Sclerotiniaceae, a family of stromatic inoperculate discomycetes. *Mycologia* 37:648–714.

Wilson, C. L. 1989. Management of disease resistance in harvested fruit and vegetables. Managing the microflora in harvested fruits and vegetables to enhance resistance. *Phytopathology* 79:1387–90.

Wisniewski, M., C. Wilson, E. Chalutz, and W. Hershberger. 1988. Biological control of postharvest diseases of fruit: Inhibition of Botrytis rot on apples by an antagonistic yeast. *Proc. Elec. Microsc. Soc. Am.* 46:290–91.

Phomopsis Rot of Apple

Phomopsis mali Roberts

Phomopsis rot was found in Yellow Newtown apples stored in Santa Cruz in 1976 (Combrink et al. 1976). The disease is thought not to be of concern to the industry because of its low incidence. It does not appear in the handbook on market diseases published in 1971 (Pierson, Ceponis, and McColloch 1971). The rot was first reported on Yellow Newtown apple in Virginia in 1912 (Roberts 1912). *Phomopsis mali* was recently reported (Rosenberger and Burr 1982) as causing a serious core rot of apples held in controlled-atmosphere storage in New York.

Symptoms. In stem-end decay, rot symptoms are first seen in the cavity as a light- to dark-brown skin discoloration (**fig. 35**). The tissue below is also discolored and softens as the lesion enlarges. The growing lesion has a well-defined margin, and the decayed area cannot be easily separated from surrounding healthy tissue. In storage near 0°C the rot does not develop for about four months. Aerial mycelium is sparse on rotted fruit and there is no spread of rot from affected to healthy fruit. On *Phomopsis*-inoculated fruit held at room temperature, black pycnidia developed in about six weeks. In core rot, the light-brown lesions may appear on any portion of the fruit after the decay has progressed from the infected carpels to the fruit surface (Rosenberger and Burr 1982).

Causal organism. The causal organism is *P. mali* Roberts; the perfect stage is *Diaporthe perniciosa* Marchal (Dunegan 1932). The fungus is easily cultured on potato-dextrose medium, where the aerial mycelium is sparse, chalk-white to gray, flocculent, and darker with age, and the substrate becomes bluish-purple. Pycnidial development is enhanced by exposure to near-ultraviolet illumination, the pycnidia measuring 0.14 to 0.22 by 0.12 to 0.16 mm. The conidia are produced on linear conidiophores that measure approximately 20 by 2 μm. Alpha spores are spindle-shaped (7 to 10 by 3 to 4 μm) and biguttulate; the beta spores are threadlike, hooked or S-shaped, attenuate, and measure 20 to 36 by 1.5 μm. The fungus grows rapidly at 20°C, covering a 90-mm Petri dish in six days; at 0°C no growth was observed (Combrink et al. 1976; Roberts 1912). A second species, *P. ambigua* (Sacc.) Trav., is reported (English 1940) to cause a decay of pear fruit in Washington and a canker disease of pears in Japan (Hiragi, Nakatani, and Sekizawa 1980).

Disease development. In the process of disease development, the inoculum is thought to come from the cankers and leaf spots on the tree (Roberts 1912; Smith 1941). Stem-end decay develops from incipient infections in the stem (Combrink et al. 1976). In core rot, there is evidence (Rosenberger and Burr 1982) that some fruit infection occurs during bloom, but it is possible, especially in cultivars with open calyx canals, that some of the decay is initiated later in the season.

Control. Some control of the decay is possible by cold storage at about 0°C. Field spraying with benomyl is reported to control Phomopsis canker on peach (Daines and Peterson 1976) and could be beneficial. Extended storage definitely is an inducement to development of Phomopsis decay as well as other fruit rots.

REFERENCES

Brown, A. E., and T. R. Swinburne. 1978. Stimulants of germination and appressorial formation by *Diaporthe perniciosa* in apple leachates. *Trans. Brit. Mycol. Soc.* 71:405–11.

Combrink, J. C., N. F. Sommer, R. H. Tyler, and R. J. Fortlage. 1976. Postharvest Phomopsis rot of apple fruits. *Plant Dis. Rep.* 60:1060–64.

Daines, R. H., and J. L. Peterson. 1976. The occurrence and control of Phomopsis fruit rot of peach. *Plant Dis. Rep.* 60:141–43.

Dunegan, J. C. 1932. The occurrence of the perfect stage of *Phomopsis mali* in the United States. *Phytopathology* 22:922–24.

English, W. H. 1940. *Taxonomic and pathogenicity studies of the fungi which cause decay of pears in Washington.* Ph.D. Thesis, Washington State University, Pullman, 270 pp.

Hiragi, T., F. Nakatani, and H. Sekizawa. 1980. [Studies on the canker of pear I. Occurrence of the disease and symptoms:. *Ann. Rept. Soc. Plant Protect. N. Japan* 31:91–92 (*Rev. Plant Pathol.* 60:5032. 1981).

Nawawi, A., and T. R. Swinburne. 1979. Observations on the infection and rotting of apple var. Bramley's Seedling by *Diaporthe perniciosa*. *Ann. Appl. Biol.* 66:245–55.

Pierson, C. F., M. J. Ceponis, and L. P. McColloch. 1971. Market diseases of apples, pears, and quinces. USDA Agric. Handbook 376, 112 pp.

Roberts, J. W. 1912. A new fungus on the apple. *Phytopathology* 2:263–64.

Rosenberger, D. A., and T. J. Burr. 1982. Fruit decays of peach and apple caused by *Phomopsis mali*. *Plant Dis.* 66:1073–75.

Smith, R. E. 1941. Diseases of fruits and nuts. *Calif. Agric. Ext. Serv. Circ.* 120, 168 pp.

Terui, M., and Y. Harada. 1968. On *Phomopsis mali* causing "jikugusare" disease of apples. *Bull. Fac. Agric. Hirosaki Univ.* 14:43–46. (*Rev. Appl. Mycol.* 46:3151. 1968).

Miscellaneous Fungal Rots of Pome Fruit

Diseases mentioned here occur in the Pacific Northwest, where many of the apples and pears sold in California originate. Some of these diseases have been reported in California but are generally of rare occurrence here.

Black rot. This disease is of worldwide distribution and is prevalent in the Atlantic Coast states (Beisel, Hendrix, and Starkey 1984; Hesler 1916; Pierson, Ceponis, and McColloch 1971; Smith and Hendrix 1984; Starkey and Hendrix 1980; Sutton 1981; Taylor 1955) but occurs rarely in the far West, including California (Anderson 1956; Pierson, Ceponis, and McColloch 1971; Smith 1941; Stillinger 1920; Zeller 1924). There is some confusion regarding the correct binomial of the causal fungus, but the name generally preferred in this country and Europe is *Botryosphaeria obtusa* (Schw.) Schoemaker (= *Physalospora obtusa* [Schw.] Cke.) (anamorph: *Sphaeropsis malorum* [Berk.] Berk.). The pathogen attacks leaves, wood, and fruit of various pome fruit species. The disease on the fruit is primarily a ripe fruit rot, although immature fruit also are somewhat susceptible. In Georgia primary infection of apple buds may occur as early as the silver-tip stage of bud development (Smith and Hendrix 1984). Later infections of the fruit commonly take place through insect injuries, various other wounds, and open calyx tubes. Typical fruit lesions are firm and brown or with concentric zones of different shades of brown. In advanced stages, the infected area is dark brown to black and covered with pycnidia (**fig. 36A**) (Pierson, Ceponis, and McColloch 1971). Arauz and Sutton (1989) have critically studied the temperature and wetness duration requirements for apple fruit and leaf infection by the pathogen.

Black rot is controlled by reducing insect and mechanical injuries to the fruit, through the use of fungicidal sprays, and by storing the fruit at temperatures below 4.4°C. In Georgia, a single spray of captafol at the silver-tip stage provided excellent control of black rot of apples (Smith and Hendrix 1984). Control measures are not required in California.

Brown rot. Fruit on the tree, or more commonly in storage, occasionally develop a dark-colored rot with little or no sporulation of the causal fungi *Monilinia fructicola* (Wint.) Honey and *M. laxa* (Aderh. & Ruhl.) Honey (**fig. 36B**). The few large sporodochia that do develop are gray to tan and resemble those of *M. fructigena*, a species not known to occur in California or other parts of the United States. Brown rot has been reported on apple, pear, and quince in California (Smith 1941) and occurs on apple and pear in the more humid regions of the Pacific Northwest (English 1940; Pierson, Ceponis, and McColloch 1971).

Cladosporium-Hormodendrum rots. The fungi *Cladosporium herbarum* Lk.: Fr., *Hormodendrum* (*Cladosporium*) *cladosporioides* Sacc., and *Alternaria alternata* (Fries) Keissler are found frequently in moldy core or core-rot types of apple fruit decay, and commonly invade apple tissues affected by soft scald. Mechanical and insect injuries in apple and pear fruit also provide common avenues of entry for these weakly pathogenic fungi; lenticel invasion apparently does not occur (English 1940; Heald and Ruehle 1931). This type of rot is usually found after fruit has been in cold storage several months. Although these rots probably occur to a limited extent in California apples and pears, their presence has not been definitely established.

Coprinus rot of pear and apple. A species of *Coprinus* was reported by Spotts, Traquair, and Peters (1981) to cause a serious loss of d'Anjou pears in controlled-atmosphere storage at Hood River, Oregon. More recently (Meheriuk and McPhee 1984) it has been reported on both pears and apples held in cold storage in British Columbia. The pathogen is cospecific with a *Coprinus* species in the *urticicola* complex that causes winter crown rot of alfalfa. In cold-stored apples and pears it causes lesions that are circular and sunken with dark brown margins and lighter centers (**fig. 36C**); the infected fruit tissue is firm and dry. Early symptoms are similar to those of bull's-eye rot. White mycelium frequently develops on fruit surfaces, fruit wraps, and trays, and causes secondary infection of adjacent healthy fruit. In vitro the fungus grows optimally at 10°C, shows slow growth at 0°C, and no growth after three weeks at 25°C. It caused extensive decay in d'Anjou pears stored in controlled atmosphere (2.5 percent oxygen, 1 percent carbon dioxide) at −1.1°C. The fungus produces sclerotia, but their role in the disease cycle has not been established. Sterol inhibitors and dithiocarbamates were active against the pathogen in vitro, and a preharvest application of ziram reduced fruit infection both in the orchard and in cold storage (Spotts, Traquair, and Peters 1981). Currently, Coprinus rot has been reported only from Oregon and British Columbia, but it poses a potential threat to the pome fruit industry in other producing states if the pathogen becomes established there.

Mucor rot. Although *Mucor piriformis* Fischer was reported to cause a decay of Washington apple in 1931 (Heald and Ruehle) and pears in 1940 (English), little attention was paid to this disease until more recent years. In 1972, a stem-end decay of cold-stored Anjou pears by this fungus was reported in British Columbia (Lopatecki and Peters), and in 1975 a similar disease outbreak occurred in the mid-Columbia area of Oregon and Washington (Bertrand and Saulie-Carter 1980). In California, Mucor rot has been reported (Michailides 1980) on peaches, nectarines, plums, and prunes, and also was found on nectarines imported from Chile; more recently it was detected on Fuji apple (Ogawa, Feliciano, Manji, and Adaskaveg 1990).

In Anjou pears, where the disease has been most prevalent, Mucor rot is commonly centered at the stem, is light to medium brown, soft, and watery, and lacks a musty odor (**fig. 36D**) (Bertrand and Saulie-Carter 1980; Lopatecki and Peters 1972). A somewhat similar rot is caused by the related fungus *Rhizopus stolonifer* (Ehr.:Fr.) Vuill. (**fig. 36D**). However, this is a relatively high-temperature organism and is of no consequence in cold-stored fruit (Pierson, Ceponis, and McColloch 1971).

Cold-stored pears infected with *M. piriformis* sometimes show a narrow band of clear, "water-soaked" tissue at the lesion margin (English 1940). The skin of decayed fruit often ruptures, with the release of juice and the enhanced development of sporangiophores. Secondary infections commonly occur at points of contact between diseased and healthy pears and sometimes from the juice that leaks from decomposing fruit (Bertrand and Saulie-Carter 1980).

In apples the symptoms are similar to those in pear, but primary infection occurs mainly at wounds, and there is relatively little secondary infection (Bertrand and Saulie-Carter 1980). The optimum temperature for growth of *M. piriformis* is 20°C, but the sporangiospores can germinate, grow, and cause decay at −1.0°C (Bertrand and Saulie-Carter 1980). The fungus is common in orchard soils and surface debris in the Hood River Valley of Oregon (Bertrand and Saulie-Carter 1980; Dobson and Spotts 1988; Spotts and Cervantes 1986); here Michailides and Spotts (1986a) recovered its two mating types and developed a technique for the production of zygospores. The function of these spores in the disease cycle has not been determined. Although not a good competitor, *M. piriformis* is able to survive and reproduce in orchard soils, especially if an appreciable amount of organic matter is present (Michailides and Ogawa 1987). In California's Central Valley, soil temperature was a more important factor than soil moisture in the survival of sporangiospores of this fungus (Michailides and Ogawa 1986, 1989). Michailides and Spotts (1986b) reported that soil is the primary source for contamination of pears by *M. piriformis*. They found that soil adhering to picking bins contains a large number of propagules of the fungus and probably serves as an important source of inoculum for postharvest infection.

Several measures, including the reduction of fruit injuries during harvesting and subsequent handling, can be used to reduce the seriousness of Mucor rot. Additional control can be obtained by using chlorine, especially in combination with surfactants, in the dump tank of the packing house (Combrink and Grobbelaar 1984; Spotts and Cervantes 1989; Spotts and Peters 1982). Storage of pears at 1.1°C in a controlled atmosphere (1 percent oxygen, 0.05 percent carbon dioxide) also reduces the incidence of Mucor rot (Chen, Spotts, and Mellenthin 1981). Additional suggestions include removal of fallen fruit from the orchard floor, avoiding harvest during wet weather, washing soil from fruit bins prior to their reaching the dump tank, prohibition against putting fallen fruit in harvest bins, and individually wrapping the harvested fruit (Bertrand and Saulie-Carter 1980; Michailides and Spotts 1986b).

Side rot. This disease, known earlier as "Sporotrichum rot," occurs in many apple-growing regions. In north-central Washington it is said (English 1944) to be the third most common rot of apples, and it also is important on pears in both Washington and Oregon (Bertrand et al. 1977; English 1940; Sugar and Powers 1986). While the rot does not ordinarily affect a high percentage of the stored fruit, it occurs quite commonly on several apple and pear cultivars. It has not as yet been reported from California.

The causal fungus, *Phialophora malorum* (Kidd and Beaum.) McColloch (McColloch 1944), was originally named *Sporotrichum malorum* Kidd and Beaumont and was also called *S. carpogenum* Newton. It produces roughly circular, slightly sunken, brown spots with distinct margins (**fig. 36E**). The fungus appears to enter fruit most commonly through punctures and lenticels (Bertrand et al. 1977; English 1940, 1944; Gardner 1929; Kidd and Beaumont 1924; Pierson, Ceponis, and McColloch 1971; Ruehle 1931) but also may infect through worm injuries, calyces, and open calyx canals. Benomyl was inactive against the causal fungus in vitro (Bertrand et al. 1977), but certain spray materials used for the control of bull's-eye rot are reported to control side rot (Pierson, Ceponis, and McColloch 1971).

Stemphylium and Pleospora rots. *Stemphylium congestum* Newton and *Pleospora fructicola* (Newt.) Ruehle sometimes produce lesions centered at skin punctures, calyces, and lenticels of stored apples (Heald and Ruehle 1931) in the state of Washington. English (1940) reported that Washington pears also were decayed by *P. fructicola* and at least one species of *Stemphylium*. The anamorph of *P. fructicola* is the form genus *Stemphylium*. In England, *Pleospora pomorum* Horne, which also has a *Stemphylium* imperfect stage, produces shallow discolored sports on apples (Kidd and Beaumont 1924). These spots later turn almost black, with submerged pseudothecia of the fungus developing in their centers. Pleospora and Stemphylium rots have not been recorded in California.

REFERENCES

Anderson, H. W. 1956. *Diseases of fruit crops*. McGraw-Hill, New York, 501 pp.

Arauz, L. F., and T. B. Sutton. 1989. Temperature and wetness duration requirements for apple infection by *Botryosphaeria obtusa*. Phytopathology 79:440–44.

Beisel, M., F. F. Hendrix, Jr., and T. E. Starkey. 1984. Natural inoculation of apple buds by *Botryosphaeria obtusa*. Phytopathology 74:335–38.

Bertrand, P. F., I. C. MacSwan, R. L. Rackham, and B. J. Moore. 1977. An outbreak of side rot in Bosc pears in Oregon. Plant Dis. Rep. 61:890–93.

Bertrand, P. F., and J. Saulie-Carter. 1980. Mucor rot of pears and apples. Oregon State Univ. Agric. Exp. Stn. Spec. Rept. 568, 21 pp.

Chen, P. M., R. A. Spotts, and W. M. Mellenthin. 1981. Stem-end decay and quality of low oxygen stored d'Anjou pears. J. Am. Soc. Hort. Sci. 106:695–98.

Combrink, J. C., and C. J. Grobbelaar. 1984. [Influence of temperature and chlorine treatments on post-harvest decay of apples caused by *Mucor piriformis*.] Hort. Sci. 1:19–20. (Rev. Plant Pathol. 64:20804. 1985.)

Dobson, R. L., and R. A. Spotts. 1988. Distribution of sporangiospores of *Mucor piriformis* in pear orchard soils. Plant Dis. 72:702–05.

English, H. 1940. *Taxonomic and pathogenicity studies of the fungi which cause decay of pears in Washington*. Ph.D. Thesis, Washington State University, Pullman, 270 pp.

———. 1944. Notes on apple rots in Washington. Plant Dis. Rep. 28:610–22.

Gardner, M. W. 1929. Sporotrichum fruit spot and surface rot of apples. Phytopathology 19:443–52.

Heald, F. D., and G. D. Ruehle. 1931. The rots of Washington apples in cold storage. Wash. Agric. Exp. Stn. Bull. 253, 48 pp.

Hesler, L. R. 1916. Black rot, leaf spot, and canker of pomaceous fruits. N. Y. (Cornell) Agric. Exp. Stn. Bull. 379:153–252.

Kidd, M. N., and A. Beaumont. 1924. Apple rot fungi in storage. Trans. Brit. Mycol. Soc. 10:98–118.

Lopatecki, L. E., and W. Peters. 1972. A rot of pears in cold storage caused by *Mucor piriformis*. Can. J. Plant Sci. 52:875–79.

McColloch, L. P. 1942. An apple rot fungus morphologically related to a human pathogen. Phytopathology 32:1094–95.

———. 1944. A study of the apple rot fungus *Phialophora malorum*. Mycologia 36:576–90.

Meheriuk, M., and W. J. McPhee. 1984. Postharvest handling of pome fruits, soft fruits, and grapes. Commun. Branch Agric. Can. Pub. 1768E, 50 pp.

Michailides, T. J. 1980. *Studies on postharvest decay of stone fruit caused by Mucor species*. M. S. thesis, University of California, Davis, 63 pp.

Michailides, T. J., and J. M. Ogawa. 1987a. Effect of soil temperature and moisture on the survival of *Mucor piriformis*. Phytopathology 77:251–56.

———. 1987b. Colonization, sporulation, and persistence of *Mucor piriformis* in unamended and amended orchard soils. *Phytopathology* 77:257-61.

———. 1989. Effects of high temperatures on the survival and pathogenicity of propagules of *Mucor piriformis*. *Phytopathology* 79:547-54.

Michailides, R. J., and R. A. Spotts. 1986a. Mating types of *Mucor piriformis* isolated from soil and pear fruit in Oregon orchards (on the life history of *Mucor piriformis*). *Mycologia* 78:772-76.

———. 1986b. Factors affecting dispersal of *Mucor piriformis* in pear orchards and into the packing house. *Plant Dis.* 70:1060-63.

Ogawa, J. M., A. J. Feliciano, B. T. Manji, and J. E. Adaskaveg. 1990. Postharvest decay organisms of Fuji apples and disease control with iprodione (abs.). *Phytopathology* 80:892.

Pierson, C. F., M. J. Ceponis, and L. P. McColloch. 1971. Market diseases of apples, pears, and quinces. *USDA-ARS Agric. Handb.* 376, 112 pp.

Ruehle, G. D. 1931. New apple-rot fungi from Washington. *Phytopathology* 21:1141-52.

Smith, M. B., and F. F. Hendrix, Jr. 1984. Primary infection of apple buds by *Botryosphaeria obtusa*. *Plant Dis.* 68:707-09.

Smith, R. E. 1941. Diseases of fruits and nuts. *Calif. Agric. Ext. Circ.* 120, 168 pp.

Spotts, R. A., and L. A. Cervantes. 1986. Populations of *Mucor piriformis* in soil of pear orchards in the Hood River Valley of Oregon. *Plant Dis.* 70:936-37.

———. 1989. Evaluation of disinfestant-flotation salt-surfactant combinations on decay fungi of pear in a model dump tank. *Phytopathology* 79:121-26.

Spotts, R. A., and B. B. Peters. 1982. Use of surfactants with chlorine to improve pear decay control. *Plant Dis.* 66:725-27.

Spotts, R. A., J. A. Traquair, and B. B. Peters. 1981. D'Anjou pear decay caused by a low temperature basidiomycete. *Plant Dis.* 65:151-53.

Starkey, T. E., and F. F. Hendrix, Jr. 1980. Reduction of substrate colonization by *Botryosphaeria obtusa*. *Plant Dis.* 65:292-94.

Stillinger, C. R. 1920. Apple black rot (*Sphaeropsis malorum* Berk.) in Oregon. *Phytopathology* 10:453-58.

Sugar, D., and K. Powers. 1986. Interactions among fungi causing postharvest decay of pear. *Plant Dis.* 70:1132-34.

Sutton, T. B. 1981. Production and dispersal of ascospores and conidia of *Physalospora obtusa* and *Botryosphaeria dothidea*. *Phytopathology* 71:584-89.

Taylor, J. 1955. Apple black rot in Georgia and its control. *Phytopathology* 45:392-98.

Zeller, S. M. 1924. *Sphaeropsis malorum* and *Myxosporium corticola* on apple and pear in Oregon. *Phytopathology* 14:329-33.

ABIOTIC DISORDERS OF APPLE IN STORAGE

Each year millions of tons of apples are stored until they can be released for consumption. Storage time ranges from a few weeks to several months or a whole year. Short-term storage relies on low temperatures while long-term holding requires not only cold storage but controlled atmospheres (CA) and special chemical treatments to prevent loss in quality. Even at the lowest temperature at which fruit can be held without freezing, apples are subject to a number of disorders; some of these are parasitic, others are nonparasitic. The parasitic are discussed under "Biotic diseases of pome fruit in storage." The nonparasitic disorders discussed here (and, for pears, in the next section) are peculiar to stored fruit and, except for Jonathan spot, rarely develop while fruit is on the tree. In this respect these diseases differ from disorders such as bitter pit, which develops both on the tree and in storage. The cause of these diseases is usually associated with the effect of storage conditions on the metabolic processes of the fruit, which continue at a slow rate even though the fruit may be held at temperatures slightly below freezing. The nomenclature of these disorders is confusing. Not only have certain disorders been given several names each, but different disorders sometimes have been given similar names. Reducing these names to synonymy is difficult because the causes of several of the disorders are either unknown or in doubt. The nomenclature and classification employed here are, for the most part, those used by Meheriuk and McPhee (1984), Pierson, Ceponis, and McColloch (1971), and Porrit, Meheriuk, and Lidster (1982). Postharvest losses and their implications are discussed by Baritelle and Gardner (1984), Capellini and Ceponis (1984), and Kelman (1984).

Fresh apples grown in North America are marketed throughout the year with decay problems that are greatly reduced when compared with those of years past. Most important has been the technology developed for cold storage. In view of the facts that low-temperature disorders present problems with some cultivars, that apples ripen twice as fast with each 4.5°C increase in temperature, and that they can be stored at or near freezing temperatures, it becomes apparent that long-term storage requires critical handling not only of temperature but also of other factors of the storage environment. The development of controlled-atmosphere (CA) storage became a landmark in the reduction of abiotic as well as biotic diseases. Also contributing to disease control were better orchard cultural practices, including more careful handling of the fruit during harvest and postharvest, and the use of chemical inhibitors of both nonparasitic and parasitic disorders. With the recent expansion of the apple industry in California, storage conditions and storage disorders have become critical areas for research. In the following discussion, the disorders are categorized alphabetically on the basis of the location of the symptoms: First come nine with external symptoms, then five with internal symptoms, then two with external and internal symptoms.

Carbon Dioxide Injury

This injury, which appears as brown, roughened, well-defined, and partly sunken lesions in the skin (**fig. 37A**) is related to the exposure of fruit to relatively high carbon dioxide concentrations. Such concentrations can lead to toxic accumulations of acetaldehyde. The damage is associated with such factors as the rapid exposure of fruit to high carbon dioxide levels before it is cooled (Meheriuk, Porritt, and Lidster 1977), low oxygen, immaturity of fruit, cultivar, and the presence of free moisture on the fruit epidermis (Lau and Looney 1977). Carbon dioxide injury is seen early in the CA storage period and does not increase with storage time.

Examples of carbon dioxide injury are as follows: Delicious apples stored in normal oxygen (21 percent) and 5 percent carbon dioxide at 0°C for six months and then held in air at 21°C for six days developed extensive death and discoloration of the skin, while Golden Delicious held at normal oxygen and 15 percent carbon dioxide at 0°C developed injury in 4½ months. Rome Beauty apples stored in polyethylene liners that accumulated 6 to 8 percent carbon dioxide during a six-month period at 0°C showed little injury (Pierson, Ceponis, and McColloch 1971).

Chemical Injury

Exposure of fruit to toxic levels of soluble and volatile forms of chemicals can result in small, dark areas in the skin as well as in the underlying tissue centered at lenticels. The lesions may become desiccated and depressed with loss of moisture. Also, circular patches of injured tissue may occur where a liquid chemical is retained at the contact points between adjacent fruit or between the fruit and the container. Postharvest chemicals that sometimes are involved in fruit injury include those that inhibit scald (diphenylamine or ethoxyquin) (**fig. 38D,E,F**), protect against fungi (sodium orthophenylphenate), control internal breakdown and bitter pit (calcium chloride or calcium nitrate) (**fig. 38B,C**), aid in fruit floation (sodium salts) (**fig. 38G**), clean fruit surfaces of debris (detergents), enhance fruit appearance (waxes), and are used as refrigerants (ammonia) (**fig. 38A**).

To prevent possible injuries from these chemicals, a few salient points are discussed here. For scald control, wettable powder formulations of diphenylamine (DPA) cause more injury than ethoxyquin. DPA preparations may break down into concentrated crystals that adhere to the fruit and induce the formation of black lesions during cold storage. On the Golden Delicious cultivar, DPA may cause a bluish discoloration. Yet DPA is more compatible than ethoxyquin with calcium salts and enhances the action of calcium in the control of internal breakdown and bitter pit. Emulsions of ethoxyquin are very effective, but after drench or immersion treatments the fruit bins must be properly drained. Otherwise the ethoxyquin may become concentrated during cold storage and induce the formation of a ring of blackened tissue, especially in the calyx or stem cavity or at the point of contact between adjacent fruit (Porritt, Meheriuk, and Lidster 1982). This injury is more common on pears than on apples. Sodium orthophenylphenate is used in a highly alkaline solution to reduce decay caused by fungi. A water rinse is required to reduce the residues below toxic levels. In the event of equipment breakdown, fruit must be manually removed and rinsed immediately with water (Pierson, Ceponis, and McColloch 1971).

Of the calcium salts used against bitter pit and internal breakdown, calcium chloride is less likely to cause injury than calcium nitrate. A high concentration of calcium salts induces plasmolysis of cells. Injury of the calyx basin sometimes results in small cracks, which are sites of infection for decay organisms. Vacuum or pressure calcium treatments can damage cultivars with open calyx tubes or permeable skin, but cultivars such as McIntosh, Delicious, Spartan, and Cox's Orange are relatively resistant to injury from this procedure. The inclusion of ethoxyqin or benomyl in the calcium treatment increases the likelihood of phytotoxicity, while DPA acts as a safener. The addition of lecithin has a synergistic effect on calcium, and xanthan gum increases the retention of calcium by increasing the viscosity of the solution applied.

Detergents and waxes are usually nontoxic at the concentrations used.

Ammonia used as a refrigerant can escape into the storage room, and concentrations of 50 to 700 mg/mL can cause fruit damage. The odor of ammonia can be detected in concentrations as low as 50 mg/mL.

Friction Marking

Skin abrasion caused during handling and transit may result in browning of the skin and underlying tissue of such fruit as apples, pears, apricots, and peaches (Mellenthin and Wang 1974; Smith 1946) (fig. 39). Friction marking of apples is uncommon unless the fruit are subjected to extensive vibration, except for light-colored cultivars such as Golden Delicious. The susceptibility of fruit to this injury increases with maturity and storage temperature (Sommer et al. 1960). Impact bruising results in activation of an enzyme that causes an oxidative reaction with tannins and other compounds (Tate, Luh, and York 1964). Careful handling and packaging in special cell packs have largely eliminated friction bruises.

Heat Injury

Exposure to hot water or hot air can injure the fruit (Porritt and Lidster 1978; Porrit, Meheriuk, and Lidster 1982) (fig. 40). Hot-water wash treatments are used before fruit is waxed and packed. Scaldlike symptoms can develop in certain apple cultivars exposed to water for 30 seconds at 55°C. Symptoms are not visible at the time of treatment but develop after several weeks of cold storage. The nonblushed portions and the calyx lobes of Delicious apples are most sensitive, and injury appears as a diffuse browning of the skin somewhat similar to that of mild sunscald. Heating of apples and pears to 38° to 40°C for several days can inhibit the softening of apples destined for long-period storage and can delay the ripening of Bartlett pears (Liu 1978).

Jonathan Spot

This disease is characterized by brownish to almost black circular skin spots 2 to 4 mm in diameter that are not consistently associated with fruit lenticels (fig. 41). On red sports the lesions may be almost black and more concentrated on the most highly colored areas. The color reaction is thought to be related to low acidity which changes the color of anthocyanin pigments from red to blue (Richmond, Dilley, and Cation 1964). The color of the spots is lighter on green or yellow apples. The lesions appear after storage for several months and may increase in size and coalesce. The disorder occurs not only on the Jonathan cultivar but also on Rome Beauty, Wealthy, Idared, Newtown, and Golden Delicious. It has been induced by subjecting fruit to ultraviolet light (Balazs and Toth 1974). The most effective control measure for Jonathan spot, in addition to harvesting mature but not ripe fruit and immediately lowering the pulp temperature to 0° to 1.1°C. is CA storage in which the carbon dioxide level is held as low as 0.7 percent (McColloch, Yeatman, and Hardenburg 1965; Pierson, Ceponis, and McColloch 1971).

Low Oxygen (Alcohol) Injury

The advantages of low oxygen (1.5 percent or slightly lower) during storage of apples have been well established (Porritt, Meheriuk, and Lidster 1982; Smock 1977). The injury from low oxygen results from the accumulation of alcohol of over 120 mg/100 g, which renders the fruit unfit for market because of unacceptable taste. Aside from flavor, irregular dark-brown water-soaked lesions that resemble scald develop in the skin and at times include subepidermal tissues (fig. 42). Under severe anaerobic conditions, the fruit cortex and core tissue may become brown, moist, and water-soaked (Fidler and North 1971). Generally, oxygen deficiency is uncommon because the gas can diffuse readily throughout the fruit, and during CA storage a safety margin of 1 percent oxygen is provided.

Scald

Scald, also known as storage scald, common scald, and superficial scald, is the most prevalent and consequently the most widely known functional disorder of stored fruit. It probably has occurred since apples were first stored in bulk because it is not strictly a cold-storage disorder, but develops in ordinary storage as well. Jones (1897) remarked on its prevalence in Vermont apples, and Jones and Orton (1898) recognized it as noparasitic in character.

Cultivars that are highly susceptible are Wagener, Stayman Winesap, Ben Davis, Cortland, Rome Beauty, Rhode Island Greening, Baldwin, York Imperial, Granny Smith, and Cleopatra. Considered less susceptible are Jonathan, Delicious, Spartan, McIntosh, Northern Spy, Golden Delicious, and Newtown.

The disorder appears after several months in cold storage and becomes more prevalent after removal of the fruit from cold storage to room temperature. In its early stages, scald appears as a diffuse browning of the fruit skin. The affected areas vary in size from small dots to large, brown, irregular and often confluent patches (**fig. 43**). At this time the flesh beneath the affected areas is normal in appearance and texture. Later, however, the flesh may become discolored for a short distance below the skin (Bain 1956; Bain and Mercer 1963). The margins of this discolored area are sharply delimited and, except in very late stages, the discoloration seldom extends very deep into the fruit. Although the edible quality of the fruit is not materially damaged by a few scalded areas, the attractiveness of the fruit is reduced and the affected areas are readily invaded by rot organisms.

This disease differs from most other functional disorders in being more prevalent on the greener or non-blush side of the fruit. Red fruit surfaces are seldom affected, and yellow areas are somewhat less affected than green ones.

Scald is thought to be caused by volatile oxidation products of a naturally occurring terpene, a-farnesene (Anet 1972a,b; Anet and Coggiolia 1974). Formerly, the injurious factor was thought to be acetaldehyde (Powers and Chestnut 1920). Cultural and environmental conditions that increase the severity of this disorder are fruit immaturity, high fruit nitrogen, low fruit calcium, warm preharvest weather, delayed cold storage, high storage temperatures, high relative humidity in storage, restricted ventilation, extended storage periods, and, in CA storage, slow oxygen reduction and high oxygen concentration (Porritt, Meheriuk, and Lidster 1982). Scald can be reduced by avoiding the above conditions, and added protection can be provided by wrapping the fruit in oil-impregnated paper (Brooks 1923; Brooks and Cooley 1924). According to Stevens and Nance (1932) the use of oiled wrappers reduced scald development in Washington-grown apples by about 80 percent.

The method now adopted for routine scald control is the use of the antioxidant diphenylamine (DPA) as a dip or ethoxyquin (6-ethoxy-1,2-dihydro-2-2-4-trimethyl-quinoline) as either a dip or impregnated into fruit wraps (Anet 1974; Dilley and Dewey 1963; Hansen and Mellenthin 1967; Hardenburg and Anderson 1965; Melville 1967; Scott and Roberts 1966).

Soft Scald

This disorder is induced during storage at temperatures of about 2.2°C or lower and is most likely to develop on fruit exposed directly to cold air discharged from the cooling units (Pierson, Ceponis, and McColloch 1971; Porritt, Meheriuk, and Lidster 1982). The symptoms are expressed between mid-November and later December as sharply defined, irregularly shaped, smooth brown areas in the fruit skin (**fig. 44**). These lesions may be small or large and may affect most of the fruit surface except that at the stem or calyx ends. The etiology of soft scald is not known, but the metabolism of fatty acids is suspected. Most susceptible to this disorder are fruit with high respiratory activity, such as those subjected to delays in postharvest cooling that enhance ripening and increase respiration. The disorder can be prevented by exposing apples to 20 to 39 percent carbon dioxide for two days during the cooling period, or reduced by heating the fruit to 38° to 42°C for 8 to 12 hours before cold storage. If the required handling procedures to control this disorder cannot be met, the fruit should be stored at temperatures above 2.5°C for the first six to eight weeks.

Sunscald

Sunscald symptoms, which appear as a bleached or bronzed area on the fruit (**fig. 45**), may not be obvious at the time of harvest but may appear after several months of cold storage. Damage may be confined to the skin or it may extend into the cortex. The affected tissue is brownish and usually firm. The bitter taste of affected fruit is caused by a high concentration of chlorogenic acid. Incidence of the disorder can be lessened by reducing exposure of fruit to sun (Pierson, Ceponis, and McColloch 1971; Porritt, Meheriuk, and Lidster 1982).

Core Browning

This disorder, also called "brown core" and "core flush," is characterized by diffuse browning of the flesh next to the fruit carpels, with no clear line of distinction between the affected tissue and the healthy core flesh (**fig. 46**). In some cultivars the first symptoms are a pinkish or yellowish discoloration. The disease appears after three to four months in cold storage and becomes more extensive upon removal of the fruit to warmer temperatures. Susceptible fruit are those picked before full maturity, those with a high nitrogen content, those from shaded parts of the tree, and those harvested after an extended period of cloudy, wet weather. Once in storage, core browning usually is associated with storage temperatures of $-0.5°$ to $3.5°C$ but may occur at temperatures as high as $4.5°C$. Lowering the storage temperature in stages can reduce the incidence of the disease on Granny Smith apples. In CA storage, carbon dioxide levels of 0 to 1 percent are used, but cases have been reported of a high incidence of the disorder in a zero carbon dioxide atmosphere. Excessive boron may cause symptoms indistinguisable from this disease.

Apples susceptible to core browning include McIntosh, Yellow Newtown, Granny Smith, Gravenstein, Grimes Golden, Baldwin, and Rhode Island Greening; those less susceptible are Delicious, Golden Delicious, Spartan, Cox's Orange, Jonathan, Rome Beauty, and Sturmer (Porritt, Meheriuk, and Lidster 1982).

Internal Browning

Apples grown in the Pajaro Valley of California were found to exhibit a tendency to develop a brown condition of the flesh when held at certain storage temperatures (**fig. 47**). Although the condition occurs to some extent in most apple cultivars grown in this valley, it is of economic importance only in Yellow Newtown (Ballard, Magness, and Hawkins 1922; Winkler 1923). It is occasionally found in Yellow Newtown grown in other states, notably Washington, Oregon, and New York, and in the Rhode Island Greening cultivar grown in New York. This disorder is common only in fruit that have been stored at $-0.5°$ to $0°C$ for some time and is first expressed as a discoloration of the tissue around the core. A cross-sectional view reveals the presence of elongated brown streaks that radiate outward from the core. The areas that first become brown are adjacent to the primary vascular bundles, the bundles themselves becoming discolored only in late stages of the disorder. Once begun, the brown discoloration spreads rapidly in the region of the secondary vascular bundles and advances toward the calyx end of the fruit before involving the pulp. In late stages, the discoloration spreads to all parts of the apple including, occasionally, the thick-walled cells of the epidermis, which develop a scalded appearance. Most frequently, however, the skin retains its natural color and luster, and the flesh remains firm.

Apparently, internal browning is caused by some peculiarity of the cultivar and to some condition peculiar to the Pajaro Valley. Here the outstanding climatic feature is the prevalence of cool foggy weather in late July and early August when the fruit is growing most rapidly. In this locality, fruit produced in the most shaded locations on the tree develop more of the disorder than do fruit located in the less shaded sites. Artificially shaded fruit develop more of the disease than do unshaded fruit. Winkler (1923) showed that enclosing the fruit in black cloth bags reduces the incidence of internal browning, which suggests that temperature rather than sunlight is the factor involved. Also, riper more mature fruit with a yellow skin color were more susceptible than less mature fruit with green skin. Regardless of the time of harvest, the onset of internal browning occurs two to three months after the fruit is stored. Storage temperature greatly influences the rate and extent of the disorder's development: fruit stored at $0°C$, for example, develop the disorder earlier and more severely than fruit stored at $4.4°C$ or above. In fact, it is said that the disease was relatively unknown before low-temperature storage was practiced (Stevens and Nance 1932).

Volatile substances are thought to be implicated in the cause of internal browning, as increased air movement, better ventilation, and the use of oiled fruit wraps provide some control. Winkler (1923) found that the essential oils, in concentrations even lower than those reported for ripening fruit by Powers and Chestnut (1920), caused a marked increase in the permeability of the cell protoplasm. This suggested that, as a result of the increased permeability of the cell, enzymes were no longer prevented from acting upon the substrate. For example, the tannins could be oxidized by oxidase and thereby produce the brown discoloration of the affected tissue (Hulme 1956).

Control is possible through prompt storage of apples at 3.3° to 5°C, and for cultivars that are sensitive to temperatures of $-0.6°$ to 0°C, CA storage should be used.

Internal Carbon Dioxide Injury

This disorder, also called brown heart, is caused by an accumulation of carbon dioxide in the storage area and is important only where aeration is restricted. In early years it caused extensive losses in apples shipped from Australia to England. During the long sea voyage, fruit in poorly aerated holds of the ship evolved sufficient carbon dioxide to cause this disorder. The susceptibility of fruit to injury from carbon dioxide is greatest at the low temperatures commonly used for storage. When the carbon dioxide content of the storage air reaches 13 percent, injury is likely to occur, and it is almost certain to occur when the content reaches 15 percent (Allen 1924; Eaves 1938; Kidd, West, and Kidd 1927; Smith 1925).

Fruit affected with carbon dioxide injury seldom exhibit external sypmtoms. The flesh of the fruit is discolored, and in mild cases the discoloration occurs in small patches scattered throughout the fruit tissue. These patches are usually associated with vascular bundles, especially the larger patches. In severe cases, large areas of the fruit are discolored, sometimes involving all the flesh around the core. The discolored areas are firm and sharply delimited from the surrounding, normal-appearing tissue. When the fruit is removed from storage, the affected regions become dry and shrunken, producing cavities lined with brown, leathery, collapsed tissue (Plagge, Maney, and Pickett 1935). Cultivars sensitive to carbon dioxide include McIntosh, Cortland, Cox's Orange, Fameous, and Northern Spy, whereas Spartan, Delicious, and Golden Delicious are more tolerant (Porritt, Meheriuk, and Lidster 1982).

Control consists of properly ventilating the storage area (Carne and Martin 1938).

Internal Breakdown

Internal breakdown, internal browning, brown dry rot, flesh collapse, Jonathan scald, Jonathan breakdown, and physiological decay are terms applied to softening and discoloration of the pulp of apple fruit under storage conditions. Some authorities (Harley and Fisher 1927) include all such disorders under the name internal breakdown, but others (Plagge and Maney 1924) recognize two principal types: mealy breakdown and soggy breakdown. Porritt, Meheriuk, and Lidster (1982) classified these conditions as senescent breakdown (**fig. 48C**), low-temperature breakdown (**fig. 48A**), and vascular breakdown (**fig. 48D**). These breakdowns are evident as browning and softening of the cortex tissue. The immediate cause of this browning appears to be associated with failure of the membranes to retain the phenolic precursors of browning within the cell vacuole.

Mealy breakdown (senescent breakdown) develops in fruit that have been held too long in storage or have been stored at too high a temperature. Believed to be caused by senescence of the fruit tissues, it develops late in the storage season. Some fruit may undergo a year or more of storage before manifesting this condition; other fruit develop it after only a few months of storage. Early harvested cultivars are more prone to develop this breakdown than are late cultivars, and large fruit are more susceptible to it than small ones. Immature or overripe fruit develop the disorder sooner than crisp-ripe fruit (Daly 1924; Plagge and Gerhardt 1930).

In mealy breakdown, the flesh of the fruit becomes brown, dry and mealy. The symptoms vary among apple cultivars. In Grimes Golden, York Imperial, and Delicious, the breakdown appears first in the area around the core; in Jonathan and McIntosh it develops first in the pulp tissue just below the skin. In the late stages of mealy breakdown, the skin cracks and the outer portions of the flesh readily break away from the underlying core tissue (**fig. 48B**). Plagge and Gerhardt (1930) considered the disorder known as Jonathan breakdown (Daly 1924) to be a type of mealy breakdown.

To prevent mealy breakdown, fruit should be picked at the proper stage of maturity and promptly stored at a temperature of 1.1° to 2.2°C (Plagge 1930). It is important to know the approximate storage life of each cultivar and to base the length of the storage period on this information. Senescent breakdown can be reduced by the use of calcium sprays, the postharvest treatment with solutions of calcium salts, or CA storage. Treatments with calcium protect the fruit by the binding of phospholipids, which prevents membrane leakage. There is some evidence that sorbitol or volatiles, such as acetaldehyde and acetate esters, could contribute to this disorder.

Soggy breakdown symptoms, like those of mealy breakdown, vary greatly among cultivars. The first symptom, which appears even before changes in color or consistency, is the development of an alcoholic flavor. Then, small brown areas develop in the cortical region of the flesh and enlarge rapidly until most of the cortical tissue is involved. In advanced stages, the soft, brown, spongy condition may extend entirely around the fruit. The inner margins of the affected area are sharply delimited and seldom include the core region. Although the skin usually does not become discolored until the final stages of soggy breakdown, it becomes lusterless, and a slight pressure on it reveals the spongy texture of the tissue beneath (Pentzer 1925; Plagge 1925, 1930).

According to Plagge and Gerhardt (1930), the disorder known as "soft scald," which is particularly severe on fruit of Jonathan, Northwestern Greening, Rome Beauty, Wealthy, and Golden Delicious, is a form of soggy breakdown, but in this section we have treated it separately. Typical soggy breakdown is most common in the cultivars most prone to develop soft scald. It is less important than apple scald because it is limited to fewer important cultivars. Common cultivars that seldom develop soggy breakdown are Winesap, Stayman Winesap, Delicious, Willow Twig, York Imperial, Baldwin, Rhode Island Greening, McIntosh, Ben Davis, Fameuse, and Mammoth Black Twig (Plagge and Maney 1924).

Soggy breakdown is also considered to be the same as "low-temperature breakdown" because it develops at 0° to 1.1°C, but seldom occurs at 2.2°C and above. According to Harding (1935), it is caused by some disturbance in the fruit incident to the interruption of respiration by storage at low temperature. The initial high respiration rate followed by cold storage results in soggy breakdown. Apples held for a few days at a moderate temperature before storage develop more soggy breakdown than fruit stored immediately after picking. Apples produced on heavily nitrated trees respire at a higher rate and are somewhat more susceptible to soggy breakdown than fruit produced on trees not so fertilized.

Soggy breakdown can be prevented by prompt storage at a temperature not lower than 2.2°C (Plagge and Maney 1928) or by holding apples in CA storage containing 20 to 30 percent carbon dioxide for two days during the cooling period. Gibberellic acid sprays or dips and increased fertilization with magnesium and potassium also have been shown to reduce this breakdown. However, treatments with calcium, except with the Jonathan cultivar, have not been beneficial.

Water Core

This disorder is discussed under "Abiotic Orchard Disorders of Pome Fruit."

Bitter Pit

This disorder is discussed under "Abiotic Orchard Disorders of Pome Fruit."

Freezing Injury

Apples occasionally are subjected to freezing injury in the spring, shortly before harvest or postharvest. Apples and pears that are frost damaged in the spring commonly develop a ring-like russet that usually surrounds the fruit (**fig 49A**). The extent of damage is determined by both temperature and duration of exposure. Cultivars differ in their susceptibility to freezing injury but usually can be supercooled to about −4°C without damage to the fruit. Freezing injury sypmtoms vary considerably depending on the severity of the exposure. Mild freezing damage of mature fruit results in a brown discoloration of the skin, often accompanied by irregular water-soaked areas (**fig. 49B**). Internal injury results in a brown discoloration of the affected tissue, which may develop cavities during subsequent cold storage. Freeze-damaged apples are apt to have a higher respiration rate, to be more susceptible to decay, to soften sooner than normal fruit, and to become mealy.

The average freezing point for apples is −2°C, with a range of −1.4° to −2.3°C. Apples usually can sustain temperatures 1° to 2°C below this freezing point for short periods without any visible injury after thawing. The amount of sugar and the number of electrolytes in fruit largely determine the freezing point. In recent years, postharvest freezing injury has been reduced by improved refrigeration facilities in storage and transit (Porritt, Meheriuk, and Lidster 1982). In CA storage, freezing injury, and low-temperature injury in cultivars such as McIntosh, can be avoided by holding the fruit at 1.7°C; with Golden Delicious and Delicious, the recommended storage temperatures are −0.5°C and 0°C, respectively.

ABIOTIC DISORDERS OF PEAR IN STORAGE

Pears are classified into two groups based on their ripening characteristics. One group includes cultivars such as d'Anjou, Winter Nelis, Packham's Triumph, and Hardy, which ripen slowly in cold storage and generally do not lose the capacity for normal ripening. These pears are subject to storage scald, for which they respond to such control measures as the use of oiled wraps or treatment with ethoxyquin. The other group includes Bartlett, Bosc, Howell, Comice, Sierra, and Flemish Beauty, which do not ripen normally at low temperatures and can lose their ability to ripen after prolonged cold storage. The only known control measure for this condition is to reduce the storage period.

As in the previous section, disorders with external symptoms are presented first, then those with internal symptoms, then those with both kinds of symptom.

Chemical Injury

For pear, chemical injury is similar to that of apple. Flotation chemicals used to increase buoyancy in pears are the highly alkaline sodium carbonate and sodium silicate, and the pH-neutral sodium sulfate. High alkalinity (pH 10 or more) can cause injury manifested as small black dots at the lenticels of pears during storage.

Friction Marking

Friction marking (**fig. 39**) lessens as the fruit mature. Pears should be packed no later than three to four weeks after harvest. Once cooled, the fruit should not be warmed before packing. The degree of brown marking of pears is determined by the concentration of phenolic substances in the skin and flesh. With fruit maturity, the phenolic concentration declines but, at least with d'Anjou pears, advanced maturity in storage is associated with an increase in the concentration of phenolic compounds. As polyphenol oxidase is most active in basic conditions, alkaline salts are not advisable for use in flotation systems (Mellenthin and Wang 1974; Tate, Luh, and York 1964). Also, Wang and Mellenthin (1974) obtained a reduction of friction discoloration with the enzyme inhibitor 2-mercaptobenzothiazole. Current control measures, however, do not include the use of chemicals. Suggested ways to minimize friction marking include the modification of handling and packing procedures, especially the use of smooth liners in bulk bins with rigid bottoms. Also, in packing sheds, reduction in the rotational speed of the cleaning brushes is beneficial as is the use of packs that prevent movement of the fruit during handling, storage, and transit.

Senescent Scald

In Bartlett, Bosc, Sierra, and Howell pears, this disorder begins as a dark brown skin discoloration in small, isolated areas, usually toward the calyx end. In advanced stages, large areas of the skin may turn brown and the fruit may fail to ripen and soften normally (**fig. 50**) (Bangerth, Dilley, and Dewey 1972).

Superficial Scald

Cultivars affected are d'Anjou, Packham's Triumph, and Winter Nelis. After long-term storage, affected pears show a diffuse brown skin discoloration, appearing first around the neck of the pear (**fig. 44**).

Carbon Dioxide Injury

Pears are less tolerant of carbon dioxide than are apples. Browning of tissues occurs in the walls of the carpels and sometimes in adjacent core and cortex tissue (**fig. 37B**). Injury to the skin has not been reported. The susceptibility of pears to carbon dioxide injury increases with fruit maturity. Fruit from trees with low vigor and those grown during especially cool seasons also are more susceptible. The Bartlett cultivar appears to be more resistant to injury than d'Anjou, Bosc, or Clapp's Favorite (Claypool 1973; Hansen 1957).

Core Breakdown

Cultivars such as Bartlett and Bosc, which do not ripen normally at low temperatures, develop core breakdown before ripening if held too long in cold storage. The disorder appears as a soft, brown, watery collapse of tissue around the core during or after ripening (**fig. 48E**). Extended cold storage can lead to the complete loss of ripening capacity. Cultivars such as d'Anjou, Winter Nelis, and Packham's Triumph are capable of ripening slowly at cold storage temperatures but may develop core breakdown while still in storage. Controlled atmosphere storage is effective in reducing core breakdown (Porritt, Meheriuk, and Lidster 1982).

Freezing Injury

The freezing points of most pear cultivars are similar to those of apples, but vary more because of a greater range of soluble solids. Bartlett pears with a soluble solids content of less than 8 percent have freezing points as high as $-1.2°C$. Freeze-damaged tissue becomes light brown, often in a pattern conforming to the shape of a pear (**fig. 49B**) (Porritt, Meheriuk, and Lidster 1982).

Bitter Pit (Anjou Pit, Cork Spot)

This disorder occurs in d'Anjou and Packham's Triumph as brown, corky lesions in the flesh toward the calyx end (**fig. 27B**). The skin surface is uneven, with dark pitlike depressions, and the fruit tend to color and soften prematurely. Control of this disorder is the same as that for bitter pit of apples (Porritt, Meheriuk, and Lidster 1982).

Premature Ripening (Pink End)

The pink coloration of the calyx end of fruit indicates premature ripening in the field and is not a postharvest-related condition (**fig. 51**). The pink coloration of fruit after canning is related to excessively high temperatures during cooking.

REFERENCES

Allen, F. W. 1924. Carbon-dioxide storage for Yellow Newtown apples. *Proc. Am. Soc. Hort. Sci.* 40:193–200.

Anet, E. F. L. J. 1972a. Superficial scald, a functional disorder of stored apples: VII. Volatile products from the autotoxidation of *a*-farnasene. *J. Sci. Food Agric.* 23:605–08.

———. 1972b. Superficial scald, a functional disorder of stored apples: IX. Effect of maturity and ventilation. *J. Sci. Food Agric.* 23:763–69.

———. 1974. Superficial scald, a functional disorder of stored apples: XI. Apple antioxidants. *J. Sci. Food Agric.* 25:299–304.

Anet, E. F. L. J., and I. M. Coggiolia. 1974. Superficial scald, a functional disorder of stored apples: X. Control of *a*-farnasene autoxidation. *J. Sci. Food Agric.* 25:293–98.

Bain, J. M. 1956. A histological study of the development of superficial scald in Granny Smith apples. *J. Hort. Sci.* 31:234–38.

Bain, J. M., and F. W. Mercer. 1963. The submicroscopic cytology of superficial scald, a physiological disease of apples. *Austral. J. Biol. Sci.* 16:422–49.

Balazs, T., and A. Toth. 1974. Jonathan spot induced by ultraviolet light. *Acta Phytopathol. Acad. Sci. Hung.* 9:179–84.

Ballard, W. S., J. R. Magness, and L. A. Hawkins. 1922. Internal browning of the Yellow Newtown apple. *U. S. Dept. Agric. Bull.* 1004, 24 pp.

Bangerth, F., D. R. Dilley, and D. H. Dewey. 1972. Effect of postharvest calcium treatments on internal breakdown and respiration of apple fruits. *J. Am. Soc. Hort. Sci.* 97:679–82.

Baritelle, J. L., and P. D. Gardner. 1984. Economic losses in food and fiber system: From the perspective of an economist. In *Postharvest pathology of fruits and vegetables: Postharvest losses in perishable crops*, 14. Univ. Calif. Agric. Exp. Stn. Div. Agric. Nat. Resour. Bull. 1914.

Brooks, C. 1923. Oiled wrappers, oils, and waxes in the control of apple scald. *J. Agric. Res.* 26:513–36.

Brooks, C., and J. S. Cooley. 1924. Oiled paper and other oiled materials in the control of scald on barrel apples. *J. Agric. Res.* 29:129–35.

Cappellini, R. A., and M. J. Ceponis. 1984. Postharvest losses in fresh fruit and vegetables. In *Postharvest pathology of fruit and vegetables: Postharvest losses in perishable crops* 24–30. Univ. Calif. Agric. Exp. Stn. Div. Agric. Nat. Resour. Bull. 1914.

Carne, W. M., and D. Martin. 1938. Apple investigations in Tasmania: Miscellaneous notes. 8. The influence of carbon dioxide concentraiton on brown heart and other disorders. *J. Counc. Sci. Indus. Res. Austral.* 11:47–60.

Claypool, L. L. 1973. Further studies on controlled atmosphere storage of "Bartlett" pears. *J. Am. Soc. Hort. Sci.* 98:289–93.

Daly, P. M. 1924. The relation of maturity to Jonathan breakdown. *Proc. Am. Soc. Hort. Sci.* 21:286–91.

Dilley, D. R., and D. H. Dewey. 1963. Dip treatment of apples in bulk boxes with diphenylamine for control of storage scald. *Quart. Bull. Mich. Agric. Exp. Stn.* 46(1):73–79.

Eaves, C. A. 1938. Physiology of apples in artificial atmospheres. *Sci. Agric.* 18:315–38.

Fidler, J. C., and C. J. North. 1971. The effect of periods of anaerobiosis on storage of apples. *J. Hort. Sci.* 46:213–21.

Hansen, E. 1957. Reactions of Anjou pears to carbon dioxide and oxygen content of the storage atmosphere. *Proc. Am. Soc. Hort. Sci.* 69:110–15.

Hansen, E., and W. M. Mellenthin. 1967. Chemical control of superficial scald on Anjou pears. *Proc. Am. Soc. Hort. Sci.* 91:860–62.

Hardenburg, R. E., and R. E. Anderson. 1965. Postharvest chemical, hot water packaging treatment to control apple scald. *Proc. Am. Soc. Hort. Sci.* 87:93–99.

Harding, P. L. 1935. Physiological behavior of Grimes Golden apples in storage. *Iowa Agric. Exp. Stn. Res. Bull.* 182:317–52.

Harley, C. P., and D. F. Fisher. 1927. The occurrence of acetaldehyde in Bartlett pears and its relation to pear scald and breakdown. *J. Agric. Res.* 35:983–93.

Hulme, A. C. 1956. Carbon dioxide injury and the presence of succinic acid in apples. *Nature* 178:218–19.

Jones, L. R. 1897. Report of the botanist. II. Apple scald. *Vermont Univ. Agric. Exp. Stn. Ann. Rept.* 10:55–59.

Jones, L. R., and W. A. Orton. 1898. Report of the botanist. II. Apple scald. *Vermont Univ. Agric. Exp. Stn. Ann. Rept.* 11:198–99.

Kelman, A. 1984. The importance of research on postharvest losses of perishable crops. In *Postharvest pathology of fruits and vegetables,* 1–3. Univ. Calif. Agric. Exp. Stn. Div. Agric. Nat. Resourc. Bull. 1914.

Kidd, F., C. West, and M. N. Kidd. 1927. Gas storage of fruit. In *Functional diseases and their control,* 29–33. Great Brit. Dept. Sci. Indus. Res. Food Invest. Bd. Spec. Rept. 30.

Lau, O., and N. E. Looney. 1977. Water dips increase carbon dioxide-associated peel injury in "Golden Delicious" apple. *HortScience* 12:503–04.

Liu, F. W. 1978. Modification of apple quality by high temperature. *J. Am. Soc. Hort. Sci.* 103:730–32.

McColloch, L. P., J. N. Yeatman, and R. E. Hardenburg. 1965. A review of literature on harvesting, handling, storage and transportation of apples. *U. S. Dept. Agric.* ARS 51–4:166–215.

Meheriuk, M., and W. J. McPhee. 1984. Postharvest handling of pome fruits, soft fruits, and grapes. Commun. Branch Agric. Can. Pub. 1768E, 50 pp.

Meheriuk, M., S. W. Porritt, and P. D. Lidster. 1977. Effects of carbon dioxide treatment on controlled atmosphere behavior of McIntosh apples. *Can. J. Plant Sci.* 54:457–60.

Mellenthin, W. M., and C. Y. Wang. 1974. Friction discoloration of d'Anjou pears in relation to fruit size, maturity, storage, and polyphenoloxidase. *HortScience* 9:592–93.

Melville, F. 1967. Ethoxyquin for the control of scald of Granny Smith apples. *J. Agric. West. Austral.* 8(1):16–20.

Pentzer, W. T. 1925. Color pigment in relation to the development of Jonathan spot. *Proc. Am. Soc. Hort. Sci.* 22:66–69.

Pierson, C. F., M. J. Ceponis, and L. P. McColloch. 1971. Market diseases of apples, pears, and quinces. USDA Agric. Handb. 376, 112 pp.

Plagge, H. H. 1925. Cold storage investigations with Wealthy apples. *Iowa State Coll. Agric. Exp. Stn. Bull.* 230:58–72.

———. 1930. A study of soggy breakdown and some related functional diseases of the apple. *Proc. Am. Soc. Hort. Sci.* 26:315–18.

Plagge, H. H., and F. Gerhardt. 1930. Acidity changes associated with the keeping quality of apples under various storage conditions. Iowa State Coll. Agric. Exp. Stn. Res. Bull. 131.

Plagge, H. H., and R. J. Maney. 1924. Apple storage investigations. I. Jonathan spot and soft scald. II. Apple scald and internal breakdown. Iowa State Coll. Agric. Exp. Stn. Bull. 222, 64 pp.

———. 1928. Soggy breakdown of apples and its control by storage temperature. Iowa State Coll. Agric. Exp. Stn. Res. Bull. 115.

Plagge, H. H., T. J. Maney, and B. S. Pickett. 1935. Functional diseases of the apple in storage. *Iowa Agric. Exp. Stn. Bull.* 329, 79 pp.

Porritt, S. W., and P. D. Lidster. 1978. The effect of pre-storage heating on ripening and senescence of apples during storage. *J. Am. Soc. Hort. Sci.* 103:584–87.

Porritt, S. W., M. Meheriuk, and P. D. Lidster. 1982. Postharvest disorders of apples and pears. *Commun. Branch Agric. Can. Pub.* 1737E, 66 pp.

Powers, F. B., and V. K. Chestnut. 1920. The odorous constituents of apples. Emanation of acetaldehyde from the ripe fruit. *J. Am. Chem. Soc.* 42:1509–26.

Richmond, A. E., D. R. Dilley, and D. H. Cation. 1964. Organic acid and pH relationships in peel tissue of apple fruits affected with Jonathan spot. *Plant Physiol.* 39:1056–60.

Scott, K. J., and E. A. Roberts. 1966. The relative effectiveness of diphenylamine and ethoxyquin in inhibiting superficial scald of Granny Smith apples. *Austral. J. Agric. Exp. Anim. Husb.* 6:445–47.

Smith, A. J. 1925. Brown heart of Australian apple shipments. Great Brit. Dept. Scientif. Indus. Res. Food Invest. Bd. Spec. Rept. 22, 28 pp.

Smith, E. 1946. Handling injuries on pears following cold storage. *Proc. Am. Soc. Hort. Sci.* 47:79–83.

Smock, R. M. 1977. Nomenclature for internal storage disorders of apples. *HortScience* 12:306–08.

Sommer, N. F., F. G. Mitchell, R. Guillou, and D. A. Luvisi. 1960. Fresh fruit temperatures and transit injury. *Proc. Am. Soc. Hort. Sci.* 76:156–62.

Stevens, N. E., and N. W. Nance. 1932. Efficiency of oiled wraps in the commercial control of apple scald. *Phytopathology* 22:603–07.

Tate, J. N., B. S. Luh, and G. K. York. 1964. Polyphenoloxidase in Bartlett pears. *J. Food Sci.* 29:829–36.

Wang, C. Y., and W. M. Mellenthin. 1974. Inhibition of friction discoloration on d'Anjou pears by 2-mercaptobenxzothiazole. *Hortscience* 9:196.

Wang, C. Y., W. M. Mellenthin, and E. Hansen. 1971. Effect of temperature on development of premature ripening of "Bartlett" pears. *J. Am. Soc. Hort. Sci.* 96:122–25.

Winkler, A. J. 1923. A study of the internal browning of Yellow Newton apples. *J. Agric. Res.* 24:165–84.

MISCELLANEOUS DISEASES OF POME FRUIT

Diseases of minor importance in California are bitter rot, black rot, fly speck, and white rot. A few other diseases have also been reported, but are considered to be insignificant in apple culture (French 1987).

Bitter rot, caused by *Glomerella cingulata* (Ston.) Spauld. & Schrenk, is more destructive than black rot or white rot. The fungus attacks healthy leaves, twigs, and fruit, and twigs killed by fireblight bacteria. Removal of infected host parts and use of protectant fungicides are suggested for disease control.

Black rot of apple, caused by *Botryosphaeria (Physalospora) obtusa* (Schw.) Cke., appears as sepal infections of blossoms, frog-eye symptoms on leaves, limb cankers, and fruit rot. Pycnidia on the infected tissue produce nonseptate, ellipsoidal to pyriform spores measuring 12 by 25 μm. Perithecia formed in dead wood and cankers produce ellipsoid ascospores (11 to 15 by 23 to 34 μm) with thickened walls. The fungus overwinters in dead bark, twigs, cankers, and mummified fruit, and spores are released and spread during rainfall. Control measures require removing inoculum sources and protecting susceptible tissues with fungicides.

Fly speck is caused by *Schizothyrium pomi* (Mont.: Fr.) von Arx (anamorph: *Zygophiala jamaicensis* Mason). There is no apparent damage to the fruit other than a reduction of its aesthetic quality. Symptoms of fly speck are distinctive in that the groups of ascocarps formed on the fruit surface are minute, sharply defined, shiny black spots. Infection is favored by heavy rainfall and warm temperatures. Fungicide treatments applied during the summer are highly effective in control of this disease.

White rot is caused by *Botrytosphaeria dothidea* (Moug.: Fr.) Ces. & de Not. (= *B. ribis* Gross. & Dug.), and fruit infection can cause heavy losses in the southeastern United States. Symptoms on the fruit are soft, watery lesions that are clear to tan in color under warm conditions, but under cooler conditions are darker tan, resembling those of the black rot disease. Branch and twig infection occurs through the lenticels and results in the formation of blisters in which pycnidia are produced. Control involves removal of the inoculum source (dead spurs, twigs, and branches) and use of protectant fungicide sprays.

Diseases not present in California but important in apple-growing areas in the eastern United States are black pox, black root rot, blister spot, blossom-end rot, blotch, Brook's spot, cedar-apple rust, Clitocybe root rot, Nectria blight, sooty blotch, thread blight, white root rot, and X-spot.

Black pox, caused by *Helminthosporium papulosum* Berg., occurs on apples and pears, affecting twigs, fruit, and foliage. The disease is severe in the mid-Atlantic and lower midwestern states. Symptoms on bark are well-defined, conical, shiny black elevations (papules); on fruit they are smooth, black, slightly sunken round spots; and on leaves they appear as light green lesions surrounded by a red halo. Spread of the disease can be prevented by using disease-free planting stock. Fungicides used for other summer diseases are effective in the control of black pox.

Black root rot, caused by *Xylaria mali* Fromme, has been reported on apples, pears, and sweet cherries. Peaches are not affected. Disease symptoms are similar to those of other wood rot diseases. The fruiting structures of the fungus are occasionally found at the base of affected trees and appear as clusters of finger-like structures that are white at first but eventually turn black.

Blister spot, caused by *Pseudomonas syringae* pv. *papulans* (Rose) Dhanvantari, is an economic problem on the cultivar Mutsu, and the disease occurs on other cultivars in several of the north-central states. The bacteria colonize the surface of leaves, blossoms, and twigs, but only the fruits are infected. Antibiotic sprays are effective in controlling this disease.

Blossom-end rot is usually caused by *Sclerotinia sclerotiorum* (Lib.) de By., but similar symptoms have been identified with infection of *Botrytis cinerea* Pers.: Fr. Fruit infection is associated with senescing flower parts. No control measures are practiced.

Blotch, caused by *Phyllosticta solitaria* Ell. & Ev., at one time was considered second only to apple scab in importance in the eastern United States, but now is rare in commercial plantings. Control measures include the use of disease-free nursery stock, the planting of resistant cultivars, and the application of fungicide sprays.

Brook's spot (Phoma fruit spot), caused by *Mycosphaerella pomi* (Pass.) Lindau, is common in the apple-production areas of the eastern United States. The disease occurs on apple and quince. Symptoms, most common on the calyx end of the fruit, occur as irregular, slightly sunken, dark green lesions, but the color varies somewhat depending on the skin color of the cultivar. The lesions can be confused with those of cork or bitter pit, but they are more shallow and the flesh is not corky. The fungus overwinters in apple leaves on the orchard floor. Both ascospores and conidia can serve as inoculum for primary infection. Protectant fungicide sprays effectively control this disease.

Cedar apple rust, caused by *Gymnosporangium juniperi-virginianae* Schw., is the most important of the pome fruit rusts (others are quince rust and hawthorn rust) found in the eastern United States. Severe infection of apple trees results in defoliation and deformation of fruit, with production of aecial and pycnial fruiting bodies. Cedar leaves are attacked, and reddish brown galls as large as 2 inches in diameter are formed with production of telial fruiting bodies. Two years are required to complete the pathogen's life cycle. Control measures involve the removal of nearby cedar trees that provide inoculum for infection of apples and the application of protective fungicides to apple orchards.

Clitocybe root rot caused by *Armillaria* (*Clitocybe*) *tabescens* (Scop.:Fr.) Dennis, Orton & Hora, is more common on peach than on apple in the eastern United States and can be confused with shoestring root rot caused by *A. mellea* (see discussion of shoestring root rot in Chapter 4).

Sooty blotch, caused by *Gloeodes pomigena* (Schw.) Colby, commonly occurs together with fly speck. There is no apparent damage to the fruit other than a reduction in its aesthetic quality. Sooty blotch symptoms appear as superficial sooty or cloudy blotches on fruits, leaves, and twigs. Infection is favored by high rainfall and warm temperatures. Fungicide treatments applied during the summer to control other diseases are effective in controlling this disease.

Nectria blight, caused by *Nectria cinnabarina* (Tode:Fr.) Fr., the perfect state of *Tubercularia vulgaris*, affects the twigs and causes them to wilt. The symptoms appear similar to those of fire blight, but during the summer bright, orange-colored sporodochia appear on the dead tissue. The disease is a minor problem and control measures are not practiced.

Thread blight, caused by *Corticium* (*Ceratobasidium*) *stevensii* Burt, is a problem only in poorly managed orchards in the southeastern United States. The signs identifying the fungus are the presence of silvery-tan rhizomorphs and white to tan sclerotia that harden and turn dark brown with age. Once established in an orchard, the disease is difficult to control.

White root rot is caused by *Scytinostroma* (*Corticium*) *galactinum* (Fr.) Donk, and is prevalent in the southeastern apple-production areas. Trees affected with this fungus die more rapidly than those affected by other root rot fungi. Forest trees and shrubs are also attacked by this fungus. Replanting infested areas has not been successful, and no resistant rootstocks are available.

X-spot, associated with *Nigrospora oryzae* (Berk. & Broome) Petch, occurs in the southern Appalachian apple-production region. This disease is characterized by lesions on the calyx end of the fruit, developing late in the season. Fungicide sprays are effective in its control.

REFERENCES

Farr, D. F., G. F. Bills, G. P. Chamuris, and A. Y. Rossman. 1989. *Fungi on plants and plant products in the United States.* APS Press, American Phytopathological Society, St. Paul, MN. 1252 pp.

French, A. M. 1987. *California plant disease host index. Part 1. Fruits and nuts.* California Department of Food and Agriculture, Sacramento. 39 pp.

Gadoury, D. M., W. E. MacHardy, and D. A. Rosenberger. 1989. Integration of pesticide application schedules for disease and insect control in apple orchards of the northeastern United States. *Plant Dis.* 73:98–104.

Jones, A. L., and T. B. Sutton. 1984. Diseases of tree fruits. *Mich. St. Univ. Coop. Ext. Serv., North Cent. Reg. Ext. Publ.* 45. 59 pp.

USDA. 1960. Index of plant diseases in the United States. Crops Research Division, *Agric. Res. Serv. Agric. Handb.* 165. 531 pp.

Chapter 3

Stone Fruit and Their Diseases

Chapter 3 Contents

Page numbers in bold indicate photographs.

INTRODUCTION	128	
The Almond	128	
Prunus dulcis (Mill.) Webb		
The Apricot	128	
Prunus armeniaca L.		
The Cherry	129	
Prunus spp.		
The Peach and Nectarine	130	
Prunus persica (L.) Batsch and		
P. persica var. *nectarina* Maxim.		
Plums and Prunes	130	
Prunus salicina Lindl. and *P. domestica* L.		
REFERENCES	131	
DISEASES OF STONE FRUIT CAUSED BY BACTERIA AND MYCOPLASMALIKE ORGANISMS	132	
Bacterial Canker and Blast of Stone Fruit	132	
Pseudomonas syringae pv. *syringae* van Hall		
Crown Gall of Stone Fruit	132	
Agrobacterium tumefaciens (Smith & Townsend) Conn		
Foamy Canker of Almond	132	403
Cause Unknown		
REFERENCE	133	
Leaf Scorch of Almond	133	403
Xylella fastidiosa Wells et al.		
REFERENCES	135	
X-Disease of Cherry and Peach	136	404
Mycoplasmalike Organism		
REFERENCES	139	
FUNGAL DISEASES OF STONE FRUIT	141	
Anthracnose of Almond	141	
Colletotrichum gloeosporioides Penz. (=*Gloeosporium amygdalinum* Brizi)		
REFERENCES	142	
Brown Rot of Stone Fruit	142	405
Monilinia laxa (Aderh. & Ruhl.) Honey		
Monilinia fructicola (Wint.) Honey		
Brown Rot: European Brown Rot	142	405
Monilinia laxa (Aderh. & Ruhl.) Honey		
Brown Rot: American Brown Rot	145	405
Monilinia fructicola (Wint.) Honey		
REFERENCES	149	
Ceratocystis (Mallet Wound) Canker of Almond and Prune	153	406
Ceratocystis fimbriata Ell. & Halst.		
REFERENCES	155	
Cercospora Leaf Spot of Almond and Stone Fruit	156	
Cercospora circumscissa Sacc.		
REFERENCES	156	
Cytospora and Rhodosticta Cankers of Stone Fruit	156	406
Cytospora leucostoma (Pers.) Sacc.		
Rhodosticta quercina Carter		
REFERENCES	160	
Dothiorella Canker of Almond	161	407
Botryosphaeria dothidea (Moug.:Fr.) Ces. and de Not.		
REFERENCES	163	
Eutypa Dieback of Apricot	164	407
Eutypa lata (Pers.:Fr.) Tul.		
REFERENCES	167	
Green-Fruit Rot of Stone Fruit	169	408
Botrytis cinerea Pers.:Fr., *Sclerotinia sclerotiorum* (Lib.) dBy., *Monilinia laxa* (Aderh. & Ruhl.) Honey, and *M. fructicola* (Wint.) Honey		
REFERENCES	170	
Hull Rot of Almond	171	409
Rhizopus spp., *Monilinia fructicola* (Wint.) Honey, and *M. laxa* (Aderh. & Ruhl.) Honey		
REFERENCES	172	
Leaf Blight of Almond	172	409
Seimatosporium lichenicola (Cda.) Shoemaker & Müller		
REFERENCES	173	
Leaf Curl of Cherry	173	409
Taphrina cerasi (Fckl.) Sadeb.		
REFERENCES	174	
Leaf Curl of Nectarine and Peach	174	410
Taphrina deformans (Berk.) Tulasne		
REFERENCES	176	
Leaf Spot of Cherry	177	411
Blumeriella jaapii (Rehm) von Arx (*Coccomyces hiemalis* Hig.)		
REFERENCES	178	
Phytophthora Pruning Wound Canker of Almond	178	411
Phytophthora syringae (Kleb.) Kleb.		
REFERENCES	180	
Phytophthora Root and Crown Rot of Stone Fruit	180	
Phytophthora spp.		
Plum Pockets	180	411
Taphrina spp.		
REFERENCES	181	

Powdery Mildew of Stone Fruit	181	412

Sphaerotheca pannosa (Wallr.:Fr.) Lév.,
Podosphaera tridactyla (Wallr.) de Bary,
P. clandestina (Wallr.:Fr.) Lév.,
P. leucotricha (E. & E.) Salm.

REFERENCES	186	
Russet Scab of Prune	187	413
Cause Unknown		
REFERENCES	188	
Rust of Stone Fruit	188	413

Tranzschelia discolor (Fuckel) Tranz. & Litv.

REFERENCES	190	
Scab of Almond and Stone Fruit	191	414

Cladosporium carpophilum Thüm.

REFERENCES	192	
Shoestring (Armillaria) Root Rot of Stone Fruit	193	

Armillaria mellea (Vahl.:Fr.) Kummer

Shothole of Stone Fruit	193	414

Wilsonomyces carpophilus (Lév.)
Adaskaveg, Ogawa, and Butler
(= *Stigmina carpophila* [Lév.] Ellis)

REFERENCES	196	
Silver Leaf of Stone Fruit	197	

Chrondrostereum purpureum (Pers.:Fr.) Pouz.

Verticillium Wilt of Stone Fruit	197	

Verticillium dahliae Kleb.

VIRUS AND VIRUSLIKE DISEASES OF STONE FRUIT	198	
Almond Union Disorders	198	415
Cause Unknown		
REFERENCE	198	
Cherry Mottle Leaf	198	416
Graft-Transmissible Pathogen		
REFERENCES	199	
Cherry Necrotic Rusty Mottle	199	416
Graft-Transmissible Pathogen		
REFERENCES	199	
Cherry Rasp Leaf	200	416
Cherry Rasp Leaf Virus (CRLV)		
REFERENCES	200	
Cherry Rusty Mottle	200	416
Graft-Transmissible Pathogen		
REFERENCES	201	
Cherry Stem Pitting	201	416
Graft-Transmissible Pathogen		
REFERENCES	202	
Peach Mosaic	202	416
Graft-Transmissible Pathogen		
REFERENCES	203	
Prune Brownline	204	417
Tomato Ringspot Virus (TmRSV)		
REFERENCES	206	
Prune Dwarf	206	417
Prune Dwarf Virus (PDV)		
REFERENCES	208	
Prunus Necrotic Ringspot	209	417
Prunus Necrotic Ringspot Virus (PNRSV)		
REFERENCES	211	
Prunus Stem Pitting (PSP)	212	418
Tomato Ringspot Virus (TmRSV)		
REFERENCES	214	
Yellow Bud Mosaic	214	418
Yellow Bud Mosaic (YBMV) Strain of Tomato Ringspot Virus (TmRSV)		
REFERENCES	216	

ABIOTIC DISORDERS OF STONE FRUIT	218	
Chilling Canker of Peach	218	418
Cause—Chilling		
REFERENCES	218	
Corky Growth on Almond Kernels	218	418
Cause Unknown		
REFERENCE	219	
Crinkle Leaf and Deep Suture of Sweet Cherry	219	419
Cause—Genetic		
REFERENCES	220	
Nectarine Pox	220	419
Cause Unknown		
REFERENCES	221	
Noninfectious Bud Failure of Almond	221	419
Cause Unknown		
REFERENCES	224	
Noninfectious Plum Shothole	225	420
Cause—Genetic		
REFERENCES	225	
Nonproductive Syndrome of Almond	225	420
Cause—Genetic		
REFERENCES	226	
Plum Rusty Blotch	226	421
Cause Unknown		
REFERENCE	226	

POSTHARVEST DISEASES OF STONE FRUIT	227	
Sweet Cherry	227	421
Apricot	231	422
Nectarine and Peach	231	421, 422
Plum	234	422
Prune	235	422
REFERENCES	236	

MISCELLANEOUS DISEASES OF STONE FRUIT	241	
REFERENCES	242	

Introduction

Stone fruit belong to the family Rosaceae. The fruit of *Prunus* species is a drupe with the seed surrounded by a hard shell or stone. The stone develops from the inner part of the ovary wall, and the soft flesh from the outer part. Blossom parts are located around the single ovary (perigynous). The blossoms are borne singly on peduncles or in clusters. Almond trees start to bloom in early February, followed by apricots in the latter part of February, then peaches and nectarines, followed by plums, prunes, and finally cherries.

Important considerations for production of stone fruit in California are pruning, fruit thinning, fertilization, pest and disease control, weed control, irrigation, harvest, packaging, postharvest decay control, storage, and shipping. These vary with each crop and also could differ depending on whether the orchard is located in the warm Sacramento and San Joaquin valleys, the cooler areas of Santa Clara Valley, or the North Bay region. The Sacramento Valley can have 16 inches or more rain annually, the southern San Joaquin Valley might have only 6 or 8 inches. Humidity and dew occur more in areas adjacent to rivers and lakes. Fog areas are usually related to proximity to rivers, lakes, and the ocean.

The Almond

Prunus dulcis (Mill.) Webb

The almond, *Prunus dulcis* (Mill.) Webb (*Amygdalus communis* L.), a native of southwest Asia, was brought to Greece and North Africa in prehistoric times. Almond growing in North America is confined almost entirely to California, on more than 423,000 acres located mainly in the Sacramento and San Joaquin valleys.

Most of the principal cultivars were developed in California in the last half of the nineteenth century from seedlings brought from Europe and northern Africa. These cultivars are Nonpareil, Ne Plus Ultra, Mission (Texas), Drake, and Peerless. In about 1938 the U.S. Department of Agriculture and the University of California Agricultural Experiment Station introduced two new cultivars, Jordanolo and Harpareil. Only the Jordanolo, favored because of its large meats, was planted extensively. Later, the Experiment Station introduced the Davey and more recently Padre, Solano, and Sonora.

Disease evaluations based on opinions and experience of experts on 11 major cultivars (Nonpareil, Merced, Mission, Peerless, Ne Plus Ultra, Butte, Carmel, Fritz, Price Cluster, Ruby, and Thompson) show some variations in disease susceptibility (Kester et al. 1980) but none with resistance that could be considered for incorporation into new cultivars.

Almonds are propagated by budding the desired cultivar onto seedlings of the bitter almond or one of the commercial cultivars. The Mission cultivar is frequently used. The peach was formerly used as a rootstock, but this practice has declined.

Almonds are self-sterile, so it is necessary to provide proper conditions for cross-pollination among cultivars. This is done by alternating one, two, or three rows of Nonpareil with adjacent rows of other cultivars such as Peerless, Ne Plus Ultra, or Mission. Because these cultivars blossom at different times, some thought is required in planting so that there is sufficient overlap in their blossoming periods. Research is in progress to develop self-fertile cultivars but as yet no completely satisfactory selections are available.

August, September, and October are the harvest months. The nuts are mechanically shaken from the tree and picked up mechanically on preleveled ground.

The Apricot

Prunus armeniaca L.

The cultivated apricot, *Prunus armeniaca* L., originated about 4,000 years ago in China near Beijing. It reached southwest Asia (Armenia) before the time of Alexander the Great. The famous "Golden Apples" of Greek mythology were actually apricots. The movements and commerce of ancient peoples resulted in the introduction of apricots into Italy by 100 B.C., into England by 1620 A.D., and into Virginia about 1629 (although the fruit did not adapt to the climate there). The Spaniards brought the apricot to Mexico in the sixteenth century. Seedlings were planted in California missions in the eighteenth century. European cultivars were introduced before 1850.

Commercial production of apricots in California was started more than 175 years ago. The first recorded production was in 1792 in an orchard near the town of Santa Clara. Today, apricot production is mainly confined to California, Washington, Utah, and Arizona.

Numerous cultivars have been grown in California, but since 1920 major commercial production has been limited to Tilton and Royal (Blenheim). Castlebright, Derby, Katy, Modesto, Patterson, and other minor cultivars make up the remainder. The Royal supposedly originated in France about 1830, and shortly thereafter the Blenheim originated in England. The Royal is highly flavored and used for canning, drying, and fresh shipment. The Tilton was originated in Kings County, California, by J. E. Tilton in 1890. It is used primarily for canning, and tends to bear alternate heavy and light crops. Moorpark has a fine quality with a plumlike taste. Derby Royal resembles Royal but sets a lighter crop and matures earlier. It was first planted near Winters, California, in 1895; next to Perfection it is the earliest commercial cultivar for fresh shipment. Perfection originated in the state of Washington in 1911.

California leads the nation in apricot production. The 23,000 acres of production represents over 90 percent of the U.S. crop. About 70 percent of the crop is used for canning, 18 percent for drying, 6 percent for freezing, and 6 percent for the fresh market. Other products made from apricots are wine, brandy, champagne, and nectar. Ground apricot pits are used to clean jet engines, and the kernel oil is used for soaps and perfume. Fresh apricots are high in vitamin A.

Apricots are usually propagated by budding the commercial cultivar on seedling peaches or apricots. Myrobalan plum is sometimes used as a rootstock if the planting is to be made in heavy, poorly drained soils. The apricot does well on the Marianna 2624 plum rootstock. Because the apricot is self-fertile, it is not necessary to alternate cultivars in the plantings. The tree starts to bear in its third year and reaches peak production by the sixth year. Some apricot trees in Solano County are still providing excellent fruit after more than 60 years. Apricots blossom in February, and harvest begins in the middle of June and continues for about a month. Trees can be pruned in late summer or fall before the leaves are shed, but pruning is usually done after leaf fall in the winter and before blossoming in the spring.

The Cherry

Prunus spp.

Ancestors of the domesticated cherries of today were native to southwestern Asia around the Caspian and Black seas. They were of three distinct kinds: sweet cherry (*Prunus avium* L.); sour cherry, (*P. cerasus* L.); and Duke cherry, which is thought to be a hybrid between sweet and sour cherry.

In California, sweet cherries are grown in the Stockton, San Jose, Hollister, and Fresno areas. In Oregon, they are grown at Hood River and The Dalles, and in Washington at Wenatchee and in the Yakima Valley. They are also grown in Montana at Flathead Lake and in Utah near Salt Lake City. The principal cultivars planted on 11,512 acres in California are Early Burlat, Larian, Van, Lambert, Black Tartarian, Royal Ann (Napoleon), Bing, Jubilee, and Rainier.

Cross-pollination among cultivars is necessary. Examples of pollenizers are Black Tartarian for Bing, Early Burlat, and Mona; and Bing for Black Tartarian, Early Burlat, Jubilee, Larian, Mona, and Van.

The predominant rootstock for sweet cherry is mahaleb, but in new orchards, mazzard is gaining in popularity. A third stock sometimes employed is Stockton Morello. Colt, a recently developed clonal rootstock with considerable resistance to Phytophthora root and crown rot, is now being used to some extent.

The harvest season starts in the middle of May in Stockton and lasts until the end of June in Hollister. This is followed by harvests in Oregon, Washington, Idaho, Montana, and Utah. About three-fourths of the crop is sold fresh and the remainder is brined or canned.

Sour cherry does best where winters are rather cold and summers are cool. Consequently, sour cherry is not grown commercially in California but is grown around the Great Lakes and in the Pacific Northwest, including Canada. The most important cultivars are Montmorency, Early Richmond, and English Morello.

The Peach and Nectarine

Prunus persica (L.) Batsch and *P. persica* var. *nectarina* Maxim.

The peach, *Prunus persica* (L.) Batsch. (*Amygdalus persica* L., *Persica vulgaris* Mill.), was introduced into Europe from Persia, which in turn acquired the peach from China where it apparently has been cultivated for some 3,000 years. (The term "peach" is derived from a Latin word meaning "Persian.") By 330 B.C. the peach had reached Greece, and during the Middle Ages its culture spread throughout Europe. By the 1570s it was in Mexico; it was introduced into Florida by the Spanish, into Louisiana by the French, and into Virginia and Massachusetts by English settlers. Today, about one-fourth of the world's peach supply is produced in the San Joaquin and Sacramento valleys of California.

California fresh peaches for market are harvested from the second week of May to late September. Of more than 125 improved cultivars in California, some of the most important are Springcrest, Merrill Gemfree, Redtop, Elegant Lady, June Lady, O'Henry, Fairtime, Carnival, Flavorcrest, and Flamecrest. Those gaining popularity are May Crest, Spring Lady, and Queencrest. Most are sold fresh; the rest are canned, frozen, or dried. Total freestone (melting flesh) acreage in California is 27,503.

Clingstone canning peaches total 35,781 acres, with an average annual production of over 400,000 tons. Harvest starts in the middle of July and extends to the beginning of September. Important cultivars, grown mostly in California, are the extra early Loadel and Carson; early Andros and Ross; late Dr. Davis and Halford; and extra late Starn and Corona. All clingstone peaches are canned. The only freestone peach commercially canned is the Fay Elberta.

The nectarine, *Prunus persica* var. *nectarina* Maxim., is identical with peach in all features except in the skin of the fruit, which is without trichomes (fuzz). Current cultivars are yellow-fleshed rather than white and are highly colored with a red blush overspreading a rich amber-yellow skin. Harvest starts in late May and extends through late September. Principal cultivars are Fantasia, Royal Giant, Flamekist, May Grand, Summer Grand, Late Le Grand, and Autumn Grand. California supplies over 97 percent of the U.S.-grown nectarines, the current production being about 184,000 tons per year. The 23,600 acres in production are mostly in the San Joaquin Valley.

Peach and nectarine are commonly propagated by budding to seedlings of peach; seedlings of Lovell and Nemaguard (which is resistant to root-knot nematodes) have been used extensively. The myrobalan plum and Marianna 2624 have been tried as rootstocks for planting in heavy, poorly aerated soils, but they are not satisfactory because of compatibility problems.

Peach and nectarine cultivars are self-fertile, so large numbers of trees of the same cultivar are planted together. The practice has considerable influence on the incidence of diseases.

Plums and Prunes

Prunus salicina Lindl. and *P. domestica* L.

Plums are native to almost all land areas of the North Temperate Zone and have been used as food since prehistoric times. Natives of certain parts of North America sometimes added the flesh of plums to venison or buffalo meat in making pemmican, a staple food. As with many other stone fruits, the cultivation of plums began in the ancient civilizations near the Caucasus bordering the Caspian Sea. But the wild progenitor of the cultivated plums of Europe is unknown. Some think that *Prunus domestica* L., the species to which European plums and prunes are assigned, actually arose as hybrids between *Prunus cerasifera* Ehrh. and *P. spinosa* L. Many plum cultivars are self-sterile.

California, with 42,752 acres, produces more than 90 percent of the total U.S. supply of plums, which are available from early May through August. All of the prominent types originated in either Japan or Europe. The Japanese types, *Prunus salicina* Lindl. (*P. triflora* Roxb.), came from stock introduced in the 1870s and are mostly early or midseason types with red or yellow skins. The European cultivars, which are descended from stock apparently introduced into America by the Pilgrims, are generally mid- or late season and always blue-to-purple in skin color. Plum cultivars are frequently classified according to skin color. Examples of red ones are Red Beaut, Casselman,

Simka, Santa Rosa, Laroda, and Late Santa Rosa; purple types are Friar, Blackamber, and Angeleno; green sorts are Kelsey and Wickson; and black-red types are El Dorado and Nubiana.

By definition, a prune is a dried plum. Certain European or "domestica" plums can be sundried or dehydrated without fermenting. The California prune is an offshoot of La Petite d'Agen, a prune-plum native to southwest France. In 1856 Jean and Pierre Pellier obtained cuttings and brought them to San Jose, where their brother Louis grafted them onto the American wild plum rootstocks. Today the d'Agen prune is known as the California or French prune and constitutes approximately 96 percent of the state's prune production. Other commercial types of importance are Imperials (Imperial Epineuse), which originated in France in 1870, Robe de Sargeant, Sugar, and Burton. California's prune orchards currently produce approximately 70 percent of the world's prune supply and 99 percent of the nation's crop. In 1979, 131,897 dried tons were produced in California on 84,337 acres (9,583 acres nonbearing); 1987 acreage was 82,032.

Full production of prunes occurs between the eighth and twelfth year after planting and continues on a commercial basis for about 30 years. Harvesting starts in mid-August and lasts for about a month. Mechanical shakers drop the fruit onto fabric catching-frames, and conveyor belts place them in bins in which they go to the dehydrator. Washed fruit are placed in wooden trays and subjected to a constant flow of warm circulating air (76.6°C) for 14 to 24 hours, at which time the moisture content has been reduced to 16 to 20 percent of the original. After sorting, the moisture content is restored at 26 to 32 percent.

Rootstocks used for cultivars of *P. salicina* are marianna plum and nematode-resistant selections of peach. For *domestica* plum trees, the myrobalan plum is the most commonly used; marianna plum and peach are also used.

REFERENCES

Aldrich, T. M., et al. 1978. Irrigation, fertilization, and soil management of prune orchards. *Univ. Calif. Div. Agr. Sci. Leaf.* 21016, 18 pp.

California Crop and Livestock Reporting Service. 1984. *California fruit and nut acreage—1983.* 28 pp.

Day, L. H. 1951. Cherry rootstocks in California. *Univ. Calif. Agr. Exp. Stn. Bull.* 725, 31 pp.

———. 1953. Roostocks for stone fruits. *Univ. Calif. Agr. Exp. Stn. Bull.* 736, 76 pp.

Foytik, Jerry. 1961. Trends and outlook, California cherry industry. *Univ. Calif. Agr. Exp. Stn. Circ.* 501, 34 pp.

Hartmann, H. T., and J. A. Beutel. 1981. Propagation of temperate-zone fruit plants. *Univ. Calif. Div. Agr. Sci. Leaf.* 21103, 61 pp.

Kester, D. E., W. C. Micke, D. Rough, D. Morrison, and R. Curtis. 1980. Almond variety evaluation. *Calif. Agr.* 34(8):4–7.

La Rue, J. H., and M. H. Gerdts. 1976. Commercial plum growing in California. *Univ. Calif. Div. Agr. Sci. Leaf.* 2458, 22 pp.

———. 1983. Growing shipping peaches and nectarines in California. *Univ. Calif. Coop. Ext. Leaf.* 2851, 26 pp.

Meith, C., W. C. Micke, D. Rough, A. D. Rizzi, and B. Teviotdale. 1983. Almond production. *Univ. Calif. Coop. Ext. Leaf.* 2463, 22 pp.

Micke, W., A. A. Hewitt, J. K. Clark, and M. Gerdts. 1980. Pruning fruit and nut trees. *Univ. Calif. Div. Agr. Sci. Leaf.* 21171, 47 pp.

Micke, W., and D. Kester, tech eds. 1981. *Almond orchard management.* Univ. Calif. Div. Agr. Sci. Priced Pub. 4092, 150 pp.

Micke, W. C., and F. G. Mitchell. 1972. Handling sweet cherries for the fresh market. *Univ. Calif. Div. Agr. Sci. Circ.* 560, 18 pp.

Micke, W. C., and K. S. Moulton. 1982. Commercial almond production in California: Information for prospective growers. *Univ. Calif. Div. Agr. Sci. Leaf.* 2824, 3 pp.

Norton, R. A., C. J. Hanson, H. J. O'Reilly, and W. H. Hart. 1963. Rootstock for peaches and nectarines in California. *Univ. Calif. Agr. Exp. Stn. Leaf.* 2157.

Norton, M. V., W. H. Krueger, and G. S. Sibbett. 1988. Growing prunes in California. *Univ. Calif. Div. Agr. Nat. Resour. Leaf.* 21465, 3 pp.

Ramos, D. E., tech ed. 1981. *Prune orchard management.* Univ. Calif. Div. Agr. Sci. Spec. Pub. 3269, 156 pp.

Reuther, W., et al. 1981. Irrigating deciduous orchards. *Univ. Calif. Div. Agr. Sci. Leaf.* 21212, 52 pp.

Ross, N., and A. D. Rizzi. 1976. Cling peach production. *Univ. Calif. Div. Agr. Sci. Leaf.* 2455, 7 pp.

Teviotdale, B. L., J. M. Ogawa, G. Nyland, and B. Kirkpatrick. 1989. Diseases. In: *Peaches, plums, and nectarines: Growing and handling for fresh market.* Chap. 17:118–29. Tech eds. J. H. LaRue and R. S. Johnson, Univ. Calif. Div. Agr. & Nat. Resour. Publ. 3331.

United States Department of Agriculture. 1976. *Virus diseases and noninfectious disorders of stone fruits in North America.* U.S.D.A. Agric. Handb. 437, 433 pp.

Westwood, M. N. 1978. *Temperate-zone pomology.* W. H. Freeman, New York, 428 pp.

Wilson, E. E., and J. M. Ogawa. 1979. *Fungal, bacterial, and certain nonparasitic diseases of fruit and nut crops in California.* Univ. Calif. Div. Agr. Sci., Berekeley, CA 190 pp.

Diseases of Stone Fruit Caused by Bacteria and Mycoplasmalike Organisms

Bacterial Canker and Blast of Stone Fruit

Pseudomonas syringae pv. *syringae* van Hall

The bacterial canker organism kills buds in the winter and blossoms in the spring, and induces canker formation on branches and trunks from late fall to spring. The bud-blight phase has been related to infections through leaf scars or buds during the fall and winter months, while the blossom-blast phase is most serious in years of cold or freezing temperatures during bloom. Associated with blossom blast has been infection of green fruits, especially in low-lying parts of the orchard. The canker phase has been considered one of the most important diseases of stone-fruit trees during the first few years after planting. Mature trees can also be attacked, although they are not as susceptible.

Partial control is achieved through soil fumigation, chemical sprays, and cultural programs such as time of pruning, rootstock selection, and protecting the blossoms from frost injury.

Details of this disease are discussed in Chapter 4.

Crown Gall of Stone Fruit

Agrobacterium tumefaciens (Smith & Townsend) Conn.

Crown gall is important in nursery trees, young orchard plantings, and mature trees. To reduce seedling infection in the nursery, the seed is soaked in a chlorine solution and nursery fields are preplant fumigated with methyl bromide-chloropicrin or a mixture of dichloropropane and dichloropropene. Trees having galls when dug are eliminated.

Root infections often occur when trees are planted in the orchard, and galls are commonly found on mature trees. Recent studies in Australia and the United States have shown that roots of nursery trees dipped in certain strains of *Agrobacterium rhizogenes* (*radiobacter*) are protected from infection. Rootstocks for stone fruit trees are in general very susceptible. Rootstocks moderately resistant are apricot (*Prunus armeniaca* cv. Royal) seedlings, mahaleb (*P. mahaleb*) seedlings, and Stockton Morello (*P. cerasus*) softwood cuttings or suckers. Of the plum rootstocks, myrobalan (*P. cerasifera*) seedlings are less susceptible than peach seedlings, while the hardwood cuttings of Myrobalan 29C and Marianna 2624 (*P. cerasifera* × *P. munsoniana*) are considered more resistant than *P. cerasifera*.

Details of this disease are discussed in Chapter 4.

Foamy Canker of Almond

Cause Unknown

This condition, first reported in 1974, occurs on a few scattered almond trees throughout the San Joaquin Valley and in Yolo and Butte counties in the Sacramento Valley. Second- and third-leaf trees of the Carmel cultivar appear most prone to foamy canker, but it also occurs on both weak and apparently healthy trees of other cultivars of all ages. The disease has not been reported elsewhere in the United States or other countries. Because of the low incidence of foamy canker, crop losses have been minimal.

Symptoms and disease development. Active symptoms of the disease first appear in late July, when copious amounts of watery, reddish gum (**Fig. 1**) drain down from affected branches and form a puddle on the ground. The scaffold crotch is where the disorder usually is initiated. The cankers then advance up into the primary and secondary limbs and down into the trunk, but not below the bud union. At the height of canker activity, a whitish foam resembling that of beer cascades down the tree; it has an alcoholic odor, suggesting fermentation of the affected tissues. The cambial region just beneath the bark is rotted, white, and mushy. The affected areas are irregularly shaped; they may completely surround healthy tissue that is still capable of developing new but short-lived shoots. The most serious aspect of the condition is the girdling and killing of the trunk and scaffold branches. From a distance, the disorder could be confused with such diseases as bacterial canker, Ceratocystis canker, aerial Phytophthora canker, or Verticillium wilt. The cankers are inactive during tree dormancy, and most of them do not reactivate.

Cause. The cause of foamy canker has not been established. A bacterium, tentatively identified as *Zymononas* sp., has been isolated from affected tissues (B. Teviotdale, personal communication), but only further research can determine if this organism is the causal agent.

Control. Control measures include pruning out diseased limbs well below the affected area and complete removal of severely cankered trees.

REFERENCE

University of California. 1985. Foamy canker. In *Integrated pest management of almond*, 104–05. Univ. of Calif., Div. Agric. Nat. Resour. Pub. 3308.

Leaf Scorch of Almond

Xylella fastidiosa Wells et al.

A leaf-scorch disease of almond was noticed in the Antelope Valley of Los Angeles County during the early 1950s. A similar disease has been seen in Contra Costa County for more than a decade and referred to as "almond decline" or "golden death." Surveys have shown that this disease occurs in Contra Costa, San Joaquin, Yolo, and 14 other counties throughout California (Sanborn et al. 1974). In a few surveyed orchards over half of the trees were affected. Almond cultivars susceptible to infection are Long IXL, IXL, Jordanolo, Nonpareil, Peerless, Mission (Texas), Ne Plus Ultra, Drake, Trembath (Baker), and Languedoc (Moller et al. 1974). Similar leaf-scorch symptoms have been observed on grape since 1892 (Pierce 1892; Houston, Esau, and Hewitt 1947), but not on other fruit crops.

Almond leaf scorch and Pierce's disease—the name given to the leaf-scorch disorder of grape—are caused by the same bacterium. This organism also causes a dwarfing disease of alfalfa. Two other stone-fruit diseases, phony peach and plum leaf scald, are caused by bacteria morphologically similar to the one causing almond leaf scorch and Pierce's disease (Raju and Wells 1986; Raju et al. 1982; Wells et al. 1981a, 1981b). The almond leaf-scorch pathogen, however, differs serologically and pathogenically from those causing phony peach and plum leaf scald (Németh 1986; Raju et al. 1982). The latter diseases occur in the southeastern United States (Raju et al. 1982), and plum leaf scald also is present in Argentina, Brazil, and Paraguay (Kitajima, Barkarcic, and Fernandez-Valiela 1975; Németh 1986). Neither phony peach nor plum leaf scald is known to occur in California.

Symptoms. Trees affected with almond leaf-scorch show a lack of terminal growth, but the most diagnostic character is the leaf scorch or scald symptom, which usually appears in June (**fig. 2 A,B,C**). The scorch of the leaf margin resembles that caused by an excess of sodium; the tissue first becomes desiccated and then gradually turns brown. The tissues bordering the scorched part turn golden yellow in some cultivars. Phytotoxins were originally suspected as the cause of leaf scorch (Lee et al. 1982), but later studies by Goodwin, DeVay, and Meredith (1988a, b) concluded that water stress resulting from xylem blockage in the node and petiole was sufficient to cause the marginal leaf necrosis. Generally, only a portion of the tree—usually the terminal branches—initially shows symptoms, but within three to eight years the entire tree may become affected. Branches exhibiting leaf-scorch symptoms are more compact and show a rough-bark symptom. Removal of the rough bark may reveal longitudinal pitting in the woody cylinder of cultivars such as Long IXL and Jordanolo. Diseased trees are stunted, less productive, and show no ability to recover (Moller et al. 1974).

Causal organism. The similarity in symptomatology between leaf scorch of almond and Pierce's disease of grape (Goheen, Nyland, and Lowe 1973; Auger, Mircetich, and Nyland 1974; Auger, Shalla, and Kado 174) prompted electron microscope studies of scorched leaves from naturally and experimentally infected almond trees. Numerous rod-shaped bacteria were observed in the xylem vessels of diseased trees. The bacterial cells had an average diameter of 0.4 µm and lengths up to 1.9 µm, with a multilayered, rippled, and convoluted wall. The similarity in topography of the cell wall between bacteria causing Pierce's disease, phony peach disease, and almond leaf scorch suggested that they might be related (Lowe, Nyland, and Mircetich 1976). Also, the green sharpshooter (leafhopper) *Draeculacephala minerva* (**Fig. 2E**) that transmits the Pierce's disease bacterium (Hewitt 1958; Hewitt et al. 1946) also transmits almond leaf scorch bacterium from infected to healthy almonds. Typical leaf symptoms of both diseases are observed within two months after exposure of the hosts to infectious leafhoppers. The almond leaf scorch bacterium can also be graft-transmitted by buds, bud chips, or stems from infected to healthy young almond trees (Mircetich et al. 1976; Mircetich et al. 1974). This evidence, as well as serological analysis, confirms that almond leaf scorch and Pierce's disease are caused by the same bacterium (Nome et al. 1980).

Pierce's disease is now known to be caused by a small, gram-negative, xylem-limited fastidious bacterium that can be grown in culture (Davis, Purcell, and Thomson 1978; Davis, Thomson, and Purcell 1980). The proposed name for this bacterium is *Xylella fastidiosa* Wells, Raju, Hung, Weisburg, Mandelco-Paul, and Brenner (Wells et al. 1987). Hopkins (1985) detected strains of the Pierce's disease bacterium and found that virulent types multiplied and moved systemically in the xylem vessels, while the avirulent strains did not move beyond the inoculated internode. Strains of the organism that could produce Pierce's disease symptoms in inoculated grapevine were cultured from American elder, Virginia creeper, peppervine, American beauty berry, and blackberry (Hopkins and Adlerz 1988).

Disease diagnosis. Goheen and Lowe (1973) detected bacteria in the xylem of grapes affected with Pierce's disease using the electron microscope. Three years later, the same technique was employed to demonstrate the presence of bacteria in almond xylem infected with the leaf-scorch pathogen (Mircetich et al. 1976). A rapid staining procedure to detect almond leaf-scorch bacteria in xylem vessels of petioles and midveins was developed by Lowe, Nyland, and Mircetich (1977). Thin sections are immersed for 15 to 20 minutes in 1 percent acid fuchsin in lactophenol, rinsed in distilled water for 1 to 2 minutes and mounted in glycerol on a glass slide. Observations using a light microscope with phase-contrast show white to faintly yellow rods and spheres embedded in a golden-yellow matrix. Later, Nome et al. (1980) showed that the enzyme-linked immunosorbent assay (ELISA) could be used to detect the Pierce's disease bacterium. Another test that can be used to diagnose ALS was originally developed for phony peach disease (Hutchins, Cochran, and Turner 1951). In this technique, two- to three-year-old affected wood is cut into slices 1 to 2 mm thick and submerged in a solution containing 500 mL absolute methyl alcohol and 5 mL concentrated hydrochloric acid. This test stains infected, woody-twig pieces with well-defined purplush red or dark pink spots or streaks (**fig. 2D**). (The success of this technique can vary; it works best with samples collected in June and with recently infected wood.)

Disease development. The development and spread of the disease is similar in almond, grape, and alfalfa. Xylem-feeding suctorial insects (certain leafhoppers [Cicadellidae] and spittlebugs [Cercopidae]) (**fig. 2E**) transmit the disease to all three hosts in the field. In grape, the bacterium is vectored by at least 21 leafhopper and spittlebug species. Common weed hosts, such as Bermudagrass, rye, fescue grasses, watergrass, blackberry, elderberry, cocklebur, and nettle, which are commonly found in areas near stream beds or ditches, can serve as reservoirs of the bacterium. As these weeds dry, the resident insect vectors move to other plant hosts, such as almond, grape, and alfalfa. Secondary spread within almond or grape plantings does not appear to be a problem. The pathogen can also be transmitted by budding and grafting, but because the disease shows such strong symptoms, grafters rarely select budwood from affected trees. Experimental transmission has been accomplished by hypodermic injection and by xylem infiltration (Mircetich et al. 1976; Davis, Purcell, and Thomson 1978; David, Thomson, and Purcell 1980).

Control. Cultural methods are most appropriate for reducing the spread of this disease. Weed hosts adjacent to orchards should be removed to reduce reservoirs of the pathogen as well as populations of the insect vectors.

Insecticides are not normally recommended, as weed hosts that harbor the vectors are usually in rather inaccessible locations. Once trees are affected, they usually become unproductive within two to three years, so they should be removed and replanted. Resistant cultivars are not available. To insure that budwood comes from trees that have never shown symptoms of the disease, propagation material should be obtained from leaf-scorch-free orchards or, ideally, from the Foundation Seed and Plant Materials Service.

Another approach to control is to suppress the pathogen by the use of either trunk injection or soil drenching with an antibiotic such as tetracycline (Hopkins and Mortensen 1971). However, trunk injection of tetracycline into almond trees with leaf-scorch symptoms has caused severe leaf chlorosis and failed to provide adequate disease control.

REFERENCES

Auger, J., S. M. Mircetich, and G. Nyland. 1974. Interrelation between bacteria causing Pierce's disease of grapevines and almond leaf scorch (abs.). *Ann. Proc. Am. Phytopathol. Soc.* 1:90.

Auger, J. G., T. A. Shalla, and C. I. Kado. 1974. Pierce's disease of grapevines: Evidence for a bacterial etiology. *Sci. Agric.* 184:1375–77.

Davis, M. J., A. H. Purcell, and S. V. Thomson. 1978. Pierce's disease of grapevines: Isolation of the causal organism. *Science* 199:75–77.

———. 1980. Isolation medium for the Pierce's disease bacterium. *Phytopathology* 70:425–29.

Davis, M. J., S. V. Thomson, and A. H. Purcell. 1980. Etiological role of the xylem-limited bacterium causing Pierce's disease in almond leaf scorch. *Phytopathology* 70:472–75.

Goheen, A. C., and S. K. Lowe. 1973. Use of electron microscopy for indexing grapevines for Pierce's disease. *Rev. Pathol. Veg. Entomol. Agric. Fr.* (Ser. IV), 279–80.

———. 1973. Association of a rickettsia-like organism with Pierce's disease of grapevines and alfalfa dwarf and heat therapy of the disease in grapevines. *Phytopathology* 63:341–45.

Goodwin, P. H., J. E. DeVay, and C. P. Meredith. 1988a. Roles of water stress and phytotoxins in the development of Pierce's disease of grapevine. *Physiolog. Molec. Plant Pathol.* 32:1–15.

———. 1988b. Physiological responses of *Vitis vinifera* cv. "Chardonnay" to infection by the Pierce's disease bacterium. *Physiolog. Molec. Plant Pathol.* 32:17–32.

Hewitt, W. B. 1958. The probable home of Pierce's disease virus. *Plant Dis. Rep.* 42:211–15.

Hewitt, W. B., B. R. Houston, N. W. Frazier, and J. H. Freitag. 1946. Leafhopper transmission of the virus causing Pierce's disease of grape and dwarf of alfalfa. *Phytopathology* 36:117–28.

Hopkins, D. L. 1985. Physiological and pathological characteristics of virulent and avirulent strains of a bacterium that causes Pierce's disease of grapevine. *Phytopathology* 75:713–17.

Hopkins, D. L., and W. C. Adlerz. 1988. Natural hosts of *Xylella fastidiosa* in Florida. *Plant Dis.* 72:429–31.

Hopkins, D. L., and J. A. Mortensen. 1971. Suppression of Pierce's disease symptoms by tetracycline antibiotics. *Plant Dis. Rep.* 55:610–12.

Houston, B. R., K. Esau, and W. B. Hewitt. 1947. The mode of vector feeding and the tissues involved in the transmission of Pierce's disease virus in grape and alfalfa. *Phytopathology* 37:247–53.

Hutchins, L. M., L. C. Cochran, and W. F. Turner. 1951. Phony. In *Virus diseases and other disorders with viruslike symptoms of stone fruits in North America.* 17–25. U.S.D.A. Handb. 10.

Kitajima, E. W., M. Bakarcic, and M. V. Fernandez-Veliela. 1975. Association of rickettsia-like bacteria with plum leaf scald disease. *Phytopathology* 65:476–78.

Lee, R. F., B. C. Raju, G. Nyland, and A. C. Goheen. 1982. Phytotoxin(s) produced in culture by the Pierce's disease bacterium. *Phytopathology* 72:886–88.

Lowe, S. K., G. Nyland, and S. M. Mircetich. 1976. The ultrastructure of the almond leaf scorch bacterium with special reference to topography of the cell wall. *Phytopathology* 66:147–51.

———. 1977. A simple and rapid staining procedure for the in situ detection of almond leaf scorch bacterium (abs.). *Proc. Am. Phytopathol. Soc.* 4:208.

Mircetich, S. M., S. K. Lowe, W. J. Moller, and G. Nyland. 1976. Etiology of almond leaf scorch disease and transmission of the causal agent. *Phytopathology* 66:17–24.

Mircetich, S. M., G. Nyland, J. Auger, and S. K. Lowe. 1974. Almond leaf scorch disease caused by bacterium vectored by leafhoppers (abs.). *Proc. Am. Phytopathol. Soc.* 1:90–91.

Moller, W. J., R. R. Sanborn, S. M. Mircetich, H. E. Williams, and J. A. Beutel. 1974. A newly recognized and serious leaf scorch disease of almond. *Plant Dis. Rep.* 58:99–101.

Németh, M. 1986. *Virus, mycoplasma and rickettsia diseases of fruit trees.* Martinus Nijhoff, Dordrecht. 841 pp.

Nome, S. F., B. C. Raju, A. C. Goheen, and D. Docampo. 1980. Enzyme-linked immunosorbent assay for Pierce's disease bacteria in plant tissues. *Phytopathology* 70:746–49.

Pierce, N. B. 1892. The California vine disease. *USDA Div. Veg. Pathol. Bull.* 2, 222 pp.

Raju, B. C., and J. M. Wells. 1986. Diseases caused by fastidious xylem-limited bacteria and strategies for management. *Plant Dis.* 70:182–86.

Raju, B. C., J. M. Wells, G. Nyland, R. H. Brlansky, and S. K. Lowe. 1982. Plum leaf scald: Isolation, culture, and pathogenicity of the causal agent. *Phytopathology* 72:1460–66.

Sanborn, R. R., S. M. Mircetich, G. Nyland, and W. J. Moller. 1974. "Golden death" a new leaf scorch threat to almond growers. *Calif. Agric.* 28(12):4–5.

Wells, J. M., B. C. Raju, H. Y. Hung, W. G. Weisburg, L. Mandelco-Paul, and D. J. Brenner. 1987. *Xylella fastidiosa* gen. nov., sp. nov.: Gram-negative, xylem-limited, fastidious plant bacteria related to *Xanthomonas* spp. *Internat. J. System Bacteriol.* 37:136–43.

Wells, J. M., B. C. Raju, G. Nyland, and S. K. Lowe. 1981a. Medium for isolation and growth of bacteria associated with plum leaf scald and phony peach diseases. *Appl. Env. Microbiol.* 42:357–63.

Wells, J. M., B. C. Raju, J. M. Thompson, and S. K. Lowe. 1981b. Etiology of phony peach and plum leaf scald diseases. *Phytopathology* 71:1156–61.

Wells, J. M., B. C. Raju, and W. G. Weisburg. 1986. *Xylemella fasidiosum*, genus novum, species novum, of the family Pseudomonadaceae: Gram-negative, xylem-limited, fastidious bacteria from plants (abs.). *Phytopathology* 76:1083.

X-Disease of Cherry and Peach

Mycoplasmalike Organism

X-disease, formerly considered a virus disease, is closely associated with a mycoplasmalike organism (Gilmer and Blodgett 1976). A variety of descriptive names have been given to the disease and include such synonyms on cherries as western X little cherry, buckskin, and wilt and decline, and on peach as yellow leaf roll, leaf casting yellows, and western X. Natural economic hosts include sweet and sour cherry, peach, and nectarine. The pathogen has been experimentally transmitted to almond, apricot, and mahaleb cherry. In contrast, certain species, such as myrobalan plum, fenzl almond, desert apricot, desert peach, David peach, black cherry, and Klamath plum, are considered nonhosts. Hosts without evident symptoms include European plum and several other *Prunus* species (Gilmer and Blodgett 1976; Gilmer, Moore, and Keitt 1954; Rawlins and Thomas 1951; Stoddard 1947; Stoddard et al. 1951). The native chokecherry is an additional host which in some areas serves as an important inoculum source.

X-disease was first reported in California in 1931 as buckskin of sweet cherry (Rawlins and Horne 1931); two years later it was reported in Connecticut on peach. Since then it has been identified in all of the major U.S. peach-producing states except South Carolina, Georgia, Arkansas, and Texas. In California, serious epidemics of the peach yellow leaf roll (PYLR) strain of X-disease have occurred several times since 1950. The incidence of X-disease on sweet cherry is more chronic and has caused severe problems for many years. In sweet cherry, particularly when grafted on mahaleb (*P. mahaleb*) rootstock, tree decline and death usually occur within one year or two following infection, whereas infected trees on mazzard (*P. avium*) rootstocks may live for many years. Often, peach trees infected with the common strain of X-disease (Green Valley strain) do not die or become unproductive for many years. However, peach trees infected with the PYLR strain usually collapse two or three years following infection.

On sweet cherry in California, two strains of the X-disease pathogen have been described as the Green Valley and Napa Valley strains. The Green Valley strain (Gold and Sylvester 1982; Purcell 1985; Suslow and Purcell 1982), now predominant in San Joaquin County, is part of a disease complex that contributes to an annual tree loss of about 10 percent. Disease outbreaks caused by the Napa Valley strain virtually eliminated commercial cherry production in Napa and Sonoma counties.

Although the first recorded outbreak of peach yellow leaf roll (PYLR) dates back to 1950 in orchards a few kilometers northeast of Marysville, California (Nyland and Schlocker 1951), Yuba County Agricultural Commissioner Arthur Worledge reported increasing numbers of sick peach trees in this district in 1948 and 1949. A 1950 survey detected 877 diseased trees; in 1951, the problem was considered so severe that an extensive survey and tree removal program was initiated. By 1969, some 9,700 trees had been removed. A sharp decrease in new disease incidence in 1953 and 1954, followed by less than 100 new infections per year between 1957 and 1964, seems to indicate that the locally adopted ordinance which required mandatory destruction of trees was beneficial (O'Reilly, Schlocker, and Rosenberg 1966). Unfortunately, the orchard inspection program was discontinued in 1971 due to lack of funds. As predicted by several scientists, in 1978–81, another epidemic of PYLR culminated in a loss of 35,000 (2.4 percent) of the bearing trees in Yuba, Sutter, and Butte counties. By 1980, in 23 selected orchards in the Yuba-Sutter-Butte county area, 10,312 diseased trees were found. Again, within the same 23 orchards, 3,653 new cases occurred the following spring (Richards 1981).

Symptoms: Sweet cherry. On mahaleb rootstock, the initial foliar symptoms include wilted, upward-folded, light-green leaves. Such symptoms first appear in late May or early June, and the trees usually collapse (**fig. 3A**) later in the summer or early the following year. Foliar symptoms of infected trees on mazzard rootstock (*P. avium*) include small leaves with an upright growth habit. The tree canopy is more sparse than that of a healthy orchard; affected trees, unlike those on mahaleb, may survive for a number of years. The terminals of affected branches are slightly rosetted and, in late summer, their leaves may be slightly undersized and bronze or rusty green. A few leaves may show a red or orange coloration along the midrib, and a late flush of growth may develop from normally dormant, terminal buds.

Fruit symptoms also vary and are dependent upon the strain of the X-disease agent and the rootstock. Infected trees on mahaleb rootstock may or may not produce symptomatic fruit. In contrast, trees on mazzard rootstock that are infected with the Green Valley strain produce small, pointed, flat-sided, pale red to greenish white, insipid-tasting fruit (fig. 3B). The pedicels are abnormally short and thick. Affected fruit are often initially confined to a few branches, while the remainder of the tree bears normal fruit. Trees on mazzard rootstock infected with the Napa Valley strain produce small but normal-shaped fruit with normal-length pedicels. Both strains cause a ragged appearance of spur buds. Infected trees on mahaleb rootstock develop smaller and weaker lateral roots and fewer feeder roots; also, a brown line develops at the graft union, and there is a decreased amount of starch in the roots and in the trunk below the union (Kirkpatrick et al. 1985a).

Symptoms: Peach. Trees infected with the common strain of the X-disease agent develop leaves at budbreak that appear normal until a month or so thereafter. However, previously affected branches may leaf out earlier than normal and produce abnormally small leaves. Before they drop prematurely, these leaves usually turn yellow and show red spots, symptoms similar to those caused by nitrogen deficiency. By mid-May or June, pale green blotches often develop in the leaf blades; these areas commonly drop out and give the foliage a tattered, shotholed appearance. These leaves commonly drop soon after symptoms appear, and the terminals of the infected shoots cease growth. The symptoms in midsummer are quite striking, with the inward-rolled leaves showing a yellow to reddish brown coloration and scorched ragged edges. Leaf casting throughout the summer is common. Affected trees show a progressive decline that often culminates with the death of small shoots and even large branches (fig. 3C).

In trees infected with the PYLR strain (fig. 3D), the infected shoots continue growth, less foliage drops prematurely than with the common strain, and there is a tendency for the affected leaves to curve down and back toward the petiole. Shoot terminals continue slow growth, and many of their leaves are retained. A prominent enlargement of the midvein and lateral veins is a key symptom for identification of the disease caused by this strain (fig. 3E). Interestingly, mechanical limb girdling can produce quite similar symptoms (Chaney et al. 1979). There is a lighter than normal set of undersized fruit on branches affected with either strain of the pathogen. The fruit initially have a prominently swollen apex, and most fall before harvest. If the fruit persist they become more pointed than normal and have shriveled nonviable seeds, internal discoloration around the pit, and a bitter unpalatable flavor (Gilmer and Blodgett 1976).

Cause. Despite considerable effort, the causal agent of X-disease has not been artificially cultured (Purcell et al. 1981). Hence, the suspected incitant, a mycoplasmalike organism (MLO), has not been confirmed experimentally. However, the constant presence of pleomorphic, mycoplasmalike bodies in the phloem tissue of diseased trees strongly implicates an MLO (Grannett and Gilmer 1971; Nasu, Jensen, and Richardson 1970).

Corroborating evidence that an MLO causes X-disease includes remission of symptoms following treatment with tetracycline (Nyland, Raju, and Lowe 1981, 1982) and graft-transmissibility of the pathogen (Hildebrand 1953). Antibodies and DNA copies of the mycoplasmalike organism have been produced (Kirkpatrick and Garrott 1984; Kirkpatrick et al. 1985b, 1987) and are currently being used in conjunction with visual symptoms for more positive disease identification. Involvement of a spiroplasma in the disease was reported (Thomson et al. 1978; Purcell et al. 1981), but was later disproved (Kloepper and Garrott 1983; Raju, Purcell, and Nyland 1984).

Several investigators (Granett and Gilmer 1971; Kloepper and Garrott 1983; McBeath, Nyland, and Spurr 1972; Nasu, Jensen, and Richardson 1970; Sinha and Chiykowski 1980), by means of electron microscopy, have studied the morphology of the MLO bodies present in the leaf sieve tubes of infected cherry and peach trees. These nonhelical, polymorphic organisms varied considerably in both size and shape; some appeared to be spherical or oval and others cylindrical or filamentous. The spherical and oval bodies had diameters of 120 to 200 nm; the elongate forms had diameters of 120 to 130 nm and lengths of 360 to 5,400 nm.

Transmission. Researchers in Washington were the first to demonstrate that leafhoppers transmit the pathogen (Anthon and Wolfe 1951; Wolfe and Anthon 1953). Since then, additional leafhopper species have been confirmed as vectors (Nielson 1962; Nielson and Jones 1954; Purcell and Elkinton 1980; Wolfe 1955). In California, Jensen and coworkers (Jensen 1953; Jensen, Frazier, and Thomas 1952; Jensen and Thomas 1955) established the vectoring ability of the leafhopper *Colladonus geminatus* (Van Duzee) for the peach yellow leaf roll strain. More recently, Suslow and Purcell (1982) showed that *Colladonus montanus* could acquire and transmit the Green Valley strain but not the PYLR strain of X-disease. This insect was caged on infected cherry trees throughout a growing season and at various times transferred to celery indicator plants. They found that *C. montanus* transmitted the X-disease MLO at rates of 0 percent (April), 5 percent (May and July), and 20 percent or higher (August and September). Evidently, the titer of the MLO in cherry is initially low but increases thereafter. Serological testing of cherry trees with X-disease supports this conclusion (Kirkpatrick et al. 1985a). In similar caging and feeding trials using peach trees infected with PYLR, however, none of the 4,500 *C. montanus* used in the test transmitted the pathogen to the indicator plants. It would appear that peach is a poor host for acquisition of the PYLR strain by this leafhopper.

The X-disease MLO is not mechanically transmissible and is heat-labile. Hildebrand (1953) reported that the agent was inactivated when infected peach buds were immersed in water at 50°C for 6 to 7 min. Gilmer, Moore, and Keitt (1954) found that the PYLR strain was inactivated in experimentally infected sweet cherry buds exposed to 37°C for 15 days or more. Nichols and Nyland (1952) reported that the PYLR MLO was considerably more tolerant of heat than the common strain.

Epidemiology. Even though *C. montanus* is the most frequently captured insect vector in cherry orchards in California, the numbers found in monitored sites show little or no correlation with disease incidence (Kirkpatrick et al. 1985a). This is not too surprising, however, as *C. montanus* prefers to feed and breed on herbaceous hosts. The main function of this leafhopper may be to establish initial infection sites in previously healthy orchards.

Another leafhopper, *Fieberiella florii* (Stal) (**fig. 3F**), is probably even more important in the spread of X-disease. This species resides on several woody perennials and is known to breed on cherry (*P. avium*). During the period from 1982 to 1985, among six monitored orchards where *F. florii* was captured, four orchards with nine or more of these leafhoppers contained at least 30 percent diseased trees. In one orchard, where only one *F. florii* was caught from 1923 to 1983, X-disease trees ranged from 8 to 11 percent during the years 1982 to 1984. However, in 1984 nine *F. florii* were trapped in this orchard, and the following year the disease incidence exceeded 30 percent (Kirkpatrick et al. 1985a).

During the winter, the common X-disease strain, but not the PYLR strain, usually dies out in the tops of peach trees (G. Nyland, unpublished). The common strain (Green Valley), however, does not die out in naturally infected cherry trees during the winter. The PYLR strain has not been identified in sweet cherry but is capable of inciting buckskin symptoms in artificially inoculated cherry trees (G. Nyland, unpublished).

Following the sharp increase in trees with PYLR symptoms in 1978, surveys in both the Sacramento and San Joaquin valleys indicated that the highest incidence of disease occurred in peach orchards adjacent to commercial pear orchards, and that the incidence generally decreased with increasing distance from pears. There was no evidence of peach-to-peach spread of PYLR (Purcell et al. 1981). These data suggest that pear orchards may be a source of PYLR vectors or even serve as a reservoir of the PYLR pathogen.

Control. Evidence accumulated to date suggests that symptomatic cherry trees, but not peach trees, serve as reservoirs of inoculum. Hence, affected cherry trees should be sprayed with an insecticide and then rogued. Historically, tree-removal programs carried out in Washington (1947 to 1951) reduced the incidence of X-disease in cherry orchards (Gilmer and Blodgett 1976). Chokecherry (*P. virginiana* var *demissa* [Torr. & Gray] Torr.), an alternate host of the X-disease MLO in some regions, has not been observed in San Joaquin

County. However, X-disease-infected chokecherry trees were recently found in the foothills of Placer County (B. C. Kirkpatrick, unpublished). In the eastern United States (Gilmer, Moore, and Keitt 1954, James and Sutton 1984; Lukens et al. 1971) and elsewhere, chokecherry eradication has provided effective X-disease control.

A cultural practice sometimes employed to control X-disease in sweet cherry involves the use of multiple-scion grafts on the disease-resistant mahaleb rootstock (Gilmer and Blodgett 1976). When individual scaffolds become diseased, they are removed. The MLO is prevented from further movement within the tree by the hypersensitive mahaleb frame. Control through the use of resistant cherry cultivars is of only limited value at present. However, some resistance is known to occur in Early Burlat, Napa Long Stem Bing, Dicke Braune Blakenburger, and Coop's Special (Gilmer and Blodgett 1976).

Injection of the antibiotic tetracycline into diseased cherry trees on mazzard or Stockton Morello (*P. cerasus* L.) rootstocks has resulted in symptom remission and the production of quality fruit (Buitendag and Bronkhorst 1983; Lee, Nyland, and Lowe 1987; Nyland, Raju, and Lowe 1982; Reil and Beutel 1976; Schreader 1985). Trees on mahaleb rootstock, however, seldom benefit from this treatment. Furthermore, this no longer is a legal application on cherry.

In conjunction with orchard sanitation, that is, removal of diseased cherry trees, *F. florii* control may also prove beneficial. Although insecticides have been tested extensively for the control of other virus and prokaryote vectors, they have largely failed to decrease the incidence of disease. However, the unique biology of *F. florii* may make this vector an exception, and controlling its nymphal and adult populations in cherry trees might significantly reduce tree-to-tree spread of the disease.

Recommendations for X-disease control are similar for peach. Some growers are continually inspecting and removing trees with PYLR and western X-disease. Other growers have injected diseased trees annually with a solution of oxytetracycline hydrochloride prepared initially for the control of pear decline (Lee, Nyland, and Lowe 1987), but this application no longer has EPA approval.

Important questions that still remain include the source of X-disease inoculum outside the orchard, the significance of disease spread within the orchard via insect vectors, the benefits of clean culture under the trees, the removal of host plants adjacent to orchards, and the use of insecticides to control the vectors.

REFERENCES

Anthon, E. W., and H. R. Wolfe. 1951. Additional vectors of western X-disease. *Plant Dis. Rep.* 35:345–46.

Buitendag, C. H., and G. J. Bronkhorst. 1983. Microinjection of citrus trees with N-pyrrolidinomethyl tetracycline (PMT) for the control of greening disease. *Citrus Subtropical Fruit J.* 592:8–10.

Chaney, D. H. 1980. *Peach yellow leaf roll: Fieldman's conference, February 19, 1980, Davis.* Sutter and Yuba Counties Farm Advisor's Office, Yuba City, Calif. 8 pp.

Chaney, D., G. Nyland, J. Beutel, and W. Moller. 1979. Yellow leaf roll of peaches. *Univ. Calif. Agr. Sci. Leaf.* 21092, 2 pp.

Gilmer, R. M., and E. C. Blodgett. 1976. X-disease. In *Virus diseases and noninfectious disorders of stone fruits in North America*, 145–55. USDA Handb. 437.

Gilmer, R. M., J. D. Moore, and G. W. Keitt. 1954. X-disease virus: I. Host range and pathogenesis in chokecherry. *Phytopathology* 44:180–85.

Gold, R. E., and E. S. Sylvester. 1982. Pathogen strains and leafhopper species as factors in the transmission of western X-disease agent under varying light and temperature conditions. *Hilgardia* 50:1–43.

Gonot, K., and A. H. Purcell. 1981. Seasonal acquisition of X-disease agent (cherry buckskin) from cherry by the leafhopper vector, *Colladonus montanus* (abs.). *Phytopathology* 71:105.

Grannett, A. L., and R. M. Gilmer. 1971. Mycoplasmas associated with X-disease in various *Prunus* species. *Phytopathology* 61:1036–37.

Hildebrand, E. M. 1953. Yellow-red or X-disease of peach. *N. Y. State Coll. Agric. Mem.* 323, 54 pp.

Jensen, D. D. 1953. Leafhopper-virus relationships of peach yellow leaf roll. *Phytopathology* 43:561–64.

Jensen, D. D., N. W. Frazier, and H. E. Thomas. 1952. Insect transmission of yellow leaf roll virus of peach. *J. Econ. Entomol.* 45:335–37.

Jensen, D. D., and H. E. Thomas. 1955. Transmission of the green valley strain of cherry buckskin virus by means of leafhoppers (abs.) *Phytopathology* 45:694.

Jones, A. L., and T. B. Sutton. 1984. Diseases of tree fruits. *N. Central Reg. Ext. Publ.* 45. Coop. Ext. Serv. Mich. State Univ. 59 pp.

Kirkpatrick, B. C., and D. G. Garrott. 1984. Detection of western X-disease mycoplasma-like organisms by enzyme-linked immunosorbent assay (abs.). *Phytopathology* 74:825.

Kirkpatrick, B. C., T. J. Morris, G. Nyland, and A. H. Purcell. 1985a. *Etiology, epidemiology and control of cherry buckskin: Final progress report 1983–1985.* USDA/ARS and Department of Plant Pathology, University of California, Davis, 128 pp.

Kirkpatrick, B. C., D. C. Stenger, T. J. Morris, and A. H. Purcell. 1985b. Detection of X-disease mycoplasma-like organisms in plant and insect hosts using cloned, disease specific DNA (abs.). *Phytopathology* 75:1351.

———. 1987. Cloning and detection of DNA from a nonculturable plant pathogenic mycoplasma-like organism. *Science* 238:197–200.

Kloepper, J. W., and D. G. Garrott. 1983. Evidence for mixed infection of spiroplasmas and nonhelical mycoplasmalike organisms in cherry with X-disease. *Phytopathology* 73:357–60.

Lee, R. F., G. Nyland, and S. K. Lowe. 1987. Chemotherapy of cherry buckskin and peach yellow leafroll diseases: An evaluation of two tetracycline formulations and methods of application. *Plant Dis.* 71:119–21.

Lukens, R. J., P. M. Miller, G. S. Walton, and S. W. Hitchcock. 1971. Incidence of X-disease of peach and eradication of chokecherry. *Plant Dis. Rep.* 55:645–47.

McBeath, J. H., G. Nyland, and A. R. Spurr. 1972. Morphology of mycoplasma-like bodies associated with peach X-disease in *Prunus persica*. *Phytopathology* 62:935–37.

Nasu, J. D., D. Jensen, and J. Richardson. 1970. Electron microscopy of mycoplasma-like bodies associated with insect and plant hosts of peach western X-disease. *Virology* 41:583–95.

Németh, M. 1986. *Virus, mycoplasma and rickettsia diseases of fruit trees.* Martinus Nijhoff, Dordrecht, 841 pp.

Nichols, C. W., and G. Nyland. 1952. Hot water treatments of some stone fruit viruses (abs.). *Phytopathology* 42:517.

Nielson, M. W. 1962. A synonymical list of leafhopper vectors of plant viruses. *USDA Agric. Res. Serv.* ARS-33-74.

Nielson, M. W., and L. L. Jones. 1954. Insect transmission of Western-X little cherry virus. *Phytopathology* 44:218–19.

Nyland, G., B. C. Raju, and S. K. Lowe. 1981. Chemical control of X-disease (abs.) *Phytopathology* 71:246.

———. 1982. Control of peach yellow leaf roll with tree injection of terramycin (abs.) *Phytopathology* 72:1005.

Nyland, G., and A. Schlocker. 1951. Yellow leaf roll of peach. *Plant Dis. Rep.* 35:33.

O'Reilly, H. J., A. Schlocker, and D. Y. Rosenberg. 1966. Deadly yellow leaf virus threatens peach industry. *Cling Peach Quart.* 4(1):22–23.

Purcell, A. 1985. Cherry buckskin update. *Univ. Calif. Coop. Ext. San Joaquin Co. Orch. Dig.*, May 1985:1–2.

Purcell, A. H., and J. S. Elkinton. 1980. A comparison of sampling methods for leafhopper vectors of X-disease in California cherry orchards. *J. Econ. Entomol.* 73:854–60.

Purcell, A. H., G. Nyland, B. C. Raju, and M. R. Heringer. 1981. Peach yellow leafroll epidemic in northern California: Effects of peach cultivar, tree age, and proximity to pear orchards. *Plant Dis.* 65:365–68.

Purcell, A. H., B. C. Raju, and G. Nyland. 1981. Transmission by injected leafhoppers of spiroplasma isolated from plants with X-disease (abs.) *Phytopathology* 71:108.

Raju, B. C., A. H. Purcell, and G. Nyland. 1984. Spiroplasmas from plants with aster yellows disease and X-disease: Isolation and transmission by leafhoppers. *Phytopathology* 74:925–31.

Rawlins, T. E., and W. T. Horne. 1931. "Buckskin", a destructive graft-infectious disease of the cherry. *Phytopathology* 21:331–35.

Rawlins, T. E., and H. E. Thomas. 1941. The buckskin disease of cherry and other stone fruits. *Phytopathology* 31:916–25.

———. 1951. Virus diseases of sweet cherry: Buckskin. In *Virus diseases and other disorders with viruslike symptoms of stone fruits in North America*, 98–102. USDA Handb. 10.

Reil, W. O., and J. A. Beutel. 1976. A pressure machine for injecting trees. *Calif. Agric.* 30(12):4–5.

Rice, R. E., and R. A. Jones. 1972. Leafhopper vectors of the western X-disease pathogen: Collections in central California. *Environ. Entomol.* 1:726–30.

Richards, B. L., and L. C. Cochran. 1957. Virus and viruslike diseases of stone fruits in Utah. *Utah Agric. Exp. Stn. Bull.* 384, 129 pp.

Richards, D. 1981. Peach yellow leaf roll update. *Cling Peach Quart.* 18(1):11.

Schreader, W. 1985. Cherry buckskin update: The UC cherry buckskin vector control program. *Univ. Calif. Coop. Ext. San Joaquin Co. Orch. Dig.*, May 1985:2–7.

Sinha, R. C., and L. N. Chiykowski. 1980. Transmission and morphological features of mycoplasmalike bodies associated with peach X-disease. *Can. J. Plant Pathol.* 2:119–24.

Stoddard, E. M. 1947. The X-disease of peach and its chemotherapy. *Conn. Agric. Exp. Stn. Bull.* 506, 19 pp.

Stoddard, E. M., E. M. Hildebrand, D. H. Palmiter, and K. G. Parker. 1951. X-disease. In *Virus diseases and other disorders with viruslike symptoms of stone fruits in Northern America*, 37–42. USDA Hand. 10.

Suslow, K. G., and A. H. Purcell. 1982. Seasonal transmission of X-disease agent from cherry by leafhopper *Colladonus montanus*. *Plant Dis.* 66:28–30.

Thomson, S. V., D. G. Garrott, B. C. Raju, M. J. Davis, A. H. Purcell, and G. Nyland. 1978. A spiroplasma consistently isolated from western X-disease infected plants. *Proc. 4th Internat. Conf. Plant Path. Bact.*: 975.

Thomson, S. V., and B. N. Wadley. 1981. Selecting for X-disease resistant sweet cherry varieties. *Phytopathology* 71:1007.

Wolfe, H. R. 1955. Transmission of the western X-disease virus by the leafhopper *Colladonus montanus* (Van D.). *Plant Dis. Rep.* 39:298–99.

Wolfe, H. R., and E. W. Anthon. 1953. Transmission of western X-disease virus from sweet and sour cherry by two species of leafhoppers. *J. Econ. Entomol.* 46:1090–92.

Fungal Diseases of Stone Fruit

Anthracnose of Almond

Colletotrichum gloeosporioides Penz.
(= *Gloeosporium amygdalinum* Brizi)

This disease was first reported on almond in California as a leaf and fruit pathogen in 1916 (Czarnecki) and again in 1925 (Taylor & Philp). The first diseased samples collected were from Napa and Alameda counties where the disease appeared mainly on the young green fruit, producing large, brown to orange, sunken spots that caused the affected fruit to shrivel and drop (Czarnecki 1916). Leaf symptoms resembling those of shothole caused by *Wilsonomyces carpophilus* (Lév.) Adaskaveg, Ogawa, and Butler or *Cercospora circumscissa* Sacc. were reported by both Czarnecki (1916) and Taylor and Philp (1925). Smith (1941) did not mention anthracnose of almond in his bulletin on "Diseases of fruits and nuts" in California, but it is listed as occurring here in "Index of plant diseases in the United States" (USDA 1960). In Europe, Israel, and South Africa, the disease causes serious crop losses and is popularly known as "gumming of the almonds," "gummosis," and "kernel rot" (Dippenaar 1931; Grosclaude 1972; Shabi and Katan 1983). As the disease appears to be serious only in areas of high rainfall, we do not expect it to become important in the semi-arid climate typical of California's Central Valley. However, it is possible that anthracnose could cause damage in years with especially heavy spring rains (Shabi and Katan 1983). Although the anthracnose disease has not been detected on almond since 1925, the disease was identified on peach and nectarine fruit in Fresno County in summer 1990 after a heavy rainfall at the end of May (A. Feliciano, personal communication).

Symptoms. The symptoms of anthracnose, as described by Dippenaar (1925) in South Africa, involve the fruit, leaves, and twigs at all stages of development. The principal damage is to the young fruit, 1 to 1.5 cm in diameter; they develop yellowish brown, sunken lesions 0.5 cm or more in diameter. The fungus penetrates deeply into the hull and shell, and finally into the kernel. Infections in young fruit rarely exude gum, but half-grown and full-grown fruit show profuse gumming. Many of the infected fruit become mummified and these usually remain on the tree throughout the winter. Twig infections result in twig die-back and shedding of the leaves. Leaf infections occur at the tips or along the margin causing water-soaked areas that die and become bleached. Affected leaves may drop off or remain hanging on the tree. In Israel (Shabi and Katan 1983), many of the twigs bearing diseased fruit die and drop their leaves, but direct twig and leaf infection has not been reported.

Of the almond cultivars grown in South Africa, Nonpareil and Jordan were more susceptible than I.X.L. and Paper Shell (Dippenaar 1931); in Israel, all three cultivars studied (Non-plus-Ultra, Poria 10, and Um-El-Fachem) were highly susceptible (Shabi and Katan 1983).

Causal organism. The fungus on almonds was first isolated in Italy and named *Gloeosporium amygdalinum* Brizi (Brizi 1986). Cultures of the anthracnose fungus isolated from almond in Israel and submitted to the Central-bureau voor Schimmel Cultures, Baarn, Netherlands by Shabi and Katan (1983) were identified as *Colletotrichum gloeosporioides* Penz. This is in agreement with the monographic treatment of the genus *Colletotrichum* by von Arx (1957) who considered *G. amygdalinum* a synonym of *C. gloeosporioides*. The fungus overwinters on the tree in mummified fruit and infected peduncles. In South Africa, it is reported to overwinter also in infected twigs. Wet weather during the growing season is conducive to fungus sporulation, dispersal, infection, and disease development. Since the Pseudoacacia tree is a host of this pathogen in Japan (S. Tsuchiya, personal communication), alternate hosts of the fungus are being examined in California.

Control. Disease control is directed at reducing the overwintering source of inoculum by the removal of mummies and infected twigs. The frequent application of fungicide sprays, using Bordeaux mixture or lime sulfur, protected new growth and resulted in increased crop yields in South Africa (Dippenaar 1931); in Israel

(Shabi and Katan 1983), the best treatment was three to five sprays of captan or its analogs captafol and folpet applied at 10-day intervals starting at petal fall.

REFERENCES

Arx, J. A. von. 1957. Die arten der Gattung *Colletotrichum* Cda. *Phytopathol. Z.* 29:413–68.

Brizi, U. 1896. Eine neue Krankheit (anthracnosis) des Mandelbaumes. *Z. PflKrankh.* 6:66–72.

Czarnecki, H. 1916. A Gloeosporium disease of the almond, probably new to America. *Phytopathology* 6:310.

Dippenaar, B. J. 1931. Anthracnose disease of almonds. *Farm. S. Africa* 6:133–34.

Grosclaude, C. 1972. L'anthracnose de l'amandier en France. *C. R. Acad. Agric. Fr.* 58:1392–95.

Shabi, E., and T. Katan. 1983. Occurrence and control of anthracnose of almond in Israel. *Plant Dis.* 67:1364–66.

Taylor, R. M., and G. L. Philp. 1925. The almond in California. *Univ. Calif. Agr. Exp. Stn. Circ.* 284.

United States Department of Agriculture, Agric. Res. Serv. 1960. *Index of plant diseases in the United States.* USDA Handb. 165, 531 pp.

Brown Rot of Stone Fruit

Monilinia laxa (Aderh. & Ruhl.) Honey
Monilinia fructicola (Wint.) Honey

Brown rot is the name given to diseases of deciduous tree fruit caused by species of *Monilinia* (= *Sclerotinia*). In 1900, Woronin showed that two types of brown rot diseases existed in Europe: one largely caused by *Sclerotinia cinerea* Bon. on stone fruit, and one largely caused by *S. fructigena* on pome fruit. It was not until almost 25 years later, however, that the common brown rot disease of stone fruit in North America was recognized as a third distinct type. Consequently, the earliest literature on the subject names either *S. fructigena* or *S. cinerea* as the causal organism of the American type of brown rot. Such misconceptions, and the tendency to name new fungus species on the basis of slight differences, led to confusion and duplication in the names of the causal organisms.

Even after the identities of the two brown rot fungi on North American stone fruit were established, few persons attempted to differentiate between the diseases they produce (Ogawa 1960). For obvious reasons it is difficult to find common names that are entirely satisfactory. In this discussion, the brown rot disease of stone fruit originally found in Europe is referred to as "European brown rot," and the disease of stone fruit originally found in North America is called "American brown rot." The disease caused by *S. fructigena* (Humphrey 1893) was authenticated in one experimental pear orchard in the eastern United States (Batra 1979) and eradicated, so is not discussed here.

Brown Rot: European Brown Rot

Monilinia laxa (Aderh. & Ruhl.) Honey

This disease, also known as "brown rot blossom blight," "apricot brown rot," and "monilinia blossom blight" (Rudolph 1925; Wilson 1953) attacks apricot, almond, cherry, domesticated plum, nectarine, peach, prune, flowering quince, and certain wild species of *Prunus*. It occurs in Australia, Europe, eastern Asia, New Zealand, North America, Manchuria, South America, and Japan (Dunegan 1958; English, Pinto de Torres, and Kirk 1969). Surveys in South America have revealed that brown rot on stone fruit in Chile (English, Moller, and Nome 1967) is caused by only *M. laxa*, and in Argentina by *M. laxa* and *M. fructicola*; whereas in Brazil it is caused entirely by *M. fructicola*. In South Africa the stone-fruit pathogen is *M. laxa*. In North America, the disease occurs in the states and provinces along the Pacific coast (Barss 1928, 1925, Evans and Owens 1941). It also occurs in some fruit-growing districts of Michigan, Wisconsin and New York (Cation, Dunegan, and Kephart 1949; Jones and Sutton 1984). It was probably introduced into California when the state's fruit industry was being established. In 1898 Bioletti (1902) found brown rot on apricot blossoms in the San Francisco Bay area. This strongly suggests that the disease was European brown rot, because severe blossom and twig blighting of apricot is one of the principal features of that disease. In 1921, Howard and Horne and in 1925 Rudolph gave similar descriptions of brown rot on apricots in San Mateo and Santa Clara counties. It was not until 1939, however, that both European and American types of brown rot were shown to occur in California (Hewitt and Leach 1939).

Symptoms. The principal feature of European brown rot is severe blossom and twig blighting (**fig. 4A**). Sudden withering of the flowers occurs during the blossoming period. For the next three or four weeks, twigs die in large numbers owing to the growth of the fungus from the infected flowers into the twigs. On apricot and prune, profuse gumming occurs with extensive twig blighting, whereas on peach and nectarine, blossom infection causes some twig blighting but in many instances only partial girdling of the twigs (Ogawa et al. 1980). Fruit infection is sporadic and usually not as serious as that caused by *Monilinia fructicola*. Loss of fruit from *M. laxa* was more common in the 1950s than in the 1980s, presumably because of poorer blossom blight control during the earlier years. Fruit rot epidemics on all stone-fruit species, including almonds, have been caused primarily by *M. fructicola*. Yet in orchards with severe *M. laxa* blossom blight, fruit rot is mostly caused by this species.

Fruit rot begins as small dark spots that expand rapidly. The affected flesh is brown, firm, and fairly dry (as contrasted with the soft watery rots caused by some other fungi). The fungus sometimes grows from infected fruit into the twigs, much as it does from infected blossoms. Somewhat the same symptom has been observed in almond, where the fungus first colonizes the inner surface of the fruit hull after it dehisces. There is evidence, however, that the twig is usually killed not by invasion of the fungus, but probably by a toxin (Mirocha and Wilson 1961).

Tufts of conidia (sporodochia) are seen on infected blossoms (**fig. 4C**) as soon as they are blighted but develop on the twigs later in the spring (**fig. 4B**). During the following winter, after the blighted parts (including mummified fruit) are wet with rain or fog, new conidial tufts develop on these structures. The sporodochia are pulverulent, ash-gray, and 1 to 2 mm in diameter; they remain on the host parts even after the rains cease and provide the inoculum for blossom infection.

Causal organism. The confused state of the taxonomy of the fungus will be discussed only briefly here. Honey (1928) erected a new genus, *Monilinia*, for the monilioid species previously placed in the genus *Sclerotinia*, but Harrison (1933) did not favor such a transfer. At present most plant pathologists accept Honey's classification and call the fungus *Monilinia laxa* (Aderh. & Ruhl.) Honey. Synonyms are *M. cinerea* Bon., *S. cinerea* (Bon.) Wr., *S. cinerea* (Bon.) Wr. f. *pruni* Wormald, and *M. oregonensis* Barss & Posey. Two biological forms are reported in Europe (Wormald 1919), one of which attacks stone fruit and the other pome fruit.

The conidia (moniliospores) of *M. laxa* are hyaline, lemon-shaped, 6.25 to 7.35 by 10.23 to 11.61 μm, and are produced in branched chains arising from a hypostroma, the whole fruiting structure being regarded as a sporodochium. Conidia are produced in spring and early summer on newly blighted blossoms, and in late winter and early spring on twigs, blossoms, and peduncles blighted the previous growing season, as well as on mummified fruit.

Apothecia of *M. laxa* are extremely rare. Aderhold and Ruhland (1905) described the species on the basis of a few apothecia. Wormald (1921) and Harrison (1934–35) described apothecial development in England, but no apothecia of this fungus have been reported since then.

Features indicating the presence of *M. laxa* in the orchard are extensive blossom infection followed by blighting of the twigs, and production of abundant ash-gray conidial tufts in early winter on the blighted blossoms and twigs. Where fruit infection by *M. laxa* occurs, the fungus produces abundant conidial tufts on the surface of the rotted fruit (**fig. 4C**) and these commonly occur in concentric zones. Although in Washington and Oregon, *M. laxa* and *M. fructicola* (the American or peach brown rot fungus) can occur simultaneously in the same orchard, in California the two species exhibit a considerable degree of host selectivity. While *M. fructicola* occurs frequently on peach and nectarine causing blossom blight and fruit rot, *M. laxa* is seldom found on these hosts (Corbin and Ogawa 1974; Ogawa, English, and Wilson 1954; Tate 1973). However, *M. laxa* occurs frequently but *M. fructicola* infrequently on apricot and almond. Both species may be found in cherry, prune, and plum orchards (Ogawa, English, and Wilson 1954).

Surveys of apricot and prune orchards during the early 1980s indicated that *M. laxa* was the prevalent species on blighted apricot twigs but *M. fructicola* was the common form on prune fruit (Michailides, Ogawa, and Opgenorth 1986).

Growth characteristics on potato-dextrose agar afford a fairly reliable method for distinguishing *M. laxa* from *M. fructicola* (Hewitt and Leach 1939). On this medium the former grows more slowly than the latter, the colonies are characteristically lobed, instead of having smooth margins with concentric rings, and there are few rather than many conidia. However, in continuous light at 21°C both organisms produce numerous conidia. Another valuable characteristic is that interaction zones do not develop between colonies of *M. laxa* isolates, whereas they do appear when isolates of *M. laxa* and *M. fructicola* are paired (Sonoda, Ogawa, and Manji 1982).

Mycelial characteristics also help to identify the two species. *M. laxa* produces geniculate (bent) germ tubes, and the hyphae seldom anastomose. *M. fructicola* produces relatively straight germ tubes, and its hyphae anastomose frequently (Hewitt and Leach 1939; Ogawa and English 1954).

Mycelial cultures (potato-dextrose agar) of both *Monilinia* species may remain viable for a one-year period in cold storage (3°C). Another technique for maintaining *M. fructicola* cultures is to grow the fungus on sterilized filter paper, which is then dried at 21°C before storage at −8°C (Baxter and Fagan 1986).

Disease development. *Monilinia laxa* survives from one season to the next in twig cankers, blighted blossoms, peduncles, and in the rotted, mummified fruit hanging in the tree. Conidia begin to develop on these parts in late December and continue to develop until April. A single blighted almond twig often bears as many as 25 sporodochia, each of which produces thousands of conidia. A high percentage of newly formed conidia is germinable, and if not exposed to direct sunlight and high temperature they retain their viability for months. An abundant and timely supply of inoculum, therefore, is strategically located in the tree when blossoms emerge in the spring (Corbin and Ogawa 1974; Corbin, Ogawa, and Schultz 1968).

The conidia are blown about by wind (Corbin and Ogawa 1974; Corbin, Ogawa, and Schultz 1968; Wilson and Baker 1946) and washed about by rain (Corbin, Ogawa, and Schultz 1968). When they lodge on susceptible tissue they germinate in two to four hours if moisture is present and temperature is favorable.

As noted earlier, *M. laxa* attacks the blossoms of almond and apricot, producing extensive flower and twig blighting. The critical period for flower infection extends from the time the unopened flowers emerge from the winter buds until the petals are shed. There is evidence that the flowers are most susceptible to infection when fully open. Though some infection through the side of the floral tube may occur, Calavan and Keitt (1948) reported that the most frequent sites of infection in cherry blossoms are the anthers, stigmas, and petals. In almond, stigma infection is most common, with anthers and petals the next most frequently infected (Ogawa and McCain 1960). In apricot and prune, the sepals are susceptible, as are other floral parts.

At ordinary springtime temperatures, three to six days elapse between blossom infection and the first evidence of necrosis. This is followed by rapid necrosis of the entire blossom. Infection and development of disease symptoms occur over a relatively wide temperature range (4° to 30°C), the optimum being about 24°C (Calavan and Keitt 1948). Low-moisture conditions limit infection; little or no infection occurs in rainless weather even if humidity is high (Weaver 1943).

Susceptibility of stone-fruit cultivars. Among almond cultivars, Drake and Jordanolo are highly susceptible to blossom infection, and Ne Plus Ultra and Texas (Mission) are moderately susceptible. Severe blossom infection is uncommon in Nonpareil, Peerless, and Davey. With the emphasis by almond growers to control blossom blight on the early pollinators (Ne Plus Ultra, Peerless, or Merced), an increase in blossom blight has been observed on the later-blooming Mission cultivar (Ogawa, Manji, and Sonoda 1985; Ogawa, Manji, and MacSwan 1986). *M. laxa* blossom and twig blight is rarely found on apple and pear, but it is common on the pomaceous flowering quince (*Chaenomeles* sp.), a popular ornamental shrub in California. Yet severe blossom blight in Bartlett pears has been found adjacent to severely diseased apricot trees.

The commercially important Royal, Blenheim, Perfection, and Derby Royal cultivars of apricot are highly susceptible to blossom infection, whereas Tilton is noticeably less susceptible (Hesse 1938). Crossa-Raynaud (1969) evaluated resistance based on the rate of canker development in young branches of apricot and almond cultivars and showed some differences.

In California, Santa Rosa and Wickson plums and Imperial and French prunes may suffer severe blossom infection by *M. laxa*. In Oregon, Italian prune is susceptible to sporadic blossom infection by both *M. laxa* and *M. fructicola*. In a few nectarine and peach orchards in California, severe blossom blight and fruit rot by *M. laxa* has occurred, but in most orchards only *M. fructicola* has been isolated.

Control. Rudolph (1925) developed a protective spray schedule that has proved relatively effective on apricots in California. The trees are sprayed with Bordeaux mixtures of 16–16–100 when the blossoms are at red bud (before the white petals are exposed) (**fig. 4H**). Where the disease has been severe two sprays are advisable, one at early red-bud stage (sepals showing) and one at full bloom. On almond, a protective spray is recommended at the early pink-bud stage of bloom (**fig. 4H**) as the sepals are relatively resistant to infection.

Later, dormant eradicative sprays were proved effective in reducing brown rot blossom blight in apricot, almond, prune, and plum (Chitzanidis 1971; Hesse 1938; Wilson and Baker 1946; Wilson 1950, 1953). The purpose of the eradicative spray is to destroy the conidial inoculum in the tree. Applications are made in late winter (January 15 to February 1) while the trees are still dormant. Monocalcium meta-arsenite penetrates into the blighted blossoms and twigs of apricot trees and kills the fungus (Wilson 1942). This treatment, however, is no longer approved. In other tests, a formulation of 79 percent sodium pentachlorophenoxide at the rate of 3 to 4 lbs. per 100 gallons was applied on apricot, almond, and prune. (This agent is no longer available for use on apricot or almond.) Later studies (Chitzanidis 1971; Sanborn, Manji, and Ogawa 1983; Ogawa, Hall, and Koepsell 1967) showed that early applications (in December), before sporodochial development, reduced the number of spore pustules formed and the ability of the spores to be wind-dispersed. The sporodochia on sprayed trees were white. The eradicant fungicides could be applied either as dilute sprays or as concentrate sprays (100 gallons of spray per acre).

On almond and apricot, benomyl suppresses sporodochial development and protects the blossoms and fruit from infection (Ramsdell and Ogawa 1973a, b). Mixtures of benomyl and oil are superior to benomyl alone. Early winter application of this combination, before sporodochia appear, suppresses sporodochial development on twigs and blighted blossoms (Ramsdell and Ogawa 1973a). A spray application of benomyl before the blossoms open provides protection of anthers and stigmas by systemic movement of the chemical (Ramsdell and Ogawa 1973b). Benomyl-resistant M. laxa was detected by Ogawa et al. (1984) in apricot orchards, but its population has not increased (Michailides, Ogawa, and Opgenorth 1986), and the isolates appeared less pathogenic than the benomyl-sensitive ones collected from a severely diseased almond orchard (Cañez and Ogawa 1985).

Captan, fixed coppers, dichlone, and coordinated mixtures of nabam plus metal salts can be applied as protectants at early and full bloom. Other effective materials are the sterol biosynthesis inhibitor (triforine) and the dicarboximides (iprodione and vinclozolin) introduced in the 1980s. Derivatives of sterol biosynthesis inhibitors (triazole, piperazine, pyrimidine, and imidazole) show activity against both brown rot *Monilinia* species in field tests on stone fruit crops in California (Ogawa, Gubler, and Manji 1988).

Only limited research has been done on the biocontrol of stone fruit brown rot, but recent research in Spain (De Cal, M. Sagasta, and Melgarejo 1988) indicates that antibiotic substances produced by *Penicillium frequentans* are highly active against M. laxa and may find a place in the control of this pathogen. In Japan, Y. Harada (personal communication) showed that *Lambertella corni-maris* Hohnel parasitizes the mycelial stroma of *Monilinia fructigena* in apple mummies found in the orchard.

Brown Rot: American Brown Rot

Monilinia fructicola (Wint.) Honey

This disease has been called "peach brown rot" and "fruit brown rot" to distinguish it from the European stone fruit brown rot, the usual designation in America being simply "brown rot." American brown rot is indigenous to North America, probably occurring on species of wild *Prunus* long before the continent was settled by Europeans. It is found in Australia, New Zealand, South Africa, Japan, Argentina, and Brazil, but is unknown in Europe (Wormald 1928).

Although this brown rot disease is present wherever stone fruit are grown in North America, the most frequent and severe outbreaks occur where rainfall and humidity are high during the growing season. It is particularly destructive in the Atlantic Coast states south of New Jersey and in the western parts of Washington and Oregon. But in central and eastern Washington, where the winters are cold and summers dry, brown rot is not a problem on stone or pome fruit. It is sporadic in California, where summers are usually rainless. Nevertheless, the disease occurs frequently enough to be a major problem (Smith and Bassett 1963; Sonoda, Ogawa, and Vertrees 1967).

American brown rot affects all species of stone fruit but is particularly severe on peach, nectarine, plum, prune, and cherry. Infection of apricot and almond blossoms is rare. Fruit rot of apricot and hull rot of almond are commonly caused by M. *fructicola* (Mirocha and Wilson 1961). The disease has been found at times on apple fruit.

Symptoms. Symptoms occurring on California peaches are included here. Blossom infection is much less prevalent in California than in some other parts of the country (Landgraf and Zehr 1982). Nevertheless, each year some peach blossoms are infected. The first indication of the disease is necrosis of the anthers; this is followed by rapid death of the floral tube, the ovary, and the peduncle. In peach, blossom blight results almost entirely from anther infections (**fig. 4H**); infected petals usually dehisce before the pathogen can enter the hypantheum. The fungus can extend into the support spur or shoot but usually does not cause extensive death of these parts, in contrast to the European brown rot fungus M. laxa.

The fungus immediately produces abundant, dusty, buff-colored tufts of conidia on the blighted blossoms and twigs. It remains alive in diseased blossoms and twigs throughout the following fall and winter and, under California conditions, sometimes produces a few new conidial tufts in the spring on these parts. Some reports indicate that the fungus fails to produce conidia on these tissues in spring (Cation and Dunegan 1949; Jenkins 1967; Norton and Ezekiel 1924; Roberts and Dunegan 1932), but sporodochia have been observed on both peduncles and blighted twigs in California.

During rainy spring weather, infection of green shoots is sometimes serious on plum trees in the eastern states, but seldom occurs on plum or other stone-fruit crops in California.

The most important aspect of American brown rot is fruit infection. While incipient infection of green fruit may occur, ripe fruit are much more susceptible to infection (Manji and Ogawa 1985, 1987). The symptoms appear as dark brown circular spots that spread rapidly over the fruit. The affected tissues remain relatively firm and dry, in contrast to the soft watery type of rotting produced by the fungus *Rhizopus*. Buff-colored masses of conidia are produced on the rotted area (**fig. 4E**).

In some years, extensive killing of twigs and even branches follows fruit infection (**fig. 4F**). Although this condition would seem to indicate that the twigs and branches are killed because they are invaded by the fungus from the infected fruit, toxins produced by the fungus may be implicated, much as they are in hull rot of almond caused by *Rhizopus* (Mirocha and Wilson 1961; Mirocha, DeVay, and Wilson 1961).

Causal organism. The incitant is *Monilinia fructicola* (Wint.) Honey. Synonyms are *Sclerotinia fructicola* (Wint.) Rehm., *S. cinerea* (Bon.) Schroeter, forma *americana* Wormald, and *S. americana* (Wormald) Norton & Ezekiel (Norton and Ezekiel 1924; Wormald 1928). Most authorities agree that the American brown rot fungus differs from any of the Sclerotinias occurring in Europe. Earlier American writers called this fungus *Sclerotinia cinerea*, *Monilia fructigena*, or *Sclerotinia fructigenum* (Cooley 1914; Dunn 1926; Hewitt and Leach 1939; Matheny 1913).

Though some workers have tried to show differences in size, the conidia (moniliospores) of M. fructicola are indistinguishable from those of M. laxa. They are produced in chains on sporodochia and average about 15 µm in length and 10µm in width. Phillips (1982) observed a conidial response to incubation temperature, with larger, higher viability, and more pathogenic spores formed at 15°C than at 25°C. Further tests (Phillips 1984) showed that conidia formed on 2 percent potato-dextrose agar were smaller and less aggressive than those produced on fresh peaches. In these tests comparisons were made among spores produced on peaches incubated at 15°, 20°, and 25°C for five days and on potato-dextrose agar for two weeks. The age of the spores was not considered, but the differences observed are considered significant in epidemiology of the disease. Furthermore, when the fungus was grown on glucose (dextrose) at a concentration of 15 percent, spore size and aggressiveness were markedly greater than on 30 percent glucose (Phillips and Margosan 1985).

The apothecia are smooth, fleshy, brown to reddish brown, cuplike structures, (**fig. 4G**) varying in size from ¼ to ¾ inch in diameter and arising by means of a stalk from the mummied fruit on the ground. Some stalks may be two inches long if they arise from mummied fruit buried that deep in the soil. The interior of the cup is lined with a hymenial layer consisting of long slender asci averaging 9 by 155 µm. The ascospores are ovoid to ellipsoid, single celled and 6 by 12 µm. Micronidia are often found in abundance on pseudosclerotia which form on rotted fruit on the ground; these are 2 to 5 µm, hyaline, spherical structures produced on short, bottle-shaped sterigmata arising from the mycelium (Ezekiel 1921, Hewitt and Leach 1939).

Production of apothecia by M. fructicola on synthetic media was achieved after exposure to cold temperatures (Terui and Harada 1966). Willets and Harada (1984) have reviewed the research efforts in Japan to produce apothecia of *Monilinia* species in culture. Critical events required for formation of *M. fructicola* apothecia were, first, the inducement of a mature stroma on potato-sucrose agar or fruit tissue (four to eight weeks at 20° to 30°C), followed by exposure to low temperatures (10° to 15°C for three to four months) for the development of apothecial initials, and, finally, the incubation at 10° to 15°C in diffuse sunlight or under daylight fluorescent illumination of intensity above 1,500 lux with a 12-hour photoperiod. These conditions closely parallel those under which stone fruit are grown in California. Apotehcia are found in California orchards at the time peach trees bloom, when mean temperatures are around 13°C and the photoperiod approaches the 12-hour requirement.

Disease development. The fungus survives the winter in several ways: (1) As mycelium in rotted fruit hanging in the tree; here conidia are produced on the surface of the fruit in spring. Bertram (1916) reported that in Vermont, conidia produced in the autumn remained viable throughout winter. When temperatures of −12° to −22°C prevailed, however, few conidia survived. Jenkins (1967) showed that in either sun or shade, the reduction in viability of conidia was less than 1 percent. (2) As mycelium in rotted fruit on the ground; on such fruit the fungus produces the typical flat sclerotial mat from which arise the apothecia. The apothecia appear and mature at the time the host blossoms in the spring. They discharge their spores into the air for a few weeks and then disintegrate. (3) As mycelium in blossom parts, peduncles, and twigs killed by the pathogen the previous year (Sutton and Clayton 1972). Sporulation on peduncles, twigs, and branch cankers occurs frequently in the eastern United States and Australia (Kable 1965a) but is less common in California. Sources of inoculum for South Carolina peach orchards were found to be nonabscised aborted fruit, infected thinned fruit on the ground, and infected fruit on wild plum trees (Landgraf and Zehr 1982).

Apothecia develop in areas where the soil is moist in the spring, but they seldom occur in California when the weather becomes dry just before and during bloom. Under such conditions these structures are found only where the soil is protected from drying by weeds or debris. Other factors affecting apothecial development are temperature and soil pH. Moderate spring temperatures (10° to 15.5°C) favor development while cold weather deters it. Ezekiel (1923) reported that soil with a pH below 7.0 favors apothecial development, whereas an alkaline soil does not.

In California, the most important primary inoculum is the conidia produced on mummies (**fig. 4G**) and blighted flowers that remain in the tree.

However, under favorable environmental conditions ascospores can be important in initiating primary infection. Ascospores are forcibly ejected into the air and are carried by air currents about the orchard. Slight disturbances in the air (which change the humidity) initiate ascospore discharge.

In the Pacific Northwest, blossom infection is often severe, causing a marked reduction in fruit production. In California, blossom infection is much less abundant. Nevertheless, conidia produced on the infected blossoms are often the only inoculum present in the orchard in midsummer when fruit infection begins. Conidia are freely disseminated by moving air, rainwater (Jenkins 1965; Kable 1965b), and insects (Tate and Ogawa 1975).

A few green fruit may become rotted in early summer. This is thought to result mainly from quiescent (incipient) infections (Tilford 1936) or insect wounds (Tate 1973), because direct infections require over 30 hours of continuous moisture. Biggs and Northover (1988b), however, have shown that young peach fruit are highly susceptible to infection, that the fruit become resistant at pit hardening, and that they become increasingly susceptible two to three weeks before full ripeness. Infected fruit sporulate and increase the inoculum supply. In California fruit infection is most common during the last four weeks before harvest. When the fruit are not picked until fully ripe (for canning or freezing), heavy losses from brown rot in the field are common. Fresh-market fruit for shipping are picked before they are fully ripe; consequently loss in the field is relatively light. Loss during shipment and in the market, however, may be heavy, as the fungus develops rapidly when packaged fruit are removed from cold storage.

Although the wounding of fruit may lead to an increase in infection, the fungus readily infects when no wound is present. It commonly enters the fruit by way of the trichome (hair) sockets—breaking of the trichomes increases the likelihood of infection (Smith 1936). Pathogenicity has been related to the production of hydrolytic enzymes (Hall 1971a,b, 1972).

Although the fungus can grow slowly at 1.7° to 4.4°C, its optimum temperature is 22.2° to 23.9 °C.

At 23°C or above it produces visible symptoms on infected fruit within two days and can completely rot the fruit in four or five days. The optimum temperatures for infection of peach and sweet cherry fruit are 22.5° to 25°C and 20° to 22.5°C, respectively (Biggs and Northover 1988a). Infection incidence increases directly with wetness duration.

Rains and accompanying high humidity favor infection. Thus the disease occurs most frequently and causes the greatest destruction in the more humid fruit-growing areas. In California, where summers are relatively rainless, an epidemic of fruit rot may follow a brief shower (Sonoda et al. 1967); nevertheless, the disease may also appear during a rainless period. Dew, which forms at night following a sharp drop in temperature, probably provides the moisture for spore germination and infection. Rot also can develop from quiescent infections (Gubler et al 1987; Jenkins and Reinganum 1965; Tate and Corbin 1978; Wade 1956).

Control. Certain cultural practices, such as the removal of mummies immediately after harvest, can reduce the number of spores available for blossom and fruit infection. Apothecial development can be reduced by cultivating the orchard just before blossom time. Baur and Huber (1941) and Huber and Baur (1939, 1941) were able to destroy many of the apothecia by dusting the orchard floor in the spring with finely ground calcium cyanamide, but they did not determine the degree of control obtainable by this procedure. No cultivars of peach are known to be highly resistant to blossom brown rot, but Bolinha has shown moderate resistance (A. J. Feliciano, J. E. Adaskaveg, and J. M. Ogawa, unpublished). The cultivars Dixon, Fortuna, Vivian, and Walton show blossom blight more often than do Halford and Stuart.

Resistance in the Brazilian peach cultivar Bolinha was reported by Feliciano, Feliciano, and Ogawa (1987). Their method of evaluation involved a comparison of lesion diameters of inoculated ripe fruit. Using the same method of evaluation, Scorza and Gilreath (1983) and Scorza and van der Zwet (1984) were able to show differences in lesion diameters of inoculated sweet cherry and peach cultivars grown in the eastern United States, but the significance of their evaluation is yet to be determined.

Chemical prevention has been widely used and found to be effective. Eradication of spore pustules with dormant sprays has not been possible, although sodium pentachlorophenoxide can help control M. *laxa* on prune. (This agent causes severe injury to peach trees and is not available for use on almond or apricot.) Fungicides for blossom protection on peach should be applied before rains when about 5 percent of the blossoms are open, and again at 70 percent bloom. Chemicals which have been recommended for blossom-blight control are sulfur, copper, captan, febram, nabam plus salts, dichlone, and benomyl. Liquid lime-sulfur applications on blossoms may result in severe damage resembling that caused by M. *fructicola*. Benomyl and thiophenate-methyl have provided effective control (Gilpatrick 1973; Ogawa, Manji, and Bose 1968; Ogawa, Wilson, and Corbin 1967; Tate et al. 1974) and can be applied as early as the pink-bud stage of bloom (**fig. 4H**). Their application at this time protects the anthers from infection. Aircraft applications of these systemic fungicides have provided excellent coverage as well as disease control (Ogawa et al. 1972; Ogawa, Manji, and Sonoda 1985).

Repeated applications of benomyl during bloom and preharvest have resulted in the selection of benomyl-resistant M. *fructicola* in Australia (Whan 1976), in the states of Michigan (Jones and Ehret 1976) and California (Ogawa et al. 1988) and M. *laxa* in California (Cañez and Ogawa 1972; Ogawa et al. 1984). Benomyl-resistant strains of M. *fructicola* have been detected in most of the major stone-fruit areas of California, and benomyl is no longer registered for control of blossom blight on sweet cherry. In an attempt to delay the development of benomyl-resistant isolates as well as to ensure disease control, the manufacturer stipulated (for California) that benomyl be used only in a mixture with another fungicide. Field tests indicate that benomyl combined with less effective compounds does not delay selection of resistant populations (Szkolnik et al. 1978). With low populations of benomyl-resistant M. *fructicola*, effective disease control can be obtained with benomyl sprays (Sonoda et al. 1983). There is laboratory evidence that benomyl-sensitive isolates tend to predominate over resistant isolates when inoculated onto benomyl-free peach fruit (Sonoda, Ogawa, and Sholberg 1982), but only through isolations from orchards without benzimidazole treatments can competitive parasitic fitness be determined. Strategies for management of fungicide treatment to prevent or delay development of resistant strains of *Monilinia* are discussed by Ritchie (1985).

The recently introduced sterol biosynthesis inhibitor (SBI) triforine and two dicarboximides, iprodione and vinclozolin, will probably replace much of the benzimidazole usage in California. Although more spray applications of these new fungicides are required for effective disease control, M. fructicola resistance to them has not been detected in California. Reese and Moore (1982) induced resistance in M. fructicola to an SBI compound, Nustar, through ultraviolet radiation of spores, but these isolates lost their resistance after three to nine transfers on a fungicide-free medium. For the dicarboximide fungicides, Sztejnberg and Jones (1978) reported induction of resistance in the laboratory, and Benes and Ritchie (1984) provided evidence of increased melanin content in the resistant isolates. Ritchie (1983) concluded that dicarboximide-resistant isolates of M. fructicola were less parasitically fit and would not be apt to rapidly increase to a dominant population level. However, iprodione-resistant *Botrytis cinerea* has become dominant in grape vineyards in France and strawberry fields in California. Thus, laboratory tests may not reliably indicate the parasitic fitness of isolates that develop under field conditions. Prochloraz is another SBI compound that shows promise for the control of brown-rot blossom blight (Dijkhuizen, Ogawa, and Manji 1983). A new class of fungicides, presently designated SCO858 by Stauffer Chemical Company, shows promise in the control of isolates resistant to both benomyl and the dicarboximides.

Protection of fruit from infection can be achieved only if fungicides are applied before free moisture occurs on the fruit. Aircraft or ground dusting is effective if done before rains (Sonoda et al. 1967; Yates, Ogawa, and Akesson 1974). Repeated ground spray or dust applications are beneficial in sprinkler-irrigated peach orchards but not in prune orchards.

Fruit with quiescent infections (**fig. 4D**) usually develop rot during the last month before harvest, regardless of the application of protective fungicides. In recent tests with green cherry fruit showing quiescent infection of M. fructicola, spray applications of iprodione, triforine, and benomyl reduced the incidence of decayed fruit at harvest (Manji and Ogawa 1987).

Eradication of incipient fruit infection on cling peaches following rains during the last three weeks before harvest was shown to be possible with ground application of liquid lime-sulfur within 37 hours from the beginning of rain (Ogawa et al. 1954). Phytotoxicity to leaves was observed in Fresno County, but not in the Sacramento Valley or as far south in the San Joaquin Valley as Stanislaus County. Benomyl was not as effective, although it significantly reduced the incidence of decay (Ogawa 1970).

Two preharvest Botran sprays applied to ripe cling peaches prevented fruit decay after mechanical harvesting and holding for processing (Ogawa, Sandeno, and Mathre 1963). Cling peaches mechanically harvested are more prone to decay than hand-harvested fruit, but may be protected by dipping in DCNA (Botran) suspensions (Ogawa, Sandeno, and Mathre 1963). Hand-harvested fruit were protected from brown rot by two preharvest applications of DCNA, a single application of benomyl (Ogawa 1970), or by heat treatment followed by a DCNA dip (Fish et al. 1968). When a disinfestant such as calcium hypochlorite was mixed with protectant fungicides such as captan or DCNA, better disease control was obtained in the orchard (Ogawa, Clason, and Corbin 1967).

Studies on nonchemical means of postharvest decay control have centered on the application of biological control agents (*Bacillus subtilus*) for controlling *Monilinia fructicola* (Pusey and Wilson 1984), but orchard trials using a commercial formulation of this organism have provided variable results (B. T. Manji and J. M. Ogawa, unpublished).

REFERENCES

Aderhold, R., and W. Ruhland. 1905. Zur Kenntnis der Obstbaum-sklerotinien. *Arb. Biol. Abt. Land- Forstw. Kaiserl. Gesundheitsamte* 4:427–42.

Barss, H. P. 1923. Brown rot and related diseases of stone fruits in Oregon. *Ore. State Coll. Agric. Exp. Stn. Circ.* 53.

———. 1925. Serious blossom blight in Pacific Northwest orchards due to species of *Monilinia* (abs.). *Phytopathology* 15:125.

Batra, L. R. 1979. First authenticated North American record of *Monilinia fructigena* with notes on related species. *Mycotaxon* 8:476–84.

Baur, K., and G. A. Huber. 1941. Effect of fertilizer materials and soil amendments on development of apothecia of *Sclerotinia fructicola*. *Phytopathology* 31:1023–30.

Baxter, L. W., Jr., and S. G. Fagan. 1986. Method for maintaining three selected fungi. *Plant Dis.* 70:499–500.

Benes, S. E., and D. F. Ritchie. 1984. Evidence for increased melanin content in dicarboximide-resistant strains of *Monilinia fructicola* (abs.) *Phytopathology* 74:877.

Bertram, H. E. 1916. A study of the brown rot fungus in northern Vermont. *Phytopathology* 6:71–78.

Biggs, A. R., and J. Northover. 1988a. Influence of temperature and wetness duration on infection of peach and sweet cherry fruits by *Monilinia fructicola*. *Phytopathology* 78:1352–53.

———. 1988b. Early and late-season susceptibility of peach fruits to *Monilinia fructicola*. *Plant Dis.* 72:1070–74.

Bioletti, F. T. 1902. Brown-rot of stone fruits. *Pacific Rural Press* 60:67, and *Calif. Univ. Agric. Exp. Stn. Rept.* 1898–1901:330–31.

Calavan, E. C., and G. W. Keitt. 1948. Blossom and spur blight (*Sclerotinia laxa*) of sour cherry. *Phytopathology* 38:857–82.

Cañez, V. M., and J. M. Ogawa. 1982. Reduced fitness of benomyl-resistant *Monilinia laxa* (abs.). *Phytopathology* 72:980.

———. 1985. Parasitic fitness of benomyl-resistant and benomyl-sensitive *Monilinia laxa* (abs.). *Phytopathology* 75:1329.

Cation, D., and J. C. Dunegan. 1949. The overwintering of *Monilinia fructicola* in twig cankers under Michigan conditions. *Plant Dis. Rep.* 33:97–98.

Cation, D., J. C. Dunegan, and J. Kephart. 1949. The occurrence of *Monilinia laxa* in Michigan. *Plant Dis. Rep.* 33:96.

Chitzanidis, A. 1971. Tests of eradicant fungicides against *Sclorotinia laxa* on sour cherry trees. *Inst. Phytopathol. Benaki Ann.* 10:119–24.

Chochriakova, T. M. 1971. Evaluation of cherry resistance to moniliosis, *Monilinia cinerea*. *Tr. Prikladnoi Bot. Genet. Selek.* 43:231–36. (English Summary)

Cooley, J. S. 1914. A study of the physiological relations of *Sclerotinia cinerea* (Bon.) Schroeter. *Ann. Missouri Bot. Gard.* 1:291–326.

Corbin, J. B., and J. M. Ogawa. 1974. Springtime dispersal patterns of *Monilinia laxa* conidia in apricot, peach, prune, and almond trees. *Can. J. Bot.* 52:167–76.

Corbin, J. B., J. M. Ogawa, and H. B. Schultz. 1968. Fluctuations in numbers of *Monilinia laxa* conidia in an apricot orchard during the 1966 season. *Phytopathology* 58:1387–94.

Crossa-Raynaud, P. H. 1969. Evaluating resistance to *Monilinia laxa* (Aderh. & Ruhl.) Honey of varieties and hybrids of apricots and almonds using mean growth rate of cankers on young branches as a criterion of susceptibility. *J. Am. Soc. Hort. Sci.* 94:282–84.

De Cal, A., E. M.-Sagasta, and P. Melgarejo. 1988. Antifungal substances produced by *Penicillium frequentans* and their relationship to the biocontrol of *Monilinia laxa*. *Phytopathology* 78:888–93.

Dijkhuizen, J. P., J. M. Ogawa, and B. T. Manji. 1983. Activity of captan and prochloraz on benomyl-sensitive and benomyl-resistant isolates of *Monilinia fructicola*. *Plant Dis.* 67:407–09.

Dunegan, J. C. 1953. Brown rot of peach. In *Plant diseases* 684–88. USDA Agricultural Yearbook.

Dunn, M. S. 1926. Effect of certain acids and their sodium salts upon the growth of *Sclerotinia cinerea*. *Amer. J. Bot.* 13:48–58.

English, H., W. J. Moller, and S. F. Nome. 1967. New records of fungus diseases of fruit crops in Chile. *Plant Dis. Rep.* 51:212–14.

English, H., A. Pinto de Torres, and J. Kirk. 1969. Reconocimiento de especies del genero *Monilinia* en frutales de carozo y en membrillo de flor en Chile. *Agric. Tecn.* 29:54–59.

Evans, A. W., and C. E. Owens. 1941. Incidence of *Sclerotinia fructicola* and *S. laxa* in sweet cherries in Oregon. *Phytopathology* 31:469–71.

Ezekiel, W. N. 1921. Some factors affecting the production of apothecia of *Sclerotinia cinerea*. *Phytopathology* 11:495–99.

———. 1923. Hydrogen-ion concentration and the development of *Sclerotinia apothecia*. *Science* 58:166.

Feliciano, A., A. J. Feliciano, and J. M. Ogawa. 1987. *Monilinia fructicola* resistance in peach cultivar Bolinha. *Phytopathology* 77:776–80.

Fish, S., et al. 1968. *Research work on brown rot of stone fruits: The brown rot research committee report, 1957–62.* Victorian Plant Research Institute, Burnley, Australia, 117 pp.

Gilpatrick, J. D. 1973. Control of brown rot of stone fruits with thiophanate-methyl and a piperazine derivative fungicide. *Plant Dis. Rep.* 57:457–59.

Gubler, W. D., B. T. Manji, J. M. Ogawa, and F. Yoshikawa. 1987. Quiescent infection of *M. fructicola* on Friar plum. *Proc. 61st Ann. West. Orch. Pest Dis. Mgmt. Conf.*: 2.

Hall, R. 1971a. Pathogenicity of *Monilinia fructicola*. I. Hydrolytic enzymes. *Phytopathol. Z.* 72:245–54.

———. 1971b. Pathogenicity of *Monilinia fructicola*. II. Penetration of peach leaf and fruit. *Phytopathol. Z.* 72:281–90.

———. 1972. Pathogenicity of *Monilinia fructicola*. III. Factors influencing lesion expansion. *Phytopathol. Z.* 73:27–38.

Harrison, T. H. 1928. Brown-rot of fruits and associated diseases in Australia. Part I. History of the disease and determination of the causal organisms. *J. Proc. Roy. Soc. New South Wales* 62:99–151.

———. 1933. Brown-rot of fruits and associated diseases of deciduous fruit trees. I. Historical review and critical remarks concerning taxonomy and nomenclature of the causal organisms. *J. Proc. Roy. Soc. New South Wales* 67:132–77.

———. 1934–35. Brown-rot of fruits and associated diseases of deciduous fruit trees. II. The apothecia of the causal organisms. *J. Proc. Roy. Soc. New South Wales* 68:154–76.

Hesse, C. O. 1938. Variation in resistance to brown-rot in apricot varieties and seedling progenies. *Proc. Am. Soc. Hort. Sci.* 36:266–68.

Heuberger, J. W. 1934. Fruit-rotting Sclerotinias. IV. A cytological study of *Sclerotinia fructicola* (Wint.) Rehm. *Maryland Agric. Exp. Stn. Bull.* 371:167–89.

Hewitt, W. B., and L. D. Leach. 1939. Brown-rot Sclerotinias occurring in California and their distribution on stone fruits. *Phytopathology* 29:337–51.

Honey, E. E. 1928. The monilioid species of *Sclerotinia*. *Mycologia* 20:127–56.

Howard, W. L., and W. T. Horne. 1921. Brown-rot of apricots. *Calif. Univ. Agric. Exp. Stn. Bull.* 326:73–88.

Huber, G. A., and K. Baur. 1939. The use of calcium cyanamide for the destruction of apothecia of *Sclerotinia fructicola*. *Phytopathology* 29:436–41.

———. 1941. Brown rot on stone fruits in western Washington. *Phytopathology* 31:718–31.

Humphrey, J. E. 1893. On *Monilinia fructigena*. *Bot. Gaz.* 8:85–93.

Jenkins, P. T. 1965. The dispersal of conidia of *Sclerotinia fructicola* (Wint.) Rehm. *Austral. J. Agric. Res.* 16:627–33.

———. 1967. The longevity of conidia of *Sclerotinia fructicola* (Wint.) Rehm. under field conditions. *Austral. J. Biol. Sci.* 21:937–45.

Jenkins, P. T., and C. Reinganum. 1965. The occurrence of quiescent infection of stone fruits caused by *Sclerotinia fructicola* (Wint.) Rehm. *Austral. J. Agric. Res.* 16:131–40.

Jones, A. L., and G. R. Ehret. 1976. Isolation and characterization of benomyl-tolerant strains of *Monilinia fructicola*. *Plant Dis. Rep.* 60:765–69.

Jones, A. L., and T. B. Sutton. 1984. Diseases of tree fruits. *N. Central Reg. Ext. Publ.* 45. Coop. Ext. Serv. Mich. State Univ. 59 pp.

Kable, P. F. 1965a. The fruit peduncle as an important overwintering site of *Monilinia fructicola* in the Murrumbidgee Irrigation Areas. *Austral. J. Exp. Agric. Anim. Husb.* 5:172–75.

———. 1965b. Air dispersal of conidia of *Monilinia fructicola* in peach orchards. *Austral. J. Exp. Agric. Anim. Husb.* 5:166–71.

———. 1970. Eradicant action of fungicides applied to dormant peach trees for control of brown rot (*Monilinia fructicola*). *J. Hort. Sci.* 45:143–52.

———. 1971. Significance of short-term latent infections in the control of brown rot in peach fruits. *Phytopathol. Z.* 70:173–76.

Landgraf, F. A., and E. I. Zehr. 1982. Inoculum sources for *Monilinia fructicola* in South Carolina peach orchards. *Phytopathology* 72:185–90.

Manji, B. T., and J. M. Ogawa. 1985. Quiescent infections and disease control in the shipping container. *Proc. 44th Ann. Conv. Natl. Peach Council* (Nashville): 23–24.

———. 1987. Quiescent infections of sweet cherry in California. *Proc. 61st Ann. West. Orch. Pest Dis. Mgmt. Conf.*: 8.

Matheny, W. A. 1913. A comparison of the American brown-rot fungus with *Sclerotinia fructigena* and *S. cinerea* of Europe. *Bot. Gaz.* 56:418–32.

Michailides, T. J., J. M. Ogawa, and D. C. Opgenorth. 1986. Distribution of *Monilinia* species and detection of benomyl-resistant isolates in prune and apricot orchards in California (abs.) *Phytopathology* 76:845.

Mirocha, C. J., J. E. DeVay, and E. E. Wilson. 1961. Role of fumaric acid in the hull rot disease of almond. *Phytopathology* 51:851–60.

Mirocha, C. J., and E. E. Wilson. 1961. Hull rot disease of almonds. *Phytopathology* 51:843–47.

Norton, J. B. S., and W. N. Ezekiel. 1924. The name of the American brown rot *Sclerotinia*. *Phytopathology* 14:31–32.

Norton, J. B. S., W. N. Ezekiel, and R. A. Jehle. 1932. Fruit-rotting *Sclerotinia*. I. Apothecia of the brown-rot fungus. *Maryland Agric. Exp. Stn. Bull.* 256.

Ogawa, J. M. 1958. The influence of emanations from fruits of *Prunus* species on spore germination of the brown-rot organisms (abs.). *Phytopathology* 48:396.

———. 1970. Brown rot control developments *Cling Peach Quart.* 7:7–9.

———. 1983. Controlling fungicide-resistant postharvest pathogens on stone fruits. *Abst. 4th Internat. Cong. Plant Pathol.* (Melbourne) 108.

Ogawa, J. M., G. W. Clason, and J. B. Corbin. 1967. Calcium hypochlorite added to selected fungicide sprays improves effectiveness in disease control (abs.). *Phytopathology* 57:824.

Ogawa, J. M., and H. English. 1954. Means of differentiating atypical isolates of *Sclerotinia laxa* and *S. fructicola* (abs.). *Phytopathology* 44:500.

———. 1960. Relative pathogenicity of two brown rot fungi, *Sclerotinia laxa* and *Sclerotinia fructicola*, on twigs and blossoms. *Phytopathology* 50:550–58.

Ogawa, J. M., H. English, W. J. Moller, B. T. Manji, D. Rough, and S. T. Koike. 1980. Brown rot of stone fruits. *Calif. Univ. Div. Agric. Sci. Leaf.* 2206, 7 pp.

Ogawa, J. M., W. H. English, and E. E. Wilson. 1954. Survey for brown rot of stone fruits in California. *Plant Dis. Rep.* 38:254–57.

Ogawa, J. M., W. D. Gubler, and B. T. Manji. 1988. Effect of sterol biosynthesis inhibitors on diseases of stone fruits and grapes in California. In *Sterol biosynthesis inhibitors*, Ed. D. Berg and M. Plempel, Chapter 9:262–87. Ellis Horwood Ltd., Chichester, England. 583 pp.

Ogawa, J. M., B. T. Manji, J. E. Adaskaveg, and T. J. Michailides. 1988. Population dynamics of benzimidazole-resistant *Monilinia* species on stone fruit trees in California. In: *Fungicide resistance in North America*. Ed. C. J. Delp, 36–39. APS Press, The American Phytopathological Society, St. Paul, Minnesota. 133 pp.

Ogawa, J. M., D. H. Hall, and P. A. Koepsell. 1967. Spread of pathogens within crops as affected by life cycle and environment. In *Air-borne microbes*, 247–67. Society for Genetic Microbiology, London.

Ogawa, J. M., B. T. Manjji, and E. Bose. 1968. Efficacy of fungicide 1991 in reducing fruit rot of stone fruit. *Plant Dis. Rep.* 52:722–26.

Ogawa, J. M., B. T. Manji, R. M. Bostock, V. M. Cañez, and E. A. Bose. 1984. Detection and characterization of benomyl-resistant *Monilinia laxa* on apricots. *Plant Dis.* 68:29–31.

Ogawa, J. M., B. T. Manji, and I. C. MacSwan. 1986. Field test procedures for evaluation of fungicides for control of *Monilinia laxa* on stone fruits. In: *Methods for evaluating pesticides for control of plant pathogens*. Ed. K. D. Hickey, 152–54. APS Press, St. Paul, 312 pp.

Ogawa, J. M., B. T. Manji, and D. J. Ravetto. 1970. Evaluation of pre-harvest benomyl applications on postharvest rot of peaches and nectarines (abs.) *Phytopathology* 60:1306.

Ogawa, J. M., B. T. Manji, and R. M. Sonoda. 1985. Management of the brown rot disease on stone fruits and almonds in California. In *Proceedings of the Brown rot of stone fruit workshop, Ames. Iowa, July 11, 1983*, 8–15. N.Y. State Agric. Exp. Stn. Geneva Spec. Rept. 55.

Ogawa, J. M., and A. H. McCain. 1960. Relations of spore moisture content to spore shape and germination reaction temperature (abs.) *Phytopathology* 50:85.

Ogawa, J. M., R. Sanborn, H. English, and E. E. Wilson. 1954. Late season protective and eradicative sprays as a means of controlling brown rot of peach fruit. *Plant Dis. Rep.* 38:869–73.

Ogawa, J. M., J. L. Sandeno, and J. H. Mathre. 1963. Comparisons in development and chemical control of decay-causing organisms on mechanical- and hand-harvested stone fruits. *Plant Dis. Rep.* 47:129–33.

Ogawa, J. M., and E. E. Wilson. 1960. Effects of the combinations of sodium pentachlorophenoxide and liquid lime-sulfur on the brown rot fungi (abs.) *Phytopathology* 50:649.

Ogawa, J. M., E. E. Wilson, and J. B. Corbin. 1967. Brown rot of cling peaches in California: Its history, life cycle and control. Department of Plant Pathology, University of California, Davis, 10 pp.

Ogawa, J. M., W. E. Yates, B. T. Manji, and R. E. Cowden. 1972. Ground and aircraft applications of thiophanate-methyl on control of stone fruit brown rot blossom blight (abs.) *Phytopathology* 62:781.

Phillips, D. J. 1982. Changes in conidia of *Monilinia fructicola* in response to incubation temperature. *Phytopathology* 72:1281–83.

———. 1984. Effect of temperature on *Monilinia fructicola* conidia produced on fresh stone fruit. *Plant Dis.* 68:610–12.

Phillips, D. J., and D. A. Margosan. 1985. Glucose concentration in growth media affects spore quality of *Monilinia fructicola* (abs.) *Phytopathology* 75:1285.

Pusey, P. L., and C. L. Wilson. 1984. Postharvest biological control of stone fruit brown rot by *Bacillus subtilis*. *Plant Dis.* 68:753–56.

Ramsdell, D. C., and J. M. Ogawa. 1973a. Reduction of *Monilinia laxa* inoculum potential in almond orchards resulting from dormant benomyl sprays. *Phytopathology* 63:830–36.

———. 1973b. Systemic activity of methyl 2-benzimidazole carbamate (MBC) in almond blossoms following pre-bloom sprays of benomyl MBC. *Phytopathology* 63:959–64.

Reese, R. L., and L. F. Moore, Jr. 1982. UV irradiation induced resistance to CGA 74251 in *Monilinia fructicola* (abs.) *Phytopathology* 72:980.

Ritchie, D. F. 1983. Mycelial growth, peach fruit-rotting capability, and sporulation of strains of *Monilinia fructicola* resistant to dichloran, iprodione, procymidone, and vinclozolin. *Phytopathology* 73:44–47.

———. 1985. Strategies for the management of fungicide resistance in *Monilinia fructicola*. In *Proceedings of the Brown Rot of Stone Fruit Workshop*, 16–19. N.Y. State Agric. Exp. Stn. Geneva Spec. Rept. 55.

Roberts, J. W., and J. C. Dunegan. 1932. Peach brown rot. *USDA Agric Bull.* 328, 59 pp.

Rudolph, B. A. 1925. Monilinia blossom blight (brown-rot) of apricots. *Calif. Univ. Agric. Exp. Stn. Bull.* 383.

Sanborn, R. R., B. T. Manji, and J. M. Ogawa. 1983. Suppression of sporodochia in *Monilinia laxa* blighted apricot twigs with thiophanate methyl plus oil (abs.) *Proc. 57th Ann. Western Orchard Pest Dis. Mgmt. Conf.*: 27–28.

Scorza, R., and L. Gilreath. 1983. Resistance to brown rot (*Monilinia fructicola* in sweet cherry (*Prunus avium* L.) (abs.) *Phytopathology* 73:969.

Scorza, R., and T. van der Swet. 1984. Evaluation of resistance to *Monilinia fructicola* in peach (abs.). *Phytopathology* 74:759.

Smith, M. A. 1936. Infection studies with *Sclerotinia fructicola* on brushed and nonbrushed peaches. *Phytopathology* 26:1056–60.

Smith, W. L., Jr., and R. D. Bassett. 1963. Hydrothermal and hygrothermal inactivation of *Monilinia fructicola* and *Rhizopus stolonifer* spores (abs.) *Phytopathology* 53:747.

Sonoda, R. M., J. M. Ogawa, T. E. Esser, and B. T. Manji. 1982. Mycelial interaction zones among single ascospore isolates of *Monilinia fructicola*. *Mycologia* 74(4):681–82.

Sonoda, R. M., J. M. Ogawa, and B. T. Manji. 1982. Use of interactions of cultures to distinguish *Monilinia laxa* from *Monilinia fructicola*. *Plant Dis.* 66:325–26.

Sonoda, R. M., J. M. Ogawa, and P. L. Sholberg. 1982. Competition between conidial isolates of benomyl resistant and benomyl sensitive *Monilinia fructicola* on peach fruit. (abs.). *Phytopathology* 72:988.

Sonoda, R. M., J. M. Ogawa, R. A. Vertrees, et al. 1967. Evaluations of the 1965 and 1966 brown-rot epidemics on cling peaches in California. Department of Plant Pathology, University of California, Davis. 25 pp.

Sonoda, R. M., J. M. Ogawa, B. T. Manji, E. Shabi, and D. Rough. 1983. Factors affecting control of blossom blight in a peach orchard with low level benomyl-resistant *Monilinia fructicola*. *Plant Dis.* 67:681–84.

Sutton, T. B., and C. N. Clayton. 1972. Role and survival of *Monilinia fructicola* in blighted peach branches. *Phytopathology* 62:1369–73.

Szkolnik, M., J. M. Ogawa, B. T. Manji, C. A. Frate, and E. A. Bose. 1978. Impact of benomyl treatments on populations of benomyl-tolerant *Monilinia fructicola* (abs.). *Phytopathol. News* 12:239.

Sztejnberg, A., and A. L. Jones. 1978. Tolerance of the brown rot fungus *Monilinia fructicola* to iprodione, vinclozolin and procymidone fungicides (abs.). *Phytopathol. News* 12:187–88.

Tate, K. G., and J. B. Corbin. 1978. Quiescent fruit infections of peach, apricot, and plum in New Zealand caused by the brown rot fungus, *Sclerotinia fructicola*. *New Zeal. J. Exp. Agri.* 6:319–25.

Tate, K. G., and J. M. Ogawa. 1975. Nitidulid beetles as vectors of *Monilinia fructicola* in California stone fruits. *Phytopathology* 65:977–83.

Tate, K. G., J. M. Ogawa, B. T. Manji, and E. Bose. 1974. Survey for benomyl-tolerant isolates of *Monilinia fructicola* and *Monilinia laxa* in stone fruit orchards of California. *Plant Dis. Rep.* 58:663–65.

Terui, M., and Y. Harada. 1966. Apothecial production of *Monilinia fructicola* on artificial media. *Trans. Mycol. Soc. Japan* 7:309–11.

Tilford, P. E. 1936. The relation of temperature to the effect of hydrogen and hydroxyl-ion concentration in *Sclerotinia fructicola* and *Fomes annosus* spore germination and growth. *Ohio Agric. Exp. Stn. Bull.* 567, 27 pp.

Tucker, D. H. Jr., and N. E. McGlohon. 1985. First report of Monilinia leaf and shoot blight of peach in Georgia. *Plant Dis.* 69:811.

Wade, G. C. 1956. Investigations on brown rot of apricots caused by *Sclerotinia fructicola* (Wint.) Rehm. I. The occurrence of latent infection in fruit. *Austral. J. Agric. Res.* 7:504–15.

Weaver, L. O. 1943. Effect of temperature and relative humidity on occurrence of blossom blight of stone fruit (abs.). *Phytopathology* 33:15.

Whan, J. H. 1976. Tolerance of *Sclerotinia fructicola* to benomyl. *Plant Dis. Rep.* 60:200–01.

Willetts, H. J., and Y. Harada. 1984. A review of apothecial production by *Monilinia* fungi in Japan. *Mycologia* 76(2):314–25.

Wilson, E. E. 1942. Experiments with arsenite sprays to eradicate *Sclerotinia laxa* in stone fruit trees as a means of controlling the brown rot disease in blossoms. *J. Agric. Res.* 64:561–94.

———. 1950. Sodium pentachlorophenate and other materials as eradicative fungicides against *Sclerotinia laxa*. *Phytopathology* 40:567–83.

———. 1953. Apricot and almond brown rot. In *Plant Diseases* 886–91. USDA Agricultural Yearbook 1953.

Wilson, E. E., and G. A. Baker. 1946. Some aspects of the aerial dissemination of spores, with special reference to conidia *Sclerotinia laxa*. *J. Agric. Res.* 72:301–27.

Winter, G. 1883. Uber einige Nordamerikanische Pilz. *Hedwigia* 22:26–72, 129–31.

Wormald, H. 1919. The brown rot disease of fruit with special reference to two biologic forms of *Monilinia cinerea*. I. *Ann. Bot.* 33:361–404.

———. 1920. The brown-rot disease of fruit trees with special reference to the two biologic forms of *Monilinia cinerea* II. *Ann. Bot.* 34:143–71.

———. 1921. On the occurrence in Britain of the ascigerous state of a brown rot fungus. *Ann. Bot.* 35:125–34.

———. 1928. Further studies of the brown rot fungi. III. Nomenclature of the American brown-rot fungi. A review of literature and critical remarks. *Trans. Mycol. Soc.* 13:194–204.

———. 1954. The brown rot diseases of fruit trees. Ministry Agric. Fisheries (Eng.) Tech. Bull. 3, 113 pp.

Yates, W. E., J. M. Ogawa, and N. B. Akesson. 1974. Spray distribution in peach orchards from helicopter and ground application. *Trans. ASAE* 17(4):633–39, 644.

Zwygart, T. 1970. Studies on host parasite interactions in Monilinia diseases of fruit trees. *Phytopathol. Z.* 68:97–130.

Ceratocystis (Mallet Wound) Canker of Almond and Prune

Ceratocystis fimbriata Ell. & Halst.

The disease of almond (*Prunus dulcis* [Mill.] Webb) often called "mallet wound canker" has been known in California for about 45 years. This canker develops on the areas of the branches struck by rubber-covered mallets during harvesting and on other areas damaged by mechanical harvesting or cultivation equipment. Extensive damage by the disease is reported on almond and prune (*P. domestica* L.), and occasionally on peach (*P. persica* [L.] Batsch.) and apricot (*P. armeniaca* L.) (DeVay et al. 1968). The cause of the disease was first reported in California in 1960 (DeVay et al.). Its increased importance was coincident with the mechanical harvesting of almond and stone-fruit crops and the resultant bark damage. In a survey in 1964, over 13,000 acres of bearing French prune trees in the Sacramento Valley were affected. In certain orchards, 100 percent of the trees were involved. On almond the disease is not as extensive, except on certain cultivars such as Mission and Nonpareil (DeVay et al. 1965). A "dry canker" of almond, not associated with harvest injuries but apparently caused by *Ceratocystis fimbriata*, has recently been reported in California by Teviotdale et al. (1989).

Symptoms. The cankers appear first as water-soaked and darkened areas which become sunken and amber colored. Gum usually forms at the margins of cankers. Infected tissues on prune trees are found to be red, (**fig. 5B**) while on almond, peach, and apricot the tissues are dark brown. The heartwood shows permeation of a brown black stain that extends longitudinally 50 cm or more past the margins of the canker in the bark. The cankers develop only in injured bark and expand throughout the year, but extend fastest during the summer months. They can eventually girdle and kill infected limbs or trunks (**fig. 5A**). A limb 4 to 6 inches in diameter can be girdled in three to four years. Infrequently, small secondary cankers develop under intact bark tissues above well-developed cankers, especially after periods of high summer temperatures (DeVay et al. 1968).

Causal organism. *Ceratocystis fimbriata* Ell. & Halst. causes the disease, although other *Ceratocystis*, *Graphium*, *Peptographium*, and *Chalaropsis* species have been isolated from freshly injured bark tissues. Carrot disks are a selective medium for *Ceratocystis fimbriata* (Parkinson 1964; Moller and DeVay 1968b). Halstead first described the fungus in 1890 on sweet potatoes and established the genus as *Ceratocystis*, although he mistook the perithecia for pycnidia. Elliott in 1923 discovered the perithecia with evanescent asci and designated the fungus *Ceratocystis fimbriata* (Ell. & Halst.) Elliott. The fungus was later transferred to *Ophiostoma* H. and P. Sydow by Melin and Nannfeldt (1934), and to *Endoconidiophora* Munch by Davidson in 1935. In 1950, Bakshi revived the generic name *Ceratocystis*. This position was maintained by Hunt (1956), who considered the other genera as synonyms. *Ceratostomella* was excluded because it contains only species with persistent asci.

Ceratocystis fimbriata is a diverse species consisting of numerous strains characterized by pathogenic specialization and variable asexual structures, but with morphological characteristics of the perithecia and ascospores quite uniform. The perithecial dimensions of nine isolates from different hosts ranged from 130 to 250 μm in diameter at the perithecial base, 250 to 900 μm in the neck length, and with a 10.8 μm neck width at the tip. The number of ostiolar hyphae ranged from 7 to 16, with a mean width of 2.2 μm (Webster and Butler 1967). The ascospores are hat-shaped, with a tapered gelatinous brim from 0.5 to 1.0 μm in width, and their dimensions are 2 by 6 μm (Webster and Butler 1967). Three types of asexual spore forms are produced. The dimensions of the cylindrical endoconidia are 2.4 by 17.4 μm; doliform endoconidia are 7.2 by 6.4 μm; the thick-walled conidia are 12.9 by 10.5 μm; mean conidiophore dimensions are 79.8 by 4.8 μm, 34.9 by 6.1 μm, and 33.8 by 4.9 μm, respectively.

On malt-extract agar, the color of the hyphae of stone-fruit isolates ranges from brown to dark olive, and perithecia are produced in clumps or in concentric rings in 6 to 12 days. Thiamine is required for perithecial production, vegetative growth being sparse in its absence. Optimum growth occurs between 24° and 27°C. The flask-shaped (tapered) endoconidiophores which give rise to cylindrical endoconidia are hyaline to subhyaline and septate, and may arise singly or in clusters from both aerial and subsurface hyphae; the truncate endoconidia are in chains containing up to 20 spores (Webster and Butler 1967).

The endoconidiophores which produce the doliform endoconidia are shorter and wider at the top. They often tend to aggregate, especially around perithecia. The spores are at first hyaline (clear) but later become subhyaline to light brown (Webster and Butler 1967).

The shapes of the thick-walled conidia vary from oval to subglobose, with smooth to rough walls. They are formed singly, or in short chains on simple or branched conidiophores, and are pale brown to olive brown. They germinate in 48 hours at 25°C (Webster and Butler 1967). The production of these spores in the vessels and their possible upward movement in the transpiration stream may explain the development of the small, secondary cankers sometimes found some distance above the initial wound canker (DeVay et al. 1968).

The various isolates of *C. fimbriata* are basically homothallic, but produce some self-sterile, cross-fertile female strains in their ascospore progeny (Webster and Butler 1967).

Disease development. The primary inoculum comes from the fungus sporulating on old diseased bark. The spores contaminate insects or are ingested and excreted by them. Vectors of particular importance are the nitidulid beetle, *Carpophilus freemani*, and a drosophilid fly, *Chymomyza procnemoides*, and to a lesser degree, *Carpophilus hemipterus*, *Litargus balteatus*, *Euzophera semifuneralis*, *Scolytus rugulosus*, *Drosophila melanogaster*, and a species of *Tarsonemus*. The pathogen could be detected on *C. fremani* up to eight days after it stopped feeding on the fungus. Pupation of contaminated larvae resulted in emerging adults still retaining the fungus. The adult insects in old bark wounds are contaminated by the fungus even in the winter months (Moller and DeVay 1968a).

On Mission almond, injuries are susceptible to infection for about 10 days but are resistant after 14 days. They are most susceptible in orchards with recent irrigation. Once infection is established, mature perithecia have been found in bark tissue four to five days after the initial bark injury. Infection usually does not occur from December through April, at which time established cankers expand very slowly (DeVay et al. 1965, 1968; Moller, DeVay, and Backman 1969). However, in summer when temperatures are higher, the cankers may enlarge at the rate of 2 to 3 inches a month. The cankers are perennial and can continue their activity year after year. Among stone fruit, almond trees appear to be the most susceptible, with apricots not far behind. Of the almond cultivars that have been studied, Mission (Texas) is most susceptible, followed by Nonpareil, Ne Plus Ultra, and Peerless. Drake appears to be highly resistant. In *Prunus domestica*, French and Imperial prunes show greatest susceptibility, with Robe de Sargent and Beauty (*P. salicina*) being resistant. Next in order of susceptibility are peach trees, where several clingstone cultivars have shown a moderate amount of infection. Artificially inoculated Bing cherry trees showed no evidence of infection, nor has the disease been observed in sweet cherry orchards. Walnut trees were moderately susceptible, but most cankers became inactive in one or two years. Apple and pear proved highly resistant (DeVay et al. 1968).

The fungus initially colonizes injured bark and exposed cambium, then invades uninjured bark tissues (except cork) and young xylem. A dark stain permeates the heartwood, but the pathogen is not present in this tissue. The fungus usually can be isolated only from the margins of the cankers (DeVay et al. 1968).

Control. The best control is to avoid bruise-type injuries. Since most of these injuries occur during mechanical harvesting, shakers that cause minimum damage should be used (**fig. 5C**). Irrigation should be avoided during the three weeks preceding harvest because it increases insect vector activity and tends to increase the susceptibility of the bark to bruise injury.

Limbs severely cankered or killed by the disease should be removed by cutting at least 6 inches below the canker margin. Small to medium-sized cankers can be excised (preferably in winter when the insect vectors are absent) by cutting away the face of the canker and a layer of the underlying wood ¼ to ½ inch thick. The diamond-shaped cut should extend at least 1 inch beyond the visible bark-canker margin. Such an eradication procedure, although difficult and expensive, is reasonably effective when properly done. The cleaned-up cankers should be examined a year later to determine the thoroughness of the eradication procedure. (DeVay et al. 1965, 1968; Moller, DeVay, and Backman 1969).

REFERENCES

Bakshi, B. K. 1950. Fungi associated with ambrosia beetles in Great Britain. *Brit. Mycol. Soc. Trans.* 33:111–20.

Davidson, R. W. 1935. Fungi causing stain in logs and lumber in the southern states, including five new species. *J. Agric. Res.* 50:789–807.

DeVay, J. E., H. English, F. L. Lukezic, and H. J. O'Reilly. 1960. Mallet wound canker of almond trees. *Calif. Agric.* 14(8):8–9.

DeVay, J. E., F. L. Lukezic, W. H. English, W. J. Moller, and B. W. Parkinson. 1965. Cotrolling Ceratocystis canker of stone fruit trees. *Calif. Agric.* 19(10):2–4.

DeVay, J. E., F. L. Lukezic, H. English, E. E. Trujillo, and W. J. Moller. 1968. Ceratocystis canker of deciduous fruit trees. *Phytopathology* 58:949–54.

Elliott, J. A. 1923. The ascigerous stage of the sweet potato black rot fungus (abs.). *Phytopathology* 13:56.

Halstead, B. D. 1890. Some fungus diseases of the sweet potato: The black rot. *New Jersey Agric. Exp. Stn. Bull.* 876:7–14.

Hunt, J. 1956. Taxonomy of the genus *Ceratocystis*. *Lloydia* 19:1–58.

Melin, E., and J. A. Nannfeldt. 1934. Researches into the blueing of ground wood pulp. *Svenska Skogsvardsforen. Tidskr.* 32:397–616.

Moller, W. J., and J. E. DeVay. 1968a. Insect transmission of *Ceratocystis fimbriata* in deciduous fruit orchards. *Phytopathology* 58:1499–1508.

———. 1968b. Carrot as a species-selective isolation medium for *Ceratocystis fimbriata*. *Phytopathology* 58:123–24.

Moller, W. J., J. E. DeVay, and P. A. Backman. 1969. Effect of some ecological factors on Ceratocystis canker on stone fruits. *Phytopathology* 59:938–42.

Parkinson, B. W. 1964. *Studies on the etiology of Ceratocystis canker of stone fruits.* M.S. Thesis, University of California, Davis, 32 pp.

Teviotdale, B. L., D. H. Harper, M. Viveros, M. Freeman, and J. Connell. 1989. Dry canker of almond. *Calif. Plant Pathol.* No. 84, June, 1989. 3 pp.

Webster, R. K., and E. E. Butler. 1967. A morphological and biological concept of the species *Ceratocystis fimbriata. Can. J. Bot.* 45:1457–68.

Cercospora Leaf Spot of Almond and Stone Fruit

Cercospora circumscissa Sacc.

Cercospora leaf spot was first noted in Southern California by N. B. Pierce in 1891 (Pierce 1891) on wild black cherry trees (*Prunus serotina*). In a nearby almond orchard he observed serious defoliation caused by the disease and also found it on apricot, peach, and nectarine trees with symptoms similar to those of the shothole disease caused by *Wilsonomyces carpophilus* (*Stigmina carpophila*). In 1941 Smith (1941) noted that the disease "is not common in California and requires no special treatment." In the United States, it is widespread on peach and is found to some extent on several other stone fruits; in Oregon it has been reported on almond (Farr et al. 1989).

Symptoms. On almond, defoliation caused by *Cercospora* leaf infection occurs in June and July with a possible decrease in fruitfulness. Leaf symptoms appear as yellowish brown spots that fall out and result in a shothole condition, which is followed by leaf fall. Infections also occur on the fruit hulls and twigs. Sporulation of the pathogen is abundant on both leaf and twig lesions. On peach the disease is expressed on the fruit as shallow, black, circular, depressed spots less than 1/8 inch in diameter; leaf infections also occur but show little or no fungal sporulation. Twigs of peaches are rarely infected. Leaves of both prune and nectarine grafted on almond stock are susceptible to the disease.

Cause. The causal fungus is *Cercospora circumscissa* Sacc. The straight or variously curved conidia are one- to seven-celled (mostly, two- to five-celled). The distal one-fourth to one-half of the conidium is usually reduced in diameter and the cells are longer than those of the proximal portion. The width of the distal end is 3 to 4 µm with the greatest breadth taken toward the base varying between 4 and 6 µm; the length ranges between 20 and 64 µm (rarely up to 106 µm) with an average of 40.6 µm.

The sporodochial size is up to 40 µm in diameter and 14 to 43 µm in height, with 20 to 50 conidiophores in one fascicle. The conidiophore fascicle pushes through the epidermis, or in some instances through the stomata.

Disease development. In California, no published information has been found on epidemiology of the disease, nor is there evidence that tests proving the pathogenicity of the suspected causal fungus have been conducted.

Control. Application of a protectant copper compound (ammoniacal solution of copper carbonate prepared by mixing 5 ounces copper carbonate and 3 pints aqua ammonia [26 degrees baumé] in 45 gal water) as soon as leaves appear followed by two to five cover sprays has been suggested (Galloway 1891), but no data have been published on this method's performance. This formulation was reported to be less phytotoxic to peach and almond than Bordeaux mixture.

REFERENCES

Farr, D. F., G. F. Bills, G. P. Chamuris, and A. Y. Rossman. 1989. *Fungi on plants and plant products in the United States.* APS Press, St. Paul, Minn. 1252 pp.

Galloway, B. T. 1891. Suggestions in regard to the treatment of *Cercospora circumscissa. J. Mycol.* 7(2):77–78.

Pierce, N. B. 1891. A disease of almond trees. *J. Mycol.* 7(2):66–77.

Smith, R. E. 1941. Diseases of fruits and nuts. *Calif. Agric. Exp. Serv. Circ.* 120. 168 pp.

Cytospora and Rhodosticta Cankers of Stone Fruit

Cytospora leucostoma (Pers.) Sacc.
Rhodosticta quercina Carter

The disease commonly known as "Cytospora canker" has also been called "perennial canker" and "peach canker." It occurs throughout the United States and southeastern Canada (Gairola and Powell 1970; Helton and Konicek 1961; Hildebrand 1947; James and Davidson 1971; Jones and Leupschen 1971; Jones and Sutton 1984; McCubbin 1918; Willison 1933) and has been reported in many other countries (Latorre 1988; Schulz 1981; Togashi 1931; Wormald 1912).

Cytospora canker has affected peach in the United States for many years. It was found in New York as early as 1900 by Stewart, Rolfs, and Hall. It also occurs on plum, prune, cherry, and apricot (Helton 1961; Lukezic, DeVay, and English 1960, 1965). It has been found on apple in New Mexico (Leonian 1921), the Pacific Northwest (Fisher and Reeves 1931), Illinois (Stevens 1919), and Michigan (Proffer and Jones 1989).

Cytospora canker occurs throughout the Central Valley of California on *Prunus domestica* plums such as President, French, and Imperial prune. It does not affect cultivars of *Prunus salicina*, nor is it a serious problem on peaches or other stone fruit in California (Lukezic, DeVay, and English 1960, 1965).

Symptoms. The most obvious symptom on prune and plum trees is the dying of branches that frequently occurs during mid- to late summer (**fig. 6A**). Upon close examination, a dying branch reveals a dark, depressed, girdling canker that is usually centered at a sunburned area. During the growing season an exudation of amber-colored gum often occurs near the canker margin. Removal of the outer bark reveals the zonate margin of active cankers. Eventually, small, olive-gray pycnidia develop just beneath the periderm (**fig. 6B**); as they enlarge they give the bark a pimpled appearance. When showers (especially in the spring) are followed by humid, but not wet, conditions amber-colored cirrhi (spore tendrils) protrude from the pycnidia (**fig. 6D**). These tendrils consist of a mass of conidia embedded in a hydrophillic matrix. Under prolonged rainy conditions, cirrhi do not form; the spore masses merely exude and are dispersed by the surface moisture. The erumpent pycnidial stromata or the succeeding ascostromata form a white apical disk that is quite evident to the naked eye and is diagnostic for Cytospora canker (**fig. 6C**). In President plum the cankers are perennial, whereas in French prune they appear to enlarge year after year only in trees that are physiologically stressed (Bertrand and English 1976a; Bertrand et al. 1976; Bertrand, English, and Carlson 1976; Lukezic, DeVay, and English 1960).

Causal organisms. Earlier investigators of the disease identified the causal organisms as *Cytospora leucostoma* Sacc. (*Valsa leucostoma* Fr.), *Cytospora cincta* Sacc., *Valsa persoonii* Nit., and *Cytospora rubescens*. Though Hildebrand (1947) considered C. cincta more aggressive than C. leucostoma on peaches in New York, Treshow and Scholes (1958) and Treshow, Richards, and Scholes (1959) in Utah identified the principal species on peach as C. rubescens; Lukezic, DeVay, and English (1960, 1965) identified two species on plums in California as C. rubescens and C. leucostoma. It is now generally accepted that only C. leucostoma is important as a pathogen of plum and prune in California (Bertrand and English 1976a; DeVay et al. 1974; Lukezic, DeVay, and English 1965).

Kern (1955) split the genus *Valsa*, placing *V. leucostoma* in the genus *Leucostoma* as *L. persoonii* (Nits.) Hoehn on the basis that the ascostroma is surrounded by a black conceptacle. The asci are clavate, and the eight ascospores are 8 to 14 by 1.5 to 3 µm, hyaline and allantoid. The pycnidial stroma is olive-brown, multilocular, convoluted, ostiolate, erumpent, and located just beneath the periderm through which it becomes erumpent. Hyaline, septate, branched conidiophores, which terminate with phialidic conidiogenous cells, line the walls of the locules. The conidia are hyaline, allantoid, 4 to 6.5 by 0.5 to 1.5 µm, and are extruded from the stroma as amber-colored tendrils or globular masses. Evidence obtained by Leonian (1923) indicates that at least some strains of the fungus are homothallic.

Kern (1955) placed *Cytospora cincta* in the genus *Leucostoma* but did not mention C. rubescens. According to Willison (1936) the ascostroma of C. cincta has a loose texture and is delimited from the host cells by a thin black zone. The ascospores are 14 to 18 by 4 to 7 µm. The pycnidiospores measure 5 to 10 by 1 to 2 µm. The perithecial stromata are intermingled with the pycnidia.

Another organism associated with severe cankers on President plum in California is *Rhodosticta quercina* Carter (Lukezic, DeVay, and English 1965). Colonies of this fungus on potato-dextrose agar grow slowly and are characterized by an entire margin, orange pycnidia, and a zonate color of dark orange in the center, white at the margin, and an orange pigment diffusing through the medium.

Mycelial growth in culture of L. persoonii is stimulated by the combination of thiamine and biotin. The response to myo-inositol is variable. Choline is also stimulatory to most isolates. Mycelial growth in culture decreased as the pH increased from 4 to 8. *Rhodosticta quercina* was not able to grow in media prepared with deionized water supplemented with myo-inositol, but on basic agar medium or PDA supplemented with myo-inositol growth was stimulated (Lukezic, DeVay, and English 1965). Myo-inositol was utilized as a carbon source by *L. persoonii* but not by *R. quercina* (Lukezic, DeVay, and English 1965).

The approximate minimum, optimum, and maximum temperatures for growth in vitro of different isolates of *C. leucostoma* are 3° to 6°, 30° to 33°, and 39°C, respectively, while for *C. cincta* they are 0° to 3°, 21° to 24°, and 30° to 36°C (Bertrand 1974; Hildebrand 1947). In a study of environmental and nutritional factors affecting pycnidiospore germination, Rohrback and Luepschen (1968) found that maximum germination occurred in 24 hours at 27°C, with an optimum pH of 5.2. A carbon source was necessary for germ tube production, and optimum germination occurred with maltose, mannitol, sucrose, sorbitol, and natural peach gum.

Disease development. Cytospora canker can be confused with bacterial canker caused by *Pseudomonas syringae*. However, the irregular, nonzonate margin of bacterial cankers is generally in sharp contrast to the U-shaped, zonate margin of active Cytospora cankers. The latter cankers also eventually develop characteristic fruiting bodies. A further difference is that bacterial cankers are principally active during the dormant season, whereas Cytospora cankers, at least in California, are usually most active during the growing season (Bertrand and English 1976a; Lukezic, DeVay, and English 1960). But *Cytospora leucostoma* is a frequent secondary invader of bark tissues killed by *P. syringae*; from this saprophytic base it sometimes further damages the tree by invading adjacent healthy bark. The secondary invasion of bacterial cankers in peach by *C. cincta* has been reported in the southeastern United States (Endert-Kirkpatrick and Ritchie 1988).

The availability of inoculum and infection sites and the predisposition of the tree are all highly important in disease development by *Cytospora*. Inoculum has been found on dead or cankered branches throughout the year (Bertrand and English 1976b; Leupschen and Rohrbach 1969). Conidia of *C. leucostoma* are largely dispersed by windblown rain, whereas ascospores are released and dispersed both during and shortly after a rainy period. The pycnidia form during the first year after infection, while the ascostromata do not mature until two or three years later (Bertrand and English 1976b). With each rain as many as 61 billion conidia are released from the pycnidia on a single tree, whereas the maximum ascospore release is only 13 million (Bertrand and English 1976b).

In many regions, common infection sites on peach and plum are winter injuries, pruning wounds, and other mechanical injuries (Helton 1961; Hildebrand 1947; Leupschen, Hetherington, and Stahl 1979; Leupschen and Rohrbach 1969). Infections of blossom and fruit peduncles, dead buds, leaf scars, and insect injuries also have been reported (DeVay et al. 1974; Dhanvantari 1978; Leupschen, Hetherington, and Stahl 1979; Lukezic, DeVay, and English 1960; Rolfs 1910; Royse and Ries 1978b; Willison 1937). In the peach orchards of southern Ontario, Canada, the principal canker pathogen is *C. cincta* (*Leucostoma cincta*), and infections reportedly occur mostly at twig nodes (TeKauz and Patrick 1974). In California, pruning cuts and leaf scars are not important infection courts, but sunburned areas and bark tissues killed by other pathogens, especially *Pseudomonas syringae*, are major infection sites (Bertrand 1974; DeVay et al. 1974). Also contributing to infection and canker development are scale insects (especially San Jose scale, *Aspidiotus perniciosus*), and the shothole borer (*Scolytus rugulosus*) (Chiarappa 1960; DeVay et al. 1974; Lukezic, DeVay, and English 1965). The latter insect may also act as a vector and carry *Cytospora* to trees under stress, which are attractive to these borers (DeVay et al. 1974).

Infection is not necessarily followed by canker development. Sunburned areas may serve as infection sites on prune but do not necessarily predispose trees to canker development. On French prunes, *C. leucostoma* cankers develop rapidly on weakened trees during summer, but not on healthy, vigorously growing trees (Bertrand 1974; Bertrand and English 1976a). The inhibition of canker development in vigorous trees appears to be associated with the wound healing process. The weakening of limbs above a large canker (as expressed by wilting and defoliation) is attributable to the plugging of the xylem vessels by gum (Banko and Helton 1974; Hampson and Sinclair 1973). Tsakadze (1959) reported that *C. leucostoma* produces a toxin in vitro that causes xylem plugging, necrosis of xylem and bark, wilting, and finally death of *Coleus* shoots. Trees subjected to moisture stress or freezing damage, as well as trees deficient in potassium or those grown on a heavy clay soil, are more susceptible to canker development (Bertrand et al. 1976; Bertrand, English, and Carlson 1976; Dhanvantari 1978; Helton 1961, 1962; Hildebrand 1947; Kable, Fliegel, and Parker 1967; Rolfs 1910). More recently it has been shown that the ring nematode, *Criconemella xenoplax*, but not the pin nematode, *Paratylenchus neoamblycephalus*, predisposes French prune to Cytospora canker (English et al. 1982).

Investigators in Idaho reported induced resistance below the infection point when multiple inoculations of *Cytospora cincta* were made on branches of prune and peach (Braun and Helton 1971; Helton and Braun 1970; Hubert and Helton 1967). Systemic resistance was also associated with localized Prunus ringspot virus infections (Helton and Hubert 1968). No explanation of the mechanism of resistance is available. Recent research (Biggs 1989; Biggs and Miles 1988) with peach in Ontario has shown that the susceptibility of wounds to infection decreases with time and that resistance is correlated with the accumulation of suberin in the infection-court tissues. The susceptibility of host tissues to *Cytospora* has been related to warm temperatures followed by freezing damage (Helton 1961; Kable, Fliegel, and Parker 1967; Rolfs 1910). Canker size also can vary with infection by different isolates of the fungus. A range in mean canker lengths from 2.0 to 7.0 cm during a four-month period following inoculation of French prune with *Cytospora* spp. has been reported (Bertrand 1974; Bertrand and English 1976a). Lukezic and DeVay (1965) found a wide range of virulence among monoascospore isolates of *Leucostoma persoonii* (*C. leucostoma*) inoculated into President and Duarte plum trees; and Wysong and Dickens (1962) reported variation in virulence among peach isolates of the fungus.

Rhodosticta quercina cankers are similar to those caused by *C. leucostoma*, except that infected tissues are reddish orange, and the erumpent stromata emerging from the bark are light red to dark orange, with an irregular outer surface, not a crustose one. No layer of callus is formed by infected trees, as is sometimes produced around cankers caused by *C. leucostoma*. Both young and mature trees of President plum are highly susceptible to *R. quercina*, a fungus erroneously identified as a strain of *Cytospora* (Chiarappa 1960; Lukezic, DeVay, and English 1965). On the other hand, Beauty and Duarte cultivars (*Prunus salicina*) are resistant to both *R. quercina* and *C. leucostoma* (Lukezic and DeVay 1964; Lukezic, DeVay, and English 1960). The mechanism of resistance of *P. salicina* to *R. quercina* is not clearly understood, although myo-inositol has been implicated (Lukezic and DeVay 1964). There appear to have been no recent outbreaks of *R. quercina*-induced cankers in California plum orchards.

Control. The apparent ineffectiveness of fungicidal sprays in controlling Cytospora canker of President plum has led to the development of other control strategies for use in California's plum and prune orchards (Bertrand, English, and Carlson 1976; DeVay et al. 1974). In President plum, whitewashing the trees in the spring reduced sunburn injury and subsequent canker development (DeVay et al. 1974). Still more effective was a modified method of pruning that reduced sunburning and *Cytospora* infection. It is not known whether these strategies would be effective also with French prune. For the latter cultivar, control of Cytospora canker is based largely on avoiding the factors that predispose the trees to this disease (Bertrand and English 1976a; Bertrand et al. 1976; Bertrand, English, and Carlson 1976). Insofar as possible, planting in heavy clay or shallow soils should be avoided. Trees should not be stressed for moisture, and steps should be taken to correct any evident potassium deficiency. Badly cankered branches should be removed by cutting several inches below the canker margin. To reduce the likelihood of sunburning and subsequent infection, heavy pruning cuts preferably should be made just above a vigorous lateral that is directed toward the south or southwest. Also, since there is evidence that the ring nematode, *Criconemella xenoplax*, predisposes prune trees to Cytospora canker, control of this common root-feeding pathogen should be considered (Bertrand 1974; English et al. 1982).

Reports from other parts of North America indicate that fall and winter spray applications of captafol and benomyl aid in the control of Cytospora canker of peach (Luepschen 1976; Northover 1976; Royse and Ries 1978b). Whether these materials could be effectively used to control Cytospora canker of California prunes and plums remains to be determined. Delayed pruning and the use of shellac on pruning cuts are recommended measures for reducing the severity of this disease on peach in Colorado (Luepschen and Rohrback 1969). No major variations in resistance have been found in peach cultivars (Luepschen et al.

1970; Scorza and Pusey 1984). The use of virus-induced resistance has not been confirmed nor is it recommended at this time (Helton and Hubert 1968). Two recent reports (Royse and Ries 1978a; Schulz 1981) suggest the possibility of biological control of Cytospora canker of stone-fruit trees.

REFERENCES

Banko, T. J., and A. W. Helton. 1974. Cytospora-induced changes in stems of *Prunus persica*. *Phytopathology* 64:899–901.

Bertrand, P. F. 1974. *Cytospora canker of French prune*. Ph.D. dissertation, Department of Plant Pathology, University of California, Davis, 113 pp.

Bertrand, P. F., and H. English. 1976a. Virulence and seasonal activity of *Cytospora leucostoma* and *C. cincta* in French prune trees in California. *Plant Dis. Rep.* 60:106–10.

———. 1976b. Release and dispersal of conidia and ascospores of *Valsa leucostoma*. *Phytopathology* 66:987–91.

Bertrand, P. F., H. English, and R. M. Carlson. 1976. Relation of soil physical and fertility properties to the occurrence of Cytospora canker in French prune orchards. *Phytopathology* 66:1321–24.

Bertrand, P. F., H. English, K. Uriu, and F. J. Schick. 1976. Late season water deficits and development of Cytospora canker in French prune. *Phytopathology* 66:1318–20.

Biggs, A. R. 1989. Integrated approach to controlling Leucostoma canker of peach in Ontario. *Plant Dis.* 73:869–74.

———. 1989. Temporal changes in the infection court after wounding of peach bark and their association with cultivar variation in infection by *Leucostoma persoonii*. *Phytopathology* 79:627–30.

Biggs, A. R., and N. W. Miles. 1988. Association of suberin formation in uninoculated wounds with susceptibility to *Leucostoma cincta* and *L. persoonii* in various peach cultivars. *Phytopathology* 78:1070–74.

Braun, J. W., and A. W. Helton. 1971. Induced resistance to Cytospora in *Prunus persica*. *Phytopathology* 61:685–87.

Chiarappa, L. 1960. Distribution and mode of spread of Cytospora canker in an orchard of the President plum variety in California. *Plant Dis. Rep.* 44:612–16.

DeVay, J. E., M. Gerts, H. English, and F. L. Lukezic. 1974. Controlling Cytospora canker in President plum orchards of California. *Calif. Agric.* 28(12):12–14.

Dhanvantari, B. N. 1978. Cold predisposition of dormant peach twigs to nodal cankers caused by *Leucostoma* spp. *Phytopathology* 68:1779–83.

Endert-Kirkpatrick, E., and D. F. Ritchie. 1988. Involvement of pH in the competition between *Cytospora cincta* and *Pseudomonas syringae* pv. *syringae*. *Phytopathology* 78:619–24.

English, H., B. F. Lownsbery, F. J. Schick, and T. Burlando. 1982. Effect of ring and pin nematodes on the development of bacterial canker and Cytospora canker in young French prune trees. *Plant Dis.* 66:114–16.

Fisher, D. F., and E. L. Reeves. 1931. A Cytospora canker of apple trees. *J. Agric. Res.* 43:431–38.

Gairola, G., and D. Powell. 1970. Cytospora peach canker in Illinois. *Plant Dis. Rep.* 54:832–35.

Hampson, M. C., and W. A. Sinclair. 1973. Xylem dysfunction in peach caused by *Cytospora leucostoma*. *Phytopathology* 63:676–81.

Helton, A. W. 1961. Low temperature injury as a contributing factor in *Cytospora* invasion of plum trees. *Plant Dis. Rep.* 45:591–97.

———. 1962. Effect of simulated freeze-cracking on invasion of dry-ice-injured stems of Stanley prune trees by naturally disseminated *Cytospora* inoculum. *Plant Dis. Rep.* 46:45–47.

Helton, A. W., and J. W. Braun. 1970. Relationship of number of *Cytospora* infections on *Prunus domestica* to rate of expansion of individual cankers. *Phytopathology* 60:1700–01.

Helton, A. W., and J. J. Hubert. 1968. Inducing systemic resistance to *Cytospora* invasion in *Prunus domestica* with localized Prunus ringspot virus infections. *Phytopathology* 58:1423–24.

Helton, A. W., and D. E. Konicek. 1961. Effect of selected *Cytospora* isolates from stone fruits on certain stone fruit varieties. *Phytopathology* 51:152–57.

Hildebrand, E. M. 1947. Perennial peach canker and the canker complex in New York, with methods of control. *Cornell Univ. Agric. Exp. Stn. Mem.* 276:3–61.

Hubert, J. J., and A. W. Helton. 1967. A translocated resistance phenomenon in *Prunus domestica* induced by initial infection with *Cytospora cincta*. *Phytopathology* 57:1094–98.

James, W. C., and T. R. Davidson. 1971. Survey of peach canker in the Niagara peninsula during 1969 and 1970. *Can. Plant Dis. Surv.* 51(4):148–53.

Jones, A. C., and N. S. Leupschen. 1971. Seasonal development of Cytospora canker in peach in Colorado. *Plant Dis. Rep.* 55:314–17.

Jones, A. L., and T. B. Sutton. 1984. Diseases of tree fruits. *N. Central Reg. Ext. Publ.* 45, Coop. Ext. Serv., Mich. State Univ. 59 pp.

Kable, P. F., P. Fliegel, and K. G. Parker. 1967. Cytospora canker of sweet cherry in New York state: Association with winter injury and pathogenicity to other species. *Plant Dis. Rep.* 51:155–57.

Kern, H. 1955. Taxonomic studies in the genus *Leucostoma*. *Michigan Acad. Sci. Arts Lett.* 40:9–22.

Latorre, G. B. 1988. Enfermedades de las plantas cultivadas. Ediciones universidad Catolica de Chile, Casilla 114-D Santiago, Chile. 307 pp.

Leonian, L. H. 1921. Studies on the Valsa apple canker in New Mexico. *Phytopathology* 11:236–43.

———. 1923. The physiology of perithecial and pycnidial formation in *Valsa leucostoma*. *Phytopathology* 13:257–72.

Luepschen, N. S. 1976. Use of benomyl sprays for suppressing Cytospora canker on artificially inoculated peach trees. *Plant Dis. Rep.* 60:477–79.

Luepschen, N. S., J. E. Hetherington, and F. J. Stahl. 1979. Cytospora canker of peach trees in Colorado: Survey of incidence, canker location and apparent infection courts. *Plant Dis. Rep.* 63:685–87.

Luepschen, N. S., A. C. Jones, K. G. Rohrbach, and L. E. Dickens. 1970. Peach varietal susceptibility to Cytospora canker. *Colorado State Univ. Exp. Stn.* PR 70–5, 2 pp.

Luepschen, N. S., and K. G. Rohrbach. 1969. Cytospora canker of peach trees: Spore availability and wound susceptibility. *Plant Dis. Rep.* 53:869–72.

Lukezic, F. L., and J. E. DeVay. 1964. Effect of myo-inositol in host tissues on the parasitism of *Prunus domestica* var. President by *Rhodosticta quercina*. *Phytopathology* 54:697–700.

———. 1965. Serological relationships between pathogenic and nonpathogenic isolates of *Leucostoma persoonii* and *Rhodosticta quercina*. *Mycologia* 57(3):442–47.

Lukezic, F. L., J. E. DeVay, and H. English. 1960. Occurrence of Cytospora canker in stone fruit trees in California (abs.). *Phytopathology* 50:84.

———. 1965. Comparative physiology and pathogenicity of *Leucostoma persoonii* and *Rhodosticta quercina*. *Phytopathology* 55:511–18.

McCubbin, W. A. 1918. Peach canker. *Can. Dept. Agric. Bull.* 37, 20 pp.

Northover, J. 1976. Protection of peach shoots against species of *Leucostoma* with benomyl and captafol. *Phytopathology* 66:1125–28.

Proffer, T. J., and A. L. Jones. 1989. A new canker disease of apple caused by *Leucostoma cincta* and other fungi associated with cankers on apple in Michigan. *Plant Dis.* 73:508–14.

Rohrbach, K. G., and N. S. Luepschen. 1968. Environmental and nutritional factors affecting pycnospore germination of *Cytospora leucostoma*. *Phytopathology* 58:1134–36.

Rolfs, F. M. 1910. Winter killing of twigs, cankers, and sun scald of peach trees. *Missouri State Exp. Stn. Bull.* 17, 101 pp.

Royse, D. J., and S. M. Ries. 1978a. The influence of fungi isolated from peach twigs on the pathogenicity of *Cytospora cincta*. *Phytopathology* 68:603–07.

———. 1978b. Detection of *Cytospora* species in twig elements of peach and its relation to the incidence of perennial canker. *Phytopathology* 68:663–67.

Schulz, U. 1981. Untersuchungen zür biologischen Bekämpfung von Cytospora-Arten. *Z. Planzenkrank. Pflanzenschutz* 88:132–41.

Scorza, R., and P. L. Pusey. 1984. A wound-freezing inoculation technique for evaluating resistance to *Cytospora leucostoma* in young peach tree. *Phytopathology* 74:569–72.

Stevens, F. L. 1919. An apple canker due to *Cytospora*. *Illinois Agric. Exp. Stn. Bull.* 217:367–79.

Stewart, F. C., F. M. Rolfs, and F. H. Hall. 1900. A fruit disease survey of western New York in 1900. *New York Agric. Exp. Stn. Bull.* 191.

TeKauz, A., and Z. A. Patrick. 1974. The role of twig infections on the incidence of perennial canker of peach. *Phytopathology* 64:683–88.

Togashi, K. 1931. Studies on the pathology of peach canker. *Imp. Coll. Agric. For. Morioka Bull.* 16, 178 pp.

Treshow, M., B. L. Richards, and J. F. Scholes. 1959. The relation of temperature to pathogenicity of *Cytospora rubescens* (abs.) *Phytopathology* 49:114.

Treshow, M., and J. F. Scholes. 1958. The taxonomy of some species of *Cytospora* found in Utah. *Utah Acad. Proc.* 35:49–51.

Tsakadze, T. A. 1959. The effect of *Cytospora leucostoma* toxin on the plant cell. *Bull. Cent. Bot. Gdn. Moscow* 35:75–77.

Wensley, R. N. 1964. Occurrence and pathogenicity of *Valsa* (*Cytospora*) species and other fungi associated with peach canker in southern Ontario. *Can. J. Bot.* 42:841–57.

Willison, R. S. 1933. Peach canker investigation. I. Some notes on incidence, contributing factors and control measures. *Sci. Agric.* 14:32–46.

———. 1936. Peach canker investigation. II. Infection studies. *J. Can. Res.* 14:27–44.

———. 1937. Peach canker investigations. III. Further notes on incidence, contributing factors, and related phenomena. *J. Can. Res.* 15:324–39.

Wormald, H. 1912. The Cytospora disease of cherry. *J. Southeastern Agric. Coll.* (Wye, Kent) 12:367–80.

Wysong, D. S., and L. S. Dickens. 1962. Variation in virulence of *Valsa leucostoma*. *Plant Dis. Rep.* 46:274–76.

Dothiorella Canker of Almond

Botryosphaeria dothidea
(Moug.: Fr.) Ces. and de Not.

A canker of the trunk and scaffold branches of almond (*Prunus dulcis*) was reported by English, Davis, and DeVay in 1966. This disease, sometimes called "band canker," was first noted in 1959 as occurring in Tehama and Stanislaus counties of California. It was found again in San Joaquin County in 1960, in Merced County in 1961, in Yolo County in 1963, and more recently in Solano County. Its occurrence in recent years has been sporadic (English, Davis, and DeVay 1975).

Symptoms. The disease occurs most frequently in vigorous Nonpareil trees 4–6 years of age. Narrow, bandlike, or irregular cankers may extend around the trunk or scaffold branches (**fig. 7B,C**). They differ from other cankers in that their greatest dimension is transverse to the long axis of the branch or trunk. The cankers appear to originate in growth cracks, usually become noticeable in summer, and are accompanied by copious gum formation.

The bark of the branch, including the cambium, is killed; these areas become noticeably sunken because of subsequent desiccation. Where a girdling canker extends to the wood, the portion of the branch or trunk above the canker dies. Although complete girdling of the trunk is relatively uncommon, half or more of its circumference may be involved. Frequently, scaffold branches are killed (**fig. 7A**). Under some conditions, minute, white spore tendrils exude from pycnidia immersed in the outer bark; they are evident on the canker surface. The sapwood beneath the killed bark, often extending longitudinally several centimeters beyond the canker margin, is discolored. In some instances, especially in very narrow cankers, the cambium is not destroyed and newly formed phloem tends to lift off the outer necrotic tissue (English, Davis, and DeVay 1975).

The cankers are active during the growing season in which they first appear, but they usually do not reactivate in subsequent seasons. Because of this feature and the fact that the disease occurs infrequently, this disorder is of relatively minor importance.

Cause. One or both of two fungi may be associated with the development of this canker disease. These are the *Dothiorella* stage of *Botryosphaeria dothidea* (Moug.: Fr.) Ces. and de Not. and *Hendersonula toruloidea* Nattrass. The latter is the cause of a branch wilt of English walnut in California and is also pathogenic to peach, apricot, and fig trees in California (Wilson 1947). It has been associated with dieback symptoms in plum, apricot, and apple trees in Egypt (Nattrass 1933).

Isolations from almond band cankers occurring in different parts of the Sacramento and San Joaquin valleys indicated that *B. dothidea* was much more common than *H. toruloidea*. Moreover, *B. dothidea* was the only fungus found at the margins of "apparently active" cankers. English, Davis, and DeVay (1975) therefore concluded this fungus was "largely responsible for this canker condition, and that *H. toruloidea* was a common secondary invader." Nevertheless, either fungus was "able to induce canker formation when mycelial inoculum was placed in cortical wounds on the cambium, or on xylem exposed by pruning." There was no evidence of a synergistic effect when wounds were inoculated simultaneously with both fungi (English, Davis, and DeVay 1966).

Since the sexual stage of *B. dothidea* was not found, identification of the almond isolate was based on asexual stage (*Dothiorella*) morphology, serology, and pathology. The almond *Dothiorella* produced erumpent, dark-walled, globose to flask-shaped, mostly nonstromatic, cortical pycnidia on inoculated almond branches. The pycnidia were 238 to 442 μm wide by 525 to 574 μm high and contained a wall layer of simple, hyaline conidiophores that measured approximately 7 to 8 by 1 to 2 μm. During humid weather, fine white cirrhi were extruded from the erumpent ostioles. The cirrhi contained both macro- and microconidia with mean dimensions, respectively, of 21.5 by 5.9 μm and 5.7 by 2.0 μm. The macroconidia were ellipsoid to fusoid, hyaline, and aseptate; with age, however, they often became biseptate, with the central cell brownish and the end cells hyaline. They germinated in four to five hours on water agar at room temperature. Germination was typically unipolar, but occasionally bipolar or lateral. During germination, aseptate spores sometimes became uni- or biseptate. Attempts to germinate the microconidia on water agar, potato-dextrose agar, or nutrient agar were negative (English, Davis, and DeVay 1975).

The relation of temperature to mycelial growth of the almond isolates agrees, in general, with that reported for *B. dothidea* (Witcher and Clayton 1963). The minimum, optimum, and maximum temperatures at which growth occurred were, respectively, 8.5°, 24° to 30°, and 31.5°C. Following a 26-day incubation period, the isolates were viable at 1.0° and 37.0°C, but no growth had occurred at these temperatures (English, Davis, and DeVay 1975). Weaver (1974, 1979) in studying isolates of *B. dothidea* from gummosis-affected peach trees in Georgia reported the optimum temperature for in vitro growth to be 28°C, with good growth at 36°C and slight growth at 38°C. He found that the optimum temperatures for conidial germination and germ-tube growth were, respectively, 25° to 35° and 30°C. It is obvious that isolates of the fungus from different hosts or different localities vary somewhat in their temperature relations.

Mycelial inoculations have caused infection and canker development when the inoculum was placed either on fresh pruning cuts or cork-borer wounds extending to the cambium. Infection also occurred when inoculum was placed over longitudinal slits through the bark to the cambium or over natural growth cracks. Little infection occurred when inoculations were made into shallow bark injuries or on the surface of uninjured bark. The largest cankers developed from spring and summer inoculations, and most ceased to enlarge in late fall or winter and seldom became reactive the following year (English, Davis, and DeVay 1975). Lenticel infection, as reported for peach (Weaver 1974, 1979) and blueberry (Milholland 1972), has not been observed in almond.

Histopathological studies (English, Davis, and DeVay 1975) of almond cankers have shown that the mycelium is present in the cortex and sieve tubes but not in the phellogen or phellem. In the xylem, the mycelium was observed in vessels, tracheids, and ray parenchyma. It was largely intracellular and appeared to penetrate the cells mostly through wall pits. Both tyloses and gum deposits were evident in the vessels of infected xylem. Similar results have been reported by Milholland (1972) for *B. dothidea* infections in blueberry.

The source of inoculum for infection of young almond orchards has not been determined. But the wide, woody-plant host range of the pathogen suggests that some unrelated host may be involved. Persian walnut orchards, known hosts of *B. dothidea*, occur in close proximity to some of the infected almond trees and could provide an inoculum source (English, Davis, and DeVay 1975). The conidia are largely rain-disseminated, but the ascospores are both waterborne and airborne (Sutton 1981; Sutton and Boyne 1983). The latter spores, in particular, could be carried by strong air currents to young almond trees from an inoculum source some distance away.

Host range and cultivar susceptibility. *Botryosphaeria dothidea* causes canker diseases in a tremendously broad range of woody plants, including shade trees, ornamental shrubs, tree fruits, and small fruits (English, Davis, and DeVay 1975; Milholland 1972; Smith 1934). It causes a serious canker and fruit-rot disease of apple in some midwestern and eastern states (Brown and Britton 1986; McGlohon 1982; Sutton 1981; Sutton and Boyne 1983) and a damaging gummosis disease of peach trees in Georgia (Britton and Hendrix 1982; McGlohon 1982; Weaver 1974, 1979). It is not known on peach in California, but early records indicate its presence on walnut, citrus, and avocado (Smith 1934), and it recently has been reported to cause a canker and fruit rot of apple (French 1987) and a blighting of pistachio shoots (Rice et al. 1985). Isolates from several of these hosts produced cankers when artificially inoculated into almond, plum, and sweet cherry (Smith 1934).

Serious outbreaks of Dothiorella canker have been encountered in almond cultivars Nonpareil and Carmel. It has been found only occasionally in cultivars Davey, Drake, and Mission. Artificial inoculations indicate that the decreasing order of susceptibility of three almond cultivars is Nonpareil, Ne Plus Ultra, and Mission (English, Davis, and DeVay 1975).

Control. The excision (bark only) of cankers during the dormant season, with or without the application of a wound disinfectant, failed to prevent further canker activity (English, Davis, and DeVay 1975). This failure, it is believed, was due to the survival of the fungus in the xylem from which it spread into the surrounding healthy tissue. Measures designed to protect the tree against infection have not been investigated. Severely cankered or killed scaffold branches should be removed by cutting a short distance below any evidence of necrosis in the bark or wood.

REFERENCES

Britton, K. O., and F. F. Hendrix. 1982. Three species of *Botryosphaeria* cause peach tree gummosis in Georgia. *Plant Dis.* 66:1120–21.

Brown, E. A., and K. O. Britton. 1986. Botryosphaeria diseases of apple and peach in the southeastern United States. *Plant Dis.* 70:480–84.

English, H., J. R. Davis, and J. E. DeVay. 1966. Dothiorella canker, a new disease of almond trees in California (abs.). *Phytopathology* 56:146.

———. 1975. Relationship of *Botryosphaeria dothidea* and *Hendersonula toruloidea* to a canker disease of almond. *Phytopathology* 65:114–22.

French, A. M. 1987. *California plant disease host index. Part I: Fruits and nuts.* Calif. Dept. Food & Agric. Div. Plant Ind., Sacramento, 39 pp.

McGlohon, N. E. 1982. *Botryosphaeria dothidea*—where will it stop. *Plant Dis.* 66:1202–03.

Milholland, R. D. 1972. Histopathology and pathogenicity of *Botryosphaeria dothidea* on blueberry stems. *Phytopathology* 62:654–60.

Nattrass, R. M. 1933. A new species of *Hendersonula* (*H. toruloidea*) on deciduous trees in Egypt. *Trans. Brit. Mycol. Soc.* 18:189–98.

Rice, R. E., J. K. Uyemoto, J. M. Ogawa, and W. M. Pemberton. 1985. New findings on pistachio problems. *Calif. Agric.* 39(1,2):15–18.

Smith, C. O. 1934. Inoculations showing the wide host range of *Botryosphaeria ribis*. *J. Agric. Res.* 49:467–76.

Sutton, T. B. 1981. Production and dispersal of ascospores and conidia of *Physalospora obtusa* and *Botryosphaeria dothidea* in apple orchards. *Phytopathology* 71:584–89.

Sutton, T. B., and J. V. Boyne. 1983. Inoculum availability and pathogenic variation in *Botryosphaeria dothidea* in apple production areas of North Carolina. *Plant Dis.* 67:503–06.

Weaver, D. J. 1974. A gummosis disease of peach trees caused by *Botryosphaeria dothidea*. *Phytopathology* 54:1429–32.

———. 1979. Role of conidia of *Botryosphaeria dothidea* in natural spread of peach tree gummosis. *Phytopathology* 69:330–34.

Wilson, E. E. 1947. The branch wilt of Persian walnut trees and its cause. *Hilgardia* 17:413–36.

Witcher, W., and C. N. Clayton. 1963. Blueberry stem blight caused by *Botryosphaeria dothidea* (*B. ribis*). *Phytopathology* 53:705–12.

Eutypa Dieback of Apricot

Eutypa lata (Pers.:Fr.) Tul.

Other names for the disease on apricot are gummosis, dieback, and Eutypa canker. It is known to occur on apricot or grapevine in Australia, New Zealand, Libya, South Africa, the United States, Canada, Mexico, Turkey, and most of the countries of central and southern Europe (Carter 1990; Carter, Bolay, and Rappaz 1983; Wilson and Ogawa 1979). The disease is found on apricot and grapevine in California and grapevine in Michigan, New York, and Washington (Carter, Bolay, and Rappaz 1983). It was first reported in Tasmania by Dowson in 1931 and in South Australia by Harris in 1932. However, Adam, Grace, and Flentje (1952) believed the disease had been present much longer in most apricot-producing areas of Australia. In the northern hemisphere, it was first discovered in California apricots in 1959 (English, Davis, and DeVay 1962; English, McNelly, and Rizzi 1960) and now is recognized as one of the most serious diseases of this crop in the central part of the state (English et al. 1963; Moller et al. 1980). Eutypa dieback, being somewhat similar in appearance to bacterial canker, probably occurred in California for many years before its true identity was established (English et al. 1963). The disease is most serious in regions with moderately high rainfall, which is conducive to the production of abundant inoculum (Carter 1957a; Ramos 1974). As many as 68 percent of the trees in some orchards are reportedly infected with this chronic, progressive, lethal disease (**fig. 8A**) (Wilson and Ogawa 1979).

Symptoms. Diagnosis of this disease was discussed by Adam, Grace, and Flentje (1952), English et al. (1963), and more recently by Carter (1968) and Moller et al. (1980). The most conspicuous symptom is the death of leaves on certain tree branches in mid- to late summer after the fruit is harvested (**fig. 8B**). The leaves on an affected branch may suddenly wilt and dry out, although they usually remain attached to the branch throughout the following winter season. The bark of such branches is dead and the wood is discolored light to dark brown. The entire limb is dry, brittle, and breaks easily when bent. Examination of the affected branch reveals a canker surrounding a pruning wound (**fig. 8C**) or other injury that has exposed the xylem. The bark in this area is usually depressed and darkened, and at times becomes swollen and cracked at the canker margin. Small brown flecks often are present in the inner bark adjacent to the canker. Globules of amber-colored gum commonly exude from the marginal region of the canker, especially during the growing season. The internal bark is light to dark brown, often zonate, and has a definite margin with the adjacent healthy tissue (**fig. 8D**). Discolored wood (xylem) may extend for a considerable distance above or below the margin of a bark canker.

The disease occurs most commonly in mature trees. Leaves on badly affected (but not girdled) branches sometimes are slightly cupped, somewhat silvered, and marginally scorched. The fruit on such branches ripen earlier and the leaves fall earlier than those on uninfected branches.

Causal organism. Known for many years only in its asexual stage (*Cytosporina* sp., syn. *Libertella blepharis* Smith), the fungus was shown by Carter in 1955 to produce a perithecial stage, which he named *Eutypa armeniacae* Hansf. & Carter (1957a). A recent revision of this genus indicates that the correct binomial for the organism is *E. lata* (Pers.:Fr.) Tul. (Rappaz 1984).

The perithecial stage was first reported in California in 1963 (English et al.), and surveys in 1965 showed this stage of the fungus to be moderately abundant in the counties of Solano, San Benito, and Santa Clara (Moller, English, and Davis 1966). This stage appears to be quite rare in California's Central Valley, where it was only recently discovered (English et al. 1983).

The asexual stage usually is found on wood exposed by a pruning cut or on decorticated wood associated with a pruning-wound infection. Pycnidia are single or closely aggregate, partly immersed or sitting almost wholly exposed on a black, irregularly pulvinate stroma which may be up to 2 mm in diameter. Exposed pycnidia possess an apical round pore; immersed pycnidia have a short neck protruding to the surface. Pycnidia may be separate or more or less confluent, or sometimes compound with irregular locules partially divided by folds of the wall. Conidiophores are hyaline, simple, straight or slightly bent, and 8 to 15 μm long by 1 μm wide. Hyaline, bent, to arcuate, filiform pycnidiospores, 18 to 25 by 1 μm are borne on the ends of the conidiophores. The pycnidiospores are extruded in cirri or as droplets from the pycnidia. Attempts to germinate or induce infection with pycnidiospores have been negative (Carter 1957a; English, Davis, and DeVay 1962). The evidence suggests that the so-called pycnidia and pycnidiospores may in fact be spermogonia and spermatia, but only further research can resolve this dilemma.

Branches on which perithecia develop have been dead for two to five years, are largely decorticated, and look as if they had been partially burned by fire (**fig. 8E**). The perithecia, produced in a black stromatic layer, are globose bodies with fleshy-membranous walls and are 450 μm in diameter. The asci, which attenuate into a stipe, are cylindric-clavate and measure approximately 4 by 30 μm. The eight one-celled ascospores are uni- to biseriate, overlapping, and allantoid. At maturity the ascospores are pale yellowish brown, slightly bent with rounded ends, and are 7 to 11 μm long by 1.5 to 2 μm wide. They readily stain with aniline blue in lactophenol (Carter 1968).

All eight ascospores are discharged at one time and aerially disseminated intact. Germination occurs at relative humidities over 90 percent (Carter 1957a).

When wet, the asci appear as a gelatinous, olivaceous gray mass in the locules of the perithecia. When cut tangentially or radially, the perithecial locules are clearly visible, especially with the aid of a hand lens (**fig. 8E**).

Potato dextrose or Czapek-Dox agars are best suited for isolation of the fungus. Cultures from infected living sapwood can be obtained from margins of the canker, with the white mycelium becoming apparent in two to three days when incubated at 15° to 25°C. Cultures from ascospores are easily obtained by washing the perithecial stromata, immersing them in water for 1 to 1½ hours, and draining the excess water. The stromata-bearing chips are then attached to the lids of sterile petri dishes containing the culture medium for 10 minutes to allow for the discharge of the ascospores onto the medium. The cultures are incubated at 15° to 25°C, and the ascospores germinate in 11.5 to 16 hours. The optimum temperature for spore germination is 22° to 25°C (Carter 1957a).

Pycnidia develop slowly on artificial media, and some isolates appear incapable of forming these structures (English et al. 1983). Cultures held in complete darkness form few if any pycnidia. These asexual bodies appear to form most abundantly in cultures on a high-glucose medium incubated in either complete light or under diurnal light conditions at a temperature of approximately 20° to 24°C. Attempts in Australia (Carter 1957a) and California to produce the perithecial stage in vitro on apricot wood and a variety of culture media have met only with failure. In these tests, however, consideration was not given to the possibility that the so-called pycnidiospores might, in fact, be spermatia. Although most of the isolates studied show little host specificity, recent research (Carter et al. 1983, 1985; Ramos, Moller and English 1975a, b) reveals a broad range of virulence among isolates from different geographic regions on different hosts. Work in California (English et al. 1983) shows that monoascospore isolates from a single ascus vary considerably in both virulence and morphology. Further studies on pathogenic variation are in progress in several countries.

Price (1973) was able to use serological methods to identify isolates from diseased apricots from Switzerland and South Africa. Cell-wall-free extracts were used in the tests. Fluorescent antibody staining techniques gave less reliable results than a gel diffusion assay.

Other hosts. Although originally thought to affect only apricot, Eutypa dieback is now known to occur in more than 60 woody plant species in some 23 families (Bolay and Carter 1985; Carter 1990; Carter,

Bolay, and Rappaz 1983). The pathogen is quite unique in that, despite its broad host range, it often does not attack species closely related to its hosts. In California, in addition to apricot, the disease has been found only on grapevine (Moller, English, and Davis 1968; Moller and Kasimatis 1978), *Ceanothus* (Moller, Ramos, and Hildreth 1971), and western chokecherry (English and Davis 1965). In other countries the fungus is reportedly pathogenic on additional fruit and nut species, such as peach, plum, almond, apple, pear, walnut, pistachio, and currant (Carter 1990; Carter, Bolay, and Rappaz 1983; Rumbos 1986). Artificial inoculations in California showed that almond and peach were slightly susceptible to an apricot isolate of the pathogen (English and Davis 1978).

Disease development. As no one has succeeded in germinating the pycnidiospores, and since the perithecial stage is rare in some areas, the source of inoculum in these regions is unclear. Carter (1957a) found the perithecial stage to be abundant in the high-rainfall areas (greater than 20 inches) of Australia, but rare in low-rainfall areas (less than 13 inches). A similar finding was reported by Ramos, Moller, and English (1975b) in California.

In Australia, apricot branches were readily infected when 100 or more ascospores were placed in wounds in the winter (Carter and Moller 1971). In California, a high incidence of infection of fresh pruning wounds was obtained with ascospore dosages of 1, 10, and 100 (Ramos, Moller, and English 1975). Under Australian conditions, wound susceptibility during dormancy declines rapidly in the first two weeks after pruning (Carter and Moller 1970). In California, Ramos, Moller, and English (1975) showed that pruning wounds made on mature trees on September 21 remained susceptible to the fungus for at least 42 days, while wounds made on March 15 became resistant within 14 days. Pruning wounds on trees held at 20°C became resistant to infection much more quickly than those maintained in a dormant condition at 3°C. High humidity around the inoculation sites hastened development of resistance. Post-inoculation temperatures between 7° and 19°C had no effect on the degree of infection, but canker size was proportional to the temperature (Moller, English, and Davis 1968). On Modesto apricot the size of pruning wounds was proportional to the degree of infection, but heartwood of large pruning wounds was not directly infected. Four to six months' incubation was required for symptom development when inoculations were made with mycelia, and longer incubations when inoculations were made with ascospores. The Patterson apricot developed fewer cankers than did Modesto. On Tilton apricot a California isolate was more virulent than the Australian isolate tested (Ramos, Moller, and English 1975).

Ramos, Moller, and English (1975b) provided evidence for long-distance dispersal of airborne ascospores from areas of high rainfall, such as Suisun and Hayward, to the interior valley apricot districts of Tracy, Patterson, and Los Banos—distances of approximately 30 to 60 miles. A correlation was established between ascospore release with rain and long-distance dissemination with prevailing winds. Somewhat similar results were obtained more recently by Petzoldt, Sall, and Moller (1983a). Limited numbers of perithecia have been found in semicoastal regions (Brentwood, San Jose, and Hollister) with a mean annual precipitation of 13 to 15 inches (Ramos 1974). More recently, mature perithecia were found for the first time in the Central Valley (Davis) on a shaded, sprinkler-irrigated grapevine with Eutypa dieback symptoms (English et al. 1983). This finding makes probable the presence of other inoculum foci in the Central Valley and helps explain the occurrence of the disease in parts of the valley 100 to 200 miles from the nearest known source of inoculum. No ascospore dispersal occurs during the dry summer months, but large numbers are discharged during certain rainy periods of fall, winter, and spring (Petzoldt, Sall, and Moller 1983b; Ramos, Moller and English 1975b). The two parameters identified as important in determining the number of ascospores discharged in a single release period are duration of surface wetness and days from previous release (Petzoldt, Sall, and Moller 1983b).

The mode of infection and pathogenesis by *E. lata* is now reasonably well understood. The fungus is unable to infect injured bark or heartwood and causes relatively little infection if placed on the cambium (English and Davis 1978; Ramos, Moller and English 1975a). However, when ascospores are deposited in a film of moisture on the surface of the sapwood of a fresh pruning wound, many are drawn by capillary action into the xylem vessels (Carter 1960). Here, after a few hours, they germinate to produce hyphae that ramify within the vessel lumens and invade adjacent cells either by direct penetration or through wall pits (Carter 1955; English and Davis 1978). The hyphae invade and kill the cambium and bark and also colonize the heartwood. In advanced stages of infection, the hyphae penetrate the secondary walls, especially those of fiber-tracheids, where they grow longitudinally within the wall and, through enzymatic action, cause the type of wood decomposition called "soft rot" (English and Davis 1978). This soft rot causes the brittleness of branches killed by the pathogen. The fungus also causes the formation of dark gum plugs in some of the tracheary elements, which undoubtedly interfere with water transport (English and Davis 1978). The darkened xylem often extends longitudinally for a considerable distance beyond the canker margin. Except in close proximity to the canker, the darkened xylem appears to be free of the pathogen. This suggests the movement of either a toxin produced by the pathogen or toxic metabolites resulting from pathogenesis. *Eutypa lata* is essentially a vascular pathogen with the additional capacity to subsequently kill the bark and cause cankers. Many infections are chronic and, over a period of years, can result in complete collapse of a tree. In some instances, however, cankers are completely arrested by some unknown defensive mechanism of the host (English and Davis 1978; English et al. 1983). There is some evidence that the incidence and severity of dieback are greater in trees grown under poor cultural conditions—moisture and nutritional stress—than in those grown under more favorable conditions (English, Davis, and DeVay 1969).

Control. In California this disease can be effectively controlled by pruning trees in July and August rather than during the dormant season (English, Davis, and DeVay 1969; Moller et al. 1980; Ramos, Moller and English 1975a). It is especially important that the shaping of young trees be done at this time. The success of this procedure is due to the following: (1) *Eutypa* ascospores are discharged only during wet weather, (2) little if any rainfall occurs in California from July through September, and (3) pruning wounds, the normal infection courts, become resistant to infection in four to eight weeks (Carter and Moller 1970; Moller and Carter 1965; Moller et al. 1980; Ramos, Moller and English 1975a). If winter pruning must be done, the wounds must be treated immediately with a benomyl suspension (Moller, Ramos, and Sanborn 1977). Other benzimidazole fungicides also are reportedly effective (Bolay 1986). Recent reports (Carter 1985, Carter and Perrin 1985) indicate that the fungicide can be effectively applied with either a manual or pneumatic-powered spraying shears. In South Africa, a homopolymeric polyvinyl acetate base provides excellent protection of wounds against infection by *E. lata* (Thomas, Matthee, and Hadlow 1978). Fungicidal sprays have been relatively ineffective (Carter 1960; Carter and Price 1977; Moller et al. 1980; Moller, Ramos, and Sanborn 1977).

Three other measures that aid in combating the disease are removal and burning of abandoned apricot trees or grapevines in the vicinity of commercial orchards; removal and burning of severely affected branches during the summer (the pruning cuts should be made at least 8 inches below the lower margin of the cankers); and maintenance of tree vigor by providing adequate irrigation and fertilization (English et al. 1963; Moller et al. 1980; Wilson and Ogawa 1979).

Biological control studies using *Fusarium lateritium* spore suspensions containing 104 macroconidia per milliliter applied to pruning cuts protected against *E. lata*. *Fusarium lateritium* is restricted to a zone of sapwood 2 cm from the pruned surface (Carter 1971; Carter and Price 1974). Furthermore, *F. lateritium* tolerates water suspensions containing 400 ppm benomyl or thiabendazole. Although *F. lateritium* shows considerable promise as a biocontrol agent, Carter (1983) recently concluded that as long as *E. lata* remains sensitive to benzimidazoles there is no immediate need to develop commercial production of *F. lateritium* for this purpose. Also, *F. lateritium* has proved ineffective in control of Eutypa dieback in grapevines (Gendloff, Ramsdell, and Burton 1983).

[For a more comprehensive discussion of *E. lata* and its pathogenicity and control, the reader is referred to the recent monograph by Carter (1990).]

REFERENCES

Adam, D. B., J. Grace, and N. T. Flentje. 1952. The "gummosis" or "dieback" disease of apricots. *J. Dept. Agric. South Austral.* 55:450–55.

Bolay, A. 1986. Comment protéger la vigne et les arbres frutiers des attaques d'eutypoise? *Rev. Suisse Vit. Arbor. Hort.* 18(1):7–13.

Bolay, A., and M. V. Carter. 1985. Newly recorded hosts of *Eutypa lata* (=*E. armeniacae*) in Australia. *Plant Protect. Quart.* (1):10–12.

Carter, M. V. 1955. Apricot gummosis—A new development. *J. Dept. Agric. South Austral.* 59:178–84.

———. 1957a. *Eutypa armeniacae* Hansf. & Carter, sp. nov., an airborne vascular pathogen of *Prunus armeniaca* L. in Southern Australia. *Austral. J. Bot.* 5:21–35.

———. 1957b. Vines aid spread of apricot "gummosis." *J. Dept. Agric. South Austral.* 60:482–83.

———. 1960. Further studies on *Eutypa armeniacae* Hansf. and Carter. *Austral. J. Agric. Sci.* 2:498–504.

———. 1968. Diagnosis of canker disease of apricot trees caused by *Eutypa armeniacae* Hansf. and Carter. *Tech. Comm. Intl. Soc. Hort. Sci.* 11(3):389–90.

———. 1971. Biological control of *Eutypa armeniacae*. *Austral. J. Exp. Agric. Anim. Husb.* 11:687–92.

———. 1983. Biological control of *Eutypa armeniacae*. 5. Guidelines for establishing routine wound protection in commercial apricot orchards. *Austral. J. Exp. Agric. Anim. Husb.* 23:429–36.

———. 1985. Evaluation of a manual spraying secateur for protecting trees and grapevines against wound-invading pathogens. *Austral. Plant Path.* 14(2):43–44. (*Rev. Plant Path.* 65:599. 1986).

———. 1990. The status of *Eutypa lata* as a pathogen. CAB International, Wallingford, Oxon., U.K. Phytopathological Paper 32. 72 pp.

Carter, M. V., A. Bolay, H. English, and I. Rumbos. 1985. Variation in the pathogenicity of *Eutypa lata* (=*E. armeniacae*). *Austral. J. Bot.* 33:361–66.

Carter, M. V., A. Bolay, and F. Rappaz. 1983. An annotated host list and bibliography of *Eutypa armeniacae*. *Rev. Plant Pathol.* 62:251–58.

Carter, M. V., A. Bolay, F. Rappaz, W. H. English, and I. Rumbos. 1983. Pathogenic variation in *Eutypa armeniacae*. *Fourth Internat. Cong. Plant Pathol. Abstr.* 199.

Carter, M. V., and W. J. Moller. 1970. Duration of susceptibility of apricot pruning wounds to infection by *Eutypa armeniacae*. *Austral. J. Agric. Res.* 21:915–20.

———. 1971. The quantity of inoculum required to infect apricot and other *Prunus* species with *Eutypa armeniacae*. *Austral. J. Exp. Agric. Anim. Husb.* 11:684–86.

Carter, M. V., and E. Perrin. 1985. A pneumatic-powered spraying secateur for use in commercial orchards and vineyards. *Austral. J. Exp. Agric.* 25:939–42.

Carter, M. V., and T. V. Price. 1974. Biological control of *Eutypa armeniacae*. II. Studies of the interaction between *E. armeniacae* and *Fusarium lateritium*, and their relative sensitivities to benzimidazole chemicals. *Austral. J. Agric. Res.* 25:105–19.

———. 1977. Explanation of the failure of commercial scale application of benomyl to protect pruned apricot trees against Eutypa dieback disease. *Austral. J. Exp. Agric. Anim. Husb.* 17:171–73.

Dowson, W. J. 1931. The die-back of apricots: Preliminary note. *Tasman. J. Agric.* 2:165.

English, H., and J. R. Davis. 1965. Apricot dieback fungus found on western chokecherry. *Plant Dis. Rep.* 49:178.

———. 1978. *Eutypa armeniacae* in apricot: Pathogenesis and induction of xylem soft rot. *Hilgardia* 46:193–204.

English, H., J. R. Davis, and J. E. DeVay. 1962. Cytosporina dieback, a new disease of apricot in North America (abs.). *Phytopathology* 52:361.

———. 1969. Factors affecting the incidence and development of Cytosporina dieback of apricot (abs.). *Phytopathology* 59:11.

English, H., J. R. Davis, J. M. Ogawa, and F. J. Schick. 1983. Variation in *Eutypa armeniacae* and discovery of its ascigerous stage in California's Central Valley (abs.). *Phytopathology* 73:958.

English, H., B. McNelly, and A. D. Rizzi. 1960. New apricot fungus disease found in Santa Clara county orchards. *Sunsweet Standard* 43(12):5.

English, H., H. J. O'Reilly, L. B. McNelly, J. E. DeVay, and A. D. Rizzi. 1963. Cytosporina dieback of apricot. *Calif. Agric.* 17(2):2–4.

Gendloff, E. H., D. C. Ramsdell, and C. Burton. 1983. Fungicidal control of *Eutypa armeniacae* infecting Concord grapevine in Michigan. *Plant Dis.* 67:754:56.

Harris, J. B. 1932. Dieback of apricot trees in the Barossa district. *J. Dept. Agric. South Austral.* 35:1394–95.

Moller, W. J., and M. V. Carter. 1965. Production and dispersal of ascospores in *Eutypa armeniacae*. *Austral. J. Biol. Sci.* 18:67–80.

Moller, W. J., H. English, and J. R. Davis. 1966. The perithecial stage of *Eutypa armeniacae* in California. *Plant Dis. Rep.* 50:53.

———. 1968. *Eutypa armeniacae* on grape in California. *Plant Dis. Rep.* 52:751.

Moller, W. J., and A. N. Kasimatis. 1978. Dieback of grapevines caused by *Eutypa armeniacae*. *Plant Dis. Rep.* 62:254–58.

Moller, W. J., A. N. Kasimatis, D. E. Ramos, W. H. English, K. W. Bowers, J. J. Kissler, D. Rough, and R. R. Sanborn. 1980. Eutypa dieback of apricot and grape in California. *Univ. Calif. Div. Agric. Sci. Leaf.* 21182, 7 pp.

Moller, W. J., D. E. Ramos, and W. R. Hildreth. 1971. Apricot pathogen associated with *Ceanothus* limb dieback in California. *Plant Dis. Rep.* 55:1006–08.

Moller, W. J., D. E. Ramos, and R. R. Sanborn. 1977. Eutypa dieback in California apricot orchards: Chemical control studies. *Plant Dis. Rep.* 61:600–04.

Petzoldt, C. H., M. A. Sall, and W. J. Moller. 1983a. Eutypa dieback of grapevines: Ascospore dispersal in California. *Am. J. Enol. Vit.* 32:265–70.

———. 1983b. Factors determining the relative number of ascospores released by *Eutypa armeniacae* in California. *Plant Dis.* 67:857–60.

Price, T. V. 1973. Serological identification of *Eutypa armeniacae*. *Austral. J. Biol. Sci.* 26:389–94.

Ramos, D. E. 1974. *Epidemiology of Eutypa (Cytosporina) dieback of apricot*. Ph.D. dissertation, Department of Plant Pathology, University of California, Davis, 64 pp.

Ramos, D. E., W. J. Moller, and H. English. 1975a. Susceptibility of apricot tree pruning wounds to infection by *Eutypa armeniacae*. *Phytopathology* 65:1359–64.

———. 1975b. Production and dispersal of ascospores of *Eutypa armeniacae* in California. *Phytopathology* 65:1364–71.

Rappaz, F. 1984. Les éspèces sanctionées du genre *Eutypa* (Diatrypaceae, Ascomycetes) étude taxonomique et nomenclaturale. *Mycotaxon* 20:567–86.

Rumbos, I. C. 1983. Eutypa canker and dieback of almond in Greece. *Zeitschr. Pflanzenkr. Pflanzensch.* 90:99–101.

———. 1986. Isolation and identification of *Eutypa lata* from *Pistacia vera* in Greece. *J. Phytopathol.* 116:352–57.

Thomas, A. C., F. N. Matthee, and J. Hadlow. 1978. Wood-rotting fungi in fruit trees and vines. III. Orchard evaluation of a number of chemical bases as wood protectants. *Decid. Fruit Grow.* 28(3):92–98.

Wilson, E. E., and J. M. Ogawa. 1979. *Fungal, bacterial, and certain nonparasitic diseases of fruit and nut crops in California.* University of California Division of Agricultural Science, Berkeley, 190 pp.

Green-Fruit Rot of Stone Fruit

Botrytis cinerea Pers.: Fr., *Sclerotinia sclerotiorum* (Lib.) dBy., *Monilinia laxa* (Aderh. & Ruhl.) Honey, and *M. fructicola* (Wint.) Honey

"Green rot," "jacket rot," "calyx rot," and "blossom rot" are names given to a disease of apricot, prune, plum, almond, and cherry (Shotwell and Ogawa 1984). Blossom rot is most common in coastal districts of California where protracted rains occur during the blossoming period. Green-fruit rot occurs both in coastal and interior areas of California. The disease is of little importance in most years, but sometimes causes severe loss of crops.

Symptoms. As the names imply, the disease affects both blossoms and green fruit (**fig. 9**). The early stage of the disease, blossom rot, becomes apparent during the latter part of the blossom stage as a withering of the calyces. As the fruit sets and starts to grow, a brown lesion commonly develops on fruit to which the calyces have adhered. The lesion expands rapidly, involving the entire young fruit.

Causal organisms. *Monilinia laxa* and *M. fructicola*, the two brown rot fungi attacking stone fruit, cause blossom blighting similar to that just described. Occasionally, these fungi attack the green fruit, and symptoms produced by them are distinguishable by development of the conidial stage on the blossoms and on the rotted fruit. The diseases produced by these fungi are discussed under "Brown Rot" in this chapter. Nixon and Curry (1910) and Smith and Smith (1911) believed that blossom and green-fruit rot is primarily caused by *Botrytis cinerea* (*Botryotinia fuckeliana* [dBy.] Whetz.) and *Sclerotinia sclerotiorum* (earlier classified as *S. libertiana* and *Whetzelinia sclerotiorum* [Lib.] Korf and Dumont [1972]). Smith (1931) later believed that the disease is caused almost entirely by *S. sclerotiorum*. Yarwood (1945, 1948), however, reported both fungi to be involved in blossom and fruit rotting, one being more common than the other in some localities. Yarwood's observations were made principally on apricots in coastal districts of California; in the Sacramento and San Joaquin valleys, blossom rotting is less common.

Botrytis cinerea Pers.:Fr. is a loosely defined species, containing forms exhibiting a wide range of variability in sclerotial size, conidial production, rate of growth, and production of aerial mycelium. Groves and Drayton (1939) obtained an apothecial stage from isolates of the *cinerea* type; hence this fungus belongs, as do other species of *Botrytis*, to the genus *Botryotinia*. Groves and Drayton did not consider a change in nomenclature to be justified, inasmuch as the morphological differences among apothecia produced by their different isolates had not been determined.

Sclerotinia sclerotiorum (Lib.) dBy. has well-defined characters (Purdy 1955). Apothecial production from sclerotia in the laboratory was described by Purdy (1956).

The name *B. cinerea* is reserved for *Botrytis* forms that produce grayish brown conidiophores and conidia, and large flat sclerotia that adhere close to the substrate. *Sclerotinia sclerotiorum* produces large, elongated, black sclerotia but no conidia. Under some conditions it produces abundant microconidia that apparently have no function in the disease.

Disease development. Sclerotia of *S. sclerotiorum* are rarely seen on infected blossoms or fruit on the tree; they form after these affected structures fall to the ground. *Botrytis cinerea* probably survives in a like manner or as mycelium in dead organic material (Ogawa and English 1960). Ascospores from apothecia (**fig. 9E**), which develop from fall until spring, are probably the only agents for primary infection by *S. sclerotiorum*. Conidia constitutes the principal inoculum for infection by *B. cinerea*; the apothecial stage is rare in the life cycle of this fungus. Both the ascospores of *S. sclerotiorum* and the conidia of *B. cinerea* are disseminated by air currents, the former being forcibly ejected from the apothecia (Leach and Hewitt 1939). Smith

(1931) mentioned the wide occurrence of *S. sclerotiorum* in the spring. Infection of green fruit occurs by growth of the mycelium from the infected calyx cup into the flesh of the fruit.

Presumably the infection pegs of both organisms invade the blossom directly from appressoria. Infection of fruit by both organisms probably results from the direct penetration of the surface by mycelium established first in the dead flower parts.

Judging from Yarwood's (1948) studies, blossom infection occurs from the time the flower is fully open until the petals are shed, and the disease develops rapidly during the latter part of the blossoming period. Fruit infection occurs from the time of fruit set until the shucks (floral tube) are shed. The relation between the presence of infected calyces and development of fruit infection is established for both *S. sclerotiorum* (Smith 1931) and *B. cinerea*. All floral parts are susceptible to infection by *B. cinereea*; the pollen grains stimulate spore germination and infection of petals (Ogawa and English 1960). Kawakubo and Nasuda (1976) found that extracts from anthers, calyces, and petals stimulate conidial germination. On apricot, stylar infection by *B. cinerea* is conducive to infection of green fruit (Ogawa and English 1960).

Prolonged rains during the blossom period favor blossom infection by both fungi. After fruit are set and begin to grow, rains that moisten the shucks (jackets) favor fruit infection. Cool weather prolongs retention of the shucks and favors fruit infection (Kawakubo and Nasuda 1976). Yarwood (1948) observed that the disease is less prevalent where the orchard floor is bare than where weeds cover it. It is a common observation that apothecia of *S. sclerotiorum* are most abundant where weeds protect the soil surface form rapid drying.

Host susceptibility. All commercial cultivars of apricot, plum, prune, almond, and cherry are susceptible to infection by these fungi. Blossoms in dense clusters, such as those produced by Ne Plus Ultra and Drake, are highly susceptible to infection. Peach and nectarine blossoms, however, are rarely affected.

Control. Clean cultivation of orchards probably is beneficial in preventing apothecial development. Forceful removal of the jackets from the young fruit might conceivably reduce infection. Some growers have reported benefits from removing the jackets by shaking the branches or by blowing them off with a helicopter. Such a practice is of limited practicality.

Early attempts to prevent green-fruit rot by spraying the fruit soon after it set were unsuccessful. Yarwood (1944, 1945, 1948) succeeded in preventing fruit and blossom infection of apricot by sprays applied during the full-bloom stage. Ferbam (ferric dimethyl-dithiocarbamate) at 1½ pounds per 100 gallons of water plus a spreader was recommended. Botran, however, injured young apricot fruit when applied at the petal-fall stage. (Ogawa, Manji, and Schreader 1975). On cherries, benomyl provides excellent control when applied at full bloom.

REFERENCES

Groves, J. W., and F. L. Drayton. 1939. The perfect stage of *Botrytis cinerea*. *Mycologia* 31:485–89.

Kawakubo, Y., and K. Nasuda. 1976. [Studies on gray mold of apricots caused by *Botrytis cinerea* II. Factors affecting occurrence of the disease.] *Bull. Fukui Agric. Exp. Stn.* 13:127–38.

Korf, R. P., and K. P. Dumont. 1972. *Whetzelinia*, a new generic name for *Sclerotinia sclerotiorum* and *S. tuberosa*. *Mycologia* 64:248–51.

Leach, L. D., and W. B. Hewitt. 1939. Forced ejection of ascospores of apothecia of *Sclerotinia* species. *Phytopathology* 29:373.

Nixon, W. H., and H. W. Curry. 1910. Disease of young apricot fruits. *Pacific Rural Press* 80:124.

Ogawa, J. M., and H. English. 1960. Blossom blight and green fruit rot of almond, apricot and plum caused by *Botrytis cinerea*. *Plant Dis. Rep.* 44:265–68.

Ogawa, J. M., B. T. Manji, and W. R. Schreader. 1975. *Monilinia* life cycle on sweet cherries and its control by overhead sprinkler fungicide applications. *Plant Dis. Rep.* 59:876–80.

Purdy, L. H. 1955. A broader concept of the species *Sclerotinia sclerotiorum* based on variability. *Phytopathology* 45:421–27.

———. 1956. Factors affecting apothecial production by *Sclerotinia sclerotiorum*. *Phytopathology* 46:409–10.

Reidemeister, W. 1909. Die Bedingungen der Sclerotien-Ringbildung von *Botrytis cinerea* auf Künstlichen Nährboden. *Ann. Mycol.* 7:19–44.

Shotwell, K. M., and J. M. Ogawa. 1984. Botrytis blossom blight and fruit rot of sweet cherry (*Prunus avium*) and their control using fungicides. (Abs.) *Phytopathology* 74:1141.

Smith, R. 1900. *Botrytis* and *Sclerotinia*: Their relation to certain plant diseases and to each other. *Bot. Gaz.* 29:369–406.

———. 1931. Life history of *Sclerotinia sclerotiorum* with reference to the green rot of apricots. *Phytopathology* 21:407–23.

Smith, R. E., and E. H. Smith. 1911. California plant diseases. *Calif. Agric. Exp. Stn. Bull.* 218:1039–1193.

Yarwood, C. E. 1944. Apricot diseases in the Santa Clara Valley in 1943. *Plant Dis. Rep.* 28:32–33.

———. 1945. Apricot diseases in coastal California in 1945. *Plant Dis. Rep.* 29:678–80.

———. 1948. Apricot jacket rot (abs.) *Phytopathology* 38:919–20.

Hull Rot of Almond

Rhizopus spp., *Monilinia fructicola* (Wint.) Honey, and *M. laxa* (Aderh. & Ruhl.) Honey

Hull rot of almond has occurred sporadically in the Sacramento and San Joaquin valleys for many years. In some years it causes some loss of the current crop and considerable reduction of fruiting wood and terminal shoots. Diseased fruit that remain on the trees harbor the navel orange worm in wintertime (Brown, Ogawa, and Gashaira 1977).

Symptoms. There are three macroscopic symptoms of the disease: (1) In late summer, after almond hulls have began to dehisce (separate at the suture), light brown spots appear on the hulls, usually along the suture. These later coalesce and the entire hull becomes wrinkled and shrunken. Mycelium, black sporangia, and sporangiophores of *Rhizopus* frequently develop on the inner surface of the separating hull and on the shell (**fig. 10A,B**). (2) Leaves on twigs nearest the affected fruit curl and dry up, but remain attached to the twig. Where rotted fruit occur on a terminal shoot, the leaves along the side of the shoot on which the fruit are attached may curl and desiccate; some of these leaves, before drying, may develop necrotic streaks along one side of the midvein. (3) Soon after the leaves wither, the supporting twig (spur or terminal shoot) becomes discolored and dies, and at times portions of branches are blighted. Where *Monilinia* is involved, the symptoms are similar, with the exception that here buff-colored masses of conidia develop both within the ruptured hull and on its outer surface (**fig. 10C**).

Causal organisms. *Rhizopus* spp., *Monilinia laxa*, and *M. fructicola* are the pathogens responsible for this disease. The species of *Rhizopus* involved are, according to the keys of Zycha (1935) and Kockova-Kratochvilova and Palkoska (1958), *R. stolonifer* (Ehr.:Fr.) Lind., *R. circinans* v. Tiegh., and *R. arrhizus* Fischer (Mirocha and Wilson 1961).

Although mycelium of the fungi may occur in fruit stems, it is not present in the twigs or leaves. With *Rhizopus* spp., the twig and leaf symptoms are caused by a toxin (fumaric acid) produced by these fungi (Mirocha, DeVay, and Wilson 1961). Tests with radiocarbon-labeled fumeric acid show that this acid is present in large amounts in hulls rotted by *R. circinans*, is transported via the xylem vessels up the twig, and accumulates in the leaves before leaf and twig symptoms appear. During symptom development, the toxin disappears and the radiocarbon is then found in the malic, citric, and tartaric acid components of the leaves and twigs. Since the symptom syndrome is the same with both *Rhizopus* and *Monilinia*, the mechanism of pathogenesis is probably the same with both pathogens.

Disease development. Fruit infection occurs shortly after the hull dehisces as the spores gain access to the moist inner hull tissue (**fig. 10D**). The disease develops during mid- to late summer before fruit harvest. It is favored by rain showers, though these are not required. Insects (nitidulid beetles) have been implicated as vectors of the causal fungi. They commonly feed in diseased hulls then carry the spores into newly dehiscing hulls.

All almond cultivars may be affected, but the disease occurs most commonly in Nonpareil and Jordanolo. Although hulls of Texas (Mission) and Ne Plus Ultra may be rotted, little blighting of the supporting twigs occurs (Browne, Ogawa, and Gashaira 1977).

Control. No control has been developed for the disease caused by *Rhizopus* spp. or *Monilinia fructicola*. Hull rot caused by *Monilinia laxa* could possibly be reduced by controlling the blossom blight caused by the same fungus. Current suggested control measures are (1) harvesting as early as possible or harvesting twice, in order to remove the fruit before the toxin is translocated into the shoots and twigs; (2) reducing or eliminating irrigations just before and during hull dehiscence to hasten the harvest and reduce the activity of the nitidulid beetles; and (3) removing the diseased and mummified fruit during the winter to reduce the source of inoculum.

REFERENCES

Browne, L. T., J. M. Ogawa, and B. Gashaira. 1977. Search continues for control of almond hull rot. *Calif. Agric.* 31(1):16–17.

Kockova-Kratochvilova, A., and V. Palkoska. 1958. A taxonomic study of the genus *Rhizopus* Ehrenberg 1820. *Preslia* 30:150–64.

Mirocha, C. J., and E. E. Wilson. 1961. Hull rot disease of almond. *Phytopathology* 51:843–47.

Mirocha, C. J., J. E. DeVay, and E. E. Wilson. 1961. Role of fumaric acid in the hull rot disease of almond. *Phytopathology* 51:851–60.

Zycha, H. 1935. Mucorineae. *Kryptogamenflora der Mark Brandenburg*, vol. 6a. Leipzig, 264 pp.

Leaf Blight of Almond

Seimatosporium lichenicola (Cda.) Shoemaker & Müller

Leaf blight first became known in 1950 when it developed in a few orchards in the Sacramento valley near Chico. It later appeared in the northern part of the San Joaquin Valley. Approximately 20 percent of the state's almond orchards are located in these two areas. The disease was more recently reported in Australia (Moller 1972).

Although leaf blight seldom destroys more than 15 to 20 percent of the leaves, repeated attacks eventually weaken the tree. An immediate reduction in productivity follows the loss of flower buds killed by the growth of the causal fungus from the leaf petiole into the supporting twig.

Symptoms and signs. Leaf blight (**fig. 11A**) is characterized by the sudden dying of leaves. Beginning in June, one or more leaves of spurs or shoots wither, turn brown, and dry out; other leaves then die throughout the summer. Though a few affected leaves fall, most do not; fragments of the petioles of such leaves remain on the tree until spring (**fig. 11B**).

No signs of the causal fungus are present on affected leaves in summer. In winter, however, the lower end of the persisting diseased petiole turns light tan, and shortly afterward small, dark fruiting bodies of the fungus develop on this area.

In late fall, buds in the axils of the affected leaves are killed by the fungus growing from the petiole into the twig. In spring, flowers are killed in the same manner as they emerge from the winter buds.

Susceptibility of cultivars. Almond cultivars vary considerably in susceptibility. Drake, Ne Plus Ultra, and Peerless are highly susceptible and Nonpareil, Texas (Mission), and IXL are moderately susceptible.

Causal organism. Only the imperfect stage of the causal fungus is known to occur on almond. This stage was originally identified as *Hendersonia rubi* West. by Ogawa, Wilson, and Engish (1959). Since that time taxonomic revisions of this and related genera have resulted in the rejection of the genus *Hendersonia* and the renaming of *H. rubi* as *Seimatosporium lichenicola* (Cda.) Shoemaker & Müller (Shoemaker and Müller 1964; Sutton 1980). Shoemaker and Müller (1964) showed that this fungus has an ascigerous stage, which they named *Clathridium corticola* (Fckl.) Shoemaker and Müller. The fungus occurs in North America and Europe on woody plant hosts such as species of *Cornus, Rubus, Rosa, Prunus,* and *Salix.*

In culture, the fungus colony consists of abundant brown subsurface hyphae and dirty-gray aerial hyphae. The conidia are produced on separate sporodochium-like structures, which at first are arranged in concentric rings but later become so numerous as to form a continuous layer. In some isolates pycnidia may be quite abundant, in others comparatively sparse. The conidia, borne on the ends of slender, tapering conidiophores, are elliptical in shape but with the basal end more pointed than the apex. Most of them are four-celled; the lower cell is subhyaline while the other cells are light brown. The wall of the conidium is slightly constricted where a septum joins it (Ogawa, Wilson, and English 1959).

On the host, the acervuli vary from 115 to 165 μm in diameter. In the early stages these structures may be uncovered, or they may possess a covering of loosely connected hyphal elements so fragile that it is seldom found to be intact after the conidia mature.

The pycnidia are ovoid in shape, with a peridium of relatively thick-walled pseudoparenchymatous cells and an ostiole. They range in diameter from 95 to 115 μm. The conidia produced in pycnidia are identical to those produced in acervuli and are more uniform in size and in number of cells than those from culture. They range from 11. to 17.7 μm in length and from 6.1 to 7.5 μm in breadth.

Disease development. The life cycle of the pathogen (that is, its imperfect stage) is completed on the tree. The fungus survives from one year to the next in affected petioles that remain in the twigs throughout the winter. It is quiescent in these parts during summer, but after rains begin in late fall it produces conidia in fruiting bodies that develop over the basal part of the petiole. Nothing is known about how and when the conidia are disseminated. They are probably washed about by rains in winter, but it is unlikely they are freely disseminated by air currents. Conidia produced in acervuli occasionally may be picked up and distributed by air, but those in pycnidia probably seldom are. In some manner, however, they succeed in reaching and infecting the petioles of the newly formed leaves. Once inside the petiole, the fungus interferes with water conduction and the leaf withers. The pathogen grows from the petiole into the twig and kills the contiguous bud in late autumn and winter. Enlargement of the twig lesion is resumed in the spring, at which time flowers are killed as they emerge from the winter buds. These small nodal lesions do not continue to enlarge from year to year.

Control. Leaf blight has been controlled by spraying the trees with either a protective fungicide or an eradicative fungicide. Captan, ziram, or dichone in water are effective protective fungicides when applied at the petal-fall stage. The eradicative treatment consisted of spraying trees in late winter before the buds begin to open, using sodium pentachlorophenoxide (Ogawa, Wilson, and English 1959). This eradicant, however, is no longer available for use on almond.

REFERENCES

Archer, W. A. 1956. Morphological characters of some Sphaeropsidales in culture. Ann. Mycol. 24:46–51.

Moller, W. J. 1972. *Hendersonia rubi*, and *Rhizopus stolonifer*, two newly recorded fungal pathogens of almond in South Australia. Search 3(3):86–87.

Ogawa, J. M., E. E. Wilson, and Harley English. 1959. The leaf blight of almond and its control. Hilgardia 28:239–54.

Shoemaker, R. A., and E. Müller. 1964. Generic correlations and concepts: *Clathridium* (=*Griphosphaeria*) and *Seimatosporium* (=*Sporocadus*). Can. J. Bot. 42:403–10.

Sutton, B. C. 1980. *The Coelomycetes*. Commonwealth Mycological Institute, Kew, England, 696 pp.

Zeller, S. M. 1925. *Coryneum ruborum* Oud., and its ascigerous stage. Mycologia 17:33–41.

———. 1927. A correction. Mycologia 19:150–51.

Leaf Curl of Cherry

Taphrina cerasi (Fckl.) Sadeb.

Cherry leaf curl, also known as "witches'-broom," occurs worldwide, having been reported from Europe, Australia, New Zealand, South Africa, Japan, and North America. In Oregon it occurs every year, but in California it has been found only in Napa County in 1929 (Mix 1949), in Berkeley in 1935 (Mix 1949), and in San Joaquin County in 1970. The disease is not of economic importance in California.

Symptoms. Symptoms on the leaves resemble the typical leaf curl found on peaches and nectarines, with curled, slightly thickened leaves that have a reddish coloration (**fig. 12**). On branches, large, broomlike tufts of shoots develop; they are distinguishable at blossom time because they have few flowers and develop a full flush of leaves earlier than healthy branches. These branches bear no fruit. The disease incidences in California in 1929 and 1970 showed only leaf symptoms.

Cause and disease development. *Taphrina cerasi* (Fckl.) Sadeb. attacks cultivated cherries (*Prunus avium* and *P. cerasus*). In California it has been observed only on *P. avium*. Ascospores, produced on leaves, attack the buds and stimulate excessive branching (a witches'-broom). The asci are hypophyllous, rarely amphigenous, clavate, rounded at the apex and provided with a stalk cell. There are eight ascospores per ascus. These spores are round, ovate, or elliptic, and often bud in the ascus. The dimensions of the clavate asci are 17 to 53 by 5 to 15 µm, the stalk cells are 5 to 26 by 4 to 12 µm and the ascospores are 3.5 to 9 by 3 to 6 µm (MacSwan and Koepsell 1974). Once infected, the branch produces a witches'-broom each year. The leaves on these branches also become infected; this indicates that the fungus is perennial within the shoot (MacSwan and Koepsell 1974). The abnormal growth of the branches is thought to be due to the production of cytokinins by the pathogen (Johnston and Trione 1974).

Control. Copper or sulfur has been recommended to prevent leaf infection (Smith 1941). When witches'-broom occurs, the supporting branch should be pruned at least 12 inches below the point of abnormal branching (MacSwan and Koepsell 1974).

REFERENCES

Johnston, T. C., and E. J. Trione. 1974. Cytokinin production by the fungi *Taprina cerasi* and *T. deformans*. *Can. J. Bot.* 52:1583–89.

MacSwan, I. C., and P. A. Koepsell. 1974. *Oregon plant disease control handbook.* Oregon State University Bookstores, Corvallis, 187 pp.

Mix, A. J. 1949. A monograph of the genus *Taphrina*. *Univ. Kansas Sci. Bull.* 33(1), 167 pp.

Smith, R. E. 1941. Diseases of fruits and nuts. *Calif. Agric. Ext. Circ.* 120, 168 pp.

Leaf Curl of Nectarine and Peach

Taphrina deformans (Berk.) Tulasne

Peach leaf curl, described as early as 1821 in England, is now practically coextensive with peach and nectarine culture (Eftimiu 1927). The disease has also been reported on almond in Europe and New Zealand and on apricot in New Zealand (Atkinson 1971), Australia (**fig. 13E**), Argentina, India, the USSR, and South Carolina, but it has not been recorded on these latter hosts in the western United States nor on trees grown in tropical climates.

In spite of control measures, the loss of peach fruit in the United States in 1921 was 254,000 bushels (Anonymous 1922), in 1922 it was 145,000 bushels (Anonymous 1923), and in 1924 it was 840,000 bushels (Anonymous 1925). Losses are negligible today with proper control.

Symptoms. In spring, reddish or yellowish areas appear on the developing leaves (**fig. 13A, B**). These areas progressively become thick and puckered, causing the leaf to curl dorsally and soon fall. Green shoots are also frequently invaded; they become thickened and distorted (**fig.13B**) and may die (**fig. 13D**). Blossoms, which are sometimes infected, shrivel and fall. Though fruit infection is rare, fruit sometimes develop irregular, somewhat elevated, wrinkled, reddish lesions (**fig. 13C**).

Cause. The causal fungus has been called *Taphrina deformans* (Berk.) Tulasne (= *Exoascus deformans* (Berk.) Fuckel) and *Exoascus amygdali* (Jaczewski) (Mix 1949). Asci of the fungus, which form mostly on the upper leaf surface, are cylindric-clavate, rounded or truncate at the apex, and provided with a stalk cell. They measure 17 to 56 by 7 to 15 μm and form a compact whitish hymenium. The eight ascospores, which frequently bud within the ascus, are round, ovate, or elliptic, and measure 3 to 7 μm in diameter. The ascus wall is unitunicate and splits at the apex to release the ascospores (Syrop and Beckett 1976). Pierce (1900) described the "vegetative" and "fruiting" hyphae of the fungus, the former being more threadlike and with longer cells than the latter. The fruiting hyphae are composed of irregular short cells, mostly located just beneath the cuticle of the upper leaf surface. The mycelium in infected shoots is largely vegetative; it is intercellular in all host tissue.

Because of differences in pathogenicity among isolates from different hosts, Schneider (1971) suggested that *T. deformans* be divided into two varieties, namely, *persicae* on peach and *amygdali* on almond. More recently Lorenz (1976a), on the basis of inoculations of 20 isolates into 12 peach-hybrid clones, concluded that *T. deformans* comprises several physiological races. More research is needed to clarify the variety-race situation in this species.

Disease development. The fungus was first believed to survive from year to year as perennial mycelium in twigs (Sadebeck 1893), but Pierce (1900) recognized that such a concept did not explain the behavior of the disease in the orchard. Caporali (1965) stated that hyphae rarely survive through January in the cortical tumors of branches. Some workers believe that the fungus survives California's hot, dry summers as ascospores. When fall rains begin, the ascospores germinate, giving rise to bud-conidia (blastospores), which were seen by most earlier workers, including Pierce (1900), and which were studied by Caporali (1965). As a result of the successive budding of the blastospores, the twigs and buds of the trees are covered by a film of spores (Fitzpatrick 1934). These blastospores were found by Lorenz (1976b) to reproduce on shoot tips and thus keep pace with the growth of the host during the entire season. Ascospores are capable of surviving several months of dry, hot weather. Mix (1924) showed that bud-conidia dried on glass slides survived for 140 days at 30°C and 315 days at lower temperatures. Fitzpatrick (1934) found that these conidia in a dry condition survived one and one-half years. Dissemination is only by water, with little evidence of lateral drift of spores from unsprayed trees to nearby sprayed trees (Wilson 1937).

Infection was observed by Fitzpatrick (1934) to occur as follows: The spore fastens itself to the leaf surface by means of a short infection thread that grows out of the spore at a position usually occupied by the daughter bud. As the infection thread presses against the leaf cuticle, the cuticle thickens at this point. The contents of the spore pass into the infection peg within the thickened cuticle, and the spore collapses. Infection may occur through either surface of the leaf. Martin (1925) claimed to have found stomatal penetration, but Fitzpatrick did not. The nuclear condition of the penetrating hyphae (dicaryon) and the cells formed from budding ascospores (both uninucleate and binucleate) is described by Caporali (1965). Lorenz (1976b) found that infection occurs only if blastospores come into contact with undifferentiated host tissue. He hypothesized that the fungus changes from the yeastlike saprophytic phase to the mycelial parasitic phase in response to some substance excreted by this tissue.

Some controversy still exists as to the exact time of the initial infection of the leaf, although Fitzpatrick (1934) showed by means of spray application that most if not all infection occurs after the leaves are fully emerged from the bud. Young leaves may be repeatedly infected by spores under favorable climatic conditions, such as the recurrent fog found in the coastal area.

Fitzpatrick (1934) stated that in eastern Canada summer infection rarely occurs. This he considered to be evidence that the current crop of ascospores does not take part in infection during the summer in which they are produced. He found that at a mean temperature of 7°C (44°F), the time between inoculation and the first recognizable symptoms was 14 days.

Symptoms on leaves do not develop until the slender hyphae move inward between the epidermal cells and reach the palisade or spongy parenchyma cells, depending on which surface infection occurs. The host cells fail to differentiate, begin to enlarge, lose their chlorophyll, and often accumulate a red pigment in large vacuoles. Thus, only host cells in direct contact with the mycelium are affected, and a single infection may involve the entire leaf or produce a sharply delineated and isolated lesion. Later, the upper surface of the swollen areas appears white (**fig. 13B**); this is due to production of asci and ascospores.

Several studies have been made of the role that plant hormones may play in the malformation of host tissue by *T. deformans* (Johnston and Trione 1974; Sommer 1961; Sziraki, Balazs, and Kiraly 1975). The fungus produces auxin and cytokinins in culture. There is evidence for increased cytokinin activity and increased levels of indole-3-acetic acid and tryptophan in curled peach leaves. It is postulated that the increased cell growth and water uptake of affected tissues is related to the higher auxin content; the stimulated cell division and abnormal growth of infected leaves could be the result of increased cytokinin activity (Sziraki, Balazs, and Kiraly 1975). Since the stimulus for hyperplasia and hypertrophy in peach leaves is not translocated to uninfected areas, this suggests that the immobile cytokinins, rather than the mobile auxins, may be responsible for the abnormal growth (Johnston and Trione 1974).

Yarwood (1941) studied the diurmal cycle of ascus maturation in April and May on ornamental peaches. Binucleate ascogenous cells ocurred between the epidermal cells and cuticle, and the ascogenous cells underwent nuclear fusion and elongation between 5 and 9 P.M. Septa were formed by midnight, and three successive nuclear divisions in the ascus were completed by 5 A.M. By 8 A.M., ascospores were delimited; discharge began in the afternoon, reaching the maximum by about 8 P.M. The ascospores were forcibly ejected as octads surrounded by ascus epiplasm.

Caporali (1965) used a medium rich in glucosides, organic and inorganic nitrogen compounds, and mineral salts to complete the life cycle of the fungus in culture. The dicaryotic mycelial threads, ascogenous cells, and asci with mature ascospores were similar to those observed on naturally infected peach leaves.

Periods of cool, wet weather at leaf-emergence favor leaf curl development. The optimum temperature for growth of the fungus in culture is about 20°C, the minimum about 8.9°C, and the maximum between 26° and 30.5°C (Lorenz 1976a). Budding of the blastospores occurs at a relative humidity of 95 percent and above. It is probable that cool weather prolongs the period of leaf susceptibility by checking leaf growth.

Control. Control of leaf curl is largely dependent on chemicals because very few peach and nectarine cultivars are resistant, and sanitation and cultural practices have proved ineffective. Many chemicals, including some of the standard fungicides, effectively control the disease if applied at the proper time. In the eastern United States, fungicides usually are applied in early spring before the leaf buds begin to open. In California, leaf curl and shothole are controlled by a single spray applied after leaf fall in autumn. Copper fungicides are commonly used because of their long-lasting residual quality. English (1958) showed that late-fall dimethyldithiocarbamate applications on peaches in California provide better control than coppers, captan, dichlone, or nabam plus coordinated salts, although all provide significant disease reduction. Captafol is superior to other chemicals tested in two other regions (MacSwan 1970; Northover 1978), but it has not been registered for use in California. In recent tests in Chile (Pinto de T. and Carreño 1988), the following experimental sterol biosynthesis inhibiting fungicides gave effective leaf curl control: flusilazole, hexaconazole, myclobutanil, penconazole, and triforine. Other compounds reportedly effective are calcium polysulfides (McCain 1983), dodine, and chlorothalonil. In orchards where a high inoculum potential exists or in years when especially heavy rains occur, two fungicide applications, one in late fall and a second at delayed dormant, are recommended.

Ritchie and Werner (1981) recently reported on the susceptibility of 67 peach and 11 nectarine cultivars to leaf curl. None are immune, but there is great variation in susceptibility. Redhaven and most cultivars derived from it are tolerant to leaf curl, whereas Redskin and cultivars derived from it are susceptible to highly susceptible. Further breeding could result in additional commercially desirable cultivars with resistance to this disease.

REFERENCES

Anonymous. 1922. Losses due to peach leaf curl in 1921. *Plant Dis. Rep. Sup.* 24:510.

———. 1923. Losses due to peach leaf curl in 1922. *Plant Dis. Rep. Sup.* 30:484–85.

———. 1924. Varietal susceptibility to peach leaf curl in 1924. *Plant Dis. Rep. Sup.* 33:96–97.

———. 1925. Losses due to peach leaf curl in 1924. *Plant Dis. Rep. Sup.* 43:404–05.

Atkinson, G. F. 1894. Leaf curl and plum pocket. *Cornell Agric. Exp. Stn. Bull.* 73.

Atkinson, J. B. 1971. *Diseases of tree fruits in New Zealand.* A. R. Sherrer, Wellington, 406 pp.

Brefeld, O. 1891. Die Formen der Ascomyceten und ihre Kulturen in Nährlösungen. *Untersuch. Gesammtgeb. Mykol.* 9:119–49.

Caporali, L. 1964. Nouvelles observations sur la biologie du *Taphrina deformans* (Berk.) Tul. *Ext. Ann. Inst. Nat. Agron.* Vol. 2.

———. 1965. Le comportement du *Taphrina deformans* (Berk.) Tul. in vitro. *Rev. Cytol. Biol. Veg.* 28:301–413.

Cation, D. 1935. One spray controls peach leaf curl. *Mich. Agric. Exp. Stn. Quart. Bull.* 18(2):86–88.

Dangeard, P. A. 1895. La production sexuelle des Ascomycetes. *Le Botaniste* 4:21–58.

Duggar, B. M. 1899. Peach leaf curl. *Cornell Agric. Exp. Stn. Bull.* 164:371–88.

Eftimiu, P. 1927. Contribution à l'étude cytologique des Exoascées. *Le Botaniste* 18:1–154.

English, H. 1958. Fall applications of ziram and ferbam effectively control peach leaf curl in California. *Plant Dis. Rep.* 42:384–86.

Fitzpatrick, R. E. 1934. The life history and parasitism of *Taphrina deformans*. *Sci. Agric.* 14(16):305–26.

———. 1935. Further studies on the parasitism of *Taphrina deformans*. *Sci. Agric.* 15:341–44.

Grady, E. E., and F. T. Wolf. 1959. The production of indole acetic acid by *Taphrina deformans* and *Dibotryon morbosum*. *Physiol. Plant.* 12:526–33.

Johnston, J. C., and E. J. Trione. 1974. Cytokinin production by the fungi *Taphrina cerasi* and *T. deformans*. *Can. J. Bot.* 52:1583–89.

Knowles, Etta. 1887. The curl of the peach leaves; a study of abnormal structure induced by *Exoascus deformans*. *Bot. Gaz.* 12:216–18.

Link, G. K. K., H. W. Wilcox, and V. Eggers. 1938. Growth tests with extracts of *Erwinia amylovora*, *Phytomonas rhizogenes*, *Taphrina cerasi*, *T. deformans*, and *Ustilago zeae*. *Phytopathology* 28:15.

Lorenz, D. H. 1976a. Untersuchungen über das Pathogenitätsverhalten von *Taphrina deformans* (Berk.) Tul. *Phytopathol. Z.* 85:333–44.

———. 1976b. Beitrage zür weiteren Kenntnis des Lebenszyklus von *Taphrina deformans* (Berk.) Tul. unter besonderer Berücksichtigung der Saprophase. *Phytopathol. Z.* 86:1–15.

MacSwan, I. 1970. Diseases of stone fruits (abs.) *Proc. 44th Ann. West. Coop. Spray Project* (Jan. 14–16, 1970, Portland):17.

Martin, E. M. 1925. Cultural and morphological studies of some species of *Taphrina*. *Phytopathology* 15:67–76.

———. 1927. Cytological studies on the Exoascaceae: *Taphrina johansonii* and *Taphrina deformans*. *J. Elisha Mitchell Sci. Soc.* 43:15–16.

———. 1940. The morphology and cytology of *Taphrina deformans*. *Am. J. Bot.* 27:743–51.

McCain, A. H. 1983. Leaf curl fungicides. *Calif. Plant Pathol.* 64:1–2.

Mix, A. J. 1924. Biological and cultural studies of *Exoascus deformans*. *Phytopathology* 14:217–33.

———. 1925. The weather and peach leaf curl in eastern Kansas in 1924. *Phytopathology* 15:244–45.

———. 1929. Further studies on Exoascaceae. *Phytopathology* 19:90.

———. 1935. Life history of *Taphrina deformans*. *Phytopathology* 25:41–66.

———. 1949. A monograph of the genus *Taphrina*. *Univ. Kansas Sci. Bull.* 33 (Part I, No. 1):3–167.

———. 1953. Differentiation of species of *Taphrina* in culture. Utilization of nitrogen compounds. *Mycologia* 45:649–70.

Northover, J. 1978. Prevention of peach leaf curl, caused by *Taphrina deformans*, with preharvest and pre-leaf fall fungicide applications. *Plant Dis. Rep.* 62:706–09.

Ogawa, J. M., I. C. MacSwan, B. T. Manji, and F. J. Schick. 1977. Evaluation of fungicides with and without adhesives for control of peach diseases under low and high rainfall. *Plant Dis. Rep.* 61:672–74.

Pierce, N. B. 1900. *Peach leaf curl: Its nature and treatment.* USDA Div. Veg. Path. Physiol. Bull. 20, 204 pp.

Pinto, de T., A., and I. Carreño. 1988. Control quimico de Monilia, oidio, cloca, y corineo en nectarinos Late Lagrand. *Agric. Tech. (Chile)* 48(2):106–10.

Ritchie, D. F., and D. J. Werner. 1981. Susceptibility and inheritance of susceptibility to peach leaf curl in peach and nectarine cultivars. *Plant Dis.* 65:731–34.

Roberts, C., and J. T. Barrett. 1944. Intercellular mycelium of *Taphrina deformans* in peach fruit. *Phytopathology* 34:977–79.

Sadebeck, R. 1893. Die parasitischen Exoasceen. *Jahrb. Hamburg Wiss. Anst.* 10:1–110.

Schneider, A. 1971. Mise en évidence de deux variétés de *Taphrina deformans* parasites l'une du pêcher, l'autre de l'amandier. *C. R. Hebdo. Séances Acad. Sci. Paris.* D273:685–88.

Selby, A. D. 1898. Preliminary report upon diseases of the peach. *Ohio Agric. Exp. Stn. Bull.* 92:226–31.

———. 1899. Variations in the amount of leaf curl of the peach (*Exoascus deformans*) in the light of weather conditions. *Proc. Assoc. Prom. Agric. Sci. Ann. Mtg.* 20:98–104.

Sommer, N. F. 1961. Production by *Taphrina deformans* of substances stimulating cell elongation and division. *Physiol. Plant.* 14:460–69.

Syrop, M., and A. Beckett. 1976. Leaf curl disease of almond caused by *Taphrina deformans*. III. Ultrastructural cytology of the pathogen. *Can. J. Bot.* 54:293–305.

Sziraki, I., E. Balazs, and Z. Kiraly. 1975. Increased levels of cytokinin and indoleacetic acid in peach leaves infected with *Taphrina deformans*. *Physiol. Plant Pathol.* 5:45–50.

Wallace, E., and H. H. Whetzel. 1910. Peach leaf curl. *Cornell Agric. Exp. Stn. Bull.* 276:157–78.

Wilson, E. E. 1937. Control of peach leaf curl by autumn applications of various fungicides. *Phytopathology* 27:110–12.

Yarwood C. E. 1941. Diurnal cycle of ascus maturation of *Taphrina deformans*. *Am. J. Bot.* 28:355–57.

Leaf Spot of Cherry

Blumeiella jaapii (Rehm) von Arx
(*Coccomyces hiemalis* Hig.)

Leaf spot does not occur in the major sweet-cherry-growing areas of California but sometimes is abundant in the coastal districts. Sour cherry, a common host in several eastern and midwestern states, is not grown commercially in California. The names "yellow leaf" and "shothole disease" have sometimes been used because the leaves turn yellow soon after symptoms develop (**fig. 14A**), and the necrotic circular lesions often drop out.

This disease was first described in Finland in 1885 (Karsten 1885), in America by Peck from New York in 1878, and in the coastal districts of California by Smith in 1941. The organism was first described as *Cylindrosporium padi* in 1884 by Karsten (1885), while in America the fungus name used was *Septoria cerasina* (Peck 1878). The ascigerous stage was found by Arthur in 1887 but he did not name it (Arthur 1887). Higgins (1913, 1914) described three species, with *Coccomyces hiemalis* occurring commonly on *Prunus cerasus*, *P. avium*, and *P. pennsylvanicum*. The other species were *C. prunophorae* on plum and *C. lutescens* on mahaleb cherry. There appears to be some question regarding the validity of the latter species as they are not currently listed as pathogens of these hosts (Farr et al. 1989).

The damage from this disease is extensive defoliation and debilitation of the tree resulting in reduction of fruit set and size. In regions with severe cold temperatures, the lack of stored sugars in the wood of affected trees causes freezing injury and tree death (Jones and Sutton 1984).

Symptoms. The characteristic symptoms on the leaves are numerous, minute, purple spots on the upper leaf surface (**fig. 14C**) that enlarge and become necrotic. The lesions may fall out, causing a shothole effect.

Affected leaves turn yellow and fall off early in the season, but frequently the area around the infected spots remains green, giving the leaf a mottled appearance. Infections may also occur on the petioles, fruit, and pedicels (**fig. 14B**) (Anderson 1956).

Cause and disease development. The name of the teleomorph is *Blumeriella jaapii* (Rehm) von Arx; the anamorph is *Phloeosporella padi* (Lib.) von Arx (=*Cylindrosporium padi* [Lib.] Karst. ex Sacc.). Infections occur through the stomata, and an intercellular, septate mycelium (stroma) is formed between the epidermis and mesophyll tissue. Acervuli are formed on the stroma, and the hyaline, elongated, curved, or flexuous conidia break through the lower epidermis as a creamy mass that is seen as a whitish speck (**fig. 14C**). The conidia are 45 to 60 by 2.4 μm with one or two septa. In the spring, apothecia arise from the same stromata in fallen leaves, and produce ascospores (30 to 50 μm long by 3.5 to 4.5 μm wide) that are forcibly ejected when the leaves are thoroughly soaked. Initial infections of unfolding leaves are from ascospores, and conidia are responsible for much of the secondary infection during rains.

Control. No highly resistant cultivars are available for either sweet or sour cherry, but, of the two, the sweet types generally have somewhat greater resistance (Sjulin, Jones, and Andersen 1989). Eradicant dinitro sprays applied on leaves on the orchard floor have been reported to reduce initial inoculum (Anderson 1956). Three protective sprays have been recommended in Oregon (MacSwan and Koepsell 1974), starting at the petal-fall stage of bloom, with such fungicides as captan, dodine, ferbam, and dichlone. Bordeaux mixture or lime sulfur applied soon after fruit set was suggested by Smith (Smith 1941).

REFERENCES

Anderson, H. W. 1956. *Diseases of fruit crops.* McGraw-Hill, New York, 501 pp.

Arthur, J. C. 1887. Report of the botanist. *N.Y. (Geneva) Agric. Exp. Stn. Rept.* 5:259–96.

———. 1888.. Report of the botanist. *N.Y. (Geneva) Exp. Stn. Rept.* 6:343–71.

Farr, D. F., G. F. Bills, G. P. Chamuris, and A. Y. Rossman. 1989. *Fungi on plants and plant products in the United States.* APS Press, St. Paul, 1252 pp.

Higgins, B. B. 1913. The perfect stage of *Cylindrosporium* on *Prunus avium. Science.* 37:637–38.

———. 1914. Contribution to the life history and physiology of *Cylindrosporium* on stone fruits. *Am. J. Bot.* 1:145–73.

Jones, A. L., and T. B. Sutton. 1984. Diseases of tree fruits. *N. Central Reg. Ext. Publ.* 45, Coop. Ext. Serv. Mich. State Univ. 59 pp.

Karsten, P. A. 1885. Symbolae ad mycoligiam Fennicam, XVI. *Medel. Soc. Fauna Flora Fenn.* 11:148–61.

Keitt, G. W. 1918. Inoculation experiments witih species of *Coccomyces* from stone fruits. *J. Agric. Res.* 13:539–69.

Keitt, G. W., E. C. Blodgett, E. E. Wilson, and R. O. Magie. 1937. The epidemiology and control of cherry leaf spot. *Wisc. Agric. Exp. Stn. Res. Bull.* 132.

MacSwan, I. C., and P. A. Koepsell. 1974. *Oregon plant disease control handbook.* Oregon State University Bookstores, Corvallis, 187 pp.

Peck, C. H. 1878. Report of the botanist. *N.Y. State Mus. Nat. Hist. Rept.* 29:29–82.

Sjulin, T. M., A. L. Jones, and R. L. Andersen. 1989. Expression of partial resistance to cherry leaf spot in cultivars of sweet, sour, duke, and European ground cherry. *Plant Dis.* 73:56–61.

Smith, R. E. 1941. Diseases of fruits and nuts. *Calif. Agric. Ext. Serv. Circ.* 120, 168 pp.

Phytophthora Pruning Wound Canker of Almond

Phytophthora syringae (Kleb.) Kleb.

Since 1982, gumming cankers in mature, bearing almond trees in California have been observed to be associated with pruning wounds made in fall and winter (Bostock and Doster 1985; Doster and Bostock 1988b). Some of the symptoms resembled those of Ceratocystis canker (DeVay et al. 1968). Yet, isolations from the cankers yielded only *Phytophthora syringae*. This fungus has also been isolated from pruning wounds on apricots and French prune in California (Doster and Bostock 1988e) and has been reported to cause pruning-wound cankers on young peach and apricot trees in New Zealand (Smith 1956).

Economic importance. Pruning-wound infections caused by *Phytophthora syringae* were found in more than 20 almond cultivars in seven orchards in the Central Valley of California (Bostock and Doster 1985). This fungus occurs in orchard soils throughout the almond producing regions of the Central Valley. The incidence of Phytophthora pruning wound canker can be quite high, especially in years with heavy precippitation during or shortly after pruning. In one surveyed orchard, 23 percent of the larger wounds were infected, with one portion of the orchard showing 50 percent infection (Doster and Bostock 1988b). The disease reduces the vigor of affected limbs and can cause a girdling and killing of branches, especially on young trees (Bostock and Doster 1985).

Symptoms. The disease is distinguished from Ceratocystis canker by its association with pruning wounds made during the cool fall and winter seasons. In Ceratocystis canker infection occurs almost entirely through injuries caused by harvest operations (DeVay et al. 1968). In two almond orchards examined in 1984, 11 and 23 percent of the pruning wounds had cankers, and these were located at heights from 0.5 m to more than 5 m (Doster and Bostock 1988b). Infections cause the profuse development of sour-smelling gum, which is extruded onto the bark surface as large amber-colored balls (**fig. 15**). With active cankers, removal of the outer bark reveals irregularly zonate areas that vary in color from greenish yellow to brown. These cankers appear in winter soon after pruning, remain active during spring, and seem to "die out" in summer. In this regard they differ from Ceratocystis cankers, which usually are perennial. The bark over old cankers appears sunken.

Causal organism. *Phytophthora syringae* (Kleb.) Kleb. is a widespread soil-inhabiting fungus commonly associated with a crown and root rot disease of stone fruits and nuts in the Central Valley of California. The pathogen can be isolated from pruning wound cankers during the winter and until mid-May, but not in June or thereafter. All isolates obtained to date from these cankers conform to the description of *P. syringae* as given by Waterhouse (1963). Oospores are formed in culture after several weeks at 5° and 9°C; (Doster and Bostock 1988a); they have an average diameter of 30 μm and are dark yellow to light brown. The antheridia are paragynous. The semipapillate sporangia are borne on short stalks, are ovoid to obpyriform, and have an average dimension of 61 by 31 μm. On synthetic media, all isolates have a typical petaloid growth pattern and colonies frequently have numerous hyphal swellings. The growth optimum is 21°C, with a minimum of 2°C and maximum of 27°C (Bostock and Doster 1985). Culture media amended with vegetable oil or made from leaves of stone fruit trees are favorable for the production of oospores. Temperatures exceeding 12°C are not conducive to oospore formation *in vitro* or in almond leaves (Doster and Bostock 1988a).

Disease development. Pruning wounds made during fall, winter, and early spring are susceptible to infection. In general, larger (20 to 40 mm diameter) pruning wounds are more susceptible than smaller ones (10 to 15 mm diameter) (Doster and Bostock 1988b). Temperature plays an important role in the resistance of pruning wounds to infection. Wounds become resistant much more rapidly at 25°C than at 6°C (Doster and Bostock 1988d). As the temperature decreases, less lignin, suberin, and ligninthioglycolic acid are detectable in almond bark wounds (Doster and Bostock 1988c,d). In orchard trees, pruning wounds made in fall and winter become decreasingly susceptible to infection with time and are almost immune after six weeks.

All the major almond cultivars, including Nonpareil, Ne Plus Ultra, Mission, and Price, are susceptible. Inoculation of fresh pruning wounds on one-year-old potted Nonpareil almond trees showed that at 6°C cankers reached a length of 10 cm within one month. They were only slightly larger at 25°C (Doster and Bostock 1988d). In excised almond shoots, *P. syringae* was virulent over a range of temperatures from 2° to 20°C (Bostock and Doster 1985). Uninjured bark was not susceptible to infection. Common winter conditions of cool temperature and rain appear to favor infection and disease development. The cankers become inactive in June, and isolations of the fungus are not possible thereafter. Since the pathogen does not appear to survive from one year to the next in cankers but to inhabit the soil and fallen leaves, it is presumed that the infective propagules are sporangia or zoospores that arise from the orchard floor. It seems probable that the inocula are zoospores, as the sporangia of *P. syringae* are considered to be nondeciduous. Neither inoculum dispersal nor the exact environmental parameters for infection have been studied.

Control. The application of fosetyl-Al to fresh pruning wounds by means of a paint brush prevented infection by inoculum consisting of either zoospores or mycelial plugs (Doster and Bostock 1988f). The protection provided by this material persisted for the three-to six-week period when untreated wounds are susceptible to infection. Cupric hydroxide was less effective than fosetyl-A1, and showed evidence of phytotoxicity. Another method of managing pruning wound canker is to prune at a time other than during the cool, rainy periods of late fall and winter (Bostock and Doster 1987). In mature orchards, pruning in early fall would take advantage of the relatively warm weather and rapid development of natural resistance. In young nonbearing orchards, delaying pruning until late winter or early spring, when trees are approaching the resistant stage, would be worth consideration.

REFERENCES

Bostock, R. M., and M. A. Doster. 1985. Association of *Phytophthora syringae* with pruning wound cankers of almond trees. *Plant Dis.* 69:568–71.

———. 1987. Managing canker diseases of almonds. *Almond Facts* 52(2):36–37.

DeVay, J. E., F. Lukezic, H. English, M. Trujillo, and W. J. Moller. 1968. Ceratocystis canker of deciduous fruit trees. *Phytopathology* 58:949–54.

Doster, M. A., and R. M. Bostock. 1988a. The effect of temperature and type of medium on oospore production by *Phytophthora syringae*. *Mycologia* 80:77–81.

———. 1988b. Incidence, distribution, and development of pruning wound cankers caused by *Phytophthora syringae* in almond orchards in California. *Phytopathology* 78:468–72.

———. 1988c. Quantification of lignin formation in almond bark in response to wounding and infection by *Phytophthora* species. *Phytopathology* 78:473–77.

———. 1988d. Effects of low temperature on resistance of almond trees to Phytophthora pruning wound cankers in relation to lignin and suberin formation in wounded bark tissue. *Phytopathology* 78:478–83.

———. 1988e. Susceptibility of almond cultivars and stone fruit species to pruning wound cankers caused by *Phytophthora syringae*. *Plant Dis.* 72:490–92.

———. 1988f. Chemical protection of almond pruning wounds from infection by *Phytophthora syringae*. *Plant Dis.* 72:492–94.

Smith, H. C. 1956. Collar-rot of apricots, peaches, and cherries. *Orchard. N. Z.* 29:22–23.

Waterhouse, G. M. 1963. Key to the species of *Phytophthora* de Bary. *Mycol. Pap.* 92. *Commonw. Mycol. Inst.*, Kew, England, 22 pp.

Phytophthora Root and Crown Rot of Stone Fruit

Phytophthora spp.

Phytophthora root and crown rot can cause severe tree losses in years of heavy rainfall, but because of constant changes in fruit culture and the introduction of new rootstocks, this disease is always a threat. In 1975, 15 to 25 percent of the mature cherry trees in San Joaquin County were killed by crown rot. Whole plantings of young and mature almond and peach trees have occasionally been killed. Least affected have been prunes.

To reduce infection of the crown, the trees are sometimes planted on soil ridges (berms) with the upper roots near the soil level. This keeps rain and irrigation water from accumulating around the crowns. Heavy clay soils should be avoided or provided with good drainage. Proper choice of rootstocks is important, but in making such choices other disease problems must also be carefully considered. Details of this disease are discussed in Chapter 4.

Plum Pockets

Taphrina spp.

"Plum pockets," "bladder plum," "mock plums," or "fools" are names given to a disease that attacks plum species. The disease occurs to some extent in California on American species of plums, but is rarely found on *Prunus domestica* (European plum) or *Prunus salicina* (Oriental plum) (Smith 1941), which are the most common species grown in California. The only definite record of its occurrence on either of these species in California was on the cultivar Burbank (*P. salicina*) in Placer County. The disease is fairly common on the native Pacific or Sierra plum (*P. subcordata*) in California, Oregon, and Colorado, and is reported on French prune and Italian prune (*P. domestica*) in Oregon (Heinis 1960; Mix 1949; Zeller 1927).

Symptoms. As the name implies, the most distinctive phase of the disease occurs on the fruit (**fig. 16**). Infected plum fruit become enlarged, bladderlike objects without pits and with leathery flesh. The surface of the fruit is covered by a whitish coating of asci that project above the cuticle. Other symptoms often encountered are curled leaves, swollen shoots, and witches'-brooms similar to those in cherry caused by *T. cerasi*.

Causal organism. Considerable confusion exists regarding the species of *Taphrina* that cause the pockets disease of plums in North America (Anderson 1956; Mix 1949). There is some question as to whether the European species *T. pruni* Tul. is present in North America despite its reported occurrence in the Northeast and in Canada (Anderson 1956; Mix 1949). *Taphrina communis* (Sadeb.) Giesenh. is found on several *Prunus* species in the eastern and midwestern states, but is not known to occur west of the Rocky Mountains. The only species positively known to infect plums in the Pacific Coast states is *T. pruni-subcordata* (Zeller) Mix (1949). Its asci are clavate, rounded or truncate at the apex, provided with a stalk cell, and with eight round, ovate, or elliptic ascospores that often bud within the ascus. Dimensions of the asci are 33 to 73 by 7 to 12 μm, and of the ascospores, 4 to 7 by 3.5 to 6 μm (Mix 1949).

Disease development. The exact cycle of this disease has not been studied in detail, but apparently infection is initiated in spring soon after the blossoms begin to open. Infection appears to be produced by the bud conidia (blastospores), which persist on the tree surfaces from one growing season to the next much like those of *T. deformans*.

Control. When necessary, a spray should be applied in spring before the blossom buds begin to open. Bordeaux mixture has been used, as have other equally persistent fungicides.

REFERENCES

Anderson, H. W. 1956. *Diseases of fruit crops*. McGraw-Hill, New York, 501 pp.

Atkinson, G. J. 1894. Leaf curl and plum pockets. *Exoascus pruni* Fckl. *Cornell Univ. Agric. Exp. Stn. Bull.* 73:329–30.

Galloway, B. J. 1889. Plum pocket. *U. S. Agric. Comm. Rept.* 1888:366–69.

Heinis, J. L. 1960. *Taphrina pruni-subcordata* (Zeller) Mix, causing witches'-broom on Italian and French prunes. *Plant Dis. Rep.* 44:630–31.

Mix, A. J. 1949. A monograph of the genus *Taphrina*. *Kansas Univ. Sci. Bull.* 33:3–67.

Smith, Ralph E. 1941. Diseases of fruit and nuts. *Calif. Agric. Exp. Stn. Ext. Serv. Circ.* 120, 168 pp.

Swingle, D. B., and H. E. Morris. 1918. Plum pocket and leaf gall of American plums. *Montana Agric. Exp. Stn. Bull.* 123:167–88.

Zeller S. M. 1927. Contribution to our knowledge of Oregon fungi—II. Mycological notes for 1925. *Mycologia* 19:130–43.

Powdery Mildew of Stone Fruit

Sphaerotheca pannosa (Wallr.:Fr.) Lév.,
Podosphaera tridactyla (Wallr.) de Bary,
P. clandestina (Wallr.:Fr.) Lév.,
P. leucotricha (E. & E.) Salm.

The typical climate in areas suited for stone-fruit production favors the attack and survival of powdery mildew fungi (Spencer 1978; Yarwood 1950, 1957). On orchard trees the primary destructive effects are on the fruit. In the absence of effective control measures, powdery mildew can spoil the fruit of apricot, cherry, nectarine, peach, and plum. In 1973 fruit infections were observed on almond and prune. On almond, the upper fruit surface appeared like "rusty spot" of peach, whereas on prune typical powdery mildew signs were present.

The susceptibility of stone fruit to powdery mildew varies greatly among cultivars. In peach, nonglandular Peak and Paloro are the most susceptible, while glandular cultivars such as Walton, Johnson, Halford, and Stuart are more resistant (Ogawa and Charles 1956). In cherry, Black Tartarian, Bing, and Chapman are most susceptible. Interplanting of susceptible cultivars for pollination increases the chance of losses. Early plum fruit infection results from their exposure to mildew on other hosts, such as nearby roses. Kelsey, Gaviota, and Wickson plums are the most susceptible plum cultivars. Mildew infection does not reduce the market quality of Beauty plum if fruit are brushed during postharvest treatment. In the spring of 1980, powdery mildew signs were found on the fruit of Black Beaut and Red Beaut plums in Fresno and Kern counties; at harvest, fruit russeting resulted in significant crop losses. In apricot, Blenheim is more susceptible than Tilton, but both can be severely affected.

Foliage infection by powdery mildew is of primary concern in nursery plantings, especially cherry. On other stone fruit, damage from infection on leaves in the nursery or orchard has not been fully assessed. It is recognized, however, that infected leaves are an important source of inoculum for fruit infection. Crop loss from the killing of buds appears to be minor, as is that from shoot blight.

Symptoms. Leaves, buds, green shoots, and fruit are commonly attacked, but flower infection is rare. As with apple, the inner scales of buds are also susceptible (Weinhold 1961a). The infection of various plant parts is largely determined by the species of the pathogen and the maturity of the host tissue. In general, younger leaves are more susceptible to *Sphaerotheca*, whereas older leaves, with the exception of sweet cherry, are more susceptible to *Podosphaera*. The first evidence of infection is a cobweblike growth on the leaf surface. Soon the fungus starts to sporulate, producing masses of white powdery spores. At this time the leaves show little change in color, but in a short time chlorosis and necrosis commonly occur. If the leaves are covered with the fungus they tend to curl upward and dehisce, but if infection is localized they usually remain on the tree. Infected shoots are commonly stunted, and infected lateral buds may be destroyed. On the twigs, mildew appears as a white, feltlike growth even in winter. On the leaves and shoots, the cleistothecia (*Podosphaera* only) first appear as yellow, then reddish brown, and finally black, spherical bodies. Cleistothecia of *Podosphaera clandestina* and *P. tridactyla* are common, but those of *Sphaerotheca pannosa* have never been found in California on fruit trees and only occasionally on rose.

Young fruit are susceptible and show typical mildew growth (**fig. 17A,B,D**). They usually are somewhat deformed, with either depressed or slightly raised areas. After the pit-hardening stage, peach and plum fruit become resistant, and the epidermis of infected areas becomes necrotic. This scabby condition remains visible on nectarine, peach, apricot, and some plum fruit even at maturity. Occasionally, infected areas on apricot and peach fruit remain green, but on Beauty plum these areas show no such effect. On sweet cherry fruit the infected areas are most noticeable near harvest time (**fig. 17F**). These areas are covered with a thin, weblike growth of whitish mycelium that makes the fruit unmarketable.

Causal organisms. Identification of the powdery mildew species on stone fruit is difficult because more than one species can occur on a single host. Furthermore, cleistothecia often are lacking or difficult to find. Zaracovitis (1965) placed the powdery mildews into three groups on the basis of rate of spore germination and appressorial formation in vitro at 21°C, in a saturated atmosphere and in the dark: (1) germination greater than 70 percent, with appressorial formation in five hours (*Uncinula necator*); (2) germination and formation of distinct appressoria in ten hours (*Erysiphe graminis*); and (3) germination less than 30 percent in five hours. Greater than 70 percent germination requires more than 10 hours (all species of *Podosphaera* and *Sphaerotheca*). The conidia of the third group germinate by producing straight, rather long germ tubes without well-differentiated appressoria. The powdery mildew fungi that attack stone fruit belong to this group.

Blumer (1967) classified the mildews on the basis of cleistothecial and conidial characteristics, and divided *P. oxyacanthae* into species such as *P. tridactyla* on apricot and *P. clandestina* on cherry. The species *P. oxyacanthae* is no longer recognized (Blumer 1967; Mukerji 1968a; Braun 1984). On samples of *Podosphaera* infected apricot, cherry, and plum collected in California, the cleistothecial appendages were located on the upper half of the fruiting body (sometimes more or less fasciculate), which indicates they belong in the species *P. tridactyla* (Blumer 1967; Braun 1984). The powdery mildews of almond and stone fruit are as follows:

Host	Pathogen
Almond (*Prunus dulcis* [Mill.] Webb)	*Podosphaera tridactyla* (Wallr.) de Bary (Weigle 1956) *Sphaerotheca pannosa* (Wallr.:Fr.) Lév. (Smith 1941)
Apricot (*P. armeniaca* L.)	*P. tridactyla* (Wallr.) de Bary (Yarwood, 1952) *S. pannosa* (Wallr.:Fr.) Lév. (Smith 1941)
Cherry (*P. avium* L.)	*P. tridactyla* (Wallr.) de Bary (Blumer 1967; English 1947)
Plum (*P. salicina* Lindl.)	*P. tridactyla* (Wallr.) de Bary (Ogawa, Hall, and Koepsell 1967) *S. pannosa* (Wallr.:Fr.) Lév. (Ogawa, Hall, and Koepsell 1967)

Prune (*P. domestica* L.)	Unidentified—California 1973
Peach (*P. persica* [L.]Batsch)	*P. clandestina* (Wallr.:Fr.) Lév. (Keil and Wilson 1961) *P. leucotricha* (E. & E.) Salm. (Manji 1972) *S. pannosa* (Wallr.) Lév. (Yarwood, 1939; Yarwood et al. 1954)

A description of *Podosphaera tridactyla* (Wallr.) de Bary (= *P. clandestina* [Wallr.]:Fr.) Lév. var. *tridactyla* (Wallr.) W. B. Cooke and *P. oxyacanthae* (DC.) de Bary var. *tridactyla* (Wallr.) Salm., as given by Mukerji (1968a), follows:

> Mycelium amphigenous, usually disappearing at maturity. Cleistothecia scattered or loosely aggregated, globose to subglobose, dark brown, 70 to 110 μm diam., cells of the peridium polygonal, up to 15 μm wide. Appendages 2–8, 1 to 8 times the diameter of the cleistothecium, usually unequal in length, springing in a cluster from the apex of the cleistothecium, more or less erect, upper half hyaline, lower half brown, broader at base, apex 3 to 6 times dicotomously branched, primary branches usually more or less elongated and sometimes slightly recurved, ultimate branches rounded, swollen and more or less knob-shaped. Ascus one per cleistothecium, globose to subglobose, 60–80 × 60–70 μm mwith no distinct stipe. Ascospores 8, ovate to ellipsoidal, 18–30 × 12–15 μm. Conidia few, formed apically in chains on septate conidiophores, hyaline, ellipsoidal, 22–45 × 14–20 μm with fibrosin granules.

The mildew on cherry leaves from Wisconsin and on cherry and plum leaves from California had appendages randomly located on the upper portion of the cleistothecium. They were not formed in a cluster at the apex of the cleistothecium as described above. However, according to the current concept of *P. tridactyla* (Braun 1984), the appendages may be either fasiculate or merely located on the upper half of the cleistothecium. This arrangement is distinctly different from that in *P. clandestina*, where the appendages are strictly radial or equatorial (Khairi and Preece 1975).

Podosphaera clandestina (Wallr.:Fr.) Lév. (= *P. oxyacanthae* [DC.] de Bary) is described by Khairi and Preece (1975) as follows:

> Mycelium aphigenous but mostly on upper surface, persistent. Cleistothecia in groups, embedded in a mass of old persistent mycelium on leaf surface, subglobose, black when mature, 86–102 μm diam. Apendages 4–12, unequal in length, 134–191 μm 1–2 times the diam. of the cleistothecia, septate, lower two thirds brown, uper third hyaline, apex 3 times dichotomously branched, equatorially or radially distributed on the cleistothecium. Ascus one per cleistothecium, ovate, 50–80 × 40–60 μm. Ascospores 8, subreniform 18 to 25 × 10 to 18 μm. Conidia are borne apically in chains on septate condiophores, with fibrosin granules, cylindrical, 18–23 × 9–10 μm.

Sphaerotheca pannosa (Wallr.:Fr.) Lév. (= *S. persicae* [Woronich.] Erikss.) is described by Mukerji (1968b) as follows:

> Mycelium persistent, on stem, thorns, calyx, fruits, patches (felt) first white, becoming gray to buff, composed of densely interwoven hyphae. Cleistothecia formed rarely, imbedded in mycelial felt, more frequently on stems, especially around thorns, globose to pyriform, 80–120 μm. diam. Peridial cells obscure, 10 μm wide. Appendages few, narrow, hyaline to slightly brown, septate, shorter than the diam. of cleistothecia. Ascus one per ascocarp, broadly oblong, subglobose to globose, 60–75 × 80–100 μm. Ascospores eight, oval, hyaline, 20–30 × 12–17 μm. Conidia in chains on long conidiophores, ellipsoidal, hyaline, 20–35 × 14–20 μm.

Disease development. Because of variations in the life cycle of the pathogens on the various stone-fruit species, the mildews on each host group are discussed separately.

On plum, as on apricot, *S. pannosa* is important on the young green fruit (Ogawa, Hall, and Koepsell 1967; Weigle and French 1956). *Sphaerotheca pannosa* does not overwinter on plum trees. At no time of the year are there signs of the fungus on the foliage or shoots of the plum.

Infection occurs shortly after the flowering period and is mostly found on fruit clusters in the interior of the tree. The higher incidence of infection in this area is probably related to the greater persistence of moisture or the higher humidity here. Mycelial growth on plum is evident in six days, and sporulation occurs in fourteen days. The spores from fourteen-day-old lesions are capable of producing the disease. The mildew spreads very quickly on the surface of plum fruit and on most cultivars causes chlorotic areas, which later turn into scabby lesions with no sign of the fungus. At the latter stage the fruit are resistant to new infection. The primary source of inoculum for both plum and apricot is rose bushes (Yarwood 1952). The fungus overwinters in the buds of rose, with about 50 percent of the terminal buds infected.

The conidia of *S. pannosa* are disseminated by wind, which results in a decreasing gradient of infection outward from the spore source. Cleistothecia are absent on plum and extremely rare on rose.

Lesions produced by *Podosphaera tridactyla* appear on apricot and plum leaves in late summer and autumn after fruit harvest (Ogawa, Hall, and Koepsell 1967). By November, cleistothecial production is heavy.

On apricot, *S. pannosa* is responsible for leaf and fruit infections (Smith 1941) in spring (**fig. 17B**), and *Podosphaera tridactyla* attacks leaves (Yarwood et al. 1954) in late summer and fall. The *S. pannosa* inoculum for infection comes from mildewed roses (**fig. 17C**), especially *Rosa banksiae* (Yarwood 1952) and Dorothy Perkins roses. In one instance (Weinhold 1961a), the only inoculum source found was a heavily infected peach tree. Secondary infection of leaves and fruit is minimal, because sporulation on the primary lesions of leaves and fruit is sparse. In California, cleistothecia of *S. pannosa* have not been found on apricot or peach and are rare on rose. Thus, the primary inoculum source appears to be the mycelium that overwinters in the dormant buds of peach and rose. The spread of mildew from peach to apricot, although suspected, is not proven. Mildew (*P. tridactyla*) infections seen after harvest occur initially on the upper surface of leaves located in the inner part of the leaf canopy. The infected leaves are not distorted nor do they drop prematurely. Abundant cleistothecial production occurs in late fall. No infection of stems or buds has been observed. The source of inoculum for infection is thought not to be infected buds, but cleistothecia, because new spring shoots do not develop mildew as they do in apple infected with *P. leucotricha*.

Though Keil and Wilson (1961) reported *P. oxycanthae* (*P. clandestina*) on peach seedlings in the eastern United States, only *S. pannosa* has been reported on peach from the western United States. The fungus *S. pannosa* overwinters as mycelium in the inner scales of the buds (Yarwood 1939) and the leaves are infected as they emerge from the bud. Secondary infection of young developing leaves occurs throughout the season, with the first symptoms becoming apparent about a month after the blooming period. Peach fruit are susceptible from the early stages of growth to about the beginning of pit hardening. According to Weinhold (1961a), neither foliage nor fruit infection appeared to be influenced by environmental conditions found in the orchard. The optimum temperature for germination of the conidia of *S. pannosa* on rose leaves is 21°C, with slight germination of 2°C and no germination at 37°C (Yarwood et al. 1954). An increase in moisture to saturation increases the germination percentage, and the presence of free moisture appears to be the primary requirement for germination (Longree 1939; Rogers 1959; Weinhold 1961a,b). The presence of sugars and amino acids in the water does not enhance germination. The resistance to infection of older leaves is thought to result from their thicker cuticle (Mence and Hildebrandt 1966; Weinhold 1961a).

Leaf infection is most apt to occur at night or in early morning, presumably because at that time the humidity is highest and the osmotic pressure of the leaves is lowest. This probably explains why mildew is more severe during overcast weather than during bright sunny days.

In New South Wales, Australia, Kable, Fried, and MacKenzie (198) reported on the spread of powdery mildew (*S. pannosa*) on peach. They found evidence of inoculum spread from rose to peach, as is the case with apricot. They indicated that the mildew from rose causes only fruit blemishes, with no secondary spread. They also determined that the isolation distance from the inoculum source to the peach orchard would be less than 60 m to hold the disease to less than 5 percent. There has not been an instance in California where *S. pannosa* infection was restricted to peach fruit.

Another fruit disorder, first reported on Elberta peach by Blodgett (1941), is "rusty spot" (**fig. 17E**). He found spots on young fruit in Idaho that resembled those left by drops of rusty water that had settled into the hairs on the fruit surface. Fruit collected by workers in Colorado showed the same condition. Sprague and Figaro (1956) were unable to detect any fungus associated with such areas. However, Daines et al. (1960) in New Jersey discovered mycelium in such lesions and found that the condition developed on peaches in the vicinity of mildew-infected apple trees. They reported that the rusty spots resembled old lesions produced by the common peach mildew fungus (*S. pannosa*) but differed in being more bare, by causing more host-cell necrosis, and by having a sparsity of fungus hyphae and sporulation on Rio Oso Gem, Goldeneast, and Summer Queen. In California, Manji (1972) reported an outbreak of rusty spot on Rio Oso Gem interplanted among Jonathan apple trees severely infected with mildew. Inoculations of peach fruit with both *S. pannosa* and mildew spores from Jonathan apple resulted in symptoms of typical peach mildew with the former and typical rusty spot with the latter.

More recently, in Illinois, Ries and Royse (1977) made microscopic examinations of rusty spot lesions and observed fungal hyphae on the peach epidermal hair, indicating the possibility that the causal agent is a fungus. They then related the incidence of rusty spot to the distance of peach trees from a powdery mildew-infected apple orchard (Ries and Royce 1978). This further supports the conclusion by Manji that rusty spot of peach is caused by the apple-mildew fungus. The optimum temperature for germination of the conidia of this mildew is 16°C.

Mildew on the foliage of cherry (*Prunus avium*) was reported by Galloway in 1889 and mentioned by Stewart in 1915, but it was not until 1947 that infection of the fruit was reported (Blumer 1967; English 1947). The pathogen (*P. tridactyla*) attacks young shoots and leaves as they emerge in the spring. The first signs of the mildew appear on the terminal leaves of the young shoots; at the time of shoot emergence the terminal growth is free of mildew. The source of inoculum is the overwintered cleistothecia (Grove 1987). Secondary infection occurs throughout the growing period, and by the time of fruit harvest numerous black cleistothecia have been formed, especially on the undersurface of the leaves. Fruit infection can be severe, especially after a period of cool, overcast weather or intermittent rain. Fruit infection does not continue to spread in storage.

Smith (1941) first reported powdery mildew on almond in the United States, in California. A few years later, Sprague (1954) recorded the disease on almond in Washington. Both Smith and Sprague considered the mildew to be *Sphaerotheca pannosa* (Wallr.:Fr.) Lév. In 1938 De Nardo reported *S. pannosa* on almond in Malta, and in 1940 Gianai reported *Phyllactinia salmonii* Blumer on almond in India. *Podosphaera tridactyla* (Wallr.) de Bary was reported on almond seedlings for the first time in California by Weigle (1956). Powdery mildew on almond is rare in California, where most of the commercial production of the United States is located. It occasionally is found on the tip leaves of young shoots of seedling almonds late in the fall.

Control. Mildew control on stone fruit has been based largely on the use of fungicides and the eradication of other-host sources of inoculum. Fungicdes have been used as either protectants or as eradicants that kill the mildew after infection. The latter fungicides are in the sterol biosynthesis-inhibitor group, such as triforine and triadimefon (Ogawa, Gubler, and Manji 1988).

For the control of mildews on stone fruit, the timing of fungicde applications appears to be very critical. On peach, Ogawa and Charles (1956) obtained excellent disease control with three spray applications, the first before any signs of mildew developed (March 24) the second at the time of dehiscence of the floral tube (April), and the third when first infections were visible and expanding (April 28). According to data of Weinhold (1961a), the third spray might have been too late because by this time the fruit could have become resistant to infection.

Liquid-lime sulfur, a mixture of lime-sulfur and wettable sulfur, or dinocap can significantly reduce shoot infection and increase the amount of marketable peach fruit. Actidione, however, causes injury to the leaves and induces the dropping of green fruit.

New classes or formulations of fungicides that are especially effective against *S. pannosa* on roses have not been tested sufficiently on peach. Fungicides of this type are benomyl, folpet, triforine, triadimefon, semicarbazone, and derivatives of cycloheximide. Several of these newer biosynthesis inhibiting compounds (namely flusilazole, hexaconazole, penconazole, myclobutanil, triadimefon, and triforine) were recently shown to be effective in controlling mildew on nectarines in Chile (Pinto de T. and Carreño 1988). Resistance of mildews to these compounds could later restrict their usage as has been reported for triadimefon on grapes (Ogawa, Gubler, and Manjii 1988). Studies comparing the performance of such compounds in controlling *Podosphaera* on stone fruit have not been made. Dinocap, benomyl, triadimefon, and sulfurs have been recommended for apple mildew control.

On apricot, sulfur fungicides are not recommended because of their phytotoxicity and, according to R.

Covey (personal communication) both dinocap and sulfur induce defoliation and fruit drop in orchards near Wenatchee, Washington. Removal of rose plantings near orchards reduces mildew infection on apricot and certain plum cultivars. The roses *Rosa banksiae* and Dorothy Perkins, both of which are very susceptible to mildew, are of greatest concern although some other cultivars are also a source of inoculum. Fruit of the plum cultivars Kelsey and Wickson are susceptible to *S. pannosa*, which can be effectively controlled by removal of nearby rose plantings. On the plum cultivars Red Beaut, Black Beaut, and Ambra, the mildew species has not been identified, but it can be controlled by the use of fungicide applications at late bloom (shuck split). On sweet cherry, as mildew symptoms develop on the leaves after fruit set and the fruit are susceptible to infection after the pit-hardening stage, control measures are started as soon as mildew symptoms appear on the inner shoots of scaffold branches. A spray of wettable sulfur, triforine, or triadimefon applied at this time prevents spread of the fungus to the maturing fruit. Vigorously growing cherry shoots, such as those on trees that have been severely pruned, are most prone to powdery mildew infection. But under high-temperature conditions (near 32°C in the month of June) new, vigorous growth of Bing cherry has been observed to be free of mildew infection.

REFERENCES

Bailey, L. H. 1949. *Manual of cultivated plants*. Macmillan, New York, 1116 pp.

Bender, C. L., and D. L. Coyier. 1984. Isolation and identification of races of *Sphaerotheca pannosa* var. *rosae*. *Phytopathology* 74:100–03.

Blodgett, E. C. 1941. Rusty spot of peach. *Plant Dis. Rep.* 25:27–28.

Blumer, S. 1967. *Echte Mehltaupilze (Erysiphaceae): Ein Bestimmungsbuch für die in Europa vorkommenden Arten*. Gustav Fischer, Jena, 436 pp.

Braun, U. 1984. Taxonomic notes on some powdery mildews (III). *Mycotaxon* 19:369–74.

———. 1987. *A monograph of the Erysiphales (Powdery mildews)*. Berlin: J. Cramer, 1987. Beiheft zur Nova Hedwigia 89. Berlin, Stuttgart. 700 pp.

Childs, J. F. L. 1940. Diurnal cycles of spore maturation in certain powdery mildews. *Phytopathology* 39:65–73.

Daines, R. H., C. M. Haenseler, E. Brennan, and I. Leone. 1960. Rusty spot of peach and its control in New Jersey. *Plant Dis. Rep.* 44:20–22.

DeNardo, A. 1938. Report on plant pathology. In *Report of the Department of Agriculture of Malta, 1936–1937*, 60–63.

English, H. 1947. Powdery mildew of cherry fruit in Washington. *Phytopathology* 37:421–24.

Galloway, B. T. 1889. The powdery mildew of cherry. *U.S. Agric. Comm. Rept. 1888*:352.

Gianai, M. A. 1940. A species of *Phyllactinia* occurring on almond (*Prunus amygdalus*). *Indian J. Agric. Sci.* 10:96–97.

Grove, G. G. 1987. Survival of *Podosphaera oxyacanthae* on senescent cherry leaves. (Abs.). *Phytopathology* 77:1239.

Kable, P. F., P. M. Fried, and D. R. MacKenzie. 1980. The spread of a powdery mildew of peach. *Phytopathology* 70:601–04.

Keil, H. L., and R. A. Wilson. 1961. Powdery mildew on peach. *Plant Dis. Rep.* 45:10–11.

Khairi, S. M., and T. F. Preece. 1975. *Podosphaera clandestina. Commonw. Mycol. Inst. Descrip. Pathogen. Fungi Bact.* 478.

Longree, K. 1939. The effect of temperature and relative humidity on the powdery mildew of roses. *Cornell Univ. Agric. Exp. Stn. Mem.* 223, 43 pp.

Manji, B. T. 1972. Apple powdery mildew on peach (abs.) *Phytopathology* 62:776.

Mence, M. J., and A. C. Hildebrandt. 1966. Resistance to powdery mildew in rose. *Ann. Appl. Biol.* 58:309–20.

Mukerji, K. G. 1968a. *Podosphaera tridactyla. Commonw. Mycol. Inst. Descrip. Pathogen. Fungi Bact.* 187.

———. 1968b. *Sphaerotheca pannosa. Commonw. Mycol. Inst. Descrip. Pathogen. Fungi Bact.* 189.

Ogawa, J. M., and F. M. Charles. 1956. Powdery mildew on peach trees. *Calif. Agric.* 10(1):7.

Ogawa, J. M., W. D. Gubler, and B. T. Manji. 1988. Effect of sterol biosynthesis inhibitors on diseases of stone fruit and grapes. In *Sterol biosynthesis inhibitors*. Eds. D. Berg and M. Plempel. Chap. 9:262–87. Ellis Horwood, Chichester, England. 583 pp.

Ogawa, J. M., D. H. Hall, and P. A. Koepsell. 1967. Spread of pathogens within crops as affected by life cycle and environment. *Symp. Soc. Gen. Micro. Airborne Microbes* 17:248–67.

Pinto de T., A., and I. Carreño. 1988. Control quimico de monilia, oidio, cloca, y corineo en nectarinos Late Legrand. *Agric. Tech. (Chile)*:48(2):106–10.

Ries, S. M., and D. J. Royse. 1977. Rusty spot of peach in Illinois. *Plant Dis. Rep.* 61:317–18.

———. 1978. Peach rusty spot epidemiology: Incidence as affected by distance from a powdery mildew-infected apple orchard. *Phytopathology* 68:896–99.

Rogers, M. N. 1959. Some effects of moisture and host plant susceptibility on the development of powdery mildew of roses, caused by *Sphaerotheca pannosa* var. *rosae*. *N. Y. (Cornell) Agric. Exp. Stn. Mem.* 363, 37 pp.

Smith, R. E. 1941. Diseases of fruits and nuts. *Calif. Agric. Ext. Serv. Circ.* 120, 168 pp.

Spencer, D. M. 1978. *The powdery mildews.* Academic Press, New York, 565 pp.

Sprague, R. 1954. Powdery mildew on almonds. *Plant Dis. Rep.* 38:695.

Sprague, R., and P. Figaro. 1956. Rusty spot, powdery mildew, and healthy skin of peach fruit compared histologically (abs.). *Phytopathology* 46:640.

Stewart, V. B. 1915. Some important leaf diseases of nursery stock. *N. Y. (Cornell) Agric. Exp. Stn. Bull.* 358:171–268.

Weigle, C. G. 1956. Powdery mildew (*Podosphaera tridactyla*) on almond. *Plant Dis. Rep.* 40:584.

Weigle, C. G., and A. M. French. 1956. Laboratory diagnosis. *Bull. Calif. Dept. Agric.* 45:186.

Weinhold, A. R. 1961a. The orchard development of peach powdery mildew. *Phytopathology* 51:478–81.

———. 1961b. Temperature and moisture requirements for germination of conidia of *Sphaerotheca pannosa* from peach. *Phytopathology* 51:699–703.

Weinhold, A. R., and H. English. 1964. Significance of morphological barriers and osmotic pressure in resistance of mature peach leaves to powdery mildew. *Phytopathology* 54:1409–14.

Yarwood, C. E. 1939. Powdery mildews of peach and rose. *Phytopathology* 29:282–84.

———. 1950. Dry weather fungi. *Calif. Agric.* 4(10):7–12.

———. 1952. Apricot powdery mildew from rose and peach. *Bull. Calif. Dept. Agric.* 41:19.

———. 1957. Powdery mildews. *Bot. Rev.* 23:235–301.

Yarwood, C. E., and A. H. McCain. 1981. Host list of powdery mildews of California. *Univ. Calif. Div. Agric. Sci. Leaf.* 2217. 16 pp.

Yarwood, C. E., S. Sidky, M. Cohen, and V. Santilli. 1954. Temperature relations of powdery mildews. *Hilgardia* 22:603–22.

Zaracovitis, C. 1965. Attempts to identify powdery mildew fungi by conidial characters. *Trans. Brit. Mycol. Soc.* 48:553–58.

Russet Scab of Prune

Cause Unknown

Russetting or "russet" of prunes was first brought to the attention of entomologists in the early 1930s. Since then it has caused severe losses in years of heavy rains during blossom period. In 1967, estimated losses in California from russet scab and associated defects (including brown rot, limb or leaf friction (**fig. 18B**), green-fruit rot, etc.) were $3 million. These included lowering of the quality of the dried fruit and the extra cost of sorting (Corbin, Lider, and Roberts 1968). Though heavy losses occur infrequently (Corbin, Lider, and Roberts 1968), some russet scab is present each year.

The original name given to this condition was "lacy scab" because of the netted pattern on the fresh fruit. Corbin, Lider, and Roberts (1968) named it "russet scab" to better describe the appearance of affected dried prunes.

Symptoms and disease development. Symptoms are first apparent on green fruit during or shortly after shuck-fall (**fig. 18A**). Wax on the surface of the fruit is lacking and the affected area is at first shiny (Ogawa, MacSwan, and Manji 1963, unpublished); such areas on mature or dried fruit may or may not constitute a major defect (**fig. 18C**). Some entomologists attributed the condition to the feeding of thrips on the surface of the fruit beneath the shuck, but others thought microorganisms must be involved. However, the effect of low temperature cannot be disregarded, since the symptoms appear typical of frost damage to apple and pear fruit. In 1963, and again in 1967, outbreaks of russet scab on prune fruit followed rainfall during the bloom period (Corbin, Lider, and Roberts 1968). Artificial overhead sprinkling of trees during the shuck-fall period also increased the incidence of russet scab (Ogawa, MacSwan, and Manji 1963). Removal of the shuck during full bloom reduced the incidence of russetting (Ogawa, MacSwan, and Manji 1963, unpublished).

Studies on russet scab during the 1982 season (Ogawa and Michailides 1982) confirmed that the disorder is correlated with a reduction of fruit wax on the skin of prunes and with shiny areas on immature fruit. Only shiny areas of considerable size resulted in russet scab symptoms on prunes after dehydration. The shiny areas were categorized as follows: 0 = no shiny areas, 1 = small shiny spots, 2 = small shiny area(s), and 3 = a large shiny area in a zone surrounding the stylar end of the fruit. Categories 0 and 1 showed no russet scab, while categories 2 and 3 developed 8 and 29 percent scabs, respectively. Furthermore, with categorization of shiny and russetted areas on fresh fruit samples at harvest, one could estimate the percent of dried fruit with russeting beyond that acceptable by inspectors.

In a 1963 trial, the fungicides captan and dichlone, applied for brown-rot blossom blight, reduced the incidence of russet scab (Ogawa, MacSwan, and Manji 1963, unpublished); this suggested that a fungus, possibly *Botrytis cinerea*, might be involved. Later studies, however, revealed no reduction in scab from the use of a benzimidazole fungicide (benomyl) that provides excellent control of *B. cinerea*. Recently, another fungicide—chlorothalonil (Bravo 500)—was shown to be effective against the disorder. The fact that several fungicides control russet scab suggests that a microorganism could be involved by inciting initial lesions which, when exposed to prolonged rains, fail to form adequate epidermal wax, as suggested by Watanabe (1969) in his study of apple fruit russet.

Control. Although the cause of this disorder is unknown, a single spray of captan or folcid applied during full bloom (**fig. 18D**) will almost eliminate it. Sprays applied during early bloom provide some reduction in russet scab while an application at or after the petal-fall stage provides no control. In field tests with French prunes during the 1986 season, Ogawa et al. (1986) showed that a chlorothalonil spray provides scab control equivalent to that obtained with captan. The sterol-biosynthesis-inhibiting and dicarboximide fungicides are relatively ineffective.

REFERENCES

Corbin, J. B., J. V. Lider, and K. O. Roberts. 1968. Controlling prune russet scab. *Calif. Agric.* 22(11):6–7.

Ogawa, J. M., R. M. Bostock, T. J. Michailides, B. T. Manji, and J. E. Adaskaveg. 1986. Control of brown rot, russet scab, and leaf rust of prunes. In *Prune research reports, 1986,* 44–47. California Prune Board.

Ogawa, J. M., and T. J. Michailides. 1982. Effect of various chemicals in controlling brown rot and russet scab of prune trees. In *Prune research reports, 1982,* 16–23. California Prune Board.

Watanabe, S. 1969. Histological studies on the cause of russet in apples. *Bull. Yamagata University* 5(4):823–89.

Rust of Stone Fruit

Tranzschelia discolor (Fuckel) Tranz. & Litv.

The rust disease considered here affects peach, nectarine, apricot, plum, prune, almond, and cherry. It is widely distributed geographically, and while it occurs more or less sporadically it can be a serious disease in years of excessive moisture during the growing season.

In the major stone-fruit production areas in the Sacramento and San Joaquin valleys, rust is observed yearly on prunes from the middle of summer to autumn, while on the other crops rust has been of little significance except on almonds in the northern Sacramento Valley in 1983 (Ogawa, Bolkan, and Krueger 1984), and throughout both the Sacramento and San Joaquin valleys in 1988. Of greatest concern during rust epidemic years is preharvest leaf infection, which results in excessive leaf fall during mechanical harvest and makes separation of the fruit from the leaves difficult. Also, early leaf fall occasionally stimulates blossoming during the winter. Fruit infections are rare and of no consequence. Major recorded outbreaks of rust have been during years with summer rains—1925, 1926, 1927, 1942, 1983 and 1988. Possible crop losses resulting from early defoliation have not been confirmed in California.

Symptoms. On peach fruit, circular spots 2 to 3 mm in diameter develop and become green and sunken when the fruit ripen (**fig. 19B**). On peach, prune, and almond leaves, bright yellow, somewhat angular spots appear (**fig. 19A**). On peach twigs, lens-shaped cracks up to 10 mm long develop in the bark; the center of the cracks is filled with a rusty-brown powdery mass of urediniospores. Lesions on the leaves are abundant and visible, but those on the twigs are few and inconspicuous (**fig. 19B**). Even in years of severe leaf infection, only about 1 to 2 percent of peach twigs develop lesions. With prune and almond, twig lesions are difficult to locate except in orchards with a history of yearly rust development. On apricot, the leaf lesions are similar to those on almond and prune.

Causal organism. The name of the causal fungus has undergone several changes. An early designation by Persoon was *Puccinia pruni-spinosae*. Arthur (1929) maintained that the name should be *Tranzschelia punctata* in recognition of the work of W. A. Tranzschel (1905) on heteroecism of the fungus. Later, Dunegan (1936) made a thorough study of the rust in the central United States and found what he considered to be two forms of *T. pruni-spinosae*. He called the form occurring on the wild *Prunus* species *T. pruni-spinosae typica* and the one on cultivated species *T. pruni-spinosae discolor*. The latter form was later raised to specific rank as *T. discolor* by American workers. In most other countries, the fungus is still called *T. pruni-spinosae*. According to the CMI descriptions (Laundon and Rainbow 1971), the teliospores of *T. pruni-spinosae* var. *discolor* are distinctive, with uneven, thicker walls, in contrast to those of *T. pruni-spinosae* var. *pruni-spinosae*, which have walls of uniform thickness. On the basis of this information and other research, the rust species found on stone fruit in California should be classified as *T. discolor*.

Tranzschel (1905) produced infection on almond and the wild *Prunus* species *P. spinosa* and *P. divaricata* with aeciospores from *Anemone divaricata*. Arthur (1929) later listed species of *Hepatica* and *Thalictrum* as alternate hosts of the fungus. Although *T. discolor* (*T. pruni-spinosae*) is macrocyclic under certain conditions, and the passage from one host to another is probably necessary to its survival in cold climates, the aecial stage produced on *Anemone*, *Hepatica*, and *Thalictrum* is apparently not necessary for its survival in warmer regions, such as the U.S. Pacific Coast. Though the aecial stage has been found on *Anemone coronaria* in California, it is comparatively rare, and the alternate host apparently plays no part in outbreaks of rust on peach in the Sacramento Valley. Goldsworthy and Smith (1931) found no aecial stage of the fungus in the Sacramento Valley during the 1925 and 1927 outbreaks. In addition, they found no teliospores on peach in that locality.

Scott and Stout (1931) reported teliospore production on prune leaves collected from the Cosumnes River district of Sacramento County. Teliospores are found to some extent on peach and are abundant on almond, prune, and *P. domestica* plum foliage. Even so, the aecial host probably plays no important part in the life cycle of the fungus in central California. Primary infection in the spring is probably initiated by urediniospores which overwinter on twigs or even on foliage, which in this locality frequently remains on the tree until spring. Cunningham (1925) in New Zealand and Perlberger (1943) in Palestine also discounted the importance of the aecial stage.

The urediniospores, measuring 12.2 to 33.6 μm in length and 9.0 to 14.4 μm in width, are heavily echinulate, except at the apex, and sharply constricted at the base. They are cinnamon brown and have two germ pores, usually located beneath the apical cap and opposite each other (Goldsworthy and Smith 1931). These spores are produced on pedicellate cells formed on a stromatic layer of pseudoparenchymatous tissue (sorus) directly beneath the epidermis of leaves and bark; on the fruit, the spore-producing layer is found at various depths beneath the epidermis. Infection by urediniospores is related to such factors as moisture, temperature, time, and spore age and maturity. Disease outbreaks in California are mostly related to the occurrence of rain during the growing season. Spores germinate only in free moisture or a saturated atmosphere; at 22°C they germinate in 12 hours and the germ tubes elongate at the rate of 1 μm or more per minute. Spores in the leaf sori are relatively short lived but remain viable longer on living leaves than on detached leaves. On living leaves, the sorus spores were found to be viable for five months but on detached leaves for only about one month. Spores less than one week old fail to germinate. The optimum temperature for spore germination is 13° to 26°C, the minimum 10°C or lower, and the maximum about 38°C. Spores collected from living leaves and frozen remain viable for more than one year.

Identity of rust species on different hosts. Dunegan (1936) reported that teliospores produced on cultivated peach in the eastern United States differ morphologically from those produced on indigenous *Prunus*. The so-called "discolor" type occurring on peach and other cultivated hosts is associated with the aecial stage on *Anenome coronaria*, whereas the "typica" form occurring on indigenous hosts is associated with the aecial state on *Hepatica* and *Anenome*. Scott and Stout (1931) produced infection of peach, apricot, almond, and prune with aeciospores from *A. coronaria*, but failed to obtain infection of any of seven cultivars of cherry.

Physiological specialization among rusts on almond, peach, and prune was shown in leaf inoculations with urediniospores (Bolkan et al. 1985). The names proposed were *Tranzchelia discolor* f. sp. *dulcis* for strains that attack almond, *T. discolor* f. sp. *persicae* for strains on peach, and *T. discolor* f. sp. *domesticae* for strains on prune. This study supported observations that peach or almond trees adjacent to severely rusted prune trees were free of disease. Other workers (Goldsworthy and Smith 1931; Smith 1947) have made similar orchard observations and have reported differences in specificity of pathogenicity (Sztejnberg 1976; Sztejnberg and Afek 1979), further supporting the natural subgrouping of rusts on stone fruit. It must be noted that in studies in California (Goldsworthy and Smith 1931) and Australia (Kable, Ellison, and Banbach 1985) cross-infection of rusts among host species has occurred but with different latent periods and symptom expression. On hosts other than the natural host, disease symptoms typical of resistance—smaller lesions with reduced sporulation—usually are observed.

Disease development. On peach in the Sacramento Valley, the fungus undergoes a period of inactivity during the dry summer months and passes this period on infected leaves and twigs. A second period of inactivity probably occurs in mid-winter. During this time the fungus survives in twig lesions, some of which developed during the previous growing season; other lesions were initiated in autumn but are not evident until spring. Barrett (1915) noted that in the moderate climate of southern California the fungus passes the winter on leaves that stay on the tree. The life cycle on almond is similar to that on peach; on prune it probably is the same although rust pustules on the twigs are rare or difficult to detect. The fungus is widespread and most important on prune in California. More recently rust on almond has become widespread. Alternate hosts appear to play only a minor role in the life cycle of the fungus in this country.

Urediniospores are the predominant if not the only inoculum for infection in the Sacramento Valley. They retain their viability for four to six weeks and are disseminated to a limited extent by wind but probably to a much greater extent by water.

The disease cycle is believed (Pierce 1894) to be as follows: In the spring, infection occurs through the undersurface of the leaves. There may be a series of leaf-infection periods, depending upon the weather. The fruit, apparently, are not susceptible to infection until they are of considerable size. Fruit lesions on peach seldom appear until late May or early June, while on other *Prunus* species lesions on the fruit are of no consequence. During most of the summer and fall, the leaves are susceptible to infection. Heavily infected leaves drop prematurely. Some time during the fall or early winter twig infection occurs, and the lesions are sometimes visible in late fall; at this time, however, the yellowish-orange sori have not produced urediniospores. There is little, if any, fungus activity in midwinter. In the spring and early summer, leaf infection is initiated by urediniospores produced on twig lesions. The incubation period in the twigs is not known; that in leaves is not known with certainty but apparently is about eight days.

Control. The application of a fungicide in early summer prevents leaf and fruit infection on peach (Linfield and Price 1983). Twig infection can be prevented by a fungicide application in mid-October. On prune the only registered material at this time is wettable sulfur, which is suggested for application before rust is evident. Recent studies with the fungicide mancozeb showed that an application before or at the onset of rust infection of the leaves provides control superior to that obtainable with wettable sulfur. A single application in the summer or early fall prevents leaf infection for two months (Michailides and Ogawa 1985). Information on physiologic specialization of stone-fruit rusts suggests that the disease on each species could be managed in accordance with its history on that crop. However, severe rust development on one stone-fruit species could be a signal for closer monitoring of rust on other species.

REFERENCES

Arthur, J. C. 1906. Cultures of Uredineae in 1905. *J. Mycol.* 12:11–27.

———. 1929. *The plant rusts (Uredinales)*. John Wiley & Sons, New York, 446 pp.

Barrett, J. T. 1915. Observations on prune rust, *Puccinia pruni-spinosae* Pers. in southern California (abs.). *Phytopathology* 5:293.

Bolkan, H. A., J. M. Ogawa, T. J. Michailides, and P. F. Kable. 1985. Physiological specialization in *Tranzschelia discolor*. *Plant Dis.* 69:485–86.

Blumer, S. 1960. [Studies on the morphology and biology of *Tranzschelia pruni-spinosae* and *Tranzschelia discolor*.] *Phytopathol. Z* 38:355–85.

Carter, M. V., W. J. Moller, and S. M. Pady. 1970. Factors affecting uredospore production and dispersal in *Tranzschelia discolor*. *Austral. J. Agric. Res.* 21:906–14.

Cunningham, G. H. 1925. *Fungous diseases of fruit trees in New Zealand*. Brett, Auckland. 382 pp.

Dunegan, J. C. 1936. The occurrence in the United States of two types of teliospores of *Tranzschelia pruni-spinosae* (abs.). *Phytopathology* 26:91.

———. 1938. The rust of stone fruits. *Phytopathology* 28:411–27.

Dunegan, J. C., and C. O. Smith. 1941. Germination experiments with uredio- and teliospores of *Tranzschelia pruni-spinosae* discolor. *Phytopathology* 31:189–91.

Duruz, W. P. 1928. Further notes regarding peach rust control. *Proc. Am. Soc. Hort. Sci.* 25:333–37.

Duruz, W. P., and M. Goldsworthy. 1928. Spraying for peach rust (A progress report). *Proc. Am. Soc. Hort. Sci.* 24:168–71.

Goldsworthy, M. C., and R. E. Smith. 1931. Studies on a rust of clingstone peach in California. *Phytopathology* 21:133–68.

Hutton, K. E. 1950. Rust of stone fruits. *Agric. Gaz. New South Wales* 61:135–38.

Jafar, H. 1958. Studies on the biology of peach rust (*Tranzschelia pruni-spinosae*) in New Zealand. *New Zeal. J. Agric. Res.* 1:642–51.

Kable, P. F., P. J. Ellison, and R. W. Banbach. 1986. Physiologic specialization of *Tranzschelia discolor* in Australia. *Plant Dis.* 70:202–04.

Laundon, G. F., and A. F. Rainbow. 1971. *Tranzschelia pruni-spinosae* var. *discolor* and *T. pruni-spinosae* var. *pruni-spinosae*. *Commonw. Mycol. Inst. Descrip. Pathogen. Fungi Bact.* 287, 288.

Linfield, C. A., and D. Price. 1983. Host range of plum anemone rust *Tranzschelia discolor*. *Trans. Brit. Mycol. Soc.* 80:19–21.

Michailides, T. J., and J. M. Ogawa. 1985. Chemical control of prune leaf rust (*Tranzschelia discolor*) in California orchards. *Plant Dis.* 70:307–09.

Ogawa, J. M., H. A. Bolkan, and W. H. Krueger. 1984. Outbreak of *Tranzschelia discolor* causing almond rust disease in northern California. *Plant Dis.* 68:351.

Perlberger, J. 1943. The rust disease of stone fruit trees in Palestine. *Rehovath Agric. Exp. Stn. Bull.* 34. (Hebrew with English summary.)

Pierce, N. B. 1894. Prune rust. *J. Mycol.* 7:354–63.

Roth, G. 1966. *Tranzschelia discolor* (Fuck.) Tranz. et Litv. on peaches in Transvaal lowveld, South Africa. *Phytopathol. Z.* 56:141–50.

Scott, C. E., and G. L. Stout. 1931. *Tranzschelia punctata* on cultivated *Anemone* in the Santa Clara Valley. *Monthly Bull. Calif. Dept. Agric.* 20(10–11):1–7.

Smith, C. O. 1947. A study of *Tranzschelia pruni-spinosae* on *Prunus* species in California. *Hilgardia* 17:251–66.

Sztejnberg, A. 1976. Physiologic specialization in rust of stone fruits. *Poljopr. znanst. Smotra (Zagreb) (Agric. Consp. Sci.)* 39:253–59.

Sztejnberg, A., and U. Afek. 1979. Physiological races of the stone fruit rust *Tranzschelia pruni-spinosae* var. *discolor* on *Anenome coronaria* plants from different sites (abs.). *Phytoparasitica* 7:51.

Thomas, H. E., R. A. Gilmer, and C. E. Scott. 1939. Rust of stone fruits. *Monthly Bull. Calif. Dept. Agric.* 28:322–27.

Tranzschel, V. 1935. [The cherry rust, *Leucotelium cerasi* (Bereng.) n. gen. n. comb (*Puccinia cerasi* Cast.) and its aecial stage.] *Riv. Pat. Veg.* 25(5–6):177–83.

Tranzschel, W. A. 1905. Bietrage zür biologie der Uredineen. *Trav. Mus. Bot. Acad. Imp. Sci. St. Petersbourg* 2:67–69.

Ward, J. R. 1955. Diseases of peaches and nectarines. *Tasman. J. Agric.* 26:52–58.

Wilson, E. E., and C. E. Scott. 1943. Prevention of three peach diseases by ferric dimethyldithiocarbamate. *Phytopathology* 33:962–63.

Scab of Almond and Stone Fruit

Cladosporium carpophilum Thüm.

A scab disease of *Prunus* spp. occurs in many parts of the world and in the warmer stone fruit areas of the United States (Anderson 1956; Jones and Sutton 1984; Keitt 1917). In California, it was first reported on almond by Smith (1924), who identified the pathogen as *C. carpophilum* Thüm. Since then, the disease has occurred sporadically on this host, generally being more severe in the Sacramento Valley than in the San Joaquin Valley (Ogawa, Nichols, and English 1955a, 1955b). It occurred in epidemic proportions in the Sacramento Valley in 1957 and 1983. These were years of unusually heavy spring and summer rains. Scab outbreaks have become more prevalent in both valleys with the change from furrow to sprinkler irrigation; with drip systems, the disease has not been a problem.

In addition to almond, scab has been reported in California on apricot, nectarine, peach, plum, and prune (English, Ogawa, and Nichols 1955; French 1987; Harris 1941). A similar scab disease of cherry, caused by *C. cerasi* Aderh. (teleomorph: *Venturia cerasi* Aderh.), occurs in the Midwest (Bensaude and Keitt 1928) but has not been reported in California. Since scab of *Prunus* in California is of importance only on almond, our discussion will be restricted largely to this host.

Symptoms. The symptoms on almond leaves, fruit, and twigs are similar to those of the scab disease of peach (**fig. 20D**). On the leaves, the first visible signs of infection are many small, indistinct, somewhat circular, greenish yellow blotches on the undersurface. The lesions later enlarge, some reaching 10 mm or more in diameter. With the production of spores, they take on an olivaceous appearance (**fig. 20B**), and eventually become yellowish brown to dark brown. On petioles and midribs they have the same color, but are more elongate. Severe infection causes the leaves to fall. Leaves with only a few lesions may remain on the tree until autumn. Premature dropping of the leaves is the most serious phase of the disease.

On the shoots, the lesions at first are indistinct, water-soaked spots. Later, the spots become necrotic and brown. In the fall and early winter, the lesions become olivaceous at the margin due to sporulation of the pathogen (fig. 20C).

On the fruit, indistinct circular olivaceous spots develop and coalesce into large irregular grayish black blotches (fig. 20A). Though lesions may occur on any part of the fruit, they are most common on the upper side.

Causal organism. The causal fungus is similar morphologically to the peach scab pathogen, which has been known for many years as *Cladosporium carpophilum* Thüm. (Keitt 1917). Hughes (1953) proposed that the name *C. carpophilum* be changed to *Fusicladium carpophilum* (Thüm.) Oudem., but this change generally has not been accepted. The teleomorph of *C. carpophilum* was found on overwintered apricot leaves in Australia and named *Venturia carpophila* sp. nov. by Fisher (1961). It is not known whether the form found on *Prunus* spp. in California has a sexual stage. In some European countries, the binomial *F. amygdali* Ducomet is used for the almond scab fungus.

Host relations. Although scab is found in California on most of the cultivated species of *Prunus*, it is not known with certainty whether the fungus on all hosts is *C. carpophilum*. There also is no information as to whether more than one pathotype of this species is present here. Research in Wisconsin (Bensaude and Keitt 1928) indicated that the form on peach and plum (*P. americana*) infected apricot, but not cherry or European plum (*P. domestica*). In California, scab occurs on most of the commercially grown almond cultivars and is especially severe on Ne Plus Ultra and Drake.

Disease development. The pathogen overwinters as mycelium in twig lesions. In the spring it resumes growth at the margin of the lesions, where it produces conidia. Lesions on new shoots develop in early spring (March), while those on fruit and leaves usually do not become evident until June.

In an environmental study of *C. carpophilum* in South Carolina, Lawrence and Zehr (1982) found that mycelial growth *in vitro* was best at 20° to 30°C. Maximum conidial germination occurred in free water, but in the absence of free moisture germination was best at 98 to 100 percent relative humidity (RH). Also, sporulation on peach twig lesions was most abundant at 98 to 100 percent RH. Sporulation on twig lesions was sparse until after overwintering. The conidia in a peach orchard were found to be both airborne and waterborne, and were most numerous in May. Forty to seventy days elapse from the time a spore lands on a peach fruit until symptoms appear (Jones and Sutton 1984). They are released in greatest abundance at low relative humidity (below 40 percent) and with exposure to infrared radiation (Gottwald 1983).

Control. Spraying almond trees with a suitable fungicide five weeks after petal fall materially reduces leaf and fruit infection (English, Ogawa, and Nichols 1955). Sprays applied at petal fall for the shothole disease give some added protection. Fungicides reported to be effective against scab are ziram, captan, wettable sulfur, dichlone, benomyl, thiophanate-methyl, and chlorothalonil. Strains of the pathogen resistant to the benzimidazoles have been found in some regions. The disease can be markedly reduced by avoiding high-angle sprinkler irrigation.

Research in Argentina has shown that the susceptibility of nectarine to scab can be significantly reduced by the foliage application of monosodium phosphate plus urea (Montero and Espósito 1984; Montero, Espósito, and Gonzalez de las Heras 1986). Resistance was correlated with a high level of sugar in the host tissue.

REFERENCES

Anderson, H. W. 1956. *Diseases of fruit crops.* McGraw-Hill, New York, 491 pp.

Bensaude, M., and G. W. Keitt. 1928. Comparative studies of certain *Cladosporium* diseases of stone fruits. *Phytopathology* 18:313–29.

English, H., J. M. Ogawa, and C. W. Nichols. 1955. Almond scab is found causing some economic loss to growers. *Almond Facts* 20(5):8–9.

Fisher, E. E. 1961. *Venturia carpophila* sp. nov., the ascigerous state of the apricot freckle fungus. *Trans. Brit. Mycol. Soc.* 44:337–42.

French, A. M. 1987. *California plant disease index. Part 1: Fruits and nuts.* Calif. Dept. Food and Agric., Div. Plant Ind., Sacramento. 39 pp.

Gottwald, T. R. 1983. Factors affecting spore liberation by *Cladosporium carpophilum*. *Phytopathology* 73:1500–05.

Harris, M. R. 1941. Unusual occurrence of scab on prunes in California. *Plant Dis. Rep.* 25:409.

Hughes, S. J. 1953. Some foliicolous hyphomycetes. *Can. J. Bot.* 31:560–76.

Jones, A. L., and T. B. Sutton. 1984. Diseases of tree fruits. *N. Central Regional Ext. Publ.* 45, Coop. Ext. Serv., Mich. State Univ., 59 pp.

Keitt, G. W. 1917. Peach scab. *USDA Bull.* 395:1–66.

Lawrence, E. G., and E. I. Zehr. 1982. Environmental effects on the development and dissemination of *Cladosporium carpophilum* on peach. *Phytopathology* 72:773–76.

Miller, R. W. 1983. Peach scab. In: *Peach growers handbook*. M. E. Ferree and P. F. Bertrand, editors. GES Handb. 1, Coop. Ext. Serv., University of Georgia, Athens, GA.

Montero, J. C., and S. M. Espósito. 1984. Modificación de la predisposición a sarna causada por *Fusicladium carpophilum* en frutos de nectarina (*Prunus persica* var.. *nectarina*). *Fitopatologia* 19(2):74–77.

Montero, J. C., S. M. Espósito, and B. Gonzalez de las Heras. 1986. Modificación de la predisposición a la sarna en frutos de nectarina mediante sustancias fertilizantes aplicadas al follaje. Influencia de la dotación de azucares en las plantas. *Boletin Tecnico Estacion Experimental de Mercedes.* 2, 18 pp. Gowland, Argentina.

Ogawa, J. M., C. W. Nichols, and H. English. 1955a. Almond scab. *Calif. Dept. Agric. Bull.* 44(2):59–62.

———. 1955b. Almond Scab. *Plant Dis. Rep.* 39:504–08.

Smith, E. H. 1924. Some diseases new to California (Abs.). *Phytopathology* 14:125.

Shoestring (Armillaria) Root Rot of Stone Fruit

Armillaria mellea (Vahl.:Fr.) Kummer

Armillaria root rot, one of the most important soil-borne diseases of stone-fruit trees, was first reported on peaches in 1891, on prunes in 1902, almonds in 1918, apricots in 1921, sweet cherries in 1926, and on Japanese plums between 1930 and 1949. *Armillaria* kills trees by attacking the roots and stem at the soil line and moving from one tree to the next by rhizomorphs or root grafts. Infected areas can be confined by trenching or by the annual injection of methyl bromide. However, methyl bromide does not completely kill the fungus in the soil, so resistant rootstocks offer the least expensive and most practical method of control. The plum rootstocks Marianna 2624, Myrobalan 29C, St. Julien, and Damson have some resistance to *Armillaria*. Marianna 2624, the most resistant stock, is widely used for plum, prune, and apricot.

Nematode-resistant rootstocks such as Nemaguard seedlings (FV 234–1), Rancho Resistant, and S–37 are susceptible to *Armillaria*. However, if nematodes are not a problem, plum rootstocks are available for peach and nectarine plantings.

Almond cultivars such as Jordanolo, Ne Plus Ultra, Peerless, and Mission may be grown on *Armillaria*-resistant Marianna 2624, but other cultivars cannot. Intergrafts of New Havens B with the Nonpareil scion are available.

Mahaleb and Stockton Morello cherries are susceptible, but Mazzard is considered moderately resistant after it is well established.

Details of the shoestring root rot are given in Chapter 4.

Shothole of Stone Fruit

Wilsonomyces carpophilus (Lév.) Adaskaveg, Ogawa, and Butler
(= *Stigmina carpophila* [Lév.] Ellis)

This disease, known as "California peach blight," "peach blight," "shothole," "Coryneum blight," and "corynosis," is particularly prevalent in the Pacific Coast states, but occurs to some extent in other western states and occasionally in midwestern and eastern states. It is also known in Canada, Australia, Africa, Asia, New Zealand, Europe, and South America.

The disease was first described in 1843 (Léveillé 1843) on peaches near Paris, and by 1864 it was in England (Berkeley 1864). From 1882 to 1889 it destroyed the Maori cultivar of peaches in New Zealand (Boucher 1901). In the United States it was first reported in Michigan in 1894 and in California in 1900 (Pierce 1900 [p. 179]; Smith 1902). California had severe outbreaks of shothole following years of economic depression in the 1930s and in years of prolonged wetness during the winter and spring months. The most recent epidemic on almond came in 1982, when yield was reduced significantly more on untreated control trees than on trees treated with fungicide.

Shothole is a major disease of peach, nectarine, apricot, and almond in California, but it has rarely been found on other stone fruit trees, except during 1982 when symptoms were observed on leaves and fruit of sweet cherry (**fig. 21F**) and plum. These infections were recorded only on trees adjacent to heavily infected peach or apricot trees. Commercial orchards of sweet cherry and plum were not affected. In the Pacific

Northwest and British Columbia, however, the disease sometimes does much damage to cherries. Other plants attacked are prune, cherry laurel (*Prunus laurocerasus*), and some species of native American cherry.

Symptoms. The lesions on twigs first appear as purplish spots 2 to 3 mm in diameter. As the lesions enlarge, they then brown, and often have a light tan center (**fig. 21C**). Small tufts of spores in the necrotic areas are visible with a hand lens. The scales of affected buds are dark brown to black, and are sometimes covered with a gummy exudate that gives them a varnished appearance. Leaf and fruit lesions begin as small purplish areas that then expand to brownish spots 3 to 10 mm in diameter (**fig. 21A**). On leaves, the lesions are often surrounded by a narrow light green to yellow zone. Many of the spots abscise to produce the symptoms of shothole. Fruit lesions (**fig. 21D**) eventually become rough and corky.

The symptoms vary on four kinds of stone fruit. On peach and nectarine, the principal symptoms are twig and bud blighting, and if the disease is not controlled on the twigs during the winter, and the spring is wet, fruit infection occurs. In apricot, bud blighting and fruit and leaf spotting are the principal symptoms (**fig. 21E**), while twig lesions rarely occur. On almond, the principal symptoms are leaf spotting and blossom and fruit infection; some twig infection occurs, but is not serious (Wilson 1937). Defoliation following leaf infection is much more common on apricot and almond than on peach and nectarine.

Causal organism. In 1843, the causal organism was described by J. H. Léveillé. Others subsequently described the organism, resulting in numerous synonyms that include *Coryneum beyerinckii* Oud. In 1901, Aderhold recognized Léveillé's earlier description and made the combination *Clasterosporium carpophilum* (Lév.) Aderh. Many researchers in North America, however, continued to use *Coryneum beyerinckii* (*C. beijerinckii*) following the publication by Smith et al. (1907) on peach blight in California.

This difference in nomenclature, as reviewed by Samuel (1927), was largely based on disease development on different host tissues. On stem tissue, the fungus produces a well-developed stroma that was interpreted as an acervulus, thus justifying its taxonomic placement in the genus *Coryneum* Nees:Fr. of the Melanconiales. While on leaf tissue, the stroma is not as well developed and the fungus was considered similar to other hyphomycetous fungi. For this reason the fungus was placed in the genus *Clasterosporium* Schw., the name researchers in Europe and Australia continued to us.

Ellis (1959), after studying hyphomycetous fungi in the genus *Clasterosporium* and other allied genera with dematiaceous phragmospores, transferred the species to the genus *Stigmina* Sacc. based on spore ontogeny and disease development. Ellis considered the conidiophore to be an annellide, with spores produced in a sporodochium similar to that of other species in the genus. After ontological and morphological studies of conidia and a critical review of the nomenclature of this fungus, Adaskaveg, Ogawa, and Butler (1990) erected the new genus *Wilsonomyces* to include this organism, since no other species is described with sporodochial conidiomata, rhexolytic separation of conidia, sympodially developing conidiogenous cells, and smooth-walled phragmospores.

The conidia are solitary, dry acrogenous, simple, thick-walled, cylindrical, clavate, ellipsoidal or fusiform, occasionally forked, rounded to acute at the apex, truncate at the base, and generally possessed of three to five longitudinal septa. Their dimensions are 20 to 90 by 7 to 16 μm, 3.5 μm at the base, and their color ranges from subhyaline to golden brown and dark olivaceous to black in mass. The conidium is divided from the conidiogenous cell by a double-walled septum, and conidial separation is by rhexolysis. The conidia are mass-produced in sporodochia that are usually punctiform, tan to olivaceous brown and then black, and are formed on leaves, twigs, flowers, and fruit of *Prunus* species (Adaskaveg, Ogawa, and Butler 1990).

Vuillemin (1888) claimed that the fungus produces a perfect stage, *Ascospora beijerinckii*, on dead leaves, but this has not been corroborated. The fungus is easily cultured, but the most rapid growth is on a basal agar medium in which betamaltose supplies the carbon, L-asparagine supplies the nitrogen, the pH is 5.4 to 5.8, and the culture is incubated at 15°C (Williams and Helton 1971).

Disease development. The pathogen survives from one season to the next in lesions on twigs and in blighted buds. Conidia are produced on twigs and in blighted buds, the latter source being particularly important in apricot, nectarine, and peach.

In peach conidia are present inside blighted buds the year round, and may be produced for at least 18 months after the buds are blighted. Under favorable conditions, a crop of conidia can be produced in six hours; in spring they form on old twig lesions and on infected blossoms and leaves (**fig. 21B**). The fungus rarely sporulates on fruit lesions. When kept dry, conidia remain viable for several months (Samuel 1927; Wilson 1937). They are not easily detached from the conidiophores by moving air but are readily detached by water and, therefore, are largely disseminated by rain. In the germination process, all cells do not germinate at one time, thus providing a mechanism for extended viability of spores during wet and dry periods. Spores can germinate almost immediately after their formation, and germination has been observed in a one-hour period. In 15 to 20 hours at 15° to 27°C, all of the spores may germinate.

In almond viable conidia were associated with dormant buds collected throughout the 1982 season in a commercial almond orchard in Merced County (Highberg and Ogawa 1986b). Viability of the spores ranged from 65 to 96 percent for samples collected throughout the dormant season. Thus as blossoms and leaves emerge in spring, the inoculum is available for infection.

Direct penetration of the leaves by germ tubes from the conidia has been observed (Aderhold 1902; Samuel 1927). The conidia germinate equally well on both leaf surfaces, but the underside is more prone to infection than the upper side (Ogawa, Yates, and Kilgore 1964). In all probability, the fungus is capable of penetrating the twigs, fruit, and blossoms in the same manner. In California, twig and bud infection of peach and apricot occurs during rainy periods at any time from autumn until spring. Blossoms and leaves are infected when they emerge or later. The incubation period of the disease is from 5 to 14 days, depending on temperature and the type of tissue infected.

Samuel (1927) found the fungus to be inter- as well as intracellular in leaves. Host cells adjacent to the mycelium die. "Shotholing" of *Prunus* leaves is caused by an abscission layer that forms around the infected area (Samuel 1927).

The spores germinate over a relatively wide temperature range, even as low as 2° to 4°C, so California winters do not limit germination. Moisture, however, is a factor in disease production: a period of at least 24 hours of continuous moisture is necessary for twig infection, the incubation period being 7 to 14 days. This is probably the reason that twig infection is not usually initiated during the first autumn rains, which commonly are of short duration. Leaf infection, however, may be initiated during such rains. When localized conditions of free moisture were produced with a moisture generator controlled by a datalogger on Nonpareil and Carmel almond trees, optimal conditions for disease development resulting in 3.1 lesions per leaf were established with a 3- or 4-second misting duration every 2.5 or 5.0 minutes in a 16-hour period with an average of 14° to 17°C (Adaskaveg, Shaw, and Ogawa 1989).

Soil conditions have little or no influence on infection, but an extremely low soil fertility might influence the tree's ability to recover from a heavy attack.

Cultivar susceptibility. Most cultivars of peach grown in California are susceptible to infection. Clingstone cultivars such as Gaume, Loadel, Palora, and Phillips Cling appear to be the most susceptible; freestone types grown in California are less affected. In California and Oregon, Elberta peach and the common cultivars of apricot (Blenheim and Tilton) are very susceptible. Among almonds, Ne Plus Ultra, Drake, Peerless, and Nonpareil are probably the most susceptible.

In spring 1986 severe shothole infection was found on Merced, Nonpareil, and Mission cultivars grown in the southern San Joaquin Valley. Infections were abundant on the leaves and fruit but not on the twigs.

Control. Common cultural practices, such as furrow irrigation, cultivation, and soil fertilization, have no apparent influence on shothole, but the recent use of high-angled sprinkler irrigation has increased the disease incidence.

Bordeaux mixture and the fixed copper materials continue to provide excellent control of shothole on dormant twigs of peach, nectarine, and apricot. Dormant sprays on almond reduce leaf infections in spring but do not provide effective control (Teviotdale et al. 1989). Copper sprays applied to apricot and almond after leaf emergence in the spring sometimes cause leaf spots, chlorosis of the leaf blade, and leaf-fall symptoms similar to those of the shothole disease. For protection of leaves, the organic fungicides—ziram, ferbam, dichone, captan, iprodione, and chlorothalonil—have proved as effective as copper fungicides and are not toxic to the foliage. Other organics that have shown promise are hexaconazole, triadimefon, and triforine (Pinto de T. and Carreño 1988). On almond, a single ziram or captan spray applied at petal fall provides significantly more kernel yield than a single application at the pink-bud stage. Chlorothalonil applied at petal fall is equivalent to sprays of ziram or captan applied at pink bud. Kernel weight and size are not affected by the treatments (Highberg and Ogawa 1986a). The effectiveness of zineb, or nabam mixed with zinc, iron, and manganese salts, is marginal (English 1958; English and Davis 1962; Ogawa et al. 1959).

While the principal damage to peach and nectarine from shothole is the killing of twigs and buds (fruit and leaf infection being much less important), the principal damage to apricot is caused by fruit and leaf infection. In addition, bud infection in apricot sometimes causes some crop loss. Almond suffers very little loss of twigs or buds, but may be damaged by defoliation following severe leaf infection (Ogawa, Teviotdale, and Rough 1983). Fruit infections occur primarily on the upper surface and may restrict the normal enlargement of the affected side (Ogawa et al. 1959). Fruit infection is of no consequence in almond unles overhead sprinkler irrigation is practiced (Aldrich, Moller, and Schulbach 1974).

The timing of protective spray applications must be varied to meet the differences in the effects of the disease on different hosts. For peach and apricot, for example, an autumn application after leaf fall but before winter rains begin is necessary to protect against twig and bud infection. No other application is usually necessary for peach, but for apricot an application at full bloom to petal fall may be necessary to prevent severe fruit and leaf infection. For almond, where protection of leaves is of prime importance, a single ziram spray applied at the end of the blossoming period (petal-fall stage) has given control throughout the remainder of the season. A spray at bud swell with sodium pentachlorophenoxide has also significantly reduced the disease, but is no longer available for use.

Applications of concentrate sprays of ziram, using 20 gallons per acre, were as effective as high-volume applications (400 gallons per acre). Helicopter spray applications give excellent disease control provided the droplet size is reduced from 450 μm to 320 μm and the forward speed does not exceed about 25 mph (Kilgore, Yates, and Ogawa 1964; Ogawa et al. 1964). Fixed-wing aircraft give little spray deposit on the under surface of leaves and afford poor control (O'Reilly 1957).

Almond hulls are used for cattle feed, so chemical residues that persist on the hulls at harvest must not exceed the legal limits.

REFERENCES

Adaskaveg, J. E., J. M. Ogawa, and E. E. Butler. 1990. Morphology and ontogeny of conidia in *Wilsonomyces carpophilus*, gen. nov. and comb. nov., causal pathogen of shot hole disease of *Prunus* species. *Mycotaxon* 37:275–90.

Adaskaveg, J. E., D. A. Shaw, and J. M. Ogawa. 1989. A mist generator and environmental monitoring system for field studies on shothole disease of almond. *Plant Dis.* 74:558–62.

Aderhold, R. 1902. Über *Clasterosporium carpophilum* (Lév.) Ader. und Beziehungen desselben zum Gummiflusse des Steinobstes. *Arb. Biol. Reichst. Forstw.* 2:515–59.

Aldrich, T., W. J. Moller, and H. Schulbach. 1974. Shothole disease control in almonds—By injecting fungicides into overhead sprinklers. *Calif. Agric.* 28(10):11.

Berkeley, M. J. 1864. (Report of peach blight in England). Note in *Gard. Chron.* 24:938.

Boucher, W. A. 1901. Peach fungus (*Clasterosporium amygdalearum*). *N. Z. Dept. Agric. 9th Rept.*:348–52.

Ellis, M. B. 1959. *Clasterosporium* and some allied Dematiaceae-Phragmosporae. III. *Mycol. Pa. (CMI)* 72:1–75.

English, H. 1958. Fall application of ziram and ferbam effectively control peach leaf curl in California. *Plant Dis. Rep.* 42:384–87.

English, H., and J. R. Davis. 1962. Efficacy of fall applications of copper and organic fungicides for the control of Coryneum blight of peach in California. *Plant Dis. Rep.* 46:688–91.

Ghanea, M. A., and P. Assadi. 1970. Shot-hole of stone fruits in Iran: *Stigmina carpophila* (Lév.) Ellis = *Coryneum beijerinckii*. Oud. *Iran. J. Plant Pathol.* 7:39–63.

Highberg, L. M., and J. M. Ogawa. 1986a. Yield reduction in almond related to incidence of shot-hole disease. *Plant Dis.* 70:825–28.

———. 1986b. Survival of shot-hole inoculum in association with dormant almond buds. *Plant Dis.* 70:828–31.

Jauch, C. 1940. A spotting of stone fruits in Argentina. *Coryneum carpophilum* (Lév.) nov. comb. *Rev. Argen. Agron.* 7:1–26.

Kilgore, W. W., W. E. Yates, and J. M. Ogawa. 1964. Evaluation of concentrate and dilute ground air-carrier and aircraft spray coverage. *Hilgardia* 35:527–36.

Léveillé. J. H. 1843. Observations sur quelques champignons de la flore des environs de Paris. *Ann. Sci. Bot.* 19:215.

Luepschen, N. S., et al. 1968. Coryneum blight infection and control studies. *Colo. State Univ. Exp. Stn. Prog. Rept.* 68–2, 2 pp.

Ogawa, J. M., H. English, W. E. Yates, and H. J. O'Reilly. 1959. Chemical control of Coryneum blight. *Almond Facts* 23(2):4.

Ogawa, J. M., B. L. Teviotdale, and D. Rough. 1983. Shot hole of stone fruits. *Univ. Calif. Div. Agric. Sci. Leaf.* 21363, 4 pp.

Ogawa, J. M., W. E. Yates, and W. W. Kilgore. 1964. Susceptibility of almond leaf to Coryneum blight, and evaluation of helicopter spray applications for disease control. *Hilgardia* 35:537–43.

O'Reilly, H. J. 1957. Relative efficiency of airplane and ground application of sprays in controlling almond shot hole disease (abs.). *Phytopathology* 47:530.

Parker, C. S. 1925. Coryneum blight of stone fruits. *Howard Univ. Rev.* 2:3–40.

Pierce, N. B. 1900. *Peach leaf curl.* USDA Div. Veg. Physiol. Pathol. Bull. 20, 204 pp.

Pinto de T., A., and I. Carreño. 1988. Control quimico de monilia, oidio, cloca y corineo en nectarinos Late Le-Grand. *Agric. Tech. (Chile)* 48(2):106–10.

Samuel, G. 1927. On the shot-hole disease caused by *Clasterosporium carpophilum* and on the "shot-hole" effect. *Ann. Bot.* 41:374–404.

Shaw, D. A., J. E. Adaskaveg, and J. M. Ogawa. 1990. Influence of wetness period and temperature on infection and development of shothole disease of almond caused by *Wilsonomyces carpophilus*. *Phytopathology* 80:749–56.

Smith, R. E. 1907. California peach blight. *Calif. Univ. Agric. Exp. Stn. Bull.* 191:73–100.

Teviotdale, B. T., M. Viveros, M. W. Freeman, and G. S. Sibbet. 1989. Effect of fungicides on shothole disease of almonds. *Calif. Agric.* 43(3):21–23.

Vuillemin, Paul. 1888. L'*Ascospora beijerinckii* et la maladie des cérisiers. *J. Bot.* 2:255–59.

Williams, R. E., and A. W. Helton. 1971. An optimum environment for culturing of *Coryneum carpophilum*. *Phytopathology* 61:829–30.

Wilson, E. E. 1937. The shot-hole disease of stone-fruit trees. *Calif. Univ. Agric. Exp. Stn. Bull.* 608:3–40.

Silver Leaf of Stone Fruit

Chondrostereum purpureum (Pers.: Fr.) Pouz.

This disease, caused by the fungus *Chondrostereum purpureum*, is important in Europe and New Zealand but is of minor importance in California, probably because of California's relatively dry climate. Only one apricot orchard in Contra Costa County and two peach orchards in Butte County have been reported to have the disease in the last few years. In these plantings, only a small area was infected and the disease did not spread to other orchards.

Details of this disease are covered in Chapter 4.

Verticillium Wilt of Stone Fruit

Verticillium dahliae Kleb.

Verticillium wilt of almond and stone fruit is caused by the fungus *Verticillium dahliae* Kleb., which was also named *V. albo-atrum* in early literature. There are two strains of *V. dahliae* on cotton in California, and both are known to attack stone-fruit trees. Because of the diversification of agriculture in California and the wide host range of the pathogen, the disease is important, especially on young plantings of stone fruit and on mature trees grown in cool climates.

When newly planted trees are affected, some branches are killed and an occasional tree is lost, but in most instances the trees recover. Interplanting with susceptible hosts such as cotton, tomato, and strawberry should be avoided. It is important that this disease not be confused with other diseases such as prune dwarf or Phytophthora root and crown rot, which may give somewhat similar symptoms.

Details of this disease are discussed in Chapter 4.

Virus and Viruslike Diseases of Stone Fruit

Almond Union Disorders

Cause Unknown

The first disorder described is almond brownline and decline (ABLD). During 1986, certain plantings of almond on the plum rootstock Marianna 2624 (*Prunus cerasifera* × *P. munsoniana*) contained declining trees. Disease symptoms comprised a lack of terminal shoot growth, giving rise to a rosette of tightly clustered leaves. On mildly affected trees, such symptoms were observed on one or more limbs. With severe disease development, no shoot growth was apparent, the leaves yellowed and drooped (**fig. 22A**), and the trees died. On all symptomatic trees, a close examination of the scion-rootstock union revealed either scattered pockets of necrotic phloem tissue and corresponding xylem pits (mild infection) or a continuous line of bark necrosis (**fig. 22B**) and xylem pits (severe infection) (**fig. 22C**). To date, ABLD has been observed on first- to fourth-leaf trees of the almond cultivars Carmel, Price, and Peerless. Disease incidence in three orchards ranged from 3 to 18 percent (J. K. Uyemoto, unpublished).

A second union disorder was observed in 1987, and principally involved the Carmel cultivar on Marianna 2624. Affected trees showed delayed budbreak and produced fewer blossoms than healthy trees. During late summer, diseased trees defoliated prematurely (**fig. 22D**) and the leaves became boat-shaped with extensive margin necrosis (**fig. 22E**). Tree death was evident. At the scion-rootstock junction, a light brown discoloration of the phloem and a corresponding mild etching of the xylem (**fig. 22F**) were evident. Disease incidence in cv. Carmel in a first- and a fourth-leaf orchard was 20 percent each. In these orchards, adjacent rows of Price/Marianna contained 1 percent diseased trees or fewer (J. K. Uyemoto, unpublished).

Cause. The causes of these union disorders are unknown, but their respective etiologies are currently under study. Graft-transmissible pathogens are suspected.

Comparison of ABLD with prune brownline (PBL). With ABLD, attempts to recover tomato ringspot virus (TmRSV) from either the soil or diseased almond and plum leaf tissues were unsuccessful. Graft inoculations with chip buds from diseased almond scions and Marianna rootstock suckers onto *P. tomentosa* (an indicator for TmRSV) were negative. The brownline symptoms described for ABLD are indistinguishable from those of PBL (described elsewhere in this chapter). However, the incitant of PBL is TmRSV, and it does not infect Marianna 2624 (Hoy and Mircetich 1984). Therefore, it seems probable that a different causal agent is involved in ABLD.

REFERENCE

Hoy, J. W., and S. M. Mircetich. 1984. Prune brownline disease: susceptibility of prune rootstocks and tomato ringspot virus detection. *Phytopathology* 74:272–76.

Cherry Mottle Leaf

Graft-Transmissible Pathogen

Cherry mottle leaf (CML) was initially described by Zeller (1934) in Oregon. Later, McLarty (1935) in British Columbia and Reeves (1935) in Washington State independently demonstrated its graft transmissibility. The disease has been observed in sweet cherry trees grown in nearly all of the western United States and Canada (Cheney and Parish 1976), including California (G. Nyland, personal communication). Severe symptoms are induced on several sweet cherry cultivars (e.g., Bing). Cherry mottle leaf infections are essentially symptomless on certain other sweet cherry cultivars and *Prunus* species. Native bitter cherry (*P. emarginata*) trees serve as a reservoir for the pathogen.

Symptoms. On highly susceptible sweet cherry cultivars, disease symptoms consist of chlorotic mottling, distortion, and puckering of the younger set of leaves (**fig. 24**). Fruit borne on symptomatic limbs are small and ripen late. Some stunting of shoot growth occurs.

Cause. Cherry mottle leaf has been graft-transmitted to healthy plants. Also, when infected budwood was exposed to a hot-water dip at 50°C for 20 minutes or to hot air at 37°C for 21 days, the disease agent survived both treatments (Cheney and Parish 1976).

Transmission. Cherry mottle leaf is easily transmitted by budding or grafting. Symptoms may appear after incubation periods of 9 to 14 days in the greenhouse or 32 days in the field. Under natural conditions, field spread has been observed in cherry orchards located near wild stands of bitter cherry trees. The vector is a scale mite, *Eriophyes inaequalis* (Cheney and Parish 1976).

Control. To control the disease, plant only certified nursery stock. Remove infected cherry trees and destroy stands of volunteer cherry and bitter cherry.

REFERENCES

Cheney, P. W., and C. L. Parish. 1976. Cherry mottle leaf. In *Virus diseases and noninfectious disorders of stone fruits in North America*. pp. 216–18. USDA Agric. Handb. 437. 433 pp.

McLarty, H. R. 1935. Cherry mottle leaf. *Northwest. Asoc. Hort. Entomol. Plant Pathol. Abstr.* 1:5.

Reeves, E. L. 1935. Mottle leaf of cherries. *Wash. State Hort. Assoc. Proc.* 31:85–89.

Zeller, S. M. 1934. Cherry mottle leaf. *Oregon State Hort. Soc. Ann. Rept.* 26:92–95.

Cherry Necrotic Rusty Mottle

Graft-Transmissible Pathogen

In 1945, Rhoads reported on the incidence of a rusty mottle disease of sweet cherry trees in orchards of Utah. The Utah rusty mottle disease was later designated necrotic rusty mottle (NRM) by Richards and Reeves (1951). In the literature and based on similarity in symptomatologies, Lambert mottle (LM) described in British Columbia (Lott 1945) and Montana (Afanasiev and Morris 1951) is reportedly closely related to NRM. Cherry bark blister, described in California by Stout (1949), is now recognized as necrotic rusty mottle. As a result of California's clean stock program, the disease here is no longer of commercial significance.

Symptoms. Leaf symptoms first appear about a month after full bloom and consist of necrotic lesions (**fig. 24**). Depending on the extent of tissue necrosis, affected leaves may abscise prematurely; also, some of the necrotic spots may drop out. Before fruit harvest, leaves that are not cast develop a yellow mottle, with patches of green tissue, and show autumnal color earlier than normal. In affected trees, buds on terminal shoots die and bark symptoms include blisters, necrotic areas, and gum pockets (common in Bing).

Cause. The disease is graft transmissible and the causal agent is eliminated from cherry budsticks subjected to a hot-water treatment at 50°C for 10 to 13 minutes or 52°C for 5 minutes (Nyland 1959). The true nature of the pathogen is unknown.

Transmission. In addition to graft-transmission, natural field spread of NRM has been observed in sweet cherry orchards in Oregon (Cameron and Moore 1985) and Utah (Wadley and Nyland 1976), and of LM in Montana (Afanasiev and Mills 1957). No vector has been found.

Control. Plant only certified nursery trees and remove diseased orchard trees.

REFERENCES

Afanasiev, M. M., and I. K. Mills. 1957. The spread of Lambert mottle virosis in Lambert and Peerless varieties of sweet cherries in Montana. *Plant Dis. Rep.* 41:517–20.

Afanasiev, M. M., and H. E. Morris. 1951 Virus diseases of sweet cherries in Montana. *Plant Dis. Rep.* 35:91.

Cameron, H. R., and D. L. Moore. 1985. Reduction in spread of necrotic rusty mottle with removal of affected trees (abs.) *Phytopathology* 75:1311.

Lott, T. B. 1946. "Lambert Mottle," a transmissible disease of sweet cherry. *Sci. Agric.* 25:260–62.

Nyland, G. 1959. Hot-water treatment of Lambert cherry budsticks infected with necrotic rusty mottle virus. *Phytopathology* 49:157–58.

Rhoads, A. S. 1945. Virus and viruslike diseases of sweet cherry in Utah, and notes on some conditions affecting various fruit crops. *Plant Dis. Rep.* 29:6–19.

Richards, B. L., and E. L. Reeves. 1951. Necrotic rusty mottle. In *Virus diseases and other disorders with viruslike symptoms in stone fruits in North America*. pp. 120–22. USDA Agric. Handb. 10.

Stout, G. L. 1949. Cherry bark blister. *Calif. Dept. Agric. Bull.* 38:257–60.

Wadley, B. N., and G. Nyland. 1976. Rusty mottle group. In *Virus diseases and noninfectious disorders of stone fruits in North America.* pp. 242–49. USDA Agric Handb. 437. 433 pp.

Cherry Rasp Leaf

Cherry Rasp Leaf Virus (CRLV)

Rasp leaf, as a virus disease of sweet cherry, was first described in Colorado (Bodine and Newton 1942). Subsequently, diseased trees were observed elsewhere in western North America (Nyland 1976). In California, cherry rasp leaf has been found in eight counties (Wagnon et al. 1968). The virus is also the incitant of the flat apple disease of apple (Parish 1977) and causes a decline in peach (Stace-Smith and Hansen 1976).

Symptoms. Characteristic disease symptoms consist of prominent enations on the underside of leaves (**fig. 23**). Affected leaves are deformed, have almost normal coloration, and are found initially on the lowest part of the tree. As the disease advances upward, additional leaves develop enations and the lower limbs exhibit sparse foliation and dieback. Fruit yield is markedly reduced. Trees infected at an early age are often killed. The incubation period varies from eight or nine months to two or three years. In California, cling peach trees growing in sites from which infected cherries had been removed became infected and developed severe symptoms within three years after planting. Symptoms consisted of severely dwarfed shoots with dark green, narrow leaves. The leaves showed abundant enations on the underside. The tree trunks and scaffolds developed extensive cankers. Transmission tests to cherry confirmed the suspected rasp leaf etiology (G. Nyland, personal communication).

Characteristics of the virus. The virus particles are isometric, about 30 nm in diameter, seedborne in dandelion(*Taraxacum officinale*) (Stace-Smith and Hansen 1976), and vectored through the soil by the dagger nematode (*Xiphinema americanum*) (Németh 1986; Nyland et al. 1969). The virus is classified in the nepovirus (*nematode polyhedral virus*) group.

Transmission. The virus is readily transmitted to healthy cherry trees by grafting chip buds from diseased sources. However, under orchard conditions, CRLV is spread by its nematode vector. Both mahaleb *Prunus mahaleb*) and mazzard (*P. avium*) are susceptible cherry rootstocks (Wagnon et al. 1968). CRLV is sap transmissible from cherry leaf extracts to herbaceous host plants, which include cucumber (*Cucumis sativus*) and *Chenopodium quinoa*. In British Columbia, natural infections (although symptomless) have been found in balsam root (*Balsamorhiza sagittata*), dandelion (*Taraxacum officinale*), and plantain (*Plantago major*) (Stace-Smith and Hansen 1976).

The virus-vector relationship and disease epidemiology are similar to those for Prunus stem pitting, described elsewhere in this chapter.

Control. Control of this disease is essentially the same as that given for Prunus stem pitting; namely: fumigation of diseased sites, control of broadleaf weeds, and planting with disease-free stock.

REFERENCES

Bodine, E. W., and J. H. Newton. 1942. The rasp leaf of cherry. *Phytopathology* 32:333–35.

Németh, M. 1986. *Virus, mycoplasma and rickettsia diseases of fruit trees.* Martinus Nijhoff, Dordrecht, 841 pp.

Nyland, G. 1976. Cherry rasp leaf. In *Virus diseases and noninfectious disorders of stone fruits in North America*, 219–21. USDA-ARS Agric. Handb. 437.

Nyland, G., B. F. Lownsbery, S. K. Lowe, and J. F. Mitchell. 1969. The transmission of cherry rasp leaf virus by *Xiphinema americanum*. *Phytopathology* 59:1111–12.

Parish, C. L. 1977. A relationship between flat apple disease and cherry rasp leaf disease. *Phytopathology* 67:982–89.

Stace-Smith, R., and A. J. Hansen. 1976. Cherry rasp leaf virus. C.M.I./A.A.B. Descrip. Plant Viruses 159, 4 pp.

Wagnon, H. K., J. A. Traylor, H. E. Williams, and A. C. Weiner. 1968. Investigations of cherry rasp leaf disease in California. *Plant Dis. Rep.* 52:618–22.

Cherry Rusty Mottle

Graft-Transmissible Pathogen

Cherry rusty mottle (CRM) and mild rusty mottle (MRM) are diseases described, respectively, in Washington (Reeves 1940) and Oregon (Zeller and Milbrath

1947). Both CRM and MRM cause similar symptoms on several cultivars of sweet and sour cherries. The causal agent or agents also can be successfully inoculated into other *Prunus* species, including peach, apricot, damson plum, flowering cherry, chokecherry, *P. mahaleb, P. tomentosa,* and *P. serotina* (Wadley and Nyland 1976). Cherry rusty mottle is frequently seen in Washington (G. I. Mink, personal communication), but only MRM is known to occur in California.

Symptoms. Symptoms of CRM first appear about two to three weeks before fruit ripening. Conspicuous chlorotic spots and rings develop first on basal leaves of water sprouts (**fig. 24**). Then, at or before fruit harvest, the older leaves within the tree canopy turn a golden-yellow to rusty bronze color with green spots (islands) and begin to defoliate, causing a 30 to 70 percent loss of leaves. Trees decline in vigor, and fruit ripening may be delayed. Symptoms of MRM are reportedly milder than those of CRM.

Cause. The diseases are graft-transmitted to healthy plants, hence the involvement of a viruslike agent is suspected. Heat treatment of MRM-infected material at 38°C for four weeks did not eliminate the pathogen (Wadley and Nyland 1976).

Transmission. The disease agent is transmitted by grafting, and inoculated trees develop symptoms after a one- to three-year incubation period. With MRM, graft transmission has occurred even without an apparent union between the infected chip bud and the recipient host tissue (Zeller and Milbrath 1947). Natural spread of CRM occurs in Washington, but the vector has not been determined (G. I. Mink, personal communication).

Control. Since the diseases are graft transmitted, it is essential that clean nursery stock be planted. In established orchards, infected trees should be rogued.

REFERENCES

Reeves, E. L. 1940. Rusty mottle, a new virosis of cherry. (abs.) *Phytopathology* 30:789.

Wadley, B. N., and G. Nyland. 1976. Rusty mottle group. In *Virus diseases and noninfectious disorders of stone fruits in North America.* pp. 242–49. USDA Agric. Handb. 437. 433 pp.

Zeller, S. M., and J. A. Milbrath. 1947. Mild rusty mottle of sweet cherry (*Prunus avium*). *Phytopathology* 37:77–84.

Cherry Stem Pitting

Graft-Transmissible Pathogen

Cherry stem pitting (CSP) is similar in symptomatology to Prunus stem pitting (PSP). However, PSP is incited by tomato ringspot virus (TmRSV) (Smith, Stouffer, and Soulen 1973), and CSP is not (Hoy 1983). The disease has been found in Santa Clara (S. M. Mircetich, unpublished) and San Joaquin counties.

Symptoms. Affected trees show delayed bud break and produce abnormally small leaves. The tree canopy is sparse (**fig. 25A**). Symptomatic trees contain small, pointed fruit, borne on a short pedicel. These fruit symptoms are similar to those caused by the X-disease mycoplasmalike organism (see "X-Disease of Peach and Cherry" elsewhere in this chapter). With advanced infection, terminal shoot growth is impaired and severe pits and grooves occur in the woody cylinder under the trunk bark (**fig. 25B**). Affected trees decline rapidly. In a 7.3-ha Bing cherry orchard, disease incidence was nearly 50 percent (Uyemoto, Luhn, and Griesbach 1987). Surveys of other cherry orchards showed a 4 to 26 per incidence (J. K. Uyemoto, unpublished).

Cause. Although a virus is suspected as being the causal agent, the viral nature of the disease is unproven. Attempts to demonstrate the presence of TmRSV in symptomatic trees have failed (Hoy 1983; Uyemoto, Luhn, and Griesbach, 1987).

Transmission. The stem pitting agent was graft-transmitted to mazzard, mahaleb (Hoy 1983), Stockton Morello, Nanking cherry, and Lovell peach (Mircetich, Moller, and Nyland 1977), suggesting that a viruslike agent is involved. Preliminary observations of CSP suggest a soilborne cause. For example, the symptoms of affected trees range in severity from mild (one-sided) to severe (entire tree). On mildly affected trees, the trunk pits and grooves are evident only directly below a symptomatic scaffold. Also, trunk pits and grooves are most prominent near the soil line (J. K. Uyemoto, unpublished).

Efforts are currently in progress to determine the etiology, epidemiology, and control of CSP.

Control. No means of control have been developed.

REFERENCES

Hoy, J. W. 1983. *Studies on the nature of the prune brownline and sweet cherry stem pitting diseases and epidemiology and control of prune brownline.* Ph.D. thesis, Universisty of California, Davis, 69 pp.

Mircetich, S. M., W. J. Moller, and G. Nyland. 1977. Prunus stem pitting in sweet cherry and other stone fruit trees in California. *Plant Dis. Rep.* 61:931–35.

Smith, S. H., R. F. Stouffer, and D. M. Soulen. 1973. Induction of stem pitting in peaches by mechanical inoculation with tomato ringspot virus. *Phytopathology* 63:1404–06.

Uyemoto, J. K., C. F. Luhn, and J. A. Griesbach. 1987. A severe outbreak of stem pitting in young trees of sweet cherry (abs.). *Phytopathology* 77:1242.

Peach Mosaic

Graft-Transmissible Pathogen

Peach mosaic is also called "Texas mosaic" because of its discovery in Texas in 1931 (Hutchins 1932). During the five years of the tree-removal program in the 1930s in southern California (Kunkel 1936), 204,193 diseased peach trees were destroyed (Cochran 1948; Pine 1976; Stout 1939). Most seriously affected were the popular freestone peach cultivars (e.g., J. H. Hale and Elberta). In North America, the disease is limited to the southwestern United States and Mexico (Bodine, 1934, 1937a,b; Bodine and Durrell 1937, 1941). Peach mosaic also has been reported in Greece, Italy, and India (Németh 1986). In California the disease has not been found north of Los Angeles County (Pine 1976; Thomas et al. 1944) and is of no economic importance today on peach or any other fruit crop. The pathogen has been transmitted to 269 cultivars and 62 numbered selections of peach and nectarine (Cochran 1949; Cochran and Pine 1953). In these studies, most freestone cultivars were severely damaged, but clingstones showed little injury. The pathogen has been found naturally in many stone-fruit species, including almond, apricot (Pine 1967b), peach, and plum (Pine and Cochran 1962), but not in commercial cherries. It has not been transmitted to mahaleb, sour, or sweet cherries.

Symptoms. Most cultivars of peach and nectarine exhibit symptoms, but there is a great variation in symptom severity. Severely damaged peaches are Elberta, Fay Elberta, J. H. Hale, Red Haven, and Rio Oso Gem; moderately damaged are Babcock, Redtop, and Merrill Fiesta; slightly damaged are Carolyn, Paloro, Peak, Springtime, and Suncrest. Among nectarines, Gower and Stanwick are severely affected, but Victoria and Gold-mine are only slightly damaged. Plum cultivars such as Becky Smith and Flaming Delicious exhibit symptoms; symptomless carriers are apricot, prune, almond, and certain plum cultivars.

The symptoms on peach are often difficult to diagnose because the pathogen exists in several forms that vary in virulence, and because peach cultivars differ in tolerance (Hutchins, Bodine, and Thornberry 1937; Hutchins et al. 1951). Symptoms expressed are color break in the blossoms, retarded foliation, mottling and deformity of the leaves, fruit deformity, and abnormal tree growth. On cultivars with large, showy, pink petals (Rio Oso Gem and Fay Elberta), color break in the petals ranges from irregular streaks, spots, and mottle to crinkling and dwarfing (**fig. 26A**). On the foliage and branches, highly susceptible cultivars exhibit leaf retardation and diminished growth in the spring and early summer. One or more branches may show such symptoms. As the season advances, the leaves may gradually enlarge, but much of the severely affected foliage often drops. Leaf markings vary considerably in size but are characteristically irregular yellow areas ranging from mere flecks to bold blotches, streaks, or vein feathering (**fig. 26B**). Some of the chlorotic areas may become necrotic and fall out, resulting in a "shothole" condition. Veinlet clearing may occur in the early season but it disappears later. With the advent of high temperatures in late summer, the yellow areas may develop normal color. Affected twigs exhibit short internodes, are stunted, and are somewhat greater in diameter than normal. Abnormal numbers of short twigs may develop at the tips of affected branches. On the fruit, about two months after set (about the time of pit hardening), round raised islands surrounded by depressed areas appear. The "bumpiness" may develop into irregular protuberances or ridges. Affected fruit are dwarfed, ripen several days later than healthy fruit, and have inferior texture and flavor.

Some plum cultivars are highly susceptible to peach mosaic, and the symptoms expressed on *Prunus domestica* and *P. salicina* are somewhat similar to those on peach. Infected trees rarely show delayed foliation, dwarfing, or fruit injury, and affected leaves are not deformed as seriously as those of peach. Natural infections of peach mosaic occur more often on apricot than on plum. Symptoms on most apricot cultivars are less severe than those on peach, but on highly sensitive cultivars the symptoms are similar to those on peach. Symptoms of peach mosaic on almond are comparatively mild. Severe outbeaks on almond have occurred only in orchards close to diseased peach orchards.

Causal agent. Because of its symptomatology and graft-transmissibility, peach mosaic for many years was assumed to be caused by a virus (Németh 1986; Pine 1976). However, since the causal agent has not yet been identified, it seems prudent to indicate merely that the disease is caused by an uncharacterized graft-transmissible pathogen. The pathogen is vectored by the eriophyid mite *Eriophyes insidiosus* Keifer and Wilson (Keifer and Wilson 1955; Wilson, Jones, and Cochran 1955). This mite was first collected from peach in southern California and subsequently from other *Prunus* species in California.

The range of the vector does not, however, coincide with the present range of the disease. The vector is found in the San Joaquin Valley (Madera County) as well as in the eastern states of Georgia, Indiana, and Mississippi where peach mosaic has never been observed. It commonly feeds and reproduces under the bud scales and is found in greatest abundance during the month of April. The disease spreads whenever foliage and the vector are present (Jones and Wilson 1951). Infections that occur before the leaves expand express symptoms in 21 to 30 days, while infections that take place later do not produce symptoms until the following spring (Hutchins, Bodine, and Thornberry 1937; Jones and Wilson 1951). Natural spread of PMV occurs from peach to peach and sometimes from peach to apricot and almond but rarely from apricot or almond to peach (Cochran and Stout 1947). The pathogen has not been transmitted to any herbaceous host and has not been purified from *Prunus* tissue. It has been transmitted by tissue grafts (Cochran and Rue 1944) but not by dodder (Bennett 1944). Contact of affected and nonaffected tissue for a period of two days is sufficient for transmission (Kunkel 1938), presumably resulting from the migration and feeding of infective mites on the healthy plant. Definable strains of the pathogen exist in nature on the basis of differential symptoms on *Prunus* species, cross-protection, and consistent recovery from natural infection (Bodine and Durrell 1941; Bodine 1942; Cochran 1938, 1944; Cochran and Hutchins 1937; Hutchins and Cochran 1940; Pine 1965).

Control. Affected trees have been removed in southern California. Selection of noninfected budwood is the only means of preventing distribution of the disease into and out of the nursery (Hutchins, Bodine, and Thornberry 1937). Kunkel (1936) found that heat-treating the budwood did not kill the pathogen. Chemotherapy of infected peach trees has been attempted on a limited scale without conclusive results (Pine 1967a).

REFERENCES

Bennett, C. W. 1944. Studies of dodder transmission of plant viruses. *Phytopathology* 34:905–32.

Bodine, E. W. 1934. Peach mosaic disease in Colorado. *Colorado Agric. Exp. Stn. Bull.* 421, 11 pp.

———. 1937a. Occurrence of peach mosaic in Colorado. *Plant Dis. Rep.* 18:123.

———. 1937b. Control of peach mosaic in Colorado (abs.). *Phytopathology* 27:954.

———. 1942. Antagonism between strains of the peach-mosaic virus in Western Colorado (abs.). *Phytopathology* 32:1.

Bodine, E. W., and L. W. Durrell. 1937. The Maynard plum—A carrier of the peach mosaic (abs.). *Phytopathology* 27:954.

———. 1941. Peach mosaic in western Colorado (abs.). *Phytopathology* 31:4.

Cation, D. 1934. Peach mosaic. *Phytopathology* 24:1381–82.

Cochran, L. C. 1938. Further studies on the host relationships of peach mosaic in southern California. *Phytopathology* 28:890–93.

———. 1944. The "Complex concept" of the peach mosaic and certain other stone fruit viruses (abs.). *Phytopathology* 34:934.

———. 1948. The virus problem. *Am. Fruit Grower* 68(11):9–11.

———. 1949. Relation of peach mosaic disease to other stone fruits. *Calif. Cultivator* 87:164–65.

Cochran, L. C., and L. M. Hutchins. 1937. Peach mosaic host relationship studies in southern California (abs.) *Phytopathology* 27:954.

Cochran, L. C., and T. S. Pine. 1953. Present status of information on host range and host reactions to peach mosaic virus. *Plant Dis. Rep.* 42:1225–28.

Cochran, L. C., and J. L. Rue. 1944. Some host-tissue relationships of the peach mosaic virus. (abs.) *Phytopathology* 34:934.

Cochran, L. C., and G. L. Stout. 1947. Studies on the natural spread of the peach mosaic virus among apricots, almonds and peaches (abs.) *Phytopathology* 37:843.

Hutchins, L. M. 1932. Peach mosaic—A new virus disease. *Science* 76(1962):123.

Hutchins, L. M., E. W. Bodine, L. C. Cochran, and G. L. Stout. 1951. Peach mosaic. In *Virus diseases and other disorders with virus-like symptoms of stone fruits in North America*, 226–36. USDA Agric. Handb. 10.

Hutchins, L. M., E. W. Bodine, and H. H. Thornberry. 1937. Peach mosaic, its identification and control. *USDA Circ.* 427, 48 pp.

Hutchins, L. M., and L. C. Cochran. 1940. Peach-mosaic virus strain studies (abs.). *Phytopathology* 30:11.

Jones, L. S., and N. S. Wilson. 1951. Peach mosaic spreads throughout the growing season. *Calif. Dept. Agric. Bull.* 40:117–18.

Keifer, H. H., and N. S. Wilson. 1955. A new species of eriophyid mite responsible for the vection of peach mosaic virus. *Univ. Calif. Dept. Agric. Bull.* 44:145–46.

Kunkel, L. O. 1936. Peach mosaic not cured by heat treatment. *Am. J. Bot.* 23:683–86.

———. 1938. Contact period in graft transmission of peach viruses. *Phytopathology* 28:491–97.

Németh, M. 1986. Peach mosaic. In *Virus, mycoplasma and rickettsia diseases of fruit trees*, 398–403. Martinus Nijhoff, Dordrecht.

Pine, T. S. 1965. Host range and strains of peach mosaic virus. *Phytopathology* 55:289–91.

———. 1967a. Reactions of peach trees and peach tree viruses to treatment with dimethylsulfoxide and other chemicals. *Phytopathology* 57:671–73.

———. 1967b. Apricot ring pox and peach mosaic viruses in apricot trees. *Phytopathology* 57:1268.

———. 1976. Peach mosaic. In *Virus diseases and noninfectious disorders of stone fruits in North America*, 61–70. USDA Agric. Handb. 437.

Pine, T. S. and L. C. Cochran. 1962. Peach mosaic virus in horticultural plum varieties. *Plant Dis. Rep.* 46:495–97.

Reed, H. S., and H. H. Thornberry. 1935. A peach tree disease recently discovered in California (abs.) *Phytopathology* 25:897.

Stout, G. L. 1939. Peach mosaic. *Calif. Dept. Agric. Bull.* 28(3):177–200.

Thomas, H. E., C. E. Scott, E. E. Wilson, and J. H. Freitag. 1944. Dissemination of peach mosaic. *Phytopathology* 34:658–61.

Wilson, N. S., L. S. Jones, and L. C. Cochran. 1955. An eriophyid mite vector of peach mosaic virus. *Plant Dis. Rep.* 39:889–92.

Prune Brownline

Tomato Ringspot Virus (TmRSV)

Prune brownline, a recently diagnosed disease of prune trees, is widespread and serious in certain areas of northern California (Hoy et al. 1984). The disease has been found in Yolo, Solano, Sutter, Yuba, Tehama, Placer, and El Dorado counties, but its occurrence in other areas of California has not been ascertained.

The disease is known to affect the *Prunus domestica* L. cultivars French prune, Empress plum, and President plum when propagated on myrobalan plum (*P. carasifera* Ehrh.) and peach (*P. persica* [L.] Batsch.) rootstocks (Hoy et al. 1984). The same disorder has recently been diagnosed on Stanley prune in New York (Cummins and Gonsalves 1986a) where for many years it was called "Stanley constriction disease," cause unknown (Parker and Gilmer 1976). This Stanley prune disorder is also present in Maryland and Pennsylvania (S. M. Mircetich, unpublished). Because of the wide geographic distribution of the causal virus and its nematode vectors, it seems likely that prune brownline will be encountered in other regions of the United States.

Symptoms. Initial symptoms of brownline-affected prune trees are yellowing of leaf margins and interveinal chlorosis accompanied by drooping and upward rolling of leaves followed by marginal necrosis of leaf lamina. The trees develop autumn colors and become defoliated much earlier than normal. They usually have a heavy set of small poor-quality fruit that drop prematurely. In older trees, poor growth of terminal shoots usually precedes leaf symptoms. As the disease progresses, affected trees show dieback of terminals, general decline, and subsequent death (**fig. 27**). The diagnostic symptom of the disease is the presence of a narrow strip of brown to dark-brown cambial tissue (brownline) at the junction of the scion and rootstock. This discoloration can be seen only by removing a section of bark at the graft union. In early stages of the disease, the brownline is not continuous around the trunk, so it is necessary to examine the union at several points. The necrosis gradually extends into the bark, causing a girdling at the union and resulting in tree death one to five years after girdling is completed. Young vigorous trees are usually subject to quicker decline and death than larger or nonvigorous trees. Affected trees commonly develop an overgrowth of the scion at the union, although sometimes younger trees die before this overgrowth has occurred. With affected Stanley prune trees, the constricted growth of the myrobalan rootstock was responsible for the name "Stanley prune constriction disease" given to this disorder years ago (Parker and Gilmer 1976).

The root system of brownline-affected trees usually is weaker than that of healthy trees on the same rootstocks. Also, in an advanced stage of the disease, myrobalan seedling rootstocks often develop a bark canker below the necrotic union. This condition usually is followed by a sudden collapse of the tree. The Myrobalan 29C rootstock in brownline-affected trees frequently develops sprouts with viruslike symptoms (mottling, chlorotic spots, and rings) in the leaves.

Cause. Studies in California (Mircetich and Hoy 1981; Hoy and Mircetich 1984; and Hoy, Mircetich, and Lownsbery 1984) and more recently in New York (Cummins and Gonsalves 1986a,b) have shown that prune brownline is caused by tomato ringspot virus (TmRSV). Isolates of TmRSV associated with peach yellow bud mosaic, cherry leaf mottle, Prunus stem pitting, and California peach stem pitting can cause the brownline disease in prune trees on peach and myrobalan rootstocks, but not in those on Marianna 2624 stocks (Hoy and Mircetich 1984). In New York, Stanley prune is susceptible to the disease when propagated on myrobalan rootstocks and some selections of *Prunus domestica* (Cummins and Gonsalves 1986a).

Tomato ringspot virus is an RNA-containing virus, with isometric particles about 25 to 28 nm in diameter. It is a member of the nepovirus (*nematode polyhedral virus*) group. In herbaceous plants the virus is readily transmissible by sap inoculation. It has a wide host range, including both woody and herbaceous plant species. The virus is transmitted by the dagger nematodes *Xiphinema americanum*, *X. californicum*, and *X. rivesi*. Adults and three larval stages of *X. americanum* can serve as vectors. Nematodes can acquire the virus in one hour and then inoculate it into plants within an hour. It is transmissible in seed of soybean, strawberry, raspberry, and tobacco. In geographic distribution, the virus appears to be restricted to the temperate regions of North America where its nematode vectors are found (Stace-Smith 1970). It is endemic in native wild hosts along the California coast. Several strains of the tomato ringspot virus that differ in host relationships, symptomatology, or serology are recognized. Three of these variants that cause disease in stone-fruit trees in California are the yellow bud mosaic strain in peach and almond, the stem pitting strain in peach and apricot, and the cherry mottle leaf strain in sweet cherry (Hoy, Mircetich, and Lownsbery 1984).

Several herbaceous plants, including bean, cowpea, tobacco, *Chenopodium amaranticolor*, and *C. quinoa*, have been of value as assay species (Hoy et al. 1984; Stace-Smith 1970). For positive identification, however, serological tests are essential. Tests that have been effective include direct and indirect enzyme-linked immunosorbent assay (D-ELISA and I-ELISA, respectively) and radio-immunosorbent assay (RISA) (Hoy and Mircetich 1984; Cummins and Gonsalves 1986b). I-ELISA and RISA are more reliable than D-ELISA for detecting TmRSV in diseased trees (Hoy and Mircetich 1984).

In tobacco sap, the virus loses infectivity after 10 minutes at about 58°C, two days at 20°C, three weeks at 4°C, or several months at $-24°C$. The dilution endpoint of sap from inoculated cucumber cotyledons or tobacco leaves is 10^{-3}; that of sap from systemically infected tobacco is less than 10^{-1} (Stace-Smith 1970). The yellow bud mosaic strain of the virus was inactivated in buds of infected peach trees exposed to 38°C for 21 days (Schlocker and Traylor 1976).

Disease development. The tomato ringspot virus is a soilborne virus that survives from year to year in infected commercial hosts, weed hosts, and in its dagger nematode (*Xiphinema* spp.) vectors. When prune or susceptible plum cultivars on myrobalan, peach, or certain *P. domestica* rootstocks are planted in infested sites, they are subject to infection with this brownline-inducing virus. The only natural mode of infection is through the feeding in the roots of viruliferous dagger nematodes. Since the virus is not present in French prune, Empress, or President plum scions, the disease apparently is not spread through scion budwood or pollen to healthy trees (Hoy et al. 1984).

Prune brownline commonly spreads within an orchard from diseased to adjacent healthy trees through migration of viruliferous nematodes. Radial spread from infected trees was found to occur commonly at the rate of one tree per year in any direction (Mircetich and Hoy 1981). From the infected roots, the virus moves upward into the trunk until it comes in contact with the scion at the graft union. Here, with a highly sensitive scion cultivar such as French prune, a hypersensitive reaction occurs that results in cell necrosis and the development of the diagnostic brownline symptom (Mircetich and Hoy 1981). Based on root-chip inoculations into myrobalan rootstocks supporting French prune scions, at least two years is required for the development of typical disease symptoms. In surveying for the disease in Stanley prune (plum), Cummins and Gonsalves (1986b) found that visual samplings ("windows" at the graft union) from all four quadrants were required to identify 90 percent of the infected trees, whereas ELISA samples from three quadrants were required for the same degree of accuracy.

Control. Partial control of prune brownline can be achieved by removing affected trees along with adjacent symptomless trees that already may have virus infection in the roots. In transporting these trees from the orchard and in all other cultural operations, care should be taken not to move soil from diseased into healthy areas. Prune or susceptible plums on Marianna 2624 rootstock should be used to replant brownline-affected areas or to plant new orchards on sites with a history of tomato ringspot virus diseases. In New York, it is recommended that own-rooted Stanley prune trees be used in establishing new plantings (Cummins and Gonsalves 1986a).

The use of a nematicidal fumigant in diseased areas will reduce the population of dagger nematodes and decrease the chances that prune brownline will continue to spread within an orchard (Hoy et al. 1984). These areas should also be kept weed-free for two years, since some orchard weeds are known to be reservoirs of tomato ringspot virus and some also serve as hosts of the nematode vectors. To reduce the likelihood of introducing the disease into new areas, use only stone-fruit trees free from virus diseases as planting stock.

REFERENCES

Cummins, J. N., and D. Gonsalves. 1986a. Constriction and decline of Stanley prune associated with tomato ringspot virus. *J. Am. Soc. Hort. Sci.* 111:315–18.

———. 1986b. Irregular distribution of tomato ringspot virus at the graft unions of naturally infected Stanley plum trees. *Plant Dis.* 70:257–58.

Hoy, J. W., and S. M. Mircetich. 1984. Prune brownline disease: Susceptibility of prune rootstocks and tomato ringspot virus detection. *Phytopathology* 74:272–76.

Hoy, J. W., S. M. Mircetich, R. S. Bethell, J. E. DeTar, and D. M. Holmberg. 1984. The cause and control of prune brownline disease. *Calif. Agric.* 38(7 & 8):12–13.

Hoy, J. W., S. M. Mircetich, and B. F. Lownsbery. 1984. Differential transmision of Prunus tomato ringspot virus strains by *Xiphinema californicum*. *Phytopathology* 74:332–35.

Mircetich, S. M., and J. W. Hoy. 1981. Brownline of prune trees, a disease associated with tomato ringspot virus infection of Myrobalan and peach rootstocks. *Phytopathology* 71:30–35.

Parker, K. G., and R. M. Gilmer. 1976. Constriction disease of Stanley prune. In *Virus diseases and noninfectious disorders of stone fruits in North America*, 268–71. USDA Handb. 437.

Schlocker, A., and J. A. Traylor. 1976. Yellow bud mosaic. In *Virus diseases and noninfectious disorders of stone fruits in North America*, 156–65. USDA Handb. 437.

Stace-Smith, R. 1970. Tomato ringspot virus. *Commonw. Mycol. Inst./Assoc. Appl. Biologists Descrip. Plant Viruses* 18, 4 pp.

Prune Dwarf

Prune Dwarf Virus (PDV)

Diseases incited by prune dwarf virus (PDV) are widespread on stone fruit and described under a variety of names: on prune, "prune dwarf," "prune mosaic," "willows," and "shoestring;" on peach, "Muir peach dwarf" and "peach stunt"; and on apricot, "gummosis." In sweet cherry, the virus may by symptomless or it may induce symptoms indistinguishable from tatter leaf (Gilmer, Nyland, and Moore 1976). In sour cherry, the disease has been called "sour cherry yellows" (Keitt and Clayton 1943), "yellow leaf," "physiological yellow leaf" (Stewart 1919), "physiological leaf drop" (Gloyer and Glastow 1928), and "boarder tree" (Keitt et al. 1951), while the initial acute symptoms have been called "chlorotic mottle" and "chlorotic ringspot" (Gilmer, Brase, and Parker 1957). It is important to note that these diseases were often misidentified as Prunus ringspot, a similar-appearing but entirely different disease (Gilmer, Nyland, and Moore 1976).

In plum and prune, where the disease was first discovered, prune dwarf is rare, although a few infected trees have been found in New York, Ontario, and British Columbia (Thomas and Hildebrand 1936). In peach, natural infection has been reported only from California (Hutchins et al. 1951; Uyemoto et al. 1989) where, in combination with Prunus necrotic ringspot virus (PNRSV), it causes a peach stunt disease in cultivars insensitive to PDV alone. In both sweet and sour cherry, the PDV is widely distributed throughout North America and occurs as well in Europe (Posnette, Cropley, and Swait 1968; Zawadaska 1965) and Australia (Pares 1967). Prune dwarf virus undoubtedly occurs in all areas where stone fruit are grown. Chronic effects of prune dwarf virus infection on sour cherry may have been observed in France as early as 1768, and in England in 1839 (Gilmer 1961b). The disease probably was introduced into North America in the sour cherry cultivars Montmorency (Gilmer 1961b). Reports of yellows disease of sour cherry, later shown by Keitt and Clayton (1939, 1943) to be caused by a virus, were observed in New York in 1919 (Stewart) and again in 1928 (Gloyer and Glastow). In 1936, Thomas and Hildebrand described prune dwarf in Italian prune in New York and Ontario and demonstrated the virus etiology of the disease. At the same time, Hutchins et al. (1951) demonstrated that Muir peach dwarf, observed in California as early as 1920, is also caused by a virus.

Symptoms. In California two-year-old peach trees in the orchard were inoculated in the fall with pure cultures of PDV and observed for symptoms over a four-year period. Mild symptoms of stunting, only in inoculated branches and on rootstock suckers, developed the following spring and persisted as mild stunting during the next four years. Meanwhile, Italian prune trees growing on inoculated peach rootstock showed severe symptoms of prune dwarf. When PDV was introduced into peach trees already infected with pure cultures of PNRSV, moderate to severe symptoms of peach stunt were produced in succeeding years (**fig. 28**). In all cases symptoms became less obvious as the season progressed, but they recurred each spring (Gilmer, Nyland, and Moore 1976).

Many of the early descriptions of the symptoms of PDV infection were complicated by the presence of other contaminating viruses such as PNRSV or green ring mottle virus (GRMV) (Keitt et al. 1951). In this discussion symptom descriptions are limited to peach, sweet cherry, and the indicator host Shiro-fugen flowering cherry.

On Shiro-fugen, gumming and necrosis occur about four to six weeks after insertion of infected (PDV) budwood or other tissue. Inoculation usually is by T-budding, and initial symptoms are localized bark necrosis and gumming at the point of contact between the diseased and healthy tissue. If the infected shoots are not removed, the virus moves downward at rates of about 2 to 5 cm per day (Hampton and Fulton 1963) until the branches and finally the entire tree are killed.

In sweet cherry, the presence of PDV is more difficult to determine, and the effect of the disease on production has not been well established. However, when Way and Gilmer (1963) pollinated healthy trees with infected pollen, fruit set was reduced by 25 to 90 percent.

A strain of the virus known to occur in California induces stunting and the formation of narrow, rough-textured leaves in sweet cherry (Nyland 1967). This is the only strain known to cause symptoms in Japanese plum (*P. salicina*). It causes sour cherry yellows and typical prune dwarf symptoms in European plum.

Causal agent. The causal virus has several synonyms (Gilmer 1963): prune dwarf virus, sour cherry yellows virus (Keitt and Clayton 1943), "yellows" strain of ringspot virus (Milbrath 1956, 1961), peach stunt virus (Milbrath 1961), Muir peach dwarf virus (Hutchins et al. 1951), and virus B (Fulton 1957a,b, 1958). It is a member of the ilarvirus group. Currently, dwarf virus (PDV) exists naturally in a wide spectrum of strains that differ in certain properties (Fulton 1957b; Waterworth and Fulton 1964), herbaceous host range (Fulton 1957a; Waterworth and Fulton 1964), rate of movement (Gilmer and Brase 1963; Milbrath 1963; Willison 1944), and symptomatology in *Prunus* (Nyland 1967; Waterworth and Fulton 1964). The sizes and shapes of PDV and PNRSV are similar, and they produce similar symptoms on *Prunus* species; they differ in herbaceous host range (Fulton 1957a) and in some properties (Fulton 1957b; Waterworth and Fulton 1964). Both viruses have many characteristics in common with nepoviruses, but the individual viruses of that group do not share common antigens. Evidence is lacking that PDV and PRSV are soilborne. Cole and Mink (1984) reported that an extract obtained from pollen taken from honeybee hives degraded PDV and necrotic ringspot virus (NRSV) into serologically undetectable units.

Prune dwarf virus particles are isometric "spheres" about 22 nm in diameter, and the ultraviolet adsorption spectra are typical of nucleoproteins (Fulton 1959). The thermal inactivation point in cucumber sap is 44°C (Fulton 1957b). The dilution endpoint for infectivity is between 10^{-2} and 10^{-3}. At a dilution of 1:80, PDV lost 50 percent of its infectivity after three hours at 24°C; no loss occurred when the virus was held for the same period at 0°C (Fulton 1957b). Prune dwarf virus was inactivated in infected stone-fruit trees by thermotherapy at 36° to 37°C for 15 days or longer (Nyland 1960).

Transmission. Prune dwarf virus is readily transmitted by budding or grafting and through seeds of several stone fruit, such as mahaleb (Cation 1949; Gilmer and Kamalsky 1962), sweet cherry (Gilmer and Kamalsky 1962), and sour cherry (Gilmer and Way 1960). The percentage of infected seeds may range from 20 to 70 percent for sour cherry (Gilmer and Way 1960) and 15 to 30 percent in mahaleb and mazzard cherries (Gilmer and Kamalsky 1962). The planting of infected rootstocks, resulting from seed transmission, accounts for the prevalence of this virus in both sweet and sour cherry. The natural spread of PDV in the orchard is more rapid in sour cherry than in sweet cherry (Gilmer 1961a). There is also a pronounced tendency for new infections to occur downwind from the source of inoculum in sour cherry orchards as well as in peach orchards containing peach stunt and Muir peach dwarf (Hutchins et al. 1951). Transmission by pollen to seed of sour cherry was proved in 1961 by Gilmer and Way (1960). Pollen transmission of PDV from tree to tree was later shown in sour cherry by George and Davidson (1963, 1964) and in sweet cherry by Gilmer (1965). Because pollen transmission is responsible for spread in the orchard, insects, especially bees, undoubtedly serve as indirect vectors. Ehlers and Moore (1957) and Williams, Traylor, and Wagnon (1963) showed that pollen from infected trees contains the PDV, and the latter group of researchers transmitted the virus to peach by inserting infected pollen beneath the bark.

Control. The first step in control of prune dwarf is to establish orchards with nursery trees developed with both the scion and rootstock free of PDV. To insure PDV-free seedling rootstocks, the foundation seed-source trees must be protected from infected pollen and indexed (Mink and Aichele 1984) to ensure freedom from the virus. Even in certified nursery trees, PDV could be present at a low level. Thus, during the first five years, orchards should be visually monitored for leaf casting symptoms within three to four weeks after petal fall, and infected trees removed. Removal of infected trees and replanting of vacant sites are suggested at tree ages of 6 to 10 years but not thereafter because of the economics of establishing fruitful trees before the entire orchard is replanted. Also, replanting in older orchards could very well result in infection of the replants when they produce blossoms.

REFERENCES

Cation, D. 1949. Transmission of cherry yellows virus complex through seeds. *Phytopathology* 39:37–40.

Cole, A., and G. I. Mink. 1984. An agent associated with bee-stored pollen that degrades intact viruses. *Phytopathology* 74:1320–24.

Ehlers, C. G., and J. D. Moore. 1957. Mechanical transmission of certain stone fruit viruses from *Prunus* pollen (abs.). *Phytopathology* 47:519–20.

Fulton, R. W. 1957a. Comparative host ranges of certain mechanically transmitted viruses of *Prunus*. *Phytopathology* 47:215–20.

———. 1957b. Properties of certain mechanically transmitted viruses of *Prunus*. *Phytopathology* 47:683–87.

———. 1958. Identity of and relationships among certain sour cherry viruses mechanically transmitted to *Prunus* species. *Virology* 6:499–511.

———. 1959. Purification of sour cherry necrotic ringspot and prune dwarf viruses. *Virology* 9:522–35.

George, J. A., and T. R. Davidson. 1963. Pollen transmission of necrotic ring spot and sour cherry yellows viruses from tree to tree. *Can. J. Plant Sci.* 43:276–88.

———. 1964. Further evidence of pollen transmission of necrotic ring spot and sour cherry yellows viruses in sour cherry. *Can. J. Plant Sci.* 44:383–84.

Gilmer, R. M. 1961a. The frequency of necrotic ring spot, sour cherry yellows, and green ring mottle viruses in naturally infected sweet and sour cherry orchard trees. *Plant Dis. Rep.* 45:612–15.

———. 1961b. A possible history of sour cherry yellows. *Phytopathology* 51:265–66.

———. 1963. Some synonyms of sour cherry yellows virus. *Phytopathology* 53:369–70.

———. 1965. Additional evidence of tree-to-tree transmission of sour cherry yellows virus by pollen. *Phytopathology* 55:482–83.

Gilmer, R. M., and K. D. Brase. 1963. Nonuniform distribution of prune dwarf virus in sweet and sour cherry trees. *Phytopathology* 53:819–21.

Gilmer, R. M., K. D. Brase, and K. G. Parker. 1957. Control of virus diseases of stone fruit nursery trees in New York. *N.Y. State Agric. Exp. Stn. (Geneva) Bull.* 779, 53 pp.

Gilmer, R. M., and L. R. Kamalsky. 1962. The incidence of necrotic ring spot and sour cherry yellows viruses in commercial mazzard and mahaleb cherry rootstocks. *Plant Dis. Rep.* 46:583–85.

Gilmer, R. M., G. Nyland, and J. D. Moore. 1976. Prune dwarf. In *Virus diseases and noninfectious disorders of stone fruits in North America*, 179–90. USDA Handb. 437.

Gilmer, R. M., and R. D. Way. 1960. Pollen transmission of necrotic ringspot and prune dwarf viruses in sour cherry. *Phytopathology* 50:624–25.

Gloyer, W. O., and H. Glastow. 1928. Defoliation of cherry trees in relation to winter injury. *N.Y. State Agric. Exp. Stn. Bull.* 555, 27 pp.

Hampton, R. E., and R. W. Fulton. 1963. Rate and pattern of prune dwarf virus movement within inoculated cherry trees. *Phytopathology* 53:998–1002.

Hutchins, L. M., C. F. Kinman, L. C. Cochran, and G. L. Stout. 1951. Muir peach dwarf. In *Virus diseases and other disorders with virus-like symptoms of stone fruits in North America,* 63–70. USDA Handb. 10.

Keitt, G. W., and C. N. Clayton. 1939. A destructive bud-transmissible disease of sour cherry in Wisconsin. *Phytopathology* 29:821–22.

———. 1943. A destructive virus disease of sour cherry. *Phytopathology* 33:449–68.

Keitt, G. W., et al. 1951. Sour cherry yellows. In *Virus diseases and other disorders with viruslike symptoms of stone fruits in North America,* 152–58. USDA Handb. 10.

Milbrath, J. A. 1956. Squash as a differential host for strains of stone fruit ringspot viruses. *Phytopathology* 46:638–39.

———. 1961. The relationship of stone fruit ringspot virus to sour cherry yellows, prune dwarf and peach stunt. *Tidsskr. Planteavl.* 65:125–33.

———. 1963. Variability in rates of movement of different strains of the virus responsible for prune dwarf. *Phytopathology.* 53:1106–08.

Mink, G. I., and M. D. Aichele. 1984. Detection of *Prunus* necrotic ringspot and prune dwarf viruses in *Prunus* seed and seedlings by enzyme-linked immunosorbent assay. *Plant Dis.* 68:378–81.

Nyland, G. 1960. Heat inactivation of stone fruit ringspot virus. *Phytopathology* 50:380–82.

———. 1967. An unusual strain of sour cherry yellows virus associated with disease symptoms in cherry, peach, and plum (abs.). *Phytopathology* 57:101.

Pares, R. D. 1967. Prune dwarf virus latent in cherries in New South Wales. *Austral. J. Sci.* 30:32–33.

Posnette, A. F., R. Cropley, and A. A. J. Swait. 1968. The incidence of virus diseases in English sweet cherry orchards and their effect on yield. *Ann. Appl. Biol.* 61:351–60.

Stewart, F. C. 1919. Notes on New York plant diseases. *N.Y. (Geneva) Agric. Exp. Stn. Bull.* 463:157–88.

Thomas, H. E., and E. M. Hildebrand. 1936. A virus disease of prune. *Phytopathology* 26:1145–48.

Uyemoto, J. K., C. F. Luhn, W. Asai, R. Beede, J. A. Beutel, and R. Fenton. 1989. Incidence of ilarviruses in young peach trees in California. *Plant Dis.* 73:217–20.

Waterworth, H. E., and R. W. Fulton. 1964. Variation among isolates of necrotic ringspot and prune dwarf viruses isolated from sour cherry. *Phytopathology* 54:1155–60.

Way, R. D., and R. M. Gilmer. 1963. Reductions in fruit sets on cherry trees pollinated with pollen from trees with sour cherry yellows. *Phytopathology* 53:399–401.

Williams, H. E., J. A. Traylor, and H. K. Wagnon. 1963. The infectious nature of pollen from certain virus-infected stone fruit trees (abs.). *Phytopathology* 53:1144.

Willison, R. S. 1944. Strains of prune dwarf. *Phytopathology* 34:1037–49.

Zawadska, B. 1965. Observations and preliminary experiments on fruit tree virus diseases in Poland. *Zast. Bilja (Belgrade)* 16:513–16.

Prunus Necrotic Ringspot

Prunus Necrotic Ringspot Virus (PNRSV)

The virus was first named peach ringspot virus (Cochran and Hutchins, 1941), but many strains of the virus exist (Fulton 1981), and, based largely on Fulton's description (Fulton 1970), it currently is referred to as Prunus necrotic ringspot virus (PNRSV). Strains of PNRSV include almond calico virus, apple mosaic virus, cherry recurrent ringspot virus, cherry rugose mosaic virus, rose mosaic virus, some isolates that incite plum line pattern, and hop mosaic virus.

The diseases caused by PNRSV are widely distributed among the *Prunus* species of fruit and nut crops (Uyemoto et al. 1989a,b). The worldwide occurrence of the virus is attributed to its relative ease of natural spread. The virus is readily pollenborne and seed-transmitted in some *Prunus* hosts. It may infect ovule parent trees and is perpetuated by common plant propagation methods. Disease synonyms include peach necrotic leaf spot, tatter leaf, lace leaf, cherry ringspot, cherry rugose mosaic (almond mosaic), almond calico, and Stecklenberger disease (Németh 1986); also, perhaps, peach mule's ear, peach willow twig, and almond infectious bud failure (Nyland 1976). Sweet cherry trees may be infected with "symptomless" virus strains that produce no acute or chronic symptoms (G. I. Mink, personal communication).

Infected trees are economically affected, as was illustrated in 14-year trial involving cling peaches that showed average annual yield losses of 3.75 tons per hectare (Schmitt, Williams, and Nyland 1977) and an annual reduction in income of $250 per hectare (Heaton, Ogawa, and Nyland 1981). Also, when PNRSV and prune dwarf virus (PDV) coinfect peach trees (to cause the disease called peach stunt (**fig. 28**), losses are greater than those caused by PNRSV alone (Schmitt, Williams, and Nyland 1977).

Symptoms. In general, PNRSV incites tissue chlorosis, necrosis, deformity, and some stunting of infected trees. Disease symptoms are usually expressed the year following infection rather than during the initial season of infection. For the most part, in succeeding years, such acute symptoms are rarely expressed in those portions of the tree that had shown initial "shock" symptoms. But certain virus strains (called recurrent ringspot strains) cause symptoms annually (Nyland, Gilmer, and Moore 1976). Symptoms induced by some of the recognized strains of PNRSV are described below.

Necrotic ringspot strains (NRSS). Typically, NRSS produce shock of acute symptoms on emerging leaves: diffuse chlorotic rings or areas, or severe chlorotic spotting, followed by recovery and the absence of disease symptoms. In peach trees, infections are displayed in the spring as retarded budbreak, death of many partially opened flowers and leaf buds, killing of previous season's twigs, or cankering and bark splitting of main scaffolds. Mild to severe gum production may accompany these effects. As the season progresses, new growth is essentially symptomless except on rapidly elongating shoots, such as growth produced immediately after a period of high temperatures, where terminal leaves exhibit necrotic spotting or shothole. With NRSS infections of cling peach trees, reduced fruit yields (up to 5 tons per acre) have been reported. Freestone cultivars are apparently more damaged than the clingstone cultivars. Almond and sweet cherry trees react to NRSS in a manner similar to that described for peach. In the nursery, leaf casting and necrosis of the cortical tissues of young shoots, or death of the shoot tip have been observed on "June budded" peach trees. This in the older literature was known as peach necrotic leafspot. The peach necrotic leafspot phase is commonly seen in nursery trees propagated in June with buds from infected trees.

With ordinary NRSS, infected trees experience the so-called shock phase (where leaves develop chlorotic rings, necrosis, and shotholing (**fig. 29A**), which is followed by a recovery phase (no obvious symptoms). However, some ringspot strains cause primary symptoms indistinguishable from those induced by the ordinary NRSS followed by chronic symptoms of specific diseases, such as almond calico or cherry rugose. These are aptly called recurrent ringspot virus strains (Nyland, Gilmer, and Moore 1976).

Cherry rugose mosaic strain (CRMS). Initial shock symptoms produced on the hosts are essentially the same as those produced by the NRSS. However, in sweet and sour cherries, acute shock symptoms are followed by "rugose mosaic" symptoms. These consist of chlorotic blotches and leaf distortion (**fig. 29B**), which may cause affected leaves to drop by late June or July. The chlorotic area can turn necrotic; also, most isolates produce some leaf enations. The fruit may be both deformed and delayed in maturity.

Infected peach trees recover from the shock phase but continue to show necrotic spotting on leaves produced during flushes of growth in the summer, and often mild mosaic symptoms on leaves produced in the spring or early summer.

In almond, diseased trees exhibit severe leaf distortion, with yellow, white, or light green mosaic or calico patterns and bud failure. With prolonged infections, this last symptom resembles that described for Drake almond virus bud failure. Isolates of mosaic virus from almond trees, when graft-inoculated to healthy cherry and peach, cause severe ringspot shock symptoms followed by rugose mosaic in cherry and leafspot symptoms in peach.

Almond calico strain. Symptomatic leaves exhibit chlorotic spots or blotches (Thomas and Rawlins 1951) that are scattered or aggregated at the tip or base (**Fig. 29C**); these symptoms are especially prominent in early spring. A secondary symptom may be observed on the shoots as a failure of the vegetative buds to grow; also, fewer flower buds develop on most cultivars. Leaf symptoms are scarce in some cultivars, especially in Mission. The large wavy leaves of Mission and the general tree appearance resemble nonproductive syndrome of almond (G. Nyland, personal communication).

In sweet cherry, the varieties Black Tartarian, Chapman, and Saylor express obvious shock symptoms of calico, but chronic infections produce only occasional symptomatic leaves scattered throughout the tree canopy.

Peach trees infected with almond calico strain develop symptoms similar to those already described for cherry rugose mosaic infections. The terminal leaves of shoots that are produced late in the season may be more erect than normal and may not have buds in their axils. This has been called "mule ear" disease.

Characteristics of the virus. Prunus necrotic ringspot virus particles are polyhedrons of various sizes and densities. In general, the virus particles average 17 to 23 nm (Fulton 1959; Shagrun 1964), but those of cherry rugose mosaic range from 21.5 to 28.7 nm. In sucrose gradient columns, three sedimenting bands are produced and these correspond to sedimentation constants of 80 to 90, 89 to 98, and 101 to 114 (Fulton 1983). The virus can be heat-inactivated in infected fruit trees by holding plants at 38°C for two to three weeks (Nyland 1960). The virus is classified in the ilarvirus (*isometric labile ringspot virus*) group.

Detection and, hence, diagnosis of PNRSV is accomplished by bioassays and serological assays. For bioassay, bud chips from the candidate tree are inserted underneath the bark of the Shiro-fugen flowering cherry (*Prunus serrulata*) indicator tree. Within four to six weeks, virus-infected bud chips induce the indicator host to produce excessive amounts of darkened gum and necrosis of the phloem and xylem (Milbrath and Zeller 1945). In contrast, healthy bud chips cause no obvious ill effect on the indicator. Serological test procedures used to identify PNRSV strains include agar gel diffusion (Fulton 1968) and enzyme-linked immunosorbent assay (ELISA) (Mink 1980; Mink and Aichele 1984). Currently, ELISA is the method of choice because it is sensitive, rapid, and reliable, and it differentiates between some of the PNRSV strains (McMorran and Cameron 1983; Mink, Cole, and Regev 1982; Mink et al. 1987).

Transmission. Prunus necrotic ringspot virus is intimately associated with pollen grains (Cole, Mink, and Regev 1982; Kelley and Cameron 1986). Under field conditions the virus can be transmitted to healthy mother trees during the pollination process (George and Davidson 1963, 1964). Also, infected seeds are produced on diseased trees (Cochran 1946, 1950) or on healthy trees pollinated with diseased pollen (George 1962; Traylor et al. 1963; Way and Gilmer 1958). The incidence of disease in seedlings from infected seeds ranges from a trace to 12 percent in cherry and peach (Cochran 1946, 1950; Wagnon et al. 1960; Mink and Aichele 1984) and up to 28 percent in almond (Williams et al. 1970).

The virus is easily transmitted by common nursery grafting techniques. The virus is also sap-transmissible to many herbaceous plants, including cucumber (*Cucumis sativus*), *Gomphrena globosa,* and *Chenopodium amaranticolor* (Fulton 1981).

Reports of possible PNRSV vectors have appeared in the literature and include the mite *Vasates jockeui* (Proeseler 1968) and the nematode *Longidorus macrosoma* (Fritzsche and Kegler 1968). Also, since pollen can contain infectious PNRSV, it is possible that honeybees serve as involuntary vectors (George and Davidson 1963; Howell and Mink 1988; Mink 1983), with the virus perhaps transmitted during their foraging process (Cole, Mink, and Regev 1982).

Control. Since infected scion buds and rootstocks can efficiently perpetuate and disseminate PNRSV, use of virus-free nursery stocks is advised. Although complete control of PNRSV is virtually impossible, practical economic control is possible if you plant trees that are free of the virus. The effects of PNRSV entering from an outside source after trees are established are considerably less than if even a few infected trees are included at planting time. The economic impact of virus infection is usually greatest when diseased scion buds or rootstocks are used to produce the nursery tree. In Washington, rugose mosaic infections of 15- to 20-year-old sweet cherry trees cause fruit to become unmarketable (G. I. Mink, personal communication).

REFERENCES

Cochran, L. C. 1946. Passage of the ring-spot virus through mazzard cherry seed. *Science* 104:269–70.

———. 1950. Passage of the ring-spot virus through peach seed (abs.). *Phytopathology* 40:964.

Cochran, L. C., and L. M. Hutchins. 1941. A severe ringspot virosis on peach (abs.). *Phytopathology* 31:860.

Cole, A., G. I. Mink, and S. Regev. 1982. Location of Prunus necrotic ringspot virus on pollen grains from infected almond and cherry trees. *Phytopathology* 72:1542–45.

Fritzsche, R., and H. Kegler. 1968. Nematoden als Vectoren von Viruskrankheiten der Obstgehötze. *Tag. Ber. D. Akad. Land. Wiss. DDR* 97:289–95.

Fulton, R. W. 1959. Purification of sour cherry necrotic ringspot and prune dwarf viruses. *Virology* 9:522–35.

———. 1968. Serology of viruses causing cherry necrotic ringspot, plum line pattern, rose mosaic, and apple mosaic. *Phytopathology* 58:635–38.

———. 1970. Prunus necrotic ringspot virus. *Commonw. Mycol. Inst./Assoc. Appl. Biol. Descrip. Plant Viruses* 5, 4 pp.

———. 1981. Ilarviruses. In *Handbook of plant virus infections and comparative diagnosis*, edited by E. Kurstak, 377–413. Elsevier/North-Holland Biomedical Press, New York.

———. 1983. Ilarvirus group. *Commonw. Mycol. Inst./Assoc. Appl. Biol. Descrip. Plant Viruses* 275, 3 pp.

George, J. A. 1962. A technique for detecting virus-infected Montmorency cherry seeds. *Can. J. Plant Sci.* 42:193–203.

George, J. A., and T. R. Davidson. 1963. Pollen transmission of necrotic ring spot and sour cherry yellows viruses from tree to tree. *Can. J. Plant Sci.* 43:276–88.

———. 1964. Further evidence of pollen transmission of necrotic ringspot and sour cherry yellows viruses in sour cherry. *Can. J. Plant Sci.* 44:383–84.

Heaton, C. R., J. M. Ogawa, and G. Nyland. 1981. Evaluating economic losses caused by pathogens of fruit and nut crops. *Plant Dis.* 65:886–88.

Howell, W. E., and G. I. Mink. 1988. Natural spread of cherry rugose mosaic disease and two Prunus necrotic ringspot virus biotypes in a central Washington sweet cherry orchard. *Plant Dis.* 72:636–40.

Jaffee, B. A., C. A. Powell, and M. A. Derr. 1986. Incidence of Prunus necrotic ringspot virus in some Pennsylvania peach orchards and nurseries. *Plant Dis.* 70:688–89.

Kelley, R. D., and H. R. Cameron. 1986. Location of prune dwarf and Prunus necrotic ringspot viruses associated with sweet cherry pollen and seed. *Phytopathology* 76:317–22.

McMorran, J. P., and H. R. Cameron. 1983. Detection of 41 isolates of necrotic ringspot, apple mosaic, and prune dwarf viruses in *Prunus* and *Malus* by enzyme-linked immunosorbent assay. *Plant Dis.* 67:536–38.

Milbrath, J. A., and S. M. Zeller. 1945. Latent viruses in stone fruits. *Science* 101:114–15.

Mink, G. I. 1980. Identification of rugose mosaic-diseased cherry trees by enzyme-linked immunosorbent assay. *Plant Dis.* 64:691–94.

———. 1983. The possible role of honeybees in long distant spread of Prunus necrotic ringspot virus from California into Washington sweet cherry orchards. In *Plant virus disease epidemiology*, edited by R. T. Plumb and J. M. Thresh, 85–91. Blackstone Press, Oxford.

Mink, G. I., and M. D. Aichele. 1984. Detection of Prunus necrotic ringspot and prune dwarf viruses in *Prunus* seed and seedlings by enzyme-linked immunosorbent assay. *Plant Dis.* 68:378–81.

Mink, G. I., A. Cole, and S. Regev. 1982. Identification and distribution of three Prunus necrotic ringspot virus serotypes in Washington sweet cherry orchards (abs.). *Phytopathology* 72:988.

Mink, G. I., W. E. Howell, A. Cole, and S. Regev. 1987. Three serotypes of Prunus necrotic ringspot virus isolated from rugose mosaic-diseased sweet cherry trees in Washington. *Plant Dis.* 71:91–93.

Németh, M. 1986. *Virus, mycoplasma and rickettsia diseases of fruit trees*. Martinus Nijhoff, Dordrecht, 841 pp.

Nyland, George. 1960. Heat inactivation of stone fruit ring spot virus. *Phytopathology* 50:380–82.

———. 1976. Almond virus bud failure. In *Virus diseases and noninfectious disorders of stone fruits in North America*, 33–42. USDA Handb. 437.

Nyland, G., R. M. Gilmer, and J. D. Moore. 1976. "Prunus" ring spot group. In *Virus diseases and noninfectious disorders of stone fruits in North America*, 104–38. USDA Handb. 437.

Proeseler, G. 1968. Übertragungsversuche mit dem latenten Prunus-virus und der Gallmilbe *Vasates fockeiu* Nal. *Phytopathol. Z.* 63:1–9.

Schmitt, R. A., H. Williams, and G. Nyland. 1977. Virus diseases can decrease peach yields. *Cling Peach Quart.* 13:17–19.

Shagrun, M. A. 1964. Cytology of Cucumis sativa L. infected with a strain of Prunus ringspot virus. Ph.D. Thesis, University of California, Davis.

Thomas, H. E., and T. E. Rawlins. 1951. Almond calico. In *Virus diseases and other disorders with viruslike symptoms of stone fruits in North America*, 189–90. USDA Agric. Handb. 10.

Traylor, J. A., H. E. Williams, J. H. Weinberger, and H. K. Wagnon. 1963. Studies on the passage of Prunus ringspot virus complex through plum seed (abs.). *Phytopathology* 53:1143.

Uyemoto, J. K., J. A. Grant, W. H. Krueger, W. H. Olson, J. W. Osgood, G. S. Sibbett, M. Viveros, and C. V. Weakley. 1989a. Survey detects viruses in almond, prune, and sweet cherry orchards. *Calif. Agric.* 43(5):14–15.

Uyemoto, J. K., C. F. Luhn, W. Asai, R. Beede, J. A. Beutel, and R. Fenton. 1989b. Incidence of ilarviruses in young peach trees in California. *Plant Dis.* 73:217–20.

Wagnon, H. K., J. A. Traylor, H. E. Williams, and J. H. Weinberger. 1960. Observations on the passage of peach necrotic leafspot and peach ringspot viruses through peach and nectarine seeds and their effects on the resulting seedlings. *Plant Dis. Rep.* 44:117–19.

Way, R. D., and R. M. Gilmer. 1958. Pollen transmission of necrotic ringspot virus in cherry. *Plant Dis. Rep.* 42:1222–24.

Williams, H. E., R. W. Jones, J. A. Traylor, and H. K. Wagnon. 1970. Passage of necrotic ringspot virus through almond seeds. *Plant Dis. Rep.* 54:822–24.

Prunus Stem Pitting (PSP)

Tomato Ringspot Virus (TmRSV)

This disease has also been referred to as peach stem pitting. Christ (1960) and Barrat (1966) reported on scion-rootstock incompatibility and wood pitting disorders, respectively, of orchard and nursery peach trees. Mircetich, Fogle, and Civerolo (1970) demonstrated the graft-transmissibility of PSP and Smith, Stouffer, and Soulen (1973) established its causal relationship with TmRSV.

In addition to peach (*Prunus persica*), other economic hosts affected by PSP include nectarine (*P. persica* var. *nectarina*), apricot (*P. armeniaca*), European plum (*P. domestica*), Japanese plum (*P. salicina*), *P. mahaleb*, and sweet (*P. avium*) and sour cherry (*P. cerasus*) (Cameron 1969; Mircetich and Fogle 1969; Mircetich, Fogle, and Barrat 1968; Mircetich, Moller, and Nyland 1977).

Symptoms. Following dormancy, trees with PSP develop leaves later than healthy trees. Also, during early summer, affected trees appear wilted, with drooping, light-green to yellow leaves. By late summer the tree canopy may turn reddish purple prior to normal autumn coloration (**fig. 30A**). Fruit production is greatly affected by reduced yield, failure to size, poor flavor, and premature drop.

The above canopy symptoms result from the abnormal enlargement that develops at the base of the trunk. Here the affected bark is thick and spongy in texture, and the underlying woody cylinder contains pits and grooves (**fig. 30B**). Depending upon the age of the infection and the cultivar involved, the disease's effect on the trunk may so weaken the tree structure that affected trees break over.

Characteristics of the virus. The virus particles are isometric, about 28 nm in diameter, seedborne in certain hosts (Mountain et al. 1983), and efficiently vectored by the nematodes *Xiphinema americanum* (Teliz, Lownsberry, and Grogan 1966), *X. californicum* (Hoy, Mircetich, and Lownsberry 1984) and *X. rivesi* (Forer, Hill, and Powell 1981). The virus tends to be restricted in its distribution in peach trees and appears to be most abundant in the bark of the below ground portion of the trunk (Bitterlin, Gonsalves, and Barrat 1988). TmRSV infects and causes economic diseases in a wide variety of fruit trees, for example prune brownline, apple union necrosis and decline, and peach yellow bud mosaic (see discussion of these diseases elsewhere in the text).

On herbaceous host plants, TmRSV induces chlorotic local lesions and systemic mottle in cucumber (*Cucumis sativus*) and necrotic local lesions and systemic apical necrosis in *Chenopodium quinoa*. Even so, and although viral antigens may be serologically detected in peach leaves (Powell 1984), direct sap-transmission of TmRSV to herbaceous hosts with tissue extracts from PSP-affected trees frequently fails (Mircetich, Fogle, and Civerolo 1970). However, TmRSV was recovered by first grafting tissue chips, preferably of roots, from PSP-affected trees onto *P. tomentosa* (Mircetich, Moller, and Nyland 1977), and then preparing buffer extracts of symptomatic *P. tomentosa* leaves and rubbing the extracts onto herbaceous host plants.

Transmission. Although scion bud transmission are possible (Mircetich, Civerolo, and Fogle 1971), orchard spread of TmRSV is solely by the nematode vector, *Xiphinema americanum* sensu lato. Disease incidence in peach and nectarine trees in some Eastern commercial orchards has reportedly ranged from 12 to over 75 percent (Mircetich and Fogle 1976), which strongly suggests the occurrence of natural spread. Before the nature of the disease was known, it was spread widely in eastern United States in nursery trees.

The epidemiology of the disease includes the following scenario. In Pennsylvania, the common dandelion (*Taraxacum officinale*) is reported to be a major reservoir of TmRSV. The virus is seedborne in this weed host, with an incidence of 7 to 35 percent infected seedlings (Mountain et al. 1983). The virus is acquired by the nematode vector, which in turn is capable of inoculating healthy dandelion seedlings or woody-plant hosts. In California, recent surveys of orchards in the Sacramento and San Joaquin valleys and the Paso Robles region revealed that average populations of *X. americanum* were 11 to 19 per 250 cm^3 of soil (McKenry and Kretsch 1987). This suggests that the nematode vector is endemic in California soils. However, as the nematode vector is unable to disperse long distances, the establishment of virus reservoirs such as infected weeds and their seeds or infected nursery trees (Stouffer and Smith 1971) must be guarded against.

Control. On known diseased sites, a complete soil fumigation is mandatory. However, as total nematode control is unrealistic, a few infected trees will likely appear in the orchard. These sites should be re-fumigated, followed by tree removal and replanting with disease-free stock. If a sod cover is desired, weed control measures should be taken against broadleaf plants.

Lastly, only clean nursery stock should be planned. This recommendation is based on extensive nursery surveys in Pennsylvania of peach trees propagated during 1968–71, which showed a PSP incidence of 15.5 percent (L. B. Forer, Bureau of Plant Industry, Pennsylvania Dept. Agriculture, personal communication). The possibility of using cross-protection as a strategy for controlling Prunus stem pitting was discussed by Bitterlin, Gonsalves, and Barrat (1988).

REFERENCES

Barrat, J. G. 1966. Problems of young peach trees. In *The peach*, 36–42. Rutgers—The State University, New Brunswick, N. J.

Bitterlin, M. W., D. Gonsalves, and J. G. Barrat. 1988. Distribution of tomato ringspot virus in peach trees: Implications for viral detection. *Plant Dis.* 72:59–63.

Cameron, H. R. 1969. Stem pitting of Italian prune. *Plant Dis. Rep.* 53:4–6.

Christ, E. G. 1960. New peach tree problem. *New Jersey Hort. Soc. News* 41:4006.

Forer, L. B., N. Hill, and C. A. Powell. 1981. *Xiphinema rivesi*, a new tomato ringspot virus vector (abs.). *Phytopathology* 71:874.

Hoy, J. W., S. M. Mircetich, and B. F. Lownsbery. 1984. Differential transmission of Prunus tomato ringspot virus strains by *Xiphinema californicum*. *Phytopathology* 74:332–35.

McKenry, M. V., and J. Kretsch. 1987. Survey of nematodes associated with almond production in California. *Plant Dis.* 71:71–73.

Mircetich, S. M., E. L. Civerolo, and H. W. Fogle. 1971. Relative differential efficiency of buds and root chips in transmitting the causal agent of peach stem pitting and incidence of necrotic ringspot virus in pitted trees. *Phytopathology* 61:1270–76.

Mircetich, S. M., and H. W. Fogle. 1969. Stem pitting in *Prunus* spp. other than peach. *Plant Dis. Rep.* 53:7–11.

———. 1976. Peach stem pitting. In *Virus diseases and noninfectious disorders of stone fruits in North America*, 77–87. USDA-ARS Agric. Handb. 437.

Mircetich, S. M., H. W. Fogle, and J. G. Barrat. 1968. Further observations on stem pitting in *Prunus*. *Plant Dis. Rep.* 52:287–91.

Mircetich, S. M., H. W. Fogle, and E. L. Civerolo. 1970. Peach stem pitting: Transmission and natural spread. *Phytopathology* 60:1329–34.

Mircetich, S. M., W. J. Moller, and G. Nyland. 1977. Prunus stem pitting in sweet cherry and other stone fruit trees in California. *Plant Dis. Rep.* 61:931–35.

Mountain, W. L., C. A. Powell, L. B. Forer, and R. F. Stouffer. 1983. Transmission of tomato ringspot virus from dandelion via seed and dagger nematodes. *Plant Dis.* 67:867–68.

Powell, C. A. 1984. Comparison of enzyme-linked immunosorbent assay procedures for detection of tomato ringspot virus in woody and herbaceous hosts. *Plant Dis.* 68:908–09.

Smith, S. H., R. F. Stouffer, and D. M. Soulen. 1973. Induction of stem pitting in peaches by mechanical inoculation with tomato ringspot virus. *Phytopathology* 63:1404–06.

Stouffer, R. F., F. H. Lewis, and D. M. Soulen. 1969. Stem pitting in commercial cherry and plum orchards in Pennsylvania. *Plant Dis. Rep.* 53:434–38.

Stouffer, R. F., and S. H. Smith. 1971. Present status of the *Prunus* stem pitting disease in the United States. *Ann. Phytopathol.* No. hors série: 109–16.

Stouffer, R. F., and J. K. Uyemoto. 1976. Association of tomato ringspot virus with apple union necrosis and decline. *Acta Horticulturae* 67:203–8.

Teliz, D., B. F. Lownsbery, and R. G. Grogan. 1966. Transmission of tomato ringspot, peach yellow bud mosaic, and grape yellow vein viruses by *Xiphinema americanum*. *Phytopathology* 56:658–63.

Yellow Bud Mosaic

Yellow Bud Mosaic (YBMV) Strain of Tomato Ringspot Virus (TmRSV)

Yellow bud mosaic, also known as "Winters disease" and "Winters peach mosaic" (Németh 1986), was found for the first time in 1936 (Thomas and Rawlins 1939) affecting peach and almond trees in the Winters area of Solano and Yolo counties. Close inspection of the area showed that the disease was present in at least 40 orchards. Since then it has been found in peach in all of the peach-growing areas in the northern counties of Placer, El Dorado, Nevada, Yuba, Napa, Butte, Sutter, and Tehama. Yellow bud mosaic is also present in San Bernardino County, but not in the southern San Joaquin Valley (Schlocker, McClain, and Hill 1958). On sweet cherry, the disease occurs in both San Joaquin and Stanislaus counties. In almond, it so far has been found only in Yolo and Contra Costa counties.

Apricot is a natural host but is not seriously damaged; symptoms are not so obvious as they are with other stone-fruit hosts. The disease has not been reported on peach outside of California, but it has been described on cherry in Oregon as Eola rasp leaf.

Yellow bud mosaic causes serious unproductiveness in peach, sweeet cherry, and almond (Weintraub 1961). In apricot (Thomas and Rawlins 1951) and Wickson plum (Wagnon and Schlocker 1953) the damage is slight.

Symptoms. On peach and nectarine the disease can be best identified during April and May (Nyland and O'Reilly 1965; Schlocker and Traylor 1976). At this time, newly infected trees display the "mosaic" or primary phase of the disease. Irregular chlorotic areas or spots and vein clearing appear on the leaf blades of recently infected shoots. The chlorotic areas usually are located along the main veins and often occur on one side of the midrib near the base of the leaf. A field diagnostic symptom is the clearing of small veins at the edge of the chlorotic areas. The affected leaf blades may be distinctly twisted or puckered. Often, the chlorotic areas will die and drop out, resulting in holes and tattering.

In the second year of infection the yellow bud or secondary phase of the disease predominates, although the mosaic phase still occurs in newly invaded tissue. The yellow bud phase is most clearly expressed in the spring, soon after the first leaves have expanded. Leaves from normal buds grow rapidly and display a healthy green color, whereas leaf growth from infected buds is extremely retarded and results in tufts of small, pale-yellow leaves (**fig. 31A**). These tufts may dry up and die, or continue to grow as small rosettes of distorted, yellow-green leaves. An individual branch on an infected tree may have both healthy and infected leaves. Many leaf tufts die after the onset of hot weather, resulting in sparsely foliated trees. Leaves and fruit often remain only on the terminal parts of branches not yet invaded by the virus. Affected trees sunburn easily and become very susceptible to wood-boring beetles.

On almond (Schlocker and Traylor 1976), yellow bud mosaic symptoms consist mainly of sparse foliation, leaf tufting and yellowing (**fig. 31A**), and lack of terminal growth. The Mission (Texas) almond is most severely affected. Sprouts from the trunk and scaffold limbs of affected trees usually show a rosette type of growth. The fruits are wrinkled and rough with abnormally thickened hulls. The reduction in yield is directly proportional to the severity of the symptoms. In trees of Nonpareil and Ne Plus Ultra, the symptoms seldom appear above the main trunk.

Affected sweet cherry trees (Schlocker and Traylor 1976) are characterized by a bare-limb appearance that starts in the lower portion of the tree and slowly advances upward as the disease kills the spurs, twigs, and small branches (**fig. 31B**). The leaves on affected spurs are small, with prominent whitish secondary veins that branch off from the midrib at almost right angles, resembling those of elm leaves. Enations develop on the underside of affected leaves, most commonly adjacent to the midrib.

Observing symptoms on diseased trees in the field is important for identifying yellow bud mosaic, but more conclusive evidence is necessary to positively identify the causal agent. Two important tools are host-range indicator plants and serology, both of which are discussed below.

Causal agent. Yellow bud mosaic is caused by the yellow bud mosaic strain (YBMV) of the tomato ringspot virus (TmRSV). The YBMV particle is spherical, with a diameter close to 25 nm (Cadman and Lister 1961, 1962; Weintraub 1961). It is classified in the nepovirus (*nematode polyhedral virus*) group. Infectivity of the sap is lost when diluted one-fifth with distilled water and heated for 10 minutes at 60°C, but not at 58°C. Infectivity is stable when sap is kept at $-10°C$ for 24 hours but it drops by 80 to 90 percent when sap is stored at this temperature for five days. Thus, the YBMV is considered to be less stable than other viruses of the ringspot type. The virus is inactivated in buds of infected peach trees exposed to 38°C for 21 days (G. Nyland and A. C. Goheen, unpublished).

Identification of the virus. Plant tissues suspected of containing the virus should be collected in early spring when the symptoms are still very distinct. As soon as possible, virus-containing sap should be extracted from the leaves by grinding them in a mortar containing phosphate buffer. The resultant sap is used to transmit the virus to herbaceous hosts by mechanical inoculation. Leaves of the herbaceous hosts should be dusted with an abrasive material such as carborundum and then inoculated by rubbing a small amount of sample sap on the leaf surface. The most useful herbaceous plants for disease diagnosis are bountiful bean, cowpea, cucumber, *Nicotiana rustica,* and *N. tabacum* (Karle 1960). Symptoms characteristic of tomato ringspot virus will develop on these hosts. The most diagnostic symptom in herbaceous hosts occurs in virus-infected cowpeas, which develop top necrosis and then collapse below the primary leaves. Symptoms on tobacco include necrotic local spots, chlorotic rings, and line patterns. On cucumber, local chlorotic spots, systemic chlorosis, and mottling develop.

Serological techniques can be used to detect and positively identify YBMV in virus-containing plant sap. Agar-gel double immunodiffusion tests using an antiserum previously prepared against YBMV have been widely used. Because virus concentrations are much higher in herbaceous plants than in *Prunus* hosts, these tests are most successful when sap from virus-infected herbaceous hosts is used (Milbrath and Reynolds 1961; Nyland and O'Reilly 1965; Ouchterlony 1962).

At present, the most reliable method for the detection of TmRSV in fruit trees is the use of immunosorbent assays, preferably I-ELISA or RISA (Hoy and Mircetich 1984).

Transmission. *Xiphinema californicum* (Lamberti and Bleve-Zacheo 1979), the dagger nematode formerly identified as *Xiphinema americanum* Cobb, is the natural vector of the virus. Generally, all active stages of the nematode are capable of transmission. The larvae lose infectivity when they molt. Adult nematodes retain their infectivity for long periods (three to eight months). The virus is transmitted by grafting and through the use of mechanically expressed sap. Peach, almond, mahaleb cherry, and mazzard rootstocks became infected when grown in infested soil. Teliz, Grogan, and Lownsbery (1966) and Teliz, Lownsbery, and Kimble (1967) succeeded in transmitting the virus to roots of Royal apricot and damson plum seedlings. In Oregon, a Lovell peach seedling bud-inoculated with Eola rasp leaf of cherry developed typical yellow bud mosaic symptoms the second season after inoculation (Milbrath and Reynolds 1961). Karle (1960) showed that transmission of YBMV does not occur in steamed soil, and found no evidence of seed transmission in the herbaceous hosts he examined. His recovery of YBMV from *Malva parviflora* growing under an infected peach tree is believed to be its first recovery from a herbaceous plant in the field. The YBMV has been found to be seedborne in chickweed.

Disease development. In California, the disease originally spread from native wild plants to fruit crops grown in infested soil. Infected trees in an orchard are not randomly scattered but are in compact groups (Thomas and Rawlins 1951), indicating that infection moves slowly from tree to tree. In some orchards the disease spreads more rapidly in the direction of irrigation water flow than at right angles to it, consistent with the involvement of nematodes (Nyland and O'Reilly 1965; Teliz, Grogan, and Lownsbery 1966). The nematode vector, *Xiphinema californicum*, probably occurs in all of California's agricultural districts, but the virus appears to have a more limited distribution.

Control measures. The primary measures for controlling yellow bud mosaic are strict adherence to quarantine regulations that restrict movement of susceptible host material and soil from infected properties and areas, and the use of propagating stock free of known viruses. To control the disease where it already occurs, infected trees must be removed and destroyed, as must healthy trees in at least two adjacent rows, and the nematode associated with the disease must be killed. If several areas of diseased trees are present in an orchard, it may be advisable to remove the block completely. Before removal, the trees should be girdled or treated with a weedkiller to help kill the roots. Upon removal of the trees, special precautions must be taken to remove the roots, and the area should be fallowed for two years before fumigation. The fallowing period must be long enough for all remaining tree roots to die, plus an additional year for the death of the root-feeding nematodes. The fallowed land must be kept free of weeds during this entire period, because many weeds are hosts for both the virus and the nematode vector. The fumigants DD, Telone, and methyl bromide are reported to be effective. If the soil is not fumigated, only nonhosts of the virus should be planted. Such crops include plums and prunes on a resistant rootstock (Marianna 2624), and probably walnuts (Nyland and O'Reilly 1965; Schlocker and Traylor 1976).

REFERENCES

Breece, J. R., and W. H. Hart. 1959. The possible association of nematodes with the spread of peach yellow bud mosaic. *Plant Dis. Rep.* 43:989–90.

Cadman, C. H., and R. M. Lister. 1961. Relationship between tomato ringspot and peach yellow bud mosaic viruses. *Phytopathology* 51:29–31.

———. 1962. Relationship and particle sizes of peach yellow bud mosaic and tomato ringspot viruses. *Plant Dis. Rep.* 46:770–71.

Hoy, J. W., and S. M. Mircetich. 1984. Prune brownline disease: Susceptibility of prune rootstocks and tomato ringspot virus detection. *Phytopathology* 74:272–76.

Karle, H. P. 1960. Studies on yellow bud mosaic virus. *Phytopathology* 50:466–72.

Lamberti, F., and J. Bleve-Zacheo. 1979. Studies on *Xiphinema americanum* sensu lato with descriptions of 15 new species (Nematoda, Longidoridae). *Nematol. Mediterr.* 7:51–06.

Matthews, R. E. F. 1970. Serological reactions. In *Plant virology*, 475–506. Academic Press, New York.

Milbrath, J. A., and J. E. Reynolds. 1961. Tomato ring spot virus isolated from Eola rasp leaf of cherry in Oregon. *Plant Dis. Rep.* 45:520–21.

Németh, M. 1986. Peach yellow bud mosaic. In *Virus, mycoplasma and rickettsia diseases of fruit trees*, 432–36. Martinus Nijhoff, Dordrecth.

Nyland, G., B. F. Lownsbery, and W. D. Gubler. 1985. Yellow bud mosaic. *Univ. Calif. Div. Agric. Nat. Resour. Leaf.* 2862.

Nyland, G., and H. J. O'Reilly. 1965. Yellow bud mosaic. (California Plant Diseases–24.) *Univ. Calif. Div. Agric. Ext. Serv.* AXT-204.

Ouchterlony, O. 1962. Diffusion-in-gel methods for immunological analysis. II. *Progr. Allergy* 6:30–154.

Schlocker, A., R. L. McClain, and J. P. Hill. 1958. Peach yellow bud mosaic. *Calif. Dept. Agric. Bull.* 47(2):183–84.

Schlocker, A., and J. A. Traylor. 1976. Yellow bud mosaic. In *Virus diseases and noninfectious disorders of stone fruits in North America*, 156–65. USDA Agric. Handb. 437.

Teliz, D., R. G. Grogan, and B. F. Lownsbery. 1966. Transmission of tomato ringspot, peach yellow bud mosaic, and grape yellow vein viruses by *Xiphinema americanum*. *Phytopathology* 56:658–63.

Teliz, D., B. F. Lownsbery, and K. A. Kimble. 1967. Transmission of peach yellow bud mosaic virus to peach, apricot, and plum by *Xiphinema americanum*. *Plant Dis. Rep.* 51:841–43.

Thomas, H. E., and T. E. Rawlins. 1939. Some mosaic diseases of Prunus species. *Hilgardia* 12:623–644.

———. 1951. Yellow bud mosaic. In *Virus diseases and other disorders with virus-like symptoms of stone fruits in North America*, 53–55. USDA Agric. Handb. 10.

Tidwell, T., and D. Mayhew. 1978. Serodiagnnosis: A tool for plant disease identification. Calif. Dept. Food Agric. Lab Serv. *Plant Pathology Rep.* 2 (Jan.).

Wagnon, H. K., and A. Schlocker. 1953. Peach yellow bud mosaic. *Calif. Dept. Agric. Bull.* 42:266–67.

Weintraub, M. 1961. Purification and electron microscopy of the peach yellow bud mosaic virus. *Phytopathology* 51:198–200.

Abiotic Disorders of Stone Fruit

Chilling Canker of Peach

Cause—Chilling

A condition that develops in peach seedlings when the plants are held at 3° to 7°C, but not when they are held at 14° to 27°C, was reported by Davis and English (1965, 1969). This condition, they concluded, was similar to the "chilling injury" that develops in certain other plants exposed to temperatures slightly above freezing. Cankers similar to those on seedlings have occasionally been observed in the field on Red Haven, Lovell, and Halford peaches.

Although the lesions produced on young plants exposed to low temperatures resemble those produced by *Pseudomonas syringae* pv. *syringae*, the bacterial canker pathogen, they do not yield that organism. Furthermore, no other organism is associated with the cankers, and attempts at graft-transmission of the disorder have failed (Davis and English 1969).

Symptoms. Plants exposed to 6°C develop water-soaked areas, usually at the nodes of the stems. Necrosis of the affected portions follows (**fig. 32**), along with gumming and a "sour-sap" smell. The lesions sometimes girdle the stem, killing the portions above (**fig. 32**).

The root systems of affected plants appear to be normal, and normal shoots later grow from the base of the trunk below the lesions.

Factors affecting development of chilling injury. Davis and English (1969) found that the condition developed when plants were exposed to temperatures of 3° to 7°C, which is well above freezing. It was, therefore, not brought on by temperatures that cause formation of ice in plant tissues. Nevertheless, as Belehradek (1957) noted, the separation between chilling injury and freezing injury is not necessarily distinct.

The minimum time for production of chilling injury at 6°C is 25 to 30 days. A longer exposure to low temperature increases the extent of injury. Factors that cause early leaf drop during chilling result in fewer cankers. Increases in day length or in nitrogen promote the maintenance of green growth and thereby increase the plant's resistance to chilling injury. Chilling injury is reduced by treating the leaves with gibberellic acid, which slows their aging, but is increased by treating the leaves with peach seed leachate, which accelerates their aging. Lipe (1966) reported that the principle in peach seed leachate that accelerates leaf aging is dormin or a material similar in its affects. Two applications of ziram or one application of decenylsuccinic acid, a treatment that Kuiper (1964) recommended for freezing protection, did not lessen chilling injury in peach seedlings. Research (Davis and English 1969) has shown that peach seedlings subjected to chilling injury are more susceptible to bacterial canker than unchilled seedlings.

REFERENCES

Belehradek, J. 1957. Physiological aspects of heat and cold. *Ann. Rev. Physiol.* 19:19–82.

Davis, J. R., and H. English. 1965. A canker condition of peach seedlings induced by chilling. *Phytopathology* 55:805–06.

———. 1969. Factors involved in the development of peach seedling cankers induced by chilling. *Phytopathology* 59:305–13.

English, H., and J. R. Davis. 1969. Effect of temperature and growth state on the susceptibility of *Prunus* spp. to *Pseudomonas syringae*. *Phytopathology* 59:1025.

Kuiper, J. C. P. 1964. Inducing resistance to freezing and desiccation in plants by decenylsuccinic acid. *Science* 146:544–46.

Lipe, W. N. 1966. *Physiolgoical studies of peach seed dormancy with reference to growth substances*. Ph.D. thesis, University of California, Davis, 112 pp.

Corky Growth on Almond Kernels

Cause Unknown

This abnormality occurs sporadically in California, and except for one almond cultivar, Jordanolo, it is of minor importance.

Symptoms. A dry corky whitish tan material is closely appressed to the surface of the kernel (**fig. 33**). A crack in the inner surface of the shell usually occurs in a position overlying the corky growth.

Cause and development. The exact cause of corky growth is not known, but it is evidently related to the growth of the almond fruit. The almond fruit grows rapidly from the time it is formed until early May, and during this time the tissues of the hull and shell are soft and the endosperm is watery. After early May, growth of the hull and shell is less rapid. Under some conditions, the inner part of the shell hardens and ceases growth while the outer part of the shell continues to grow slowly. Probably this is the cause of cracks that develop in the inner part of the shell. It is during this period that the corky growth develops (Kester 1957). The growth is produced by the extrusion of callus tissue from the cracks. This tissue becomes closely appressed to the surface of the kernel.

Contributing factors. Factors contributing to corky growth are those that favor rapid early growth of the fruit. Two of the more important factors are crop size and root troubles that exert a girdling affect on the tree. Light crops, induced by frost or inadequate pollination, are also closely related to the incidence of corky growth. The susceptible Jordanolo cultivar blossoms early when frosts are likely and when trees of other (pollenizing) cultivars are not yet in bloom.

Experiments show that the removal of a strip of bark from around almond branches will increase both the size of the fruit and the amount of corky growth.

Control. The surest way to avoid corky growth is to avoid planting Jordanolo. Jordanolo also has another defect (noninfectious bud failure) that limits its value, so it is not being recommended for new plantings. When Jordanolo trees occur in an almond orchard, the best ways to prevent corky growth are to protect the trees against frost, provide adequate pollenizers, and correct any adverse soil conditions.

REFERENCE

Kester, D. E. 1957. Corky growth of kernels is recurring Jordanolo problem. *Almond Facts* 1957:8–9.

Crinkle Leaf and Deep Suture of Sweet Cherry

Cause—Genetic

The disorder "crinkle leaf" has been called "curly leaf," "wild trees," "male trees," "red bud," "maple leaf," and "unproductive cherry," while synonyms of "deep suture" are "long leaf" and "rough leaf." Crinkle leaf was first described by Kinman (1930) and deep suture by Reeves (1943). The most susceptible sweet cherry cultivars are Black Tartarian and Bing. Neither disorder occurs to any extent on Lambert or Napoleon.

These disorders generally are seed-transmitted in a small percentage of cases, but in some instances the percentage is large (Kerr 1963; Rhoads 1945). A small percentage of mazzard seedlings may show deep suture symptoms. The distribution patterns of both disorders coincide with cherry culture and few if any orchards are free of the diseases (Posnette 1951; Blodgett and Nyland 1976).

Crinkle and deep suture are considered two the most important diseases of sweet cherries in California because of their common occurrence, the unproductiveness of affected trees, and the low quality of the fruit (Nyland and O'Reilly 1961). Crinkle can reduce the crop by 50 percent or more, and deep suture can cause total rejection of the fruit at the packing house. The crop losses from these two disorders are probably greater than those from all of the sweet cherry virus diseases.

Symptoms. Both disorders can be recognized by their leaf and fruit symptoms. Furthermore, both can occur on the same tree, sometimes even on different parts of the same branch. In the crinkle leaf disorder, the leaves often have deep indentations or sinuses, and the color of affected areas is usually light green, silvery, and mottled. The leaf margin is irregular, crinkled, and distorted (**fig. 34A**). In severely affected trees of the cultivars Bing and Black Tartarian, the foliage often appears wilted, and the total leaf area is markedly reduced.

Fruit production is greatly reduced on trees that are moderately to severely affected with crinkle leaf. This unfruitfulness is related, in large part, to the development of small, defective blossoms. The peduncles of affected blossoms are abnormally short, and the pistils are short and slender, with a tendency to discolor even before the blossoms open. The fruit that set are small and pointed (**fig. 34B**), and often have a raised suture;

they tend to hang with their long axis at an angle to the peduncle. The flower buds nearest the base of the preceding year's growth often fail to open fully; they turn a reddish brown, become dry, and some may remain on the spur for several weeks.

In deep suture, the leaves can vary in size and shape from almost normal to a long, narrow, and straplike form. They are thick, leathery, rugose, and darker green than normal; the leaf margin is less serrated than that of healthy foliage. Affected fruit develop a pronounced suture depression (**fig. 34C**), but are normally rounded at the stylar end. They usually are located close to symptomatic leaves. Fruit symptoms alone usually are not diagnostic since abortive doubling that results from other causes may also yield deeply cleft fruit.

Cause. Both crinkle leaf and deep suture are regarded as genetic. There is no evidence that they are caused by infectious agents. The disease factor does not move past a graft or bud union. For crinkle leaf, a single recessive controlling gene is postulated for which certain cultivars prone to the disorder are heterozygous. Presumably, certain environmental stress conditions could induce a mutation in this gene and result in symptom expression (Kerr 1963).

Control. Since both crinkle leaf and deep suture are regarded as mutations, mother trees used for budwood should be inspected carefully for even traces of these disorders. Disease-free budwood sources are now available for Bing and some other cultivars (Nyland and O'Reilly 1961). Even when budwood is obtained from registered or certified sources, some of the buds occasionally produce shoots with crinkle leaf. All crinkle-affected trees in the nursery row should be rogued out before the trees become dormant. If diseased trees are planted in a commercial orchard, branches with crinkle should be topworked while the trees are young, preferably before five or six years of age (Brooks and Hewitt 1949; Nyland and O'Reilly 1961). If affected branches are too large to bud, they can be cut back severely in late May or early June to stimulate sucker shoots that can be budded the same season. If either crinkle leaf or deep suture is found on a tree, the affected limbs should be removed or topworked. Nonsusceptible cultivars are Lambert and Napoleon.

Climate and other environmental factors influence the incidence of both disorders. Bing clone 260 trees planted near Davis, California, developed more crinkle leaf than similar trees planted at Corvallis, Oregon. Also, new shoots that grew on trees pruned heavily in August or early September developed less crinkle and deep suture than those growing on trees that were dormant-pruned (just before budbreak) for two or three years (Blodgett and Nyland 1976).

REFERENCES

Blodgett, E. C., and G. Nyland. 1976. Sweet cherry crinkle leaf, deep suture, and variegation. In *Virus diseases and noninfectious disorders of stone fruits in North America*, 306–13. USDA Agric. Handb. 437.

Brooks, R. M., and W. B. Hewitt. 1949. Occurrence of certain diseases in sweet cherry seedlings propagated on Stockton Morello rootstock. *Am. Soc. Hort. Sci.* 54:149–53.

Kerr, E. A. 1963. Inheritance of crinkle, variegation and albinism in sweet cherry. *Can. J. Bot.* 41:1395–1404.

Kinman, C. F. 1930. A study of some unproductive cherry trees in California. *J. Agric. Res.* 41:327–35.

Nyland, G., and H. J. O'Reilly. 1961. Cherry crinkle and deep suture diseases. *Univ. Calif. Agric. Exp. Serv.* OSA#113.

Posnette, A. F. 1951. Virus diseases of cherry in England. I. Survey of diseases present. *J. Hort. Sci.* 29:44–58.

Reeves, E. L. 1943. Virus diseases of fruit trees in Washington. *Wash. State Dept. Agric. Bull.* 1

Rhoads, A. S. 1945. Virus and virus-like diseases of sweet cherry in Utah; and notes on some conditions affecting various fruit crops. *Plant Dis. Rep.* 29:6–19.

Zeller, S. M., and A. W. Evans. 1941. Vein clearing, a transmissible disease of *Prunus*. *Phytopathology* 31:463–67.

Nectarine Pox

Cause Unknown

Several types of skin disfiguration of nectarine fruit have been found occasionally in California orchards for a number of years. The disorder is sometimes restricted to a single tree; at other times an entire planting may be affected. Nectarine growers observed "rashes" and "splitting" of the fruit surface as early as 1964, and they reported high incidences of the problem in 1977, 1978, and 1979. A survey in 1979 by the California Department of Food and Agriculture (Matsumoto, Pool, and Shaver 1979) revealed that this skin disfiguration condition was present in both commercial orchards and backyard trees. It was found throughout the San Joaquin Valley but was most severe in Fresno County; it also was present in Santa Barbara County. Two orchards in Fresno County with nearly 50 percent of the fruit disfigured were removed by the growers; in another orchard, essentially all of the fruit were reportedly disfigured one season, but not at all in subsequent years. This same disorder is reported to occur in the nectarine-growing areas of West Virginia on the Fantasia, Sunglow, and Nectared cultivars, with losses up to 15 percent. Researchers have reported seeing nectarine "pox" in Australia and France, and in Chile the disorder results in a large number of culls, again on the same cultivars as those showing symptoms in California.

The nectarine cultivars most susceptible to skin disfiguration are Firebrite, Fairlane, Fantasia, May Red, May Grand, and Royal Giant. In California the disorder appears on fruit early in their development and usually occurs to some extent each year in orchards with a history of the problem. Nectarine pox symptoms have not been observed on peach, although there is report of a "nonvirus" wart condition of peach fruit.

Symptoms. The symptoms expressed fit into several categories—scabbing, cracking, pimpling, and red spotting (**fig. 35**)—and the whole complex has been termed "nectarine pox." The oldest of these skin problems is "black scab," which has been seen for over 20 years on Late LeGrand and similar late-ripening cultivars. This type is characterized as dark, scabby patches with diameters that range from $1/16$ to $1/2$ inch. The second skin problem, and the one that caused the most general concern in 1977–79, was a pimpled condition with red or acne-like pimples. The affected areas were raised red bumps and liquid-filled subepidermal pimples that occurred on early- or late-maturing yellow-skin cultivars such as Spring Red and Fantasia. The third type of pox is the most serious and occurs mostly on May Red and May Grand. Severe pimpling and tough skin result in fruit cracking when it enlarges rapidly just before harvest. This condition can be evident before thinning. A fourth type, seen mostly on Royal Giant, consists of red pimples and small black scabby areas on the same fruit (Beutel 1979).

Cause. Considerable effort has been made to determine the cause(s) of pox. There is no evidence that bacteria, fungi, or viruses are involved. Herbicides have been suggested as a possible cause, but in Chile the disorder occurs in orchards on virgin soils where no herbicide applications have been made. Soil treatments with boron and trunk injections with tetracycline have not significantly reduced disease incidence or severity (Manji and Ogawa 1986).

Control. No control is known for this disorder. Some believe that the dormant application of certain fungicides tends to reduce its severity. However, there is no experimental evidence to support this observation.

In an attempt to ascertain whether a pesticide might cause pox, an experiment plot involving applications of many of common fungicides, insecticides, and herbicides was established in a block of Fantasia nectarines in Fresno County. The fungicides were benomyl, captan, copper, sulfur, and ziram. No pox developed in any of the pesticide treatments or in the untreated control (J. M. Ogawa and B. T. Manji, unpublished). In a somewhat similar test in New Zealand, neither Fantasia trees that received four applications of the fungicide triforine nor the unsprayed control trees developed the disorder (K. G. Tate, unpublished). These results suggest, but do not prove, that pesticides do not cause nectarine pox.

REFERENCES

Beutel, J. A. 1979. *Report on nectarine pox.* U.C. Committee on Nectarine Pox, University of California, Davis, 3 pp.

Manji, B. T., and J. M. Ogawa. 1986. Study of nectarine pox. *Res. Rept. 61st Ann. West. Orch. Pest Dis. Mgmt. Conf.:* 15–16.

Matsumoto, T., R. Pool, and R. Shaver. 1979. Scab on nectarines. *Calif. Dept. Food Agric. Plant Pathology Rept.:* 87–88.

Noninfectious Bud Failure of Almond

Cause Unknown

A disorder involving the shattering of leaf buds of Nonpareil and Peerless almonds and variously known as "bud failure," "crazy top," and "noninfectious bud failure" was first reported in California in 1939 (Milbrath 1939; Wilson 1951). During the past 25 years, this disorder has become so common on several additional cultivars, including Jordanolo (Wilson 1950, 1952), Harpareil, Merced, Harvey, Jubilee, Arboleda, and possibly Carmel, that it restricts their commercial use (Browne, Gerdts, and Yeary 1975; Gerdts, et al. 1975; Kester 1969, 1976; Kester and Jones 1970; Kester and Micke 1986; Statewide Integrated Pest Management Project 1985). Nonpareil and Peerless are old established cultivars, Jordanolo and Harpareil resulted from crosses Nonpareil × Harriott made by the U.S. Department of Agriculture and the University of California Agricultural Experiment Station, and Merced and Carmel are chance seedlings. The disorder is not known to occur in the older cultivars Drake, I.X.L., and Ne Plus Ultra, and is rare in Mission.

In addition to the noninfectious bud failure disorder, an infectious bud failure disease of almond also occurs in California (Nyland 1976). The latter disorder can be induced by two strains (almond calico and cherry rugose mosaic) of *Prunus* necrotic ringspot virus. A third virus or virus strain may also be involved in this disorder. Infectious bud failure occurs on several almond cultivars and is widely distributed in California's Central Valley, but its economic importance is relatively minor compared to that of noninfectious bud failure. (See Nyland [1976] and Statewide Integrated Pest Management Project [1985] for a more thorough discussion of infectious [virus] bud failure of almond.)

Symptoms. As the name implies, the principal symptom of the noninfectious disorder is the failure of buds to grow. This failure occurs only in leaf buds, not in flower buds. The bearing surface may be drastically reduced in severe cases of bud failure.

The failure of leaf buds to grow results in branches that are entirely or largely devoid of foliage. Buds that do grow develop into vigorous shoots. Not infrequently, the terminal bud on a shoot will fail while a lateral bud grows into a shoot. Repetitions of such an occurrence may result in radical changes in the direction of growth. This symptom has been given the name *crazy top* (**fig. 36A**).

A second symptom is the development of bands of rough bark resulting from the necrosis of patches of the outer bark of affected branches. As the branch develops, the phellogen activity in the cortex beneath the necrotic areas causes the phelloderm to crack, producing a band of rough bark around the branch (**fig. 36B**).

Such cortical necrosis is common in Jordanolo and Peerless and generally infrequent in Nonpareil, but occurs commonly in seedlings of both Nonpareil and Jordanolo. Few leaf buds remain viable in areas of cortical necrosis, although an occasional flower bud may open and set fruit there.

Cause. Experiments to test the transmissibility of noninfectious bud failure (BF) have yielded negative results, even though the experiments were extended for several years (Kester and Wilson 1961; Wilson and Schein 1956; Wilson and Stout 1944). Moreover, observations in orchards where the grower had grafted over bud failure–affected trees with scions from healthy trees showed that the scions usually grew normally for years, whereas branches of the original tree continued to develop bud failure. It was concluded, therefore, that the disorder was not caused by a transmissible entity, such as a virus, and it was named "noninfectious bud failure." In a more recent study (Fenton, Kuniyuki, and Kester 1988b), no evidence could be found of a viroid association with the disorder.

All almond trees appear to be susceptible to bud failure, but do not necessarily develop symptoms (Kester and Micke 1986). The factor imparting susceptibility (BF-potential) is of unknown nature, but it is inherited from its parents at a specific level. It can be transmitted through seed from either the pistillate (Wilson 1954; Wilson and Schein 1956) or the pollen parent (Kester 1968a,b).

It is important to understand the difference between *BF-potential* (susceptibility) and *BF-expression* (symptoms) (Kester, 1970). The development of symptoms "is the endpoint of a process in which BF-potential changes with age and is perpetuated to the next scion generation at the same level as it was in the bud used for propagation" (Kester and Micke 1986). Symptoms are not expressed in most seedlings unless one or both parents have a high BF-potential.

However, in a process of change associated with growth, vegetative propagation and exposure to high but sublethal temperatures, some plants of the clone eventually reach a threshold at which BF symptoms begin to appear. The probability for BF to develop in individual trees depends upon (a) BF-potential as determined by the previous propagation history of the bud source, (b) growth and environmental conditions in the nursery and (c) growth and environmental conditions in the orchard. Because of the variation in number of scion generations and the growth and environmental conditions associated with nursery and orchard operations, much variation has developed in BF-potential of different trees and/or sources with time (Kester and Micke 1986).

In sensitive cultivars it seems almost inevitable that BF trees will eventually appear within any propagation source of the cultivar. A mathematical model to describe the BF variability pattern within populations of almond trees over time was developed by Fenton, Kester, and Kuniyuki (1988). The rate of BF symptom development was shown to be directly proportional to accumulated temperatures above 28°C. There was a suggestion, too, that bud failure is closely associated with and dependent upon the process of growth.

As mentioned above, the environment plays an important role in the BF syndrome, and there is increasing evidence that the dominant environmental factor is high temperature (Fenton, Kester, and Kuniyuki 1988; Hellali, Kester, and Martin 1979; Kester and Asay 1978 a, b; Kester and Hellali 1972; Kester, Hellali, and Asay 1975; Kester and Micke 1986). Experiments in the orchard (Hellali, Lin, and Kester 1978; Kester and Asay 1978a), the greenhouse (Kester, Hellali, and Asay 1976), and the growth chamber (Hellali and Kester 1979) indicate that high temperature is extremely important in the development of BF symptoms. Some of this research also indicates that high temperature in the nursery or the orchard can increase the BF-susceptibility of almond selections. The role of other environmental factors in the etiology of BF has not been investigated, but observations suggest that moisture stress may be involved (Kester and Micke 1986). On the basis of current knowledge, noninfectious bud failure appears to be a high-temperature–induced disorder with an important heritable (genetic) component.

Two studies have attempted to determine the effect of temperature on cultured almond cells derived from tree sources with and without noninfectious bud failure (Fenton, Kester, and Liu 1988; Fenton, Kuniyuki, and Kester 1988a). Specific growth rates of the cells differed with the BF-susceptibility (potential) of the source plant, the cell line, the age of cell lines, and the temperature. The growth rate decreased more sharply at temperatures above the optimum for the cell lines derived from BF sources. Cells from a high BF-susceptibility source survived better at temperatures of 40° and 45°C than cells from a low BF-susceptibility source. Cells from the high BF-susceptibility source synthesized protein more slowly at 20° and 25°C than cells from the low source. Significantly more free proline was associated with cells from the high BF-susceptibility source.

In an earlier growth-chamber study (Hellali and Kester 1979; Hellali, Kester, and Ryugo 1971), high levels of abscisic acid were detected in the buds of normal almond plants exposed to high temperature (43 ± 2°C). Buds on plants from a BF-source tree showed much less abscisic acid under the same conditions, although there was more before the beginning of high-temperature exposure. Little significant effect of high temperature on the levels of gibberellic acid-like substances was detected. Only further research can elucidate how biochemical changes associated with high temperature are related to the bud failure syndrome.

Control. The only sure way of avoiding noninfectious bud failure is to avoid planting cultivars prone to this disorder (Kester and Asay 1975). The following cultivars, among other minor selections, have shown no evidence of BF: Davey, Drake, I.X.L., Ne Plus Ultra, Butte, Padre, Mono, and Tokyo (Kester and Jones 1970; Kester and Micke 1986; Kester et al. 1969).

If cultivars susceptible to the disorder are used, selecting low BF-potential sources for propagation is essential. There are three general types of source material: orchard sources, orchard-progeny sources, and single-tree sources. Each of these source types has advantages and disadvantages, and none can guarantee complete freedom from bud failure. (See Kester and Micke [1986] for a thorough discussion of this propagation source question.)

The only known "cure" for bud failure is to topwork or remove affected trees. Bud-failure trees usually are not replaced unless they are younger than 10 years of age and have more than 25 to 30 percent reduction in yield. Severely affected young trees can be topworked either to a cultivar that is not prone to the disorder or to the same cultivar from a source with low BF-potential. Source orchards in the cooler part of the Central Valley generally have lower BF-potentials than those in the warmer regions (Kester and Asay 1978a). In the warmer parts of the Central Valley, young trees that show symptoms during their first few years in the orchard and severely affected older trees (under 10 years of age) should be topworked or removed and replanted (Kester and Micke 1986). Since observations suggest that moisture stress may be conducive to bud failure, good cultural practices should always be employed (Kester and Micke 1986).

REFERENCES

Browne, L, T., M. Gerdts, and E. A. Yeary. 1975. Bud failure in almonds: Replacing bud failure trees. *Calif. Agric.* 29(3):15.

Fenton, C. A. L., D. E. Kester, and A. H. Kuniyuki. 1988. Models for noninfectious bud failure in almond. *Phytopathology* 78:139–43.

Fenton, C. A. L., D. E.. Kester, and L. Liu. 1988. The temperature response of cultured almond cells derived from tree sources with and without non-infectious bud failure. I. Specific growth rate. *Plant Sci.* 56:239–43.

Fenton, C. A. L., A. H. Kuniyuki, and D. E. Kester. 1988a. The temperature response of cultured almond cells derived from tree sources with and without non-infectious bud failure. II. Heat shock. *Plant Sci.* 56:245–51.

———. 1988b. Search for a viroid etiology for noninfectious bud failure in almond. *HortScience* 23:1050–53.

Gerdts, M., W. C. Micke, D. Rough, K. W. Hench, L. T. Browne, and G. S. Sibbett. 1975. Bud failure in almonds: Almond yield reduction. *Calif. Agric.* 29(3):14.

Hellali, R., and D. E. Kester. 1979. High temperature induced bud-failure symptoms in vegetative buds of almond plants in growth chambers. *J. Am. Soc. Hort. Sci.* 104:375–78.

Hellali, R., D. E. Kester, and G. C. Martin. 1979. Seasonal changes in the physiology of vegetative buds of normal and bud-failure affected "Nonpareil" almond trees. *J. Am. Soc. Hort. Sci.* 104:371–75.

Hellali, R., D. E. Kester, and K. Ryugo. 1971. Noninfectious bud failure in almonds. Observations on the vegetative growth patterns and concurrent changes in growth regulating substances. *Hort. Sci.* 6(3) Sec. 2:33.

Hellali, R., J. Lin, and D. E. Kester. 1978. Morphology of noninfectious bud-failure symptoms in vegetative buds of almond (*Prunus amygdalus* Batsch.) *J. Am. Soc. Hort. Sci.* 104:459–64.

Kester, D. E. 1968a. Noninfectious bud-failure, a nontransmissible inherited disorder in almond. I. Pattern of phenotype inheritance. *Proc. Am. Soc. Hort. Sci.* 92:7–15.

———. 1968b. Noninfectious bud-failure, a nontransmissible inherited disorder in almond. II. Progeny tests for bud-failure. *Proc. Am. Soc. Hort. Sci.* 92:16–26.

———. 1969. Noninfectious bud failure in almonds in California. I. The nature and origin. *Calif. Agric.* 23(2):12–14.

———. 1970. Noninfectious bud failure in almond, a nontransmissible inherited disorder. III. Variability in BF potential within plants. *J. Am. Soc. Hort. Sci.* 95:162–65.

———. 1976. Noninfectious bud failure in almond, pp. 278–82. In *Virus diseases and noninfecttious disorders of stone fruits in North America.* USDA Agric. Handb. 437.

Kester, D. E., and R. N. Asay. 1975. Bud failure in almonds: Selection for freedom from bud failure. *Calif. Agric.* 29(3):13.

———. 1978a. Variability in noninfectious bud failure of "Nonpareil" almond. I. Location and environment. *J. Am. Soc. Hort. Sci.* 103:377–82.

———. 1978b. Variability in noninfectious bud failure of "Nonpareil" almond. II. Propagation source. *J. Am. Soc. Hort. Sci.* 103:429–32.

Kester, D. E., and R. Hellali. 1972. Variations in the distribution of noninfectious bud-failure (BF) in almonds as a function of temperature and growth. *Hort. Sci.* 7:(3) Sec. 2:18.

Kester, D. E., R. Hellali, and R. N. Asay. 1975. Bud failure in almonds: Variability of bud failure in Nonpareil almonds. *Calif. Agric.* 29(3):12.

———. 1976. Temperature sensitivity of a "genetic disorder" in clonally propagated cultivars of almond. *Hort. Sci.* 11:55–57.

Kester, D. E., and R. W. Jones. 1970. Noninfectious bud-failure from breeding programs of almond (*Prunus amygdalus* Batsch.). *J. Am. Soc. Hort. Sci.* 95:492–96.

Kester, D. E., and W. C. Micke. 1986. *Noninfectious bud-failure in almond—An update.* Pomology Department, University of California, Davis. 9 pp.

Kester, D. E., A. D. Rizzi, H. E. Williams, and R. W. Jones. 1969. Noninfectious bud failure in almond varieties. II. Identification and control of bud failure in almond varieties. *Calif. Agric.* 23(12):14–16.

Kester, D. E., and E. E. Wilson. 1961. Bud-failure in almonds. *Calif. Agric.* 15(6):5–7.

Milbrath, D. G. 1939. Almond disease. *Bull. Calif. Dept. Agric.* 28(10):572–73.

Nyland, G. 1976. Almond virus bud failure. In *Virus diseases and noninfectious disorders of stone fruits in North America*, pp. 33–42. USDA Agric. Handb. 437.

Statewide Integrated Pest Management Project. 1985. *Integrated pest management for almonds.* Univ. Calif. Div. Agric. Nat. Resour. Publ. 3308. 156 pp.

Stout, G. L., and E. E. Wilson. 1947. Studies of a bud failure condition in almond trees. *Phytopathology* (abs.). 37:364.

Wilson, E. E. 1950. Observations on the bud-failure disorder in Jordanolo, a new variety of almond. *Phytopathology* (abs.). 40:970.

———. 1952. Development of bud failure in the Jordanolo variety of almond. *Phytopathology* (abs.). 42:520.

———. 1954. Seed transmission tests of almond bud failure disorders. *Phytopathology* (abs.). 44:510.

Wilson, E. E., and R. D. Schein. 1956. The nature and development of noninfectious bud failure of almonds. *Hilgardia* 24:519–42.

Wilson, E. E., and G. L. Stout. 1944. A bud failure disorder in almond trees. *Bull. Calif. Dept. Agric.* 33:60–64.

———. 1951. Almond bud failure. In *Virus diseases and other disorders with viruslike symptoms of stone fruits in North America*, pp. 205–07. USDA Agric. Handb. 10.

Noninfectious Plum Shothole

Cause—Genetic

Noninfectious plum shothole is a genetic disorder that occurs occasionally in western North America. It can be mistaken for shothole, caused by the fungus *Wilsonomyces carpophilus*, or Italian prune leaf spot (Pine and Welsh 1976), which is also a genetic disorder. The noninfectious shothole disorder of plum was first described on Beaty plum, a hybrid of *Prunus augustifolia* × *P. munsoniana* (Smith and Cochran 1943), and is known to occur on native trees of *P. munsoniana*. The condition, however, is not limited to plums carrying genes of this species. Noninfectious plum shothole is seedborne and bud-perpetuated.

Symptoms. On Japanese plums (*P. salicina*) the spring growth is usually normal, but as the season progresses the older leaves at the bases of shoots develop translucent flecks that rapidly enlarge. These spots turn brown or purple-brown and separate from the healthy tissue, resulting in a shothole condition (**fig. 37**). The symptoms move progressively toward the shoot apex, and by midsummer the foliage becomes riddled with holes. The spots and holes range in size from pinpoints to 1 to 2 mm in diameter, their size usually being inversely proportional to their number in an affected leaf. The spots are generally circular but may become irregularly shaped as they coalesce.

Remarks. This disorder is not considered economically important except when confused with the fungal shothole disease, which sometimes requires chemical control. *Wilsonomyces carpophilus* can attack plums, but infection has been observed only in very wet years on plum trees located adjacent to heavily diseased (*W. carpophilus*) peach or apricot trees. The Italian prune leaf spot is not restricted to the Italian prune (*P. domestica*) and is serious in Oregon, Washington, Idaho, and Utah (Richards 1952). It has also been observed on Stanley prune and Beaty plum (Smith and Cochran 1943). No report of this disorder in California is known.

Control. Heat therapy of affected plum clones has not alleviated this disorder. It is suggested that budwood for plums be obtained from foundation programs where trees have been selected for their trueness to type and freedom from known diseases.

REFERENCES

Helton, A. W. 1960. Common disorders in the prune orchards of Idaho. *Idaho Agric. Res. Prog. Rept.* 44, 10 pp.

Pine, R. S. and M. F. Welsh. 1976. Noninfectious plum shot hole. In *Virus diseases and noninfectious disorders of stone fruits in North America*, pp. 285–87. USDA Agric. Handb. No. 437.

Richards, B. L. 1952. Leaf spot, most destructive disease of Italian prune in Utah. *Utah Farm Home Sci.* 13:27.

Smith, C. O., and L. C. Cochran. 1943. A noninfectious heritable leaf-spot and shot-hole disease of Beaty plum. *Phytopathology* 33:1101–03.

Nonproductive Syndrome of Almond

Cause—Genetic

On the Mission cultivar of almond, nonproductive syndrome is commonly referred to as "bull mission" because of the large, vigorous trees that are nonproductive. The condition is typically found in several consecutive trees in an orchard row, indicating that it probably originates in the nursery. Crop yields of affected trees range from very low to near normal, but unlike the nuts normally produced by Mission trees, these have softer shells like Nonpareil. This nonproductive syndrome has also been found to some extent in other almond cultivars, such as Nonpareil, Carmel, and Fritz. In addition to their reduced productivity, affected trees are relatively ineffective as pollinators for other cultivars.

Symptoms. The severity of the symptoms varies among affected trees and in different portions of the same tree. In the Mission cultivar, the reduced crop provides the tree with great vigor (**fig. 38A**) and with large narrow leaves, and the fruit appear wrinkled (**fig. 38B**) like those on trees affected with tomato ring spot virus. At harvest, the nuts' shells are soft with a prominent wing extending along the suture, instead of hard and without a wing, as is characteristic of this cultivar (**fig. 38C**). Nuts on affected trees tend to have abnormal shapes; they may be longer and narrower than normal nuts, or shorter and plumper. In some instances, the kernels have a bitter flavor. Except for the longer shoot growth, this disorder is not easy to detect in winter.

Cause. The cause of this disorder appears to be genetic, in that it is bud-perpetuated and nontransmissible. If the mutations that cause nonproductive syndrome occur in the plant used for propagation, offspring trees are likely to have the disorder. Diseased trees show such abnormalities in the reproductive system as defective pistils, ovules, and pollen grains, and an abnormally high drop of flowers and nuts. However, not all of these abnormalities occur on every affected tree. Care must be exercised in the diagnosis of nonproductive syndrome, because certain abiotic factors and at least one virus can induce somewhat similar symptoms.

Control. The control of this condition starts with a nursery that selects budwood from trees that combine high yields with typical cultivar characteristics. This does not insure that offspring trees will be free of the disorder, but it increases the likelihood that they will be healthy. In the orchard, if an affected branch is observed it should be removed. Trees with abnormal vigor and markedly reduced yields should be replaced or grafted with budwood from source trees free of nonproductive syndrome.

REFERENCE

University of California. 1985. *Integrated pest management for almonds.* Univ. Calif. Div. Agric. Nat. Resour. Publ. 3308. 122 pp.

Plum Rusty Blotch

Cause Unknown

Plum rusty blotch was described as a new disease of Japanese plums (*Prunus salicina*) in California in 1960 (Pine and Cochran 1960). It occurs in all of the plum-growing areas of California and in Washington, but has not been reported in other areas of the United States. Somewhat similar disorders have been reported in Argentina and in Israel (Pine 1976). The disease is most commonly encountered in the Santa Rosa cultivar, but has been found in eleven other plum cultivars containing genes of *P. salicina*. It also has been experimentally induced in Abundance and Beauty plums (Pine 1976). It has been found in trees on their own roots and in those on apricot, peach, and plum (marianna and myrobalan) rootstocks (Pine 1976). The disorder has complicated the development of disease-free nursery stock in conjunction with California's clean stock program.

Symptoms. In the spring, affected leaves developed a chlorosis that begins at the base of the blade and progresses toward the apex along one or both margins (**fig. 39**). The chlorosis may be continuous or may take the form of chlorotic blotches. Within a month or so the blotches become reddish brown and the affected areas develop numerous distinct red spots. These spots eventually become necrotic and drop out to produce a shothole appearance. Affected leaves are markedly reduced in size and tend to become misshapen as the season progresses. By midsummer all of the leaves on an affected branch usually show some damage. These branches produce relatively few blossoms and almost no fruit (Pine 1976). Reduced fruiting, however, is not always correlated with leaf symptoms.

Cause. Evidence suggesting that rusty blotch is caused by a transmissible agent was obtained some years ago by Pine and Cochran (1960) and by Pine (1965, 1976). Other unpublished research, however, has failed to corroborate these results. The current opinion of most agriculturists familiar with the disorder is that it is an expression of a genetic abnormality. A thorough etiological investigation of this disease is urgently needed.

Control. Seven weeks of 38°C heat treatment will not eliminate plum rusty blotch in buds from affected trees (G. Nyland, unpublished). Removing affected branches does not prevent the appearance of the disorder in other portions of a tree. Damage from this disease can be minimized by the selection of propagative budwood from high-yielding trees with healthy foliage (Pine 1976).

REFERENCES

Pine, T. S. 1965. Plum rusty blotch in California. *Plant Dis. Rep.* 49:109–10.

———. 1976. Plum rusty blotch. In *Virus diseases and noninfectious disorders of stone fruits in North America*, pp. 294–296. USDA Agric. Handb. No. 437. 433 pp.

Pine, T. S., and L. C. Cochran. 1960. Plum rusty blotch—A transmissible disorder found in southern California. Plant Dis. Rep. 44:87–88.

Postharvest Diseases of Stone Fruit

The production and sale of California's agricultural products are directed toward local, national, and international markets. Successful marketing of fresh and processed fruit and nut crops requires delivery of high-quality, decay-free products from the grower and packer to the wholesaler, retailer, and consumer. The reduction of postharvest food losses on a worldwide basis has been identified as an important goal by national and international organizations because of losses amounting to 10 to 30 percent of the total yield of crops (Kelman 1984; Baritelle and Gardner 1984). Postharvest losses of California fruit and nut crops have been reduced through (1) methods used in the orchard to control quiescent infection and contamination of fruit, (2) proper harvest and handling procedures, (3) postharvest fungicide treatments, and (4) improved packaging, storage, and transportation methods (Ogawa and Manji 1984; Spotts 1984).

Yet greater emphasis on control of postharvest decay is needed because of increases in production acreages, diversity of crops, changes in cultural practices, longer harvest seasons, new cultivars, more distant markets, and longer storage periods. Postharvest fungicide treatments effectively reduce fruit decay (Eckert 1969; Eckert and Ogawa 1988), but with possible new restrictions being placed on the use of some of these chemicals and the development of *Monilinia* and *Botrytis* strains that are resistant to the benzimidazoles (Ogawa and Manji 1985), research priority on decay control without the use of chemicals is demanded. Furthermore, since the early 1980s consumers have demanded more fresh plant products in their diet, with increasing emphasis on crops grown organically (free of certain pesticide or other chemical applications). In order to meet the demands of consumers for quality products without decay, efforts are being made to ensure minimal pesticide residues on crops through use of other management procedures (O'Reilly 1947; Pusey et al. 1988). Yet residues of fungicides in safe quantities would be more desirable than the possibility of decay from fungi that produce mycotoxins known to be carcinogenic (Phillips 1984).

To present a clearer picture of the requirements for control of postharvest decay, this section will discuss specific crops and their decay. Some information on preharvest conditions is repeated from earlier pages to emphasize that effective postharvest decay control demands considerable attention to preharvest orchard operations (Conway 1984).

Crops such as apricot, peach, and prune are frequently dried and, currently, some sweet cherries and nectarines are also being dried. These fruits are fully ripened before drying and, except for prune and sweet cherry, are held for 24 to 48 hours at ambient temperatures or in an ethylene atmosphere to hasten ripening. Fresh-market fruit are hand-harvested in California to reduce bruising and mechanical injuries, which become sites of infection for pathogens. Decay control on mechanically harvested fruit for canning is more difficult (Ogawa, Sandeno, and Mathre 1963; Burton and Brown 1984). Problems of decay have also been observed on nut crops such as almond and walnut during the drying period before hulling, but not during cold storage of the shelled product.

Sweet Cherry

Loss from fruit decay during transit and storage has been a limiting factor in the shipment of sweet cherries to both local and distant markets, especially those to the eastern United States and other countries. Losses during 1936–42 in New York City averaged 2.8 percent per car lot, with 72 percent of the product showing decay that ranged from a trace to 4 percent, and 10.2 percent of the lots showing 5 to 9 percent decay (Cappellini and Ceponis 1984). A recent article (Ceponis et al. 1987) estimated retail losses of sweet cherries from the western United States in New York at 5.7 percent.

The more common pathogens causing decay are *Monilinia fructicola* (Wint.) Honey, *Rhizopus stolonifer* (Ehr.:Fr.) Vuill., *Botrytis cinerea* Pers: Fr., *Sclorotinia sclerotiorum* (Lib.) dBy., *Penicillium expansum* Lk., *P. italicum* Wehmer, *Alternaria alternata* (Fr.) Keissl. (=*A. tenuis* Nees), *Monilinia laxa* (Aderh. & Ruhl.) Honey, *Aspergillus niger* v. Tiegh., *Cladosporium herbarum* Lk.: Fr., and *Pullularia* sp. (Bratley 1931; English 1945; Harvey, Smith, and Kaufman 1972; Wiant and Bratley 1948) (fig. 40A).

The relative importance of different postharvest decay pathogens may vary from year to year because of environmental conditions. Warm weather in winter promotes formation of double fruit and spurs (two fruit joined together, with one undeveloped) which are sites for *A. alternata* and *R. stolonifer* infection. *Alternaria* infection occurs on injured fruit in the field before harvest, while injuries to cherries during picking and packing provide avenues for infection by *R. stolonifer*. Rains during the blossoming period favor infection of blossoms by *B. cinerea*, *M. laxa*, *M. fructicola*, and *S. sclerotiorum*, and evidence points to the likelihood of quiescent fruit infections by *B. cinerea* and *Monilinia*. Rains just before harvest induce cracking of the fruit, and such injuries are susceptible to fungus pathogens as well as to the growth of certain bacteria and yeasts (Brooks and Fisher 1916, 1924; English 1945; Gerhardt, English, and Smith 1945).

Seasonal and varietal factors related to fruit decay have been observed in California during the last 20 years. For the earliest harvested sweet cherry cultivar, Burlat, decay problems have been limited, probably because of a relative lack of pathogens in the field at that time, and also because the fruit is flown to market and quickly consumed. Also, the floral tubes of Burlat dehisce and fall off before the parts senesce, reducing the chances of infection from pathogens such as *Botrytis*. Chapman and Black Tartarian cherries have soft flesh, are easily bruised, and develop rot quickly. In addition, consumer demand for them is not as great as for Burlat or the later-harvested Bing. The Napoleon and Rainier cultivars are mainly used for processing, and the fruit is placed in a bisulfite brine immediately after harvest. In early harvested cultivars the pathogens are primarily *Botrytis* and *Monilinia*; under the higher temperatures of midseason, *Rhizopus* becomes more important, especially on ripe and overripe fruit.

During mid and late harvest *Alternaria* begins to predominate, especially on fruit held for an extended period. The growth of *Penicillium*, bacteria, and yeasts on injured or bruised fruit becomes more important with extended storage periods.

Botrytis decay. Postharvest Botrytis decay (gray-mold rot) of fruit is becoming more common and can be traced to the incidence of blossom blight. Blossoms are especially susceptible at about full bloom to shuck-fall when the floral parts are senescing. Infection of a single blossom can result in cluster rot through contact of diseased with healthy blossoms, or through infection from spores, masses of which are quickly produced on blighted blossoms. Study of an epidemic of sweet cherry blossom blight in 1983, when rainfall during bloom was almost three times normal and the average daily high temperature was 18°C, showed *B. cinerea* and *Monilinia* sp. sporulating on 42 and 60 percent of the necrotic styles and 51 and 35 percent of the necrotic hypanthia, respectively, when the sampled necrotic tissues were incubated. At harvest time, the two pathogens were still sporulating on 23 and 1 percent, respectively, of persistent blighted blossoms (Shotwell and Ogawa 1984). Green fruit are occasionally infected from contact with blighted blossoms of infected floral shucks. Most of these infections are quiescent at first but become active and cause extensive decay as the fruit mature. In general, quiescent infections are not common in California because of the relatively warm, dry climate during blossoming (Manji and Ogawa 1985). Mummified green fruit produce large numbers of spores. One infected fruit can cause decay of the other fruit in the cluster. Botrytis decay on ripening fruit appears as a firm, light tan area with or without typical sporulation of the fungus. It can be mistaken for *Monilinia* infection if not observed closely enought to note the lack of distinctive spore clusters. A detailed description of *Botrytis* is given above in the section "Green-Fruit Rot of Stone Fruit."

Botrytis is spread from one fruit to another by contact and by airborne spores during harvest and packing operations. In the hand sorting operation, some Botrytis fruit infections are evident as light brown areas but without typical sporulation—quiescent infections cannot be seen by the unaided eye. Spores of *Botrytis* germinate even at near-freezing temperatures, but the optimum temperature is 20°C (Togashi 1949). On fruit with quiescent infections, decay and sporulation can develop in storage at 4.4°C. At 20°C, symptoms appear within 24 hours and heavy sporulation is evident after 48 hours.

Monilinia decay. Monilinia decay (brown rot), like Botrytis decay, is produced by infection with spores from mummies of green and ripe fruit as well as from blighted blossoms. The overwintering spore sources of *Monilinia* are the blighted blossoms, twigs, and mummies on the tree. The first *Monilinia*-blighted blossoms produce spores in time to infect other blossoms. Healthy blossoms in contact with infected blossoms frequently become infected (Brooks and Fisher 1916, 1924; Manji and Ogawa 1985). *Monilinia* quiescent infections do not appear to be as common as those caused by *Botrytis*. Green fruit become infected and produce masses of spores that are readily dispersed by water and wind. During fruit ripening, cluster rotting occurs through contact of healthy and mummified fruit or blighted blossoms. In 1987, when there was little evidence of blossom blight, there was also a low incidence of Monilinia and Botrytis fruit rot at harvest.

The presence of mummified fruit is probably the reason most fruit are contaminated before harvest; during harvest the chance of contamination is even greater. *Monilinia* can infect uninjured fruit surface, but infects injured surfaces more readily. Decay can develop at storage temperatures below 5°C, but the optimum is around 25°C. Symptoms on the fruit are quite similar to those produced by *Botrytis*. The color of the sporulating surface varies from almost white to gray. An entire fruit can rot from a single infection in 48 hours at room temperature. The decayed flesh is firm to the touch and does not separate easily from the healthy portion. Partially mummified fruit tissue appears dark because of the presence of fungus pseudosclerotia.

Rhizopus decay. This decay is primarily a postharvest problem of ripe fruit, as green fruit are not attacked by this fungus. Initial infections usually occur on injured or bruised ripe fruit. Impact injury can occur during harvest, sorting, sizing, packaging, storage, and transit. *Rhizopus* spores are practically omnipresent and are the main infection propagules. At storage temperatures below 4.4°C, *Rhizopus* is not a problem, but at 27°C (its optimum growth temperature) complete fruit decay can occur in 24 hours. Aerial hyphae develop during this period and sporulation occurs within 30 hours. Rhizopus rot is serious primarily because the mycelium spreads from diseased to healthy fruit (nesting), and can grow on fruit and containers soiled by juice released from decaying fruit.

The pathogen, *R. stolonifer* (Ehr.: Fr.) Vuill., on an artificial medium such as potato-dextrose agar, is a rapid-growing fungus with white, coarse mycelium; its optimum growth temperature is 27°C (maximum 32.2°C and minimum 5°C) (Pierson 1966). At high humidity, aerial hyphae are abundant; at low humidity they are sparse. The sporangia are at first white, next gray, and then black. They range from 150 to 350 μm in diameter; the columella is broad, hemispheric, about 70 to 90 μm high, and about 70 to 250 μm wide. The sporangiospores are released by disintegration of the outer sporangial wall and are unequal, irregular, round or oval, angular, striated, and about 7 μm wide and 10 to 15 μm long. They germinate readily at 7.2°C but not at 4.4°C (Matsumoto and Sommer 1967; Pierson 1966). The zygospores are black, about 150 to 200 μm in diameter, and are formed on suspensors. The fungus is heterothallic (Zycha, Siepmann, and Linnemann 1969). Zygospore germination has seldom been observed and does not play a role in postharvest infections.

Rhizopus rot in fruit packages is characterized by a mass of white mycelium, with long stolons extending to and infecting adjacent healthy fruit. Black sporangia occur abundantly near the edge of the container because of the higher oxygen level there. In the presence of this fungus, macroscopic evaluation of other molds is difficult if not impossible. Decaying fruit on the orchard floor are covered with black sporangia, but the sporangiophores are inconspicuous and there is little evidence of aerial mycelium. Infection of fruit on the tree is rare.

Penicillium decay. Decay by *Penicillium italicum* and *P. expansum* occurs most often on fruit that have been stored for extended periods or fruit that are soft or overripe. These fungi cause a watery rot like that of *Rhizopus*. The best feature for identification is the typical mass of bluish green spores on the surface of the lesion. Infection occurs most commonly in bruised, double, or spurred fruit.

Alternaria decay. This decay develops commonly in bruised, double, or spurred fruit as a dark, olive-green depressed area. Under the microscope, *Alternaria alternata* can be identified by its spores, which are typically muriform (with cross and longitudinal walls) and catenulate (in chains). Infection is more common on ripe or overripe fruit and on fruit harvested during the latter part of the season. *Alternaria* can be controlled by a postharvest iprodione treatment.

Other decays. Bacteria and yeasts can make fruit slimy and sticky, and they develop quickly on fruit infected by other organisms. *Pullularia* occurs commonly on rain-damaged fruit as a shallow, sticky rot.

Control. Control of the fungal rots begins in the orchard. Sprays are applied to control blossom infections by *Botrytis* and *Monilinia*. Benzimidazole compounds are used (Ogawa, Manji and Bose 1968), and they also prevent quiescent infections of fruit by *Botrytis* and *Monilinia*. Fungicides such as copper and the carbamates are effective against *Monilinia* but less effective against *Botrytis*. Spraying twice during the three weeks before harvest with either captan, wettable sulfur, or DCNA (2,6-dichloro-4-nitro-aniline), however, has failed to reduce fruit infection. Infection through cracks caused by rain (Gerhardt, English, and Smith 1945) are common; combating such infections is probably not worthwhile because cracked fruit cannot be marketed. Preharvest DCNA sprays provide excellent postharvest control of *Rhizopus*, and sprays of benomyl are effective against *Monilinia*, *Botrytis*, and *Penicillium* provided the fruit are not washed before packaging (Ogawa, Lyda, and Weber 1961; Ogawa, Manji, and Bose 1968). For fruit that is mechanically sorted for size, spraying with DCNA plus benomyl or with DCNA plus captan during stem cutting and sizing provides control of all common molds except *Alternaria*, while a combination of DCNA plus iprodione controls the primary pathogens as well as *Alternaria*. A residue tolerance for iprodione in postharvest treatments on peaches, plums, nectarines, and sweet cherries was established for the United States in 1989. Reduction of cherry decay by washing in water has been suggested (Yarwood and Harvey 1952), but unless protectant chemicals are used in the water the decay increases. Chlorine washes before or during stem cutting and sizing have provided added control of molds, including *Alternaria*, and have also reduced bacteria and yeast populations. Some *Alternaria* control has been obtained with dehydroacetic acid (Young and Beneke 1952).

One of the best means of reducing decay in packages is to prevent injury of the fruit before packaging. Injured areas become pitted (sunken) and are sites for fungal infection. The removal of decaying fruit, doubles, and mechanically damaged fruit delays mold development, and rapid cooling of fruit to a flesh temperature of 0° to 5°C delays the rate of disease development.

Packaging of sweet cherries to prevent pressure bruising during storage and transportation remains one of the critical areas for study because such injuries are avenues of fungal infection (Ogawa et al. 1972).

Modifying the storage atmosphere with carbon dioxide levels up to 20 percent reduces mold development on sweet cherries (English and Gerhardt 1942, Gerhardt, English, and Smith 1945; Gerhardt and Ryall, 1939; Schomer and Olson 1964; Wells and Uota 1970). Temperatures near freezing provide the best decay control. However, the length of exposure to carbon dioxide needed to kill germinated spores precludes the commercial application of such a treatment (Bussel, Sommer, and Kosuge 1969; Bussel et al. 1969). In the Pacific Northwest, fruit is placed in polyliners (plastic bags) (Gerhardt, Schomer, and Wright 1956, 1957; Pierson 1958) to prevent moisture loss and to build up the carbon dioxide level (thereby reducing the rate of ripening and consequently the amount of decay). Ultraviolet radiation has proved ineffective in the control of fruit decay in sweet cherry (English and Gerhardt 1946).

A summary of the stepwise procedure used in California for the control of postharvest decay starts with blossom sprays of benomyl, iprodione, or triforine to reduce inoculum produced on blighted blossoms, which contaminates healthy fruit and sometimes causes quiescent infections. Preharvest sprays are generally not applied unless rains are forecast or occur during the last month before harvest. If fruit are severely cracked following a rain, effective decay control is not possible. The ripe fruit are hand picked into padded buckets to prevent bruising, dumped into large wooden bins (500-pound capacity), and quickly trucked to the packing shed. Here they are dumped into chlorinated water (50 to 100 mg/mL before entering the cluster-cutting and sizing operation. The protectant fungicides applied are either thiophanate-methyl or iprodione combined with DCNA (Shotwell and Ogawa 1984). The purpose of this spray treatment is to protect the fruit from decay caused by *Monilinia*, *Rhizopus*, *Alternaria*, and *Botrytis*, and also to lubricate the fruit and equipment during cluster-cutting and fruit-sizing. The treated fruit are placed in bins and cooled to near 0°C in refrigerated chambers or with a forced-air system; they are then hand sorted to remove decay and other defects and packed in boxes holding 8.2 kg (18 pounds) of fruit. Hydrocooled fruit are dip-treated in a fungicide suspension. The packaged fruit are immediately cooled to below 5°C before being shipped in refrigerated trucks or nonrefrigerated aircraft.

Apricot

Decays cause losses in fruit destined for fresh market and for canning, freezing, and drying. The principal apricot cultivars in California are Blenheim (Royal) for fresh market and drying and Tilton for canning. For extra early shipment, Castlebright, Katy, Derby Royal, and Perfection (harvest beginning in mid-May) are used. The canning season in the San Joaquin and Santa Clara valleys ends in mid-July.

Fruit for the fresh market is usually handpicked and placed in wooden boxes holding about 50 pounds. Some fruit for processing is mechanically harvested and placed in bins holding about 800 pounds. Fruit for drying is placed in 50-pound boxes and ripened at ambient temperatures.

The most common decay pathogens are *Monilinia laxa* and *Rhizopus stolonifer* (Ogawa, Manji, and Bose 1974; Ogawa, Sandeno, and Mathre 1963; Stanton and Ogawa 1961). *Monilinia fructicola* is selectively becoming more common with the use of benzimidazole fungicides (Michailides, Ogawa, and Opgenorth 1987). *Alternaria alternata* (**fig. 40D**), *Rhizopus arrhizus*, and *R. circinans* are also fairly common, but the importance of their role in decay has not been fully assessed. Other pathogens are *Botrytis cinerea* and *Penicillium expansum*, which occur on fruit that is held in cold storage for extended periods. *Gilbertella persicaria, Aspergillus* spp., *Mucor* spp., and *Cladosporium* sp. occur occasionally on ripening fruit.

Studies by Ogawa et al. (1974) and Strand et al (1981) have shown that *Rhizopus arrhizus* and *R. stolonifer* are sometimes involved in a serious softening of canned apricots (**fig. 40I**). If even a small amount of decayed fruit tissue is incorporated into a can of otherwise healthy fruit, the decay can cause a progressive and unacceptable maceration of the canned product. Evidence suggests that this softening is caused by heat-stable pectolytic enzymes produced in infected fruit before canning. A 6-minute treatment at 75°C before canning significantly reduces subsequent softening of the canned fruit.

Control programs for *Monilinia* have been directed at reducing the inoculum in the field by controlling both blossom blight and fruit rot. Quiescent infections of green fruit by *Monilinia* have been detected in California (Wade 1956).

Captan sprays during the last three weeks before harvest, when followed by a postharvest wax treatment, can cause superficial blemishes on the fruit epidermis (Chastagner and Ogawa 1976). Preharvest sprays with benomyl and wettable sulfur do not cause this problem after the wax treatment.

Preharvest or postharveset DCNA applications provide excellent control for *Rhizopus stolonifer* but not *R. arrhizus* or other *Rhizopus* species (Ogawa, Manji, and Bose 1974), therefore it is important to store fruit below 5°C. When the combination treatment of benomyl (Ogawa, Manji, and Bose 1968) and DCNA (Ogawa 1965b; Ogawa et al. 1964; Stanton and Ogawa 1961) is used, the pathogens still to be controlled are *R. arrhizus, G. persicaria, Aspergillus, Mucor,* and *Alternaria*. For *Alternaria* the most promising fungicidal treatment is iprodione.

Postharvest wax-fungicide combinations are applied with a special horsehair brush to prevent fruit damage. The fruit are sized, then sorted, and packed loosely. Some packers vibrate the boxes (tight-fill) to prevent fruit bruising during transit (Mitchell et al. 1968). Polyethylene wrappers (Boyes 1955) are used in Italy, but not in California.

Nectarine and Peach

Of the annual shipment of 36 million bushels of fresh fruit to New York City between 1935 and 1942, over $1 million worth was lost annually to decay (Wiant and Bratley 1948). However, as a result of improved control measures, fresh fruit can now be shipped economically even to distant markets for as long as a month after harvest. Freestone peaches for processing are harvested when mature and canned or frozen after being held for ripening at 20°C and a relative humidity over 85 percent for one to five days (Ogawa et al. 1971); without chemical treatment, serious decay would result within 24 to 48 hours. Mechanically harvested ripe cling peaches are more prone to decay than hand-harvested fruit (Ogawa, Sandeno, and Mathre 1963), but since the fruit are processed within 12 to 18 hours of picking, decay is not a real threat.

Postharvest decay pathogens are primarily (*Monilinia fructicola* and *Rhizopus stolonifer*, with some lots showing considerable *R. arrhizus* and *Gilbertella persicaria* (Anderson 1925; Harter and Weimer 1922; Harvey, Smith, and Kaufman 1972; Ogawa 1965a; Ogawa et al. 1964; Pryor 1949; Smith 1941). A special report on disorders of peach and nectarine shipments to the New York markets, 1977–85, indicated the importance, also, of decay caused by *Rhizopus*, and *Penicillium* (Ceponis et al. 1987) (**fig. 40F**). Decay from *Penicillium* and *Botrytis* infection occurs only in fruit held for extended periods, although it appears that Botrytis decay is becoming somewhat more prevalent than in earlier years. Infection by yeasts has been reported (Burton and Wright 1969; Smith 1960) but is rare in fruit produced in the western states. In general, fruit decays are more prevalent during the latter part of the harvest season. This may be attributable to increases in fungal inoculum in the field or to more moisture in the environment.

While fruit of all nectarine and peach cultivars are susceptible to fungal decay, cultivars such as the Kirkman Gem peach that tend to develop suture cracks are most subject to decay. Cultivars that develop serious blossom blight from *Monilinia* are most likely to show fruit decay. Infection of harvested fruit is most common at the stem end where the epidermis is often torn during harvest. Control of insects such as Oriental fruit moth and peach twig borer is essential in order to eliminate larval holes where nitidulid beetles contaminated with fungus spores can enter (Ogawa 1965a; Tate 1973).

Monilinia fructicola is the most common decay pathogen on peach and nectarine and causes both blossom blight and fruit rot (**fig. 40E**). *Monilinia laxa* is less common on peach in California (Ogawa 1965b) but more common on peach in Oregon. Neither species presents a problem in the peach-growing areas of Washington's Yakima Valley (Pierson et al. 1958).

One can expect almost all fruit in the orchard to be contaminated if fungus-infected, mummified fruit develop during the growing season, and serious decay will result if the harvested fruit are held at room temperature. Fruit held below 4.4°C develop decay more slowly. In the packed product, decay starts most readily if the fruit touch each other or if they are placed in a plastic pack with the stem end down; the stem end is where the highest humidity occurs and where epidermal injury is most common.

Injured fruit contaminated by spores of *M. fructicola* may develop visible decay within 24 hours at 22.2°C, and sporulation of the fungus will occur in 30 hours. The decaying tissue is at first light brown but later turns gray or black. The rotted tissue, which is firm and difficult to separate from the healthy tissue, extends into the mesocarp in a conical pattern. *Monilinia* is identified by the presence of lemon-shaped spores, and species differentiation is by mycelial characteristics on a potato-dextrose-agar medium.

Rhizopus stolonifer spreads quickly from fruit to fruit causing the tissue to turn mushy and leaky in the boxes (Anderson 1925; Brooks and Cooley 1928). Affected portions can easily be separated from the healthy tissue. Infection initially occurs only on injured fruit surfaces, so it is rarely found in fruit on the tree but is abundant on fallen fruit. Such fruit are commonly affected with both *Rhizopus* species and *Gilbertella persicaria* (McClure and Smith 1959) (**fig. 40B**).

At low humidity, *Rhizopus* produces abundant sporangia on short aerial hyphae. In fruit packages, where the humidity is high, the mycelium is abundant but spores are sparse. Though one cannot distinguish *R. stolonifer* from *R. arrhizus* in the field, they can be differentiated by incubating cultures at 36°C: *R. stolonifer* will not grow but *R. arrhizus* will. *Rhizopus* sporangia are first white, then black, and finally gray when the sporangial wall disintegrates. With *Gilbertella*, droplets of shiny black spore suspensions appear on the sporangia. When dry, these black spore masses cling to the surface of the split sporangial wall.

Rhizopus and *Gilbertella* are widely distributed in California orchards (Butler, Ogawa, and Shalla 1960; Ogawa et al. 1963). Neither organism attacks green immature fruit, but partially ripened, slightly yellow fruit are highly susceptible. Spores of *Rhizopus* are ubiquitous and can remain alive under extreme conditions of dessication and high temperature.

Rhizopus arrhizus Fischer is differentiated from *R. stolonifer* by its ability to grow at temperatures of 36°C to 40°C and by its smaller spores (5 to 7 μm). On decayed fruit, the mycelial mass develops less sporangia and less turf. When *R. oryzae* is present it can be identified by its spore dimensions of 7 to 9 μm and stouter stolons (Zycha, Siepmann, and Linnemann

1969). One of the best identifying tests for *R. arrhizus* is that it will grow in a medium containing 4 ppm of 2,6-dichloro-4-nitroaniline (DCNA), while *R. stolonifer* will not (Ogawa et al. 1963). Thus far only one isolate of *R. stolonifer* tolerant to DCNA has been found in orchards; one isolate of *R. arrhizus* sensitive to that chemical has also been reported (Weber and Ogawa 1965).

Gilbertella persicaria (Eddy) Hesseltine is distinguished from *Rhizopus* by having several hyaline appendages at both ends of the sporangiospores. The sporangia are often borne in a circinate fashion and eventually split into two equal halves. The fungus is heterothallic (Butler, Ogawa, and Shalla 1960). The temperature range for sporangiospore germination, mycelial growth in culture, and fruit rot is 4° to 39°C, the optimum being 30° to 33°C. Characteristic growth and sporulation occur on peach, nectarine, plum, apricot, and other crops, such as strawberry, tomato, and sweet potato. Maximum decay of peaches occurs at 30° to 33°C, with little or none developing at 18° or 39°C (Ogawa et al. 1964).

Gilbertella growing on a synthetic medium containing DCNA gave rise to a variant that could tolerate a high DCNA concentration. On mycelial-inoculated peach fruit, a spray of 1,000 ppm DCNA suppressed the nonvariant but not the variant. The variant remained stable during successive transfers in agar culture and on peach fruit (Ogawa, Ramsey, and Moore 1963).

Mucor piriformis Fischer (**fig. 40C**) was first isolated in California in 1977 when an unusual amount of decay developed during refrigerated transit of fresh-market peaches from California and fresh-market nectarines from Chile to California. *Mucor piriformis* is a cold-tolerant fungus (grows at 0°C) and a relative of *Rhizopus* and *Gilbertella* in the Zygomycetes. In common with these genera, it causes a soft rot in stone fruit (Smith, Moline, and Johnson 1979). Mucor rot is a serious problem during cold storage of apples and pears in Oregon and Washington (Bertrand and Saulie-Carter 1980). *Mucor piriformis* is very sensitive to high temperatures (above 27°C), but very tolerant of freezing temperatures. Thus, under the high summer temperatures common to San Joaquin Valley orchards, the soil populations of *M. piriformis* are low and postharvest decay losses almost nonexistent (Michailides 1984; Michailides and Ogawa 1987a,b). Yet in 1989, plum fruit harvested in the southern San Joaquin Valley and held in cold storage for extended periods developed extensive *Mucor* decay, indicating the serious nature of this pathogen even in California (J. M. Ogawa and B. T. Manji, unpublished).

Control. Postharvest treatments depend on whether fruit is destined for fresh market or processing. For fresh market, the main goal is to extend the decay-free life of the mature fruit; for processing, the aim is to ripen the fruit rapidly without decay.

In the 1950s, sulfur (Eaks, Eckert, and Roistacher 1958; M. A. Smith 1930; W. L. Smith 1962b, 1971) and captan dusts were widely used as pre- and postharvest protectants against fungus pathogens. Also used, but with limited success, were cold-water sprays containing 100 to 125 ppm chlorine (McClure 1958; Van Blaricom 1959). Dipping fruit in 0.1 to 1 percent sodium orthophenylphenate or the antibiotic mycostatin provided better decay control than did chlorine treatments (DiMarco and Davis 1957). In 1960, DCNA was introduced for the control of *Rhizopus* and *Monilinia* and it continues to be the key chemical for control of both fungi (Capellini and Stretch 1962; Daines 1965; Dewey and McLean 1962; Johnson 1969; Ogawa et al. 1963, 1964; Ogawa and Uyemoto 1962). Mixtures of DCNA with captan (Ogawa et al. 1971) give excellent decay control (Ogawa et al. 1964) but DCNA with benomyl is even better (Ogawa, Lyda, and Weber 1961). Waxes containing these chemicals provide better distribution and increased chemical residues on fruit (Ogawa 1965b) and reduction in fruit shriveling (Kraght 1966; Sommer and Mitchell 1959; Wells 1972). Recently it was shown that established infections of *M. fructicola* and *R. stolonifer* can be suppressed by the proper application of DCNA plus benomyl in a water-soluble wax (Manji, Ogawa, and Chastagner 1974); these chemicals were found to penetrate into the mesocarp tissues (Luepschen 1964; Ravetto and Ogawa 1972).

Studies show that DCNA induces lysis of *Rhizopus* germ tubes (Ogawa and Mathre 1964). It is active against *R. stolonifer* at 2 ppm and against *Monilinia fructicola* at 4 ppm. Benomyl does not prevent germination of *Monilinia* spores, but it suppresses growth of the organism at 0.1 ppm. Neither fungicide is effective against *Alternaria* or *Aspergillus*. Tolerant isolates of *R. stolonifer*, *Botrytis cinerea*, and *G. persicaria* have been produced in a medium containing DCNA (Ogawa, Ramsey, and Moore 1963; Webster, Ogawa, and Bose

1970; Webster, Ogawa, and Moore 1968), but in only one instance has DCNA-tolerant *R. stolonifer* been found in the field, and this was in a peach orchard not previously sprayed with DCNA. Benomyl-tolerant *Monilinia* was not found in culture or in nature (Tate et al. 1974) until 1978, when it was detected on fresh-market peaches.

Hot-water dips at 32°C for 18 to 90 seconds (Jenkins, Peggie, and Tindale 1961; Miller and Tanaka 1963; Richards, Wadley, and Barlow 1953; Smith 1962, 1971; Smith, Haller, and McClure 1956; Wells 1970, 1972; Wells and Gerdts 1971; Wells and Harvey 1970) suppress development of *Monilinia* and *Rhizopus*. Such treatments would therefore be beneficial for fruit with incipient infections. High-temperature storage (24 hours at 40°C) of fruit picked four days before canning-ripeness provides a quality canned product without rot problems from *Monilinia* or *Rhizopus* (Tindale et al. 1968). The quick cooling of fruit exposed to high temperatures is essential to prevent the loss of dessert quality.

Other decay control measures that have not proved as effective as the chemical treatments mentioned above or the regulation of temperatures are gamma radiation, polyethylene liners (Luvisi and Sommer 1960; Beraha 1959), pre- and postharvest calcium treatments (Conway, Greene, and Hickey 1987), and treatment with ozone (Spalding 1966), ammonia (Eaks, Eckert, and Roistacher 1958), alcohols (Ogawa and Lyda 1960), or acetaldehydes (Aharoni and Stadelbacher 1973). Also, the use of a modified atmosphere containing as much as 40 percent carbon dioxide at storage temperatures has reduced the incidence of decay (O'Reilly 1947). Recently, Wilson, Franklin, and Otto (1987) showed that several volatile compounds emitted from stone fruit are either fungistatic or fungicidal to *M. fructicola* and *Botrytis cinerea*. Further research is necessary to ascertain the possible practical application of this information. Showing some promise is the biological control of *Monilinia fructicola* on peaches by means of the bacterium *Bacillus subtilis* and a mixture of this organism and dicloran for the control of both *Monilinia* and *R. stolonifer* (Pusey and Wilson 1984; Pusey et al. 1986; McKeen, Reilly, and Pusey 1986; Pusey and Hotchkiss 1987; Pusey, et al. 1988; Wilson and Wisniewski 1989).

Treatments of fruit for the fresh market often involves the use of chlorine in hydrocooling, followed by a spray of triforine and then an application of DCNA and benomyl in wax applied as a spray and uniformly distributed by brushing. Iprodione alone or in combination with DCNA is now available as a treatment for the control of *Monilinia*, *Rhizopus*, *Botrytis*, and *Alternaria*. An alternative experimental fungicide is SC-0858 (Stauffer Chemical Company), which provides both protective and suppressive action against *M. fructicola* on peach (Osorio et al. 1987). Fruit for canning can be dipped in DCNA, thereby providing fungicidal residues of 20 ppm or even higher (Ogawa et al. 1964, 1971). Subsequent washing by dipping the fruit in lye before canning removes the DCNA residue.

Fungicide applications during bloom, preharvest, and postharvest provide good decay control (Ogawa, Manji, and Ravetto 1970). Equally important, however, are discarding injured and diseased fruit, packaging only quality fruit of uniform maturity, and storage of fruit at a temperature near 0°C.

Plum

Monilinia fructicola and *Rhizopus stolonifer* are the fungi most commonly associated with the decay of harvested plums (Creelman 1962; Richards, Wadley, and Barlow 1953; R. E. Smith 1941). Alternaria rot develops on injured fruit, especially those mechanically harvested. Decay caused by *Penicillium*, *Mucor*, and *Aspergillus* is minimal, yet significant amounts have developed in a few lots.

Plums are hand harvested and placed in 40-pound wooden boxes or 800-pound bins for hauling to the packing shed, where they are either cooled or immediately packaged by the tight-fill method (Mitchell et al. 1968) after a fungicide treatment. Harvest extends from May to September in California.

The symptoms caused by each of the involved pathogens are similar to those previously described for other stone fruit. *Monilinia* inoculum comes from decayed fruit that were injured and infected earlier. Blossom blight from *Monilinia* is not so severe as on other stone fruit because the peduncles of blighted blossoms dehisce as the fungus progresses in this tissue.

Control of decay has been difficult because the fruit epidermis is smooth and waxy, making effective chemical application difficult. Until the introduction of DCNA, control was largely a matter of slowing the rate of decay through cold storage and modified atmospheres (Allen 1948; Allen, Pentzer, and Bratley

1944; Ceponis and Friedman 1957; Couey 1960, 1965; Pentzer and Allen 1937). Chemical control tests show that a DCNA spray alone does not leave enough fungicidal residue for adequate control, but the addition of fruit waxes to the fungicide results in larger deposits and better control (Phillips and Uota 1971; Wells 1973; Wells and Harvey 1970). Fungicides currently used in postharvest treatments are iprodione or triforine in combination with DCNA.

Prune

Pathogens causing decay of fresh, mechanically harvested prunes are similar to those causing decay of other stone fruit, while those attacking dehydrated prunes are heat tolerant or are favored by a substrate rich in sugars. Prunes are harvested when the soluble solids content is at a minimum of 24 or 25 percent. After drying reduces moisture to an average of 20 percent, the dried prunes range from 38.4 to 51 percent reducing sugars and 0.6 to 5.5 percent sucrose, levels unsuitable for the growth of most molds.

The decay of fresh prunes held in bins more than one day after harvest can be serious. Molds on dehydrated prunes held in bulk bins are most critical in fruit invaded by insect pests which increase the moisture level (El-Behadli and Ogawa 1974).

Molds developing on fresh prunes are *Monilinia fructicola*, *M. laxa*, *Rhizopus stolonifer*, *Alternaria alternata*, *Aspergillus*, *Cladosporium*, *Chrysosporium*, *Mucor*, and *Penicillium* (El-Behadli and Ogawa 1974) (**fig. 40G**). On dried fruit (Tanaka and Miller 1963a), the organisms most commonly isolated are *Aspergillus glaucus* (*A. chevalieri* [El-Behadli and Ogawa 1974]) followed by *A. niger*, *A. repens*, *A. amstelodami*, *Penicillium* spp., *Alternaria*, *Monilinia*, *Chaetomella*, and *Mucor*. Yeasts isolated most frequently (Tanaka and Miller 1963a) are *Saccharomyces rouxii*, followed by *S. mellis*, *Torulopsis magnoliae*, *T. stellata*, *Candida krusei*, *Trichosporon behrendii*, *Pichia fermentans*, *P. membranaefaciens*, *C. chalmersi*, *S. rosei*, *S. cerevisiae*, *Sporobolomyces roseus*, and *C. parapsilosis*. Tanaka and Miller (1963a,b) examined postharvest samples obtained from bulk-stored, packaged, and processed prunes, while samples studied by El-Behadli and Ogawa (1974) were from the orchard and bulk storage bins.

Most prunes are mechanically harvested with shakers and the fruit are caught in frames attached to the shakers or on canvas laid on the ground. Before processing, the fruit are held in bins of about 1,000 pounds capacity. During harvest and dehydration, little sorting is done to remove decaying fruit. Mechanical injuries do occur and are sites for infection by *Monilinia*, *Aspergillus*, and *Alternaria*. The dehydration process (Gentry, Miller, and Claypool 1965; Rodda and Gentry 1969) involves moving single layers of fruit in trays through a treatment tunnel with a flow of hot air parallel to or against the trays for 12 to 20 hours. The maximum temperature in the counter-flow tunnel is approximately 70.4°C, while in the parallel-flow tunnel the warmest air at the front end of the tunnel is about 90°C. The fruit itself remains well below 70.4°C to prevent caramelization of sugars. Phaff et al. (1946) reported that no viable yeasts or molds remain on prunes at the end of commercial dehydration (Tanaka and Miller 1963a). El-Behadli and Ogawa (1974) showed that dehydration does not kill spores of *Aspergillus chevalieri* (*A. glaucus* group), and found this fungus to be the most common mold in bulk bins.

A description (Raper and Fennell 1965) of *Aspergillus chevalieri*, a mold that produces mycotoxins, is:

> *Aspergillus chevalieri* (Mangin) Thom and Church, synonyms *Eurotium chevalieri* Mangin and *A. allocutus* Batista and Maia, belongs to the *A. glaucus* group. On Czapek's solution agar with 20 percent sucrose it grows best at 30°C or above. The mycelium is spreading, plain to somewhat wrinkled in the central area of the colony with abundant conidial heads in shades of sage-green to andover-green or slate-olive. Yellow cleistothecia are enmeshed in orange-red hyphae at the agar surface. Conidiophores are 700 to 850 μm in length, with almost globose vesicular apices, 25 to 35 μm in diameter. Sterigmata are in a single series, closely packed, 5 to 7 μm by 3.0 to 3.5 μm. Conidia are ovate to elliptical with ends often flattened, spinulose, and 4.5 to 5.5 μm long. Cleistothecia (100 to 140 μm in diameter) are abundant and closely enmeshed in a felt of orange-red encrusted hyphae. Asci are 9 to 10 μm long; ascospores are lenticular, 4.6 to 5.0 by 3.4–3.8 μm, with walls smooth to slightly roughened and with prominent equatorial crests that are thin and often recurved.

Studies have been undertaken to find ways to prevent mold development in dry- and wet-packaged fruit

(McBean and Pitt 1965; Nury, Miller, and Brekke 1960), to determine the heat resistance of xerophilous fungi (Pitt and Christian 1968, 1970; Tanaka and Miller 1963b), and to evaluate the effect of fruit moisture and relative humidity on spoilage (El-Behadli and Ogawa 1974; Miller, Smith, and Sisler 1959). Chemicals showing the greatest promise on dehydrated fruit are ethylene oxide, propylene oxide, and sorbic acid and its salts (Nury, Miller, and Brekke 1960). One treatment is the exposure of dried prunes to 2 percent sorbic acid at 80.2°C during hydration, except for repeated turning of bins containing fruit to equilibrate the moisture level between fruits.

Potassium sorbate is applied to prevent spoilage of the final product, but no real effort has been made to control decay in bulk bins just before or after dehydration, except for repeated turning of bins containing dried fruit to equilibrate the moisture level between fruits.

Fresh prunes can be held in cold storage with no internal breakdown for two or three weeks. In recent experiments, the amount of decay during storage and the reduction of molds by chemical treatment were related to the method of harvest and the temperature of incubation (Michailides, Ogawa, and Sholberg 1987). Untreated fruit held at 4°C started to develop mold on the seventh day, while treated (chlorine, etaconazole, or K sorbate) fruit did not show mold development until the tenth day. No differences were found in the amount of decay on fruit washed (potable water) and incubated at 4°, 20°, or 28°C. Prunes mechanically harvested showed less decay than those hand harvested from the ground. A captan dip treatment controlled the molds on prunes, and the dehydration process, involving moist atmospheres and high temperatures, essentially eliminated all residues of the chemical (Archer and Corbin 1969, 1970).

On dried prunes, a deterioration of unknown cause, characterized by macerated, wet, sticky areas on the fruit surface and by skin that tends to slip with slightest pressure during inspection, has been described (**fig. 40H**). Studies by Sholberg and Ogawa (1983a) indicated that this condition most often is the result of fresh-fruit infection by the fungus *Rhizopus stolonifer*, although other fungi, such as *Aspergillus*, *Penicillium*, and *Monilinia*, could cause part of the problem, commonly referred to as "box rot." Eliminated as the cause of this condition were fruit of different sizes and soluble solids content, sunburned fruit, and fruit injured by rough handling. Significant amounts of this dried fruit deterioration were first noticed when *Rhizopus* was allowed to infect and incubate for a period of more than 24 hours at 28°C, the temperature commonly found in fruit bins before dehydration. To reduce this deterioration, one can dry the fresh prunes within 24 hours of harvest or keep the fruit in cold storage at temperatures below 5°C if they cannot be dried within 24 hours. If no cold storage is available, fruit harvested in the morning should be dried after fruit harvested in the afternoon. Other suggestions are to use a protective fungicide (potassium sorbate) dip or expose the fruit to sulfur dioxide gas to kill the fungal spores. The latter, however, causes a golden coloration of the dried prunes (Sholberg and Ogawa 1983a,b).

In recent years, prunes for fresh market have been hand picked, packaged without postharvest treatments, and shipped with few to no decay problems to both domesstic and international markets.

REFERENCES

Aharoni, Y., and G. J. Stadelbacher. 1973. The toxicity of acetaldehyde vapors to postharvest pathogens of fruits and vegetables. *Phytopathology* 63:544–45.

Allen, F. W. 1948. Storage and shipping of plums. *Calif. Fruit Grape Grower* 2(6):12–14.

Allen, F. W., W. T. Pentzer, and C. O. Bratley. 1944. Carbon dioxide investigations: Dry ice as a supplement to refrigeration of plums in transit. *Am. Soc. Hort. Sci. Proc.* 44:141–47.

Anderson, H. W. 1925. Rhizopus rot of peaches. *Phytopathology* 5:122–24.

Archer, T. E., and J. B. Corbin. 1969. The site and fate of captan residues from dipping prunes prior to commercial dehydration. *Food Technol.* 23(2):101.

———. 1970. Captan residues on dipped prune fruits as affected by different adjuvants. *Food TEchnol.* 24:710–12.

Baraha, L. 1959. Effects of gamma radiation on brown rot and Rhizopus rot of peaches and the causal organisms. *Phytopathology* 49:354–57.

Baritelle, J. L., and P. D. Gardener. 1984. Economic losses in the food and fiber system: From the perspective of an economist. In *Postharvest pathology of fruits and vegetables: Postharvest losses in perishable crops,* 4–10. Univ. Calif. Agric. Exp. Stn. Pub. NE–87.

Bertrand, P., and J. Saulie-Carter. 1980. Mucor rot of pears and apples. *Oregon State Univ. Corvallis Spec. Rep.* 568, 21 pp.

Boyes, W. W. 1955. Effect of polyethylene (polythene) wrappers on the keeping quality of apricots, peaches, and plums. *Farming South Africa* 30(346):13–19.

Bratley, C. O. 1931. Decay of sweet cherries from California. *Plant Dis. Rep.* 15:73–74.

Brooks, C., and J. S. Cooley. 1928. Time-temperature relations in different types of peach rot infection. *J. Agric. Res.* 37:507–43.

Brooks, C., and D. F. Fisher. 1916. Brown rot of prunes and cherries in the Pacific Northwest. *USDA Bull.* 368, 10 pp.

———. 1924. Prune and cherry brown rot investigations in the Pacific Northwest. *USDA Bull.* 1252, 22 pp.

Burton, C. L., and G. K. Brown. 1984. Quality retention strategies for mechanically harvested fresh-market fruit. In *Postharvest pathology of fruits and vegetables: Postharvest losses in perishable crops*, 42–49. Univ. Calif. Agric. Exp. Stn. Pub. NE-87.

Burton, C. L., and W. R. Wright. 1969. Sour rot disease of peaches on the market. *Plant Dis. Rep.* 53:580–82.

Bussel, J., P. M. Buckley, N. F. Sommer, and T. Kosuge. 1969. Ultrastructural changes in *Rhizopus stolonifer* sporangiospores in response to anaerobiosis. *J. Bacteriol.* 98:774–83.

Bussel, J., N. F. Sommer, and T. Kosuge. 1969. Effect of anaerobiosis upon germination and survival of *Rhizopus stolonifer* sporangiospores. *Phytopathology* 59:946–52.

Butler, E. E., J. M. Ogawa, and T. Shalla. 1960. Notes on *Gilbertella persicaria* from California. *Bull. Torrey Bot. Club.* 87:397–401.

Capellini, R. A., and M. J. Ceponis. 1984. Postharvest losses in fresh fruits and vegetables. In *Postharvest pathology of fruits and vegetables: Postharvest losses in perishable crops*, 24–30. Univ. Calif. Agric. Exp. Stn. Pub. NE-87.

Capellini, R. A., and A. W. Stretch. 1962. Control of postharvest decay of peaches. *Plant Dis. Rep.* 46:31–33.

Ceponis, M. J., R. A. Capellini, and G. W. Lighnter. 1987. Disorders in sweet cherry and strawberry shipments to the New York market, 1972–1984. *Plant Dis.* 71:472–75.

Ceponis, M. J., R. A. Capellini, J. M. Wells, and G. W. Lightner. 1987. Disorders in plum, peach, and nectarine shipments to the New York market, 1972–1985. *Plant Dis.* 71:947–52.

Ceponis, M. J., and B. A. Friedman. 1957. Effect of bruising injury and storage temperature upon decay and discoloration of fresh, Idaho-grown Italian prunes on the New York City Market. *Plant Dis. Rep.* 41:491–92.

Chastagner, G. A., and J. M. Ogawa. 1976. Injury of stone fruits by preharvest captan sprays followed by postharvest treatments. *Phytopathology* 66:924–27.

Conway, W. S. 1984. Preharvest factors affecting postharvest losses from disease. In *Postharvest pathology of fruits and vegetables: Postharvest losses in perishable crops*, 11–16. Univ. Calif. Agric. Exp. Stn. Pub. NE-87.

Conway, W. S., G. M. Greene II, and K. D. Hickey. 1987. Effects of preharvest and postharvest calcium treatments of peaches on decay caused by *Monilinia fructicola*. *Plant Dis.* 71:1084–86.

Couey, H. M. 1960. Effect of temperature and modified atmosphere on the storage life, ripening behavior, and dessert quality of El Dorado plums. *Am. Soc. Hort. Sci. Proc.* 75:207–15.

———. 1965. Modified atmosphere storage of Nubiana plums. *Am. Soc. Hort. Sci. Proc.* 86:166–68.

Creelman, D. W. 1962. Summary of the prevalence of plant diseases in Canada in 1961, IV. Diseases of fruit crops: Plum. *Can. Plant Dis. Survey* 42(2):80.

Daines, R. H. 1965. 2,6-dichloro-4-nitroaniline used in orchard sprays, and dump tank, the wet brusher, and the hydrocooler for control of Rhizopus rot of harvested peaches. *Plant Dis. Rep.* 49:300–04.

Dewey, D. H., and D. C. McLean. 1962. Post-harvest treatment with 2,6-dichloro-4-nitroaniline for fruit rot control on fresh market peaches. *Mich. Agric. Exp. Stn. Quart. Bull.* 44:679–83.

DiMarco, G. R., and B. H. Davis. 1957. Prevention of decay of peaches with post-harvest treatments. *Plant Dis. Rep.* 41:284–88.

Eaks, I. L., J. W. Eckert, and C. N. Roistacher. 1958. Ammonia gas fumigation for control of Rhizopus rot of peaches. *Plant Dis. Rep.* 42:846–48.

Eckert, J. W. 1969. Chemical treatments for control of postharvest diseases. *World Rev. Pest Control* 8(3):116–37.

Eckert, J. W., and J. M. Ogawa. 1985. The chemical control of postharvest diseases: Subtropical and tropical fruits. *Annu. Rev. Phytopathol.* 23:421–54.

———. 1988. The chemical control of postharvest diseases: Deciduous fruits, berries, vegetables and root/tuber crops. *Annu. Rev. Phytopathol.* 26:433–69.

El-Behadli, A., and J. M. Ogawa. 1974. Occurrence and survival of molds on fruit in California prune orchards and dehydrators (abs.). *Proc. Am. Phytopathol. Soc.* 1:143.

English, H. 1945. Flungi isolated from moldy sweet cherries in the Pacific Northwest. *Plant Dis. Rep.* 29:559–66.

English, H., and F. Gerhardt. 1942. Effect of carbon dioxide and temperature on the decay of sweet cherries under simulated transit conditions. *Am. Soc. Hort. Sci. Proc.* 40:172–76.

———. 1946. Effect of ultraviolet radiation on the viability of fungus spores and on the development of decay in sweet cherries. *Phytopathology* 36:100–11.

Gentry, J. P., M. W. Miller, and L. L. Claypool. 1965. Engineering and fruit quality aspects of prune dehydration in parallel- and counter-flow tunnels. *Food Technol.* 19(9):121–25.

Gerhardt, F., H. English, and E. Smith. 1945. Cracking and decay of Bing cherries as related to the presence of moisture on the surface of the fruit. *Am. Soc. Hort. Sci. Proc.* 46:191–98.

Gerhardt, F., and L. A. Ryall. 1939. The storage of sweet cherries as influenced by carbon dioxide and volatile fungicides. *USDA Tech. Bull.* 631.

Gerhardt, F., H. A. Schomer, and T. R. Wright. 1956. Sealed film lug liners for packing Bing cherries. *U.S. Agric. Mkt. Serv.* AMS 121, 8 pp.

———. 1957. Film lug liners lengthen market life of sweet cherries. *U.S. Agric. Mkt. Serv.* AMS 177, 3 pp.

Harter, L. L., and J. L. Weimer. 1922. Decay of various vegetables and fruits by different species of *Rhizopus*. *Phytopathology* 12:205–12.

Harvey, J. M., W. L. Smith, and J. Kaufman. 1972. Market diseases of stone fruits: Cherries, peaches, nectarines, apricots, and plums. *USDA Handb.* 414, 64 pp.

Helgeson, J. P. 1989. Management of disease resistance in harvested fruits and vegetables. Postharvest resistance through breeding and biotechnology. *Phytopathology* 79:1375–77.

Jenkins, P. T., I. D. Peggie, and G. B. Tindale. 1961. Brown rot in canning peaches: A heat therapy control. *Plant Pathol. Waite Agric. Res. Inst., Adelaide Conf.* 1(7):2.

Johnson, J. D. 1969. New techniques battle postharvest peach diseases. *Texas Agric. Progress* 15(1):24–25.

Kelman, A. 1984. The importance of research on postharvest losses in perishable crops. In *Postharvest pathology of fruits and vegetables: Postharvest losses in perishable crops.* 1–3. Univ. Calif. Agric. Exp. Stn. Publ. NE–87.

Kraght, A. J. 1966. Waxing peaches with the consumer in mind. *Produce Mkt.* 9(2):20–21.

Luepschen, N. 1964. Effectiveness of 2,6-dichloro-4-nitroaniline impregnated peach wraps in reducing Rhizopus decay losses. *Phytopathology* 54:1219–22.

Luvisi, D. A., and N. F. Sommer. 1960. Polyethylene liners and fungicides for peaches and nectarines. *Am. Soc. Hort. Sci. Proc.* 76:146–55.

Manji, B. T., and J. M. Ogawa. 1985. Quiescent infections and disease control in the shipping container. *Proc. 44th Ann. Conv. Natl. Peach Council*, Nashville, Feb. 20–23:23–34.

Manji, B. T., J. M. Ogawa, and G. A. Chastagner. 1974. Suppression of established infections in nectarine and peach with postharvest fungicide treatment (abs.). *Proc. Am. Phytopathol. Soc.* 1:43.

Matsumoto, T. T., and N. F. Sommer. 1967. Sensitivity of *Rhizopus stolonifer* to chilling. *Phytopathology* 57:881–84.

McBean, D. McG., and J. I. Pitt. 1965. Preservation of high-moisture prunes in plastic pouches. *CSIRO Food Pres. Quart.* 25(2):27–32.

McClure, T. T. 1958. Brown and Rhizopus rots of peaches as affected by hydrocooling, fungicide, and temperature. *Phytopathology* 48:322–23.

McClure, T. T., and W. L. Smith, Jr. 1959. Postharvest decay of peaches as affected by temperatures after hydrocooling in water or Dowicide A solutions. *Phytopathology* 49:472–74.

McKeen, C. D., C. C. Reilly, and P. L. Pusey. 1986. Production and partial characterization of antifungal substances antagonistic to *Monilinia fructicola* from *Bacillus subtilis*. *Phytopathology* 76:136–39.

Michailides, T. J. 1984. *Studies on the ecology and epidemiology of the postharvest fruit pathogen* Mucor piriformis *Fischer*. Ph.D. Thesis, University of California, Davis, 88 pp.

Michailides, T. J., and J. M. Ogawa. 1987a. Effect of soil temperature and moisture on the survival of *Mucor piriformis*. *Phytopathology* 77:251–56.

———. 1987b. Colonization, sporulation, and persistence of *Mucor piriformis* in unamended and amended orchard soils. *Phytopathology* 77:257–61.

———. 1989. Effects of high temperatures on the survival and pathogenicity of propagules of *Mucor piriformis*. *Phytopathology* 79:547–54.

Michailides, T. J., J. M. Ogawa, and D. C. Opgenorth. 1987. Shift in *Monilinia* spp. and distribution of isolates sensitive and resistant to benomyl in California prune and apricot orchards. *Plant Dis.* 71:893–96.

Michailides, T. J., J. M. Ogawa, and P. L. Sholberg. 1987. Chemical control of fungi causing decay of fresh prunes during storage. *Plant Dis.* 71:14–17.

Miller, M. W., and H. Tanaka. 1963. Microbial spoilage of dried prunes. III. Relation of equilibrium relative humidity to potential spoilage. *Hilgardia* 34:183–90.

Miller, W. H., W. L. Smith, Jr., and H. D. Sisler. 1959. Temperature effects on *Rhizopus stolonifer* and *Monilinia fructicola* spores on potato dextrose agar (abs.). *Phytopathology* 49:546.

Mitchell, F. G., N. F. Sommer, J. P. Gentry, et al. 1968. Tight-fill fruit packing. *Univ. Calif. Agric. Ext. Serv. Circ.* 548.

Nury, F. S., M. W. Miller, and J. E. Brekke. 1960. Preservative effect of some antimicrobial agents on high-moisture dried fruits. *Food Technol.* 14(2):113–15.

Ogawa, J. M. 1965a. Postharvest diseases of peaches and their control. *Proc. Peach Cong. Verona* (July 20–24.), 9 pp.

———. 1965b. Control of pre- and postharvest fruit decays in relation to residues of 2,6-dichloro-4-nitroaniline (DCNA). *Symposium on DCNA, Kalamazoo, Mich.* Nov. 3–5.

Ogawa, J. M., E. Bose, T. B. Manji, and W. R. Schreader. 1972. Bruising of sweet cherries resulting in internal browning and increased susceptibility to fungi. *Phytopathology* 62:579–80.

Ogawa, J. M., G. A. Boyak, J. L. Sandeno, and J. H. Mathre. 1964. Control of post-harvest fruit decays in relation to residues of 2,6-dichloro-4-nitroaniline and Difolatan. *Hilgardia* 35(14):365–73.

Ogawa, J. M., E. E. Butler, and S. D. Lyda. 1961. Relation of temperature to spore germination, growth, and disease development by *Gilbertella persicaria*. *Phytopathology* 51:572–74.

Ogawa, J. M., S. Leonard, B. T. Manji, E. Bose, and C. J. Moore. 1971. Monilinia and Rhizopus decay control during controlled ripening of freestone peaches for canning. *J. Food Sci.* 36:331–34.

Ogawa, J. M., and S. D. Lyda. 1960. Effect of alcohols on spores of *Sclerotinia fructicola* and other peach fruit-rotting fungi in California. *Phytopathology* 50:790–92.

Ogawa, J. M., S. D. Lyda, and D. J. Weber. 1961. 2,6-dichloro-4-nitroaniline effective against Rhizopus fruit rot of sweet cherries. *Plant Dis. Rep.* 45:636–38.

Ogawa, J. M., and B. T. Manji. 1984. Control of postharvest diseases by chemical and physical means. In *Postharvest pathology of fruits and vegetables: Postharvest losses in perishable crops*. 55–66. Univ. Calif. Div. Agric. Expt. Stn. Publ. NE-87.

———. 1985. Coping with fungicide ineffectiveness against postharvest decay pathogens. *Proc. 44th Ann. Conv. Natl. Peach Council* (Nashville, Feb. 20–23):21–22.

Ogawa, J. M., B. T. Manji, and E. Bose. 1968. Efficacy of fungicide 1991 in reducing fruit rot of stone fruits. *Plant Dis. Rep.* 52:722–26.

———. 1974. Molds on fresh market Blenheim apricots controlled with fungicide mixtures (abs.) *Proc. Am. Phytopathol. Soc.* 1:41–42.

Ogawa, J. M., B. T. Manji, and D. J. Ravetto. 1970. Evaluation of preharvest benomyl applications on postharvest Monilinia rot of peaches and nectarines (abs.). *Phytopathology* 60:1306.

Ogawa, J. M., and J. H. Mathre. 1964. Lysis of germinating sporangiospores of *Rhizopus stolonifer* by 2,6-dichloro-4-nitroaniline (DCNA)—A basis for control of fruit rot (abs.). *10th Internat. Bot. Cong.*:438.

Ogawa, J. M., J. H. Mathre, D. J. Weber, and S. D. Lyda. 1963. Effects of 2,6-dichloro-4-nitroaniline on *Rhizopus* species and its comparison with other fungicides on control of Rhizopus rot of peaches. *Phytopathology* 53:950–55.

Ogawa, J. M., R. H. Ramsey, and C. J. Moore. 1963. Behavior of variants of *Gilbertella persicaria* arising in medium containing 2,6-dichloro-4-nitroaniline. *Phytopathology* 53:97–100.

Ogawa, J. M., J. Rumsey, B. T. Manji, G. Tate, J. Toyoda, E. Bose, and L. Dugger. 1974. Implication and chemical testing of two Rhizopus fungi in softening of canned apricots. *Calif. Agric.* 28(7):6–7.

Ogawa, J. M., J. L. Sandeno, and J. H. Mathre. 1963. Comparisons in development and chemical control of decay-causing organisms on mechanical and hand-harvested stone fruits. *Plant Dis. Rep.* 47:129–33.

Ogawa, J. M., and J. K. Uyemoto. 1962. Effectiveness of 2,6-dichloro-4-nitroaniline on development of Rhizopus rot of peach fruits at various temperatures (abs.) *Phytopathology* 52:23.

O'Reilly, H. J. 1947. Peach storage in modified atmospheres. *Proc. Am. Soc. Hort. Sci.* 49:99

Osorio, J. M., J. M. Ogawa, A. J. Feliciano, and B. T. Manji. 1987. Efficacy of experimental fungicide SC-0858 in control of brown rot diseases of stone fruits (abs.). *Phytopathology* 77:1241.

Pentzer, W. T., and F. W. Allen. 1937. Ripening and breakdown of plum as influenced by storage temperatures. *Am. Soc. Hort. Sci. Proc.* 44:148–56.

Phaff, H. J., E. M. Mrak, R. Alleman, and R. Whelton. 1946. Microbiology of prunes during handling and drying. *Fruit Prod. J. Am. Food Manuf.* 25:140–41.

Phillips, D. J. 1984. Mycotoxins as a postharvest problem. In *Postharvest pathology of fruits and vegetables: Postharvest losses in perishable crops*. 50–54. Univ. Calif. Div. Agric. Exp. Stn. Publ. NE-87

Phillips, D. J., and M. Uota. 1971. Postharvest treatments to control brown rot on plums. *Blue Anchor* 48(3):20–21.

Pierson, C. F. 1958. Fungicides for reduction of postharvest decay of sweet cherries. *Wash. State. Hort. Assoc. Proc.* 54:115–16.

———. 1966. Effect of temperature on growth of *Rhizopus stolonifer* on peaches and agar. *Phytopathology* 56:276–78.

Pierson, C. F., A. M. Neubert, M. A. Smith, E. R. Wolford, and M. Thomson. 1958. Studies on Rhizopus rot of cannery peaches in the state of Washington. *Wash. State Hort. Assoc. Proc.* 54:179–82.

Pitt, J. I., and J. H. B. Christian. 1968. Water relations of xerophillic fungi isolated from prunes. *Appl. Microbiol.* 16(12):1853–58.

———. 1970. Heat resistance of xerophilic fungi based on microscopical asssessment of spore survival. *Appl. Microbiol.* 20(5):682–86.

Pryor, D. E. 1949. Reduction of post-harvest spoilage in fresh fruits and vegetables destined for long distance shipment. *Food Technol.* 4(2):57–62.

Pusey, P. L., and M. W. Hotchkiss. 1987. Application and efficacy of *Bacillus subtilis* for brown rot control on commercial peach packing operations (abs.). *Phytopathology* 77:1776.

Pusey, P. L., M. W. Hotchkiss, H. T. Dulmage, R. A. Baumgardner, E. I. Zehr, et al. 1988. Pilot tests for commercial production and application of *Bacillus subtilis* (B-3) for postharvest control of peach brown rot. *Plant Dis.* 72:622–26.

Pusey, P. L., and C. L. Wilson. 1984. Postharvest biological control of stone fruit brown rot by *Bacillus subtilis*. *Plant Dis.* 68:753–56.

Pusey, P. L., C. L. Wilson, M. W. Hotchkiss, and J. D. Franklin. 1986. Compatibility of *Bacillus subtilis* for postharvest control of peach brown rot with commercial fruit waxes, dicloran, and cold-storage conditions. *Plant Dis.* 70:587–90.

Raper, K. B., and D. I. Fennell. 1965. *The genus Aspergillus*. Williams & Wilkins, Baltimore, 686 pp.

Ravetto, D. J., and J. M. Ogawa. 1972. Penetration of peach fruit by benomyl and 2,6-dichloro-4-nitroaniline fungicides (abs.). *Phytopathology* 62:784.

Richards, B. L., B. N. Wadley, and J. C. Barlow. 1953. Diseases of Italian prune in Utah. *Farm Home Sci.* 14:34–36.

Rodda, E. D., and J. P. Gentry. 1969. New concepts in fruit dehydrator construction. *Am. Soc. Agric. Eng.* 12(4):540–41.

Schomer, H. A., and K. L. Olson. 1964. Storage of sweet cherries in controlled atmospheres. *U.S. Agric. Mkt. Serv.* AMS 529, 7 pp.

Sholberg, P. L., and J. M. Ogawa. 1983a. Fungus causes deterioration of dried prunes. *Calif. Agric.* 37(3–4):27–28.

———. 1983b. Relation of postharvest decay fungi to the slipskin maceration disorder of dried French prunes. *Phytopathology* 73:708–13.

Shotwell, K. M., and J. M. Ogawa. 1984. Botrytis blossom blight and fruit rot of sweet cherry (*Prunus avium*) and their control using fungicides (abs.). *Phytopathology* 74:1141.

Smith, M. A. 1930. Sulfur dust for the control of brown rot of peaches in storage (abs.). *Phytopathology* 20:122–23.

———. 1960. Etiology of sour pit of peaches. *Appl. Microbiol.* 8(4):256–61.

Smith, R. E. 1941. *Diseases of fruits and nuts.* Calif. Agric. Ext. Serv. Circ. 120, 168 pp.

Smith, W. L., Jr. 1962. Reduction of postharvest brown rot and Rhizopus decay of eastern peaches with hot water. *Plant Dis. Rep.* 46:861–65.

———. 1962. Chemical treatments to reduce postharvest spoilage of fruits and vegetables. *Bot. Rev.* 28:411–45.

———. 1971. Control of brown rot and Rhizopus rot of inoculated peaches with hot water or hot chemical suspensions. *Plant Dis. Rep.* 55:228–30.

Smith, W. L., Jr., M. H. Haller, and T. T. McClure. 1956. Postharvest treatments for reduction of brown and Rhizopus rots of peaches. *Phytopathology* 46:261–64.

Smith, W. L., Jr., H. E. Moline, and K. S. Johnson. 1979. Studies with *Mucor* species causing postharvest decay of fresh produce. *Phytopathology* 69:865–69.

Smith, W. L., Jr., R. W. Penny, and R. Grossman. 1972. Control of postharvest brown rot of sweet cherries and peaches with chemical and heat treatments. *USDA-ARS Mktg. Res. Rep.* 979.

Sommer, N. F. 1989. Management of disease resistance in harvested fruits and vegetables. Manipulating the postharvest environment to enhance or maintain resistance. *Phytopathology* 79:1377–80.

Sommer, N. F., and F. G. Mitchell. 1959. Fruit shrivel control in peaches and nectarines. *Am. Soc. Hort. Sci.* 74:199–205.

Spalding, D. H. 1966. Appearance and decay of strawberries, peaches, and lettuce treated with ozone. *USDA-ARS Mktg. Res. Rept.* 756, 11 pp.

Spotts, R. A. 1984. Environmental modification for control of postharvest decay. In *Postharvest pathology of fruits and vegetables: Postharvest losses in perishable crops.* 67–72. Univ. Calif. Div. Agric. Exp. Stn. Publ. NE-87.

Stanton, T. H., and J. M. Ogawa. 1961. Susceptibility of detached Royal apricot fruits to various fungus pathogens. *Plant Dis. Rep.* 45:659.

Strand, L. L., J. M. Ogawa, E. Bose, and J. W. Rumsey, 1981. Bimodal heat stability curves of fungal pectolytic enzymes and their implication for softening of canned apricots. *J. Food Sci.* 46:498–500, 505.

Tanaka, H., and M. W. Miller. 1963a. Microbial spoilage of dried prunes. I. Yeasts and molds associated with spoiled dried prunes. *Hilgardia* 34(6):167–70.

———. 1963b. Microbial spoilage of dried prunes. II. Studies of the osmophilic nature of spoilage organisms. *Hilgardia* 34(6):171–81.

Tate, K. G. 1973. *Nitidulid beetles as vectors of* Monilinia fructicola *in stone fruits.* Ph.D. dissertation, Department of Plant Pathology, University of California, Davis.

Tate, K. G., J. M. Ogawa, B. T. Manji, and E. Bose. 1974. Survey for benomyl tolerant isolates of *Monilinia fructicola* and *M. laxa* in stone fruit orchards of California. *Plant Dis. Rep.* 58:663–65.

Tindale, G. B., I. D. Peggie, P. T. Jenkins, et al. 1968. Postharvest control. In *Report of Research work on brown rot of stone fruit—The Brown Rot Research Committee 1957–1962.* (Austral.) Government Printer, Melbourne, 117 pp.

Togashi, K. 1949. *Biological characters of plant pathogens: Temperature relations.* Meibundo, Tokyo, 478 pp.

Van Blaricom, L. O. 1959. The effect on decay of adding various reagents to the water for hydrocooling peaches. *So. Car. Agric. Exp. Stn. Circ.* 124.

Wade, G. C. 1956. Investigations on brown rot of apricots caused by *Sclorotinia fructicola* (Wint.) Rehm. I. The occurrence of latent infection in fruit. *Austral. J. Agric. Res.* 7:504–15.

Weber, D. J., and J. M. Ogawa. 1965. The mode of action of 2,6-dichloro-4-nitroaniline in *Rhizopus arrhizus. Phytopathology* 55:159–65.

Webster, R. K., J. M. Ogawa, and E. Bose. 1970. Tolerance of *Botrytis cinerea* to 2,6-dichloro-4-nitroaniline. *Phytopathology* 60:1489–92.

Webster, R. K., J. M. Ogawa, and C. J. Moore. 1968. The occurrence and behavior of variants of *Rhizopus stolonifer* tolerant to 2,6-dichloro-4-nitroaniline. *Phytopathology* 58:997–1003.

Wells, J. M. 1970. Postharvest hot water and fungicide treatments for reduction of decay of California peaches, plums, and nectarines. *USDA Agric. Market. Res. Rept.* 908, 10 pp.

———. 1972. Heated wax-emulsions with benomyl and 2,6-dichloro-4-nitroaniline for control of postharvest decay of peaches and nectarines. *Phytopathology* 62:129–33.

———. 1973. Postharvest wax fungicide treatments of nectarines, peaches and plums for: Reducing decay, reducing moisture loss, and enhancing external appearance. *USDA Mktg. Res. Rept.* 981.

Wells, J. M., and M.H. Gerdts. 1971. Pre- and postharvest benomyl treatments for control of brown rot of nectarines in California. *Plant Dis. Rep.* 55:69–72.

Wells, J. M., and J. M. Harvey. 1970. Combination heat and 2,6-dichloro-4-nitroaniline treatments for control of Rhizopus and brown rot of peaches, plums and nectarines. Phytopathology 60:116–20.

Wells, J. M., and M. Uota. 1970. Germination and growth of five decay fungi in low oxygen and high carbon dioxide atmosphere. Phytopathology 60:50–53.

Wiant, J. S., and C. O. Bratley. 1948. Spoilage of fresh fruits and vegetables in rail shipments unloaded at New York Citiy 1935–1942. USDA Circ. 773, 62 pp.

Wilson, C. L. 1989. Management of disease resistance in harvested fruits and vegetables. Managing the microflora of harvested fruits and vegetables to enhance resistance. Phytopathology 79:1387–90.

Wilson, C. L., J. D. Franklin, and B. E. Otto. 1987. Fruit volatiles inhibitory to Monilinia fructicola and Botrytis cinerea. Plant Dis. 71:316–19.

Wilson, C. L., and M. E. Wisniewski. 1989. Biological control of postharvest diseases of fruits and vegetables: An emerging technology. Ann. Rev. Phytopathol. 27:425–41.

Yarwood, C. E., and H. T. Harvey. 1952. Reduction of cherry decay by washing. Plant Dis. Rep. 36:389.

Young, W. J., and E. S. Beneke. 1952. Treatments to prevent fruit storage rot (abs.). Phytopathology 42:24.

Zycha, H., R. Siepmann, and G. Linnemann. 1969.Mucorales. Verlag Von J. Cramer, Lehre, 355 pp.

MISCELLANEOUS DISEASES OF STONE FRUIT

Minor fungal diseases in California are Gibberella bud rot and shoot blight (*Gibberella baccata* [Wallr.] Sacc.) of peach; damping-off and root rot (*Rhizoctonia solani* Kuehn), of cherry, peach, and European plum; fruit rots of peach (*Candida albicans* [Robin] Berkh., *Fusarium avenaceum* [Fr.] Sacc., *F. oxysporum* Schlecht., *F. solani* [Mart.] Appel & Wr., and *Trichothecium roseum* L.); fruit rot of apricot and plum (*Lambertella pruni* Whetzel); Fusarium root rot of peach (*Fusarium roseum* L.); leaf blight of peach (*Fabraea maculata* Atk.); Phloeosporella shothole of apricot (*Phloeosporella padi* [Lib.] Arx); Phyllosticta leaf spot of apricot (*Phyllosticta circumcissa* Cke.); Pythium root rot of almond and cherry (*Pythium* sp.); trunk and limb gall of apricot (*Monochaetia rosenwaldia* Khazanoff); and trunk rot of apricot (*Heterobasidion annosum* [Fr.] Bres).

Viruslike diseases encountered occasionally in California stone fruit orchards are apricot ring pox, peach blotch, peach wart, and prune diamond canker. Although these diseases have commonly been attributed to viruses, their graft-transmissible causal agents have never been identified.

Two parasitic seed plants seen rarely on California stone fruit trees are leafy mistletoe (*Phoradendron flavescens* [Pursh] Nutt.) on almond and American plum (*Prunus americana* Marsh and *P. mahaleb* L.), and European mistletoe (*Viscum album* var. *album* L.) on almond.

Several stone fruit diseases not known to occur in California are important in other regions of the United States, especially in the fruit-growing areas east of the Rocky Mountains. One of the principal reasons for their importance in the eastern states is the prevalence there of warm, humid weather during the summer, whereas in California the summers are hot and dry. However, the increasing use of sprinkler irrigation in California orchards might enable some of these "foreign" diseases to become established here and, thereby, seriously complicate the disease control program. The diseases not present in California but important in other states, their causal agents, and their approximate geographic distribution are as follows:

Diseases caused by bacteria and mycoplasmalike organisms. Bacterial spot (*Xanthomonas campestris* pv. *pruni* [E. F. Smith] Dye) attacks all cultivated stone fruits, but is especially serious on apricot, peach, and Japanese plum. It is widespread east of the Rocky Mountains. Phony peach, caused by an unnamed, fastidious, xylem-limited bacterium, is serious on peach and is found to a limited extend on some other species of *Prunus*. It is present in the southern states, from North Carolina to Texas. Peach rosette, caused by an unidentified mycoplasmalike organism, is especially prevalent on peach but occurs to some extent on several other species of stone fruit. It is of greatest importance in the southern peach-growing regions.

Diseases caused by fungi. Anthracnose or bitter rot (*Glomerella cingulata* [Ston.] Spauld. & Schrenk) is a disease of peach of moderate importance, especially in the peach-growing areas of the southern and southeastern states. The pathogen was detected on harvested peaches and nectarines by A. Feliciano (personal communication) during summer 1990 in Fresno County. Black knot of plum and cherry (*Apiosporina morbosa* [Schwein.:Fr.] Arx [=*Dibotryon morbosum* {Schwein.:Fr.} Theiss. & Syd.]) is serious on plum and sometimes on cherry in the more northerly fruit-growing regions of the eastern and central states. Clitocybe root rot (*Armillaria* [*Clitocybe*] *tabescens* [Scop.:Fr.] Dennis, Orton, & Hora) is common on peach and occasional on other stone and pome fruit in the more southerly states east of the Rocky Mountains. Fungal gummosis is a branch and trunk disease of peach in the southeastern United States that is caused by three species of *Botryospaeria* (*B. dothidea* [Moug.:Fr.] Ces. & DeNot., *B. obtusa* [Schwein.] Shoemaker and *B. rhodina* [Cooke] Arx). In California, Dothiorella canker of almond is caused by *B. dothidia* (see Chapter 3). Fusicoccum canker (*Fusicoccum amygdali* Del.) is a moderately serious disease of peach that occurs principally in the Atlantic Coast region from Massachusetts south. It has not been reported on other stone fruits in this country, but is known to occur on almond in France. A disorder of great concern in the southeastern states is peach tree short life. It is a disease complex involving such factors as freeze injury, infection by *Pseudomonas syringae* pv. *syringae*, and the presence of ring nematodes in the soil. It is similar in many respects to the bacterial canker disease of peach and other stone fruits in California (see discussion of this disease in Chapter 4).

REFERENCES

Farr, D. F., G. F. Bills, G. P. Chamuris, and A. Y. Rossman. 1989. *Fungi on plants and plant products in the United States.* APS Press, American Phytopathological Soc., St. Paul, MN. 1252 pp.

French, A. M. 1987. *California plant disease host index. Part 1: Fruits and nuts.* Calif. Dept. Food and Agric., Sacramento, CA. 39 pp.

Jones, A. L., and T. B. Sutton. 1984. *Diseases of tree fruits.* Mich. State Univ. Coop. Ext. Serv., N. Cent. Reg. Ext. Publ. 45. 59 pp.

Myers, S. C., ed. 1989. *Peach production handbook.* Cooperative Ext. Service, Univ. Georgia Coll. Agric., Athens. 221 pp.

Németh, N. 1986. *Virus, mycoplasma, and rickettsia diseases of fruit trees.* Martinus Nijhoff Publishers. 841 pp.

Pine, T. S., R. M. Gilmer, J. D. Moore, G. Nyland, and M. F. Welsh, eds. 1976. *Virus diseases and noninfectious disorders of stone fruits in North America.* Agric. Handb. 437. ARS-USDA, Washington, D. C. 433 pp.

Chapter 4

DISEASES ATTACKING SEVERAL GENERA OF FRUIT AND NUT TREES

4

Page numbers in bold indicate photographs.

BACTERIAL DISEASES OF FRUIT AND NUT TREES	246	
Bacterial Canker and Blast of Fruit and Nut Trees	246	**423, 424**
Pathovars of *Pseudomonas syringae* van Hall		
REFERENCES	254	
Crown Gall of Fruit and Nut Crops	257	**425**
Agrobacterium tumefaciens (Smith & Townsend) Conn		
REFERENCES	261	
FUNGAL DISEASES OF FRUIT AND NUT TREES	263	
Phytophthora Root and Crown Rot of Fruit and Nut Trees	263	**425**
Phytophthora spp.		
REFERENCES	271	
Postharvest Decay: Mycotoxins	274	
Miscellaneous Fungi		
REFERENCES	275	
Replant Diseases of Fruit Trees	276	**426**
Various Causal Factors		
REFERENCES	279	
Shoestring (Armillaria) Root Rot of Fruit and Nut Trees	280	**427**
Armillaria mellea (Vahl:Fr.) Kummer		
REFERENCES	283	
Silver-Leaf Disease of Fruit and Nut Trees	285	**428**
Chondrostereum purpureum (Pers.:Fr.) Pouz.		
REFERENCES	286	
Verticillium Wilt of Fruit and Nut Trees	287	**428**
Verticillium dahliae Kleb.		
REFERENCES	291	
Wood Decay of Fruit and Nut Trees	292	**429, 430**
Basidiomycetous Fungi		
REFERENCES	295	

Bacterial Diseases of Fruit and Nut Trees

Bacterial Canker and Blast of Fruit and Nut Trees

Pathovars of *Pseudomonas syringae* van Hall

The disease variously known as "bacterial canker," "bacterial gummosis," and "bacterial blast" attacks most cultivated species of *Prunus* and a few species of *Pyrus* and *Malus*. It has been found on these hosts in many parts of the world. It is particularly severe on stone fruit in England, Europe, New Zealand, Chile, South Africa, and the southeastern and Pacific Coast regions of the United States (Cameron 1962b; Dye 1954a; Klement 1977; Nyczepir et al. 1983; Weaver, Wehunt, and Dowler 1974; Wilson and Ogawa 1979). The blossom-blast phase of the disease occasionally occurs on pear in the coastal regions and the Sierra foothills of California and on almond and stone fruit in the interior valleys.

Symptoms on stone fruit and almond (Cameron 1962b; Wilson and Ogawa 1979). The most destructive form of the disease occurs as roughly elliptical cankers on trunks and scaffold branches, or as complete necrosis of one or more branches or the entire tree (**fig. 1E,I**). In most *Prunus* species, gum is exuded profusely near the canker margin (**fig. 1G**). Under the outer bark, pale, water-soaked reddish brown streaks and flecks are common in the phloem (**fig. 1H**), extending from the upper and lower apices of the canker. When complete necrosis of a branch or the trunk occurs, a watery exudate commonly appears (**fig. 1F**), the sour smell of which accounts for the term "sour-sap." The necrosis may extend down the trunk but seldom very far below ground. Thus, although the aboveground parts of the tree may be killed, the basal portion of the trunk and the roots often remain alive and frequently send up new shoots (**fig. 1J**). Individually affected branches may fail to start growth in the spring, or they may produce flowers and leaves and then die later in the spring. Killed areas are sometimes invaded by *Cytospora leucostoma* Sacc., which may kill branches or the trunk later in the season (**fig. 1J**).

The dormant or semidormant buds of infected *Prunus* species may fail to grow in the spring (**fig. 1D**). They are brown and often covered with gum. Such symptoms are similar to those caused by the shothole fungus *Wilsonomyces carpophilus* (*Stigmina carpophila*) on apricot and peach. The blossoms of stonefruit and especially almond may suddenly wither and turn dark brown (**fig. 1A**). Necrosis may extend into the supporting branch to form a small canker that may exude gum.

The pathogen sometimes attacks the young green fruit of cherry (**fig. 1C**) and other stone fruit. In such cases, it forms small, dark brown, sunken lesions that may severely malform the fruit at maturity (Cameron 1962b; Wilson and Ogawa 1979).

Roughly circular to somewhat angular lesions 2 to 4 mm in diameter may develop on the leaves in early spring (**fig. 1B**). A yellowish halo usually surrounds the brown necrotic center of the lesions. Eventually, many of the infected areas drop out to produce a shothole appearance. A sudden withering of the ends of the leafy terminals sometimes occurs in early spring. In California, this phase of the disease appears to be most common in plum, especially in the Duarte, Santa Rosa, and Wickson cultivars, but it seldom is of economic importance. In England, however, the more favorable climate occasionally results in serious damage (Crosse 1954; Wilson and Ogawa 1979).

Symptoms on pear (Cameron 1962b; Dye 1956a; Wilson and Ogawa 1979). On this host the most damaging phase of the disease is the blasting of blossom and leaf clusters in the early spring. Brown to black necrotic areas develop on the receptacle and other flower parts and on pedicels and fruit-cluster bases. If the fruit buds are infected during the green-tip or tight-cluster stage, they stop growing and drop prematurely. The affected blossom and leaf clusters turn dark brown, but necrosis seldom extends far into the spur or supporting branch. This and the lack of bacterial exudate on the blossom parts distinguishes bacterial blast from fireblight. The invasion and killing of flowering spurs can lead to a shortage of fruiting sites the following year. When spurs or their supporting branches are invaded, the periderm layer of affected areas separates from the underlying tissue and, upon drying, develops a papery appearance. This is an important diagnostic feature. Blast usually is most abundant on the lower limbs.

The infected areas on pear fruit appear as black depressed spots that may cause marked fruit malformation. On leaves, a red ring frequently surrounds the infected area. Later this area becomes dry and falls away, leaving a small hole.

A blighting of dormant buds similar to that in stone fruit occurs infrequently in pears. The thickened ends of spurs that have borne fruit may become infected, and a section through such tissue reveals a watersoaked area extending back from the fruit-stem scars (Cameron 1962b; Crosse 1966; Wilson and Ogawa 1979).

Branch cankers may occur on certain pear cultivars and result mostly from the progression of the necrosis from infected blossoms and spurs. Other cankers, however, may be centered at nodes—infection apparently occurring through dormant buds or leaf-scars—or at pruning wounds (Cameron 1962b; McKeen 1955; Wilson 1936). In active cankers, the cortex is light tan or buff, and brownish streaks may extend upward and downward through the cortex and outer phloem for several inches beyond the outwardly visible canker. The infected tissue is spongy, cracked, and sometimes loosened. The branch may be weakened because the diseased area is commonly confined to the outer cortex, but usually the branch will not be killed. In California the cankers are active in late fall, winter, and early spring (Wilson 1936).

Causal organisms. Griffin in 1911 was the first person in the United States to demonstrate that a bacterium, *Pseudomonas cerasus* Griffin, was the cause of a gummosis spur- and bud-killing disease of cherry. This western Oregon report was followed by a more extensive study by Barss (1918), who showed that a gummosis and canker disorder of peach and prune was caused by the same organism. In the same year, Barrett first reported a similar bacterial disease of apricot in California. Wormald described two organisms that caused canker and shoot blight of cherry and plum in England. One he named *P. prunicola*, the other *P. morsprunorum* (Wilson and Ogawa 1979). Subsequently, Wilson (1931) showed that *P. prunicola* was similar to the organism causing bacterial canker of stone fruit in California and to the one described by Griffin as *P. cerasus*. More recently in France, Prunier, Luisetti, and Gardan (1970) described *P. mors-prunorum* f. sp. *persicae*, a new biotype pathogenic only to peach. Also in France, Gardan et al. (1973) reported that both *P. syringae* and *P. viridiflava* (Burkholder) Dowson cause cankers in apricot. The pseudomonads reported to cause canker diseases in pome fruits include *P. syringae*, *P. papulans*, and *P. eriobotryae* (see Chapter 2).

Pseudomonas syringae is the name of a fluorescent plant pathogen originally isolated from lilac, but subsequently found on a wide range of annual and perennial host plants (Canfield, Baca, and Moore 1986; Opgenorth et al. 1983). Many species related to *P. syringae* have also been isolated and identified. Most are limited in their host range to a single genus or family of plants, and most are characterized by the production of a green fluorescent pigment.

In accordance with a recent listing of the pathovars of *P. syringae* (Dye et al. 1980), the pseudomonads causing canker and blast of stone and pome fruit are as follows: *P. syringae* pv. *syringae* on stone fruit, including almond, and on apple and pear, worldwide; *P. syringae* pv. *morsprunorum* on stone fruit, mainly in Great Britain, Europe, the northeastern United States, and adjacent Canada; *P. syringae* pv. *persicae* on peach in France; *P. syringae* pv. *papulans* on apple and *Malus pumila* in the eastern United States and Ontario, Canada; and *P. syringae* pv. *eriobotryae* on loquat in Japan, New Zealand, and California. In Chile, *P. cichorii* is associated with a "gummy spot" disorder of nectarine fruit (Pinto and Carreño 1983). Two species of questionable validity are *P. viridiflava* (Burkholder) Dowson on apricot in France and *P. amygdali* Psallidas and Panagopoulos on almond in Afghanistan, Greece, and Turkey. In California, only pv. *syringae* is of commercial importance.

Pseudomonas syringae pv. *syringae* is a gram-negative rod, 0.7 to 1.2 by 1.5 to 3.0 μm, with one or more polar flagella, occurring singly, in pairs, or in short chains. It is a strict aerobe, oxidase-negative, and capable of using a large number of compounds as energy sources. Most isolates are positive for gelatin liquefaction and aesculin hydrolysis but negative for tyrosinase activity and tartrate utilization. On certain media, such as King's medium B, the organism produces a green fluorescent, water-soluble pigment (King, Ward, and Raney 1954). Optimum temperature for growth is 25° to 30°C, but some strains are able to reproduce at temperatures as low as 4°C. Most isolates induce a hypersensitive reaction in tobacco leaves, produce the toxin syringomycin, and can serve as ice-nucleating agents. Physiological tests, serology, and syringomycin production have aided in distinguishing pv. *syringae* from pv. *morsprunorum*, the other commonly occurring biotype on stone fruit (Latorre and Jones 1979; Seemüller and Arnold 1978). Pathogenicity to Lovell peach seedlings can be used to separate these two pathovars from most other species of *Pseudomonas* (Otta and English 1971). A bioassay using peach seedling cotyletons to detect pathogenicity among pathovars of *P. syringae* and strains of pv. *syringae* was developed by Endert and Ritchie (1984b).

Disease cycle. The bacteria, when not functioning as pathogens within the infected plant, live as omnipresent epiphytes on the aerial parts of the host (Crosse 1966; English and Davis 1960; Latorre and Jones 1979; Wilson 1934). The pathogen has also been isolated from the aerial parts of several woody nonhost plants, from plant debris, and from the surface of healthy weeds in the orchard (English and Davis 1960, Latorre and Jones 1979b; Waissbluth and Latorre 1978). Whether or not the bacteria on nonhost plants and orchard debris play a role in the disease cycle has not been determined. Under favorable conditions, the pathogen appears able to multiply epiphytically on its host and then to be disseminated largely by rain to suitable infection courts (Crosse 1966; Latorre et al. 1985; Wimalajeewa and Flett 1985).

For the canker phase of the disease, leaf scars are thought to be the principal sites of infection (**fig. 1D**). There is strong evidence for this mode of penetration in peach with pv. *syringae* in California (Davis and English 1969) and New Zealand (Dye 1954a), with pv. *persicae* in France (Vigouroux 1976), and in cherry with pv. *morprunorum* in England (Crosse 1966). However, Endert and Ritchie (1984a) obtained negative results from bud and leaf scar inoculation of peach in North Carolina. Studies in Oregon indicate that the dormant buds of sweet cherry are important infection sites for the *syringae* pathovar (Cameron 1962a), and research in California (H. English and J. R. Davis, unpublished) has demonstrated bud infection and subsequent canker development in both sweet cherry and almond. Bud infection in cherry and peach is also reported from France (Prunier and Luisetti 1983). Pruning wounds and other injuries can also serve as penetration sites (Crosse 1966; Dowler and Petersen 1966; Dye 1954a; Klement 1977; Otta and English 1970), but there is no evidence that wounds play an important role in the disease syndrome in California.

During the cool moist weather of spring, infection may occur in opening blossom and leaf buds and in leaves, blossoms, green shoots, fruit and fruit stems; this gives rise to the blast phase of the disease (**fig. 1A,B**) (Crosse 1966; Dye 1954a; English, Ogawa, and Davis 1957; Wilson and Hewitt 1939). Blast infection is thought to occur mainly through stomata or injuries caused by frost (Crosse 1966). Cankers induced by infection in late autumn and winter develop during tree dormancy and early spring; the tree then becomes resistant to infection and canker development until late fall (Cameron 1962b; Crosse 1966; Davis and English 1969; Wilson 1939). Most cankers are annual rather than perennial, the infected tissues being effectively walled off by callus during mid to late spring (Crosse 1966; Wilson 1939). Most bacteria die in arrested cankers during the summer (Endert-Kirkpatrick and Ritchie 1988; English et al. 1980), but several workers (Crosse 1966; Endert and Ritchie 1984a; McKeen 1955; Otta and English 1970; Wilson 1939) have reported finding viable bacteria in or adjacent to such cankers during this period. They indicated that some of these summer-holdover cankers may become active again in late fall. Blast infections sometimes give rise to small cankers that apparently do not reactivate during the following dormant period. Cameron (1970) has reported the systemic occurrence of pv. *syringae* in apparently healthy cherry trees in Oregon, and speculated that these internal bacteria give rise to the canker disease.

Hattingh, Roos, and Mansvelt (1989), in recent South African research, provide additional evidence supporting the systemicity of both pv. *syringae* and pv. *morsprunorum* in stone fruit. They believe that systemic invasion plays an extremely important role in the disease cycle under their environmental conditions. They also showed for the first time that pv. *syringae* can be seedborne in peach seeds. Although limited attempts to demonstrate systemic invasion by pv. *syringae* in California stone fruit have yielded negative results, the serious nature of this disease here would appear to justify a more thorough investigation.

Factors related to canker development. Bacterial canker exemplifies the important role that environmental factors, both physical and biological, can play in the development of a plant disease. The factors related to the canker phase have been investigated in California and elsewhere for many years. Those pertinent to stone fruit are briefly discussed here.

Moisture. Crosse (1966), in reference to bacterial canker, stated that "rain provides simultaneously for the mobilization, distribution, and penetration of inoculum, and is the most important environmental factor in infections of stone-fruit trees." Since the pathogen is an epiphyte on aerial parts of the tree, moisture also is necessary for the buildup of inoculum (Latorre et al. 1985; Wimalajeewa and Flett 1985). The importance of quantity or periodicity of rainfall in disease development has not been ascertained. Only limited information is available on the relation of soil moisture to the disease, but Wilson (1939) reported that variations in soil moisture above the permanent wilting point did not affect progress of the disease in plum trees. Inoculation of trees held at moisture levels below this point failed to induce cankers.

In more recent research, H. English and F. J. Schick (unpublished) found that French prune trees severely stressed for moisture during the growing season were no more susceptible to canker development (by artificial inoculation) during the subsequent dormant season that were unstressed trees. They also found that containerized trees whose roots were immersed in water for two months during the dormant season developed slightly smaller cankers from dormant-season inoculations than did trees whose roots were not immersed. In an experiment with young peach trees in sandy soil, fall irrigation early in October increased the severity of bacterial canker (English et al. 1980). Whether this increased moisture had a direct effect on susceptibility or an indirect effect by increasing the population of plant parasitic nematodes or decreasing soil nutrients was not determined.

Temperature. The geographic distribution and severity of bacterial canker is due, at least in part, to temperature. Since pathogenesis is largely associated with tree dormancy, the disease generally is of minor importance where winter temperatures are too low for bacteria to multiply. In a study of dormant-tree inoculations, Wilson (1939) found that cankers were initially larger at 21° to 23°C than at lower temperatures. However, if trees were held until after dormancy was broken, a temperature of 15.5°C was more favorable for canker extension than either higher or lower temperatures. In another experiment with dormant, artificially inoculated cherry trees, cankers developed much more rapidly on the south side of the branch than on the north side. Otta and English (1970) found a higher incidence of cankers on the south than on the north side of French prune trunks. In both of these cases, temperature appears to have been at least a contributory factor. Dye (1957), in a test with growing, artificially inoculated peach seedlings, reported an optimum temperature of 18.2°C for stem infection. In a California study (H. English and J. R. Davis, unpublished), 10 isolates of pv. *syringae* were inoculated into excised peach shoots and incubated 14 days at 12°, 21°, and 30°C; with five of the isolates, maximum infection occurred at 12°C, and with the other five the maximum was at 21°C. In a test with growing peach seedlings, the organism was more virulent at 12°C than at either 7° or 28°C (English and Davis 1969).

Recent research in both Europe (Klement et al. 1984) and the United States indicates that an interaction between *P. syringae* and freeze injury may be important in the bacterial canker syndrome. Klement, Rozsnyay, and Arsenijevic (1974) and Weaver (1978), working respectively with apricot and peach, found that artificially inoculated (pv. *syringae*) cut shoots had to be subjected to subfreezing temperatures in order to develop typical bark cankers. Similar results were obtained by Vigouroux (1979) with inoculations of pv. *persicae* into excised peach shoots. In 1981 Weaver, Gonzales, and English reported that ice-nucleating, syringomycin-positive strains of pv. *syringae* could cause cankers in growing peach seedlings, in dormant potted peach trees, and in dormant excised peach shoots either frozen (−10°C) or not frozen. The cankers were significantly larger in the frozen shoots than in those not frozen (unpublished). Cankers in twigs subjected to either −2°C or −6°C were no larger than those in nonfrozen twigs. Recent work in France (Prunier and Luisetti 1983) implicated the combination of subfreezing temperatures and ice-nucleating strains of *P. syringae* in bud infection of cherry and peach. In a study of leaf infection in sour cherry, Süle and Seemüller (1987) reported that leaves became infected only when ice previously had been formed in their intercellular spaces. Freezing predisposed the leaves to infection only if inoculation with *P. syringae* pv. *syringae* occurred within 20 minutes of thawing. In a somewhat similar study, Vigouroux (1989) showed that the ingress and spread of *P. syringae* pv. *persicae* in artificially inoculated stems of peach and apricot was promoted by frost-related water-soaking of the tissues. In California, the role of subfreezing temperatures and ice nucleation in pathogenesis remains to be determined.

Research has shown that peach seedlings subjected to chilling injury conditions (25 to 30 days at 6°C) are more susceptible to pv. *syringae* than unchilled seedlings (English and Davis 1969). Defoliating the seedlings before chilling reduced their susceptibility both to chilling injury and to infection. Chilling injury may or may not play a role in canker development under orchard conditions.

Nutrition and soil pH. British research (Crosse 1966) showed no consistent relation between nutrition and canker (pv. *mors-prunorum*) development in plum trees. Wilson (1939) found no evidence that nitrogen increased the resistance of trees to bacterial canker, but it appeared that affected trees recovered better with additional nitrogen. In a California test with peaches in sandy soil, the trees in low-nitrogen plots had significantly more disease than those given adequate nitrogen (English et al. 1961). There was even less disease when adequate nitrogen was combined with phosphorus and potassium. Adding large amounts of lime, magnesium oxide, or dolomite to the soil as a supplement to the nitrogen did not enhance canker control. Daniell and Chandler (1976) reported that peach seedlings grown in liquid culture deficient in iron were smaller and more susceptible to bacterial canker than those not deprived of iron.

In a recent French study (Vigouroux, Berger, and Bussi 1987), induced infection was correlated with high levels of potassium and low levels of calcium in the branch cortex of young peach trees. A high gravel:soil ratio also increased the susceptibility of the trees to infection.

Evidence shows that low soil pH is conducive to the development of bacterial canker (Vigouroux and Huguet 1977; Weaver and Wehunt 1975) and also peach tree short life (McGlohon 1982), which has bacterial canker as one of its components. In California, bacterial canker of peach and almond is usually most severe in soils with a low pH, but this relationship has not been established in other *Prunus* species. Attempts to reduce the severity of bacterial canker or peach tree short life by raising the soil pH with lime or other alkaline materials have met with variable results. Weaver and Wehunt (1975) reduced the susceptibility of peach seedlings to bacterial canker by raising the soil pH from 5.6 to 6.4 or 7.2 with calcium carbonate; McGlohon (1982) reported that adding lime to the acid soils of the southeastern United States decreases losses from peach tree short life. Other researchers (English et al. 1961; Vigouroux and Huguet 1977; Wehunt, Horton, and Prince 1980) had less success with lime in reducing the severity of these disorders. There is conflicting evidence regarding the effect of soil pH on the population of *Criconemella xenoplax*, a nematode that predisposes stone fruit to bacterial canker and peach to peach tree short life (English et al. 1982; Lownsbery et al. 1977; Weaver, Wehunt, and Dowler 1974; Wehunt, Weaver, and Doud 1976).

Nematodes. The severity of bacterial canker in California and peach tree short life in the southeastern United States is markedly increased by the ring nematode, *Criconemella xenoplax* (English et al. 1982; Lownsbery et al. 1973, 1977; Mojtahedi and Lownsbery 1975; Nyczepir et al. 1983; Reilly et al. 1986; Weaver, Wehunt, and Dowler 1974; Zehr, Miller, and Smith 1976). It is not known whether other nematodes predispose trees to these disorders, but evidence to date indicates that neither *Paratylenchus neoamblycephalus* Geraert (English et al. 1982) nor *Tylenchorhynchus claytonii* Steiner (Nyczepir et al. 1983; Wehunt, Horton, and Prince 1980) is involved. How ring nematodes predispose stone-fruit trees to bacterial canker by feeding on roots has not been determined. Compared to trees in fumigated soil, Carter (1976) found elevated levels of indoleacetic acid in the vascular cambium of peach trees in unfumigated soil containing *C. xenoplax*. He suggested that an imbalance of growth hormones causes the early breaking of dormancy and the resultant predisposition of trees to cold injury. It is possible, at least under some conditions, that the predisposed tissues interacting with ice nucleation induced by *P. syringae* could be involved in the bacterial canker syndrome.

Time of pruning. There appears to be no consistent correlation between the time of pruning and the development of bacterial canker. In South Carolina and Georgia, peach trees pruned in autumn had more bacterial canker than those pruned in late winter or spring (Chandler and Daniell 1976; Dowler and Petersen 1966). Another Georgia test, however, found no correlation between time of pruning and death of peach trees from this disease (Weaver, Wehunt, and Dowler 1974). But in general, peach tree short life—largely the result of winter injury and bacterial canker—in the southeastern states is more damaging in trees pruned before January than in those pruned later (Ritchie and Clayton 1981).

In California, French prune trees pruned and inoculated in early November developed smaller cankers than those pruned and inoculated in late December or mid-March (Otta and English 1970). Another French prune experiment (English et al. 1974) found no correlation between pruning in autumn, winter, or spring and the severity of bacterial canker. In young California apricot trees, however, a five-year pruning test showed that trees pruned in November and December had significantly more infection than those pruned in January, February, or March. Tree mortality decreased from 22 percent in the November and December prunings to 7 percent in the March pruning (D. E. Ramos, H. English, and W. J. Moller, unpublished). In Hungary, apricot trees pruned in winter are reported to suffer more damage from bacterial canker than those pruned in spring (Klement 1977).

Rootstocks and cultivars. That the rootstock and the method of budding have an effect on the severity of bacterial canker has been observed for many years in England and the United States (Cameron 1962b, 1971; Day 1947, 1951; English et al. 1980, 1983; Grubb 1944; Montgomery, Moore, and Hoblyn 1943). In England, the budding of susceptible plum cultivars on the scaffolds of Myrobalan B has minimized the damage. In California, plums on Lovell peach root and French prune on both Lovell and Nemaguard root are less affected by bacterial canker than those on Myrobalan 29C or Marianna 2624. In Oregon, Italian prune develops less disease on Lovell than on plum rootstocks. California studies indicate that apricot is damaged most on plum rootstock and least on peach. In England, Oregon, and California, scaffold budding of sweet cherry on mazzard rootstocks, as opposed to low budding, has resulted in less disease. Scaffold budding on mahaleb rootstock also is effective. Sweet cherry appears to suffer more from canker when propagated on Stockton Morello rootstock than on mazzard. In the southeastern United States, the disease has been minimized by propagating peach on either Lovell or selection NA 8 peach rootstock (Weaver, Doud, and Wehunt 1979; Zehr, Miller, and Smith 1976). With the Redhaven scion budded on different rootstocks, the natural incidence of bacterial canker was positively correlated with disease incidence in the artificially inoculated rootstock selections (Weaver, Doud, and Wehunt 1979).

Although some information is available on the comparative susceptibility to bacterial canker of stone-fruit cultivars, the marketplace usually determines cultivar selection. Among California plums, Wilson (1939) reported that President and Duarte were highly susceptible and Beauty and Kelsey were highly resistant. In general the European type plums were more susceptible to the disease than were the Japanese types. In England, Victoria is considered to be one of the most susceptible cultivars (Montgomery, Moore, and Hoblyn 1943). In sweet cherry, where both pv. *syringae* and pv. *mors-prunorum* may be important, there is little agreement on cultivar susceptibility, although in Oregon, Corum is thought to be most resistant (Allen and Dirks 1978; Cameron 1962b, 1971). In almond the blast phase is usually more severe in early blooming cultivars such as Ne Plus Ultra than in the later blooming types. This is probably because frost damage (a predisposing factor for blast) is generally more severe in the early cultivars. There is no reliable information on the relative susceptibility of cultivars of almond, apricot, and peach to the canker phase of the disease.

Toxin production. Many plant-pathogenic strains of *P. syringae* produce a broad-spectrum biocide known as syringomycin that is thought to play a role in pathogenesis (DeVay et al. 1968; Gross and DeVay 1977; Paynter and Alconero 1979; Sinden, DeVay, and Backman 1971). A second biocide, syringotoxin, has been found in cultures of a citrus isolate of this organism (DeVay, Gonzalez, and Wakeman 1978; Gross and DeVay 1977). Both of these toxins are antibiotic in vitro to several fungi, including the commonly used assay fungus *Geotrichum candidum* Link ex Fries. The toxins induce symptoms in some of their hosts that resemble those caused by *P. syringae*; some evidence indicates that non-toxin-forming strains are not pathogenic. Other research, at least with syringomycin-producing isolates, does not show a correlation between toxin production and virulence (Baigent, DeVay, and Starr 1963; Latorre and Jones 1979; Otta and English 1971; Seemüller and Arnold 1978; Weaver, Gonzalez, and English 1981). Evidence (Latorre and Jones 1979la; Paynter and Alconero 1979; Seemüller and Arnold 1978) indicates that pv. *mors-prunorum* does not produce syringomycin or syringotoxin, but one report (Filipek and Powell 1971) indicated the in vitro production of a different biocide by this pathovar. Only further research can elucidate the role of these toxins in pathogenesis in *Prunus* species.

Other factors. In California and the southeastern United States (Ritchie and Clayton 1981; Weaver, Dowler, and Nesmith 1976) bacterial canker and peach tree short life are more common on land where peaches have been grown previously than on newly cleared land. This increase in disease could be related to a higher population of ring nematodes in the old orchard site (Weaver, Wehunt, and Dowler 1974), the release of a toxic substance from dead peach roots (Chandler and Daniell 1974), or some unknown factor. The physical characteristics of the soil also appear to have a bearing on the incidence and severity of bacterial canker. In California, trees on light, sandy soil are almost invariably damaged more by the disease than those on heavier soils (English et al. 1980). French workers (Vigouroux and Huguet 1977) also reported a correlation between soil type and susceptibility to bacterial canker: potted peach trees on a sandy-gravelly soil (pH 5.1) were damaged more by inoculations with pv. *persicae* than trees on a calcareous loam-clay soil (pH 8.3); susceptibility was not markedly reduced by raising the pH of the acidic soil to 7.6. Observations in California (English et al. 1980) indicate that peach and prune trees in shallow soils above a hardpan have less vigor and are more subject to bacterial canker than trees on deeper soils. Other research (English and Davis 1969) suggests that growth regulators may be involved in host-plant reaction to this disease. Chilled and unchilled peach seedlings sprayed with a solution of gibberellic acid prior to inoculation with pv. *syringae* were more resistant to canker development than seedlings sprayed with water. Only further research can determine the role, if any, of this or other growth regulators in the development of bacterial canker in nature.

Factors related to blast development. Relatively little is known of the factors related to the development of the blast phase of the disease. Cold, wet weather during the period from bud break through bloom is thought to be the most important predisposing factor. Blast usually is most severe in the lower parts of the tree and in the lowest and, usually, coldest part of an orchard. It also is generally more prevalent in almond—and in the earliest blooming cultivars—than in the later-blooming species of *Prunus*. Blast of almond has been significantly reduced by protecting trees in early spring against frost (English et al. 1980). In another experiment, cut blossoming shoots of almond (Ne Plus Ultra and Nonpareil) subjected to −4°C for two hours were more susceptible to blast (spray-inoculated with pv. *syringae*) than unfrosted shoots. In another test, cut, leafing-out shoots of French prune were atomized with a suspension of an ice-nucleating isolate of pv. *syringae* either before or after being subjected to a temperature of −5°C; other shoots were inoculated but not chilled. Damage to the leaf clusters was significantly greater in the treatment in which the shoots were inoculated and then chilled. It appears that the interaction of ice-nucleating bacteria and freezing temperatures is an important factor in the blast syndrome.

Control of canker. Although no single practice will adequately control the canker phase of this disease, a number of measures can markedly reduce its severity:

Site selection. Orchard sites should not have acidic or sandy soils or shallow soils above a hardpan (English et al. 1980). Also, if possible, it is best to select new land or a site in which stone fruits have not been grown in recent years (Weaver, Dowler, and Nesmith 1976; Weaver, Wehunt, and Dowler 1974).

Vigor. Tree vigor should be maintained by using proper fertilization and irrigation practices. In a California test with young peach trees on sandy soil, there was a decreasing order of canker severity in trees that received, respectively, low nitrogen, high nitrogen, and high nitrogen plus potassium and phosphorus (English et al. 1961). The liming of acid soils in the southeastern United States has resulted in increased vigor and resistance to bacterial canker and peach tree short life (Ritchie and Clayton 1981; Weaver and Wehunt 1975). In California, however, adding lime or magnesium oxide to an acid soil did not reduce the severity of bacterial canker in young peach trees (English et al. 1961).

Rootstock and cultivar selection. As pointed out earlier, the severity of bacterial canker is influenced by both scion cultivar and rootstock. Where feasible, resistant or tolerant cultivars should be selected, and rootstocks that are resistant or that contribute to the resistance of the scion also should be employed. In sweet cherry, plum, and prune, the budding of susceptible cultivars on the scaffold branches of resistant rootstocks has been a beneficial practice (Cameron 1962b; Day 1951; English et al. 1980, 1983).

Bactericidal sprays. Although experiments to control bacterial canker with bactericidal sprays have been conducted in many countries, success has been limited. In Michigan, epiphytic populations of *P. syringae* pv. *morsprunorum* on Montmorency cherry were reduced by spraying with fixed copper materials in spring and early summer (Olson and Jones 1983). In Oregon sweet cherries, the "dead-bud" phase of the disease has been effectively controlled by two autumn applications of Bordeaux mixture (Cameron 1960, 1962b); streptomycin sprays were ineffective. Crosse (1962) reported that two to three fall applications of either streptomycin or Bordeaux mixture provided effective control of the canker phase in sweet cherry in England. Moore (1946), however, found no evidence that fall applications of Bordeaux mixture reduced the incidence of canker in plum trees. In France (Prunier, Luisetti, and Gardan 1974) autumn applications of copper compounds were ineffective in controlling the disease in peach, but sprays of either oxytetracycline or kanamycin gave good control. In New Zealand, fall and winter applications of copper materials and zinc sulfate were reported to control bacterial canker of apricot (D. W. Wilson 1958; Young and Paulin 1982). Dye (1954a) found streptomycin sulfate to be far superior to Bordeaux mixture in protecting wounded peach trees from infection when artificially inoculated. Two New Zealand reports (Dye 1958; Young 1977), however, indicate the occurrence of streptomycin-resistant variants of *P. syringae*.

In California, spray applications of Bordeaux mixture and streptomycin have provided either no control or only partial control of canker (English et al. 1974, 1980; English and Hansen 1954). In tests with plums, one or two applications of Bordeaux mixture at leaf fall reduced the number of cankers by 25 to 50 percent. Control was not enhanced by an additional application immediately after dormant pruning. With French prune, neither two leaf-fall applications nor a single bud-break application of either Bordeaux mixture or streptomycin significantly reduced canker severity. In another French prune test, Bordeaux mixture applied three times (at leaf fall, immediately after pruning, and at bud break) also failed to control the disease. In peach, Bordeaux mixture applied at early and late leaf fall moderately reduced twig infection and tree mortality. Streptomycin applied at the same time was ineffective. Control was not enhanced by spraying with a defoliant before the leaf-fall applications of streptomycin or Bordeaux mixture. No canker control was obtained when streptomycin and Bordeaux mixture were applied at bud swell and pink bud. Spraying peach or French prune trees with whitewash in either fall or early spring also has yielded negative results. In Australia multiple applications of copper sprays to sweet cherry during fall and winter prevented infection of scaffold branches (D. L. S. Wimalajeewa, personal communication).

Soil fumigation. Preplant soil fumigation treatments with materials such as chloropicrin, methyl bromide, and mixtures containing 1, 3-dichloropropene have yielded reduced severity of canker in young trees and increased tree growth (Davis and English 1969; English and DeVay 1964; English et al. 1961, 1980, 1983; Zehr and Golden 1986). The decrease in canker severity appears to result largely from a reduction in the amount of infection. Preplant soil fumigation has been used effectively for several years to reduce the damage to peach, almond, and French prune trees in California's Central Valley. However, in a Sonoma County test neither pre- nor postplant fumigation proved beneficial. The benefit from preplant fumigation usually lasts for two to four years, but in one test French prune trees were protected for at least seven years. Before dibromochloropropane was prohibited as a soil fumigant, this compound was employed as a postplant treatment every two to three years to maintain tree vigor and reduce susceptibility to bacterial canker. Fumigation appears to succeed largely by suppressing the ring nematode, *Criconemella xenoplax* (English et al. 1980, 1982; Lownsbery et al. 1977). Peach tree short life (of which bacterial canker is an important component) is also controlled, at least in part, by soil fumigation (Ritchie and Clayton 1981; Zehr, Lewis, and Gambrell 1982). Research (Ritchie 1984, 1989) indicates that fenamiphos may be a suitable replacement for the postplant fumigant dibromochloropropane in the control of peach tree short life. Whether this material can be used effectively to control bacterial canker in California remains to be determined.

Pruning. In California there appears to be no consistent correlation between the time of pruning and the incidence and severity of bacterial canker in stone-fruit trees. Young apricot trees seem to suffer greatest damage from canker when pruned in November or December and least damage when pruned in March. Pruning tests in French prune and observations in peach have failed to show a relationship between time of pruning and canker development. In the southeastern United States, peach trees usually are damaged more by bacterial canker and peach tree short life when pruned in fall or early winter than in late winter or early spring (Chandler and Daniell 1976; Ritchie and Clayton 1981). No information is available on the relationship of pruning to canker development in other species of *Prunus*.

Control of blast. Little information is available on the control of the spring blast phase of the disease in stone fruits. The common relationship between spring frost and the severity of blast suggests that frost-control measures should be of value. Also, because of both the pathogenic and ice-nucleating capacities of the pathogen, reducing populations of the organism on trees in late winter and early spring by either chemical or biological means would appear to offer promise. However, an attempt to control almond blast by bud-swell and popcorn applications or a leaf-fall application of either Bordeaux mixture or streptomycin was totally unsuccessful. The control of pear blast with early spring applications of these same materials (Bethel et al. 1977; Dye 1956a) suggests the need for further trials with *Prunus* species.

REFERENCES

Allen, W. R., and V. A. Dirks. 1978. Bacterial canker of the sweet cherry in the Niagara Peninsula of Ontario: *Pseudomonas* species involved and cultivar susceptibility. *Can. J. Plant. Sci.* 58:363–69.

Baigent, N. L., J. E. DeVay, and M. P. Starr. 1963. Bacteriophages of *Pseudomonas syringae*. *N. Z. J. Sci.* 6:75–100.

Barrett, J. T. 1918. Bacterial gummosis of apricots—Preliminary report. *Calif. State Comm. Hort. Monthly Bull.* 7:137–40.

Barss, H. P. 1918. Bacterial gummosis of stone fruits. *Calif. State Comm. Hort. Monthly Bull.* 7:121–36.

Bethel, R. S., J. M. Ogawa, W. H. English, R. R. Hansen, B. T. Manji, and F. J. Schick. 1977. Copper-streptomycin sprays control pear blossom blast. *Calif. Agric.* 31(6):7–9.

Cameron, H. R. 1960. Death of dormant buds in sweet cherry. *Plant Dis. Rep.* 44:139–43.

———. 1962a. Mode of infection of sweet cherry by *Pseudomonas syringae*. *Phytopathology* 52:917–21.

———. 1962b. *Diseases of deciduous fruit trees incited by Pseudomonas syringae van Hall.* Ore. Agr. Exp. Stn. Tech. Bull., 66 pp.

———. 1970. *Pseudomonas* content of cherry trees. *Phytopathology* 60:1343–46.

———. 1971. Effect of root or trunk stock on susceptibility of orchard trees to *Pseudomonas syringae*. *Plant Dis. Rep.* 55:421–23.

Canfield, M. L., S. Baca, and L. W. Moore. 1986. Isolation of *Pseudomonas syringae* from 40 cultivars of diseased woody plants with tip dieback in Pacific Northwest nurseries. *Plant Dis.* 70:647–50.

Carter, G. E., Jr. 1976. Effect of soil fumigation and pruning date on the indoleacetic acid content of peach trees in a short life site. *Hort. Sci.* 11:594–95.

Chandler, W. A., and J. W. Daniell. 1974. Effect of leachates from peach soil and roots on bacterial canker and growth of peach seedlings. *Phytopathology* 64:1281–84.

———. 1976. Relation of pruning time and inoculation with *Pseudomonas syringae* van Hall to short life of peach trees grown in old peach land. *Hort. Sci.* 11:103–04.

Crosse, J. E. 1954. Bacterial canker, leaf spot, and shoot wilt of cherry and plum. *East Malling Res. Stn. Ann. Rept.* 1953:202–07.

———. 1962. Antibiotic sprays for the control of cherry bacteriosis: The bactericidal activity of streptomycin and Bordeaux mixture on the plant surfaces. In *Antibiotics in agriculture*, 86–100. Proc. Univ. Nottingham Ninth Easter School Agric. Science.

———. 1966. Epidemiological relations of pseudomonad pathogens of deciduous fruit trees. *Ann. Rev. Phytopathol.* 4:291–310.

Daniell, J. W., and W. A. Chandler. 1976. The effect of iron on growth and bacterial canker susceptibility of peach seedings. *Hort. Sci.* 22:402–03.

Davis, J. R., and H. English. 1969. Factors related to the development of bacterial canker in peach. *Phytopathology* 59:588–95.

Day, L. H. 1947. The influence of rootstock on the occurrence and severity of bacterial canker, *Pseudomonas cerasi*, of stone fruits. *Am. Soc. Hort. Sci. Proc.* 50:100–02.

———. 1951. Cherry rootstocks in California. *Calif. Agric. Exp. Stn. Bull.* 725, 31 pp.

DeVay, J. E., C. F. Gonzalez, and R. J. Wakeman. 1978. Comparison of the biocidal activities of syringomycin and syringotoxin and the characterization of isolates of *Pseudomonas syringae* from citrus fruits. *Proc. 4th Internat. Conf. Plant Path. Bact.* 2:643–51.

DeVay, J. E., F. L. Lukezic, S. L. Sinden, H. English, and D. L. Coplin. 1968. A biocide produced by pathogenic isolates of *Pseudomonas syringae* and its possible role in the bacterial canker disease of peach trees. *Phytopathology* 58:95–101.

Dowler, W. M., and D. H. Petersen. 1966. Induction of bacterial canker of peach in the field. *Phytopathology* 56:989–90.

Dye, D. W. 1954a. Blast of stone fruit in New Zeland. *N. Z. J. Sci. Tech.* 35:451–61.

———. 1954b. Preliminary field trials to control blast of stone fruit (*Pseudomonas syringae* van Hall). *N. Z. J. Sci. Tech.* 36:331–34.

———. 1956a. Suggestions for controlling blast of stone fruit. *Orch. N. Z.* 29(4):2–3.

———. 1956b. Blast of pear. *Orch. N. Z.* 29(7):5, 7.

———. 1957. The effect of temperature on infection by *Pseudomonas syringae* van Hall. *N. Z. J. Sci. Tech.* 38:500–05.

———. 1958. Development of streptomycin-resistant variants of *Pseudomonas syringae* van Hall in culture and in peach seedlings. *N. Z. J. Agri. Res.* 1:44–50.

Dye, D. W., J. F. Bradbury, M. Goto, A. C. Hayward, R. A. Lelliott, and M. N. Schroth. 1980. International standards for naming pathovars of phytopathogenic bacteria and a list of pathovar names and pathotype strains. *Rev. Plant Pathol.* 59(4):153–60.

Endert, E., and D. F. Ritchie. 1984a. Overwintering and survival of *Pseudomonas syringae* pv. *syringae* and symptom development in peach trees. *Plant Dis.* 68:468–70.

———. 1984b. Detection of pathogenicity, measurement of virulence, and determination of strain variation in *Pseudomonas* pv. *syringae*. *Plant Dis.* 68:677–80.

———. 1988. Involvement of pH in the competition between *Cytospora cincta* and *Pseudomonas syringae* pv. *syringae*. *Phytopathology* 78:619–24.

English, H., and J. R. Davis. 1960. The source of inoculum for bacterial canker and blast of stone fruits (abs.). *Phytopathology* 50:634.

———. 1969. Effect of temperature and growth state on the susceptibility of *Prunus* spp. to *Pseudomonas syringae*. *Phytopathology* 59:1025.

English, H., and J. E. DeVay. 1964. Influence of soil fumigation on growth and canker resistance of young fruit trees in California. *Down to Earth* 20(3):6–8.

English, H., J. E. DeVay, O. Lilleland, and J. R. Davis. 1961. Effect of certain soil treatments on the development of bacterial canker in peach trees (abs.). *Phytopathology* 51:65.

English, H., J. E. DeVay, J. M. Ogawa, and B. F. Lownsbery. 1980. Bacterial canker and blast of deciduous fruits. *Univ. Calif. Div. Agric. Sci. Leaf.* 2155, 7 pp.

English, H., J. E. DeVay, F. L. Schick, and B. F. Lownsbery. 1974. Factors affecting development of bacterial canker in prune trees (abs.). *Proc. Am. Phytopathol. Soc.* 1:72.

———. 1983. Reducing bacterial canker damage in French prunes. *Calif. Agric.* 37(5 and 6):10–11.

English, H., and C. J. Hansen. 1954. Experiments on the control of bacterial canker of stone fruit trees (abs.). *Phytopathology* 44:487.

English, H., B. F. Lownsbery, F. J. Schick, and Tom Burlando. 1982. Effect of ring and pin nematodes on the development of bacterial canker and Cytospora canker in young French prune trees. *Plant Dis.* 66:114–16.

English, H., J. M. Ogawa, and J. R. Davis. 1957. Bacterial blast, a newly recognized disease of almonds in California (abs.). *Phytopathology* 47:520.

Ercolani, G. L., and A. Ghaffer. 1985. Outbreaks and new diseases. Afghanistan. Bacterial canker and gummosis of stone fruit. *FAO Plant Protect. Bull.* 33(1):37–39.

Filipek, D. M., and D. Powell. 1971. Antibiotic production by *Pseudomonas morsprunorum*. *Phytopathology* 61:892.

Gardan, L., J. P. Prunier, J. Luisetti, and J. J. Bezelgues. 1973. Etude sur les bactérioses des arbres fruitiers. VII. Responsabilité de divers *Pseudomonas* dan les dépérissement bactérien de l'abricotier en France. *Rev. Zool. Agric. Path. Veg.* 72:112–20.

Griffin, F. L. 1911. A bacterial gummosis of cherries. *Science* 34:615–16.

Gross, D., and J. E. DeVay. 1977. Production and purification of syringomycin, a phytotoxin produced by *Pseudomonas syringae*. *Physiol. Plant Pathol.* 11:13–28.

Grubb, N. H. 1944. The comparative susceptibility of high- and low-worked cherry trees in the nursery to bacterial canker. *Rept. E. Malling Res. Stn.* 1943:43–44.

Hattingh, M. J., I. M. M. Roos, and E. L. Mansvelt. 1989. Infection and systemic invasion of deciduous fruit trees by *Pseudomonas syringae* in South Africa. *Plant Dis.* 73:784–89.

King, E. O., M. K. Ward, and D. E. Raney. 1954. Two simple media for the demonstration of pyocyanin and fluorescein. *J. Lab. Clin. Med.* 44:301–07.

Klement, Z. 1977. Bacterial canker and dieback disease of apricots (*Pseudomonas syringae* van Hall). *EPPO Bull.* 7:57–68.

Klement, Z., D. S. Rozsnyay, and M. Arsenijevic. 1974. Apoplexy of apricots. III. Relationship of winter frost and the bacterial canker and dieback of apricots. *Acta Phytopathol. Acad. Scient. Hung.* 9:35–45.

Klement, Z., D. S. Rozsnyay, E. Balo, M. Panczel, and Gy. Prilesky. 1984. The effect of cold on development of bacterial canker in apricot trees infected with *Pseudomonas syringae* pv. *syringae*. *Phys. Plant Pathol.* 24:237–46.

Latorre, B. A. 1980. Cancer bacterial: Una importante enfermedad de los guindos y otros frutales. *Rev. Frut.* 1(1):9–10.

Latorre, B. A., J. A. Gonzalez, J. E. Cox, and F. Vial. 1985. Isolation of *Pseudomonas syringae* pv. *syringae* from cankers and effect of free moisture on its epiphytic populations on sweet cherry. *Plant Dis.* 69:409–12.

Latorre, B. A., and A. L. Jones. 1979a. *Pseudomonas morsprunorum* the cause of bacterial canker of sour cherry in Michigan and its epiphytic association with *P. syringae*. *Phytopathology* 69:335–39.

———. 1979. Evaluation of weeds and plant refuse as potential sources of inoculum of *Pseudomonas syringae* in bacterial canker of cherry. *Phytopathology* 69:1122–25.

Lownsbery, B. F., H. English, E. H. Moody, and F. J. Schick. 1973. *Criconemoides xenoplax* experimentally associated with a disease of peach. *Phytopathology* 63:994–97.

Lownsbery, B. F., H. English, G. R. Noel, and F. J. Schick. 1977. Influence of Nemaguard and Lovell rootstocks and *Macroposthonia xenoplax* on bacterial canker of peach. *J. Nematol.* 9:221–24.

Luisetti, J., et al. 1976. *Le dépérissement bactérien du pêcher.* Institut National Reserches Agronomiques Paris, 60 pp.

McGlohon, N. E. 1982. Management practices that are controlling peach diseases. *Plant Dis.* 66:7.

McKeen, W. E. 1955. Pear blast on Vancouver Island. *Phytopathology* 45:629–32.

Mojtahedi, H., and B. F. Lownsbery. 1975. Pathogenicity of *Criconemoides xenoplax* to prune and plum rootstocks. *J. Nematol.* 7:114–19.

Montgomery H. G. S., M. H. Moore, and T. N. Hoblyn. 1943. A field trial of measures designed for the control of bacterial canker of Victoria plum trees. *Rept. E. Malling Res. Stn.* 1942:53–61.

Moore, M. H. 1946. Bacterial canker and leaf spot of plum and cherry. A summary of present knowledge of control measures in Britain. *Rept. E. Malling Res. Stn.* 1945:134–37.

Nyczepir, A. P., E. I. Zehr, S. A. Lewis, and D. C. Harshman. 1983. Short life of peach trees induced by *Criconemella xenoplax*. *Plant Dis.* 67:507–08.

Olson, B. D., and A. L. Jones. 1983. Reduction of *Pseudomonas morsprunorum* on Montmorency sour cherry with copper and dynamics of the copper residues. *Phytopathology* 73:1520–25.

Opgenorth, D. C., M. Lai, M. Sorrell, and J. B. White. 1983. Pseudomonas canker of kiwifruit. *Plant Dis.* 67:1283–84.

Otta, J. D., and H. English. 1970. Epidemiology of the bacterial canker disease of French prune. *Plant Dis. Rep.* 54:332–36.

———. 1971. Serology and pathology of *Pseudomonas syringae*. *Phytopathology* 61:443–52.

Paynter, V. A., and R. Alconero. 1979. A specific fluorescent antibody for detection of syringomycin in infected peach tree tissues. *Phytopathology* 69:493–96.

Pinto de T., A., and I. Carreño I. 1983. *Pseudomonas cichorii* (Swingle) Stapp, asociado a "manchas gomosas" en frutos de nectarino Armking; localidad de Padre Hurtado, Chile. *Agric. Tech. (Chile)* 43(3):283–84.

Prunier, J. P., and J. Luisetti. 1983. Phytopathogenic pseudomonads behavior in natural conditions in connection with fruit tree infections. *Proc. 4th Internat. Cong. Plant Pathl. Absts. Papers*: 36.

Prunier, J. P., J. Luisetti, and L. Gardan. 1970. Etudes sur les bactérioses des arbres fruitiers. II. Caractérization d'un *Pseudomonas* nonfluorescent agent d'une bactériose novelle du pêcher. *Ann. Phytopathol.* 2:181–97.

———. 1974. Etudes sur les bactérioses des arbres fruitiers. Essais de lutte chimique contre le dépérissement bactérien du pêcher en France. *Phytiat.-Phytopharm.* 23:71–87.

Reilly, C. C., A. P. Nyczepir, R. R. Sharpe, W. R. Okie, and P. L. Pusie. 1986. Short life of peach trees as related to tree physiology, environment, pathogens, and cultural practices. *Plant Dis.* 70:538–41.

Ritchie, D. F. 1984. Control of *Criconemella xenoplax* and *Meloidogyne incognita* and improved peach tree survival following multiple fall applications of fenamiphos. *Plant Dis.* 68:477–80.

———. 1989. Improved peach tree longevity with use of fenamiphos in peach tree short-life locations. *Plant Dis.* 73:160–63.

Ritchie, D. F., and C. N. Clayton. 1981. Peach tree short life: A complex of interacting factors. *Plant Dis.* 65:462–69.

Roos, I. M. M., and M. J. Hattingh. 1983. Fluorescent pseudomonads associated with bacterial canker of stone fruit in South Africa. *Plant Dis.* 67:1267–69.

Seemüller, E., and M. Arnold. 1978. Pathogenicity, syringomycin production and other characteristics of *Pseudomonas* strains isolated from deciduous fruit trees. *Proc. 14th Internat. Conf. Plant Path. Bact.* 2:703–10.

Sinden, S. L., J. E. DeVay, and P. A. Backman. 1971. Studies on the mode of action and biogenesis of the phytotoxin syringomycin. *Physiol. Plant Pathol.* 1:199–213.

Snyder, R. L., K. T. Pau U, and J. F. Thompson. 1987. Passive frost protection of trees and vines. *Univ. Calif. Div. Agric. Nat. Resour. Leaf.* 21429. 7 pp.

Süle, S., and E. Seemüller. 1987. The role of ice formation in the infection of sour cherry leaves by *Pseudomonas syringae* pv. *syringae*. *Phytopathology* 77:173–77.

Vigouroux, A. 1976. Observations de contaminations tardives des plaies pétiolares du pêcher par *Pseudomonas mors-prunorum* f. sp. *persicae* Prunier, Luisetti, Gardan. *Ann. Phytopathol.* 8:111–15.

———. 1979. Incidence des basses températures sur la sensibilité du pêcher au dépérissement bactérien (*Pseudomonas mors-prunorum* f. sp. *persicae*). *Ann. Phytopathol.* 11:231–39.

———. 1989. Ingress and spread of *Pseudomonas* in stems of peach and apricot promoted by frost-related watersoaking of tissues. *Plant Dis.* 73:854–55.

Vigouroux, A., J. F. Berger, and C. Bussi. 1987. La sensibilité du pêcher au dépérissement bactérien en France: Incidence de certaines caractéristiques du sol et de l'irrigation. Relations avec la nutrition. *Agronomie* 7:483–95.

Vigouroux, A., and C. Huguet. 1977. Influence du substrat de culture sur la sensibilité du pêcher au dépérissement bactérien (*Pseudomonas mors-prunorum* f. sp. *persicae*. *C. R. Séances Acad. Agric. France* 63:1095–1103.

Waissbluth, M. E., and B. A. Latorre. 1978. Source and seasonal development of inoculum for pear blast in Chile. *Plant Dis. Rep.* 62:651–55.

Weaver, D. J. 1978. Interaction of *Pseudomonas syringae* and freezing in bacterial canker on excised peach trees. *Phytopathology* 68:1460–63.

Weaver, D. J., S. L. Doud, and E. J. Wehunt. 1979. Evaluation of peach seedling rootstocks for susceptibility to bacterial canker caused by *Pseudomonas syringae*. *Plant Dis. Rep.* 63:364–67.

Weaver, D. J., W. M. Dowler, and W. C. Nesmith. 1976. Association between elemental content of dormant peach trees and susceptibility to short life. *J. Am. Soc. Hort. Sci.* 101:486–89.

Weaver, D. J., C. F. Gonzales, and H. English. 1981. Ice nucleation by *Pseudomonas syringae* associated with canker production in peach. *Phytopathology* 721:109–10.

Weaver, D. J., and E. J. Wehunt. 1975. Effect of soil pH on susceptibility of peach to *Pseudomonas syringae*. *Phytopathology* 65:784–89.

Weaver, D. J., E. J. Wehunt, and W. M. Dowler. 1974. Association of tree site, *Pseudomonas syringae*, *Criconemoides xenoplax*, and pruning date with short life of peach trees in Georgia. *Plant Dis. Rep.* 58:76–79.

Wehunt, E. J., B. D. Horton, and V. E. Prince. 1980. Effects of nematicides, lime, and herbicides on peach tree short life in Georgia. *J. Nematol.* 12:183–89.

Wehunt, E. J., D. J. Weaver, and S. L. Doud. 1976. Effects of peach rootstock and lime on *Criconemoides xenoplax*. *J. Nematol.* 8:304–05.

Wilson, D. W. 1958. Stone-fruit blast experiment. *Orch. N. Z.* 31(1):32.

Wilson, E. E. 1931. A comparison of *Pseudomonas prunicola* with a canker-producing bacterium of stone fruit trees in California. *Phytopathology* 21:1153–61.

———. 1934. A bacterial canker of pear trees new to California. *Phytopathology* 24:534–37.

———. 1936. Symptomatic and etiological relations of the canker and blossom blast of *Pyrus* and the bacterial canker of *Prunus*. *Hilgardia* 10:213–40.

———. 1939. Factors affecting development of the bacterial canker of stone fruits. *Hilgardia* 12:259–98.

Wilson, E. E., and W. B. Hewitt. 1939. Host organs attacked by bacterial canker of stone fruits. *Hilgardia* 12:249–55.

Wilson, E. E., and J. M. Ogawa. 1979. *Fungal, bacterial, and certain nonparasitic diseases of fruit and nut crops in California*. University of California Division of Agricultural Sciences, Berkeley, 190 pp.

Wimalajeewa, D. L. S., and J. D. Flett. 1985. A study of populations of *Pseudomonas syringae* pv. *syringae* on stone fruits in Victoria. *Plant Pathol.* 34:248–54.

Wormald, H. 1928. Bacterial diseases of stone fruit trees in Britain. I. Preliminary note on bacteriosis in plum and cherry trees. *East Malling Res. Stn. Rept.* (Suppl. II) 1926–27:121–27.

———. 1930. Bacterial diseases of stone fruit trees in Britain. II. Bacterial shoot wilt of plum trees. *Ann. Appl. Biol.* 17:725–44.

———. 1931. Bacterial diseases of stone fruit trees in Britain. III. The symptoms of bacterial canker of plum trees. *J. Pom. Hort. Sci.* 9:239–56.

———. 1932. Bacterial diseases of stone fruit trees in Britain. IV. The organism causing bacterial canker of plum trees. *Brit. Mycol. Soc. Trans.* 17:157–69.

Wormald, H., and R. J. Garner. 1938. Manurial trial on nursery trees with reference to effect on plum to bacterial canker. *East Malling Res. Stn. Ann. Rept.* 1937:194–97.

Young, J. M. 1977. Resistance to streptomycin in *Pseudomonas syringae* from apricot. *N. Z. J. Agric. Res.* 20:249–51.

———. 1978. Survival of bacteria on *Prunus* leaves. *Proc. 14th Internat. Conf. Plant Pathol. Bact.* 2:779–86.

Young, J. M., and R. N. Paulin. 1982. A spray programme to control bacterial blast of apricot. *Orch. N. Z.* 55:413.

Zehr, E. I., and J. K. Golden. 1986. Strip and broadcast treatments of dichloropropene compared for controlling *Criconemella xenoplax* and short life in a peach orchard. *Plant Dis.* 70:1064–66.

Zehr, E. I., S. A. Lewis, and C. E. Gambrell. 1982. Effectiveness of certain nematicides for control of *Macroposthonia xenoplax* and short life of peach trees. *Plant Dis.* 66:225–28.

Zehr, E. I., R. W. Miller, and F. H. Smith. 1976. Soil fumigation and peach rootstocks for protection against peach tree short life. *Phytopathology* 66:689–94.

Crown Gall of Fruit and Nut Crops

Agrobacterium tumefaciens
(Smith & Townsend) Conn

Crown gall, also known as "root tumors" and "plant cancer," attacks a wide range (142 genera) of dicotyledonous hosts including stone fruit, pome fruit, grape, and brambles (Elliott 1951; Smith 1907; Smith, Brown, and McCulloch 1912). The importance of the disease in California was reported as early as 1892 by Wickson and Woodworth.

The disease occurs worldwide, but is especially prevalent in temperate regions. It is probably responsible for the loss of more nursery trees than any other disease. The economic loss sustained after planting is difficult to evaluate, but in 1963 alone, it was estimated to cause $7 million loss in California on 14 major crops (University of California 1965).

Symptoms. The galls occur on roots, crown, stems, and leaves, but most typically at the juncture of large roots and the trunk (**fig. 2**). The smooth, young galls enlarge rapidly and, on perennial hosts, become firm, woody tumors with irregular surfaces. The galls vary in size from nearly microscopic to 12 inches or more in diameter. The exterior color matches that of the host bark and the interior color that of the normal wood of the host. Eventually, the galls become necrotic and portions slough or break off. Secondary invasions result in masses of galls that often girdle the roots and crown. Secondary wood decay organisms quickly enter these injuries, causing further debilitation of the host. *Armillaria* may also enter the gall tissue. Galls commonly form on aerial parts of quince, brambles, and grapes.

Although crown gall is very important on young nursery trees, methods for detecting incipient infections have not been developed.

Cause. The causal organism is *Agrobacterium tumefaciens* (Smith and Townsend) Conn, formerly *Bacterium tumefaciens*. The nomenclature of *Agrobacterium* has been in a confused state for many years (Kerr and Brisbane 1983). In 1970, Keane, Kerr, and New proposed that the three pathogenic species of *Agrobacterium*—*A. tumefaciens*, *A. rhizogenes*, and *A. rubi*—be treated as varieties of *A. radiobacter*. This plan had to be abandoned, however, when it was established (Kerr and Brisbane 1983) that *A. radiobacter* is synonymous with the type species *A. tumefaciens*. The confusion in nomenclature was the result of basing speciation on pathogenicity, but the genes for pathogenicity are carried on large plasmids that can be transferred from one bacterial cell to another. Such plasmids are called *Ti* (tumor-inducing) and *Ri* (root-inducing) plasmids. Hence, as Kerr and Brisbane (1983) pointed out, one species could be changed into another by transfer of a plasmid, hardly a stable basis for speciation. Biotypes carrying the *Ti* plasmid cause crown gall; those with the *Ri* plasmid cause the disease known as hairy root.

There are three basic chromosomal forms of *Agrobacterium* that may be elevated to the rank of species (Holmes and Roberts 1981): *A. tumefaciens*, *A. rhizogenes*, and *A. rubi*. Revised descriptions of the three species were given by Holmes and Roberts (1981). Tests to distinguish the species and additional information on the complex problem of the nomenclature of *Agrobacterium* are given by Kerr and Brisbane (1983).

The *A. tumefaciens* bacteria form circular, smooth, creamy, shiny colonies on rich media. They are encapsulated, bacilliform rods, 0.4 to 0.8 by 0.8 to 3.0 μm, motile with 5 to 11 peritrichous flagella, gram-negative, and oxidase-positive. Their optiumum growth temperature is 25° to 30°C, with a generation time of about 60 minutes. The bacterium synthesizes b-glucoside-3-dehydrogenase, catalase, oxidase, urease, nitrate reductase, and b-galactosidase. On basal medium, acid is reduced from arabinose, cellobiose, fructose, galactose, lactose, melezitose, salicin, erythritol, ethanol, and dulcitol (Breed, Murray, and Smith 1957). Cultures kept in selective media or in water at room temperature retain their pathogenicity for over 20 years (C. I. Kado, unpublished), and do so indefinitely in a lyophilized state. Hildebrand (1941) found the organism not to be a good soil inhabitant, but other researchers (Dickey 1961; Schroth, Thompson, and Hildebrand 1965) considered the organism successful in this respect.

Visually distinguishing crown-gall tissue from callus (wound overgrowth) tissue usually is not difficult; however, in some instances it is almost impossible. A positive diagnosis can be made (1) by isolating the pathogen, (2) by identifying a specific opine in the suspected gall, and (3) by using T-DNA as a probe (Kerr and Brisbane 1983). The pathogenic species of *Agrobacterium* can be differentiated and distinguished from nonpathogenic forms by use of selective media (Kado and Heskett 1970, Kerr and Brisbane 1983; New and Kerr 1972a; Schroth, Thompson, and Hildebrand 1965). Soil population studies of a pathogenic isolate of *A. tumefaciens* on a selective medium excluded over 99 percent of the common soil microorganisms but recovered only 38 percent of the pathogens introduced (Schroth, Thompson, and Hildebrand 1965). Schroth et al. (1971) detected the pathogen in 18 out of 28 soils tested—in many instances where no host plants were grown. In addition to selective media, physiological tests are of great value in differentiating the pathogenic species of *Agrobacterium* (Kerr and Brisbane 1983). Another test that aids in distinguishing *A. tumefaciens* from *A. rhizogenes* is the inoculation of disks of carrot or other fleshy roots (Ark and Schroth 1958; Kerr and Brisbane 1983). In this test, biotypes carrying the *Ti* plasmid cause gall formation, whereas those with the *Ri* plasmid induce the production of roots.

Disease development. *Agrobacterium tumefaciens* survives from one year to the next in galls and in the soil. The greatest incidence of the disease occurs in nurseries, with infection occurring at the time the seeds germinate, when liners and nursery trees are harvested, and during storage in heel-in beds. On apple, the union produced from root-scion grafting is readily infected. The bacteria can infect nonmechanically injured roots of peach, with most infection apparently occurring through the lenticels. Shipment of affected nursery stock is probably the most important means of long-distance dissemination. Local dissemination is accomplished by water, movement of soil, cultivation, or by wind. Bacteria are released from galls when the galls are wet or are disintegrating.

In established orchards, the bacteria most commonly enter through growth cracks or wounds caused by hoeing, disking, and removal of suckers (Ross et al. 1970). On such crops as pea, mung bean, tomato, and barley, Schroth and Ting (1966) showed that *A. tumefaciens* is attracted to root hairs and to zones of root elongation. Broken root hairs also appear to be a major site where the bacteria accumulate. The acidity of the medium also affects attraction sites. At pH 5.4, the bacteria accumulate at cut ends of roots; at pH 6.0, they accumulate at root hairs and zones of elongation.

Because the development of galls depends upon plant growth, these overgrowths appear only during the growing season. Riker (1934) reported that incubation periods in the field range from 2 to 11 weeks depending on the time of year. Such variations reflect the influence of temperature on gall formation: the longest incubation periods occur in the spring and the shortest ones in midsummer. Still longer incubation periods are possible because infection in autumn will not result in gall formation until growth starts in spring.

The bacterium is invasive, spreading intercellularly from the site of initial invasion. During infection, cells adjacent to the invaded areas begin dividing actively. As abnormal division continues in an undifferentiated manner, a tumorlike growth becomes visible. Such neoplasms develop into massive tumors that are called crown galls.

The cause of cellular transformation by *A. tumefaciens* has been studied for over 70 years. According to Nitsch (1964), it is believed that auxin and phytokinin (cytokinin) are needed for the inception of crown-gall tumors. The tumors generally contain more auxin than do corresponding normal tissues; they also contain water-soluble phytokinin (Braun 1962). Tumor cells also have unusually high concentrations of free lysopene and, especially, free proline (Hochster 1964). Different hypotheses propose several tumor-inducing agents: auxin- or cytokinin-like substances, proteinaceous material, ribonucleic acid, deoxyribonucleic acid, bacteriophage, and toxins. More recent research has been concerned with nucleic acids (Drlica and Kado 1974; Kado and Heskett 1970), possible virus carried by the inciting bacterium, and with plasmids. It now is generally accepted that the abnormal growth pattern of crown-gall cells is the direct consequence of specific tumor genes (Drummond 1979; Kahl and Schall 1982). These genes are transferred into plant nuclei by *A. tumefaciens*, which contains natural gene vectors in the form of large *Ti* (tumor-inducing) plasmids. These plasmids carry the tumor-inducing genes and can transfer them to the host cell nucleus.

The transferred DNA also determines the synthesis in transformed tissue of novel amino-acid compounds, called opines, which can serve as the nitrogen or carbon source for the oncogenic bacteria. An avirulent strain of *A. tumefaciens* can be made virulent by transferring to it the *Ti* plasmid from an oncogenic strain through conjugation or transformation.

The bacteria are intercellular but may exist to some extent within the vessels (Robinson and Walkden 1923). They are especially abundant in hyperplastic islands near the periphery of active galls. Both hypertrophy and hyperplasia of cells occur. Smith's view that "tumor strands" give rise to secondary galls at some distance from the point of infection has been considerably modified by Riker (1923a).

A soil temperature of 22°C appears most favorable for gall formation; the maximum appears to be near 30°C. Soil moisture of 60 percent is most favorable for gall formation at all temperatures (Riker 1926). Fertility, tilth, and other soil characteristics are of minor importance in the development of this disease. Infestation by the pathogen is not limited to certain soil types to any noticeable extent. Siegler (1938) found gall formation more abundant at pH 6.8 than at 5.0. This may account for the greater severity of the disease in some soils than in others.

Host susceptibility. Almond and peach roots are the most susceptible of the stone fruits. C. O. Smith (1917) found both bitter and sweet almonds to be highly susceptible, as well as the peach commonly used for rootstocks. Japanese plums are more susceptible than the European sorts. Because of susceptibility to crown gall, propagation of walnuts on *Juglans regia* stock was abandoned in favor of *J. californica* or *J. hindsii* stock.

Control. Cultural techniques followed by some nurseries involve soaking the seeds in sodium hypochlorite and stratifying the seeds directly in the planting rows instead of putting them under refrigeration (Ark 1941; Ross et al. 1970; Siegler and Bowman 1940). At harvest, diseased nursery trees are discarded and apparently healthy trees are planted soon after delivery, with limited use of the heeling-in procedure (placement in clean wood shavings). Little benefit has resulted from fertilization or irrigation, but the use of sulfur to lower the pH of neutral or alkaline soils has been suggested.

Crown-gall resistance has been shown only for black walnut. DeVay et al. (1965) attempted to increase resistance in peach, almond, and cherry seedlings by thermal neutron irradiation of seeds. They obtained some resistance in irradiated as well as nonirradiated peach and almond seedlings, but none in cherry seedlings.

Soil fumigation with methyl bromide-chloropicrin reduces the bacterial population, but does not effectively control the disease (Deep and Young 1965; Deep, McNeilan, and MacSwan 1968; Dickey 1962; Munnecke and Ferguson 1960). Furthermore, fumigants do not kill bacteria in the gall tissues. Treatments to kill the galls in the root-crown region, thereby preventing further girdling, has received much attention. Ark (1941, 1942) and Ark and Scott (1951) obtained good results with a sodium dinitro-o-cresol-methanol mixture (no longer used) on almond. Streptomycin or tetracycline failed as 60-minute dips.

Bacticin, a hydrocarbon containing 2,4-xylenol and metacresol, has produced positive results (Schroth and Hildebrand 1968). On crops such as peach, cherry, almond, pear, and walnut (Ross et al. 1970) this treatment is suggested for use during the first three years of tree growth. The crown area of the plant is exposed by hosing away the soil so as not to injure the host tissue. The dry gall is treated once with the chemical and left exposed for several weeks. The gall dies within a three- to four-month period. Galls over 4½ inches in diameter may require two applications. Except for walnuts, removal of any part of the gall before treatment is not necessary. Bacticin penetrates the diseased but not the healthy tissue, and therefore does not retard the formation of new callus. Such a treatment has prevented the regrowth of tumors. Though effective when applied to tumors, bacticin cannot be used as a dip for nursery stock because of its phytotoxic effects.

Biological control of crown gall has been under investigation by many researchers for several years. New and Kerr (1972a,b) used a nonpathogenic isolate of *A. radiobacter* var. *radiobacter* biotype (biovar) 2 (strain 84, now called K84) to reduce gall incidence in seedlings which grew from seeds that were inoculated with both the pathogen and the nonpathogen. In inoculation tests on pinto bean leaves (Lippincott and Lippincott 1969), the ratio of pathogenic to non-pathogenic populations required for control was 1:1. The prevention of crown gall on *Prunus* roots with strain K84 has been confirmed by Schroth and Moller (1976), Moore (1977), Moore and Warren (1979), and others. The mechanism of action has been related to the presence of bacteriophage (Leff and Beardsley 1970), the exclusion of the pathogen from the infection or attachment site (Lippincott and Lippincott 1969), and the production of an antibiotic by the strain used (Kerr 1980). Evidence (Kerr 1980; Kerr and Brisbane 1983; Kerr and Htay 1974; Moore 1977) indicates that protection is afforded, at least in part, through the production by strain K84 of an antibiotic bacteriocin called agrocin 84, a substance belonging to a new group of highly specific antibiotics known as nucleotide bacteriocins. The production of agrocin 84 is coded for by a plasmid, an extrachromosomal loop of DNA (Kerr 1980). The effectiveness of disease control with strain K84 probably is attributable to the unique infection process of *A. tumefaciens*. Pathogenic strains sensitive to K84 are prevented from transferring the Ti plasmid to the wounded host, apparently because the agrocin 84 produced by the antagonist either kills the pathogen or prevents its attachment to the host receptor site (Moore and Warren 1979).

Strain K84 is now used commercially for the control of crown gall on a variety of crops in the United States, Australia, and several other countries (Kerr 1980; Moore and Cooksey 1981; Moore and Warren 1979). Its failure in some instances is due at least in part to the prevalence of resistant strains of *A. tumefaciens* (Moore and Warren 1979, Süle and Kado 1980). Biotype 2 strains are most susceptible. Less sensitive biotype 1 or 3 strains occur in some geographic areas (Vidaver 1982); such strains appear associated with certain crops and not with others (Kerr 1980; Moore and Cooksey 1981; Moore and Warren 1979). Strain K84 was ineffective against isolates from grape and several other hosts in Europe and failed to adequately control crown gall on mazzard cherry rootstocks in California, apple seedlings in Washington, and rose plants in Pennsylvania and Texas (Alconero 1980; Moore and Warren 1979; Schroth and Hancock 1981). Süle and Kado (1980) showed that exposure of *A. tumefaciens* to agrocin 84 yields many virulent mutants resistant to this bacteriocin. This finding complicates current concepts regarding the use of K84 against crown gall.

A genetically engineered derivative of K84, *K1026*, recently developed in Australia (Jones and Kerr 1989), was found to be equal to the parent strain in its control of crown gall. In this new strain, conjugative transfer of the agrocin 84 plasmid to the pathogen, which causes the recipient cells to become insensitive to K84, is blocked. Jones and Kerr recommend that as soon as K1026 is registered for commercial use, it be employed as a replacement for K84 in the biocontrol of crown gall.

Recommendations (Moore and Warren 1979; Vidaver 1982) are: (1) use a high concentration of K84 (or, presumably K1026) cells—10^8 to 10^9 colony-forming units per milliliter (a turbid suspension); (2) treat seed, bare roots, cuttings, or aerial grafts before presumptive exposure to the pathogen, since only preventive treatment is effective; (3) dip or spray plant parts with K84 (spraying is recommended if planting stock is contaminated with other pathogens); and (4) protect treated plants from sunlight and drying to prevent death of the K84 inoculum. It must be emphasized that the use of K84 will not stop latent infections, presumably because the pathogen is already in a protected site. And the use of the bacterium will not substitute for proper sanitation and nursery management.

An effective control strategy for use with *Prunus* nursery stock has been developed recently in Oregon (Moore and Allen 1986). The incidence of crown gall in the pruned roots of dormant seedlings of three *Prunus* species was markedly reduced by heating the root systems to 18° to 25°C for one to three weeks. The reduction was even greater when the seedlings were inoculated with K84 before exposure to heat. Holding the heat boxes at 2° to 4°C during the temperature treatments prevented the buds from breaking dormancy. The success of the heat treatment is attributed to the rapid callusing of the root pruning cuts at the higher temperatures.

REFERENCES

Alconero, R. 1980. Crown gall of peaches from Maryland, South Carolina, and Tennessee, and problems with biological control. *Plant Dis.* 64:835–38.

Ark, P. A. 1941. Chemical eradication of crown gall on almond trees. *Phytopathology* 31:956–57.

———. 1942. Crown-gall of deciduous fruit trees and its control. *Blue Anchor* 19(1):16–19, 37.

Ark, P. A., and M. N. Schroth. 1958. Use of slices of carrot and other fleshy roots to detect crown gall bacteria in soil. *Plant Dis. Rep.* 42:1279–81.

Ark, P. A., and C. E. Scott. 1951. Elimination of crown gall: Treating small galls on young trees with elgetol-methanol mixture assures control in almond, peach, walnut orchard. *Calif. Agric.* 5(7):3.

Bernaerts, M., and J. DeLey. 1963. A biochemical test for crown gall bacteria. *Nature* 197:406–07.

Braun, A. C. 1962. Tumor inception and development in the crown gall disease. *Ann. Rev. Plant Physiol.* 13:533–58.

Breed, R. S., E. G. D. Murray, and N. R. Smith. 1957. *Bergey's manual of determinative bacteriology*, 7th ed. Balliere, Tindall and Cox, London.

Deep, I. W., R. A. McNeilan, and I. C. MacSwan. 1968. Soil fumigants tested for control of crown gall. *Plant Dis. Rep.* 52:102–05.

Deep, I. W., and R. A. Young. 1965. The role of preplanting treatments with chemicals in increasing the incidence of crown gall. *Phytopathology* 55:212–16.

DeVay, J. E., G. Nyland, W. H. English, F. J. Schick, and G. D. Barbe. 1965. Effects of thermal neutron irradiation on the frequency of crown gall and bacterial canker resistance in seedlings of *Prunus* rootstocks. *Radiation Bot.* 5:197–204.

DeVay, J. E., and W. C. Schnathorst. 1963. Single-cell isolation and preservation of bacterial cultures. *Nature* 199:775–77.

Dickey, R. S. 1961. Relation of some edaphic factors to *Agrobacterium tumefaciens*. *Phytopathology* 51:607–14.

———. 1962. Efficacy of five fumigants for the control of *Agrobacterium tumefaciens* at various depths in the soil. *Plant Dis. Rep.* 46:73–76.

Drlica, K. A., and C. I. Kado. 1974. Quantitative estimation of *Agrobacterium tumefaciens* DNA in crown gall tumor cells. *Proc. Nat. Acad. Sci.* 71:3677–81.

Drummond, M. 1979. Crown gall disease. *Nature* 281(5730):343–47.

Dye, D. W. 1952. The effects of chemicals and antibiotic substances on crown-gall (*Agrobacterium tumefaciens* [Smith and Townsend] Conn). Part IV. *N. Z. J. Sci. Tech.* 33:104–08.

Dye, D. W., P. B. Hutchinson, and A. Hastings. 1950. Effect of chemicals and antibiotic substances on crown-gall (*Agrobacterium tumefaciens* [Smith and Townsend] Conn). Part I. Colchicine and penicillin. *N. Z. J. Sci. Tech.* 31:31–39.

Elliott, C. 1951. *Manual of bacterial plant pathogens*, 2d ed. Chronica Botanica, Waltham, Mass., 186 pp.

Hildebrand, E. M. 1941. On the longevity of the crown gall organism in soil. *Plant Dis. Rep.* 25:200–02.

Hochster, R. M. 1964. Perspectives on the biochemistry of *Agrobacterium tumefaciens*. In *Proceedings of the Conference on Abnormal Growth in Plants*, edited by J. E. DeVay and E. E. Wilson, 26–35. University of California, Berkeley.

Holmes, B., and P. Roberts. 1981. The classification, identification, and nomenclature of *Agrobacteria*, incorporating revised descriptions for each of *Agrobacterium tumefaciens* (Smith and Townsend) Conn 1942, *A. rhizogenes*

(Riker et al.) Conn 1942, and *A. rubi* (Hildebrand) Starr and Weiss 1943. *J. Appl. Bacteriol.* 50:443–67.

Jones, D. A., and A. Kerr. 1989. *Agrobacterium radiobacter* strain K1026, a genetically engineered derivative of strain K84, for biological control of crown gall. *Plant Dis.* 73:15–18.

Kado, C. I., and M. G. Heskett. 1970. Selective media for isolation of *Agrobacterium, Corynebacterium, Erwinia, Pseudomonas,* and *Xanthomonas. Phytopathology* 60:969–76.

Kado, C. I., M. G. Heskett, and R. A. Langley. 1972. Studies on *Agrobacterium tumefaciens*: Characterization of strains 1D135 and B6, and analysis of the bacterial chromosomes, transfer RNA and ribosomes for tumor inducing ability. *Physiol. Plant Pathol.* 2:47–57.

Kahl, G., and J. S. Schall. 1982. *Molecular biology of plant tumors.* Academic Press, New York, 615 pp.

Keane, P. J., A. Kerr, and P. B. New. 1970. Crown gall of stone fruit. II. Identification and nomenclature of *Agrobacterium* isolates. *Austral. J. Biol. Sci.* 23:585–95.

Kerr, A. 1969. Crown gall on stone fruit. I. Isolation of *Agrobacterium tumefaciens* and related species. *Austral. J. Biol. Sci.* 22:111–16.

———. 1971. Acquisition of virulence by nonpathogenic isolates of *Agrobacterium radiobacter. Physiol. Plant Pathol.* 1:241.

———. 1972. Biological control of crown gall: Seed inoculation. *J. Appl. Bacteriol.* 35:493–97.

———. 1980. Biological control of crown gall through production of agrocin 84. *Plant Dis.* 64:24–25, 28–30.

Kerr, A., and P. G. Brisbane. 1983. *Agrobacterium.* In *Plant bacterial diseases: A diagnostic guide,* edited by P. C. Fahy and G. H. Persley, 27–43. Academic Press, Sydney.

Kerr, A., and A. Htay. 1974. Biological control of crown gall through bacteriocin production. *Physiol. Plant Pathol.* 4:37–44.

Leff, J., and R. E. Beardsley. 1970. Action tumorigène de l'acide nucléique d'un bactériophage présent dans les cultures due tissu tumoral de Tournesol (*Helianthus annuus*). *C. R. hebd. Séanc. Acad. Sci.* 270:2505.

Lippincott, J. A., and B. B. Lippincott. 1969. Bacterial attachment to a specific wound site as an essential stage in tumor initiation by *Agrobacterium tumefaciens. J. Bacteriol.* 97:620–28.

———. 1969. Tumor growth complementation among strains of *Agrobacterium. J. Bacteriol.* 99:496–502.

Meith, C., and W. J. Moller, 1978. Poria wood rot of deciduous fruit and nut trees. *Univ. Calif. Div. Agric. Sci. Leaf.* 21033. 3 pp.

Moore, L. W. 1977. Prevention of crown gall on *Prunus* roots by bacterial antagonists. *Phytopathology* 67:139–44.

Moore, L. W., and J. Allen. 1986. Controlled heating of root-pruned dormant *Prunus* spp. seedlings before transplanting to prevent crown gall. *Plant Dis.* 70:532–36.

Moore, L. W., and D. A. Cooksey. 1981. Biology of *Agrobacterium tumefaciens*: Plant interactions. In *Biology of the Rhizobiaceae,* edited by K. L. Giles and A. G. Atherly, 15–46. Academic Press, New York.

Moore, L. W., and G. Warren. 1979. *Agrobacterium radiobacter* strain 84 and biological control of crown gall. *Ann. Rev. Phytopathol.* 17:163–79.

Munnecke, D. E., and J. Ferguson. 1960. Effect of soil fungicides upon soil-borne plant pathogenic bacteria and soil nitrogen. *Plant Dis. Rep.* 44:552–55.

New, P. B., and A. Kerr. 1972a. A selective medium for *Agrobacterium radiobacter* biotype 2. *J. Appl. Bacteriol.* 34:233.

———. 1972b. Biological control of crown gall: Field measurements and glasshouse experiments. *J. Appl. Bacteriol.* 35:279–87.

Nitsch, J. P. 1964. Endogenous auxins and phytokinins in abnormal plant tissues. In *Proceedings of the Conference on Abnormal Growth in Plants,* edited by J. E. DeVay and E. E. Wilson, 15–20. University of California, Berkeley.

Panagopoulos, C. G., and P. G. Psallidas. 1973. Characteristics of Greek isolates of *Agrobacterium tumefaciens* (E. F. Smith and Townsend) Conn. *J. Appl. Bacteriol.* 36:233–40.

Pullman, G. S., J. E. DeVay, C. L. Elmore, and W. H. Hart. 1984. Soil solarization. *Univ. Calif. Div. Agric. Nat. Resour. Leaf.* 21377. 7 pp.

Riker, A. J. 1923a. Some relations of the crown gall organism to its host tissue. *J. Agric. Res.* 25:119–32.

———. 1923b. Some morphological responses of the host tissue to the crown gall organism. *J. Agric. Res.* 26:425–36.

———. 1925a. Crown gall in relation to nursery stock. *Science* 62:184–85.

———. 1925b. Second report of progress on studies of crown gall in relation to nursery stock. *Phytopathology* 15:805–6.

———. 1926. Studies on the influence of some environmental factors on the development of crown gall. *J. Agric. Res.* 32:83–96.

———. 1927. Cytological studies of crown gall. *Am. J. Bot.* 14:25–37.

———. 1934. Seasonal development of hairy-root and crown gall, and wound overgrowth of apple trees in the nursery. *J. Agric. Res.* 48:887–912.

Riker, A. J., and G. W. Keitt. 1926. Studies of crown gall and wound overgrowth on apple nursery stock. *Phytopathology* 16:765–808.

Riker, A. J., W. M. Banfield, W. H. Wright, G. W. Keitt, and H. E. Sagen. 1930. Studies on infectious hairy root of nursery apple trees. *J. Agric. Res.* 41:507–40.

Robinson, W., and H. Walkden. 1923. A critical study of crown gall. *Ann. Bot.* 37:299–324.

Ross, N., M. N. Schroth, R. Sanborn, H. J. O'Reilly, and J. P. Thompson. 1970. Reducing loss from crown gall disease. *Calif. Agric. Exp. Stn. Bull.* 845, 10 pp.

Schroth, M. N., and J. G. Hancock. 1981. Selected topics in biological control. *Ann. Rev. Microbiol.* 35:453–76.

Schroth, M. N., and D. C. Hildebrand. 1968. A chemotherapeutic treatment for selectively eradicating crown gall and olive knot neoplasms. *Phytopathology* 58:848–54.

Schroth, M. N., and W. J. Moller. 1976. Crown gall controlled in the field with a nonpathogenic bacterium. *Plant Dis. Rep.* 60:275–78.

Schroth, M. N., J. P. Thompson, and D. C. Hildebrand. 1965. Isolation of *Agrobacterium tumefaciens–A. radiobacter* group from soil. *Phytopathology* 55:645–47.

Schroth, M. N., and W. P. Ting. 1966. Attraction of *Agrobacterium* spp. to roots (abs.). *Phytopathology* 56:899–900.

Schroth, M. N., A. R. Weinhold, A. H. McCain, D. C. Hildebrand, and N. Ross. 1971. Biology and control of *Agrobacterium tumefaciens*. *Hilgardia* 40:537–52.

Siegler, E. A. 1938. Relations between crown gall and pH of the soil. *Phytopathology* 28:858–59.

Siegler, E. A., and J. J. Bowman. 1940. Crown gall of peaches in the nursery. *Phytopathology* 30:417–26.

Smith, C. O. 1917. Comparative resistance of *Prunus* to crown gall. *Am. Naturalist* 51:47–60.

———. 1944. A method of inoculating peach seedlings with crown gall without using punctures. *Phytopathology* 34:764–65.

Smith, E. F. 1907. A plant tumor of bacterial origin. *Science* 25:671–73.

———. 1917. Mechanisms of tumor growth in crown gall. *J. Agric. Res.* 8:165–88.

Smith, E. F., N. A. Brown, and L. McCulloch. 1912. The structure and development of crown gall, a plant cancer. *USDA Bur. Plant Indus. Bull.* 225.

Smith, E. F., N. A. Brown, and C. O. Townsend. 1911. Crown gall of plants: Its cause and remedy. *USDA Bur. Plant Indus. Bull.* 213.

Stonier, L. 1960. *Agrobacterium tumefaciens* II. Production of an antibiotic substance. *J. Bacteriol.* 79:880–98.

Stonier, T., J. McSharry, and T. Speitel. 1967. *Agrobacterium tumefaciens* Conn. IV. Bacteriophage PB2, and its inhibitory effect on tumor induction. *J. Virol.* 1:268–73.

Stroun, M., P. Anker, P. Gahan, A. Rossier, and H. Greppen. 1971. *Agrobacterium tumefaciens* ribonucleic acid synthesis in tomato cells and crown gall induction. *J. Bacteriol.* 106:634–39.

Süle, S., and C. I. Kado. 1980. Agrocin resistance in virulent derivatives of *Agrobacterium tumefaciens* harboring the *pTi* plasmid. *Physiol. Plant Pathol.* 17:347–56.

University of California Plant Pathology Statewide Conference on Plant Disease Losses Committee. 1965. *Estimates of crop losses and disease-control costs in California, 1963.* Univ. Calif. Agric. Exp. Stn. Agric. Ext. Serv., 102 pp.

Vidaver, A. K. 1982. Biological control of plant pathogens with prokaryotes. In *Phytopathogenic prokaryotes. vol. 2* edited by M. S. Mount and G. H. Lacy, 387–97. Academic Press, New York.

Wickson, E. J., and C. W. Woodworth. 1892. Root knots on fruit trees and vines. *Calif. Agric. Exp. Stn. Bull.* 99, 4 pp.

Wood, H. N. 1964. Cell membranes as regulators of biosynthetic metabolism and the biochemistry of a mitotic triggering substance in plant tumor cells. In *Proceedings of the Conference on Abnormal Growth in Plants*, edited by J. E. DeVay and E. E. Wilson, 20–26. University of California, Berkeley.

Fungal Diseases of Fruit and Nut Trees

Phytopthora Root and Crown Rot of Fruit and Nut Trees

Phytophthora spp.

Most deciduous fruit and nut species are affected by a disease known as "crown rot," "collar rot," "Phytophthora collar rot," "Phytophthora trunk rot," and, in the earlier literature, "Pythiacystis canker." All species of stone fruit, pome fruit, and walnut are more or less affected.

The root and crown rot disease occurs in many fruit-growing areas of the world, and it seems certain that it has been present since orchard trees were first cultivated. Early horticultural records give accounts of such a disease. Baines (1939), for example, cited a description in 1858 of "collar injury" of apple in the midwestern states. Although at that time the malady was believed to be caused by winter injury to the trunk, a means of avoiding it—that of double-working the trees—had been successfully employed.

Root and crown rot has been a source of major loss to Pacific Coast orchards for many years. The first reference to it in California was made in 1912, followed by reports of extensive damage and death of fruit and nut trees in nurseries and orchards in 1914, 1921, 1927, and 1941 (Day 1953; Smith 1941). The cause was attributed to excessive soil moisture and was referred to as "wet feet" or "sour sap" because of the dead roots and the characteristic odor of the decayed tissues. In California, root and crown rot kills more almond, apple, cherry, nectarine, and peach trees than perhaps any other disease, and has long been a concern to walnut growers. On pear, plum, and prune, however, the occurrence of the disease is sporadic and the damage relatively minor.

The increase in Phytophthora root and crown rot in California sweet cherry orchards has been related to such practices as changing rootstocks from the tolerant mazzard to the susceptible mahaleb, moving the pathogen on farm equipment from one orchard to the next, planting orchards on marginal soils with poor drainage, and the frequency and amount of irrigation water applied (Mircetich and Browne 1987).

Symptoms. The disease is troublesome both in the nursery and in the orchard (**fig. 3A**), and fruit infection sometimes occurs in sprinkler-irrigated orchards (**fig. 3C**). In the nursery the principal loss occurs when the young trees are dug, tied into bundles, and the bundles heeled-in (the roots and crown buried in sawdust or soil) pending their sale. Under such conditions, cankers (which in stone-fruit trees exude copious gum and in pome-fruit or walnut trees exude a watery or slimy material) develop at various places along the trunk. In the orchard, cankers usually are centered at the crown or lower part of the trunk (**fig. 3B**), but may (depending on the species of host) extend some distance up the trunk and down into the large roots.

The affected bark is brown and dead, gum-soaked in stone-fruit trees, and moist or slimy in pome-fruit trees. A shallow layer of outer sapwood underlying the cankers may be discolored, but the discoloration does not extend very far beyond the upper and lower limits of the necrotic area in the bark.

These symptoms are frequently not apparent without close examination of the crown of the tree. Thus the first indication of infection may be failure of the tree to start growth in the spring, or sparse light-green or bronze foliage after the tree starts growth and subsequent rapid decline and death of the tree.

Somewhat similar disease symptoms are caused by *Armillaria mellea*, *Rosellinia necatrix*, certain soilborne viruses, various plant pathogenic nematodes (Mircetich and Browne 1987), and by some of the wood-decay organisms discussed elsewhere in this chapter.

Phytophthora cactorum on pear does not confine its attack to the crown and large roots but may infect the small feeder roots (McIntosh 1960; McIntosh and O'Reilly 1963). The newly affected bark develops various shades of light and dark brown and has a marbled appearance. There is usually a well-defined margin between the diseased and healthy bark. Diseased trees occur at random throughout the planting, evidence that the disease is not transmitted from one affected tree to another. When young vigorously growing trees become infected, their leaves develop a purplish cast in autumn. Infected mature trees with poorly developed chlorotic foliage produce little or no new growth, and bear small, highly colored fruit. The relationship of this type of infection to the pear decline complex was investigated by McIntosh (1960). Pythiaceous fungi were found not to be the primary cause of pear decline in California (Nichols et al. 1964).

Information on the pathological histology of the disease is meager. Baines (1939) reported that the protoplasts of infected cells, particularly the parenchymatous cells, are disorganized and the cell walls disrupted. The mycelium is both intercellular and intracellular. According to Cooper (1928), however, the mycelium is intercellular, but haustoria penetrate into the adjacent cells.

A different type of Phytophthora disorder was reported recently by Bostock and Doster (1985, 1987) and Doster and Bostock (1988b)—the association of *P. syringae* with pruning wound cankers in California almond trees. This disease is discussed under the title "Phytophthora Pruning Wound Canker of Almond" in Chapter 3.

Causal organisms. Although it is generally believed that species of *Phytophthora* are involved in most crown-rot situations, it would be premature to say that other agents cannot produce similar symptoms. For example, Magness (1929) believed that various factors, including low temperature and excessive soil moisture, might cause collar-rot symptoms on apple trees.

The frequent association of *Phytophthora* spp. with crown-rot symptoms in a variety of crops and at various locations strongly suggests that these fungi are the primary cause of the disease. Though studies on this genus were started by Rosenbaum in 1917, and extended by Leonian (1925, 1936), Leonian and Geer (1929), and Tucker (1931, 1933), the status of the species is still in a state of change. Earlier workers employed morphological characters of the sexual and asexual structures in classifying species. Tucker (1931) considered these of limited value and believed that cultural and environmental criteria offer better means for identification. In a recent study (Bielenin et al. 1988) protein electrophoresis was found to be an aid in distinguishing the species and subgroups within *Phytophthora* species encountered on deciduous fruit crops. Most investigators agree that *P. cactorum* (Leb. & Cohn) Schroet. is frequently the cause of the disease in walnut and apple (Barrett 1917, 1928; Braun and Schwinn 1963; Dunegan 1935; McIntosh 1953, 1959; McIntosh and Mellor 1953; Sewell and Wilson 1959; Welsh 1942). McIntosh (1964) found it to be widely distributed in the irrigated soils of the apple-growing areas of British Columbia, but it was not found in virgin soils or in nonirrigated cultivated soils. Young and Milbrath (1959) found that *P. syringae* Kleban attacks nursery stock of peach, apricot, cherry, and crabapple in Oregon. Both *P. cactorum* and *P. cinnamomi* Rands attack pear in Oregon (Cameron 1962). Mircetich and Keil (1970) implicated *P. cinnamomi* in root rot of peach in Maryland and southern Pennsylvania.

Extensive isolations from California fruit orchards since then have revealed the following association by host. Peach and almond: Sixteen different *Phytophthora* species including *P. cactorum, P. cambivora, P. citricola, P. cryptogea, P. drechsleri, P. megasperma, P. citrophthora, P. syringae*, and eight apparently different, unidentified *Phytophthora* species. Of these 16, *P. syringae, P. cactorum, P. cambivora, P. citricola, P. megasperma*, and two of the unidentified species were able to induce extensive cankers in artificially inoculated stems of one-year-old Nemaguard, Lovell, and Elberta peach seedlings. Apricot: At least eight different *Phytophthora* species are capable of causing root and crown rot of apricot and apricot rootstocks. Among this group, based on pathogenicity tests, are *P. cinnamomi, P. megasperma, P. cambivora, P. syringae*, and four unidentified but different *Phytophthora* species (Mircetich 1982). Sweet cherry: Thirteen different *Phytophthora* species are associated with and capable of causing different degrees of root and crown rot of sweet cherry rootstocks. Of primary importance are *P. megasperma, P. cambivora*, and *P. drechsleri*; other species found to be pathogenic on mahaleb and mazzard cherry rootstocks are *P. cactorum, P. citricola, P. citrophthora, P. cryptogea, P. syringae*, and five unidentified but apparently different *Phytophthora* species. Walnut: *Phytophthora cactorum, P. cinnamomi, P. megasperma*, and *Phytophthora* spp. (Mircetich and Matheron 1976, 1983). Apple: *Phytophthora cactorum, P. cambivora, P. megasperma, P. cryptogea*, and several apparently different but unidentified *Phytophthora* species (Mircetich and Browne 1985).

Variation in their morphological and pathogenic characteristics has contributed greatly to the problem of identifying species of *Phytophthora*. Information on variation was reviewed by Erwin et al. in 1963. A key to the *Phytophthora* species is found in papers by Waterhouse (1956, 1963).

The following account is based on Blackwell's (1943) description of the life cycle of *P. cactorum*: Aerial sporangiophores, bearing sympodially a succession of sporangia, 36 by 28 μm in dimensions, are produced on the infected host tissue when the atmosphere is moist. Under favorable conditions the sporangium produces zoospores which emerge into a vesicle at the apex of the sporangium. The pear-shaped, uninucleate, biflagellate zoospores are liberated when the wall of the vesicle breaks. Production of sporangia and liberation of zoospores may occur within six or seven hours. Under unfavorable conditions the sporangium may develop into a structure Blackwell calls a conidium. Once the sporangium develops into a conidium it does not produce zoospores, but germinates by means of a germ tube that usually emerges from the apex of the structure. If kept dry it will remain dormant for months and then germinate when moisture is present.

Blackwell describes another resting structure produced by some strains of *P. cactorum* that she calls a chlamydospore. The chlamydospore is produced terminally, rarely intercalary, on the mycelium, and is easily confused with an oospore.

The oogonium and antheridium, sexual reproductive structures of the fungus, are produced on separate hyphal branches. The oogonia vary from 25 to 40 μm (avg. 33 μm) in diameter, and the oblong antheridia average about 13 to 14 μm in greatest dimension. Blackwell (1943) reported that the antheridia of *P. cactorum* are largely paragynous. The oogonium and antheridium are multinucleate, and the nuclei are arranged evenly throughout the protoplasm; eventually, however, one oogonial nucleus becomes most prominent and with a portion of surrounding protoplasm forms an oosphere, which is separated from the rest of the nuclei by a thin periplasm. Upon its contact with the oogonium wall the antheridium forms a fertilization tube, which appears to be an extension of the walls of the antheridium and oogonium. This tube presses into the oogonium and finally gives way, liberating a nucleus that fertilizes the oosphere. The resulting oospore undergoes a period of dormancy, apparently comparable to the dormant period of seeds. During this period, which may last several months, it must be in a moist environment to survive. At the end of dormancy the oospore germinates by one to three germ tubes.

Various methods of isolating *Phytophthora* from soil have been described (Banihashemi 1970; Borecki and Millikan 1969; Nichols et al. 1964; Tsao 1960, 1970). Improved techniques and selective media required for isolation and detection of *Phytophthora* spp. are discussed by Mircetich and Browne (1987). They emphasize the relative ease of isolation from newly infected plant tissues, but isolation from dried plant tissues, roots, or bark is difficult because the tissue often harbors secondary organisms. A common bait technique involves the use of pear fruit (unblemished, firm Bartlett cv. preferred) that are partially immersed in about 500 mL of soil, with 1 cm of free water above the soil surface. The fruit are held at 18° to 20°C for 48 to 72 hours, rinsed in sterile distilled water, and incubated for an additional two to four days before isolations are made onto a selective medium. The selective medium most commonly used is called 3-P (corn meal agar amended with Pimaricin 100 ppm, Penicillin 50 ppm, and Polymixin 50 ppm). A medium commonly used for the isolation of *Phytophthora* spp. from infected tree tissues is P10VP (Tsao and Ocana 1969), which has been modified to contain Pimaricin 5 mg/L, vancomycin hydrochloride 300 mg/L, PCNB 25 mg/L and cornmeal agar 17 g/L. Another isolation medium is PARP (Kannwischer and Mitchell 1978), modified to contain Pimaricin 5 mg/L, Ampicillin 250 mg/L, Rifampicin 10 mg/L, PCNB 25 mg/L, and cornmeal agar 17 g/L. The inoculated plates are incubated at 18°C and observed periodically for *Phytophthora* development. Of the latter two media, the PARP medium is less expensive; both media permit growth of soilborne *Pythium* spp. and *Mortierella* spp., but the addition of hymexazole to the medium suppresses their development (Tsao and Guy 1977; Tsao 1983).

Disease development. *Phytophthora* spp. can live in the soil without a host plant for extended periods, probably obtaining nutriment from dead plant material. In nursery salesyards, where bundles of nursery trees are heeled-in pending sale, the disease may become epidemic, indicating that the fungus finds conditions favorable for both multiplication and infection. In such situations it is not known whether the fungus increases only on the host tissue, or on both the host tissue and dead organic matter in the soil. Seemingly, the most effective way the fungus could increase would be by producing sporangia and liberating zoospores. Also, oospores are commonly produced by some *Phytophthora* species in infected host tissue. Some investigators believe that sporangia may be produced on fallen fruit or on dead organic debris in the soil and, consequently, that zoospores play an important part in disseminating the fungus and in the buildup of infection in the orchard.

Field observations and laboratory studies in California by Mircetich, Browne, and others (Mircetich and Browne 1987) indicate that the incidence and severity of Phytophthora root and crown rot symptoms in fruit and nut orchards are influenced by (1) the *Phytophthora* species present in the soil, (2) epidemiological factors, including soil moisture conditions, and (3) the relative resistance of the fruit-tree species or cultivar and its rootstock.

The disease is more likely to occur when the soil remains wet for extended periods. Welsh (1942) found that colony growth of *P. cactorum* in vitro ceased when the relative humidity fell below 90 percent and that at relative humidity levels below 60 percent the fungus soon died. Welsh also found that the disease in apple trees was favored by high soil-moisture and temperature between 23° and 32°C. In *P. cactorum*-infested soils approaching saturation, the disease increased with successive increases in soil temperature between these limits. At low soil-moisture levels, disease development did not increase as soil temperature increased. At 61 percent saturation, for example, disease development was low at all temperatures between 23° and 32°C. Cardinal temperatures for growth of the fungus in culture are: minimum 4° to 7°C, optimum about 27°C, maximum 32°C. The fungus survives only a few days at 32°C.

Welsh (1942) speculated that high soil moisture favors development of Phytophthora crown rot because it impairs the vigor of the tree by injuring the root system. Low soil moisture, he believed, is unfavorable to development of the disease because the vegetative phase of the fungus cannot survive under such conditions. Only the resting structures of the fungus would survive, and these would be unable to germinate and develop sporangia at low moisture levels.

The effect of high soil moisture on production, liberation, and dissemination of zoospores has been explored to a limited extent. Sporangial production and zoospore maturation are favored by high humidity, and certainly the mobility of the zoospores would depend upon free soil water. In addition, there is evidence that spattering of some part of the fungus, possibly zoospores, leads to infection of the stems of young nursery trees that are heeled-in in the nursery yard. In such trees individual cankers often develop both on the base of the trunk in contact with the soil and a foot or more above the soil level. Baines (1939) described infection of apple trees in which cankers developed on the trunk well above the soil level. More recently, Bostock and Doster (1985, 1987) and Doster and Bostock (1988b) studied the epidemiology of pruning-wound cankers in California almond trees caused by *P. syringae*. Most infection occurred during wet periods in late fall and early winter. The fungus probably reaches the pruning-wound infection courts in spattered water during rainy weather. (This disease is discussed fully in Chapter 3.)

Phytophthora crown-rot epidemics in stone-fruit trees occur in California in years when rainfall is plentiful in fall, winter, and spring and temperatures are moderate. Smith and Smith (1925) mentioned the heavy losses that occurred in nursery trees in California during the winter of 1921–22, a season of excessive rainfall beginning in early November. Apparently, infection occurred in the nursery row before the trees were dug for sale. Thousands of trees were lost during January and February. The greatest losses occurred among apricot and peach trees that had been heeled-in in outdoor soil trenches. Many almond, plum, apple, and pear trees were also lost.

Critical studies on the effects of soil moisture and temperature on the severity of root and crown rot were summarized by Mircetich and Browne (1987).

Soil moisture. Both incidence and severity of the disease are closely related to soil moisture conditions (Browne 1984; Mircetich and Keil 1970; Mircetich and Matheron 1976; Mircetich 1975; Welsh 1942). Field surveys indicate that water-saturated soil associated with poor surface or subsoil drainage or improper irrigation management favors disease development, and greenhouse and laboratory experiments confirm these findings. Experiments on peach seedlings with *P. cinnamomi* showed more disease on trees subjected to weekly, 48-hour periods of water saturation than when soil was watered only when dry and seedlings were wilted (Mircetich and Keil 1970). On mahaleb and mazzard cherry seedlings, simulated irrigations to provide soil moisture levels below saturation or interrupted flooding once every two weeks for four hours resulted in 20 to 60 percent root rot with *P. cambivora*. The results with other species were as follows: *P. cryptogea*, no crown rot; *P. megasperma*, negligible root rot; *P. dreschsleri*, insignificant growth reduction. However, with interrupted flood irrigations of 48 hours, all four *Phytophthora* species caused severe root rot (Wilcox and Mircetich 1985a,b).

In another experiment (using only *P. cryptogea* and *P. megasperma*), where the duration of biweekly intervals of flood irrigation was lengthened from 4 to 48 hours, less disease developed on mazzard seedlings than on mahaleb seedlings (Wilcox and Mircetich 1985c), confirming earlier field observations. Neither *P. cryptogea* nor *P. megasperma* was capable of inducing disease in the absence of flood irrigation. The increase in disease from flooding (soil-water saturation) was related to decreased oxygen tension and increased susceptibility of cherry roots (Wilcox and Mircetich 1985c).

On Red Delicious apple seedlings, long periods of flood irrigation caused more disease with *P. cryptogea* than with *P. cactorum* or *P. cambivora* (Browne 1984). On Paradox walnut rootstock, careful soil-water management was found to reduce tree losses in orchard sites infested with *P. citricola*, *P. cactorum*, *P. citrophthora*, *P. cryptogea*, and *P. megasperma* but not with *P. cinnamomi*. With the *J. hindsii* rootstock, the avoidance of prolonged or repeated soil-water saturation around tree trunks should minimize losses caused by infection by *P. cryptogea* and *P. citrophthora*, but not those caused by *P. citricola* and *P. cinnamomi*. The formation of sporangia, the release of motile zoospores, and the movement of zoospores are favored by moisture levels from field capacity to saturation, but the moisture requirement for sporangial formation differs among *Phytophthora* species (Duniway 1983). Similar information has been developed on the effect of irrigation schedules on the development of *Phytophthora* root rot and fruit rot of tomato (Hoy, Ogawa, and Duniway 1984). Thus, the role of soil moisture, especially between field capacity and saturation, governs the life cycle of *Phytophthora* spp. as well as the development of plant roots, both of which are critical to infection and disease development.

Soil temperature. Differences in temperature influence the life cycle of *Phytophthora* species and, hence, the severity of disease development (Duniway 1983; Wilcox and Mircetich 1987). In artificially infested soil, *P. syringae* caused 100 percent infection and death of peach seedlings at 10°C but no infection at 20°C; *P. cactorum* caused severe crown rot in apple seedlings at 7° to 26°C but not at 29°C; *P. cryptogea* caused severe root rot in apple seedlings at 26°C but only slight root rot at 29°C; and *P. megasperma* caused 80 and 97 percent root rot of mahaleb seedlings at 10° and 20°C, respectively, but no root rot at 25°C (Browne 1984; Wilcox and Mircetich 1987). Thus, crops grown in different climatic regions could be affected differently, even when grown on the same rootstock and under similar soil-moisture conditions.

Among most fruit and nut tree hosts, young nonbearing trees seem to be somewhat less susceptible to infection than older bearing trees (McIntosh 1959; Welsh 1942).

Susceptibility of different crops and rootstocks. Knowledge of the relative resistance of rootstocks to *Phytophthora* spp. is inadequate because of differences in the reports by various researchers. These differences can be related to variations in the techniques used as well as to seasonal variations in the susceptibility of rootstocks (Jeffers et al. 1981; Jeffers and Aldwinckle 1986). Some of these variables are discussed below for each crop.

Apple. With Phytophthora crown rot, as with other root and crown diseases, susceptibility of the aboveground parts of the tree may or may not determine the severity of the disease in the orchard. Since it is the lower trunk of the tree (the *crown*) that is attacked by *P. cactorum* and since, in trees propagated by budding, the lower trunk may be rootstock tissue, the susceptibility of the rootstock determines the severity of the disease in the orchard. In trees propagated by grafting the scion of the commercial variety onto a root from a different cultivar or species, the lower trunk is tissue of the scion. Consequently, the susceptibility of the commercial cultivar will be important in determining the severity of the disease.

Baines (1939) reported that collar rot of apple in Indiana was largely confined to Grimes Golden. He said little about the rootstock in general use at that time, but alluded to seedlings of French crab that, he intimated, were resistant to infection by *P. cactorum*. It is clear, however, that the site of infection was not in rootstock tissue but in the trunk of the commercial cultivar and thus it can be assumed that the rootstock was either resistant or escaped infection because of some special attribute. Baines noted that the disease was occasionally found in Baldwin, Tompkins King, Roxbury Russett, Rhode Island Greening, Esopus Spitzenberg, and Hubbardston.

Welsh (1942) described a situation in British Columbia in which the fungus attacked the trees at or slightly below soil level. He noted that the rootstocks were seedlings of various apple cultivars, and while he did not mention the method of propagation the trees probably were budded into the seedling trees a few inches above the soil. Consequently, when they were transplanted to the orchard, the union between rootstock and scion of some trees would be above the soil level, while with others the union would be at or below soil level. In such a situation, the susceptibility of the commercial cultivar would influence the incidence of infection in trees planted with the bud union at or near the soil level. Susceptibility of the rootstock will be the more influential factor in trees planted with the bud union well above the soil level.

Aldwinckle et al. (1975) reported the isolation of only *P. cactorum* from two orchard locations in the state of New York and, using 15 isolates, identified four pathogenicity groups on the basis of differential interactions with excised twigs of 31 apple cultivars. Later, *Phytophthora megasperma* and *Pythium irregulare* were also isolated and shown to be pathogenic (Jeffers et al. 1981, 1982). Studies by Jeffers and Aldwinckle (1986) extended the list of pathogens to include *P. cambivora* and *P. cryptogea*. Meanwhile, in North Carolina both *P. cactorum* and *P. cambivora* were isolated from Malling, Malling-Merton, and seedling apple rootstocks (Julis, Clayton, and Sutton 1978).

In California Mircetich and Browne (1987) noted a higher incidence of Phytophthora root and crown rot on MM106 than on MM11, M26, or M9 rootstocks in some commercial orchards infested with *P. cactorum*, but in several other orchards infested by other *Phytophthora* species they saw no differences in incidence or disease severity among the different rootstocks. Another finding was that California-grown MM11 was more resistant than MM106, MM104, MM109, M7, or standard apple seedlings to *P. cambivora*. All of the rootstocks tested appeared to be most susceptible when inoculated with *P. cryptogea* in March and *P. cambivora* in May, while all were least susceptible to *P. cryptogea* in November and to *P. cambivora* in February.

Seasonal susceptibility of apple and variation in the activity of the fungi in soil have also been studied to provide better timing of fungicide applications for control and improved evaluation of rootstocks for resistance (Borecki and Millikan 1969; Gates and Millikan 1972; McIntosh 1964; Sewell and Wilson 1973).

Pear. McIntosh (1959) cited d'Anjou as the pear most often affected by Phytophthora crown rot in British Columbia. Bartlett was sometimes affected. In experiments in Oregon, Bartlett and Winter Nelis proved susceptible (Cameron 1962) whereas Old Home, Old Home × Farmingdale, and *Pyrus calleryana* "were not severely damaged." Critical studies on Phytophthora resistance among various pear rootstocks have not been conducted in California.

Stone fruit. Smith and Smith (1925) noted epidemic infection of apricot and peach nursery trees by *P. citrophora* during 1921–22. Almond and plum trees were also highly susceptible to infection. It has become more and more evident that the incidence of disease under one set of conditions may lead to erroneous conclusions regarding the inherent susceptibility of different host plants. One reason for this is that it may be that certain hosts or rootstocks merely escape infection while others do not. Another reason is the existence of pathogenic races of the fungus.

Mahaleb seedling rootstock for sweet cherry was susceptible to severe root and crown rot as well as stem cankers when inoculated with *P. megasperma* and *P. cambivora*, but no stem cankers developed when it was inoculated with *P. drechsleri*. *Phytophthora drechsleri* caused decay of feeder roots and growth stunting resulted. Field isolations from natural infections supported these data. Mazzard cherry rootstocks were more resistant than mahaleb to these pathogens (Mircetich, Matheron, and Schreader 1974). In lathhouse and greenhouse experiments using stem inoculations of 6- to 12-month-old cherry rootstocks, mazzard and Stockton Morello were more resistant than mahaleb to *P. cambivora* and *P. megasperma*; mahaleb also was more susceptible than other rootstocks, such as Colt, Vladimir, Falstran M × M2, M × M38, M × M60, and M × M97, to *P. cambivora*. In greenhouse tests, using artificially infested soil, mazzard rootstock was more resistant than mahaleb to both root and crown rot caused by *P. cambivora* and *P. megasperma*, but the two stocks were equally susceptible to *P. cryptogea*. Mazzard was also more resistant than maheleb to crown rot caused by *P. cinnamomi* and *P. citricola*, but no differences were evident in their resistance to root rot caused by these two species (Wilcox and Mircetich 1985a,b,c). Similar *Phytophthora* species were found to be involved with root and crown rot of cherry trees in New York (Wilcox et al. 1985).

In Mississippi, isolations from peach trees with Phytophthora root rot and stem canker detected *Phytophthora cinnamomi*, *P. cactorum*, and, for the first time, *P. nicotianae* var. *parasitica* (Haygood, Groves, and Ridings 1986). In inoculation tests, all species were pathogenic.

On almond in California, the root and crown rot phase was predominantly caused by *P. megasperma* and *P. drechsleri*. But *P. syringae* caused most extensive trunk cankers on almond rootstock, with the next, in decreasing order of susceptibility, being almond × peach hybrid, Nemaguard, and Lovell (Mircetich, Moller, and Chaney 1974).

Among rootstocks used for apricot, prune, and plum, myrobalan plum was found to be more resistant than either Lovell or Nemaguard peach seedlings to root and crown rot caused by *P. megasperma* and *P. drechsleri* when evaluated in artificially infested soil in greenhouse experiments; both peach seedlings were equally susceptible to the two *Phytophthora* species (Mircetich 1981). In the same test, Lovell seedlings were more susceptible than Nemaguard seedlings to *P. cryptogea* (Mircetich 1984). In growth chamber studies, myrobalan seedlings were found to be more resistant than Lovell peach or Tilton apricot seedlings to *P. cinnamomi* and *P. megasperma* when inoculations were made on the stem. In the same tests, Nemaguard peach seedlings were more resistant than Marianna 2624 clonal rootstock, apricot seedlings, or Lovell peach seedlings. Also, Lovell peach seedlings were more resistant than apricot seedlings to *P. syringae*.

Walnut. Barrett (1928) listed the southern California black walnut (*Juglans californica*) as somewhat more susceptible than the northern California black walnut (*J. hindsii*) to attack by *P. cactorum*. The Persian (English) walnut (*J. regia*) is quite resistant. Black walnut species, especially *J. hindsii*, have been widely used in California as rootstocks for the commercial Persian walnut cultivars. More recently the Paradox hybrid (*J.*

hindsii × *J. regia*) has come into use as a rootstock, but its susceptibility to *Phytophthora* has not been fully explored. After infecting the black walnut rootstock the fungus sometimes invades the trunk of the Persian walnut above the union. Nevertheless, the susceptibility of the rootstock is more important than that of the commercial cultivar in determining the severity of the disease in the walnut orchard.

In addition to *J. hindsii*, *J. californica*, and *J. regia*, Smith and Barrett (1931) found that *P. cactorum* produced symptoms on *J. nigra*, *J. mandschurica*, *J. sieboldiana*, *J. major*, *J. pyriformis*, and a Paradox hybrid from the cross *J. californica* × *J. regia*.

Mircetich and Matheron (1983) repeatedly isolated *P. cactorum*, *P. citricola*, *P. cinnamomi*, *P. citrophthora*, *P. megasperma*, *P. cryptogea*, and four other different but unidentified *Phytophthora* species from symptomatic trees in commercial orchards. In inoculation tests, *P. citricola*, *P. cactorum*, *P. cinnamomi*, and *P. citrophthora* were more virulent than *P. megasperma* or *Phytophthora* sp. (isolate 1029) to seedlings of *J. hindsii* used as rootstocks. Seedlings of Paradox rootstock were significantly more resistant than those of *J. hindsii* to *P. cactorum*, *P. citrophthora*, *P. megasperma*, and *Phytophthora* sp. (1029), whereas seedlings of *J. regia* were as susceptible as those of *J. hindsii* to *P. cactorum*, *P. cinnamomi*, *P. megasperma*, and *P. citricola*.

Control. Measures recommended for the control of the disease in orchard trees can be divided into four categories: orchard site selection; use of a resistant understock for commercial cultivars; irrigation methods, drainage measures, and planting procedures; and fungicide application.

The selection of the orchard site is important because once the soil is infested with *Phytophthora*, eradication is difficult if not impossible, and chemical control is extremely costly. Avoid any site with a history of *Phytophthora* on tree crops, and consider that sites can become infested if infected or infested plant material or contaminated irrigation or runoff water are used (Mircetich et al. 1985), and if infested soil on farm machinery is introduced. Soil fumigation with methyl bromide is helpful, but its effect is short lived, and complete eradication of the pathogen is not possible.

The rootstock is most important when planting on sites with a history of Phytophthora root and crown rot. A considerable degree of resistance to various *Phytophthora* species is available, but none of the stone-fruit rootstocks is immune. Of the rootstocks for sweet cherry, both mazzard and Stockton Morello are more resistant than mahaleb and such others as Colt, Falstran, Vladimir, M × M2, M × M60, M × M97, M × M46, M × M39, and M × M14 (Mircetich and Matheron 1981). For apricot, prune, and plum trees, myrobalan rootstock is more resistant than Marianna 2624, Nemaguard, or Lovell peach seedlings (Mircetich 1981, 1982). For apple propagation, there is considerable disagreement as to the relative resistance of the various rootstocks. However, the M2, M4, M9, M26 and Antonovka seedling stocks have been reported to be more resistant to *P. cactorum* than MM106 or MM104. For *Phytophthora* species in general, the MM111 and M7 rootstocks have been reported relatively trouble-free in some regions but not in others. Resistance ratings are affected by the methods of evaluation used, the presence of different *Phytophthora* biotypes, the cultural conditions of the test plot, and the climatic variations in different regions. To ensure the benefits derived from resistant rootstocks, infection should be prevented at the salesyard by heeling-in nursery trees in fresh peat moss or fresh sawdust. Also, by protecting the heeled-in trees from rain, accumulation of the moisture needed for the pathogen's activities can be prevented.

Careful soil-water management should limit prolonged, repeated periods of soil saturation, especially in the vicinity of tree crowns. Planting trees on berms can reduce the incidence of disease (Baines 1939; Braucher 1930; Smith 1955), but even more important is the internal and surface water drainage in orchards. Also, graft unions should not be in contact with the soil, regardless of the rootstock used. Ellis, Ferree, and Madden (1986), however, reported that placing the union of susceptible rootstocks above or below the soil line appeared to have no effect on collar rot of Golden Delicious apple trees on MM106 rootstock with or without an East Malling 9 interstem. These results are contrary to the general observations of other investigators.

Protective fungicides, especially copper compounds, have been applied to the crown region with variable results. Copper crystals hung at the intake of pumps for sprinkler irrigation have provided control of Phytophthora fruit decay in orchards in Washington (Magness 1929). Maneb and captafol have been recommended (McIntosh 1971) for crown-rot control, and fenaminosulf, ethyl phosphites, and acylalanines have been tested with variable results. The ethyl phosphite and acylalanine groups are known not only for their protective action but also for their curative properties, as they are absorbed by the foliage and shoots and are translocated both acropetally and basipetally (ethyl phosphites) or only acropetally (the acylalanines) (Cohen and Coffey 1986; Schwinn 1983). Matheron and Mircetich (1985b) showed that monthly soil drenches of metalaxyl or fosetyl-Al reduced Phytophthora root and crown rot in seedlings of *Juglans hindsii* and Paradox walnut rootstocks. They also showed that stems sprayed with either fungicide were protected against infection for at least 54 days. In another recent study (Bielenin and Jones 1988), metalaxyl applied as a soil drench was found to be more effective than fosetyl-Al applied as a foliar spray in preventing root rot and mortality of *Prunus mahaleb* seedlings growing in a potting medium infested with *Phytophthora megasperma*, *P. cryptogea*, or *P. cambivora*. Acylalanine (metalaxyl), however, induces resistance in some target pathogens and is degraded by soil microorganisms (Cohen and Coffey 1986). Because most fungicides have a short half-life in wet soil, critical studies on the period of host susceptibility are needed (Erwin et al. 1963; McIntosh 1964; McIntosh and O'Reilly 1963). On apples in Missouri, inner bark tissues are highly susceptible to *Phytophthora* colonization at or near the blossoming period (Gates and Millikan 1972), so this may be the proper time to provide protective measures.

Recent control strategies involve some of the above chemical control measures in combination with either bacterial antagonists (Utkhede 1983a,b, 1984a,b) or the incorporation of a 1:1 mixture of composted hardwood bark and field soil into the planting hole (Ellis, Ferree, and Madden 1986). Doster and Bostock (1988c) recently obtained excellent control of almond pruning-wound cankers caused by *P. syringae* by applying fosetyl-Al to the fresh wounds.

REFERENCES

Aldwinckle, H. S., F. J. Polach, W. T. Molin, and R. C. Pearson. 1975. Pathogenicity of *Phytophthora cactorum* isolates from New York apple trees and other sources. *Phytopathology* 65:989–94.

Aljibury, F. K., J. W. Biggar, R. L. Branson, et al. 1981. Drip irrigation management. *Univ. Calif. Div. Agric. Sci. Leaf.* 21259. 39 pp.

Baines, R. C. 1939. Phytophthora trunk canker or collar rot of apple trees. *J. Agric. Res.* 59:159–94.

Banihashemi, Z. 1970. A new technique for isolation of *Phytophthora* and *Pythium* species from soil. *Plant Dis. Rep.* 54:251–62.

Barrett, J. T. 1917. Pythiacystis related to *Phytophthora* (abs.). *Phytopathology* 7:150–51.

———. 1928. *Phytophthora* in relation to crown rot of walnut (abs.). *Phytopathology* 18:948–49.

———. 1948. Induced oospore production in the genus *Phytophthora* (abs.). *Phytopathology* 38:2.

Bielenin, A., S. N. Jeffers, W. F. Wilcox, and A. L. Jones. 1988. Separation by protein electrophoresis of six species of *Phytophthora* associated with deciduous fruit crops. *Phytopathology* 78:1402–08.

Bielenin, A., and A. L. Jones. 1988. Efficacy of sprays of fosetyl-Al and drenches of metalaxyl for the control of Phytophthora root and crown rot of cherry. *Plant Dis.* 72:477–80.

Bingham, F. T., and G. A. Zentmyer. 1954. Relation of hydrogen ion concentration of nutrient solution to Phytophthora root rot of avocado seedlings. *Phytopathology* 44:611–14.

Blackwell, E. 1943. The life history of *Phytophthora cactorum* (Leb. & Cohn) Schroet. *Trans. Brit. Mycol. Soc.* 26:71–89.

Borecki, Z., and D. R. Millikan. 1969. A rapid method for determining the pathogenicity and factors associated with the pathogenicity of *Phytophthora cactorum*. *Phytopathology* 59:247–48.

Bostock, R. M., and M. A. Doster. 1985. Association of *Phytophthora syringae* with pruning wound cankers of almond trees. *Plant Dis.* 69:568–71.

———. 1987. Managing canker diseases of almonds. *Almond Facts* 52(2):36–37.

Braucher, O. L. 1930. Field trials in crown rot control: That success is based on keeping the tree crown dry. *Diamond Walnut News* 12(4):9.

Braun, H., and F. J. Schwinn. 1963. Fortgeführte Untersuchungen über den Erreger der Kragenfäule des Apfelbaumes (*Phytophthora cactorum*). *Phytopathol. Z.* 47:327–70.

Browne, G. T. 1984. *Effects of flood duration, temperature, and moisture extremes in soil on root and crown rot of apple caused by three* Phytophthora *species*. M. S. thesis, University of California, Davis, 69 pp.

Cameron, H. R. 1962. Susceptibility of pear roots to *Phytophthora*. *Phytopathology* 52:1295–97.

Cohen, Y., and M. D. Coffey. 1986. Systemic fungicides and the control of Oomycetes. *Ann. Rev. Phytopathol.* 24:311–38.

Cooper, D. 1928. Paragynous antheridia of *Phytophthora* spp. (abs.) *Phytopathology* 18:149.

Day, L. H. 1953. *Rootstocks for stone fruits*. Calif. Agric. Exp. Stn. Bull. 736, 76 pp.

Doster, M. A., and R. M. Bostock. 1988a. Incidence, distribution and development of pruning wound cankers caused by *Phytophthora syringae* in almond orchards in California. *Phytopathology* 78:468–72.

———. 1988b. Chemical protection of almond pruning wounds from infection by *Phytophthora syringae*. *Plant Dis.* 72:492–94.

Dunegan, J. C. 1935. Phytophthora disease of peach seedlings. *Phytopathology* 25:800–09.

Duniway, J. M. 1983. Role of physical factors in the development of Phytophthora diseases. In *Phytophthora: Its biology, taxonomy, ecology, and pathology*, edited by D. C. Erwin, 175–87. American Phytopathological Society, St. Paul. 392 pp.

Ellis, M. A., D. C. Ferree, and L. V. Madden. 1986. Evaluation of metalaxyl and captafol soil drenches composted with wood bark soil amendments and graft union placement on control of apple collar rot. *Plant Dis.* 70:24–26.

Erwin, D. C., G. A. Zentmyer, J. Galindo, and J. S. Niederhauser. 1963. Variation in the genus *Phytophthora*. *Ann. Rev. Phytopathol.* 53:375–96.

Fitzpatrick, R. E., F. C. Mellor, and M. F. Welsh. 1944. Crown rot of apple trees in British Columbia—Rootstock and scion resistance trials. *Sci. Agric.* 24:533–41.

Gates, J. E., and D. F. Millikan. 1972. Seasonal fluctuations in susceptibility of the inner bark tissues of apple to colonization by the collar rot fungus *Phytophthora cactorum*. *Phytoprotection* 53:76–81.

Haygood, R. A., C. H. Groves, and W. H. Ridings. 1986. Phytophthora root rot and stem canker of peach trees in Mississippi. *Plant Dis.* 70:866–68.

Houten, J. G. ten. 1958. Resistance trials against collar rot of apples caused by *Phytophthora cactorum*. *Tijdschr. Plantenziekten* 65:422–31.

Hoy, M. W., J. M. Ogawa, and J. M. Duniway. 1984. Effects of irrigation on buckeye rot of tomato fruit caused by *Phytophthora parasitica*. *Phytopathology* 74:474–78.

Hunt, J. F. 1921. Pythiacystis "brown rot" affecting deciduous trees. *Calif. Dept. Agric. Monthly Bull.* 10:143–45.

Jeffers, S. N., and H. S. Aldwinckle. 1986. Seasonal variation in extent of colonization of two apple rootstocks by five species of *Phytophthora*. *Plant Dis.* 70:941–45.

Jeffers, S. N., H. S. Adwinckle, T. J. Burr, and P. A. Arneson. 1981. Excised twig assay for the study of apple tree crown rot pathogens in vitro. *Plant Dis.* 65:823–25.

———. 1982. *Phytophthora* and *Pythium* species associated with crown rot in New York apple orchards. *Phytopathology* 72:533–38.

Jullis, A. J., C. N. Clayton, and T. B. Sutton. 1978. Detection and distribution of *Phytophthora cactorum* and *P. cambivora* on apple rootstocks. *Plant Dis. Rep.* 62:516–20.

Kannwischer, M. E., and D. J. Mitchell. 1978. The influence of fungicide on the epidemiology of black shank of tobacco. *Phytopathology* 68:1760–65.

Leonian, L. H. 1925. Physiological studies on the genus *Phytophthora*. *Am. J. Bot.* 12:444–98.

———. 1936. Control of sexual reproduction in *Phytophthora cactorum*. *Am. J. Bot.* 23:188–90.

Leonian, L. H., and H. L. Geer. 1929. Comparative values of the size of *Phytophthora sporangia* obtained under standard conditions. *J. Agric. Res.* 39:293–311.

Magness, J. R. 1929. Collar rot of apple trees. *Wash. Agric. Exp. Stn. Bull.* 236, 19 pp.

Matheron, M. E., and S. M. Mircetich. 1985a. Seasonal variation in susceptibility of *Juglans hindsii* and Paradox rootstocks of English walnut trees to *Phytophthora citricola*. *Phytopathology* 75:970–72.

———. 1985b. Control of Phytophthora root and crown rot and trunk canker in walnut with metalaxyl and fosetyl-A1. *Plant Dis.* 69:1042–43.

McIntosh, D. J. 1953. A trunk and crown rot of sweet cherry in British Columbia. *Phytopathology* 43:402–03.

———. 1959. Collar rot of pear trees in British Columbia. *Phytopathology* 49:795–97.

———. 1960. The infection of pear rootlets by *Phytophthora cactorum*. *Plant Dis. Rep.* 44:262–64.

———. 1964. *Phytophthora* spp. in soils of the Okanagan and Similkameen Valleys of British Columbia. *Can. J. Bot.* 42:1411–15.

———. 1971. Dilution plates used to evaluate initial and residual toxicity of fungicides in soils to zoospores of *Phytophthora cactorum*, the cause of collar rot in apple trees. *Plant Dis. Rep.* 55:213–16.

McIntosh, D. L., and I. C. MacSwan. 1966. The occurrence of collar rot caused by *Phytophthora cactorum* in a planting of apple trees aged 1 to 7 years. *Plant Dis. Rep.* 50:267–70.

McIntosh, D. L., and F. C. Mellor. 1953. Crown rot of fruit trees in British Columbia. II. Rootstock and scion resistance trials of apples, pears, and stone fruits. *Can. J. Agric. Sci.* 33:615–19.

McIntosh, D. L., and H. J. O'Reilly. 1963. Inducing infection of pear rootlets by *Phytophthora cactorum* (abs.) *Phytopathology* 53:1447.

Mircetich, S. M. 1975. Phytophthora root and crown rot of orchard trees. *Calif. Plant Pathol.* 24:1–2.

———. 1981. Rootstock diseases of prune trees. In *Prune orchard management*. edited by D. E. Ramos, 107–17. University of California, Berkeley, Agricultural Science Publications.

———. 1982. Phytophthora root and crown rot of apricot trees. *Acta Hort.* 21:272–76.

———. 1984. Phytophthora root and crown rot and stem canker of peach trees. *Proc. 83d Natl. Peach Council Conv.*: 17–21.

Mircetich, S. M., and G. T. Browne. 1985. Phytophthora root and crown rot of apple trees. *Proc. Oregon Hort. Soc.* 76:44–54.

———. 1987. Phytophthora root and crown rot of deciduous fruit trees: Progress and problems in etiology, epidemiology and control. In N. E. Louney (Ed.), *Proceedings of the Summerland Research Station commemorative symposium: Challenges and opportunities in fruit production, protection, and utilization research,* 64–95. Summerland Research Station, Agriculture Canada, Summerland, B.C. 111 pp.

Mircetich, S. M., G. T. Browne, W. Krueger, and W. Schreader. 1985. *Phytophthora* spp. isolated from surface water-irrigation sources in California (abs.). *Phytopathology* 75:1346–47.

Mircetich, S. M., and H. L. Keil. 1970. *Phytophthora cinnamomi* root rot and stem canker of peach trees. *Phytopathology* 60:1376–82.

Mircetich, S. M., and M. E. Matheron. 1976. Phytophthora root and crown rot of cherry trees. *Phytopathology* 66:549–58.

———. 1981. Differential resistance of various cherry rootstocks to *Phytophthora* spp. (abs.). *Phytopathology* 71:243.

———. 1983. Phytophthora root and crown rot of walnut trees. *Phytopathology* 73:1481–88.

Mircetich, S. M., M. E. Matheron, and W. R. Schreader. 1974. Sweet cherry root rot and trunk canker caused by *Phytophthora* spp. (abs.). *Proc. Am. Phytopathol. Soc.* 1:58.

Mircetich, S. M., W. J. Moller, and D. H. Chaney. 1974. Phytophthora crown rot and trunk canker of almond trees (abs.). *Proc. Am. Phytopathol. Soc.* 1:58.

Nichols, C. W., S. M. Garnsey, R. L. Rackham, S. M. Gotan, and C. N. Mahannah. 1964. Pythiaceous fungi and plant-parasitic nematodes in California orchards. I. Occurrence and pathogenicity of Pythiaceous fungi in orchards soils. *Hilgardia* 35:577–602.

Ohr, H. D., G. A. Zentmyer, E. C. Pond, and L. J. Klure. 1980. Plants in California susceptible to *Phytophthora cinnamomi*. *Univ. Calif. Div. Agric. Sci. Leaf.* 21178. 12 pp.

Reuther, W., J. A. Beutel, A. W. Marsh, et al. 1981. Irrigating deciduous orchards. *Univ. Calif. Div. Agric. Sci. Leaf.* 21212. 52 pp.

Rosenbaum, J. 1917. Studies of the genus *Phytophthora*. *J. Agric. Res.* 8:233–76.

Schwinn, F. J. 1983. New developments in chemical control of *Phytophthora*. In *Phytophthora: Its biology, taxonomy, ecology, and pathology*, edited by D. C. Erwin et al., 327–34. American Phytopathological Society, St. Paul.

Sewell, G. W. F., and J. F. Wilson. 1959. Resistance trials of some apple rootstock varieties to *Phytophthora cactorum* (L. & C.) Schroet. *J. Hort. Sci.* 34:51–58.

———. 1973. Phytophthora collar rot of apple: Seasonal effects on infection and disease development. *Ann. Appl. Biol.* 74:149–58.

Sewell, G. W. F., J. F. Wilson, and J. T. Dakwa. 1974. Seasonal variations in the activity in soil of *Phytophthora cactorum*, *P. syringae*, and *P. citricola* in relation to collar rot disease of apple. *Ann. Appl. Biol.* 76:179–86.

Smith, C. O., and J. T. Barrett. 1931. Crown rot of *Juglans* in California. *J. Agric. Res.* 43:885–904.

Smith, E. H. 1915. Pythiaceous infection of deciduous nursery stock. *Phytopathology* 5:317–22.

Smith, E. H., and J. F. Hunt. 1922. Pythiacystis blighting of nursery stock. *Calif. Agric. Exp. Ann. Rept.* 1920–21:63.

Smith, H. C. 1955. Collar-rot and crown rot of apple trees. *Orch. N. Z.* 28(10):16, 17, 19, 21.

———. 1956. Collar rot of apricots, peaches and cherries. *Orch. N. Z.* 29(9):22, 23, 25.

Smith, R. E. 1941. *Diseases of fruits and nuts.* Calif. Agric. Ext. Serv. Circ. 120. 167 pp.

Smith, R. E., and E. H. Smith. 1925. Further studies on Pythiaceous infection of deciduous fruit trees in California. *Phytopathology* 15:389–404.

Tsao, P. H. 1960. A serial dilution end-point method for estimating disease potentials of citrus *Phytophthoras* in soil. *Phytopathology* 50:717–24.

———. 1970. Selective media for isolation of pathogenic fungi. *Ann. Rev. Phytopathol.* 8:157–86.

———. 1983. Factors affecting isolation and quantification of *Phytophthora* from soils. In *Phytophthora: Its biology, taxonomy, ecology and pathology*, edited by D. C. Erwin et al., 219–36. American Phytopathological Society, St. Paul.

Tsao, P. H., and S. O. Guy. 1977. Inhibition of *Mortierella* and *Pythium* in a *Phytophthora*-isolation medium containing hymexasol. *Phytopathology* 67:796–801.

Tsao, P. H., and J. M. Menyonga. 1966. Response of *Phytophthora* spp. and soil microflora in the pimaricin-vancomycin medium (abs.). *Phytopathology* 56:152.

Tsao, P. H., and G. Ocana. 1969. Selective isolation of species of *Phytophthora* from natural soils on an improved antibiotic medium. *Nature* 223:636–38.

Tucker, C. M. 1931. *Taxonomy of the genus* Phytophthora *de Bary*. Missouri Univ. Agric. Exp. Stn. Res. Bull. 153, 208 pp.

———. 1933. *The distribution of the genus* Phytophthora. Missouri Univ. Agric. Exp. Stn. Res. Bull. 184, 80 pp.

Utkhede, R. S. 1983a. Inhibition of *Phytophthora cactorum* by a bacterial antagonist. *Can. J. Bot.* 61:3343–48.

———. 1983b. Inhibition of *Phytophthora cactorum* by bacterial isolates and effects of chemical fungicides on their growth and antagonism. *Z. Pflanzenkrank. Pflanzenschutz.* 90:140–45.

———. 1984a. Antagonism of isolates of *Bacillus subtilis* to *Phytophthora cactorum*. *Can. J. Bot.* 62:1032–35.

———. 1984b. Effect of nitrogen fertilizers and wood compost on the incidence of apple crown rot in British Columbia. *Can. J. Plant Pathol.* 62:329–32.

Waterhouse, G. M. 1956. *The genus* Phytophthora. Commonwealth Mycological Institute, Kew, England, 120 pp.

———. 1963. Key to the species of *Phytophthora* de Bary. Commonw. Mycol. Papers 92, 22 pp.

Welsh, M. F. 1942. Studies of crown rot of apple trees. *Can. J. Res.* 20:457–90.

Wilcox, W. F., S. N. Jeffers, J. E. K. Hayes, and H. S. Aldwinckle. 1985. Phytophthora species causing root and crown rot of cherry trees in New York (abs.). *Phytopathology* 75:1347.

Wilcox, W. F., and S. M. Mircetich. 1985a. Pathogenicity and relative virulence of seven *Phytophthora* spp. on mahaleb and mazzard cherry. *Phytopathology* 75:221–26.

———. 1985b. Influence of soil water matric potential on the development of Phytophthora root and crown rot of mahaleb cherry. *Phytopathology* 75:648–53.

———. 1985c. Effects of flooding duration on the development of Phytophthora root and crown rots of cherry. *Phytopathology* 75:1451–55.

———. 1987. Lack of host specificity among isolates of *Phytophthora megasperma*. *Phytopathology* 77:1132–37.

Young, R. A., and J. A. Milbrath. 1959. A stem canker disease of fruit tree nursery stock caused by *Phytophthora syringae* (abs.). *Phytopathology* 49:114.

Zentmyer, G. A. 1980. *Phytophthora cinnamomi* and the diseases it causes.. (Phytopathological Monograph 10.) American Phytopathological Society, St. Paul, 96 pp.

Postharvest Decay: Mycotoxins

Miscellaneous Fungi

Mycotoxins are extracellular metabolites that are produced by certain fungi and adversely affect plants and animals. In plants, shoot blight associated with the almond hull rot disease is attributed to fumaric acid, a compound nontoxic to humans and animals, produced by *Rhizopus* species. The term "mycotoxin" now is generally used only for those fungal toxins that are injurious to humans and animals. Some toxins in various food products can cause cancer and other disorders when ingested by animals and humans. Mycotoxins are produced only when these fungi grow in plant products under favorable environmental parameters of moisture and temperature. Mycotoxins are of some concern on pome and stone fruits stored for extended periods, but are of greater concern on nuts and dried fruits. Postharvest decays caused by mycotoxin-producing fungi can be prevented or reduced through proper handling of the products (Phillips 1984).

Symptoms. Infections and decay caused by the mycotoxin-producing fungi are no different from those caused by nontoxigenic fungi. Furthermore, mycotoxin-producing fungi appear to be as common in nature as are other fungi, including nontoxigenic strains of the toxin-producing species. One cannot easily identify crops affected with mycotoxin-producing strains of fungi. Special assays can indicate the presence of mycotoxins in crops.

Causal organisms. Fungal pathogens that produce mycotoxins are ubiquitous, and occur with other nontoxigenic fungi on the surfaces of fruit and nut crops. Some of the more common genera that include toxigenic species are *Alternaria*, *Aspergillus*, *Fusarium*, *Penicillium*, and *Trichoderma*. *Aspergillus* has been reported to produce toxins on kernels of almond, filbert, pecan, pistachio, and walnut, and on dried figs and prunes (Buchanan, Sommer, and Fortlage 1975; El-Behadli 1975; Kader et al. 1980). *Penicillium* produces a toxin on fresh apples and pears (Scott et al. 1972). Toxins of *Fusarium* and *Alternaria* have not been reported on commercial crops of fruit and nuts. *Aspergillus flavus* produces aflatoxin, a carcinogen that can be found in fig fruit and in kernels of almond and pistachio if they are not properly handled. Postharvest toxigenic pathogens discussed elsewhere in this volume include *Alternaria alternata* on fig, pome fruit, and stone fruit; *Aspergillus amstelodami*, *A. glaucus*, and *A. repens* on prune; *A. niger* on sweet cherry, fig, pomegranate, and prune; *Cladosporium herbarum* on sweet cherry and fig; *Fusarium moniliforme* and *F. solani* on fig; and *Penicillium expansum* on feijoa, pomegranate, pome fruit, and stone fruit.

Disease and mycotoxin development. Fungi that produce mycotoxins are found naturally in orchards as contaminants of fresh and dried fruit and nut crops. Yet mycotoxins are seldom detected on commercial produce because of the market's high quality standards (King, Miller, and Eldridge 1970; Miller and Tanaka 1963). Most of these toxigenic fungi develop on low-quality produce that usually is not marketed. Most are saprophytes or weak facultative parasites that can grow only on injured, senescing, or dead tissue. A few, however, are parasites with the ability to attack living tissue (English, Ryall, and Smith 1946; Phillips 1984). On nuts and dried fruits, environmental factors of greatest influence in mycotoxin production are temperature and available moisture (Cook and Papendick 1978). For almond and pistachio kernels, the cold storage moisture levels at which fungi cannot grow are 5 and 7 percent, respectively, but at high storage temperatures (room temperature) these moisture levels could be sufficient to promote fungal spore germination and mycelial growth. Also, insects feeding on dried prunes have been shown to increase the water present on the fruit, helping thus to establish fungal growth (El-Behadli 1975; Wells and Payne 1975). The mere isolation of these mycotoxin-producing fungi from crops does not necessarily mean that they are dangerous or toxic to humans or animals. Evidence shows that when some of the fungi are allowed to grow in the presence of other natural microflora, little or no mycotoxin is produced, whereas pure cultures grown under optimal conditions produce high levels of toxin (Phillips et al. 1979).

Control. Few mycotoxins are reported on fresh fruit, so the control of the fungi has been mostly limited to field practices that provide high quality crops relatively free of blemishes and other injuries that promote fungal decay (English, Ryall, and Smith 1946; King, Miller, and Eldridge 1970; Phillips 1984). For dried products, high quality means freedom from insect injuries—the common sites of fungal infection—and the establishment of handling and storage conditions that prevent fungal growth (dried fruit with water activity below 0.70) and involve the removal of damaged products that could otherwise serve as sites for fungal infection and that may already contain mycotoxins. Temperatures up to 80°C are used to dry fruit, and only heat-tolerant fungi such as *Aspergillus* and *Rhizopus* survive these high temperatures. Peaches, apricots, and some of the prune plums and raisin grapes are treated with sulfur dioxide before drying and that kills contaminating fungi, including *Aspergillus* and *Rhizopus*, and protects the fruit from infection during storage. Chemical preservatives, such as sorbic acid, also may be used to reduce the potential fungal decay of dried fruit. Although efforts are constantly being made to reduce the use of pesticides for the control of plant pathogens, some of the toxins produced by decay fungi may pose greater health threats than the chemicals used to control them.

Efforts to detoxify mycotoxins in cotton seed and corn by ammoniation (Park et al. 1988) have proven effective, and the by-products of ammoniation are now being studied for their potentially harmful health effects.

Strategies for the reduction and elimination of aflatoxin—caused by the soilborne fungi *Aspergillus flavus* and *A. parasiticus*—on food crops have been redirected to studies involving biocompetitive agents (Ashworth, Schroeder, and Langely 1965; Cleveland and Cotty 1989; Cotty 1989; Kimura and Hirano 1988) and the enhancement of crop resistance using naturally occurring resistant germplasm as well as genetic engineering (Widstrom 1987; Zumo and Scott 1989).

REFERENCES

Ashworth, L. J., H. W. Schroeder, and B. C. Langley. 1965. Aflatoxins: Environmental factors governing occurrence in spanish peanuts. *Science* 148:1228.

Buchanan, J. R., N. E. Sommer, and R. J. Fortlage. 1975. *Aspergillus flavus* infection and aflatoxin production in fig fruits. *Appl. Microbiol.* 30:238–41.

Cleveland, T. E., and P. J. Cotty. 1989. Reduced pectinase activity of *Aspergillus flavus* is associated with reduced virulence in cotton (abs.). *Phytopathology* 79:1208.

Cook, R. J., and R. I. Papendick. 1978. Role of water potential in microbial growth and development of plant disease with special reference to postharvest pathology. *HortScience* 19:559–69.

Cotty, P. J. 1989. Prevention of aflatoxin contamination with strains of *Aspergillus flavus* (abs.). *Phytopathology* 79:1153.

El-Behadli, A. H. 1975. *Mold contamination and infection of prunes and their control.* Ph.D. dissertation, University of California, Davis. 80 pp.

English, H., A. I. Ryall, and E. Smith. 1946. Blue mold decay of Delicious apples in relation to handling practices. *U.S. Dept. Agric. Cir.* 751.

Kader, A. A., J. M. Labavitch, F. G. Mitchell, and N. F. Sommer. 1980. Quality and safety of pistachio nuts as influenced by postharvest handling procedures. In *Calif. Pistachio Assoc. Annu. Rep.* 1980, pp. 44–52.

Kimura, N., and S. Hirano. 1988. Inhibitory strains of *Bacillus subtilis* for growth and aflatoxin-production of aflatoxigenic fungi. *Agric. Biol. Chem.* 52:1173.

King, A. D., Jr., M. J. Miller, and L. C. Eldridge. 1970. Almond harvesting, processing, and microbial flora. *Appl. Microbiol.* 20:208–14.

Miller, M. W., and H. Tanaka. 1963. Microbial spoilage of dried prunes. III. Relation of equilibrium relative humidity to potential spoilage. *Hilgardia* 34:183–90.

Park, D. L., L. S. Lee, R. L. Price, and A. E. Pohland. 1988. Review of the decontamination of aflatoxins by ammoniation: Current status and regulation. *J. Assoc. Off. Anal. Chem.* 71:587–96.

Phillips, D. J. 1984. Mycotoxins as a postharvest problem. In *Postharvest pathology of fruits and vegetables: Postharvest losses in perishable crops.* ed. H. E. Moline, 50–54. Univ. Calif. Agr. Exp. Stn. Bull. 1914. 80 pp.

Phillips, D. J., B. Mackey, W. R. Ellis, and T. H. Hansen. 1979. Occurrence and interaction of *Aspergillus flavus* with other fungi on almond. *Phytopathology* 69:829–31.

Scott, P. M., W. E. Miles, P. Toft, and J. G. Dube. 1972. Occurrence of patulin in apple juice. *J. Agric. Food Chem.* 20:450–51.

Sommer, N. F., J. R. Buchanan, and R. E. Rice. 1980. Aflatoxin—Some problems and perspectives. *Calif. Pistachio Assoc. Annu. Rep.* 1980. p. 53.

Wells, J. M., and J. A. Payne. 1975. Toxigenic *Aspergillus* and *Penicillium* isolates from weevil-damaged chestnuts. *Appl. Microbiol.* 30:536-40.

Widstrom, N. W. 1987. Breeding strategies to control aflatoxin contamination of maize through host plant resistance. In *"Aflatoxin in maize: A proceedings of the workshop"*, eds. M. S. Zuber, E. B. Lillehoj, B. L. Renfro. CIMMYT, Mexico, D. F., 212.

Wyllie, T. D., and L. G. Morehouse, eds. 1977. *Mycotoxic fungi, mycotoxins, mycotoxicoses—an encyclopedic handbook. vol. 1. Mycotoxic fungi and chemistry of mycotoxins.* Marcell Dekker, Inc., N. Y. 538 pp.

Zumo, N., and G. E. Scott. 1989. Evaluation of field inoculation techniques for screening maize genotypes against kernel infection by *Aspergillus flavus* in Mississippi. *Plant Dis.* 73:313.

Replant Diseases of Fruit Trees

Various Causal Factors

In many parts of the world, the replanting of apple or peach on land previously planted to the same type of tree results in poor growth and sometimes death of the newly planted trees. In the United States this has usually been referred to as the "replant problem." Two types of replant disease have been recognized in Europe and to some extent in this country (Mai and Abawi 1981; Savory 1966): a "specific replant disease" limited to one crop or to closely related crops, and a "nonspecific replant disease" affecting several fruit-tree crops, and often correlated with high populations of plant-parasitic nematodes. Apple, cherry, peach, and citrus are reported to be severely affected by specific replant diseases. Less severely affected are pear and plum (Savory 1966).

Although many of the replant diseases of fruit crops appear to fall nicely into either the specific or nonspecific category, some do not, or at least opinions differ among investigators (Jackson 1973; Sewell 1979, 1981). Also, several investigators (Bunemann and Jensen 1970; Hoestra 1968; Savory 1966) have reported that, with apple, a "nonspecific replant disease" soil could become "specific" after cropping for a short time (Jaffee, Abawi, and Mai 1982). In New York, the apple replant disorder has specific and nonspecific characteristics (Jaffee, Abawi, and Mai 1982; Mai and Abawi 1981).

Because of this rather confused situation, this discussion will not attempt to strictly segregate the replant diseases of fruit crops into specific and nonspecific categories, but will deal more with the implication of various known or suspected causes identified with this complex disease syndrome.

In the specific replant problem, as indicated by Koch (1955) in Ontario, Canada, apricot and sour cherry grow normally after peach, while peach does not; peach grows well after apple or plum; apple grows well after cherry but not after apple. In California, Proebsting and Gilmore (1941) reported that sweet cherry and myrobalan plum grow well on old peach soils. So do almond and apricot unless they are on peach rootstock. The replant problem in California has attracted attention only in peach (Hine 1961a,b; Proebsting 1950; Proebsting and Gilmore 1941) and in recent years its importance appears to have diminished, possibly because of extensive soil fumigation (Lembright 1976). Most of the worldwide research on replant diseases of fruit crops has been devoted to apple, peach, and citrus, with a limited amount on other stone fruit and pear (Mai and Abawi 1981; Patrick, Toussoun, and Koch 1964; Rice 1974; Savory 1966).

Symptoms. Unfortunately, there are no truly diagnostic symptoms of replant diseases of fruit trees. In general, affected trees show subnormal growth (**fig. 4A**), shortened internodes, and abnormally small leaves. In newly established replant orchards, tree growth is characteristically uneven and there may be some mortality of young trees (**fig. 4B**) (Covey, Benson, and Haglund 1979; Mai and Abawi 1981). In peach, an interveinal chlorosis has been reported by Koch (1955). The aboveground effects usually are more pronounced in younger than in older trees. According to Sewell (1981), "the apple replant disease is an ill-defined 'poor growth phenomenon' with no diagnostic symptoms."

The root system of affected trees often shows a variety of symptoms. The system is reduced in size, with darkening of some of the roots and a killing and distortion of many of the feeder rootlets. In apple, these rootlets sometimes show a witches'-broom morphology. Root symptoms are more evident in young trees but also occur in mature orchards.

Causes. The nature of the replant diseases of fruit trees has been and continues to be an extremely perplexing question. Mai and Abawi (1981) stated, "although the etiology of replant diseases is incompletely understood, most investigators have concluded that replant diseases are caused by nematodes and several other soilborne organisms." The following are considered possible etiological factors in the fruit-tree replant disease syndrome.

Nematodes. Although research indicates that a variety of soilborne microorganisms probably are involved in replant disorders (Harr and Klingler 1976; Jaffee 1981; Mai and Abawi 1978; Savory 1966; Sewell 1981), there is increasing evidence that certain plant-parasitic nematodes play an important role in the root injury associated with many of these problems (Harr and Klingler 1976; Mai and Abawi 1981). These nematodes include the endoparasitic root-lesion nematodes (*Pratylenchus* spp.), which are damaging to both pome and stone fruit especially in plantings on coarse-textured soil; the ectoparasitic dagger nematodes (*Xiphinema* spp.), which in large numbers cause unthriftiness in fruit trees and, in addition, are vectors of several economically important virus diseases; the ring nematodes (*Criconemella* spp.), which are implicated in the peach-tree short-life disorder in the southeastern United States and in the bacterial canker problem of stone fruit in California; and the endoparasitic root-knot nematodes (*Meloidogyne* spp.), which are especially important on stone fruit in the warm-temperate regions of the world (Chitwood 1949; Colbran 1953; Harr and Klingler 1976; Hoestra 1968; Jaffee 1981; Mai and Abawi 1981).

Although nematodes apparently are the most important factor in the replant problem of apple in New York (Mai and Abawi 1981) and Australia (Colbran 1953), they appear to play little if any role in this disorder in either Washington (Covey et al. 1984) or England (Sewell 1981). Likewise, no evidence was obtained by Proebsting (1950) and Proebsting and Gilmore (1941) that nematodes are responsible for the peach replant problem in California. In the cherry replant disease in Switzerland, Harr and Klingler (1976) reported that the most important causal factor in sandy soil was a root-lesion nematode (*P. penetrans*), whereas in heavy soil the fungus *Thielaviopsis basicola* was the important incitant. They also found a synergistic relationship between these two organisms in this disease complex. It is possible that plant-parasitic nematodes may interact with rhizosphere microorganisms, both parasitic and nonparasitic, in the replant disease syndrome (Jaffee 1981; Mai and Abawi 1981).

Fungi. Several researchers have presented evidence supporting the hypothesis that certain replant diseases are caused by soilborne fungi. Savage (1953), in Georgia, suggested that *Clitocybe tabescens* was the cause of the peach replant problem. Peterson's studies (1961), however, failed to corroborate this finding. Hine (1961b) attributed the poor growth of replanted peach trees in California to *Pythium ultimum, Fusarium* sp., and *Rhizoctonia* sp., and Hendrix, Powell, and Owens (1966) attributed the decline of peach trees in Georgia to feeder-root necrosis caused by *Pythium* spp. Mircetich (1971), however, found no relationship between peach-tree decline and the occurrence of *P. ultimum, P. irregulare,* and *P. vexans* in peach roots and orchard soil in the mid-Atlantic and southeastern states. Sewell (1981) presented circumstantial and experimental evidence implicating *Pythium* spp. in the apple replant disease in England. Harr and Klingler (1976) implicated both *Thielaviopsis basicola* and a root-lesion nematode in the cherry replant disease in Switzerland, and Taylor (1978) reported the involvement of several basidiomycetous fungi in a cherry decline and replant disease in New Zealand. These studies, and the beneficial results from fungicidal soil fumigants in some instances (Covey et al. 1984; Hoestra 1968; Sewell and White 1979), strongly support a role for pathogenic soilborne fungi in replant diseases.

Toxins. Proebsting and Gilmore (1941), Proebsting (1950), and Havis and Gilkeson (1947) indicated that decomposing peach roots contain a material toxic to young peach trees. Likewise, Fastabend (1955) and Borner (1959) reported that apple-root residues, when added to the soil, are toxic to young apple trees. Borner suggested that the phenolic compound phlorizin or its breakdown products were probably the toxins involved. He later reported two other phytotoxins (patulin and an unidentified phenolic compound) synthesized by *Penicillium expansum* in soils supplemented with apple leaf and root residues (Patrick, Toussoun, and Koch 1964).

Havis and Gilkeson (1947) and Proebsting (1950) reported that peach roots added to the soil were not always toxic to peach trees growing therein. Proebsting explained the exceptions as indicating that the phytotoxin from peach roots forms only when certain soil microorganisms decompose the roots. Later work by Patrick (1955) and Wensley (1953) tended to substantiate this theory. Patrick found that decomposition products of amygdalin (benzaldehyde and hydrogen cyanide), as well as compounds produced by the decomposition of peach roots, were toxic to rootlets of growing peach trees. Hine (1961a), however, did not believe that the breakdown of amygdalin to hydrogen cyanide contributed to the peach replant problem in California. In a fairly recent study in Japan, Mizutani et al. (1979) reported that condensed tannins appeared to be the main root-growth inhibiting fractions obtained in peach-root extracts. Concentrations of these materials were greater under anaerobic than aerobic conditions, and their presence was thought to play a role in the peach replant problem.

Gilmore (1959, 1963), in pot experiments, found that peach-root bark somewhat reduced growth of peach seedlings and that peach-root wood (previously found fairly harmless) greatly stunted the seedlings. The stunting effect could be offset to some extent by adding adequate nitrogen to the soil.

Nutrient imbalance. The deficiency of certain nutrients in old orchard soils has been considered a possible cause of replant diseases. However, Proebsting (1950) and Proebsting and Gilmore (1941) obtained no improvement in replant peach trees by adding large amounts of nitrogen, phosphorus, potassium, manganese, zinc, or copper. Havis et al. (1958) reported no response from adding nitrogen, phosphorous, potassium, dolomitic limestone, or trace elements to old peach soil. Fastabend (1955) found that apple trees in old apple soil did not respond to the addition of copper, boron, and certain other elements. Hewetson (1953), however, obtained a response in peach trees from the application of 23–21–27 NPK fertilizer. Also, Upshall and Ruhnke (1935) reported that the growth of pear, mazzard and mahaleb cherries, and myrobalan plum on old peach land was poor except where adequate nitrogen was present. Savory (1966), after reviewing the literature on the subject, concluded that "it would seem most unlikely that a nutrient deficiency or imbalance can be the primary cause of specific replant diseases."

Viruses. Certain soilborne, nematode-vectored viruses, such as tomato ringspot virus, could be involved in a type of fruit-tree replant problem. Strains of this virus cause several important diseases of fruit crops—apple union necrosis and decline, yellow bud mosaic of peach, cherry, and almond, brownline of prune, and Prunus stem pitting. After removal of infected trees, portions of the root system remaining in the soil could be a source of inoculum for infection of replants as well as sustenance for the nematode vectors. All of the known virus-induced "replant" diseases of fruit trees would appear to fall into the nonspecific category.

Other factors. Several investigators (Hoestra 1968; Jaffee 1981, Jaffee, Abawi, and Mai, 1982; Savory 1966; Utkhede, Li, and Owen 1987) have hypothesized that nonparasitic rhizosphere organisms (possibly bacteria, actinomycetes, or fungi) are involved in the fruit-tree replant problem. Exactly how these organisms might induce a replant disease has not been shown. It is possible, however, that the rhizosphere organisms might act on root exudates to produce allelopathic (phytotoxic) compounds (Jaffee 1982; Rice 1974). Much additional research is needed to clarify the possible role that nonparasitic rhizosphere organisms play in replant diseases. In Washington, the retardation of apple-tree growth has been related to arsenic residues in the soil that date back to the time when lead arsenate was commonly used for insect control (Benson 1976; Benson, Covey, and Haglund 1978). But the current replant problem in Washington apples is thought to be specific in nature, of unknown but apparently biotic etiology, and with nematodes playing no more than a minor role (Covey, Benson, and Haglund 1979; Covey et al. 1984).

Control. The effects of replant diseases have been measured by the use of such parameters as the annual increase in circumference of the tree trunk, the length of the current year's shoots, and crop production. Preplant fumigation with a broad-spectrum biocide such as chloropicrin or methyl bromide has been most successful in controlling replant disorders of both pome and stone fruit. In the control of the nonspecific problem in fruit crops, presumably caused mainly by nematodes, preplant fumigation with either 1,3-dichloropropene or ethylene dibromide has reduced nematode populations and increased tree growth and crop yield, especially on coarse-textured soil (Bunemann and Jensen 1970; Mai and Abawi 1981). There is some evidence that nonfumigant nematicides may also be effective, but more research on these materials is needed (Mai and Abawi 1981).

For control of the specific replant disease of apple in Washington, chloropicrin and methyl bromide soil fumigation and a formaldehyde soil drench were effective, whereas nematicidal rates of 1,3-dichloropropene failed (Covey, Benson, and Haglund 1979; Covey et al. 1984). Somewhat similar results were obtained in

England by Sewell and White (1979). An apple-replant-problem site benefited from soil treatment with chloropicrin, gamma radiation, and heat (75°C for 30 minutes) (Jaffee, Abawi, and Mai 1982). Here, too, a nematicide (DD) was relatively ineffective.

In British Columbia, Yorston (1987) showed that the addition of 11–55–0 NPK fertilizer to soil treated with formaldehyde (formalin) increased apple shoot growth at one site (Kelowna) but not at another (Summerland). Increased shoot growth resulted from adding ammonium phosphate to the soil as well as from a soil drench containing *Enterobacter aerogenes*.

Although soil treatment with broad-spectrum biocides has been effective in most replant-disease situations where it has been tested, it has not helped in all cases. Jackson (1973) planted apple and cherry rootstocks on apple land that was conducive to the apple replant disease. Preplant soil fumigation with chloropicrin greatly increased the growth of apple but not of cherry. Also, in some cases soil fumigation has adversely affected the growth of fruit trees. Repeated fumigation of peach nursery soils in Pennsylvania with broad-spectrum biocides has resulted in poor seedling growth, even though the initial treatments with the same fumigants enhanced seedling growth. This injurious effect has usually been attributed to either the elimination of mycorrhizal fungi or the direct toxicity of biocide residues (Mai and Abawi 1981).

A nonchemical strategy that has had partial success in controlling replant diseases is to grow a crop or crops not susceptible to the disease in question for several years (Savory 1966). Much additional research is needed on the control of replant diseases of fruit crops, but this can best be done only after more is known of their etiology.

REFERENCES

Benson, N. R. 1976. Retardation of apple tree growth by soil arsenic residues. *J. Am. Soc. Hort. Sci.* 101:251–53.

Benson, N. R., R. P. Covey Jr., and W. A. Haglund. 1978. The apple replant problem in Washington State. *J. Am. Soc. Hort. Sci.* 103:156–58.

Blake, M. A. 1947. The growing of peach after peach was an old world problem in 1768. *New Jersey State Hort. Soc. News.* 28:1910.

Borner, H. 1959. The apple replant problem. I. The excretion of phlorizin from apple root residues and its role in the soil sickness problem. *Contrib. Boyce Thompson Inst.* 20:39–56.

Bunemann, G., and A. M. Jensen. 1970. Replant problem in quartz sand. *HortScience* 5:478–79.

Chitwood, B. G. 1949. Ring nematode (Criconematinae) a possible factor in decline and replanting problems of peach orchards. *Proc. Helminthol. Soc. Wash. D.C.* 16:6–7.

Colbran, R. C. 1953. Problems in tree replacement. I. The root lesion nematode *Pratylenchus coffeae* Zimmerman as a factor in the growth of replanted trees in apple orchards. *Austral. Agric. Res.* 4:384–89.

Covey, R. P., N. R. Benson, and W. A. Haglund. 1979. Effect of soil fumigation on the apple replant disease in Washington. *Phytopathology* 69:684–86.

Covey, R. P., B. L. Koch, H. J. Larsen, and W. A. Haglund. 1984. Control of apple replant disease with formaldehyde in Washington. *Plant Dis.* 68:981–83.

Fastabend, H. 1955. Über die Ursachen der Bodenmüdigkeit in Obstbaumschulen Landwirtschaft-Angewandte Wissenschaft. In *Sonderheft Gartenbau IV*, 94. Landwirtschaftsverlag, Hiltrup/Mainster.

Gilmore, A. E. 1959. Growth of replanted peach seedlings in pots. *Proc. Am. Soc. Hort. Sci.* 80:204–06.

———. 1963. Pot experiments related to the peach replant problem. *Hilgardia* 34:63–78.

Harr, J., and J. Klingler. 1976. Einfacher und kombinierter Effeckt von *Pratylenchus penetrans* und *Thielaviopsis basicola* auf das Wachstum von Kirschen-Stecklingen. *Z. Pflanzenkrank. Pflanzensch.* 83:615–19.

Havis, L., and A. L. Gilkeson. 1947. Toxicity of peach roots. *Proc. Am. Soc. Hort. Sci.* 50:206.

Havis, L., H. F. Morris, R. Manning, and J. E. Demmon. 1958. Repsonses of replanted peach trees to soil treatments in field tests in Texas. *Proc. Am. Soc. Hort. Sci.* 71:67–76.

Hendrix, F. F., Jr., W. M. Powell, and J. H. Owens. 1966. Relation of root necrosis caused by *Pythium* species to peach tree decline. *Phytopathology* 56:1229–32.

Hewetson, F. N. 1953. Re-establishing the peach orchard: The influence of various nutrient solutions and fertilizers on the growth and development of one-year peach trees. *Proc. Am. Soc. Hort. Sci.* 69:122–25.

Hine, R. B. 1961a. The role of amygdalin breakdown in the peach replant problem. *Phytopathology* 51:10–13.

———. 1961b. The role of fungi in the peach replant problem. *Plant Dis. Rep.* 45:462–65.

Hoestra, H. 1968. *Replant diseases of apple in The Netherlands.* Mededelingen Landbouwhogeschappen, Wageningen, 105 pp.

Jackson, J. E. 1973. Effects of soil fumigation on the growth of apple and cherry rootstocks on land previously cropped with apples. *Ann. Appl. Biol.* 74:99–104.

Jaffee, B. A. 1981. *Etiology of an apple replant disease.* Ph.D. thesis, Cornell University, 65 pp.

Jaffee, B. A., G. S. Abawi, and W. F. Mai. 1982. Role of soil microflora and *Pratylenchus penetrans* in an apple replant disease. *Phytopathology* 72:247–51.

Johanson, F. D. 1950. A preliminary report on the incidence of two types of plant parasitic nematodes on peaches in Connecticut, *Storrs Agric. Exp. Stn. Bull.* 10.

Koch, L. W. 1955. The peach replant problem in Ontario. *Can. J. Agric.* 33:450–60.

Lembright, H. W. 1976. Solutions to problems associated with the replanting of peaches. *Down to Earth* 32:12–21.

Mai, W. F., and G. S. Abawi. 1978. Determining the cause and extent of apple, cherry, and pear replant diseases under controlled conditions. *Phytopathology* 68:1540–44.

———. 1981. Controlling replant diseases of pome and stone fruits in Northeastern United States by preplant fumigation. *Plant Dis.* 65:859–64.

McCalla, J. M., and F. A. Haskins. 1964. Phytotoxic substances from soil microorganisms and crop residues. *Bot. Rev.* 28:181–207.

Mircetich, S. M. 1971. The role of *Pythium* in feeder roots of diseased and symptomless peach trees and in orchard soils in peach tree decline. *Phytopathology* 61:357–60.

Mizutani, F., H. Itamura, A. Sugiura, and T. Tomana. 1979. [Studies on the soil sickness problem for peach trees II. Condensed tannins as growth inhibitors from peach roots.] *J. Jap. Soc. Hort. Sci.* 48:279–87. (*Rev. Plant Pathol.* 59:4689. 1980.)

Parker, K. G., W. F. Mai, G. H. Oberly, K. D. Brase, and K. D. Hickey. 1966. Combating replant problems in orchards. *N.Y. Agric. Exp. Stn. Bull.* 1169, 19 pp.

Patrick, Z. A. 1955. The peach replant problem in Ontario. II. Toxic substances from microbial decomposition of products of peach root residue. *Can. J. Agric.* 33:461–86.

Patrick, Z. A., T. A. Toussoun, and L. W. Koch. 1964. Effect of crop-residue decomposition products on plant roots. *Ann. Rev. Phytopathol.* 2:267–92.

Patrick, Z. A., T. A. Toussoun and W. C. Snyder. 1963. Phytotoxic substances in arable soils associated with decomposition of plant residues. *Phytopathology* 53:152–61.

Petersen, D. H. 1961. The pathogenic relationship of *Clitocybe tabescens* to peach trees. *Phytopathology* 51:819–23.

Prince, V. E., L. Havis, and L. E. Scott. 1955. Effect of soil treatment in a greenhouse study of the peach replant problem. *Proc. Am. Soc. Hort. Sci.* 65:139–48.

Proebsting, E. L. 1950. A case history of a "peach replant" situation. *Proc. Am. Soc. Hort. Sci.* 38:21–25.

Proebsting, E. L., and A. E. Gilmore. 1941. The relation of peach root toxicity to re-establishing of peach orchards. *Proc. Am. Soc. Hort. Sci.* 38:21–25.

Rice, E. L. 1974. *Allelopathy*. Academic Press, New York, 353 pp.

Savage, E. F. 1953. Root rot of peaches. *Am. Fruit Grower* 73:1642–43.

Savory, B. M. 1966. Specific replant diseases causing root necrosis and growth depression in perennial fruit and plantation crops. *Commonw. Bur. Hortic. Kent., Res. Rev. No.* 1, 64 pp.

Sewell, G. W. F. 1979. Reappraisal of the nature of the "specific replant disease" of apple. *Rev. Plant Pathol.* 58:209–11.

———. 1981. Effects of *Pythium* species on the growth of apple and their possible causal role in apple replant disease. *Ann. Appl. Biol.* 97:31–42.

Sewell, G. W. F., and G. C. White. 1979. The effects of formalin and other soil treatments on the replant disease of apple. *J. Hortic. Sci.* 54:333–35.

Taylor, J. B. 1978. The source of infections by basidiomycete fungi causing a decline and replant disease in Central Otago, New Zealand. In *Microbiol ecology*, ed. M. W. Loutit and J. A. R. Miles, 346–49. Springer-Verlag, New York.

Upshall, W. H., and G. N. Ruhnke. 1935. Growth of fruit tree stocks as influenced by a previous crop of peach trees. *Sci. Agric.* 16:16–20.

Utkhede, R., T. Li, and G. Owen. 1987. Replant problem: Apple. In *Research Reports, 61st Annual Western Orchard Pest and Disease Management Conference, Portland,* 5.

Ward, G. M., and A. B. Durkee. 1956. The peach replant problem in Ontario. III. Amygdalin content of the peach tree tissues. *Can. J. Bot.* 34:419–22.

Wensley, R. N. 1953. Preliminary microbiological investigations of the peach replanting problem (abs.) *Can. Phytopathol. Soc. Proc.* 21:19.

———. 1956. The peach replant problem. IV. Fungi associated with replant failure and their importance in fumigated and nonfumigated soil. *Can. J. Bot.* 34:967–82.

Yorston, J. 1987. Apple replant disease. In *Research reports, 61st Annual Western Orchard Pest and Disease Management Conference, Portland,* 4.

Shoestring (Armillaria) Root Rot of Fruit and Nut Trees

Armillaria mellea (Vahl:Fr.) Kummer

Shoestring root rot, also known as "Armillaria root rot," "mushroom root rot," "shoestring fungus rot," and "oak root fungus disease," occurs on a wide range of hosts including woody and herbaceous plants throughout both temperate and tropical regions (Smith 1941; Thomas, Wilhelm, and MacLean 1953). According to Gardner and Raabe (1963) the disease was first recognized by Hartig on pine trees in Germany in 1873. In the United States the disease is of greatest importance on fruit trees west of the Rocky Mountains. In California it was first recorded on lemon, then on grape in Sonoma County in 1881, and at Oakville in Napa Valley in 1887. It was of great concern to walnut growers in Santa Barbara County in 1886–88, and was reported on peach and prune in Santa Clara County in 1901. A 1945 survey showed that 34 of 65 orchards in Santa Clara County had one or more infected sites (Schneider, Bodine, and Thomas 1945). The importance of shoestring root rot is emphasized by the host list compiled by Raabe, who also included the years and countries in which the disease was first reported (Raabe 1962). Once established, *Armillaria* is extremely difficult to eradicate from the orchard.

Symptoms. External symptoms of the disease are poor shoot growth, premature yellowing and dropping of leaves, and eventual death of the tree. Sudden collapse of the tree can occur during the dry summer months, an occurrence that can be confused with rodent injury and Phytophthora root rot. Symptoms usually appear on one or two limbs, then in two or three years spread throughout the tree. The roots (up to and including the crown) are killed. Between the brown dead bark and the wood are white to slightly yellow, flat, fan-shaped, feltlike plaques of mycelium (**fig. 5E**). Dark brown, somewhat shiny, rootlike mycelial strands (rhizomorphs) arise from the diseased areas and extend along the surface of the bark (**fig. 5D**) and a short distance into the soil. Rhizomorphs are not so consistently present as to be a ready means of diagnosis, especially on crops like grapes. Sporophores of the fungus commonly occur in clusters about the base of the diseased trees between October and February (**fig. 5A,B,C**). The decayed bark is not sour or putrid but has a sharp, rather agreeable mushroom odor.

Invasion by another fungus, a species of *Oxyporus* (*Poria*) (DeVay et al. 1968), is often confused with that of *Armillaria*. *Oxyporus* produces somewhat more wefty or cottony plaques, irregular in shape and not confined between the wood and bark but occurring throughout the bark and into the wood. It is primarily a woodrotting fungus and usually can be distinguished from *Armillaria* by its powdery mycelium and the soft, crumbling condition of the wood it attacks, as contrasted to the firm condition of the wood attacked by *Armillaria*. Furthermore, this fungus does not develop rhizomorphs, nor does it have the fresh mushroom odor produced by *Armillaria*. Also, its sporophores are totally different from those of *Armillaria*.

Causal organism. The rhizomorphic stage of *Armillaria*, first described as *Rhizomorpha subterranea* Pers. and *R. subcorticalis* Pers., was later shown (Hartig 1873,

1874) to give rise to the mushroom stage, then known as *Agaricus melleus* but now named *Armillaria* (*Armillariella*) *mellea* (Vahl:Fr.) Kummer. This fungus is bifactorially heterothallic, with a diploid vegetative phase.

The characteristic rhizomorphs of the fungus are not always produced on tree-hosts and have never been observed on grape. In culture, rhizomorph formation occurs when ethanol and related compounds are present in the medium (Weinhold and Garraway 1963, 1966; Garraway and Weinhold 1968). Certain naturally occurring materials in woody plants have also been shown to stimulate rhizomorph formation. Ethanol in glucose media was found to inhibit production of phenols, which are thought to prevent thallus growth (Vance and Garraway 1973).

The mushrooms, which vary in size, are usually honey-colored and have a delicate annulus (ring) on a stout stem. They are borne in clusters at the base of the tree trunk (**fig. 5A,B,C**). The caps are from 3 to 15 cm wide; the spores are colorless, elliptical or round, 7 to 10 μm in dimension and borne on white or dull-white gills. Wolpert (1924) found that *Armillaria* grows at a range of pH between 2.9 and 7.4, about 5.0 being optimum (for germination of spores, however, the optimum is 6.5). The fungus first destroys the cell-wall cellulose and associated pentosans and later the lignin. Lanphere (1934) found that rhizomorphs of *A. mellea* produce the enzymes amylase, catalase, inulase, lipase, pepsin, and sucrase.

Clones of the fungus collected from various areas of California exhibit different degrees of pathogenicity on Elberta peach in the orchard (Wilbur, Munnecke, and Darley 1972) and on plants in containers (Raabe 1967). In California, the biological species of *Armillaria* isolated are in Group VI, *A. mellea* sensu stricto (J. E. Adaskaveg, personal communication), while Proffer, Jones, and Ehret (1987) suggested that in orchard crops biologic species Group I (*A. ostoyae*) and Group III should also be considered as pathogens.

Disease development. The pathogen survives from one year to the next in the soil on roots of affected hosts. Unassociated with the host, it may survive for several years on woody materials in the soil. The rhizomorphs are the principal agents of infection. Under California conditions the basidiospores, being both comparatively rare and certainly more evanescent than the rhizomorphs, probably play no part in infection. Rhizomorphs are moderately abundant and are produced whenever the fungus finds conditions favorable for growth. The fungus is disseminated on infected material by orchard workers and by the movement of surface water. The basidiospores are doubtless carried considerable distances in air currents. In the soil, the fungus spreads by growing out along the roots of affected trees; if such roots touch those of healthy trees, the fungus may move into the latter. The rhizomorphs do not extend for great distances in the soil, as was formerly believed. British work (Morrison 1976, 1982) indicates that the formation and growth of rhizomorphs occur mostly in the upper 30 to 35 cm of soil and that their growth is enhanced by organic matter. No wounds are necessary for infection: Thomas (1934) found the rhizomorphs capable of penetrating directly through the bark of roots. Infection occurs and the disease develops at an optimum soil temperature of 17° to 24°C. Such temperatures occur during spring and early summer in California (Bliss 1946).

Thomas (1934), dealing primarily with the early stages of infection, found some evidence of destruction of the suberized cells of the bark and death of cells of the cortex ahead of the invading rhizomorph. After gaining entrance, the rhizomorph grows rapidly and kills the cells. The cellulose of cells is destroyed, followed later by destruction of lignin, where this is present. In resistant hosts the invaded regions are frequently walled off by a phelloderm.

Trees usually die in one to three years, depending on their susceptibility. Death is hastened by summer heat and lack of soil moisture.

Host susceptibility. The pathogen has an extremely wide host range (Raabe 1962, 1979). Apple, cherry, almond, olive, peach, prune, plum, and apricot are highly susceptible. The pear and fig are comparatively resistant. The fairly high resistance of the California black walnut (*Juglans hindsii*) and Paradox hybrid has led to their wide use as rootstocks for Persian (English) walnut (*J. regia*), whose roots are highly susceptible. Marianna 2624 and certain selections of myrobalan plum exhibit resistance.

Control. Soil fumigation and resistant rootstocks have been used with moderate success (O'Reilly 1963; Sisson, Lider, and Kasimatis 1978). The value of such control measures as destruction of the mushrooms, use of nitrogen to stimulate growth of affected trees, and care not to introduce the fungus, is difficult to evaluate.

The fumigation of soil before planting trees eliminates most of the *Armillaria* fungus, but spot treatment is necessary later to disinfest areas where the fungus was not killed. Most important in fumigation is low soil moisture (Bliss 1951; Thomas and Lawyer 1939) and proper application of the fumigants—carbon disulfide or methyl bromide. In preparation for fumigation, diseased trees should be removed during the winter months and, along with them, all roots greater than 1½ inches in diameter in the top 2 feet of soil. Diseased roots and stumps should be burned in place. It is necessary to remove partially affected trees and the adjacent apparently unaffected trees surrounding the margin of the *Armillaria* site. After tree removal, an entire summer without irrigation is needed for the soil to dry out. Planting a deep-rooted, unirrigated cover crop such as mustard, sudangrass, or safflower will help reduce the soil moisture. In heavy soils the moisture should be tested to a depth of at least 5 to 6 feet. Fumigation is done in the fall while the soil is still warm and low in moisture. The soil should be prepared for fumigation by cultivation to break up the clods; if a plow pan or hardpan is present, subsoiling or deep chiseling is necessary. After the final cultivation and before fumigation, the top 6 to 8 inches of soil should be moistened lightly as an aid to penetration by the soil-fumigation equipment and to help prevent escape of the fumigant.

The equipment used for applying carbon disulfide depends upon the size of the area to be treated. For small areas, hand equipment is most feasible. A self-measuring, force-fed hand applicator is used to inject the chemical 6 to 8 inches deep at 18-inch intervals in each direction. To obtain maximum soil diffusion of the gas, the injection sites are staggered so that the pattern of injection is diagonal or diamond-shaped. Immediately after injection the soil should be firmly tamped to prevent escape of the fumes. Carbon disulfide is injurious to tree roots, so treatment should not be made nearer uninfected trees than the extension of overhead branches—in apricot and peach this is 8 to 10 feet; in walnut, 15 to 25 feet. After treating the area, a light irrigation should be applied sufficient to wet the soil to a depth of about 3 inches.

Larger areas require tractor-operated equipment. The applicator consists of three or more chisels, 18 inches apart, to inject the fumigant 8 inches deep in sandy soil and 6 inches in heavier soils. If using carbon disulfide, a minimum of 302 gallons (3,025 pounds) per acre usually is applied. After application, a drag and a heavy steel roller are used to seal the soil surface. Planting in treated soil should come no sooner than six to eight weeks after the treatment.

Carbon disulfide is both explosive and toxic. The gas is 2½ times as heavy as air. Safety rules are: (1) store in a cool place and use at temperatures below 29.4°C; (2) do not light matches or smoke near the chemical; (3) never use a hammer and chisel to loosen the bung—use a wrench made to open barrels, and then allow the gas pressure to equalize slowly; (4) prevent accumulation of static electricity by grounding the barrels when they are in a truck during removal of the liquid.

Methyl bromide has done as well as carbon disulfide in light sandy soil when applied under a plastic tarp (Kolbezen and Abu-El-Haj 1972a,b; LaRue et al. 1962). The chemical is injected as deep as possible with chisel applicators, then the treated area is covered with a gas proof plastic sheet. With clay soils, the dosage should be increased. The plastic cover is removed after two weeks and the treated area is aerated for at least one month before trees are planted.

Complete eradication is rarely achieved and localized follow up treatment may be necessary. Neither fumigant treatment is successful if the soil is wet or if extensive impenetrable clay layers are present at depths reached by the roots. Methyl bromide is now used almost exclusively in California because of its ease of handling, but it is toxic and must be handled with caution.

Control of *Armillaria* by soil fumigation is apparently related to weakening of the pathogen followed by the increase in population of *Trichoderma viride*, which is antagonistic toward the pathogen (Bliss 1951; Garrett 1958; Munnecke, Kolbezen, and Wilbur, 1970, 1973; Munnecke et al. 1981; Ohr, Munnecke, and Bricker 1973; Saksena 1960).

A new approach to control of this disease through eradication of the pathogen in the stumps and roots of infected pine trees was explored in the Pacific Northwest (Filip and Roth 1977). Fumigants were injected into either the infected stump or the soil immediately adjacent to the stump. All of the fumigants tested—carbon disulfide, chloropicrin, methyl bromide, sodium N-methyldithiocarbamate, and Vorlex—were highly effective in eradicating the fungus from the stumps and tap root. It was not determined whether the pathogen survived in the smaller, more distant feeder roots. This type of treatment may show promise in the control of shoestring root rot of orchard crops.

Resistant rootstocks offer a less expensive and more practical method of coping with this disease (Raabe 1962, 1979a,b). Resistant rootstocks are plum (Marianna 2624), persimmon, fig, northern California black and Paradox hybrid walnuts, and domestic French pear. Moderately resistant are plum (Myrobalan 29-C), mazzard cherry, St. Julien and damson plum (*P. institia*), and apple. Seedlings of peach, apricot, and almond are susceptible, as are mahaleb and Stockton Morello cherries and all rootstocks of grape and quince. The Oriental pear rootstocks are, as a group, more susceptible than domestic French rootstocks. Havens 2B almond interstock is used with Nonpareil almond, a cultivar that is not compatible with the resistant plum rootstock Marianna 2624. Almond cultivars compatible with Marianna 2624 are Jordanolo, Ne Plus Ultra, and Mission (Texas) (Chester, Hanson, and Panetsos 1962; Day 1947; O'Reilly 1963; Proffer, Jones, and Perry 1988).

REFERENCES

Bliss, D. E. 1944. Controlling Armillaria root rot in citrus. *Univ. Calif. Agric. Exp. Stn. Lithoprint* 50.

———. 1946. The relation of soil temperature to the development of Armillaria root rot. *Phytopathology* 36:302–18.

———. 1951. The destruction of *Armillaria mellea* in citrus soils. *Phytopathology* 41:665–83.

Campbell, A. H. 1934. Zone lines in plant tissues. II. The black lines formed by *Armillaria mellea* (Vahl.) Quel. *Ann. Appl. Biol.* 21:1–22.

Chester, D. E., C. J. Hansen, and C. Panetsos. 1962. Plum rootstocks for almonds. *Calif. Agric.* 16(6):10–11.

Day, L. H. 1947. Apple, quince, and pear rootstocks in California. *Calif. Agric. Exp. Stn. Bull.* 700.

DeVay, J. E., S. L. Sinden, F. L. Lukezic, L. F. Werenfels, and P. A. Backman. 1968. Poria root and crown rot of cherry trees. *Phytopathology* 58:1239–41.

Filip, G. M., and L. F. Roth. 1977. Stump injections with soil fumigants to eradicate *Armillaria mellea* from young-growth Ponderosa pine killed by root rot. *Can. J. For Res.* 7:226–31.

Gardner, M. W., and R. D. Raabe. 1963. Early references to Armillaria root rot in California. *Plant Dis. Rep.* 47:413–15.

Garraway, M. O., and A. R. Weinhold. 1968. Influence of ethanol on the distribution of glucose-^{14}C assimilated by *Armillaria mellea*. *Phytopathology* 58:1652–57.

Garrett, S. D. 1956. Rhizomorph behavior in *Armillaria mellea* (Vahl.) Quel. II. Logistics of infection. *Ann. Bot. London* 20:193–209.

———. 1958. Inoculum potential as a factor limiting action by *Trichoderma viride* Fr. on *Armillaria mellea* (Fr.) Quel. *Trans. Brit. Mycol. Soc.* 41:157–64.

Hartig, R. 1873. Vorlaufig Mittheilung über den Parasitismus von *Agaricus melleus* und desen Rhizomorphen. *Bot. Z.* 31:295–97.

———. 1874. *Wichtige krankheiten der Waldbaume: Beitrage zur Mycologie und Phytopathologie für Botaniker und Forstmanner.* Berlin, 127 pp.

Kolbezen, M. J., and F. J. Abu-El-Haj. 1972a. Fumigation with methyl bromide. I. Apparatus for controlled concentration, continuous flow laboratory procedures. *Pestic. Sci.* 3:67–71.

———. 1972b. Fumigation with methyl bromide. II. Equipment and methods for sampling and analyzing deep field soil atmospheres. *Pestic. Sci.* 3:73–80.

Lanphere, W. M. 1934. Enzymes of the rhizomorphs of *Armillaria mellea*. *Phytopathology* 24:1244–49.

LaRue, J. H., A. O. Paulus, W. D. Wilbur, H. J. O'Reilly, and E. F. Darley. 1962. Armillaria root rot fungus controlled with methyl bromide soil fumigation. *Calif. Agric.* 16(8):8–9.

Morrison, D. J. 1976. Vertical distribution of *Armillaria mellea* rhizomorphs in soil. *Trans. Brit. Mycol. Soc.* 66:393–99.

———. 1982. Effects of soil organic matter on rhizomorph growth by *Armillaria mellea*. *Trans. Brit. Mycol. Soc.* 78:201–02.

Munnecke, D. E., M. J. Kolbezen, and W. D. Wilbur. 1970. Dosage response of *Armillaria mellea* to methyl bromide. *Phytopathology* 60:992–93.

———. 1973. Effect of methyl bromide or carbon disulfide on *Armillaria* and *Trichoderma* growing on agar medium and relation to survival of *Armillaria* in soil following fumigation. *Phytopathology* 63:1352–57.

Munnecke, D. E., M. J. Kolbezen, W. D. Wilbur, and H. D. Ohr. 1981. Interactions involved in controlling *Armillaria mellea*. *Plant Dis.* 65:384–89.

Ohr, H. D., D. E. Munnecke, and J. L. Bricker. 1973. The interaction of *Armillaria mellea* and *Trichoderma* spp. as modified by methyl bromide. *Phytopathology* 63:965–73.

O'Reilly, H. J. 1963. Armillaria root rot of deciduous fruits, nuts, and grapevines. *Calif. Agric. Exp. Stn. Cir.* 525, 15 pp.

Piper, C. V., and S. W. Fletcher. 1903. Root diseases of fruit and other trees caused by toadstools. *Wash. State Coll. Agric. Exp. Stn. Bull.* 59, 14 pp.

Proffer, T. J., A. L. Jones, and R. L. Perry. 1988. Testing cherry rootstocks for resistance to infection by species of *Armillaria*. *Plant Dis.* 72:488–90.

Raabe, R. D. 1962. Host list of the root rot fungus, *Armillaria mellea*. *Hilgardia* 33(2):25–88.

———. 1967. Variation in pathogenicity and virulence in *Armillaria mellea*. *Phytopathology* 57:73–75.

———. 1979a. Some previously unreported hosts of *Armillaria mellea* in California, III. *Plant Dis. Rep.* 63:494–95.

———. 1979b. Resistance or susceptibility of certain plants to Armillaria root rot. *Univ. Calif. Div. Agric. Sci. Leaf.* 2591. 11 pp.

Saksena, S. B. 1960. Effect of carbon disulphide fumigation on *Trichoderma viride* and other soil fungi. *Trans. Brit. Mycol. Soc.* 43:111–16.

Schneider, H., E. W. Bodine, and H. E. Thomas. 1945. Armillaria root rot in the Santa Clara Valley of California. *Plant Dis. Rep.* 29:495–96.

Sisson, R. L., L. A. Lider, and A. N. Kasimatis. 1978. Some economic aspects of vineyard site preplant soil fumigation under California north coast conditions. *Am. J. Enol. Vit.* 29:97–101.

Smith, R. E. 1941. Diseases of fruits and nuts. *Calif. Agric. Exp. Stn. Serv. Circ.* 120, 168 pp.

Thomas, H. E. 1934. Studies on *Armillaria mellea* (Vahl.) Quel., infection, parasitism, and host resistance. *J. Agric. Res.* 48:187–218.

———. 1938. Suggestions for combating oak root fungus. *Almond Facts* 2(10/11):4.

Thomas, H. E., and L. O. Lawyer. 1939. The use of carbon disulfide in the control of Armillaria root rot. *Phytopathology* 29:827–28.

Thomas, H. E., H. E. Thomas, C. Roberts, and A. Amstutz. 1948. Rootstock susceptibility to *Armillaria mellea*. *Phytopathology* 38:152–54.

Thomas, H. E., S. Wilhelm, and N. A. MacLean. 1953. Two root rots of fruit trees. In *Plant diseases*, 702–05. U.S. Department of Agriculture Yearbook 1953.

Vance, C. P., and M. O. Garraway. 1973. Growth stimulation of *Armillaria mellea* by ethanol and other alcohols in relation to phenol concentration. *Phytopathology* 63:743–48.

Weinhold, A. R., and M. O. Garraway. 1963. Rhizomorph production in *Armillaria mellea* induced by ethanol and related compounds. *Science* 142:1065–66.

———. 1966. Nitrogen and carbon nutrition of *Armillaria mellea* in relation to growth promoting effects of ethanol. *Phytopathology* 56:108–12.

Wilbur, W., D. E. Munnecke, and E. F. Darley. 1972. Seasonal development of Armillaria root rot of peach as influenced by fungal isolates. *Phytopathology* 62:567–70.

Wolpert, F. S. 1924. Studies in the physiology of the fungi. XVIII. The growth of certain wood-destroying fungi in relation to the H-ion concentration of the media. *Ann. Missouri Bot. Garden* 11:43–97.

Silver-Leaf Disease of Fruit and Nut Trees

Chondrostereum purpureum (Pers.:Fr.) Pouz.

Silver leaf or silver blight is rarely found in California, although it is of some importance on apple in Washington (Hord and Sprague 1950), on apple and stone fruit in Wisconsin (Setliff and Wade 1973), and on sour cherry in Oregon (Williams and Cameron 1956). In temperate climates it is known to affect crops such as pome fruit, stone fruit, currant, chestnut, and gooseberry. Wild hosts are willow, lilac, sycamore, rhododendron, poplar, birch, and oak. It was first reported by Prillieux in France in 1885 and by Sorauer in Germany in 1886 and by Aderhold in 1895. The disease has also been reported from Chile, Japan, India, England, Australia, New Zealand, South Africa, and Canada. In Chile, France, and New Zealand it is of major economic importance.

Symptoms. The leaves of affected trees become silvery in appearance (**fig. 6A,B**) because the upper epidermis separates from the palisade layer and a layer of air enters between interfering with the normal reflection of light. Severely affected leaves may turn necrotic before normal leaf fall (Percival 1902). Some plants such as hawthorn, rhododendron, rose, beech, and birch are killed without any silvering of the leaves (Brooks and Bailey 1919). Discoloration of the heartwood is one of the most characteristic symptoms (**fig. 6C**) and after the death of the tree or branches leathery basidiocarps occur on the bark surface (**fig. 6B**). The upper portions of the basidiocarps frequently extend out from the host in the form of a small shelf.

Causal organism. *Chondrostereum (Stereum) purpureum* (Pers.: Fr.) Pouz., in the family Stereaceae, forms a thin leathery sporophore 2 to 8 cm in width, which is brown or brownish purple but fades with age (**fig. 6B**). It appears as a small circular resupinate or profusely imbricate structure with a hairy upper surface. The hymenium is smooth, lilac to purplish in color. The basidospores are hyaline, oval, apiculate at one end, and 6 to 8 by 3 to 4 μm in dimension. One of the characteristic histological features of this fungus is the presence of vesiculose, subhymenial cystidia, 15 to 30 μm long and 12 to 25 μm wide, extending above the basidia. The sporophores can develop during mild or warm moist weather at any time of the year, but are produced most abundantly in autumn (Dye 1972). The spores are forcibly abjected from the basidia within a few hours after the onset of rain at temperatures that range from 4° to 21°C. They may be shed in daylight and at night. The same spoprophores produce repeated crops of spores during mild, wet weather for one to two years. The spores are entirely wind-disseminated (Dye 1974).

Disease development. In order for infection to occur, the fungus requires a fresh wound on the trunk or branches deep enough to expose the wood. Spores deposited on the wood during moist weather germinate in place or are drawn into the xylem vessels where they germinate to produce hyphae, which spread in the living wood and kill the tissues as they advance. The wood is most susceptible during the first week after wounding, but infection is rare after one month. The invaded wood becomes dark brown and develops a gummy substance in the vessels that serves as a barrier against further spread of the fungus (Anonymous 1966; Brooks and Brenchley 1931a; Brooks and Moore 1926; Brooks and Storey 1923). Instances of tree "recovery" are reported (Grosjean 1951, 1952; Overholtz 1939), but such trees are not immune to reinfection (Dye 1972). There is evidence in cherry of tree-to-tree spread through natural root grafts.

Although the fungus cannot be isolated from the silvered leaves, substantial evidence exists to prove that *C. purpureum* causes the leaf symptoms. Bedford and Pickering (1910) produced silver-leaf symptoms by inoculating trees with sporophores of *C. purpureum*. Injecting trees with filtered extracts of the fungus induced the silver symptom in apple, cherry, and plum trees (Bishop 1979; Brooks and Brenchley 1931b; Miyairi et al. 1977). Naef-Roth, Kern, and Toth (1963) reported that a similar symptom can be produced in plum trees by using a phytolysin, a high-molecular weight, water-soluble compound prepared from cultures of the fungus.

Research in England and New Zealand has shown that trees are much less susceptible to infection in summer and fall than in winter and spring (Anonymous 1966; Dye 1972). Beever (1970) found that nitrogen and carbohydrate levels in xylem sap from peach trees were highest in late winter and early spring, which best supported in vitro growth of *C. purpureum*. He postulated that seasonal variations in the susceptibility of stone fruit to silver leaf are due to the differential ability of their xylem saps to support growth of the pathogen. Polish research (Bielenin and Malewski 1982) also has shown a correlation between the sugar content of xylem sap and growth of *C. purpureum*.

It has been suspected that a silver-leaf disorder on apricot in California is caused by *C. purpureum*, and the fungus has been implicated definitely in an eight-year-old Sacramento Valley cling peach orchard (Chaney, Moller, and Gotan 1973). The disease was later found in another peach orchard about 30 miles away. Both orchards were located near rivers where alternate hosts of the fungus, such as poplar (*Populus* sp.), were present.

Control. Control is difficult due in part to the pathogen having a wide host range of perennial plants, inoculum being produced over a long period, and the virtual impossibility of protecting all wounded surfaces. Therefore, chemotherapeutic treatments have been suggested. Iron sulfate was used by Brooks in 1911 and oxyquinoline sulfate was used by Grosjean in 1951 and Pratella and Kovaca in 1960, both with variable results. In more recent studies (Dye 1971, 1981; Wicks, Volle, and Lee 1983), paints or sprays containing captafol applied immediately to pruning wounds provided effective disease control. Silver leaf also has been controlled by applying spore suspensions of the biocontrol agent *Trichoderma viride* Pers. ex. Fr. to fresh pruning wounds (Corke 1980; Dubos and Ricard 1974; Grosclaude, Dubos, and Ricard 1974; Grosclaude, Ricard, and Dubos 1973). In other studies (Corke 1974, 1980), trunk implantation of this fungus has significantly reduced the severity of the disease.

Further precautions suggested are the use of clean nursery stock, removal of infected parts, burning of all infected material immediately after removal, heading back the rootstock to the scion bud in late rather than early spring, and pruning, where possible, during summer. Cultivar resistance has been reported for crops such as apple and plum (Anonymous 1966; Bennett 1962; Brooks and Brenchley 1931a; Grosclaude 1971), and some rootstocks impart resistance to the scion (Anonymous 1966). Also advisable is the removal of alternate hosts, such as poplar and willow, from the vicinity of the orchard (Anonymous 1966; Dye 1972).

REFERENCES

Aderhold, R. 1895. Notizen über einige im vorigen Sommer beobachtete Pflanzenkrankheiten. *Z. Pflanzenkrankh.* 5:86–90.

Anonymous. 1966. Silver-leaf disease of fruit trees. *Min. Agric. Fish. Food Great Brit. Adv. Leaf.* 246, 7 pp.

Bedford, D., and S. U. Pickering. 1910. Silver-leaf disease. *Woburn Exp. Fruit Farm 12th Rept.:* 1–34.

Beever, D. J. 1970. The relationship between nutrients in extracted xylem sap and the susceptibility of fruit trees to silver-leaf disease caused by *Stereum purpureum* (Pers.) Fr. *Ann. Appl. Biol.* 65:85–92.

Bennett, M. 1962. An approach to the chemotherapy of silver-leaf disease (*Stereum purpureum*) of plum trees. *Ann. Appl. Biol.* 50:515–24.

———. 1962. Susceptibility of Victoria plum trees to different isolates of *Stereum purpureum J. Hort. Sci.* 37:235–38.

Bielenin, A., and W. Malewski. 1982. Changes in the content of sugars in sour cherry xylem sap and their effect on the growth of *Stereum purpureum* (Pers.) Fr. fungus. *Fruit Sci. Rept.* 9(2):83–89.

Bishop, G. C. 1979. Infection of cherry trees and production of a toxin that causes foliar silvering by different isolates of *Chondrostereum purpureum*. *Austral. J. Agric. Res.* 30:659–65.

Brooks, F. T. 1911. Silver-leaf disease. I. *J. Agric. Sci.* 4:133–44.

———. 1913. Silver-leaf disease. II. *J. Agric. Sci.* 5:288–308.

———. 1928. *Plant diseases.* Oxford University Press, London, 386 pp.

Brooks, F. T., and M. A. Bailey. 1919. Silver-leaf disease. III. *J. Agric. Sci.* 9:189.

Brooks, F. T., and G. H. Brenchley. 1929. Injection experiments on plum trees in relation to *Stereum purpureum* and silver-leaf disease. *New Phytol.* 28:218.

———. 1931a. Silver-leaf disease VI. *J. Pomol. Hort. Sci.* 9:1–15.

———. 1931b. Further injection experiments in relation to *Stereum purpureum* (abs.). *New Phytol.* 30:128.

Brooks, F. T., and W. C. Moore. 1926. Silver-leaf disease. V. *J. Pomol. Hort. Sci.* 5:61–97.

Brooks, F. T., and H. H. Storey. 1923. Silver-leaf disease. IV. *J. Pomol. Hort. Sci.* 3:117–41.

Chaney, D. H., W. J. Moller, and S. M. Gotan. 1973. Silver leaf of peach in California. *Plant Dis. Rep.* 57:192.

Corke, A. T. K. 1974. The prospect of biotherapy of trees infected by silver leaf. *J. Hort. Sci.* 49:391–94.

———. 1980. Biological control of tree diseases. *Rept. Long Ash. Res. Stn.* 1979:190–98.

Cunningham, G. H. 1923. Silver blight, *Stereum purpureum*, its appearance, cause and preventative treatment. *N. Z. J. Agric.* 24:276–83.

Dubos, B., and J. L. Ricard. 1974. Curative treatment of peach trees against silver leaf disease (*Stereum purpureum*) with *Trichoderma viride* preparations. *Plant Dis. Rep.* 58:147–50.

Dye. M. H. 1971. Wound protection on deciduous fruit trees. *N. Z. J. Agric. Res.* 14:526–34.

———. 1972. Silver-leaf disease of fruit trees. *Min. Agric. Fish. N. Z. Bull.* 104, 20 pp.

———. 1974. Basidiocarp development and spore release of *Stereum purpureum* in the field. *N. Z. J. Agric. Res.* 17:93–100.

———. 1981. Protection of fruit-tree pruning wounds against silver-leaf disease by immediate spray application of captafol. *N. Z. J. Agric. Res.* 9:97–100.

Grosclaude, C. 1971. Silver-leaf disease of fruit trees. VIII. Contribution to the study of varietal resistance in plum. *Ann. Phytopathol.* 3(3):283–98.

Grosclaude, C., B. Dubos, and J. L. Ricard. 1974. Antagonism between ungerminated spores of *Trichoderma viride* and *Stereum purpureum*. *Plant Dis. Rep.* 58:71–74.

Grosclaude, C., J. Ricard, and B. Dubos. 1973. Inoculation of *Trichoderma viride* spores via pruning shears for biological control of *Stereum purpureum* on plum tree wounds. *Plant Dis. Rep.* 57:25–28.

Grosjean, J. 1952. Natuurlijk herstel van loodglansziekte. *Tijdschr. Plziekt.* 58:109–20.

———. 1951. Investigations on the possibility of silver-leaf disease control by the bore-hole method. *Tijdschr. Plziekt.* 57:103–08. (English summary.)

Heimann, M. 1962. [Spread of silver leaf by nursery material]. In *Mitt. biol BdAnst. Berl.* 108, 187 pp. 1963. *Rev. Appl. Mycol.* 43:2491g. 1964.

Hord, H. H. V., and R. Sprague. 1950. Silver-leaf disease of apple in Washington. *Wash. Agric. Exp. Stn. Circ.* 119, 5 pp.

Miyairi, K., K. Fujita, T. Okuno, and K. Sawai. 1977. A toxic protein causative of silver-leaf disease symptoms on apple trees. *Agric. Biol. Chem.* 41:1897–1902.

Naef-Roth, S., H. Kern, and A. Toth. 1963. On the pathogenesis of parasitogenic and physiological silver leaf of stone fruit. *Phytopathol. Z.* 48:232–39. (English summary.)

Overholtz, L. O. 1939. The genus *Stereum* in Pennsylvania. *Bull. Torrey Bot. Club* 66:515–37.

Percival, J. 1902. Silver-leaf disease. *J. Linn. Soc. Bot.* 35:390–95.

Pratella, G., and A. Kovaca. 1960. Systemic activity of 8-oxyquinoline sulphate in young peach plants. *Ann. Sper. Agr.* 14:171–83. (English summary.)

Prillieux, E. 1885. Le plomb des arbres fruitiers. *Bull. Seances Soc. Nat. Agr. France.*

Setliff, E. C., and E. K. Wade. 1973. *Stereum purpureum* associated with sudden decline and death of apple trees in Wisconsin. *Plant Dis. Rep.* 57:473–74.

Sorauer, P. 1886. Milchglanz. In *Handbuch der Pflanzenkrankheiten*, 2 Auf. 1:141; 3 Auf. 1:285–86. 1909.

Wicks, T. J., D. Volle, and T. C. Lee. 1983. Effect of fungicides on infection of apricot and cherry pruning wounds inoculated with *Chondrostereum purpureum*. *Austral J. Exp. Agric. Anim. Husb.* 23:91–94.

Williams, H. E., and H. R. Cameron. 1956. Silver leaf of Montmorency sour cherry in Oregon. *Plant Dis. Rep.* 40:954–56.

Verticillium Wilt of Fruit and Nut Trees

Verticillium dahliae Kleb.

Popularly known as "black heart" in trees of certain cultivated fruit, this disease has also been called "verticilliosis," "Verticillium hadromycosis," and "vascular wilt." It occurs in many parts of the world and causes much economic loss. According to Rudolph (1931) the disease was observed in California as early as the first decade of this century. Observers noticed that in orchards of young apricot, peach, plum, and almond where tomatoes had been interplanted, a disease known as "black heart" frequently developed. The condition was first considered to be associated with the heavy irrigation of tomatoes, which led to souring of the soil caused by rotting tomatoes left on the ground after harvest.

However, in 1916 Czarnecki first presented proof of the fungal origin of the disease, and in 1923 she stated that it was caused by a fungus belonging to the genus *Verticillium*. According to Rudolph (1931), as early as 1879 Reinke and Berthold (1879) noted that a disease of Irish potato in Germany was caused by a fungus named *Verticillium albo-atrum*. *Verticillium* was later associated with diseases of many plant species. In 1957, Englehard published a host index of plants affected by *Verticillium albo-atrum* (including *V. dahliae*). Besides the plants named above, the cultivated tree fruit and nuts affected are sweet and sour cherry, nectarine, pome fruit, citrus, avocado, mango, Japanese persimmon, Persian walnut, pistachio, and olive. In addition, stone fruit used as rootstocks such as myrobalan plum, mahaleb cherry, *Prunus mume*, and *P. davidiana* are susceptible. The disease also occurs on brambles (especially raspberry), strawberry, melons, cotton, and many other cultivated plants (Carter 1938; Dufrenoy and Dufrenoy 1927; Haenseler 1928; Mills 1932; Presley 1950; Waggoner 1956). Strains of the fungus that attack pome fruit in Europe are not known to occur in North America.

Symptoms. The following symptoms are largely those on apricot and almond. The initial external evidence is a sudden wilting of the foliage on one or more branches in the summer. The leaves turn a dull grayish tan, wither, and infrequently fall from the tree (**fig. 7A**). Sometimes the lower leaves wither and fall first, and development of leaf symptoms progresses up the branch.

On affected sweet cherry, the first symptom may be a withering of the leaves on one or more spurs on the one-year-old wood of the leaders. The trees become unthrifty, the remaining leaves are off-color and small, and the fruit is abnormally small (McKeen 1943).

Very young fruit trees are those generally developing the most severe cases of hadromycosis. Although very young apricot and almond trees may be killed outright by the fungus, older trees are not—in fact, the defoliated twigs and branches show no other external symptoms. In the cool California coastal areas, some of the youngest twigs die, but the older wood remains normal-looking throughout the summer and again puts out foliage the following year. But the terminal growth is subnormal, and the affected branches may again be attacked by the fungus and defoliated, severely stunting their growth. The primary effect of the disease on trees several years old is generally poor growth and low productivity.

When the bark of affected branches is removed, a portion of the underlying sapwood is found to be brown to black (**fig. 7B**). This symptom has led to the name "black heart" for the disease in apricot and almond. In olive, however, the wood of affected branches is rarely off-color, although the symptoms with respect to the withering of foliage are the same as in apricot and almond.

Causal organism. As noted earlier, Reinke and Berthold (1879) described a disease in Irish potato caused by a fungus that they named *Verticillium albo-atrum*. In 1913, Klebahn isolated a fungus from a diseased dahlia and named it *Verticillium dahliae*. The difference between *V. albo-atrum* and *V. dahliae* was said to be that the latter species produces microsclerotia, whereas the former does not. According to Rudolph (1931), whether or not the fungus described by Reinke and Berthold produced microsclerotia depends upon interpretation of the description and drawings of those authors. Rudolph gave an extensive review of the literature up to 1930 pertaining to the two species. He cited the work of Bewley (1922), who isolated six strains of a fungus he regarded as *V. albo-atrum* from plants suffering from hadromycosis. Some of the strains produced no microsclerotia; others produced microsclerotia in abundance. Rudolph concluded that the fungus isolated by Klebahn from dahlia was *V. albo-atrum*, and went on to say that morphological, physiological, and genetic studies and cross-inoculation tests of different forms of *Verticillium* were needed.

Although many workers have accepted *Verticillium albo-atrum* Reinke & Berth. as the name for the pathogen that produces hadromycosis in perennial as well as annual hosts (Presley 1941, 1950), others have maintained that *Verticillium dahliae* Kleb. is the correct name (Reiss 1969). The difference of opinion hinges on the question of whether or not Reinke and Berthold observed true microsclerotia. Though these workers claimed that the microsclerotia of the organism gave rise to the hyaline mycelium when placed in a moist environment, others believe that their drawings do not represent true microsclerotia. Rather, they believe that the structures Reinke and Berthold (1879) illustrated were composed of dark-celled, torulose, resting mycelia. Schnathorst (1965) reviewed this question in the light of studies by himself and others.

The conditions under which the fungus is grown may affect its production of microsclerotia. Wilhelm (1948) found that at temperatures of 10° to 20°C the fungus produced numerous microsclerotia, whereas at 25° to 31°C it produced a creamy white growth and only sparse microsclerotia. He concluded that the differences in temperature at which various studies are made could lead to the morphological differences on the basis of which Klebahn separated *V. dahliae* and *V. albo-atrum*. But others take a different view. Studies such as those of Berkeley, Maden, and Willison (1931), Hall (1969), and Reiss (1969) appear to show certain fundamental distinctions between isolates purported to be *V. dahliae* and *V. albo-atrum*. For example, Berkeley and his coworkers claimed that the latter is much more pathogenic to tomato, potato, and cucumber than the former. Hall found that the two "species" were clearly distinguishable by the patterns of their buffer-soluble proteins. Ludbrook (1933) found that *V. albo-atrum* grew at the same rate as *V. dahliae* at 22°C but only about one-fourth as fast at 28°C. These two organisms now are generally regarded as distinct species, and until recently only *V. dahliae* was known to occur in California. Both the highly virulent T-1 strain and less virulent SS-4 strain identified in cotton (Schnathorst and Mathre 1966) attack fruit and nut crops here. The recent discovery (Erwin et al. 1990) of a high-temperature strain of *V. albo-atrum* in alfalfa fields in southern California indicates that the Verticillium wilt problem in California crops will probably become even more complicated than it is at present.

Verticillium dahliae, in culture, initially produces white colonies composed entirely of hyaline mycelium. Some of the hyphae gradually darken (resting mycelium) and form dark, thick-walled microsclerotia by budding. The conidiophores are slender, erect, septate, and hyaline, in contrast to those of *V. albo-atrum*, which become distinctly darkened toward their base at maturity. The conidiophores typically branch verticillately (in whorls) to form two or three phialides at each locus. The conidia (phialospores) are ovoid to ellipsoid, hyaline, one-celled, borne singly or in small, moist clusters apically, and measure 4 to 11 by 2 to 14 µm. They do not become uniseptate as do some of the conidia of *V. albo-atrum*. The optimum growth temperatures for most isolates of *V. dahliae* and *V. albo-atrum* are, respectively, 24.0°C and 20.0° to 22.5°C. *V. dahliae* can grow at 30°C whereas *V. albo-atrum* cannot (Schnathorst 1973). In an in vitro test (Ragazzi and Parrini 1982), an olive isolate of *V. dahliae* grew best at pH 6.45 and 25°C. Microsclerotia formed at 5° to 30°C.

Disease development. The fungus lives from one year to the next in the soil. Wilhelm (1950a) found that it occurred to a depth of 30 to 36 inches, with the greatest concentration in the upper 6 to 12 inches. Small numbers of microsclerotia have been found in a pistachio orchard at a soil depth of 4 feet (Ashworth and Gaona 1982). In *Verticillium*-infested plant debris such as that of potato and tomato, the fungus can survive for at least a year and in some instances for several years. It has been found to persist in soil (probably as microsclerotia) for an equal length of time depending upon conditions (McKeen 1943). Nadakavikaren and Horner (1961) found that the microsclerotia "survived poorly" at temperatures above 25°C, and that in soils that were flooded, the microsclerotia population was reduced rapidly. Survival was highest at 5° to 15°C and 50 to 75 percent moisture-holding capacity. At a favorable soil moisture, the microsclerotia could survive several months even though soil temperatures rose to 30°C. Many of the microsclerotia remain dormant until stimulated to germinate by root exudates. The fungus is a soil invader rather than a true soil inhabitant.

The fungus can be disseminated in various ways; for example, it can be carried to uninfested areas in diseased plants. The role of the tomato in introducing the fungus into apricot orchards was noted by Rudolph (1931) in California as early as 1931; this has been found to occur in other states (Parker 1959). Verticillium hadromycosis in olive and pistachio has followed their planting in soils in which cotton was grown in the San Joaquin Valley of California. Wilhelm (1950a) and others have suggested that microsclerotia and *Verticillium*-infested plant debris may be blown about by wind, thereby infesting new areas. The conidia are short lived and apparently play little if any role in dissemination and only a minor role in infection.

Parker (1959) reviewed the spread of the fungus both in infected buds used to propagate perennials and on knives used in propagation. The fungus was reported to be spread from diseased to healthy maple trees by pruning.

McKeen (1943) found that the fungus isolated from such diverse plants as barberry, peach, rose, potato, muskmelon, and chrysanthemum differed only slightly in morphology and pathogencity. Under greenhouse conditions and at optimum soil moisture, Verticillium wilt was severe at 21° to 27°C, with a sharp peak at 24°C. This coincided with the optimum temperature for growth of the fungus in culture.

McKeen found the fungus to persist and remain aggressive in both sandy and clay soils and in soils that were cropped or remained fallow. The growing of an immune crop had no apparent effect on the survival of the fungus. Low soil-moisture conditions did, however, reduce the aggressiveness of the fungus. According to Parker (1959), others have found that infection is less in dry than in moist soils. Huisman and Ashworth (1976) found that once soils were infested, the rate of decrease in inoculum, even in the presence of immune crops, was very low. Their results agree with other findings that *Verticillium dahliae* persists in soils for many years and can induce severe wilt in susceptible crops even after 6 to 12 years of nonhost cropping.

Wilhelm (1950b) reported that Verticillium wilt of various crops occurs in California soils that range in pH from 4.6 to 6.7. He quoted the studies of Van der Meer in Holland, who also found that the disease attacks plants growing in soils with a pH of 4.4 to 6.6. A recent study by Ashworth, Gaona, and Surber (1985) showed that a high incidence of Verticillium wilt in pistachio is correlated with potassium deficiency in the leaves and with poor tree growth.

The disease develops quite characteristically in stone fruit and nut trees. That is, symptoms appear early in the growing season but seldom first appear in late summer, although they intensify during that time; this coincides with the development of the fungus in olive. In early summer the fungus can be readily isolated from diseased olive stem tissue, whereas in late summer and autumn it appears to be nonviable in this tissue (Wilhelm and Taylor 1965). However, the fungus survives from one season to the next in roots, and disease recurrence could, presumably, result either from renewed activity of the mycelium in the roots or from new root infections by microsclerotia in the soil. Somewhat the same course of development occurs in apricot and almond (Rudolph 1931; Taylor and Flentje 1968).

In apricot, Harrison and Clare (1970) found that as the infected wood becomes darkened the hyphae within it are darkened and inactivated. Extracts from the darkened wood inhibit spore germination and hyphal growth. Gums and tyloses in the vessels prevent upward movement of conidia in these elements. Trees recover from the disease when new wood forms around the discolored wood containing the inactivated hypahe of *V. dahliae*.

Although many studies on the biochemistry of pathogenesis in Verticillium wilt have been made, understanding of the disease syndrome is incomplete. Substances produced in vitro that are suspected of playing a role in pathogenesis include pectolytic and cellulolytic enzymes, a lipopolysaccharide, high- and low-molecular-weight proteins, and growth regulators such as indole-3-acetic acid and ethylene (Howell 1973).

Control. The most effective means of reducing infection by *V. dahliae* is to avoid planting fruit trees in soils recently cropped with susceptible plants such as cotton, tomatoes, and potatoes. Interplanting in young orchards with susceptible crops also should be avoided. If suspect soils cannot be avoided, it is advisable to delay planting trees for several years, allowing the soil to lie fallow and cultivating it to prevent the growth of susceptible weed hosts. Parker (1959), reviewing the work of others, reported that *Bacillus vulgatus* and soil-inhabiting fungi belonging to the genera *Gliocladium* and *Blastomyces* are antagonistic to *V. dahliae*. Another antagonistic fungus, *Talaromyces flavus*, produces a metabolite (glucose oxidase) that inhibits mycelial growth and germination of microsclerotia of the pathogen (Kim, Fravel, and Papavizas 1988). Allowing the soil to remain fallow would give such microorganisms a chance to reduce the inoculum of *V. dahliae*. Only trees known to be free of infection by *V. dahliae* should be planted. Proper fertilization and irrigation of the orchard better enables the trees to recover from an attack by the fungus. According to Parker (1959) there is some evidence that fertilization with nitrogen and potassium reduces the disease in susceptible plants. Ashworth, Gaona, and Surber (1985) recently showed that disease incidence and severity in pistachio are reduced by soil applications of potassium. Yearly applications of sulfur are said to decrease, and lime applications to increase, Verticillium wilt of eggplant (Haenseler 1928).

There is little that can be done in the way of selecting resistant fruit cultivars. Not only are most edible cultivars of stone fruit susceptible, but the rootstock used in propagating these varieties is susceptible to infection, thereby giving the fungus access to the wood of the scion. In olive some success has been achieved by using the clonally propagated rootstock Oblonga; in pistachio the highly resistant *Pistacia integerrima* is now used extensively and successfully as a rootstock.

In attempts to find chemicals that can be used to control the disease, LeTourneau, McLean, and Guthrie (1957) tested phenols and quinones in culture, and Waggoner (1956) applied growth regulators such as 2,4-D to potato plants.

Wilhelm (1951) found that large amounts of blood meal, fish meal, cottonseed meal, and ammonium sulfate applied to the soil substantially reduce the inoculum potential of the fungus, as did Fermate and Dithane Z78. Wilhelm and Ferguson (1953) found that chloropicrin applied to the soil at the rate 2 to 2.5 mL per square foot at a depth of 12 inches is effective in destroying *V. dahliae* inoculum. Chloropicrin diffuses in all directions, whereas allyl bromide and 55 percent chlorobromopropene, though effective, diffuse laterally and downward but not upward. Sod culture in an olive orchard offered some control (Ashworth and Wilhelm 1971).

More recent research by Ashworth and Zimmerman (1976) showed that preplant soil fumigation with a 2:1 mixture of chloropicrin and methyl bromide reduced the inoculum density and the number of pistachio trees killed by *V. dahliae* over a two-year period. Soil solarization by means of a polyethylene mulch in a young pistachio orchard also reduced inoculum density and disease incidence (Ashworth and Gaona 1982). In neither treatment, however, was the pathogen completely eliminated and, over time, orchards so treated have continued to suffer heavy wilt losses. Attempts to protect healthy trees from infection by trunk injections of benomyl have also been unsuccessful (Ashworth et al. 1982).

REFERENCES

Ashworth, L. J., Jr. 1985. *Verticillium* resistant rootstock research. *Calif. Pistachio Ind. Ann. Rept.* 1984–85:56–57.

Ashworth, L. J., Jr., and S. A. Gaona. 1982. Evaluation of clear polyethylene mulch for controlling Verticillium wilt in established pistachio nut groves. *Phytopathology* 72:243–46.

Ashworth, L. J., Jr., S. A. Gaona, and E. Surber. 1985. Verticillium wilt of pistachio: The influence of potassium nutrition on susceptibility to infection by *Verticillium dahliae*. *Phytopathology* 75:1091–93.

Ashworth, L. J., Jr., D. P. Morgan, S. A. Gaona, and A. H. McCain. 1982. Polyethylene tarping controls Verticillium wilt in pistachios. *Calif. Agric.* 36(5–6):17–18.

Ashworth, L. J., Jr., and S. Wilhelm. 1971. Verticillium wilt of olive. Sod culture offers a promise of control. *Calif. Olive Ind. News* 25.

Ashworth, L. J., Jr., and G. Zimmerman. 1976. Verticillium wilt of the pistachio nut tree: Occurrence in California and control by soil fumigation. *Phytopathology* 66:1449–51.

Berkeley, G. H., G. O. Maden, and R. S. Willison. 1931. Verticillium wilt in Ontario. *Sci. Agric.* 11:739–59.

Bewley, W. F. 1922. "Sleepy disease" or wilt of the tomato. *Ann. Appl. Biol.* 9:116–34.

Carter, J. C. 1938. Verticillium wilt of woody plants in Illinois. *Plant Dis. Rep.* 22:253–54.

Czarnecki, H. 1923. Studies on the so-called black heart disease of apricot. *Phytopathology* 13:216–24.

Dufrenoy, J., and M. L Dufrenoy. 1927. Hadromycosis. *Ann. Epiphyt.* 13:195–212.

Englehard, A. W. 1957. Host index of *Verticillium albo-atrum* Reinke & Berth. (including *Verticillium dahliae* Kleb.). *Plant Dis. Rep. Suppl.* 244:23–49.

Erwin, D. C., R. A. Kahn, A. Howell, A. Baameur, and S. B. Orloff. 1990. Verticillium wilt found in southern California alfalfa. *Calif. Agric.* 43(5):12–14.

Haenseler, C. M. 1928. Effect of soil reaction on Verticillium wilt of eggplant. *New Jersey Agric. Exp. Stn. Ann. Rept.* 41:267–73.

Hall, R. 1969. *Verticillium albo-atrum* and *V. dahliae* distinguished by acrylamide gel-electrophoresis of protein. *Can. J. Bot.* 47:2110–11.

Harrison, A. F., and B. G. Clare. 1970. Host reactions involved in the recovery of apricot trees from Verticillium wilt. *Austral. J. Biol. Sci.* 23:1027–32.

Howell, C. R. 1973. Pathogenicity and host-parasite relationships. *Proc. Work Conf. Nat. Cotton Pathol. Res. Lab.* (College Stn., Text. Aug. 30–Sept. 1, 1971) ARS-S-19:42–46.

Huisman, O. C., and L. J. Ashworth, Jr. 1976. Influence of crop rotation on survival of *Verticillium albo-atrum* in soils. *Phytopathology* 66:978–81.

Kim, K. K., D. R. Fravel, and G. C. Papavizas. 1988. Identification of a metabolite produced by *Talaromyces flavus* as glucose oxidase and its role in the biocontrol of *Verticillium dahliae*. *Phytopathology* 78:488–92.

Klebahn, H. 1913. Beitrage zur Kenntnis der fungi imperfecti. I. Eine Verticillium-krankheit auf Dahlien. *Mycol. Centralbl.* 3(2):49–66.

Le Tourneau, D., J. G. McLean, and J. Guthrie. 1957. The effect of phenols and quinones on the growth in vitro of *Verticillium albo-atrum*. *Phytopathology* 47:602–06.

Ludbrook, M. V. 1933. Pathogenicity and environmental studies on Verticillium hadromycosis. *Phytopathology* 23:117–54.

McCain, A. H., R. D. Raabe, and S. Wilhelm. 1981. Plants resistant or susceptible to Verticillium wilt. *Univ. Calif. Div. Agric. Sci. Leaf.* 2703. 10 pp.

McIntosh, D. L. 1954. Verticillium wilt of sweet cherry in British Columbia. *Plant Dis. Rep.* 38:74–75.

McKeen, C. D. 1943. A study of some factors affecting the pathogenicity of *Verticillium albo-atrum* R. & B. *Can. J. Res. Sect. C.* 21:95–117.

Mills, W. D. 1932. Occurrence of Verticillium wilt of peach in New York. *Plant Dis. Rep.* 16:132–33.

Nadakavikaren, M. J., and C. E. Horner. 1961. Influence of soil moisture and temperature on survival of *Verticillium microsclerotia* (abs.). *Phytopathology* 51:66.

Parker, K. G. 1959. Verticillium hadromycosis of deciduous tree fruits. *Plant Dis. Rep. Supp.* 255:39–61.

Presley, J. T. 1941. Saltants from a monosporic culture of *Verticillium albo-atrum*. *Phytopathology* 31:1135–39.

———. 1950. Verticillium wilt of cotton with particular emphasis on variation of the causal organism. *Phytopathology* 40:497–511.

Ragazzi, A., and C. Parrini. 1982. Caratteristiche colturali di *Verticillium dahliae* Kleb., isolato da olivo. [Cultural characters of *Verticillium dahliae* Kleb., isolated from olive.] *Riv. Patol. Veg.* 18:129–41. (*Rev. Plant Pathol.* 62:1590. 1983.)

Reinke, J., and G. Berthold. 1879. Die Zersetzung der Kartoffel durch Pilze. *Untersuch. Bot. Lab. Univ. Göttingen* 1, 100 pp.

Reiss, J. 1969. Beitrag zum Problem der systematischen Abgrenzung von *Verticillium albo-atrum* Rke. et Berth. und *Verticillium dahliae* Kleb. *Z. Pflanzenkrank. Pflanzenpathol. Pflanzenschutz* 75:480–84.

Rudolph, B. A. 1931. Verticillium hadromycosis. *Hilgardia* 5:197–353.

Schnathorst, W. C. 1965. Origin of new growth in dormant microsclerotial masses of *Verticillium albo-atrum*. *Mycologia* 57:343–51.

———. 1973. Nomenclature and physiology of *Verticillium* species, with emphasis on the *V. albo-atrum* versus *V. dahliae* controversy. *Proc. Work Conf. Nat. Cotton Pathol. Res. Lab.* (College Stn., Texas, Aug. 30–Sept. 1, 1971) ARS-S-19:1–19.

Schnathorst, W. C., and D. E. Mathre. 1966. Host range and differentiation of a severe form of *Verticillium albo-atrum* in cotton. *Phytopathology* 56:1155–61.

Taylor, J. B., and N. J. Flentje. 1968. Infection, recovery from infection and resistance of apricot trees to *Verticillium albo-atrum*. *New Zeal. J. Bot.* 6:417–26.

Waggoner, P. E. 1956. Chemotherapy of Verticillium wilt of potatoes in Connecticut. *Am. Potato J.* 33:223–25.

Wilhelm, S. 1948. The effect of temperature on the taxonomic characters of *Verticillium albo-trum* Reinke & Berth. (abs.). *Phytopathology* 38:919.

———. 1950a. Vertical distribution of *Verticillium albo-atrum* in soils. *Phytopathology* 40:368–76.

———. 1950b. Verticillium wilt in acid soils. *Phytopathology* 40:776–77.

———. 1951. Effect of various soil amendments on the inoculum potential of the Verticillium wilt fungus. *Phytopathology* 41:684–90.

Wilhelm, S., and J. Ferguson. 1953. Soil fumigation against *Verticillium albo-atrum*. *Phytopathology* 43:593–96.

Wilhelm, S., and J. B. Taylor. 1965. Control of Verticillium wilt of olive through natural recovery and resistance. *Phytopathology* 55:310–16.

Wood Decay of Fruit and Nut Trees

Basidiomycetous Fungi

Wood-rot disorders occur in fruit and nut orchards throughout California. The fungi causing these disorders are primarily fungi in the Basiomycotina, and information available on them as pathogens of fruit and nut trees mostly includes only mycological descriptions (Overholts 1953; Smith, Smith, and Weber 1979; Juelich and Stalpers 1980; Miller 1980; Gilbertson and Ryvarden 1986, 1987) and reports noting their occurrence (Anonymous 1961). Several detailed surveys of the wood-decay fungi of apple trees have been conducted in Washington (Dilley and Covey 1980; Helton and Dilbeck 1984), Minnesota (Eide and Christensen 1940; Bergdahl and French 1985), and on deciduous fruit and nut trees in California (Adaskaveg and Ogawa 1990). Other surveys and reports have been made of these fungi on fruit and nut crops in Europe (D'Aulerio, Dallavalle, and Menz 1985; Plank 1981; Penrose 1979), England (Wormald 1955), and India (Chohan, Kang, and Rattan 1984). In California, no specific studies on crop losses caused by individual species of these fungi have been conducted.

Wood-decay fungi enter trees primarily through wounds. These fungi utilize wood for energy and thus weaken the strength properties of the wood in advanced stages of decay. Some species actively invade cambial tissue and kill the host rapidly. Certain species of *Armillaria* function in this manner (Raabe 1967). Other wood-decay fungi, such as *Chondrostereum purpureum* (Pers.:Fr.) Pouz., release toxins that affect the foliage of the infected tree (Cunningham 1925; Miyairi et al. 1977; Bishop 1979). These wood-decay fungi cause shoestring root rot and silver-leaf disease, respectively, and are considered in detail elsewhere in this chapter. Most wood-rotting fungi of fruit and nut trees, however, have been implicated in more subtle diseases affecting only the structural wood of the tree. Some have been associated with the general decline and dieback of fruit trees (Setliff and Wade 1973; Alconero et al. 1978; Doepel, McLean, and Goss 1979; Covey et al. 1981; Dilley and Covey 1981).

Symptoms. The symptoms of trees associated with wood-decay fungi can vary among different hosts and decay fungi. Host tissues usually respond to wounds and infection by microorganisms. Immediately after an injury, tissues become discolored from host reactions to environmental and biological factors associated with the wound (Shigo and Hillis 1973; Hillis 1977; Shortle and Cowling 1978; Bauch 1984). Discolored wood of angiosperms contains phenolic compounds, gums, and other occlusions that occupy the cell lumens of the injured areas and provide a chemical barrier for protection against microorganisms (Shigo and Hillis 1973).

Decayed wood can be characterized and placed into two major groups, brown wood rots and white wood rots. Fungi that cause brown rots primarily degrade cell-wall polysaccharides, leaving lignin unchanged or slightly modified. The decayed wood is brown, dry, and crumbly, with both longitudinal and transverse cracking. Strength properties of wood decayed by brown-rot fungi are significantly reduced or lost entirely in a relatively short time. Mycelium of these fungi can grow extensively within cracks of decayed wood forming mycelial mats. Living trees infected with brown-rot fungi can have the older sapwood and heartwood, if present, substantially degraded. These tissues collapse, leaving the trunk or main branches hollow. The hollow regions are often utilized by nesting birds.

On living fruit and nut trees, fungi that cause white rots are more common than those that cause brown rots. White-rot fungi degrade all major structural components of wood, namely cellulose, hemicellulose, and lignin. White-rotted wood is moist, soft or spongy, and bleaches to a lighter color. The strength properties of the affected wood are reduced only after extensive decay. These losses often do not occur as quickly as in wood decayed by brown-rot fungi. Fungi that cause white rot, such as *Ganoderma* spp., as well as fungi that cause brown rot, can also infect older sapwood and heartwood in living trees. Decayed wood may also contain dark lines. These structures are not clearly understood for all types of decay but have been described as host reactions to decay fungi (Kile and Wade 1975; Dilley and Covey 1981), mycelial interactions between genetically distinct isolates of the same species, or interactions between different xylophilic fungal species (Rayner and Todd 1979).

Other symptoms induced by wood-decay fungi include limb breakage (**fig. 8D,F**) during fruit production and uprooted trees (**fig. 8B**) during wind storms or mechanical harvesting, both associated with extensively decayed wood in scaffold branches or roots, respectively. More general symptoms, such as dieback (**fig. 8E**), early decline, and short-life, have also been associated with wood decay fungi.

Fruiting bodies of wood-decay fungi can develop on decayed wood of living trees throughout the year depending on the fungal species and environmental conditions in the fruit and nut growing areas of California. Fruiting can occur on the soil surface from infected roots, on the lower trunk of the tree (**fig. 8A,C**), or on the main scaffold branches (**fig. 8F**). Basidiocarps (fruiting bodies of basidiomycetous fungi) can vary from soft and fleshy mushrooms to tough and leathery brackets or conks that may be annual or perennial and are found either solitary or in clusters. Fruiting body development occurs from the mycelium within extensively decayed wood which may or may not be associated with wounds.

Causal organisms. Fungi that cause extensive wood decay of fruit and nut trees are mostly in the division Basidiomycotina, class Hymenomycetes. Formerly these fungi were classified in the class Basidiomycetes. Although a few ascomycetous fungi have been reported to cause wood decay, this discussion will be limited to basidiomycetous fungi. Wood-decay fungi that have been recorded or observed on fruit and nut trees in California, Oregon, and Washington are presented in table 2. Fungi in the Basidiomycotina have specialized cells called basidia. In the Hymenomycetes, these cells compose a fertile tissue layer, referred to as the hymenium, which produces sexual spores called basidiospores. Basidia of the hymenial tissue can be arranged in numerous ways depending on the fruiting body of the fungus. These cells can line inner walls of tubes as in species of the Polyporaceae, Hymenochaetaceae, and Ganodermataceae; compose the surface of lamellae or gills as in the Agaricales; or compose the surface of flat or resupinate fruiting bodies of species in several other families. Basidiospores are usually disseminated by air currents and may or may not germinate on a suitable substrate. In heterothallic species, basidiospores germinate to form a homokaryotic mycelium that fuses with compatible mating types to form a dikaryotic mycelium. In other species the nuclear condition is not known and germinating basidiospores do not require other mating types; presumably these fungi are homothallic. In species of *Armillaria*, however, nuclei fuse shortly after hyphae fusion to form a diploid mycelium. Wood-decay fungi colonize woody substrates by releasing enzymes and other compounds from the mycelium that degrade structural components of the wood. After substrate colonization, fruiting bodies can form under suitable environmental and physiological conditions. Basidiospores are again produced, thus completing the life cycle. Aerially disseminated asexual spores are not commonly produced by these fungi.

Epidemiology and disease development. Wood-decay fungi enter trees primarily through wounds. Prior to infection by these fungi, however, injured wood is colonized by a succession of other microorganisms such as bacteria, yeast, and hyphomycetes (Shigo and Hillis 1973; Blanchette 1979a). In nature, wood-decay fungi function as secondary invaders of trees, but they are independently capable of decaying wood in the absence of other organisms. Some fungi can function as primary invaders by spreading vegetatively through natural host root grafts (*Armillaria* and *Ganoderma* spp.). Once established in the wood, these fungi can extensively decay the inner sapwood and heartwood of the tree. Some decay fungi primarily colonize and decay the roots and lower portion of the trunk, and are known as root- and butt-rot fungi. Other species grow in the upper trunk and scaffold branches. They colonize older sapwood and heartwood that function in the structural support of the tree. Although true heartwood may not be formed in apple trees (Kile and Wade 1974), and probably not in other fruit and nut trees under commercial production, wood-decay fungi that attack these trees have traditionally been referred to as heart-rot fungi. The term sapwood fungi sometimes has been used to describe these fungi, but this term has also been used when referring to fungi that decay wood of dead trees.

Wood-decay fungi break down cellulose and hemicellulose into simple sugar by enzymatic and nonenzymatic chemical reactions. Wood decay decreases the strength properties of wood. Affected branches or entire trees are weakened and often break from strong winds, heavy crops, harvesting with mechanical shakers, or as a result of contact with other farm equipment. Other physiological effects of decay fungi on host trees have not been extensively studied. Secondary metabolites or toxins may be produced, as in the decay caused by *Chondrostereum purpureum*. Toxins produced by this fungus affect the foliage of the tree, causing silver leaf disease. Host nutrition as related to the susceptibility of apple trees to *Trametes versicolor* (L.:Fr.) Pil. has been studied (Wade, 1968) and is discussed under "Sappy Bark of Apple" in Chapter 2. Although wood-decay fungi have been associated with fruit-tree decline by numerous researchers (DeVay et al. 1968; Kile and Wade 1974; Alconero et al. 1978; Covey et al. 1981; Dilley and Covey 1981), little information is available on their effect on crop yield, tree vigor, or tree life span.

Several species of wood-decay fungi are associated with certain injuries found in orchards in California. Sunburned areas on scaffolds are commonly attacked by *Schizophyllum commune* Fr. (**fig. 9C**). This fungus has also been associated with injuries caused by chemicals, such as those that can be caused by the application of dormant oil sprays. *Trametes versicolor* and *T. hirsuta* (Wulf.:Fr.) Pilat (**fig. 9B**) are commonly associated with pruning wounds, especially those on the main scaffold branches and trunk. *Oxyporus latemarginatus* (Dur. and Mont. ex. Mont.) Donk (**fig. 8A,B**) (formerly known as *Poria ambigua*), *O. similis* (Bres.) Ryv., and species of *Ganoderma*, namely *G. lucidum* (W. Curt.:Fr.) Karst, and *G. brownii* (Murr.) Gilbn. (**fig. 9A**), are primarily restricted to the lower butt and roots of infected trees. These fungi commonly enter the tree through injuries that expose the wood of the trunk, such as wounds caused by mechanical harvesters and cultivation equipment. In the case of *O. latemarginatus*, infection often is associated with the Phytophthora root and crown rot disease. The fruiting bodies of these species appear to be most common in orchards with sprinkler irrigation where the trunk and tree base are kept moist for long periods. *Fomitopsis cajanderi* (Karst.) Kotl. et Pouz. and species of *Phellinus* are associated with pruning wounds and trunk injuries (**fig. 10**). Zeller (1926) indicated that *F. cajanderi* (*Trametes subrosea* Weir) was a major cause of heart rot in peach and prune trees in the Pacific Coast states. Another fungus, *Laetiporus sulphureus* (Bull.: Fr.) Murr. (**fig. 8F**), is widespread in species of *Prunus* in California; it is commonly found in scaffold branches and main trunks, and probably enters through large pruning wounds.

Control. Information on the control of disorders caused by wood-decay fungi in fruit and nut trees is limited. Control of these disorders traditionally involves cultural practices that promote tree vigor. This includes good site and cultivar selection, proper irrigation and fertilization programs, and orchard management practices that minimize tree wounding. Chemical and mechanical methods of weed control can cause injuries to roots and the lower part of the trunk, and mechanical harvesters may cause substantial trunk damage when misused. Large pruning wounds are also common points of infection for wood-decay fungi. Therefore, when possible, pruning should be done when branches are small. Dead branches or trees should be removed and destroyed to prevent the formation of new inoculum. Removal of large living branches should be done with cuts adjacent to the supporting branch or trunk. Stub cuts or horizontal cut surfaces that allow water accumulation should be avoided (Thomas, Mathee, and Hadlow 1978; Covey and Dilley 1981). Thomas, Mathee, and Hadlow (1978) observed that callus formed more rapidly in wounds made during the spring (after fruit set) and winter than in other periods of the year. Kile (1976) indicated that extensive hyphal penetration by *T. versicolor* occurred following summer inoculation and the fungus was capable of infecting 30-day-old wounds. In contrast, Gendle et al. (1983) observed a greater natural resistance to damage by *T. versicolor* in the growing season than in the dormant season. Aging of wounds prior to inoculation, however, reduced the growth of this fungus regardless of the season (Kile 1976). In vitro wood-decay studies indicated that discolored wood, formed in response to wounding, is more resistant to decay by isolates of *T. versicolor* than normal sapwood (Kile and Wade, 1975).

Pruning should be done when inoculum is sparse or absent. Unfortunately, wood-decay fungi have different peak-sporulation periods depending on the species and geographical location in respect to temperature and moisture. Haard and Kramer (1970), Rockett and Kramer (1974), and McCraken (1978) determined the periodicity of spore release in numerous lignicolous Basidiomycetes. In California, where moisture is not a limiting factor in irrigated orchards, some species can produce basidiocarps year around (i.e., *Ganoderma* sp. and *Trametes* sp.). In other species, like *Armillaria mellea*, fruiting only occurs during the cooler, winter months, providing moisture is present. Additionally, once basidiocarps are formed, relative humidity and temperature also are limiting factors for sporulation (McCracken 1978). Further studies to determine sporulation periods for wood-decay fungi in California are required to determine optimum pruning times to minimize infection.

Tree paints have been used to protect wounds. However, laboratory studies indicate that asphalt paints are not fungitoxic and do not provide a barrier to several wood-decay fungi unless amended with copper naphthenate (Dooley 1980) or other fungitoxicants (May and Palmer 1959). Also, field trials showed that bituminous compounds do not protect wounds (Thomas, Mathee, and Hadlow 1978). Covey and Dilley (1981) observed decay in apple trees in Washington associated with both painted and unpainted wounds. Wound protection depends on complete coverage and maintenance of an intact protective layer. Although paints may promote callus formation, cracking of the sealant may expose unprotected wood and allow invasion by wood-decay fungi. Polyvinyl acetate was found to provide a flexible, protective layer on apple tree wounds (Thomas, Mathee, and Hadlow 1978). However, this material has not been tested under California conditions.

Systemic fungicides have not been evaluated extensively against wood-decaying Basidiomycetes. In field trials with apple, Gendle et al. (1983) reported that furmecyclox and fenpropimorph provide some wound protection and triadimenol is very effective against isolates of *T. versicolor*. Since triadimenol moves and persists well within wood, this compound may be of value in commercial orchards.

REFERENCES
(Numerals at left are cited in Table 2.)

1. Adaskaveg, J. E., and J. M. Ogawa. 1990. Wood decay pathology of fruit and nut trees in California. *Plant Dis.* 74:341–52.
2. Alconero, R., R. Campbell, J. Brittain, and B. Brown. 1978. Factors contributing to early decline of peaches in South Carolina. *Plant Dis. Rep.* 62:914–18.
3. Anonymous. 1961. *Index of plant diseases in the United States.* U.S. Dept. Agric. Handbook 165, 531 pp.
4. Bauch, J. 1984. Discoloration in the wood of living and cut trees. *I.A.W.A. Bull.* 5:92–98.
5. Bergdahl, D. R., and D. W. French. 1985. Association of wood decay fungi with decline and mortality of apple trees in Minnesota. *Plant Dis.* 69:887–90.
6. Bishop, B. C. Infection of cherry trees and production of a toxin that caused foliar silvering by different isolates of *Chondrosterium purpureum*. *Austr. J. Agric. Res.* 30:659–65.
7. Blanchette, R. A. 1979. A study of progressive stages of discoloration and decay in *Malus* using scanning electron microscopy. *Can. J. For. Res.* 9:464–69.
8. Bonar, L. 1964. *Polyporus oleaea* on olive in California. *Plant Dis. Rep.* 48:70.
9. Covey, R. P., and M. A. Dilley. 1981. Heart rot and wood decay of apple. *The Bad Apple* 2(1):32–37.
10. Covey, R. P., H. J. Larson, T. J. Fitzgerald, and M. A. Dilley. 1981. *Coriolus versicolor* infection of young apple trees in Washington State. *Plant Dis.* 65:280.
11. Chohan, J. S., I. S. Kang, and G. S. Rattan. 1984. Control of root rot and sapwood rot of peaches (Flordasun cultivar) caused by *Polyporus palustrus*, *Ganoderma lucidum*, associated with *Schizophyllum commune*. *Internat. J. Trop. Plant Dis.* 2:49–54.
12. Cunningham, G. H. 1925. *Fungous diseases of fruit trees in New Zealand.* Brett, Auckland, 293 pp.
13. Darbyshire, B., G. C. Wade, and K. C. Marshall. 1969. In vitro studies of the role of nitrogen and sugars on the susceptibility of apple wood to decay by *Trametes versicolor*. *Phytopathology* 59:98–102.
14. D'Aulerio, A. Z., E. Dallavalle, and M. Menz. 1985. Distribution and pathogenic aspects of wood destroying fungi on trees in the commune of Bologna, Italy. *Micol. Ital.* 14:39–47.
15. DeVay, J. E., S. L. Sinden, F. L. Lukezic, L. F. Werenfels, and P. A. Backman. 1968. Poria root and crown rot of cherry trees. *Phytopathology* 58:1239–41.
16. Dilley, M. C., and R. P. Covey. 1980. Survey of wood decay and associated hymenomycetes in central Washington apple orchards. *Plant Dis.* 64:560–61.
17. ———. 1981. Association of *Coriolus versicolor* with a dieback disease of apple trees in Washington State. *Plant Dis.* 65:77–78.
18. Doepel, R. F., G. D. McLean, and O. M. Goss. 1979. Canning peach decline in Western Australia. I. Associ-

ation between trunk cankers, trunk pruning wounds, and crotch angles of scaffold limbs. *Austral. J. Agric. Res.* 30:1089–1100.

19. Dooley, H. L. 1980. Methods for evaluating fungal inhibition and barrier action of tree wound paints. *Plant Dis.* 64:465–68.
20. Eide, C. J., and C. M. Christensen. 1940. Wood decay in apple trees in Minnesota. *Phytopathology* 30:936–44.
20a. Farr, D. F., G. F. Bills, G. P. Chamuris, and A. Y. Rossman. 1989. *Fungi on plants and plant products in the United States.* American Phytopathological Society, St. Paul, MN. 1252 pp.
21. French, A. M. 1987. California plant disease host index: Part 1. Fruits and nuts. California Department of Food and Agriculture, 39 pp.
22. Gendle, P., D. R. Clifford, P. C. Mercer, and S. A. Kirk. 1983. Movement, persistence, and performance of fungitoxicants applied as pruning wound treatments on apple trees. *Ann. Appl. Biol.* 102:281–92.
23. Gilbertson, R. L. 1979. The genus *Phellinus* (Aphyllophorales: Hymenochaetaceae) in western North America. *Mycotaxon* 9:51–89.
24. Gilbertson, R. L., and L. Ryvarden. 1986. *Polypores of North America, vol. 1.* Ryv. Press, Oslo, 300 pp.
25. ———. 1987. *Polypores of North America, vol. 2: Fungiflora.* Oslo, pp. 434–885.
26. Haard, R. T., and C. L. Kramer. 1970. Periodicity of spore discharge in the Hymenomycetes. *Mycologia* 62:1145–69.
27. Helton, A. W., and R. Dilbeck. 1984. Wood decay fungi in fruit trees. *Proc. Wash. State Hort. Assoc.* 80:171–75.
28. Hillis, W. E. 1977. Secondary changes in wood. *Rec. Advances Phytochem.* 11:247–309.
29. Juelich, W., and J. A. Stalpers. 1980. *The resupinate non-poroid Aphyllophorales of the temperate northern hemisphere.* North-Holland, Amsterdam, 335 pp.
30. Kile, G. A. 1976. The effect of season of pruning and of time since pruning upon changes in apple sapwood and its susceptibility to invasion by *Trametes versicolor. Phytopathol. Z.* 87:231–40.
31. Kile, G. A., and G. C. Wade. 1974. *Trametes versicolor* on apple. I. Host-pathogen relationship. *Phytopathol. Z.* 81:328–38.
32. ———. 1975. *Trametes versicolor* on apple. II. Host reaction to wounding and fungal infection and its influence on susceptibility to *T. versicolor. Phytopathol. Z* 86:1–24.
33. May, C., and J. G. Palmer. 1959. Effects of asphalt varnish-fungicide mixtures on growth in pure culture of some fungi that cause decay in trees. *Plant Dis. Rep.* 43:955–59.
34. McCracken, F. I. 1978. Spore release of some decay fungi of southern hardwoods. *Can. J. Bot.* 56:426–31.
35. Miller, O. K. 1980 *Mushrooms of North America.* E. P. Dutton, New York, 360 pp.
36. Miyairi, K., K. Fujita, T. Okuno, and K. Sawai. 1977. A toxic protein causative of silver-leaf disease symptoms on apple trees. *Agric. Biol. Chem.* 41:1897–1902.
37. Overholts, L. O. 1953. *Polyporaceae of the United States, Alaska, and Canada.* University of Michigan Press, Ann Arbor, 466 pp.
38. Penrose, L. J. 1979. Wood rot of fruit trees (*Trametes versicolor, Schizophyllum commune,* and *Pycnoporus coccineus*). Plant Dis. Bull. New South Wales Dept. Agric., 6 pp.
39. Plank, S. 1981. Contribution to the knowledge of wood destroying fungi of Greece. *Ann. Inst. Phytopathol. Benaki.* 12:244–52.
40. Raabe, R. D. 1967. Variation in pathogenicity and virulence in *Armillariella. Phytopathology* 57:73–75.
41. Rayner, A. D. N., and N. K. Todd. 1979. Population and community structures and dynamics of fungi in decaying wood. *Advances Bot. Rev.* 7:333–420.
42. Rockett, T. R., and C. L. Kramer. 1974. Periodicity and total spore production by lignicolous basidiomycetes. *Mycologia* 66:817–29.
43. Setliff, E. C., and E. K. Wade. 1973. *Stereum purpureum* associated with sudden decline and death of apple trees in Wisconsin. *Plant Dis. Rep.* 57:473–74.
44. Shaw, Charles G. 1958. Host fungus index for the Pacific Northwest I. Hosts. Wash. Agric. Exp. Stn. Circ. 335, 127 pp.
45. Shigo, A. L., and W. E. Hillis. 1973. Heartrot, discolored wood and microorganisms in living trees. *Ann. Rev. Phytopathol.* 11:197–222.
46. Shortle, W. C., and E. B. Cowling. 1978. Development of discoloration, decay and microorganisms following wounding of sweetgum and yellow-poplar trees. *Phytopathology* 68:609–16.
47. Smith, A. H., H. V. Smith, and N. S. Weber. 1979. *How to key the gilled mushrooms.* W. C. Brown, Dubuque, 334 pp.
48. Thomas, A. C., F. N. Mathee, and J. Hadlow. 1978. Wood-rotting fungi in fruit trees and vines. III. Orchard evaluation of a number of chemical bases as wood protectants. *Decid. Fruit Grower* 28:92–98.
49. Wade, G. C. 1968. The influence of mineral nutrition on the susceptibility of apple trees to infection by *Trametes (Polystictus) versicolor. Austral. J. Exp. Agric. Anim. Prod.* 8:436–39.
50. Wilson, E. E., and J. M. Ogawa. 1979. *Fungal, bacterial, and certain nonparasitic diseases of fruit and nut crops in California.* Agricultural Science Publications, University of California, 190 pp.
51. Wormald, H. 1955. *Diseases of fruit and hops.* C. Lockwood and Son, London, 325 pp.
52. Zeller, S. M. 1936. Brown-pocket heartrot of stone fruit trees caused by *Trametes subrosea* Weir. *J. Agric. Res.* 33:687–93.

Table 2. Common wood decay fungi of selected fruit and nut tree species in California, Oregon, and Washington

Fungus	Occurrence	Host*	HA†	Decay‡	Source§
Abortiporus biennis (Bull.:Fr.) Sing.	CA,OR,WA	2,6–8,18	1,2	W	8,24,44,50
Antrodia malicola (Burt. and Curt.) Donk.	WA	7	2	B	24,44
Armillaria spp.	CA;OR;WA	1–18; 3,6,7,12,13,14,18; 3,6,7,13,14,16,18	1,2	W	3,16,21,44,50
Armillaria mellea (Vahl:Fr.) Kummer s. str.	CA	10,14	1,2	W	1
Bjerkandera adusta (Willd.:Fr.) Karst.	WA	6,7	1,2	W	16,44
Ceriporia purpurea (Fr.) Donk.	WA	6	2	W	24,44
C. spissa (Schw.:Fr.) Rajch.	CA	14	2	W	1
Cerrena unicolor (Bull.:Fr.) Murr.	WA,OR	16	1	W	24,44
Chondrostereum purpureum (Pers.:Fr.) Pouz.	CA;OR;WA	11,13,14,16; 7,16,18;7,13,16,19	1,2	W	3,16,21,44,50
Coltricia perennis (L.:Fr.) Murr.	WA	7	3	NS	16
Coprinus spp.	CA	16	1,2	B	1
Coriolopsis gallica (Fr.) Ryv.	OR;WA	16;7	2	W	3,16,44
Cyphellopsis anomala (Pers.:Fr.) Donk	WA	3	3	NS	20a,44
Daedalea quercina Fr.	CA	6	1,2	B	3,21,24
Daedaleopsis confragosa (Bolt.:Fr.) Schroet.	CA;WA	6;16	1,2	W	3,21,24
Dendrophora albobadia (Schw.:Fr.) Boidin (syn. *Peniophora*)	CA	10	2	W	1
D. erumpens (E.A. Burt) G.P. Chamuris (syn *Peniophora*)	OR;WA	7,16;16	2	W	44
Flammulina velutipes (Fr.) Karst.	WA	7	1,2	W	16
Fomes fomentarius (L.:Fr.) Kickx.	OR,WA	7,13,16,18	1,2	W	18,24,44
Fomitopsis cajanderi (Karst.) Kotl. et Pouz. (fig. 10)	CA;OR;WA	13,16;1,3,13,14,16; 1,3,12,13,16	1,2	B	1,3,24,44,51
F. pinicola (Schwartz:Fr.) Karst.	OR;WA	7,13,14,16;7,12	1,2	B	3,24,44
F. rosea (Alb. et Schw.:Fr.) Karst.	OR;WA	1,4,14,16;1,4,16	1,2	B	3,24
Ganoderma annularis (Fr.) Gilbn.	CA	10,14	1,2	W	1
G. applanatum (Pers.) Pat.	CA;OR;WA	7,14;2,4,13,16; 7,16,18	1,2	W	1,3,24,44
G. brownii (Murr.) Gilbn. (fig. 9A)	CA	10,14	1,2	W	1
G. lucidum (W.Curt.:Fr.) Karst.	CA	7,10,12,14,16	1,2	W	1,24
Gloeophyllum saepiarum (Fr.) Karst.	OR;WA	7,13,14,16;12,13,16	2	B	3,24,44
G. trabeum (Fr.) Murr.	OR;WA	7,16;16	2	B	3,44
Gloeoporus dichorus (Fr.)Bres.	CA	7,11	1	W	1,24
Hyphoderma mutatum (Pk.) Donk	WA	3	2	W	20a,44
H. puberum (Fr.) Wallr.	CA	10	2	W	1
H. setigerum (Fr.:Fr.) Donk	WA	3	2	W	20a,44
Hyphodermella corrugata (Fr.) J. Erikss. & Ryv.	WA	16	2	W	44
Hyphodontia aspera (Fr.) J. Erikss.	CA	10	2	W	1
Inonotus cuticularis (Bull.:Fr.) Karst. (fig. 10)	CA	5	1	W	1
I. dryophilus (Berk.) Murr.	CA,OR	16	1	W	3,24
I. rickii (Pat.) D. Reid	CA	5	1	W	1
Irpex lacteus (Fr.:Fr.) Fr.	CA,OR,WA	12,16	1,2	W	3,24,44

Footnotes at end of table, page 299.

Continued on next page.

Table 2. Common wood decay fungi of selected fruit and nut tree species in California, Oregon, and Washington (Continued)

Fungus	Occurrence	Host*	HA†	Decay‡	Source§
Laetiporus sulphureus (Bull.:Fr.) Murr. (fig. 8F)	CA;OR,WA	2,6,10,13,16; 6,10,16	1,2	B	1,3,21,24
Lenzites betulina (Fr.) Fr.	CA	7,10,12	1,2	W	1
Maireina marginata (McAlpine) W. B. Cooke	OR;WA	7,10,14;7	1,2	W	20a,44
Merulius sp.	WA	7	3	NS	16
Naematoloma fasciculare (Huds.:Fr.) Karst.	WA	7	2	NS	16
Oligoporus tephorleucus (Fr.) Gilbn. & Ryv. (syn. *Postia*)	OR	12	2	B	3
Omphalotus olearius (DC:Fr.) Singer	CA	8	1,2	W	1
Oxyporus corticola (Fr.) Ryv.	CA;OR;WA	14,16;12,16;16	2	W	1,25,44
O. latemarginatus (Dur. & Mont. ex. Mont.) Donk (fig. 8A,B)	CA	7,10,12,16	1,2	W	1,3,15,16,21
O. populinus (Fr.) Donk	OR,WA	16	1,2	W	3
O. similis (Bres.) Ryv.	CA	7,10,14	1,2	W	1
Panellus serotinus (Fr.) Kuhn	OR,WA	7	2	W	44
Peniophora cinera (Pers.:Fr.) Cke.	WA	3	2	W	44
P. nuda (Fr.:Fr.) Bres.	WA	3	2	W	20a,44
Perenniporia fraxinophila (Pk.) Ryv.	CA	16	1	W	3
P. medulla-panis (Jacq.:Fr.) Donk (fig. 8C,D,E)	CA	16	1	W	1
Phanerochaete velutina (Fr.) Karst.	CA;WA	14;3	2	W	1
Phellinus ferreus (Pers.) Bourd. & Galz.	CA,WA;OR	3;2,3,16	1,2	W	3,23,25,44
P. ferruginosus (Schard.:Fr.) Bourd. & Galz.	CA,OR;WA	3,16;7,16	1,2	W	16,23,25,44
P. gilvus (Schw.) Pat.	CA	10,14,16	1,2	W	1,23,25
P. igniarius (L.:Fr.) Quél.	CA,OR;WA	7,16,18;7,16	1	W	3,23,25,44
P. pomaceus (Pers.:S.F. Gray) Maire	CA;OR;WA	10;13,16;3,16	1,2	W	1,3,23,25,44
P. robustus (Karst.) Bourd. & Galz. (fig. 10)	CA	7,10,13	1	W	1
P. texanus (Murr.) A. Ames	CA	13	1	W	1
Phlebia concentrica (Cke. & Ell.) Kropp & Nakasone	WA	3	2	W	20a,44
P. merismoides (Fr.:Fr.) Fr.	CA	3	2	W	20a
P. radiata Fr.	CA	7	1,2	W	1
P. rufa (Fr.) M. P. Christ.	CA	10	2	W	1
Pholiota sp.	CA	16	1	NS	1
P. adiposa (Fr.:Fr.) Kummer	WA	7	1,2	W	16
P. squarrosa (Mull.:Fr.) Kummer	WA	7	1,2	W	16
Pleurotus sp.	CA,WA	7	1,2	W	1,20a,44
P. ostreatus (Fr.) Kummer (fig. 10)	CA	5,6,8	1,2	W	1,21
Plicaturepsis crispa (Pers.:Fr.) D. Reid	OR	13	2	W	20a,44
Polyporus elegans Bull.:Fr.	WA	16	2	W	25,37,44
P. varius Fr.	CA,WA	16	2	W	44
Pycnoporus cinnabarinus (Jacq.:Fr.) Karst.	CA	16	2	W	3,25
Schizophyllum commune Fr. (Fig. 9C)	CA;OR;WA	2,5–7,9–12,14; 2,7,16;7,11–14,16	1,2	W	1,3,16,21,44

Footnotes at end of table, page 299.

Continued on next page.

Table 2. Common wood decay fungi of selected fruit and
nut tree species in California, Oregon, and Washington (Continued)

Fungus	Occurrence	Host*	HA†	Decay‡	Source§
Schizopora flavipora (Cke.) Ryv.	CA	10	2	W	1,25
Sistotrema brinkmannii (Bres.) J. Erikss.	CA	14	2	W	1
Stereum spp.	OR;WA	3,6,16;7	3	W	21,44
S. hirsutum (Willd.:Fr.) S.F. Gray	CA;OR;WA	9,10;2,10,13,14,18;3	1,2	W	1,21,44
Trametes hirsuta (Wulf.:Fr.) Pilát (fig. 9B)	CA;OR;WA	7,11,14,16; 3,6,7,12–14,16,18; 6,7,12,14,16,18	1,2	W	1,3,21,25,44
T. ochracea (Pers.) Gilbn. & Ryv.	CA,WA	7	1,2	W	1,15
T. pubescens (Schum.:Fr.) Pilát	CA,WA;OR	7;12	1,2	W	16,25,44
T. versicolor (L.:Fr.) Pilát	CA;OR;WA	7,8,10–14,18; 2,3,7,10, 12–14,16,18; 6,7,12,14,16	1,2	W	1,3,16,21,25,44
Tremella mesenterica Retz:Fr.	WA	3	2	W	20a,44
Trichaptum biforme (Fr. in Klotzsch) Ryv.	OR	18	2	W	25,44
Tyromyces chioneus (Fr.) Karst.	OR,WA	7,13,16	2	W	25,44
T. galactinus (Berk.) Lowe	OR	13,16	2	W	25,44

*Hosts included: (1) *Carya illinoensis* (Wang.) K. Koch (pecan); (2) *Castanea* spp. (chesnut); (3) *Corylus* spp. (filbert); (4) *Diospyros* spp. (persimmon); (5) *Ficus carica* L. (fig); (6) *Juglans* spp. (walnut); (7) *Malus* spp. (apple); (8) *Olea* spp. (olive); (9) *Pistacia vera* L. (pistachio); (10) *Prunus amygdalus* Batsch. (almond); (11) *P. armeniaca* L. (apricot); (12) *P. avium* L. (cherry); (13) *P. domestica* L. and *P. americana* L. (prune, plum); (14) *P. persica* (L.) Batsch. (peach); (15) *P. salicina* Lindl. (Japanese plum); (16) *Prunus* spp.; (17) *Punica granatum* L. (pomegranate); (18) *Pyrus* spp. (pear); and (19) *Cydonia oblonga* Mill (quince). Host numbers separated by semicolons correspond to occurrence by state.
†Host association (HA): 1, living trees; 2, dead wood; 3, not specified.
‡Type of decay: white rot (W), brown rot (B), and not specified (NS).
§Information obtained from specified sources in the preceding References list, as indicated by the numbers cited.

Chapter 5

The English Walnut and Its Diseases

Page numbers in bold indicate photographs.

INTRODUCTION	304	
REFERENCES	304	

Bacterial Diseases of English Walnut — 305

Blight of English Walnut — 305, **431**
Xanthomonas campestris pv. *juglandis* (Pierce) Dye
REFERENCES — 308

Crown Gall of English Walnut — 309
Agrobacterium tumefaciens (Smith & Townsend) Conn

Deep-Bark Canker of English Walnut — 309, **431**
Erwinia rubrifaciens Wilson, Zeitoun & Fredrickson
REFERENCES — 311

Shallow-Bark Canker of English Walnut — 312, **432**
Erwinia nigrifluens Wilson, Starr & Berger
REFERENCES — 313

Fungal Diseases of English Walnut — 313

Anthracnose of English Walnut — 313
Gnomonia leptostyla (Fr.) Ces. and DeN.
REFERENCES — 313

Branch Wilt of English Walnut — 314, **432**
Hendersonula toruloidea Nattrass
REFERENCES — 315

Downy Leaf Spot of English Walnut — 315, **432**
Microstroma juglandis (Bereng.) Sacc.
REFERENCES — 316

Kernel Mold of English Walnut — 316, **432**
Miscellaneous Fungi
REFERENCES — 316

Melaxuma Canker and Twig Blight of English Walnut — 316
Dothiorella gregaria Sacc.
REFERENCES — 317

Phytophthora Root and Crown Rot of English Walnut — 317
Phytophthora spp.
REFERENCES — 318

Shoestring (Armillaria) Root Rot of English Walnut — 318
Armillaria mellea (Vahl.:Fr.) Kummer

Miscellaneous Fungal Diseases of English Walnut — 318

Cytospora canker, Dematophora root rot, Diplodia canker and dieback, Fusarium root rot, wood rot, Ascochyta ring spot, Botryosphaeria dieback, Cercospora leaf spot, Cylindrosporium leaf spot, Marssonina leaf spot, Melanconis canker and dieback, Nectria canker, Phloeospora leaf spot, Phyllosticta leaf spot, Phymatotrichum root rot.
REFERENCES — 318

Virus Diseases of English Walnut — 319

Blackline of English Walnut — 319, **433**
Cherry Leafroll Virus (CLRV-W)
REFERENCES — 321

Abiotic Diseases of English Walnut — 322

Shell Perforation of English Walnut — 322
Cause Unknown
REFERENCES — 322

Sunburn of English Walnut — 322, **433**
Direct Exposure to Sun's Rays
REFERENCES — 323

Introduction

The English walnut (*Juglans regia* L.), also known as the Persian walnut, apparently originated in the Caucasus and Turkestan areas of Persia. In ancient times, trees of the species were taken to Greece and later to Rome, where the tree became known as "Jovis Glans" (Jupiter's acorn), from which the genus name *Juglans* is derived. In relatively early times, trees of *J. regia* were taken to England and then to America, where it was called the English walnut to distinguish it from the native American species.

The native American species are called black walnut because of the color of the hard, thick shell. The American black walnut, *J. nigra*, is native to the eastern and midwestern United States, while the species found in California are *J. hindsii* (northern California black walnut) and *J. californica* (southern California black walnut). Two other species of black walnut, *J. rupestris* and *J. major*, are native to the United States and are found in Arizona, New Mexico, and Texas.

Although trees of *J. regia* were grown in the eastern United States, commercial cultivation has become almost exclusively confined to the West Coast, largely in California (Batchelor 1921; Hendricks, Gripp, and Ramos 1980; Lelong 1896; Ramos 1985) and Oregon (Lewis 1906). The California English walnut industry owes its origin to the efforts of Joseph Sexton of Santa Barbara, who propagated the Santa Barbara soft-shell type of nut from seeds brought from Chile, and to Felix Gallet of Nevada City, who introduced varieties from France (Franquette, etc.). One of the earliest orchards in California was the Kellog orchard near Napa, planted in 1846. In 1987 California cultivated 196,439 acres of English walnuts and produced about 95 percent of the walnuts grown in the United States, approximately one-third of the world's supply. Other producing countries are France, Italy, Romania, China, Turkey, Yugoslavia, Bulgaria, and Hungary. California's center of walnut acreage has moved from southern California to the San Joaquin Valley. Tulare, Stanislaus, and San Joaquin counties account for about 40 percent and Sutter, Tehama, and Butte counties about 19 percent of the crop. Most acres are planted with the Hartley cultivar (58,586), followed by Payne (36,130), Franquette (35,791), Eureka (20,664) and Ashley (15,065). Other cultivars with 1,000 acres or more are Mayette, Nugget, Blackmer, Poe, Waterloo, and Concord. English walnut trees are long lived, and if grown in good soil and well cared for they may live for 200 years.

Trees are propagated by budding a patch of bark with a leaf-bud on the trunk of a young seedling tree suitable for a rootstock. Sometimes grafting is employed. Although seedlings of the English walnut were formerly used, these were abandoned in favor of seedlings of *J. hindsii*, the northern California black walnut, partly because the *J. regia* seedlings are highly susceptible to shoestring (Armillaria) root rot, are more difficult to propagate, and do not do well on marginal soil types. The southern California black walnut tends to sucker profusely at the base of the trunk. In recent years, Paradox hybrid (*J. regia* × *J. hindsii*) seedlings have been used to some extent as a rootstock because of their rapid growth and moderately high resistance to shoestring root rot. One drawback is the high degree of variability in growth among seedlings.

The pistillate and pollen-bearing flowers of English walnut are produced in different locations on the same plant. The former arise from terminal buds on the ends of shoots produced during the current year; the latter arise as long, pendulous catkins from naked buds in the axils of leaves of shoots produced the previous year. Catkins develop in the spring (April-May) shortly before the pistillate flowers emerge, and pollen from the catkins is transferred by wind to the stigma (Smith, Smith, and Ramsey 1912; Wood 1934). As shall be discussed, these features must be taken into consideration in controlling walnut blight. Mature nuts are mechanically shaken and gathered during September and October.

REFERENCES

Batchelor, L. D. 1921. Walnut culture in California. *Calif. Agric. Exp. Stn. Bull.* 332:142–217.

Forde, H. I., and W. H. Griggs. 1981. Pollination and blooming habits of walnuts. *Univ. Calif. Div. Agric. Sci. Leaf.* 2753, 9 pp.

Hendricks, L. C., R. H. Gripp, and D. E. Ramos. 1980. Walnut production in California. *Univ. Calif. Div. Agric. Sci. Leaf.* 2984, 22 pp.

Lelong, B. M. 1896. *California walnut industry.* Report of the State Board of Horticulture for 1895–96, 44 pp.

Lewis, C. I. 1906. The walnut in Oregon. *Oregon Agric. Exp. Stn. Bull.* 92, 43 pp.

Ramos, D. E., ed. 1985. Walnut orchard management. *Univ. Calif. Coop. Ext. Pub.* 21410, 178 pp.

Sibbett, G. S., D. E. Ramos, and L. C. Brown. 1980. Establishing a new walnut orchard. Univ. Calif. Div. Agric. Sci. Leaf. 21157, 19 pp.

Smith, R. E., C. O. Smith, and H. J. Ramsey. 1912. Walnut culture in California: Walnut blight. *Calif. Agric. Exp. Stn. Bull.* 231.

Wood, M. N. 1934. Pollination and blooming habits of the Persian walnut in California. USDA Tech. Bull. 387, 56 pp.

Bacterial Diseases of English Walnut

Blight of English Walnut

Xanthomonas campestris pv. *juglandis* (Pierce) Dye

The disease known as "walnut bacteriosis," "walnut blight," "black spot," and "black plague" affects English (Persian) walnut (*Juglans regia*), but rarely occurs on English walnut hybrids (Paradox seedlings). The black walnut species (*Juglans nigra, J. californica,* and *J. hindsii*) have been infected artificially but are seldom affected in nature.

Although first observed in southern California in 1891, the first published reference to the disease was in 1894 (Galloway). Oregon was the second state to report it (Lewis 1906; Barss 1928). Thereafter it was reported in New Zealand (1900), Russia (1904), Canada (1911), Tasmania (1912), Australia (1914), Mexico (1915), Chile (1917), the eastern United States (1917), South Africa (1918), Italy (1923), Holland (1923), Switzerland (1924), England (1927), France (1931), Romania (1940), and the West Indies (1943) (McMurran 1917, Wormald 1931). In California, Pierce (1901) was the first to direct attention to walnut blight in Los Angeles County, and in 1896 was credited as the first to prove that the incitant is a bacterium. By 1903 the disease had become so destructive that the California Walnut Growers Association offered a reward of $20,000 to anyone who could find a practical method of control, but no successful claimant for the prize appeared (Smith, Smith, and Ramsey 1912). During the same year, which was shortly after the establishment of the first Division of Plant Pathology at the University of California, Berkeley, the association voted funds for the University to investigate the disease. In 1905, the California State Legislature appropriated $4,000. Later, plant pathologists at the Plant Disease Laboratory at Whittier provided new insights on the disease (Smith 1922, 1931), but had little success in finding a satisfactory method of control. In 1912, C. O. Smith estimated that for the previous 10 years the average annual loss from the disease was 50 percent of the crop. In the Pacific Northwest, losses ranging from 40 to 60 percent were not uncommon, according to Miller (1934). It was not until 1928 that progress was made toward a satisfactory control program. In Oregon, the 1933 loss estimate was 35 percent, but in 1950, only 4 percent; this was probably related to effective spraying.

Symptoms. The bacteria are known to attack the catkins, fruit, green shoots, leaves, and buds (Ark and Scott 1950; Clayton 1921; Miller 1934; Miller and Bollen 1946; Pierce 1901; Rudolph 1933; Smith 1922).

Catkins, although not as susceptible as the fruit, became infected as soon as they break dormancy. Symptoms are commonly expressed after the catkins have elongated. A single floret or all the florets on one side or at the top or bottom of the catkin may show symptoms (**fig. 1B**). Infected florets first appear water-soaked and wilted, and later turn black. Similarly, the rachis may be attacked at any point along its length. Distortions and deformities in the catkins result from the killing of local areas.

Fruit infection accounts for most of the economic loss (**fig. 1A**). The fruit may become infected at any time after its formation and up until harvest; this long period of susceptibility is one of the chief obstacles to control of the disease. The first disease symptom on the fruit usually appears in the bracts, bracteoles, or involucres at the apical (blossom) end as a tiny black spot (Miller and Bollen 1946). The spot enlarges rapidly and may then involve the entire young fruit. After the stigma shrivels, new infection at the blossom end usually ceases, but lateral or side infections can occur. Lesions produced on partly grown nuts enlarge rapidly, and the somewhat shrunken and depressed areas rupture and exude a black shiny liquid (white specks when dry) containing many bacteria. During rains this material spatters onto healthy nuts and shoots. Nuts infected before, during, or immediately after the pollination period (1/8 to 1/2 inch diameter) almost invariably are shed from the tree. Nuts attacked shortly before the shell hardens tend to remain on the tree. The kernel at this time may be destroyed or blackened and shriveled. These nuts are called "blanks" or sometimes "blows." The shell finally gains a toughness impenetrable to the bacteria, but a substance formed in the lesion can cause discoloration or even a partial shriveling of the kernel, and the shell may be so badly stained that the nut becomes a cull. Even after the shell becomes impenetrable to the bacteria or to byproducts of the lesion, it can still be stained.

The mere blackening or shriveling of the kernel is not always attributable to the walnut blight disease. Drought may cause shriveling, and sunburn may do likewise. But sunburned husks are dry and leathery while diseased husks are soft and water-soaked. Also, sunburned husks show symptoms only on exposed sides of the fruit.

Infection of green shoots is possible only at an early stage in their development. Infection is most common at or near the extreme tip of the shoot, and the entire end may then be killed back a few inches. Often the infection is localized and the shoot grows away from the lesion. If the localized lesions are deep-seated, the bacteria have a better chance for survival in winter. The dying tissue sometimes exudes a mixture of decomposed cellular products and bacterial slime. When dry, it is a frosty, whitish, flaky precipitate.

All parts of the leaf may be attacked, including the parenchyma tissue of the leaflet, the midrib, lateral veins, veinlets, rachis, and petiole. Typical disease spots are dark brown with yellowish green margins; they are usually only a few millimeters in diameter but several may coalesce to form a single large spot. Severe infections of tender young leaflets result in the killing of considerable tissue, which causes the malformation of leaflets as they develop. Defoliation is rare, although killing the rachis or petiole can lead to the death of the leaflet or leaf.

Causal organism. The pathogen is now classified as *Xanthomonas campestris* pv. *juglandis* (Pierce) Dye. Previously, the genus ranged from *Pseudomonas, Bacillus,* and *Phytomonas,* to *Bacterium.* The bacterium is a rod (0.5 to 0.7 by 1.1 to 3.8 μm), gram negative, and capsulated, occurs singly or in pairs, and lacks endospores. The organism is motile by means of a long single polar flagellum (Miller and Bollen 1946). Acid, but no gas, is formed from the common diagnostic sugars, mannitol, glycerol, and starch (Burkholder 1932; Galloway 1894). The colony is pale yellow on nutrient dextrose agar. The optimum temperature for growth is 28° to 32°C, the minimum about 1°C and the maximum 35° to 37°C. The thermal death point is approximately 53°C. The optimum pH for growth is 6.4 to 7.2. Enzymes produced by the bacteria in culture are diastase, rennin, proteolase, and pectinase, but not cellulase (Burkholder 1932; Miller and Bollen 1946). The pathogen retains its virulence in culture at room temperature for over three years without repassage through the host (Miller and Bollen 1946). Mulrean and Schroth (1981a) recently developed a semiselective medium containing brilliant cresyl blue that has proved useful in the isolation of this organism from infested walnut buds and catkins.

Disease development. The pathogen survives from one year to the next in twig lesions, affected buds (Miller and Bollen 1946), and nuts that remain on the tree. Between fall and spring, Smith (1922) was able to isolate it from the diseased epidermis, wood, and pith of shoots, and from infected nuts hanging in the tree. Pollen from diseased catkins was also shown to carry the bacteria. In isolations from August through February, Ark (1944) found that 15 percent of the healthy catkins and 10 to 26 percent of the blighted leaf buds contained living bacteria. These results were confirmed by Mulrean and Schroth (1981a,b, 1982), who found sizable percentages of overwintered buds and catkins with epiphytic and internal populations of the pathogen. However, they were unable to isolate the organism from overwintered branch cankers.

In a study of potential sources of inoculum in Chilean walnut orchards, Esterio and Latorre (1982) recovered the pathogen from symptomless buds, catkins, leaves, nuts, and twigs, from blighted branches, mummy nuts, and refuse, and from symptomless weeds. As the infested buds and catkins begin to grow in the spring, the bacteria regenerate and infect the surrounding healthy tissue. There they multiply and become included in an exudate that is spread by rain, insects, and windborne pollen to other susceptible plant tissues. Reportedly, bacteria in this surface exudate can remain viable for several weeks, but with exposure to direct sunlight they die within a few days. When contaminated pollen falls upon the stigmatic surface of the pistillate flower, the bacteria invade this tissue. Miller and Bollen (1946) believed that since distribution of infected nuts in the tree is not uniform, the severity of infection is related to moisture, not to time of pollen shedding. The point generally agreed upon is that the first symptoms appear on the fruit at the blossom end.

Smith (1922) recovered the pathogen in sterilized inoculated soil after 18 days at 20°C, but it survived only 6 to 9 days in unsterilized soil. Rudolph (1943) believed that erinose mites (small arthropods that produce galls on leaves of the black walnut) carry the bacteria from one place to another. Flies, aphids, and other insects that feed upon the bacterial exudate probably pick up the bacteria on their feet and transfer them to susceptible parts of the tree (Smith, Hunt, and Nixon 1913). Although certain insects probably do carry the bacteria, they have not yet been shown to play an important part in pathogen dissemination.

Stomata are probably the principal infection courts on leaves and fruit (Miller and Bollen 1946). Insect punctures and mechanical injuries have also been suspected as infection courts but not proved as such. Fruit and leaf infections can occur at any time in early spring. In California, the disease is most prevalent in years when rains and fog occur early in the growing season (Smith, Smith, and Ramsey 1912). It is particularly severe near the coast and in areas of the Central Valley exposed to ocean influences (e.g., San Joaquin County).

Free moisture is the most important environmental factor required for the initiation and development of blight. This was clearly stated by Pierce in 1896: ". . . moisture of the atmosphere is favorable to the disease, and dry atmosphere is unfavorable to it." The presence of moisture for only 5 to 15 minutes is followed by stomatal infection in laboratory tests (Miller and Bollen 1946). So long as the temperature ranges between 5° and 27°C, bacterial infection of nuts is favored. The time required for lesion development is four days at 27°C and eight days at 15°C. Inoculated leaves develop lesions in six days at 21°C and thirteen days at 16°C.

Cultivar susceptibility. The cultivars reported most susceptible to infection are Payne, Ashley, Santa Barbara Soft Shell, Gustine, Marchetti, and Sunland, while Amigo, Chico, Mayette, Poe, Hartley, Serr, and Concord are less susceptible. Eureka, Waterloo, Franquette, San Jose, Pedro, Vina, and Tehama are least affected. Early blooming cultivars are more susceptible because climatic conditions favor infection at that time. For example, Eureka was once classified as highly resistant to immune in southern California, but when exposed to wet weather during bloom it was highly susceptible; in Oregon it is regarded as a commercial failure because of its susceptibility to blight (Miller et al. 1940; Miller and Bollen 1946). Thus, so-called resistance could very well be related to blooming and foliation in dry weather.

Control. Little can be done in the way of fertilization, irrigation, or cultivation to control walnut blight. Removing affected shoots and fruit would probably help reduce the disease, but such an operation is impractical. The use of late-blooming varieties would be beneficial. Sprinkler irrigation that wets the foliage and nuts should be avoided.

Chemical sprays and dusts are the most effective controls for blight (Scott 1948). Rudolph (1933) found that two applications of 16–8–100 Bordeaux mixture were effective. The first, a prebloom spray, was applied when few or no nuts had appeared and the catkins were out but not shedding pollen. Though he stated that this spray timing was "absolutely indispensable," it is no longer used today. The second spray was applied when the fruit was the size of a pea or small olive. Other investigators later tried copper-lime dust and proprietary copper sprays (Rudolph 1940); still others used sprays containing penicillin (Rudolph 1946) and 50 ppm streptomycin or streptomycin-pyrophyllite dusts (Ark 1955; Miller 1959). If there was a high incidence of disease the previous year, the first spray was applied during catkin elongation and the second spray when 25 to 75 percent of the pistillate flowers were in bloom. However, if the disease was not severe, only one spray was applied when 25 to 75 percent of the pistillate flowers were showing. Bordeaux mixture (16–10–100) has been used before the pistillate flowers appear, but sprays during bloom require a weaker mixture (4–2–100). Fixed-copper materials also are effective (Pinto and Carreño 1988; Teviotdale, Schroth, and Mulrean 1985) when used in accordance with the manufacturer's directions.

Current control recommendations call for the application of one or several copper-containing sprays such as Bordeaux mixture or the fixed coppers. In general, one to three treatments are sufficient in the San Joaquin Valley's dried areas and three or more in regions of greater rainfall. The first application should be made no later than the first appearance of the pistillate bloom (**fig. 1C**). Earlier treatments may be of benefit in some years or in some orchards. Good coverage of the expanding tissues is essential, and a 7- to 14-day interval, according to weather and plant growth, is suggested for additional treatments. The effectiveness of dust applications has not been established (Mulrean and Schroth 1981b; Olsen et al. 1976; Teviotdale, Schroth, and Mulrean 1985).

REFERENCES

Anderson, H. W. 1950. Bacterial blight of Persian walnut in Illinois. *Plant Dis. Rep.* 34:352.

Ark, P. A. 1944. Pollen as a source of walnut bacterial blight infection. *Phytopathology* 34:330–34.

———. 1955. Use of streptomycin-pyrophyllite dust against pear blight and walnut blight. *Plant Dis. Rep.* 39:926–28.

Ark, P. A., and C. E. Scott. 1950. Walnut blight—Symptoms and control. *Diamond Walnut News* 32(2):6–7.

Barss, H. P. 1928. Bacterial blight of walnuts. *Oregon Ext. Circ.* 239, 15 pp.

Burkholder, W. H. 1932. Carbohydrate fermentation by certain closely related species of the genus *Phytomonas*. *Phytopathology* 22:699–707.

Clayton, C. O. 1921. Some studies relating to infection and resistance to walnut blight. *Calif. Dept. Agric. Monthly Bull.* 10:367–71.

Esterio, M. A., and B. A. Latorre. 1982. Potential sources of inoculum of *Xanthomonas juglandis* in walnut blight outbreaks. *J. Hort. Sci.* 57:69–72.

Galloway, B. T. 1894. *Walnut disease*. USDA Report 1893, 272 pp.

Lewis, C. I. 1906. The walnut in Oregon. *Ore. Agric. Exp. Stn. Bull.* 92, 43 pp.

McMurran, S. M. 1917. Walnut blight in the Eastern United States. *USDA Bur. Plant Indus. Bull.* 611, 7 pp.

Miller, P. W. 1934. Walnut blight and its control in the Pacific Northwest. *USDA Circ.* 331, 13 pp.

———. 1959. A preliminary report on the comparative efficacy of copper-lime and agrimycin dust mixtures for the control of walnut blight in Oregon. *Plant Dis. Rep.* 43:401–02.

Miller, P. W., and W. B. Bollen. 1946. *Walnut bacteriosis and its control*. Ore. Agric. Exp. Stn. Tech. Bull. 9, 107 pp.

Miller, P. W., W. B. Bollen, J. E. Simmons, H. N. Gross, and H. P. Barss. 1940. The pathogen of filbert bacteriosis compared with *Phytomonas juglandis*, the cause of walnut blight. *Phytopathology* 30:713–33.

Miller, P. W., and B. G. Thompson. 1935. Walnut and filbert blight and insect pests and their control. *Ore. Agric. Exp. Stn. Tech. Bull.* 476, 16 pp.

Mulrean, E. N., and M. N. Schroth. 1981a. A semiselective medium for the isolation of *Xanthomonas campestris* pv. *juglandis* from walnut buds and catkins. *Phytopathology* 71:336–39.

———. 1981b. Bacterial blight on Persian walnuts. *Calif. Agric.* 35(9/10):11–13.

———. 1982. Ecology of *Xanthomonas campestris* pv. *juglandis* on Persian (English) walnuts. *Phytopathology* 72:434–38.

Olsen, W. H., W. J. Moller, L. B. Fitch, and R. B. Jeter. 1976. Walnut blight control. *Calif. Agric.* 30(5):10–13.

Pierce, N. B. 1896. Bacteriosis of the walnut. *Pacific Rural Press* 57(25):387.

———. 1901. Walnut bacteriosis. *Bot. Gaz.* 31:272–73.

Pinto de T., A. and I. Carreño I. 1988. Control quimico de pesta negra del nogal. *Agric. Tech.* (Chile) 48(3):258–61.

Ruldolph, B. A. 1933. Bacteriosis (blight) of the English walnut in California and its control. *Calif. Agric. Exp. Stn. Bull.* 564:3–88.

———. 1940. A blight control spray, red cuprous oxide. *Diamond Walnut News* 22(2):4–6.

———. 1943. The walnut erinose mite a carrier of walnut blight. *Diamond Walnut News* 25(6):4–5.

———. 1946. Attempts to control bacterial blights of pear and walnut with *Penicillin*. *Phytopathology* 36:717–25.

Scott, C. E. 1948. Does it pay to spray for blight? *Diamond Walnut News* 30(2):11.

Smith, C. O. 1922. Some studies relating to infection and resistance to walnut blight, *Pseudomonas juglandis* (abs.). *Phytopathology* 12:106.

———. 1931. Pathogenicity of *Bacillus amylovorus* on species of *Juglans*. *Phytopathology* 21:219–23.

Smith, R. E., J. F. Hunt, and W. H. Nixon. 1913. Spraying walnut trees for blight and aphis control. *Calif. Agric. Exp. Stn. Circ.* 107, 8 pp.

Smith, R. E., C. O. Smith, and H. J. Ramsey. 1912. Walnut culture in California: Walnut blight. *Calif. Agric. Exp. Stn. Bull.* 231:119–398.

Teviotdale, B. L., M. N. Schroth, and E. N. Mulrean. 1985. Bark, fruit, and foliage diseases. In *Walnut orchard management*, edited by D. E. Ramos, 153–57. Univ. Calif. Coop. Ext. Publ. 21410.

Wormald, H., and J. B. Hammond. 1931. The distribution of bacterial blight of walnut. *Gard. Chron.* 90(2348):476–77.

Crown Gall of English Walnut

Agrobacterium tumefaciens
(Smith & Townsend) Conn

Susceptibility to crown gall is considered one of the reasons why California black walnut rootstock has partly replaced both English and Paradox hybrid as rootstocks for commercial cultivars.

A chemotherapy treatment involving a mixture of 2,4-xylenol and metacresol (Gallex) has been used effectively to eradicate tumors on infected trees. Growers surgically remove the galls and paint the cuts with Gallex. It generally is inadvisable to use the treatment if half or more of the tree's crown tissue is galled. A recently developed biocontrol treatment involving the use of strain 84 of *Agrobacterium rhizogenes (radiobacter)* has proved effective in preventing infection of walnut seedlings by the crown-gall pathogen. A preplant dip of the root system in a suspension of this antibiotic organism protects wounded roots from infection. *Agrobacterium rhizogenes* 84 is available commercially under the names Galltrol-A and Norbac. It functions only as a preventative measure and has no curative action against existing galls. The effectiveness of these treatments may encourage greater use of Paradox hybrid rootstocks in soils where Phytophthora root and crown rot is a problem. A general discussion of crown gall appears in Chapter 4.

Deep-Bark Canker of English Walnut

Erwinia rubrifaciens
Wilson, Zeitoun & Fredrickson

In 1967, a then-undescribed disease of English walnut trees was identified (Wilson, Zeitoun, and Fredrickson 1967) in the San Joaquin Valley of central California. Although the symptoms had been noted as early as 1962, they were not recognized as symptomatic of a distinct disease, since they occurred in trees that were also affected by the shallow-bark canker disease caused by *Erwinia nigrifluens* (Schaad and Wilson 1971b; Wilson, Starr, and Berger 1957). Later, characteristic symptoms of the disease were found in trees free of shallow-bark canker, and a bacterial organism differing in important characteristics from those of *E. nigrifluens* was obtained from the diseased tissue (Wilson, Zeitoun, and Fredrickson 1967; Zeitoun and Wilson 1966, 1969). This disease originally was called "phloem canker" to differentiate it from shallow-bark canker.

So far, deep-bark canker has been an important disease only on the Hartley cultivar in the San Joaquin and Sacramento valleys. Although Hartley is grown to some extent in northern coastal districts (Santa Clara, Napa, and Sonoma valleys, and near Hollister), the disease has not been found there. However, Hartley trees with the disease were found 20 miles from the coast in southern California's Ojai Valley.

The importance of the disease was shown by surveys conducted in the Sacramento Valley on 3,680 trees in 27 orchards over a period of four years. In 1968, only 1.7 percent of the trees showed disease; in 1969, the prevalence was 4 percent; in 1970, it was 9 percent; and in 1971, it was 11.6 percent. This would project an average increase of 3 percent per year (Gardner and Kado 1972). In some orchards in the San Joaquin Valley, 70 percent of the trees were infected (Schaad and Wilson 1970c). The disease is not known to kill trees, but in combination with diseases such as Phytophthora root and crown rot or crown gall it may contribute to their gradual decline and eventual death (Teviotdale, Schroth, and Mulrean 1985) (**fig. 2A**).

Symptoms. Dark brown to black streaks of varying widths extend through the inner bark of the trunk and scaffold branches of affected trees. Because these streaks occur in the region of the phloem, the disease was named "phloem canker" to distinguish it from the previously described bark canker, the diseased areas of which occur almost exclusively in the outer bark just below the phelloderm. Individual discolored streaks of deep-bark (phloem) canker commonly merge to form necrotic areas that may extend several feet up the trunk and into the scaffold branches (**fig. 2B**).

Another characteristic internal symptom is the occurrence of numerous small, roughly circular discolored spots in the outer xylem immediately beneath the cankered areas in the bark. These spots are no more than 1 to 2 mm in diameter.

A symptom visible on the outside of the branch or trunk is a crack in the phelloderm that exudes a dark reddish brown, often slimy substance (**fig. 2B**). This material, which is produced abundantly in the summer, runs down the surface of the bark and dries, leaving discolored deposits somewhat like those on trees affected by shallow-bark canker.

Causal organism. The pathogen is a peritrichous, gram-negative, rod-shaped organism 0.37 to 0.6 μm wide and 0.92 to 1.52 μm long. It grows well at temperatures of 24° to 39°C, with an optimum of 30° to 33°C. Its pH optimum runs from 6.0 to 7.2. It utilizes arabinose, glucose, glycerol, mannose, and sucrose as carbon sources, but apparently not lactose or xylose. On YDC agar (yeast extract, dextrose, calcium carbonate agar) the organism grows profusely and produces a red pigment that diffuses through the medium surrounding the colonies. In other diagnostic cultural tests, the organism differs significantly from *Erwinia nigrifluens* and *E. rhapontici*, the latter being the only other known *Erwinia* species to produce a red pigment on the YDC medium. A bacteriophage obtained from the exudate from cankers produced by *E. nigrifluens* was found to have lysed that organism but not the deep-bark canker pathogen. Concentrated suspensions of the phage placed in contact with the latter pathogen did, however, modify the cell wall of the bacterium to such an extent that it became spherical in shape (Schaad et al. 1973). Such a phenomenon is known to occur with other bacteria and has been termed "lysis from without" and "abortive infection."

The phloem canker pathogen differs serologically from other *Erwinia* species, including *E. nigrifluens* (Wilson, Zeitoun, and Fredrickson 1967; Zeitoun and Wilson 1966). In view of the differences in cultural and serological relationships, the phloem canker organism was considered an unidentified species and was named *Erwinia rubrifaciens* n. sp. (red-producing *Erwinia*). Azad and Kado (1984) reported that there was no correlation between pathogenicity of the organism and the tobacco hypersensitivity reaction.

Hosts. *Erwinia rubrifaciens* has been identified only on English walnut (*Juglans regia*). Among walnut cultivars, Hartley is most commonly affected, and is the only one on which the disease is of economic importance. When interplanted with Hartley, the Franquette, Eureka, Ashley, and Payne are sometimes attacked, and artificial inoculations indicate that Gustine and Howe are somewhat susceptible (Kado et al. 1977). Although black walnuts (*J. hindsii* and *J. nigra*) are considered nonsusceptible, cankers originating in the trunk of a Hartley sometimes extend several inches into the tissues of the black walnut rootstock.

Hartley originated in the Napa Valley of California and was introduced in 1925. Its parentage is unknown, except that in 1909 it is known to have been produced from a seedling grown from seeds planted in 1892. The tree comes into leaf after Payne and is a pollinator for Payne, and is one of the most popular commercial cultivars in California.

Disease development. The pathogen survives from year to year in infected trees and is released from them in large numbers through cracks in the bark (**fig. 2C**). The slimy exudate is most abundant in summer and early autumn, when it runs down the surface of the bark and dries. The bacteria may remain viable in the dried exudate for as long as 123 days. Windblown rains disseminate the bacteria as far as 20 feet (Schaad and Wilson 1971a).

Under winter conditions the pathogen can survive as long as three months in infested soil. Whether infested soil constitutes an important overwintering site for the bacteria has not been determined (Schaad and Wilson 1970c).

The bacteria gain entrance to the inner bark through breaks in the outer bark. Growth cracks, which occur frequently in the trunk and scaffold branches of Hartley, as well as pruning cuts and holes made by woodpeckers and oystershell scale are avenues of entry. Injuries caused by mechanical harvesting provide infection courts that appear to be extremely important in disease epidemiology (Kado and Gardner 1977; Teviotdale, Schroth, and Mulrean 1985). These wounds may remain susceptible to infection for as long as 10 days. The bacteria are restricted to nonfunctional phloem and ray parenchyma, but adjacent conducting phloem is adversely affected. This impairs the movement of sugars, which results in a gradual weakening of scaffold branches and sometimes the entire tree.

Disease activity, in the form of systemic canker extension and exudation of bacteria from infected areas of the tree, is greatest from April to October. This, in turn, corresponds to the relatively high temperature requirement of the bacterium (optimum temperature 30° to 33°C). Increases in canker length are most rapid during the summer months, with the rate of extension corresponding to the increase in average temperature from spring to summer. Infections do not occur in the winter (Schaad et al. 1973).

Prevalence of the disease has differed greatly even within the Central Valley, being much more widespread in the southern half (San Joaquin Valley) than in the northern half (Sacramento Valley) (Schaad and Wilson 1971a). One possible factor contributing to this phenomenon is the relative youth of Hartley trees in the north compared to those in the south. Another is temperature: minimum and maximum air temperatures in the summer are higher in the central San Joaquin Valley than at Davis in the southern Sacramento Valley. In comparative tests, cankers from inoculations developed more rapidly in the San Joaquin Valley than at Davis. Unfavorable growing conditions such as poor soil or chronic water stress that decrease tree vigor favor canker development (Teviotdale, Schroth, and Mulrean 1985; Teviotdale and Sibbett 1982).

The pathogen invades the sieve tubes of the nonfunctional secondary phloem, a development that appears to be characteristic of *Juglans regia* (Schaad and Wilson 1970a,b). As the nonfunctional secondary phloem elements become involved, the bacteria pass into the laterally oriented parenchyma cells. Invaded adjacent sieve tubes and parenchyma merge and produce bands of discolored tissue. Because the sieve plate pores are large enough to allow passage of the bacterial cells, vertical invasion of the sieve elements readily follows. This results in long streaks and bands of diseased tissue, sometimes extending several feet up and down the inner bark outside of the cambium and functional phloem (Gardner and Kado 1973).

In addition to the vertical movement of the pathogen, there is a lateral movement inward from the secondary nonfunctional phloem. This occurs along the ray elements and terminates in the outer xylem. If the bark is pulled away, the surface of the outer xylem (wood tissue) is found to be covered with many dark, roughly circular pits 1 to 2 mm in diameter. The inner surface of the bark overlying this area is covered with small protrusions that correspond to the pits. The pits are, therefore, the termini of the infected ray elements. No further bacterial invasion of the xylem has been found—apparently, vertical movement of bacteria in the tree occurs only in the secondary nonfunctional phloem elements.

Control. When cankers are small, surgical removal is possible, but removing all of the cankered tissue of an infected tree is impractical. Chemical control measures have not proved effective. Applications of various copper materials, antibiotics, and household bleach (sodium hypochlorite) have failed to give satisfactory control, even though some of these preparations killed the pathogen in the exudate. Reducing injuries from mechanical harvesting appears to be one of the most effective control measures. Gardner and Kado (1972) suggested that if few trees are infected in the orchard, they should be harvested last to minimize the risk of mechanical spread on the shaker. Decontamination of harvester pads is recommended when bringing machinery into an orchard that does not have deep-bark canker. Good water management appears to be extremely important in combating this disease (Teviotdale, Schroth, and Mulrean 1985; Teviotdale and Sibbett 1982). In a three-year test in Tulare County, a midwinter irrigation significantly reduced the percentage of trees with active cankers (Teviotdale and Sibbett 1982). With good water status restored, diseased trees often recover.

REFERENCES

Azad, H. R., and C. I. Kado. 1984. Relation of tobacco hypersensitivity to pathogenicity of *Erwinia rubrifaciens*. Phytopathology 74:61–64.

Gardner, J. M., and C. I. Kado. 1972. Deep bark canker. Diamond Walnut News (5):20–21.

———. 1973. Evidence for systemic movement of *Erwinia rubrifaciens* in Persian walnuts by the use of double-antibiotic markers. Phytopathology 63:1085–86.

Kado, C. I., J. C. L ra, W. J. Moller, and D. E. Ramos. 1977. An assessment of the susceptibility of various walnut cultivars to deep bark canker. J. Am. Soc. Hort. Sci. 102:698–702.

Kado, C. I., and J. M. Gardner. 1977. Transmission of deep bark canker of walnuts by the mechanical harvester. Plant Dis. Rep. 61:321–25.

Schaad, N. W., M. G. Heskett, J. M. Gardner, and C. I. Kado. 1973. Influence of inoculum dosage, time after wounding, and season on infection of Persian walnut trees by *Erwinia rubrifaciens*. Phytopathology 63:327–29.

Schaad, N. W., and E. E. Wilson. 1970a. Structure and seasonal development of secondary phloem of *Juglans regia*. Can. J. Botany 48:1049–53.

———. 1970b. Pathological anatomy of the bacterial phloem canker diseases of *Juglans regia*. Can. J. Botany 48:1055–60.

———. 1970c. Survival of *Erwinia rubrifaciens* in soil. *Phytopathology* 60:557–58.

———. 1971a. Bacterial phloem canker of Persian walnut: Development and control factors. *Calif. Agric.* 25(4):4–7.

———. 1971b. The ecology of *Erwinia rubrifaciens* and the development of phloem canker of Persian walnut. *Ann. Appl. Biol.* 69:125–36.

Teviotdale, B. L., M. N. Schroth, and E. N. Mulrean. 1985. Bark, fruit, and foliage diseases. In *Walnut orchard management*, edited by D. E. Ramos, 153–57. Univ. Calif. Coop. Ext. Publ. 21410.

Teviotdale, B. L., and G. S. Sibbett. 1982. Midwinter irrigation can reduce deep bark canker of walnuts. *Calif. Agric.* 36(5/6):6–7.

Wilson, E. E., M. P. Starr, and J. A. Berger. 1957. Bark canker, a bacterial disease of Persian walnut trees. *Phytopathology* 47:669–73.

Wilson, E. E., F. M. Zeitoun, and D. L. Fredrickson. 1967. Bacterial phloem canker, a new disease of Persian walnut trees. *Phytopathology* 57:618–21.

Zeitoun, F. M., and E. E. Wilson. 1966. Serological comparisons of *Erwinia nigrifluens* with certain other *Erwinia* species. *Phytopathology* 56:1381–85.

———. 1969. The relation of bacteriophage to the walnut-tree pathogens *Erwinia nigrifluens* and *Erwinia rubrifaciens*. *Phytopathology* 59:756–61.

Shallow-Bark Canker of English Walnut

Erwinia nigrifluens Wilson, Starr & Berger

Bark canker (shallow-bark canker) was discovered in the Sacramento Valley in 1955. Few cases have been found in other areas of the state, and none has been reported elsewhere in the world.

Judging from the extensive nature of the cankers, one might expect the disease to be capable of serious damage to the tree. Although the trunks of some trees are completely encircled by the cankers, few trees show evidence of serious injury, apparently because the bark is killed to the cambium in only a few places (Wilson, Starr, and Berger 1957).

Symptoms. Irregular dark brown necrotic areas develop in the bark of the trunk and scaffold branches (**fig. 3B**). These originate as small, roughly circular spots in the cortical tissue just beneath the corky outer layer of bark (periderm). Extensive cankers are formed by the enlargement and coalescence of these spots. The outline of the canker is not visible from the surface, but its presence can be detected by a dark, watery exudate that stains the affected trunk or limb (**fig. 3A**). In general the canker is relatively shallow, extending only to one-fourth or one-third the depth of the bark. Occasionally a break occurs in the cortex underlying the infected area, and the bark adjacent to the break becomes involved to a greater depth.

Susceptibility of walnut species and cultivars. Shallow-bark canker has been found on the Hartley, Mayette, Payne, Mammmoth, Meyland, and Myrtleford cultivars of English walnut (*Juglans regia*). In one orchard the disease was found on the trunk of a Paradox hybrid (*J. regia* × *J. hindsii*). Northern California black walnut (*J. hindsii*) appears to be resistant if not immune to the disease. Franquette trees growing near diseased trees of other cultivars are seldom affected. Inoculation tests indicated, however, that certain clones of this cultivar are more susceptible than others.

Causal organism. The pathogen is a bacterium of the genus *Erwinia*. It resembles *E. amylovora* in some features, but differs in important characteristics; therefore, it has been given the designation *Erwinia nigrifluens* Wilson, Starr & Berger. *Erwinia nigrifluens* is a peritrichous, gram-negative, rod-shaped organism ranging from 0.8 to 1.3 μm wide by 1.6 to 3.5 μm long. The rods are slightly pointed at the ends and can be somewhat curved. There are at least five flagella, and they are two to three times the length of the cell. No endospores or capsules are formed. The bacterium grows and produces acid, but no gas, in media containing the following carbon sources: glucose, fructose, galactose, mannose, ribose, xylose, arabinose, rhamnose, sucrose, cellobiose, and raffinose. It does not utilize ammonium chloride as the sole nitrogen source, but can use amino acids as the sole source of both carbon and nitrogen. Optimum temperature for bacterial growth in culture is about 28°C; it grows slowly or not at all at or above 37°C.

Disease development. The bacterium survives from one year to the next in the bark canker and escapes from cankers in the dark, watery exudate. The manner in which the bacteria disseminate has not been determined, nor has the most common mode of entry into the host. In one orchard, the holes produced in the bark by birds (sapsuckers) were found to be affected. In other orchards, however, such injuries were not avenues of entry. The cankers spread rapidly in summer and are essentially quiescent during the fall and winter. Serological and bacteriophage studies are reviewed in the section on deep-bark canker (Zeitoun and Wilson 1966, 1969).

Control. No measures for preventing infection have been devised. However, the following method effectively prevents extension of established cankers. In spring, shave off the corky outer bark overlying the diseased area with a carpenter's drawknife. A few weeks after this treatment the necrotic bark will have dried out, and the bacteria will die. The canker margins are usually well defined, so one can easily determine the limit of the infection and thus remove only the periderm that overlies diseased bark (Wilson, Starr, and Berger 1957).

REFERENCES

Wilson, E. E., M. P. Starr, and Joyce A. Berger. 1957. Bark canker, a bacterial disease of the Persian walnut tree. *Phytopathology* 47:669–73.

Zeitoun, F. M., and E. E. Wilson. 1966. Serological comparison of *Erwinia nigrifluens* with certain other *Erwinia* species. *Phytopathology* 56:1381–85.

———. 1969. The relation of bacteriophage to the walnut-tree pathogens *Erwinia nigrifluens* and *Erwinia rubrifaciens*. *Phytopathology* 59:756–61.

Fungal Diseases of English Walnut

Anthracnose of English Walnut

Gnomonia leptostyla (Fr.) Ces. and DeN.

This disease affects eastern black walnut (*Juglans nigra*), northern California black walnut (*J. hindsii*), southern California black walnut (*J. californica*), English walnut (*J. regia*), and butternut (*J. cinerea*) of the eastern United States. It is rarely important in the English walnut orchards of the Pacific Northwest (Miller, Schuster, and Stephenson 1945) and is not known to occur in California. The disease causes serious damage to walnuts in regions with frequent rains during the growing season.

Symptoms. Circular reddish brown to grayish brown spots (1/16 to 3/4 inch in diameter) with grayish centers develop on the leaves. Oval or irregularly circular, sunken, light grayish brown spots with dark brown margins develop on green shoots. On the nuts, the disease is characterized by circular, depressed necrotic lesions in the hull. If infection occurs early in the season the infected nuts may drop or, if they remain on the tree, become malformed. Yield reduction is the most serious effect of this disease.

Causal organism. The causal organism is the fungus *Gnomonia leptostyla* (Fr.) Ces. and DeN.; the imperfect stage is *Marssonina juglandis* (Lib.) Magn.

Disease development. The fungus overwinters in infected leaves and nuts on the ground and in cankers on twigs infected the previous year. The ascigerous stage is produced on these parts. In the spring, the ascospores ejected from the perithecia are carried by air to the new growth where they produce the initial infection of the season. Conidia produced in the acervuli on old twig lesions may also serve as primary inoculum. Conidia produced in old and new lesions constitute the principal inoculum for summer infection.

Control. Control measures are rarely necessary for this disease in the United States. In France, where walnut anthracnose sometimes causes severe leaf and fruit infection of both wild and cultivated walnuts, spraying with Bordeaux mixture effectively reduces the amount of infection (Gard 1928). Rankin (1932) reported a reduction of the disease on eastern black walnut using three treatments of Bordeaux when leaves are unfolding, when leaves have reached mature size, and about two weeks after the second application. Other effective fungicides are benomyl, chlorothalonil, cupric hydroxide, and dodine; benomyl apparently provides the best control (Berry 1977).

REFERENCES

Berry, F. H. 1977. Control of walnut anthracnose with fungicides in a black walnut plantation. *Plant Dis. Rep.* 61:378–79.

Gard, M. 1928. Développement des maladies cryptogamique sur les noyers en 1926. *Ann. Epiphytes* 14:152–62.

Hammond, J. B. 1931. Some diseases in walnuts. *Ann. Rept. East Malling Res. Stn.* 1928, 1929, and 1930 supp. 2:143–49.

Miller, P. W., C. E. Schuster, and R. E. Stephenson. 1945. Diseases of walnuts in the Pacific Northwest and their control. *Ore. Agric. Exp. Stn. Bull.* 435, 42 pp. (reprinted 1947).

Rankin, W. H. 1932. Spraying for leaf diseases of shade trees. *Proc. Eighth Ann. Meeting Nat. Shade Tree Conf.*: 64–69.

Branch Wilt of English Walnut

Hendersonula toruloidea Nattrass

The fungus causing branch wilt is found on English (Persian) walnut (*Juglans regia*), lemon (*Citrus limonia*), grapefruit (*C. grandis*), orange (*C. sinensis*), European chestnut (*Castanea sativa*), fig (*Ficus carica*), almond (*Prunus dulcis*), and poplar (*Populus* sp.) (Bates 1937; English, Davis, and DeVay 1975; Ogawa 1954; Paxton, Wilson, and Davis 1964; Warner 1952). It has also been found on dead branches of the black walnut (*Juglans hindsii*). In Egypt, Nattrass (1933) isolated the causal fungus from apple, apricot, and peach trees thought to have been injured by excessive soil moisture. The fungus was first isolated in California by Fawcett (1936, pp. 218, 222).

In California the disease has not been found on walnut in coastal areas, but it is widely distributed throughout the Sacramento and San Joaquin valleys and in interior districts of southern California (Wilson 1947). Branch wilt is a major walnut disease in Tulare County, but it has been of secondary significance in other localities.

Symptoms. During July and August, the leaves on certain branches wither, turn deep brown, and dry up but remain attached to the twigs (**fig. 4A**). The symptoms that conclusively identify branch wilt occur in the branches; the outer layer of bark (periderm) loosens in certain areas and the underlying cortex is covered by a black powdery material composed of black one-celled spores (arthrospores) of the causal fungus (**fig. 4B**). The cortex underlying the spore mass is brown, and the wood is dark gray to black (Wilson 1945, 1947) (**fig. 4C**). Once established in a branch, the fungus may spread to other branches and eventually to the trunk.

Causal organism. The conidial (arthrospore) stage was described as *Exosporina fawcetti* (Wilson 1947), but a few years later a pycnidial stage was found (Wilson 1949) that proved to be identical with *Hendersonula toruloidea* Nattrass (Nattrass 1933). Until the time a perithecial stage is found, the latter is the correct binomial, though this fungus probably was first described as *Torula dimidiata* Penzig (1887), an incorrect generic classification.

Arthrospores are produced by the progressive segmentation of closely packed hyphae arising from a slight hypostroma. The structure, therefore, is a sporodochium. The pycnidia are produced in a stroma, which Nattrass (1933) believed developed from the conidium-bearing hyphae. One to six long-necked pycnidia are partially immersed in the stroma. The pycnidiospores (5.3 by 14.6 μm), which are produced on short stalks arising from the base of the pycnidium, are one-celled and hyaline as long as they remain in the pycnidium. But a few days after being extruded from the pycnidia, they become three-celled, and the central cell turns dark brown.

Disease development. The fungus undergoes a period of inactivity in the form of mycelium and arthrospores in and on infected branches. These spores constitute the chief inoculum for infection, since pycnidiospore development is comparatively rare. Arthrospores are produced in dense layers beneath the periderm of infected branches. They are capable of withstanding long periods of hot dry weather. Inasmuch as arthrospores are only 4.6 to 7.3 μm in diameter and are produced in powdery masses, they are suited for wind dissemination, but they may also be washed about by rain. Sunburn cracks and mechanical injuries are common avenues of entry into the branch; the fungus is not known to penetrate uninjured bark (Paxton and Wilson 1965). Inoculation of the cracks or cuts in the branches probably occurs during the winter, but infection does not occur until spring or early summer. The most active extension of the fungus through the branches occurs in midsummer.

The incubation period is greatly affected by temperature. Inoculations produced large cankers within 10 to 15 days during the summer, whereas inoculations in winter or early spring did not produce cankers for several months. High temperatures can cause sunburning and allow entry of the fungus—the absence of serious branch wilt in coastal areas is believed to be the result of lower summer temperatures. Temperatures below 25°C are less favorable to the fungus than are temperatures between 25° and 35°C, the optimum being somewhere between 30° and 33°C. The vigor of the tree has a marked influence on its susceptibility to the disease (Sommer 1955). Unfavorable soil fertility and soil moisture and the presence of root or crown diseases that lower the tree's vigor increase its susceptibility. Also, heavy mite populations increase the susceptibility of trees to sunburn and subsequent infection by the branch-wilt pathogen.

The pathological histology of the disease has not been thoroughly studied, but the mycelium of the fungus is known to invade most elements of the xylem, causing the formation of tyloses and a dark gummy substance. The mycelium also resides in the ray cells and most elements of the phloem and cortex.

Control. Franquette, Mayette, Eureka, and Meyland are most susceptible. Payne is somewhat less susceptible, and Concord apparently is fairly resistant, though it is not well adapted to Central Valley conditions. Cultural practices involving proper irrigation and fertilization have essentially eliminated this disease problem.

Affected branches must be removed and destroyed. One or two applications of Bordeaux mixture in winter will reduce infection (Wilson 1950), although spraying walnut trees is expensive. With good culture, spraying should not be necessary.

REFERENCES

Bates, G. R. 1937. Disease of citrus fruits in southern Rhodesia. *Mazoe Citrus Exp. Stn. Ann. Rept. Pub.* 6:173–208.

Calavan, E. C., and J. M. Wallace. 1948. Exosporina branch blight of grapefruit in southern California (abs.). *Phytopathology* 38:913.

English, H., J. R. Davis, and J. E. DeVay. 1975. Relationship of *Botryosphaeria dothidea* and *Hendersonula toruloidea* to a canker disease of almond. *Phytopathology* 65:115–22.

Fawcett, H. S. 1936. *Citrus diseases and their control.* McGraw-Hill, New York, 656 pp.

Nattrass, R. M. 1933. A new species of *Hendersonula* (*H. toruloidea*) on deciduous trees in Egypt. *Brit. Mycol. Soc. Trans.* 18:189–98.

Ogawa, J. M. 1954. The occurrence of *Hendersonula toruloidea* Nattrass on *Populus* species in California. *Plant Dis. Rep.* 38:238.

Paxton, J. D., and E. E. Wilson. 1965. Anatomical and physiological aspects of branch wilt disease of Persian walnut. *Phytopathology* 55:21–26.

Paxton, J. D., E. E. Wilson, and J. R. Davis. 1964. Branch wilt of fig caused by *Hendersonula toruloidea*. *Plant Dis. Rep.* 48:142.

Penzig, O. 1887. *Studi botanici sugli agrumi e sulle piante afini: Memoria premiata dal R. Ministro dell' Agricultura.* Tipografia Eredi Botta, Rome, 590 pp.

Sommer, N. F. 1955. Infection of the Persian walnut tree by *Hendersonula toruloidea* Nattrass. Ph.D. dissertation, Department of Plant Pathology, University of California, Davis, 82 pp.

Warner, R. M. 1952. Some observations on branch wilt on figs. *Proc. Sixth Ann. Res. Conf. Calif. Fig Inst.*: 24–25.

Wilson, E. E. 1945. A wilt disease of Persian walnuts in California. *Plant Dis. Rep.* 29:614–15.

———. 1947. The branch wilt of Persian walnut trees and its cause. *Hilgardia* 17:413–36.

———. 1949. The pycnidial stage of the walnut branch wilt fungus, *Exosporina fawcetti*. *Phytopathology* 39:340–46.

———. 1950. Studies on control of walnut branch wilt. *Diamond Walnut News* 32(4):6–9.

Downy Leaf Spot of English Walnut

Microstroma juglandis (Bereng.) Sacc.

This disease is of minor importance in the walnut-growing areas of the Pacific Coast states. During wet years it is widespread in California's Central Valley. All species of *Juglans* are susceptible.

Symptoms. White or pale yellow downy spots develop on the underside of leaves (**fig. 5**). On fruit, the lesions are roughly circular with a downy texture; they are extended by the increase of fruit size until they reach 1 inch or more in diameter. The downy mat of fungus is commonly confined to the periphery of the lesions. The surface of the hull may be slightly indented at the lesion's periphery, but the fruit is not otherwise malformed.

Cause. The fungus causing this disease is *Microstroma juglandis* (Bereng.) Sacc. Pires (1928) believed this fungus to be a Basidiomycete, but Wolf (1927) and Karakulin (1923) placed it in Melanconiaceae of the Fungi Imperfecti. The fungus now is placed in the order Basidiales of the Basidiomycetes (Farr et al. 1989). *Microstroma juglandis* var. *robustum* is said to cause a disease of pecan catkins.

Little is known about the life history of the fungus or the development of the disease. It is more common in wet years and in orchards with poor air circulation (J. K. Hasey, Personal communication 1986).

Control. Control of downy leaf spot is seldom necessary. The disease has not occurred in orchards regularly sprayed with Bordeaux mixture for control of bacterial blight. Downy leaf spot also can be controlled with sprays of benomyl.

REFERENCES

Farr, D. F., G. F. Bills, G. P. Chamuris, and A. Y. Rossman. 1989. *Fungi on plants and plant products in the United States.* APS Press, American Phytopath. Soc., St. Paul, MN. 1252 pp.

Karakulin, B. P. 1923. [On the question of the systematic position of fungi belonging to the type of Exobasidiopsis.] *Not. Syst. Inst. Cryt. Hort. Bot. Petropol.* 2(7):101–08.

Miller, P. W., C. E. Schuster, and R. E. Stephenson. 1945. Diseases of the walnut in the Pacific Northwest and their control. *Ore. Agric. Exp. Stn. Bull.* 435, 42 pp. (reprinted 1947).

Pires, V. A. 1928. Concerning the morphology of Microstroma and the taxonomic position of the genus. *Am. J. Botany* 15:132–40.

Wolf, F. A. 1927. The morphology and systematic position of the fungus, *Microstroma juglandis* (Bereng.) Sacc. *J. Elisha Mitchell Sci. Soc.* 43:97–99.

Kernel Mold of English Walnut

Miscellaneous Fungi

Under certain conditions several fungi are able to penetrate the walnut shell and cause a moldy condition of the kernel (**fig. 6**). This disorder varies in severity from year to year depending on environmental factors. The invading fungi commonly coat the kernels with grayish white mycelium and, depending on the species involved, also may form masses of greenish blue or grayish black spores. Infection renders nuts unmarketable.

The weakly pathogenic fungi that cause kernel mold belong to such genera as *Aspergillus*, *Penicillium*, *Alternaria*, and *Rhizopus*. These organisms may invade the nuts as early as midsummer if the hulls (husks) have been damaged by sunburn, drought, or husk fly. The damaged hulls adhere to the shell and allow the molds that colonize this injured tissue to penetrate into the kernel, usually through the soft suture.

Abnormally hot and dry summers cause the hulls to shrivel around the shell instead of cracking normally, and provide conditions that tend to promote kernel infection. Kernel mold also is directly correlated with the length of time nuts remain on the tree after hull split. In addition, infection increases when nuts remain too long on rain-soaked ground. The incidence of kernel mold can be reduced by providing cultural conditions that minimize sunburn and shriveling of the hull. Harvest should not be delayed, and should occur as close as possible to hull split. A spray of the growth regulator ethephon when the packing tissue around the kernel turns brown makes it possible to harvest virtually the entire crop earlier than normal. Use of this material in recent years has resulted in a marked improvement in kernel quality. Drying the nuts quickly after harvest also helps reduce kernel mold.

REFERENCES

Hendricks, L. C., R. H. Gripp, an D. E. Ramos. 1980. Walnut production in California. *Univ. Calif. Div. Agric. Sci. Leaf.* 2984, 22 pp.

Mircetich, J. M., W. J. Moller, and B. Teviotdale. 1982. Diseases. In *Integrated pest management for walnuts*, 62–65. Univ. Calif. Div. Agric. Sci. Publ. 3270.

Olson, W. H., and W. W. Coates. 1985. Maturation, harvesting, and nut quality. In *Walnut orchard management*, edited by D. E. Ramos, 172–74. Univ. Calif. Coop. Ext. Publ. 21410.

Smith, R. E. 1941. *Diseases of fruits and nuts.* Calif. Agric. Ext. Serv. Circ. 120. 168 pp.

Melaxuma Canker and Twig Blight of English Walnut

Dothiorella gregaria Sacc.

This newly described disease attracted considerable attention in Santa Barbara and Ventura counties early in the 20th century (Fawcett 1915). It has not been identified with certainty in the important walnut-producing areas of California's Central Valley, and was not encountered anywhere in the state in the most recent walnut disease survey conducted by the California Department of Food and Agriculture. Where outbreaks occur, large branches of the walnut tree may succumb or many small terminal branches may be killed. The symptoms that develop in tree crotches resemble those of deep-bark canker caused by *Erwinia rubrifaciens*.

Symptoms. The most conspicuous aspect of the disease, and the one that has given it the name "melaxuma," is the exudation of a dark, watery material from the surface of cankers located mostly at the crotches of large limbs. The bark of such cankers is moist and dark brown to black; the underlying sapwood is also dark. Occasionally, the disease develops on the outermost branches, and is first noticeable in midsummer when the leaves on such branches suddenly wither. This phase of the disease resembles the branch-wilt disease caused by *Hendersonula toruloidea*. However, melaxuma twig blight progresses more slowly than branch wilt and seldom extends downward into large branches.

Causal organism. The fungus *Dorthiorella gregaria* Sacc. was shown by Fawcett (1915) to be the cause of the disease. He isolated it both from the cankers on large limbs and from the small blighted twigs. On the latter, the fungus often fruits abundantly, producing subcuticular pycnidia either singly or in groups on a basal stroma. The spores are oblong-fusoid, 20 to 26 by 5 to 7 µm in dimension, and nonseptate until they germinate, at which time they develop a single septum. The fungus is said to be the pycnidial stage of *Physalospora gregaria* Sacc. (Grove 1935).

The same fungus occurs on other hosts, particularly members of *Cornus*, *Populus*, and *Salix*. Fawcett (1915) found it on the arroyo willow (*Salix lasiolepis*) in Santa Barbara County and believed this to be an inoculum source for adjacent walnut orchards.

Disease development. Cankers commonly occur at the crotches of large limbs, indicating that such places are particularly favorable for infection or for growth of the fungus after infection. Other probable infection sites are injuries of various sorts, including those caused by harvesting tools and cracks resulting from wind and other natural agencies. Small, weakened twigs, especially on the inside of the tree, are sometimes attacked, with the fungus passing into the supporting branch to form a canker (Fawcett 1915).

Though the height of disease activity, as judged by the enlargement of cankers and wilting of leaves on affected branches, occurs in summer, infection probably occurs in winter or no later than in spring.

Control. The removal of wilted branches and excision of cankers on larger limbs are beneficial measures (Fawcett 1915; Smith 1941). The painting of wounds with Bordeaux paste, however, has been recommended (Fawcett 1915). Experience in some areas indicates that inadequate soil moisture may predispose trees to infection.

REFERENCES

Fawcett, H. S. 1915. Melaxuma of the walnut, "*Juglans regia*". *Calif. Agric. Exp. Stn. Bull.* 261:133–48.

Grove, W. B. 1935. *British stem- and leaf-fungi*, vol. 1. Cambridge University Press, London, 240 pp.

Smith, R. E. 1941. *Diseases of fruits and nuts.* Calif. Agric. Exp. Stn. Circ. 120, 168 pp.

Phytophthora Root and Crown Rot of English Walnut

Phytophthora spp.

Some 13 different *Phytophthora* species (*P. citricola*, *P. cinnamomi*, *P. cactorum*, *P. megasperma*, *P. cryptogea*, *P. drechsleri*, *P. citrophthora*, *P. parasitica*, and five unidentified species) are associated with walnut trees in California affected by root and crown rot. Research (Mircetich and Matheron 1983; Mircetich, Matheron, and Teviotdale 1985) has shown that each of these 13 *Phytophthora* species can infect and cause root or crown rot on rootstocks used for walnuts in California. More than one species is often found in the same commercial orchard. The incidence and severity of the disease are influenced by the *Phytophthora* species present, soil moisture, temperature, and the relative resistance of the rootstock. Paradox rootstock is more resistant to several *Phytophthora* species than *Juglans hindsii* or *J. regia* rootstocks, making Paradox use advisable in sites with a history of Phytophthora root and crown rot. The graft union of planted trees should be above the soil line. Careful soil-water management and improved drainage also help reduce losses from this disease.

Details concerning this disease are discussed in Chapter 4.

REFERENCES

Mircetich, S. M., and M. E. Matheron. 1983. Phytophthora root and crown rot of walnut trees. *Phytopathology* 73:1481–88.

Mircetich, S. M., M. E. Matheron, and B. L. Teviotdale. 1985. Armillaria and Phytophthora root and crown rot diseases. In *Walnut orchard management*, edited by D. E. Ramos, 130–41. Univ. Calif. Coop. Ext. Publ. 21410.

Shoestring (Armiilaria) Root Rot of English Walnut

Armillaria mellea (Vahl:Fr.) Kummer

This disease, a general discussion of which appears in Chapter 4, is an example of a disease controlled with resistant rootstocks. English walnut on its own rootstock cannot survive in *Armillaria*-infested soil, but on resistant northern California black walnut (*J. hindsii*) rootstock the trees can be planted in infested areas where other tree fruit could not survive. Paradox hybrid rootstock, which approaches the northern California black walnut rootstock in resistance to *A. mellea*, is now in extensive use. Occasionally, walnut trees on *J. hindsii* rootstock are severely affected with shoestring root rot, presumably because of an extremely virulent strain of *Armillaria*, variability of the black walnut seedlings, or unfavorable soil conditions. In heavy, poorly drained soils, Paradox hybrid performs better than northern California black walnut rootstock (W. O. Reil, personal communication).

5 MISCELLANEOUS FUNGAL DISEASES OF ENGLISH WALNUT

Diseases of minor importance reported on English walnut in California are Cytospora canker (*Cytospora* sp.), Dematophora root rot (*Rosellinia necatrix* [Hartig] Berl.), Diplodia canker and dieback (*Diplodia juglandis* [Fr.] Fr.), Fusarium root rot (*Gibberella baccata* [Wallr.] Sacc., anamorph = *Fusarium lateritium* Nees), and wood rot (*Laetiporus sulphureus* [Bull.:Fr.] Murr., *Polyporus* sp., and *Schizophyllum commune* Fr.).

Diseases of English walnut not known in California but present in other parts of the United States are Ascochyta ring spot (*Ascochyta juglandis* Bolstshauser), Oregon, Washington; Botryosphaeria dieback (*Botryosphaeria querqum* [Schwein.] Sacc.), Florida; Cercospora leaf spot (*Cercospora* sp.), Florida; Cylindrosporium leaf spot (*Cylindrosporium juglandis* F. A. Wolf, telemorph = *Mycosphaerella juglandis* K. J. Kessler), Alabama, North Carolina, Oklahoma; Marssonina leaf spot (*Marssonina juglandis* [Lib.] Magnus), Oklahoma; Melanconis canker and dieback (*Melanconis juglandis* [Ellis & Everh.] Graves), Connecticut, New Jersey; Nectria canker (*Nectria galligena* Bres.), North Carolina; Nectria canker (*Nectria* sp.), New York; Phloeospora leaf spot (*Phloeospora multimaculans* Heald & F. A. Wolf), Oklahoma, Texas; Phyllosticta leaf spot (*Phyllosticta juglandis* [DC.] Sacc.), Georgia, Indiana, Oregon, Washington; Phymatotrichum root rot (*Phymatotrichopsis omnivora* [Dugg.] Hennebert), Texas; and wood rot (*Trametes hirsuta* [Wulfen:Fr.] Quel.), Oregon.

REFERENCES

Farr, D. F., G. F. Bills, G. P. Chamuris, and A. Y. Rossman. 1989. *Fungi on plants and plant products in the United States.* APS Press, American Phytopathological Soc., St. Paul, MN. 1252 pp.

French, A. M. 1987. *California plant disease host index. Part 1. Fruits and nuts.* Calif. Dept. Food and Agric., Sacramento, CA. 39 pp.

Virus Diseases of English Walnut

Blackline of English Walnut

Cherry Leafroll Virus (CLRV-W)

The infectious disease known as blackline or sometimes as girdle of walnut causes the death of many trees in California and Oregon. It was first observed in Oregon in 1924 and in California's Contra Costa County in 1929. The disease is quite common in some areas but rare or absent in others. For example, it is said to affect about 9 percent of the walnut trees in parts of Oregon's Willamette Valley, and is responsible for most of the decline and death of walnuts there (Miller 1942; Rawlings, Painter, and Miller 1950). In California, blackline occurs as far north as Tehama County and as far south as Tulare and Kings counties, but the highest incidence and greatest damage occur in coastal walnut-growing areas and in the southern Sacramento and northern San Joaquin valleys (Reil et al. 1985). Disease incidence in surveyed orchards has ranged from a few trees to as much as 80 percent of the orchard. The disease occurs in orchards ranging in age from 5 to over 70 years, but its incidence is slight in trees below bearing age.

Besides Oregon and California, blackline has been observed on grafted English walnut trees in England, France, and Hungary (Delbos, Bonnet, and Dunez 1984; Massalski and Cooper 1984; Mircetich, Rowhani, and Ramos 1985). In California, only English walnut growing on northern California black walnut (*Juglans hindsii*) rootstock, Paradox hybrid, or on rootstock hybrids of *J. hindsii* and other black walnuts is affected. Blackline is now considered a limiting factor in walnut production in some areas and a potential threat to the industry in others where it has not been observed or is extremely rare.

For 60 years investigators suggested a number of noninfectious causes of blackline, with scion-rootstock incompatibility as the most common. The recent exhaustive studies by Mircetich and coworkers (Mircetich, DeZoeten, and Lauritus 1980; Mircetich and Rowhani 1984; Mircetich, Rowhani, and Cucuzza 1982; Mircetich, Rowhani, and Ramos 1985; Mircetich, Sanborn, and Ramos 1978), however, have conclusively established that the disease is caused by a highly infectious virus.

Symptoms. The first outward sign of the disease is the gradual decline in tree vigor (**fig. 7A**). This is accompanied by poor terminal growth, yellowing and drooping of leaves, and premature defoliation, particularly at the top of the tree. As the disease progresses there is a dieback of terminal shoots and a profuse development of sucker shoots from the rootstock. Although this sprouting is a strong indicator of blackline, by itself it is not diagnostic. Positive diagnosis requires careful examination of the graft union. Blackline-affected trees usually show small holes or vertical cracks in the bark at the union. Removal of a small patch of bark at this location may reveal a narrow strip of darkened cambium and phloem—black line—at the junction of rootstock and scion. This necrotic strip varies in length depending on the stage of disease development. In early stages the blackline may be only an inch or two in length, but later it usually extends completely around the trunk. This necrosis causes a break in the continuity between xylem and phloem tissues of the scion above and the rootstock below. This girdles the trunk and results in the death of the scion usually two to six years after the girdle is complete. In English walnut on *J. hindsii*, this necrotic girdle is a narrow strip at the union (**fig. 7B**); in trees on Paradox hybrid, however, bark necrosis may extend downward below the union (**fig. 7C**), sometimes as far as the ground. Accordingly, trees on Paradox rootstock suffer quicker decline than those on *J. hindsii*. In only a few instances in California has the English walnut–*J. regia* combination been observed to develop blackline symptoms (Mircetich, Rowhani, and Ramos 1985). In Europe, however, a strain of the virus is reported to induce a variety of symptoms in certain English cultivars on *J. regia:* leaf spots, stunted growth and dieback of terminal shoots, poor nut yield, deformed nuts, necrotic kernels, and severe tree decline (Mircetich, Rowhani, and Ramos 1985).

Cause. The cause of walnut blackline remained obscure from the time of its discovery in 1924 until 1978, when Mircetich, Sanborn, and Ramos showed that it was caused by a transmissible agent that spreads naturally in California's walnut orchards. In succeeding studies, Mircetich and his coworkers (Mircetich, DeZoeten, and Lauritus 1980; Mircetich and Rowhani 1984; Mircetich, Rowhani, and Cucuzza 1982; Mircetich, Sanborn, and Ramos 1980) established the agent as a strain of the cherry leafroll virus (CLRV-W). This sap-transmissible nepovirus was not previously known to occur in California. The virus particles are isometric and approximately 26 nm in diameter (DeZoeten, Lauritus, and Mircetich 1982). In walnuts and inoculated herbaceous plants, virus particles were found in cell wall protrusions of the vascular parenchyma and in the cytoplasm of both parenchyma and phloem elements. In serological tests, the walnut blackline virus reacted positively with antisera prepared against strains of the cherry leafroll virus from rhubarb, dogwood, and golden elderberry (DeZoeten, Lauritus, and Mircetich 1982; Mircetich, DeZoeten, and Lauritus 1980).

The cherry leafroll virus comprises a number of strains that attack a wide variety of woody and herbaceous plants (Cropley and Tomlinson 1971). Among its hosts are cherry and other *Prunus* species, dogwood, birch, elberberry, raspberry, blackberry, olive, privet, rhubarb, and walnut. The walnut blackline strain of the virus that occurs in California is transmissible to sweet cherry but has not been found on this host in nature. The virus may have been brought to the United States from Europe in infected walnut seed. In the United States the cherry leafroll virus is currently known only in walnut, dogwood, and golden elderberry. In recent pathogenicity tests, three CLRV-W isolates from walnut (one from California and two from Europe) caused blackline symptoms at the English walnut–*Juglans hindsii* graft union. Trees inoculated with strains of the virus from cherry and golden elderbery remained healthy and virus-free (Rowhani and Mircetich 1988).

Disease development. The blackline virus can be transmitted from diseased to healthy trees or introduced into noninfested areas by budwood or graftwood from infected English walnut trees. Seed from both symptomless and blackline-affected English walnuts may be infected with the virus. Incidence of infected seed is generally greatest in trees with blackline at the graft union and ranges from a few to more than 70 percent infected nuts. English walnut seedlings developing from infected seed may be symptomless carriers of the virus and thus may serve to introduce the disease into uninfested areas (Mircetich, Rowhani, and Ramos 1985).

The blackline virus is also transmitted by infected pollen in the normal pollination process (Mircetich, Rowhani, and Cucuzza 1982; Mircetich, Rowhani, and Ramos 1985). Using naturally infected pollen (as determined by ELISA tests) to pollinate virus-free pistillate flowers artificially resulted in 4 to 14 percent nut infection. There was evidence, also, that the virus moved from the infected pistillate flowers into the spurs and, a year later, into the catkins. This research clearly demonstrated the role of infected pollen in the natural spread of the blackline virus. Viral spread from an infected to a noninfected cultivar depends, however, on the simultaneous shedding of pollen by the infected tree and the presence of pistillate flowers on the healthy tree. For example, there is little or no evidence of disease spread from infected Payne, which blooms early, to Franquette, which blooms late. In diseased orchards with a single cultivar or with two cultivars with a bloom overlap, the annual increase in disease incidence may be as high as 20 percent. From 70 to 90 percent of newly infected trees appear to be adjacent to previously diseased trees (Mircetich, Rowhani, and Ramos 1985).

The virus spreads through English walnut relatively slowly; the vertical spread in 20 inoculated cultivars was found to range from 2 to 32 inches per year depending on the cultivar, climatic conditions, and physiological state of the trees. The upward and downward spread was at approximately the same rate. Lateral spread of the virus in inoculated trees ranged from 2 to 5 inches per year (Mircetich, Rowhani, and Ramos 1985).

Although *Xiphinema* species are known to transmit certain strains of the cherry leafroll virus to several plants in Europe, it appears unlikely that they play an important role in the transmission of walnut blackline in California. The reasons for this are that (1) the walnut rootstocks (*J. hindsii* and Paradox hybrid) commonly used in California are immune to the virus, (2) the virus has a ready means of transmission through infected pollen, and (3) only two of the nematode species (*X. diversicaudatum* and *X. vuittenezi*) reported to transmit the virus in Europe occur in California, and here they are rare and not known to be present in walnut-producing areas. However, if *J. regia* rootstocks come into more general use and if nematode vectors become established in commercial walnut areas, nematodes could play a significant role in the disease syndrome.

The disease has been observed in English walnut trees propagated on wingnut (*Pterocarya stenoptera*), on seven species of *Juglans*, and on two hybrid stocks, Paradox (*J. hindsii* × *J. regia*), and Royal (*J. hindsii* × *J. nigra*) (Mircetich, Rowhani, and Ramos 1985). Field observations indicate that the majority of English walnut cultivars propagated on *J. hindsii* or Paradox rootstock in California are subject to infection and development of blackline at the union. Little is known about the effects of the blackline virus upon English cultivars grown on *J. regia* seedling rootstocks. Inoculation tests with the virus show that some seedlings develop chlorotic rings and spots, necrosis, and distortion of the leaves, as well as stunted growth and dieback of terminal shoots, while other seedlings of the same population remain symptomless and vigorous (Mircetich, Rowhani, and Ramos 1985).

Control. Since walnut blackline can be transmitted both by budding and grafting and by natural pollination, it is a most difficult disease to control. Only graftwood or budwood from virus-free English walnut trees should be used; preferably this wood should be taken from trees growing where the disease is not known to occur. Also, since blackline is both pollen-borne and seedborne, precautions should be taken to avoid using pollen and seed from virus-infected trees. In orchards with trees on Paradox or *J. hindsii* rootstocks and where blackline incidence is relatively low, it may be advisable either to graft the rootstock with virus-free scionwood or to rogue infected trees and replace them with healthy stock. Where feasible, regrafted or replanted cultivars should have a pistillate bloom schedule that does not coincide or overlap with the shedding of pollen on the infected cultivars. Since English walnuts on *J. regia* seedlings rarely develop blackline symptoms, a number of orchards have recently been established on English rootstocks with no evident problems. The following factors, however, dictate caution: (1) too little is known about different English walnut cultivars' and seedling selections' tolerance to the virus, (2) *J. hindsii* and Paradox hybrid are more resistant than *J. regia* to Phytophthora and Armillaria root and crown rot, (3) English walnut seedlings are less tolerant of toxic salts than the other two stocks, and (4) trees on *J. regia*, being symptomless carriers of the virus could pose a serious threat to healthy trees on *J. hindsii* and Paradox rootstocks.

A modified enzyme-linked immunosorbent assay called Indirect-ELISA (I-ELISA) is an efficient, rapid, and reliable procedure for detecting the blackline virus in nursery trees, English walnut seed and seedlings, and graftwood and budwood. This test could be used effectively to index propagation material for the presence of the blackline virus when and if a need arises for these control measures (Mircetich, Rowhani, and Ramos 1985; Reil et al. 1985).

REFERENCES

Cropley, R., and J. A. Tomlinson. 1971. Cherry leaf roll virus. *Commonw. Mycol. Inst./Assoc. Appl. Biol. Descrip. Plant Viruses* 80, 4 pp.

Delbos, R., A. Bonnet, and J. Dunez. 1984. Le virus de l'enroulement des feuilles du cerisier, largement répandu en France sur noyer, est-il à l'origine de l'incompatibilité de greffage du noyer *Juglans regia* sur *Juglans nigra*. *Agronomie* 4:333–39.

DeZoeten, G. A., J. A. Lauritus, and S. M. Mircetich. 1982. Cytopathology and properties of cherry leaf roll virus associated with walnut blackline disease. *Phytopathology* 72:1261–65.

Massalski, P. R., and J. I. Cooper. 1984. The location of virus-like particles in the male gametophyte of birch, walnut and cherry naturally infected with cherry leaf roll virus and its relevance to vertical transmission of the virus. *Plant Pathol.* 33:255–62.

Miller P. W. 1942. A report of progress on studies of the cause of the decline and death of walnuts in Oregon. *Ore. State Hort. Soc. Rept.* 1941:124–26.

Mircetich, S. M., G. A. DeZoeten, and J. A. Lauritus. 1980. Etiology and natural spread of blackline disease of English walnut trees. *Acta Phytopathol. Acad. Sci. Hung.* 15:147–51.

Mircetich, S. M., and A. Rowhani. 1984. The relationship of cherry leafroll virus and blackline disease of English walnut trees. *Phytopathology* 74:423–28.

Mircetich, S. M., A. Rowhani, and J. Cucuzza. 1982. Seed and pollen transmission of cherry leafroll virus (CLRV-W), the causal agent of the blackline disease of English walnut trees (abs.). *Phytopathology* 72:988.

Mircetich, S. M., A. Rowhani, and D. E. Ramos. 1985. Blackline disease. In *Walnut orchard management*, edited by D. E. Ramos, 142–152. Univ. Calif. Coop. Ext. Publ. 21410.

Mircetich, S. M., R. R. Sanborn, and D. E. Ramos. 1978. Walnut blackline disease: Graft transmission and natural spread (abs.). *Phytopathol. News* 12:226.

———. 1980. Natural spread, graft-transmission, and possible etiology of walnut blackline disease. *Phytopathology* 70:962–68.

Rawlings, C. O., J. H. Painter, and P. W. Miller. 1950. Importance of black line in certain Oregon walnut orchards. *Ore. State Hort. Soc. Rept.* 1950:150–51.

Reil, W. O., G. A. Rowe, D. E. Ramos, and S. M. Mircetich. 1985. Incidence of walnut blackline diseases in California's commercial orchards. *Calif. Agric.* 39(9/10):21–24.

Rowhani, A., and S. M. Mircetich. 1988. Pathogenicity on walnut and serological comparisons of cherry leafroll virus strains. *Phytopathology* 78:817–20.

Rowhani, A., S. M. Mircetich, R. J. Shepherd, and J. D. Cucuzza. 1985. Serological detection of cherry leafroll virus in English walnut trees. *Phytopathology* 75:48–52.

Abiotic Diseases of English Walnut

Shell Perforation of English Walnut

Cause Unknown

This malformation (Miller, Schuster, and Stephenson 1945) consists of circular or irregular holes extending through the shell and occurring any place in the shell, but usually near the apex. Affected nuts are not marketable as such, but must be cracked and sold as shelled nut meats.

Some workers have attributed the perforations to feeding by the walnut aphid, *Chromaphis juglandicola* Kalt, but the insect is not the only suspected cause of the disorder. Trees of certain cultivars are more prone to produce perforated nuts than those of other cultivars. Trees in low vigor are said to have the disorder more than highly vigorous trees.

Little is known about prevention. Top-working trees that produce abnormal numbers of perforated nuts with a cultivar such as Franquette is one way of reducing the problem.

REFERENCE

Miller, P. W., C. E. Schuster, and R. E. Stephenson. 1945. Diseases of the walnut in the Pacific Northwest and their control. *Ore. Agric. Exp. Stn. Bull.* 435, 42 pp. (reprinted 1947).

Sunburn of English Walnut

Direct Exposure to Sun's Rays

Sunburn of the hull occurs under high-temperature conditions, often commencing with the first heat wave of summer, when nuts are exposed to the direct rays of the sun. The exposed hull surface first shows yellowish brown spots that develop into dark brown, leathery lesions by mid or late summer (**fig. 8**). Sunburned nuts have shriveled, darkened kernels that are totally unmarketable (**fig. 8**).

A number of factors contribute to the development of sunburn. Honeydew excreted by walnut aphids is important on Payne and possibly on other early cultivars because it kills and blackens the epidermal cells of the hull, making the tissue more heat-absorbent. Sunburn may also result from inadequate winter pruning, especially in highly fruitful cultivars. Poor shoot growth and moisture stress also contribute to the incidence and severity of sunburn.

Sunburn can be reduced if growers follow good cultural practices and use a summer covercrop. Installation and frequent use of sprinkler irrigation during hot weather also helps. Another aid in reducing sunburn is the application of one or two whitewash sprays beginning in June when the nuts are approaching full size. However, under extreme conditions sunburning will occur even with this treatment.

In addition to the nuts, the bark of young tree trunks and the exposed limbs on older trees is subject to sunburn. The sunburned bark is bleached and cracked and susceptible to infection by *Hendersonula toruloidea*, the branch wilt fungus. The damage may be confused with deep-bark canker, but the latter is not confined to areas exposed to the sun. The trunks of young trees can be protected by painting them with either whitewash or a white water-based paint. Maintaining good tree vigor through adequate irrigation and fertilization minimizes the sunburning of branches in older trees.

REFERENCES

Hendricks, L. C., R. H. Gripp, and D. E. Ramos. 1980. Walnut production in California. *Univ. Calif. Div. Agric. Sci. Leaf.* 2984, 22 pp.

Mircetich, J. M., W. J. Moller, and B. Teviotdale. 1982. Diseases. In *Integrated pest management for walnuts*, 62–65. *Univ. Calif. Div. Agric. Sci. Publ.* 3270.

Smith, R. E. 1941. *Diseases of fruits and nuts. Calif. Agric. Ext. Serv. Circ.* 120, 168 pp.

Chapter 6

THE PISTACHIO AND ITS DISEASES

6

Chapter 6 Contents

Page numbers in bold indicate photographs.

INTRODUCTION	328	
REFERENCES	328	

FUNGAL DISEASES OF PISTACHIO	329	
Botryosphaeria Panicle and Shoot Blight of Pistachio	329	**434**
Botryosphaeria dothidea (Moug.:Fr.) Ces. & de Not.		
REFERENCES	331	
Botrytis Blossom and Shoot Blight of Pistachio	331	**434**
Botrytis cinerea Pers.		
REFERENCES	332	
Verticillium Wilt of Pistachio	332	**434**
Verticillium dahliae Kleb.		
REFERENCES	333	

MISCELLANEOUS DISEASES OF PISTACHIO	334
Alternaria leaf and fruit spot	334
REFERENCES	334
Septoria leaf spot	334
REFERENCES	335
Eutypa dieback	335
REFERENCE	335
Leaf rust	335
REFERENCE	335
Camarosporium shoot and panicle blight, Phytophthora root and crown rot, nut rots, root rots, sapwood rots	335
REFERENCES	335

Introduction

The pistachio nut tree (*Pistacia vera* L.) is native to Asia and Asia Minor. It was introduced into Mediterranean Europe during the Christian era, and thereafter into Tunisia, Australia, and the United States (Butterfield 1937–38). In 1902 the U.S. Department of Agriculture Plant Introduction Station at Chico, California, started evaluating the suitability of pistachio cultivars; they selected the Kerman cultivar for commercial test orchards in the Sacramento Valley. Today, Kerman is the only cultivar grown commercially in California. It is also grown in Arizona and Texas, in areas where winter temperatures approximate 1,000 hours below 7°C—the conditions required to break the rest period for normal growth and fruiting (Maranto and Crane 1982, 1988). All pistachio species are dioecious, and one male tree (cv. Peters) is planted to every 10 to 12 female trees. Alternate bearing or "off-year" production is of great concern, but is difficult to correct because pistachio trees do not respond to conventional pruning procedures used to stimulate the growth of lateral buds. On pistachio trees, only a few lateral, vegetative buds form, and apical dominance persists. In California about 40,000 acres of pistachio trees are in production, with an estimated annual harvest of 62 million pounds. The nonbearing acreage is estimated to be about 14,000 acres.

The tree is in the family Anacardiaceae, which includes mango, cashew nut, sumac, poison oak, and the ornamental pepper tree. Most of the original pistachio plantings in California were on seedling rootstocks of *P. atlantica* or *P. terebinthus*. Some plantings now are on *P. integerrima* rootstock, which is considered to be resistant to Verticillium wilt (Maranto and Crane 1982) and is sold as Pioneer Gold.

No serious diseases of pistachio have been reported in Turkey or Iran, two other countries where pistachio production is important. In those countries, trees have been known to attain a trunk girth of six feet and remain productive for as long as 700 years. In California, however, many orchards have been seriously threatened by Verticillium wilt. Recent orchard surveys have revealed other diseases, such as Botrytis blossom and shoot blight, Botryosphaeria panicle and shoot blight, and powdery mildew. In addition, the Peters pollenizer has a leaf scorch of unknown etiology that develops in late summer and fall. Another condition, epicarp lesion of the developing fruit, has been determined to be caused by punctures of sucking insects and not by a plant pathogen (Bolkan et al. 1984).

References

Bolkan, H. A., J. M. Ogawa, R. E. Rice, R. M. Bostock, and J. C. Crane. 1984. Leaffooted bug (Hemiptera: Coreidae) and epicarp lesion of pistachio fruits. *J. Econ. Entomol.* 77:1163–65.

Butterfield, H. M. 1937–38. History of deciduous fruits in California. *Blue Anchor*, vols. 14 and 15, 38 pp.

Crane, J. C., and J. Maranto. 1988. Pistachio production *Univ. Calif. Div. Agric. Nat. Resour. Publ.* 2279, 15 pp. (revised).

Maranto, J., and J. C. Crane, 1982. Pistachio production. *Univ. Calif. Div. Agric. Sci. Leaf.* 2279, 17 pp.

Whitehouse, W. E. 1957. A new crop for the western United States. *Econ. Bot.* 11:281–321.

Fungal Diseases of Pistachio

Botryosphaeria Panicle and Shoot Blight of Pistachio

Botryosphaeria dothidea
(Moug.:Fr.) Ces. & de Not.

In the summer of 1984 a diseases of unknown etiology was detected in an orchard in the northern Sacramento Valley near Chico (Rice et al. 1985). A 1985 survey of the major pistachio production areas in California indicated that this disease was prevalent in the northern counties of Butte, Tehama, and Glenn and sporadic in the southern San Joaquin Valley counties of Merced, Madera, Kings, and Kern (Michailides and Ogawa 1986a). Examination of fruit clusters revealed disease symptoms on most of the clusters on the lower parts of the tree during midsummer, with the infection spreading to clusters at the top of the tree by harvest time. In some orchards, the estimated crop losses resulting from reduction in the quality and quantity of nuts ranged from 25 to 50 percent. The disease was most severe in orchards with high-angle sprinkler irrigation systems that consistently wetted the lower part of the trees (Michailides, Ogawa, and Olson 1986). The only other country in which this disease has been reported is Italy (Corazza, Chilosi, and Avanzato 1987), but it also was recently identified in samples sent to California from Greece (A. J. Feliciano, personal communication).

Symptoms. The disease is characterized by blighted fruit clusters (**fig. 1B**), catkins, shoots, individual leaves, and buds. The emerging shoots are highly susceptible to infection and remain susceptible throughout the growing season. The cankers that develop on shoots may girdle and kill the terminal portion while those on the rachis may girdle and blight part or all of the fruit cluster. Infected areas are covered with characteristic black pycnidia (**fig. 1C**) by late August to mid-September (Michailides and Ogawa 1986a).

On newly developed shoots, the first symptom observed is leaf wilting. These leaves are light green at first and later turn light brown. Characteristically, the blighted leaves remain attached to the wilted shoots, the bark and cambium of which become black, and grayish white mycelium of the fungus forms within the pith area. The initial cankers that extend from blighted shoots into supporting branches have a diffuse margin, but on two-year-old or older wood they have a sharp margin. Usually, shoot blight is related to bud infection during the winter (Michailides and Ogawa 1986a).

Single-leaflet infections are common and usually result in blighting of the leaflet and sometimes of the entire leaf. In such cases, the petiole and midrib become black, while the leaf blade remains light brown. In sprinkler-irrigated orchards, leaf-blade infections begin as numerous small, angular, dark brown lesions, 2 to 4 mm in diameter (**fig. 1A**), which later coalesce to produce blighting (Michailides and Ogawa 1986a).

Newly developed panicles (female flower clusters) may shrivel by the beginning of May. This is because the flower buds become infected during the winter and the twig becomes partially girdled. After the onset of sprinkler irrigation, entire fruit clusters or parts of clusters become blighted; in some instances, infections start at the base of the cluster stem (rachis) as black, longitudinal lesions that may eventually girdle the rachis. Adjacent clusters and their supporting shoots may also be invaded and sometimes killed. The shriveled rachis turns brownish black, the affected fruit become externally and internally wrinkled, and the kernels fail to fully develop (Michailides, Ogawa, and Olson 1986).

Fruit infections begin as small, freckle-like, circular or irregular black lesions on the green epicarp, and then enlarge to coalesce and crack (**fig. 1A**). These symptoms should not be confused with those on the petioles, stems, and fruit caused by the citrus flat mite (*Brevipalpus lewisi* McGregor). This mite forms dark, roughened, scablike blotches on the surface tissue (Rice 1980; Rice et al. 1985). In the summer or early fall, the *Botryosphaeria*-infected fruit turn gray and are covered with black pycnidia embedded in the epicarp. Fruit infections can progress into the rachis and, at times, hyaline gum exudes from the juncture of the rachis branches. Infection of fruit after hull split can result in penetration of the fungus into the shell cavity and the kernel (Michailides and Ogawa 1986a).

Infected buds and leaf and bud scars appear black, and gum exudes from the scars or at the base of dead buds. Cankers progress both basipetally and acropetally from the infection points (Michailides and Ogawa 1986a).

Causal organism. In California *Botryosphaeria dothidea* (Moug.:Fr.) Ces & de Not. was reported in 1966 and 1975 (English, Davis, and DeVay) on almond (*Prunus dulcis*) cv. Nonpareil, causing a bandlike canker on the trunk and scaffold branches of vigorous young trees. It was found to cause panicle and shoot blight of pistachio in 1984. The same organism, under the synonym *B. ribis* Gross & Dugg., was previously reported to cause branch and trunk cankers on a variety of woody plants, including citrus and avocado (Smith 1934).

Botryosphaeria was placed in the Botryosphaeriaceae, a family in the order Dothiorales, by von Arx and Müller (1954). They reduced many of the species, including *B. dothidea* and *B. ribis*, to synonymy (Commonwealth Mycological Institute 1973). In *B. dothidea* the locules of the ascocarp are 170 to 250 µm in diameter, ostiolate, and darker around the neck. The asci are interspersed among filiform paraphyses, clavate, 100 to 110 by 16 to 20 µm, eight-spored, and bitunicate. The ascospores are irregularly biseriate, hyaline, unicellular, ovoid, and measure 17 to 23 by 7 to 10 µm. The sexual fruiting structures have not been found on pistachio in California, but do occur on giant sequoia and coast redwood (Worrall, Correll, and McCain 1986). They also are found on peach in the eastern United States (Weaver 1974). The pycnidia are solitary or botryose, stromatic, globose, and 150 to 250 µm in diameter, with papillate ostioles darker around the neck region. The pycnidial wall is many cells thick and composed of outer sclerotized cells and inner, thin-walled cells that line the entire cavity. The conidiogenous cells are holoblastic, hyaline, and arise from the inner lining of the pycnidial cavity. The macroconidia are hyaline, fusoid, 17 to 25 by 5 to 7 µm; the microconidia (spermatia) are hyaline, allantoid, and measure 2 to 3 by 1 µm (Smith 1934). The minimum, optimum, and maximum temperatures for growth and conidial production are 10°, 28°, and 32° to 35°C, respectively.

The pathogen is plurivorous, particularly on tree crops, and also causes rots in fruit. No physiologic specialization has been reported, and cross-inoculations among different hosts suggest that different pathogenic strains do not occur (Smith 1934).

Disease development. The spring and summer sources of *B. dothidea* inoculum are pycnidia on rachises hanging on the tree from previous crops, on dead shoots, and on newly infected plant parts. Infected rachises may remain on the tree and provide inoculum for at least two years (Michailides and Ogawa 1986a, 1987c). In addition, the fungus remains viable in twig cankers for at least four years (Michailides and Ogawa 1987a). The infection of leaves, rachises, and shoots occurs through the stomates, and that of fruit occurs through the lenticels.

The conidia of *B. dothidea* are dispersed by water, especially from high-angle sprinkler irrigation systems. In an orchard experiment, the pycnidia continued to yield spores after four repeated simulations of this type of irrigation. The optimum temperature for disease development approximates that for growth of the fungus—27° to 33°C. Thus summer temperatures in the pistachio-growing areas of the central valleys in California are ideal for infection and disease development provided that free moisture is present on the host. Both sprinkler irrigation and summer rain trigger disease development.

In the eastern United States where temperature is high and rainfall abundant during the summer months, *Botryosphaeria* is a major pathogen on peach (Britton and Hendrix 1986).

Control. In a commercial pistachio orchard with high-angle sprinklers, there was a clear separation in disease intensity between the part of the tree wetted by sprinklers and the part that remained dry. Lowering the angle of the sprinklers resulted in effective control of the disease during the 1985 season when summer rains were absent. A single application of fungicides during bloom effectively controlled the disease in an orchard with high sprinkler irrigation. Of the fungicides tested, captafol provided the best control, followed by chlorothalonil and benomyl. Data on the weight of the healthy nuts harvested indicated that trees sprayed with each of the three fungicides yielded more than double the harvest of unsprayed trees. Only captafol (no longer available) provided an increase in harvest weight when low-angle sprinklers were used (Michailides and Ogawa 1986b). In a 1987 experiment (orchard irrigated with high-angle sprinklers), the percentage of infected fruit was significantly reduced with sprays of captafol, captan, chlorothalonil, iprodione, and SC-0858 (Michailides and Ogawa, unpublished). Captafol was more effective than the other fungicides in reducing spore germination in vitro. Multiple summer applications of a fixed copper spray gave some promise of reducing fruit infection (Olsen et al. 1989–90).

Another potential control measure is the use of disease-resistant cultivars. A recent report from Italy indicates the possible resistance of several cultivars of *Pistacia vera* (Aeqina, Baglio, and Sfax) and of the seedlings of *P. terebinthus* and *P. atlantica* (Corazza, Avanzato, and Chilosi 1987).

REFERENCES

Arx, J. A. von, and E. Müller. 1954. Die Gattungen der Amerosporen Pyrenomyceten. *Beit. Kryptogamenflora Schweiz.* 11(1):1–434.

Britton, K. O., and F. F. Hendrix. 1986. Population dynamics of *Botryosphaeria* spp. in peach gummosis cankers. *Plant Dis.* 70:134–36.

Commonwealth Mycological Institute. 1973. *Botryosphaeria ribis. Commonw. Mycol. Inst., Descrip. Pathogen. Fungi Bact.* 395.

Corazza, L., D. Avanzato, and G. Chilosi. 1987. Some phytopathological problems of pistachio nut in Italy. *Proc. 7th Congress Medit. Phytopath. Union.* pp. 207–08.

Corazza, L., G. Chilosi, and D. Avanzato. 1987. Un disseccamento dei rami di pistacchio causato da *Botryosphaeria ribis. Ann. Inst. Speriment. Patol. Veg.* 1:95–8.

English, H., J. R. Davis, and J. E. DeVay. 1966. Dothiorella canker, a new disease of almond trees in California (abs.). *Phytopathology* 56:146.

———. 1975. Relationship of *Botryosphaeria dothidea* and *Hendersonula toruloidea* to a canker disease of almond. *Phytopathology* 65:114–22.

Fawcett, H. L. 1915. Melaxuma of the walnut, *Juglans regia* (A preliminary report). *Calif. Agric. Exp. Stn. Bull.* 261:133–48.

Michailides, T. J., and J. M. Ogawa. 1986a. Sources of inoculum, epidemiology, and control of Botryosphaeria shoot and panicle blight of pistachio. *Calif. Pistachio Ind. Ann. Rept.* 1985–1986:87–91.

———. 1986b. Epidemiology and control of Botryosphaeria panicle and shoot blight of pistachios (abs.). *Phytopathology* 76:1106.

———. 1987a. New findings on the epidemiology and control of Botryosphaeria panicle and shoot blight of pistachio. *Calif. Pistachio Ind. Ann. Rept.* 1986–1987:99–107.

———. 1987b. New pathogens of pistachio and other hosts of *Botryosphaeria dothidea* in California. *Calif. Pistachio Ind. Ann. Rept.* 1986–1987:115–118.

———. 1987c. Retention of panicles and petioles of pistachio infected by *Botryosphaeria dothidea* (abs.). *Phytopathology* 77:1770.

Michailides, T. J., J. M. Ogawa, and B. Olson. 1986. Shoot and panicle blight of pistachio caused by *Botryosphaeria dothidea*. *Calif. Pistachio Ind. Ann. Rept.* 1985–1986:96–101.

Olson, W. H., J. M. Ogawa, M. R. Montgomery, Y. Fujii, and A. J. Feliciano. 1989–90. Evaluation of multiple applications of fixed copper and chlorothalonil sprays on control of Botryosphaeria panicle and twig blight of pistachio. *Calif. Pistachio Ind. Ann. Rept.* 1989–90:96, 97.

Punithalingam, E., and P. Holliday. 1973. *Botryosphaeria ribis*, C.M.I. *Descript. Path. Fung. Bact.* 395. Commonwealth Agric. Bur.

Rice, R. E. 1980. The citrus flat mite on pistachios. *Calif. Pistachio Assn. Ann. Rept.*, 20–21.

Rice, R. E., J. K. Uyemoto, J. M. Ogawa, and W. M. Pemberton. 1985. New findings on pistachio problems. *Calif. Agric.* 39(1,2):15–18.

Smith, C. O. 1934. Inoculations showing the wide host range of *Botryosphaeria ribis*. *J. Agric. Res.* 49:467–76.

Weaver, D. J. 1974. A gummosis disease of peach trees caused by *Botryosphaeria dothidea*. *Phytopathology* 64:1429–32.

Worrall, J. J., J. C. Correll, and A. H. McCain. 1986. Pathogenicity and teliomorph-anamorph connection of *Botryosphaeria dothidea* on *Sequoiadendron giganteum* and *Sequoia sempervirens*. *Plant Dis.* 70:757–59.

Botrytis Blossom and Shoot Blight of Pistachio

Botrytis cinerea Pers.

In the spring of 1983, after heavy, prolonged rains and cool weather, a disease characterized by blighted shoots was observed initially in Solano County (Sacramento Valley) and subsequently in Madera County (San Joaquin Valley) (Bolkan, Ogawa, and Teranishi 1984) and Butte, Glenn, Fresno, and Yolo counties. During the following two years, with no rains during the bloom period, the incidence of this condition, caused by the *Botrytis* fungus, was considered insignificant.

Symptoms. The catkins are more severely attacked than the female flower clusters. Once these floral parts are attacked, the fungus progresses into the twigs to cause cankers and shoot blight (**fig. 2**). Cankers initiated from blighted male inflorescences can cause shoot blight even a year after the initial infection. Partial girdling of the shoots or twigs results in a sudden shriveling of the leaves from water stress. Petiole infections result in death of individual leaves and frequently in shoot blight. Catkins of the cultivars 02–16 and 02–18 are more susceptible to *Botrytis* infection than those of Peters (Michailides and Ogawa 1986a,b).

Causal organism. *Botrytis cinerea* Pers., the causal fungus, is discussed fully under "Gray Mold Rot of Apple and Pear" in Chapter 2.

Host relations and cultivar susceptibility. In a test plot in Butte County, the catkins of the male pollenizer trees were more susceptible (88 percent infection) to the disease than the blossoms of the female Kerman trees (28 percent infection). Similar differences in susceptibility between the male and female flowers were noted in test plots in Yolo and Madera counties. The cause of the higher incidence of *Botrytis* infection of catkins could be that the pollen grains may serve as a substrate for germination and colonization of the fungus (Borecka and Millikan 1973; Ogawa and English 1960). Inoculum for *Botrytis* infection appears to be ubiquitous in orchards regardless of whether they are strip-sprayed with herbicides—with turf in the centers providing crop residues for colonization by the fungus—or clean cultivated. The fungus is continuously present on floral parts from the time of flowering to nut harvest (Bolkan, Ogawa, and Manji 1984). New conidiophores and conidia of the pathogen develop during late February and early March on shoots blighted during the previous spring (Michailides, Ogawa, and Teranishi, unpublished). In addition, the fungus produces sclerotia on infected shoots pruned and left on the orchard floor. Their role in the epidemiology of the disease has not been investigated. The effect of *Botrytis* infection on pollen viability has not been evaluated.

Control. In a field plot established in Madera County, when the female Kerman cultivar was partially in bloom and the male Peters was in full bloom, benomyl (a single spray application) was the most effective fungicide tested; this material has now been approved for registration. However, captafol and copper hydroxide also significantly reduced floral infection and shoot blight.

REFERENCES

Bolkan, H. A., J. M. Ogawa, and B. T. Manji. 1984. Effect of fungicides on mycoflora and epicarp lesion of pistachio nuts grown in California. *Calif. Pistachio Ind. Ann. Rept.* 1983–1984:74–77.

Bolkan, H. A., J. M. Ogawa, and H. R. Teranishi. 1984. Shoot blight of pistachio caused by *Botrytis cinerea*. *Plant Dis.* 68:163–65.

Borecka, H., and D. F. Millikan. 1973. Stimulatory effect of pollen and pistillate parts of some horticultural species upon the germination of *Botrytis cinerea* spores. *Phytopathology* 63:1431–32.

Michailides, T. J., and J. M. Ogawa. 1987a. Botrytis shoot blight of pistachio: New findings. *Calif. Pistachio Ind. Ann. Rept.* 1986–1987:108–12.

———. 1987b. Survival of *Botrytis cinerea* in twig cankers initiated from infected male inflorescences of pistachio (abs.). *Phytopathology* 77:1241.

Ogawa, J. M., and H. English. 1960. Blossom blight and green fruit rot of almond, apricot and plum caused by *Botrytis cinerea*. *Plant Dis. Rep.* 44:265–68.

Verticillium Wilt of Pistachio

Verticillium dahliae Kleb.

A rootstock (*Pistacia atlantica*) susceptible to Verticillium wilt was used to establish the pistachio industry in California. Within a decade, beginning in 1976 when pistachio plantings totaled only 4,350 acres, there was an eightfold increase in acreage. Verticillium wilt was first reported on pistachio in California in 1950 (Snyder, Hansen, and Wilhelm 1950) and emerged as a major disease problem in Kern County in 1976. The disease occurred in plantings in virgin land and in sites that had been farmed since 1968, but had been planted to no more than three cotton crops (Ashworth and Zimmerman 1976). The only other known report of Verticillium wilt on pistachio is from Greece (Thanassoulopoulos and Kitsos 1972). The rootstocks used in California were seedlings of *P. atlantica* because studies had shown that this rootstock, in combination with the Kerman cultivar, consistently outyielded trees on other rootstocks, which included *P. terebinthus* but not *P. integerrima*. In 1982, the *P. integerrima* rootstock, resistant to Verticillium wilt and sold as Pioneer Gold, was introduced, and now it is planted where trees have died from Verticillium wilt and in new plantings, especially in Kern County. The disease remains prevalent in the southern San Joaquin Valley, with a few isolated spots in the central San Joaquin Valley (Madera County) on trees established on *P. atlantica* rootstock. The disease has not been detected in pistachio on *P. atlantica* rootstocks in the Sacramento Valley.

Symptoms. The typical symptoms of Verticillium wilt on fruit crops (Chapter 4) also occur on pistachio (**fig. 3A**), with autumn discoloration of the xylem tissue in the trunk and scaffolds (**fig. 3B**). Initial symptoms, however, are expressed on the leaves in the spring. The fungus spreads rapidly in infected trees, often killing them during the year of infection, whereas in stone fruit the branches and scaffolds may be killed, but seldom the whole tree. Thus, the primary concern of pistachio growers is not just the initial decrease in production but the loss of trees and the possible continued spread of the fungus within the orchard.

Causal organism. The mild SS-4 and severe T-1 cotton strains of *Verticillium dahliae* appear to be equally involved in Verticillium wilt of pistachio (Schnathorst 1982). The causal fungus is described under "Verticillium Wilt of Fruit and Nut Trees" in Chapter 4.

Species and cultivar susceptibility. Both the Kerman (female) and Peters (male) cultivars are grown commercially on rootstocks, so the susceptibility of the rootstock to *Verticillium* is of prime importance in pistachio production. Raabe and Wilhelm (1978) determined the susceptibility of various species and selections of *Pistacia* to *Verticillium*. The roots of seedlings were dipped in a suspension of *Verticillium* conidia, and symptoms first appeared after three to four weeks; after six weeks most plants showed severe symptoms. Final data on disease incidence and severity were recorded after ten weeks. The selection PI 246341 of *P. terebinthus* and a selection of *P. integerrima* were rated tolerant on the basis of seedling survival in this greenhouse test and on freedom from symptoms after 14 years' growth in the field; *P. atlantica*, *P. mutica*, and *P. chinensis* were rated as susceptible. Later studies by Ashworth (1983–84, 1984–85) showed that during the second year of seedling growth in soil heavily infested with *V. dahliae*, plants homotypic of *P. atlantica* and *P. chinensis* were highly susceptible to the fungus. Seedlings of *P. vera* were less susceptible, while those of *P. integerrima*, *P. terebinthus*, and hybrids of *P. terebinthus* × *P. integerrima* were the least susceptible. In the most recent study (Schnathorst 1988), only *P. integerrima* was found to be resistant to the SS-4 and T-1 strains of *V. dahliae*.

In orchards on virgin land, with only trace amounts of *Verticillium* microsclerotia (0.02 to 0.05 per gram of air-dry soil), losses of 0.6 to 1.9 percent occurred. But in orchards on land previously planted to cotton, losses usually were 10 percent or more, and microsclerotia counts ranged from 1 to 2 per gram of air-dry soil.

Control. To provide reasonable freedom from Verticillium wilt problems in the San Joaquin Valley, the first suggestion is to plant trees in virgin soil (Ashworth and Zimmerman 1976; Huisman and Ashworth 1974, 1976; Huisman and Anderson 1980). In soils previously cropped to cotton, soil fumigation with a 2:1 chloropicrin–methyl bromide mixture effectively reduces microsclerotia to trace levels and reduces tree mortality to about 1 percent over a two-year period. This loss is similar to that found in virgin soil plantings (Ashworth and Zimmerman 1976). Microsclerotia populations in the orchard can be reduced by soil solarization using a polyethylene mulch (Ashworth and Gaona 1982). Where trees have been killed by *Verticillium*, planting a resistant seedling rootstock (*P. integerrima*) is suggested (Schnathorst 1988). Such plantings have provided healthy trees that have produced normal crops for more than 20 years (Maranto 1980; Maranto and Crane 1982).

Recently, the role of potassium nutrition in the susceptibility of pistachio to infection by *Verticillium* was investigated (Ashworth, Gaona, and Surber 1985a,b). The results indicated that a high incidence of disease was associated with unthrifty trees having low levels of leaf potassium (0.5 to 0.8 percent), and disease was less in thrifty trees with leaf potassium of 1 percent or more. Furthermore, when potassium was added to the soil, disease incidence was proportionately lower. Yet the authors noted that this "appears to be a klendusic response of actively growing trees in low-inoculum soils." The authors did not have appropriate controls to determine whether adding potassium to deficient trees would provide them with an escape mechanism against *Verticillium* infection (Ashworth, Ganoa, and Surber 1985a,b).

REFERENCES

Ashworth, L. J., Jr. 1983–84. Verticillium wilt control. *Calif. Pistachio Ind. Ann. Rept.* 1983–1984:55–57.

———. 1984–85. Verticillium resistant rootstock research. *Calif. Pistachio Ind. Ann. Rept.* 1984–85:56–57.

Ashworth, L. J., Jr., and S. A. Gaona. 1982. Evaluation of clear polyethylene mulch for controlling Verticillium wilt in established pistachio nut groves. *Phytopathology* 72:243–46. (Errata Fig. 1 published in *Phytopathology* 73:1142. 1983.)

Ashworth, L. J., Jr., S. A. Gaona, and E. Surber. 1985a. Nutritional diseases of pistachio trees: Potassium and phosphorus deficiencies and chloride and boron toxicities. *Phytopathology* 75:1084–91.

———. 1985b. Verticillium wilt of pistachio: The influence of potassium nutrition on susceptibility to infection by *Verticillium dahliae*. *Phytopathology* 75:1091–93.

Ashworth, L. J., Jr., and G. Zimmerman. 1976. Verticillium wilt of the pistachio nut tree: Occurrence in California and control by soil fumigation. *Phytopathology* 66:1449–51.

Huisman, O. C., and G. Anderson. 1980. Pistachio root growth and distribution in relation to the Verticillium wilt fungus. *California Pistachio Assoc. Ann. Rept.*:35–41.

Huisman, O. C., and L. J. Ashworth, Jr. 1974. Quantitative assessment of *Verticillium albo-atrum* in field soils: Procedural and substrate improvements. *Phytopathology* 64:1043–44.

———. 1976. Influence of crop rotation on survival of *Verticillium albo-atrum* in soils. *Phytopathology* 66:978–81.

Maranto, J. 1980. Pistachio rootstock update. *Calif. Pistachio Assoc. Ann. Rept.*:43.

Maranto, J., and J. C. Crane. 1982. Pistachio production. Univ. Calif. Div. Agric. Sci. Leaflet 2279, 17 pp.

Raabe, R. D., and S. Wilhelm. 1978. Susceptibility of several *Pistacia* spp. to *Verticillium albo-atrum*. *Plant Dis. Rep.* 52:672–73.

Schnathorst, W. C. 1982. The relation of *Verticillium dahliae* strains and cotton plantings to the epidemic of wilt disease in pistachio nut trees. *Phytopathology* (Abs.) 72:960.

———. 1988. Control of Verticillium wilt of pistachio nut trees with a resistant rootstock and the comparative susceptibility of *Pistacia* species to *Verticillium dahliae*. *Phytopathology* (abs.). 78:1546.

Snyder, W. C., H. N. Hansen, and S. Wilhelm. 1950. New Hosts of *Verticillium albo-atrum*. *Plant Dis. Rep.* 34:26–27.

Thanassoulopoulos, C. C., and G. T. Kitsos. 1972. Verticillium wilt in Greece. *Plant Dis. Rep.* 56:264–67.

MISCELLANEOUS DISEASES OF PISTACHIO

Alternaria leaf and fruit spot. This disease, caused by *Alternaria alternata* (Fries) Keissler, has been reported in Egypt (Wasfy, Ibrahim, and Elarosi 1974), Italy (Corazza and Avanzato 1986), and California (Michailides and Ogawa 1986). In Egypt, crop losses of 75 percent have been reported, but in California the effect of the disease on crop production has not been assessed. The symptoms include dark-bordered foliar spots of irregular size that can eventually cause complete blighting of leaflets. The *Alternaria* fungus is widespread in California pistachio orchards (Bolkan, Ogawa, and Manji 1984). Field isolations and laboratory inoculation studies (Michailides and Ogawa 1986) have established the pathogenicity of the *Alternaria* isolates and have shown that they induce disease symptoms typical of those reported from Italy and Egypt. Experiments in Italy (Corazza and Avanzato 1986) showed variability in the susceptibility of pistachio cultivars from the United States, with Kerman and Chico recorded as moderately resistant, Red Aleppo as moderately susceptible, and Bronte as susceptible; seedlings of both *P. terebinthus* and *P. atlantica* were rated resistant. Napoletana (Bianca), the main cultivar grown in Italy, was rated highly susceptible. No other information on disease control measures has been found.

REFERENCES

Bolkan, H. A., J. M. Ogawa, and B. T. Manji. 1984. Effect of fungicides on mycoflora and epicarp lesion of pistachio nuts grown in California *Calif. Pistach. Ind. Ann. Rept. Crop Year* 1983–1984. pp. 74–7.

Corazza, L., and D. Avanzato. 1986. *Alternaria alternata* (Fries) Keissler su pistachio in Italia. *Info. Agrar,* 42(25):73–5.

Michailides, T. J., and J. M. Ogawa. 1986. *Alternaria* sp. A pathogen of pistachio? *Calif. Pistach Ind. Ann. Rept. Crop Year* 1985–1986. pp. 117–18.

Wasfy, E. H., I. A. Ibrahim, and H. M. Elarosi. 1974. New Alternaria disease of pistachio in Egypt. *Phytopath. Medit.* 13:110–11.

Septoria leaf spot. This disease, caused by *Septoria pistaciarum* Caracc. (teleomorph *Mycosphaerella pistacearum* Chitzanidis), has been reported in Italy (Pupillo and Di Caro 1952) and Greece (Zachos and Tzavella-Klonari 1971). In the United States, it was first reported in Texas (Maas, van der Zwet, and Madden 1971) and more recently in Arizona (Young and Michailides 1989). Defoliation, common in the Mediterranean countries, was also observed in Texas but not in Arizona. In Arizona, the disease was found on the *Pistachia vera* cultivars Kerman and Peters but not on *P. atlantica* Desf. or *P. terebinthus* L. In Texas, *P. vera* cultivars (Kerman, Lassen, Red Aleppo, Trabonella, Sfax, and Bronte) were very susceptible, whereas *P. chinensis*, *P. atlantica*, and *P. terebinthus* were only moderately susceptible. A somewhat similar leaf-spot disease induced by *S. pistacina* All. causes serious damage to pistachios in Turkey (Dinc, Goksedef, and Turan 1980).

REFERENCES

Dinc, N. M. O. Goksedef, and K. Turan. 1979. [Research on the bio-ecology and control of *Septoria pistacina* All., attacking pistachio trees in Gaziantep province.] *Rev. Plant Pathol.* 59:1426. 1980.

Maas, J. L., T. van der Zwet, and G. Madden. 1971. A severe Septoria leaf spot of pistachio nut trees new to the United States. *Plant Dis. Rep.* 55:72–76.

Pupillo, M., and S. Di Caro. 1952. Alcune osservazioni sulla Septoria del pistachio. *Ann. Sper. Agr. N. S.* 6(3):623–34.

Young, D. Y., and T. J. Michailides. 1989. First report of Septoria leaf spot of pistachio in Arizona. *Plant Dis.* 73:775.

Zachos, D. G., and K. Tzavella-Klonari. 1971. On the development and structure of the spermogonia and ascocarps of *Mycosphaerella pistacearum* Chitzanidis. *Ann. Inst. Phytopath. Benaki N. S.* 10:217–21.

Eutypa dieback. A branch dieback of pistachio trees caused by *Eutypa lata* (Pers.:Fr.) Tul. has recently been reported in Greece (Rumbos 1986). This disease has not been detected on pistachio in California, but since the causal fungus induces a serious dieback disease of apricot and grape, which frequently are grown close to pistachio plantings, it poses a potential threat to the pistachio industry (see discussion of Eutypa dieback of apricot in Chapter 3).

REFERENCE

Rumbos, I. C. 1986. Isolation and identification of *Eutypa lata* from *Pistacia vera* in Greece. *J. Phytopathol.* 116:352–57.

Leaf rust. This disease, caused by *Pileolaria terebinthi* (DC) Cast., is one of the major diseases of pistachio in Italy (Corazza and Avanzato 1985). The disease is widespread in the Mediterranean region, and in Egypt a 60 percent reduction in crop production has been reported. This disease has not been reported from the United States. The Kerman cultivar is considered moderately susceptible, Red Aleppo and Napoletana highly susceptible, and seedlings of *P. terebinthus* and *P. atlantica* moderately resistant and resistant, respectively.

REFERENCE

Corazza, L., and D. Avanzato. 1985. Alcune considerazioni sulla ruggine del pistacchio in Italia. *Ann. Inst. Speriment. Patol. Veg.* 10:39–42.

Other diseases. Another disease of importance on pistachio in the Mediterranean region, but as yet undetected in the United States, is a shoot and panicle blight caused by *Camarosporium pistaciae* Zachos, Tzavella-Klonari, & Roubos. A second disease of some consequence in that region that is found sparingly in California is Phytophthora root and crown rot caused by several species of *Phytophthora* (French 1987; Pontikis 1977). Several other disorders and their causal agents reported in California (French 1987) are nut rots (*Alternaria* sp., *Penicillium* sp., and *Rhizopus* sp.), root rots (*Armillaria mellea* Vahl:Fr., *Fusarium oxysporum* Schlecht., *Fusarium* sp., *Pythium* sp., and *Phymatotrichopsis* [*Phymatotrichum*] *omnivora* [Duggar] Hennebert), and sapwood rots (*Pleurotus ostreatus* [Fr.] Kummer and *Schizophyllum commune* Fr.).

REFERENCES

French, A. M. 1987. *California plant disease host index. Part 1: Fruits and nuts.* Calif. Dept. Food & Agric., Sacramento, CA. 39 pp.

Pontikis, K. A. 1977. [Contribution to studies on the resistance of some hybrids and species of the genus *Pistacia* (as rootstocks) to *Phytophthora* spp.]. *Rev. Plant Pathol.* 57:1426. 1978.

Zachos, D. G., and I. Roubos. 1977. Recherches sur la biologie du champiqnon *Camarosporium pistaciae* Zachos, Tzavella-Klonari, & Roubos. *Ann. Inst. Phytopath. Benaki* 11:346–53.

Chapter 7

The Olive
And Its
Diseases

7

Page numbers in bold indicate photographs.

INTRODUCTION	340	
REFERENCES	340	

BACTERIAL DISEASES OF OLIVE	341	
Olive Knot	341	**435**
Pseudomonas syringae pv. *savastanoi* (Smith) Young et al.		
REFERENCES	343	

FUNGAL DISEASES OF OLIVE	344	
Leaf and Fruit Spot of Olive	344	
Mycocentrospora cladosporioides (Sacc.) P. Costa ex Deighton		
REFERENCES	344	
Leaf Spot of Olive	344	**435**
Spilocaea oleaginea (Cast.) Hughes		
REFERENCES	346	
Shoestring (Armillaria) Root Rot of Olive	347	
Armillaria mellea (Vahl.:Fr.) Kummer		
Verticillium Wilt of Olive	347	**435**
Verticillium dahliae Kleb.		
REFERENCES	347	
Miscellaneous Fungal Diseases of Olive	347	
REFERENCES	348	

VIRUS AND VIRUSLIKE DISEASES OF OLIVE	348	
Sickle Leaf	348	**435**
Graft-Transmissible Pathogen		
REFERENCES	348	

ABIOTIC DISEASES OF OLIVE	349	
Shotberry of Olive	349	**435**
Soft Nose of Olive	349	**435**
Split-Pit of Olive	349	
REFERENCES	349	

INRODUCTION

The family Oleaceae, to which the edible olive, *Olea europaea* L. belongs, contains the following genera: *Fraxinus*, ash; *Syringa*, lilac; *Ligustrum*, privet; *Forsythia*, forsythia; *Jasminum*, jasmine; and *Forestiera*, California wild olive.

Other species of *Olea* are *O. ferruginosa*, *O. verucosa*, and *O. chrysophylla*. The fruit of these species is not edible.

The olive originated in southwestern Asia, probably Syria, many centuries ago and was gradually introduced into all the countries surrounding the Mediterranean Sea. Today, about 99 percent of the world's supply of olives is produced in Spain, Italy, Greece, Portugal, Turkey, Tunisia, France, Morocco, Algeria, Syria, Yugoslavia, Jordan, Cyprus, Israel, Libya, and Egypt. The remainder is produced in the United States, Argentina, Chile, China, Peru, Mexico, South Africa, Australia, and Japan.

Olives were introduced into California around 1769 when seeds were brought to Mission San Diego from Mexico by Father Junipero Serra and Don Jose de Galvez. After 1850 the number of trees increased rapidly until, by 1910, they numbered almost a million. Between 1928 and 1950, acreage dropped from 29,000 to 26,826, but between 1950 and 1957 it increased to 31,000 acres. By 1987 it had once more increased to a total of 33,175 acres.

Presently, 98 percent of the commercial plantings are located in the San Joaquin and Sacramento valleys. Although olives were planted extensively in the coastal areas of California, they were not as successful there as in the interior valleys.

Principal varieties grown in California are Mission (probably a seedling of the Spanish cultivar Cornicabra), Manzanillo, Sevillano, Ascolano, and Barouni. Most of the olive oil produced in the state comes from fruit of the first two cultivars; the fruit of the other three cultivars is used almost exclusively for pickling.

Attempts to use other species of *Olea* as rootstocks for *O. europaea* have generally proved unsuccessful. Except for Sevillano grafted onto *O. chrysophylla*, there is excessive overgrowth at the union and the trees develop many abortive shotberries or yellow leaves, and then die.

The method of propagation in some olive-growing countries is to saw off swellings (ovuli), which are common on the trunks of certain cultivars, and then plant them. Such swellings contain dormant buds and adventitious roots that grow when the ovuli are planted.

Another method of propagation sometimes employed is to plant large pieces of green wood (truncheons) horizontally 3 to 6 inches deep in the soil. From these arise shoots that become rooted and develop into trees.

More common methods of propagation are grafting seedlings with scions of the desired cultivar, self-rooting one- or two-year-old terminal growth with leaves, and self-rooting hardwood cuttings of branches several years old and 1 to 3 inches in diameter.

Where it is desirable to change the cultivar of a mature olive tree, scions of the desired cultivar can be grafted into the primary (scaffold) branches of the tree between early March and late April.

REFERENCES

Hartmann, H. T. 1976. Olive production in California. *Univ. Calif. Div. Agric. Nat. Res. Leaf.* 2474.

Hartmann, H. T., K. W. Opitz, and J. A. Beutel. 1980. Olive production in California. *Univ. Calif. Div. Agric. Sci. Leaf.* 2474. 64 pp.

Bacterial Diseases of Olive

Olive Knot

Pseudomonas syringae pv. *savastanoi*
(Smith) Young et al.

The disease "olive knot" or "olive tubercle" has been known since ancient times. It was spread around the world with the olive host and now occurs in all olive-producing regions. It probably was not introduced into California by the Spanish settlers, since the first olives grown in the state were produced from seeds. Between 1850 and 1900, however, the vegetative parts of many Spanish and Italian cultivars were brought into the state, and the disease was probably introduced during this time. It was present in California in 1898, according to Bioletti (1898).

According to Wilson and Magie (1963), studies on the cause of olive knot were reported in 1886 by Archangeli, who described the disease thoroughly and gave the name *Bacterium oleae* to an organism that he did not, however, consider to be the causal agent. Between 1887 and 1889, Savastano proved that the disease was caused by a bacterium, and this was soon confirmed by Cavara. Savastano called the organism *Bacillus oleae-tuberculosis*, but neither he nor Cavara described it adequately. The bacterium, which later became known as *Bacillus oleae* (Arch.) Trevisan, was described by Berlese in 1905 as having a yellow color in culture. In 1908, Smith published results of a careful study of the cause of olive knot and gave the name *Bacterium savastanoi* to the bacterium that he found to be the causal agent (Smith 1920).

Symptoms. The disease reduces production of the olive crop by defoliation and by killing twigs and branches. Moderately infected trees produce considerably less fruit than those with light infections (Schroth, Osgood, and Miller 1973). The reduction in yield is also associated with a substantial decrease in fruit size and reportedly with a decrease in oil and protein content (Osman, Tarabeih, and Michail 1980). Fruit from diseased trees contain compounds that impart bitter, salty, sour, or rancid tastes (Schroth, Hildebrand, and O'Reilly 1963).

If branches become infected in late winter, water-soaked lysigenous cavities 2 to 5 mm in length often develop. The cells of the host are collapsed in these cavities, and bacterial masses occur in the affected tissue. When the host starts growth, a knot (**fig. 1**) develops around this cavity because of the proliferation of tissue at its periphery. Hypertrophy and hyperplasia of cells occur. The cells nearest the center of the knot differentiate into xylem; those near the periphery differentiate into parenchyma and phloem. The elements of such tissues are arranged in a disorderly manner. The bacteria are found in sutures formed by the folding of the proliferating tissues. Smith (1920) claimed that metastasis (development of tubercles at a distance from the point of inoculation) occurred. There is evidence that secondary knots may sometimes develop ½ to 1 inch from the point of the primary infection when the lysigenous cavities extend to this distance, but seldom, if ever, are knots produced farther from the infection court.

Cause. Formerly known as *Bacterium savastanoi* E. F. Smith and *Phytomonas savastanoi*, the pathogen causing this disease is now classified as *Pseudomonas syringae* pv. *savastanoi* (Smith) Young et al. The organism producing tumors on oleander (*Nerium oleander*) similar to those on olive was found by C. O. Smith (1928), Adams and Pugsley (1934), and Pinckard (1935) to be similar to the olive pathogen in morphology and physiology. These workers and others, however, reported that while the oleander pathogen readily produces tumors on olive, the olive pathogen does not infect oleander. Wilson and Magie (1963) reported that certain isolates of *P. savastanoi* from olive do not infect the oleander, but others produce tumors on this host similar to those produced by the oleander pathogen. Furthermore, cultural characteristics, bacteriophage relations, and physiological and serological reactions are similar to the bacteria from the two hosts.

Domenico (1969) later reported evidence that under natural conditions the olive pathogen will spread to and infect oleander, but the frequency of infection is low. This frequency is in character with results of artificial inoculations, where not more than 20 percent of the isolates from olive infect oleander. Later, unpublished evidence indicates that while the pathogen from oleander can utilize sucrose as a carbon source, the pathogen from olive cannot, since it lacks the ability to produce the enzyme sucrase (invertase). The question remains whether the olive isolate's low pathogenic affinity for oleander and its inability to produce sucrase are stable characters. Despite the differences found, there seems to be little justification for regarding the oleander pathogen as a distinct species. The 1984 edition of Bergey's Manual of Systematic Bacteriology (Krieg 1984) does not recognize this pathogen as a biotype distinct from P. syringae pv. savastanoi. Janse (1982), however, proposed that pv. savastanoi be changed to a subspecies within which the pathogens of olive, oleander, and ash would become, respectively, pv. oleae, pv nerii, and pv. fraxini. He maintained that these pathovars can be distinguished on the basis of pathogenicity and host range.

Pseudomonas syringae pv. savastanoi is a gram-negative motile rod (0.4 to 0.8 by 1.2 to 3.3 μm) with one to four polar flagella. The colonies are white and smooth with wavy or entire margins. It produces acid but no gas from glucose, galactose, and certain other carbon sources. Starch is hydrolyzed. It creates a fluorescent pigment on certain media. A semiselective medium for the isolation of this pathogen has recently been developed (Surico and Lavermicocca 1989).

Other recent research (Comai and Kosuge 1980; Comai, Surico, an Kosuge 1982) has shown that tumors in olive and oleander are induced by indole-3-acetic acid (IAA). Two of the 12 strains of P. syringae pv. savastanoi that were studied contain a plasmid that codes for IAA production; in the remaining strains the IAA genes are located on the chromosome.

Disease development. Pseudomonas savastanoi survives from one season to the next in the knots, where it is produced in large numbers at all times of the year. It exudes to the surface during rains and is readily washed about (Horne, Parker, and Daines 1912). If not exposed to direct sunlight, the bacteria in the ooze may resist desiccation for several hours or even days (Smith 1920). There is some evidence (Smith 1920) that the bacteria are windborne, for short distances at least, with water droplets as the likely medium. Birds have been suspected of carrying the bacteria on their feet. In Italy, the olive fly (Dacus olea) is said to spread the bacteria. This insect is not found in California.

Ercolani (1978, 1979, 1983) found the pathogen to be a common component of the phylloplane of the olive in Italy. Populations of the organism did not differ significantly on leaves of the same age in different parts of the tree regardless of the presence of fruit. The organism was especially abundant on leaves in April and October, suggesting that the phylloplane of the host may be an important source of inoculum.

There is some evidence (E. E. Wilson, unpublished) that the pathogen survives only a matter of days in the soil.

Scars formed by the abscission of leaves are the most common avenues for entry of the bacteria into the twigs (Hewitt 1939; Horne, Parker, and Daines 1912) (**fig. 1**). Thus excessive defoliation caused by the leaf spot disease (Spilocaea oleaginea) is important in increasing the sites for bacterial infection. Blossom scars may become infected if rains occur at the end of the blossoming period. Other infection courts are pruning wounds or other injuries such as those made by hail and frost, and cracks made by adventitious shoots as they emerge from the branch. Because of their profuse production of suckers, cultivars such as Sevillano, Nevadillo, and Ascolano are prone to branch and trunk infection. Severe multiple infection of branches occurred in California in 1932 and again in 1950 through bark cracks caused by freezing injury.

Natural infection occurs during California's rainy season from late October to early June, but knots do not develop until tree growth starts in the spring. Knot development is rapid from May through June. If infection occurs in the fall, the time between infection and the first visible evidence of the knot is several months, but it is only 10 to 14 days when infection occurs in spring. The tree must be growing before the knot will develop. Systemic invasion is uncommon in olive but common in oleander (Wilson and Magie 1964).

Studies on the pathological anatomy of olive knots reveal both hypertrophy and hyperplasia of affected cells. The cambium produces disarranged and malformed tracheary elements and an abundance of radially elongated parenchyma cells instead of secondary xylem and phloem elements. Dissolution of some of the tumor cells results in fissures that harbor the pathogen and through which the bacteria migrate to the surface of the knot (Fudl-Allah and Baraka 1974; Osman, Tarabeih, and Michail 1980).

The pathogen can grow and reproduce at a wide range of temperatures, 23° to 24°C being the optimum. The development of water-soaked areas in branches following inoculation in winter indicates that the bacteria are capable of producing infection at fairly low temperatures (5° to 10°C). Temperatures below the optimum for growth are therefore not a limiting factor in infection. Moisture, however, is a limiting factor—infection rarely if ever occurs during the dry season. Rains followed by periods of high humidity are favorable to infection of such natural courts as leaf and blossom scars.

Soil fertility and soil moisture may indirectly affect the susceptibility of the host to olive knot. Anything that causes the trees to lose leaves during the rainy season will increase the likelihood of infection.

Control. Control begins with reducing the sources of inoculum by removing the knots. This may involve cutting off much of the branch system of badly infected trees. The work should be done in midsummer to avoid infection of cuts. Irrigation and fertilization to prevent the unseasonal dropping of leaves will help reduce the amount of infection. Protection against frost damage, if practical, is also helpful. Furthermore, in diseased orchards pruning should be done in summer rather than winter.

Chemicals such as phenol derivatives and kerosene painted over the knots kill the bacteria therein. It is difficult, however, to destroy the bacteria in the bark at the base of the knot without seriously injuring the branch. For this reason the Elgetol-methanol mixture that is effective against crown gall is not successful against olive knot. Schroth and Hildebrand (1968) provided the first successful chemotherapeutic treatment to selectively destroy knot tissue, a paraffin-oil-water emulsion containing 1,2,3,4-xylenol and m-cresol; the proprietary product is called Gallex.

Sprays of Bordeaux mixture have reduced olive-knot infection when applied in late fall, winter, and early spring (Wilson 1935). Although the best control was obtained with several applications, in one year a 90 percent reduction in disease incidence was achieved with a single application in early February. Control is complicated by the fact that leaf fall, although normally occurring in its greatest amount during spring, may produce infectible leaf scars at almost any time of the year. A postharvest (mid-November) spray of Bordeaux mixture helps control defoliation caused by the leaf-spot fungus and reduces bacterial infection at other infection sites. Slight copper injury from Bordeaux sprays has been reported in areas of low rainfall (Wilson 1935).

REFERENCES

Adams, D. B., and A. J. Pugsley. 1934.. A bacterial canker disease of oleander. *J. Dept. Agric. Victoria Austral.* 32:309–11.

Bioletti, F. T. 1898. The olive knot. *Calif. Agric. Exp. Stn. Bull.* 120, 11 pp.

Comai, L., and T. Kosuge. 1980. Involvement of plasmid deoxyribonucleic acid in indoleacetic acid synthesis in *Pseudomonas savastanoi*. *J. Bacteriol.* 143:950–57.

Comai, L., G. Surico, and T. Kosuge. 1980. Relation of plasmid DNA to indoleacetic acid production in different strains of *Pseudomonas syringae* pv. *savastanoi*. *J. Gen. Microbiol.* 128:2157–63.

Domenico, J. 1969. Relative infectivity of *Pseudomonas savastanoi* from olive and oleander plants (abs.). *Phytopathology* 59:11.

Ercolani, G. L. 1978. *Pseudomonas savastanoi* and other bacteria colonizing the surface of olive leaves in the field. *J. Gen. Microbiol.* 109:245–57.

———. 1979. Distribuzione di *Pseudomonas savastanoi* sulle foglie dell'olivo. *Phytopath. Mediter.* 18:85–88. (*Rev. Plant Pathol.* 62:4728. 1983.)

———. 1983. Variability among isolates of *Pseudomonas syringae* pv. *savastanoi* from the phylloplane of olive. *J. Gen. Microbiol.* 129:901–16.

Fudl-Allah, A. E., and M. M. Baraka. 1974. Pathological anatomy of olive tumors caused by *Pseudomonas savastanoi*. *Libyan J. Agric.* 3:105–08. (*Rev. Plant Pathol.* 56:5127. 1977.)

Hewitt, W. B. 1939. Leaf-scar infection in relation to the olive knot disease. *Hilgardia* 12:41–66.

Horne, W. J., W. B. Parker, and L. L. Daines. 1912. The method of spread of the olive knot disease. *Phytopathology* 2:101–05.

Janse, J. D. 1982. *Pseudomonas syringae* subsp. *savastanoi* (ex Smith) subsp. nov., nom. rev., the bacterium causing excrescences on Oleaceae and *Nerium oleander* L. *Internat. J. Syst. Bacteriol.* 32:166–69. (*Rev. Plant Pathol.* 61:6209. 1982.)

Krieg, N. R., ed. 1984. *Bergey's manual of systematic bacteriology*, vol. 1. Williams & Wilkins, Baltimore, 964 pp.

Osman, W. A., A. M. Tarabeih, and S. H. Michail. 1980. Studies on olive knot disease in Iraq with reference to response of different cultivars. *Mesopot. J. Agric.* 15:245–61. (*Rev. Plant Pathol.* 60:5514. 1981.)

Pinckard, J. A. 1935. Physiological studies of several pathogenic bacteria that induce cell stimulation in plants. *J. Agric. Res.* 50:933–52.

Schroth, M. N., and D. C. Hildebrand. 1968. A chemotherapeutic treatment for selectively eradicating crown gall and olive knot neoplasms. *Phytopathology* 58:848–54.

Schroth, M. N., D. C. Hildebrand, and H. J. O'Reily. 1963. Off-flavor of olives from trees with olive knot tumors. *Phytopathology* 53:524–25.

Schroth, M. N., J. W. Osgood, and T. D. Miller. 1973. Quantitative assessment of the effect of the olive knot disease on olive yield and quality. *Phytopathology* 63:1064–65.

Smith, C. O. 1928. Oleander bacteriosis in California. *Phytopathology* 18:503–08.

Smith, E. F. 1908. Recent studies of the olive-tubercle organism. *USDA Bur. Plant Ind. Bull.* 131 (4):25–42.

———. 1920. *Bacterial diseases of plants.* W. B. Saunders, Philadelphia, pp. 389–412.

Surico, G., and P. Lavermicocca. 1989. A semiselective medium for the isolation of *Pseudomonas syringae* pv. *savastanoi. Phytopathology* 79:185–90.

Wilson, E. E. 1935. The olive knot disease: Its inception, development and control. *Hilgardia* 9:233–64.

Wilson, E. E., and A. R. Magie. 1963. Physiological, serological, and pathological evidence that *Pseudomonas tonelliana* is identical with *Pseudomonas savastanoi. Phytopathology* 53:653–59.

———. 1964. Systemic invasion of the host plant by the tumor-inducing bacterium, *Pseudomonas savastanoi. Phytopathology* 54:576–79.

Fungal Diseases of Olive

Leaf and Fruit Spot of Olive

Mycocentrospora cladosporioides (Sacc.) P. Costa ex Deighton

This disease, affecting the leaves and fruit of olive, was reported by Hansen and Rawlins in 1944, but it has not been reported in California since that time.

Indistinct dark areas occur on the lower surface of affected leaves, which may fall. The conidial stage of the causal fungus is found on these spots. Fruit that remain on the tree during winter develop dark, circular spots. Uninfected areas of the fruit may remain green instead of turning black like normal fruit. The mycelium of the fungus is present between fruit cells, but seldom enters them. Tufts of conidiophores and conidia are produced on the surface of the lesions.

The causal fungus was identified as *Cercospora cladosporioides* Sacc. by Govi in 1952; it now is known as *Mycocentrospora cladosporioides* (Sacc.) P. Costa ex Deighton (Farr et al. 1989).

REFERENCES

Farr, D. F., G. F. Bills, G. P. Chamuris, and A. Y. Rossman. 1989. *Fungi on plants and plant products in the United States.* APS Press, American Phytopath. Soc., St. Paul, MN. 1252 pp.

Govi, G. 1952. La Cercosporiosi o "piombatura" dell'olivo. *Ann. Sper. Agrar.* 6:69–80.

Hansen, H. N., and T. E. Rawlins. 1944. Cercospora fruit and leaf spot of olive. *Phytopathology* 34:257–59.

Leaf Spot of Olive

Spilocaea oleaginea (Cast.) Hughes

Leaf spot of olive, also called "bird's-eye spot" and "peacock spot," is well known in all Mediterranean olive-growing countries, where it has received the attention of plant pathologists since the middle of the last century. In recent years the disease has been recognized as a serious problem in China (Chen and Zhang 1983; Chen, Zhang, and Zhang 1981). In 1845, Castagne described the disease in France and named the causal fungus *Cycloconium oleaginum.* In 1891, Boyer in France published the results of a careful study on the disease and gave a detailed description of the causal fungus. A few years later, Brizi (1899) investigated the morphology and life history of *C. oleaginum* and illustrated the changes it produces in the olive leaf. Brizi apparently was the first to show that the disease could be prevented by spraying with Bordeaux mixture. In the early part of this century, Ducomet (1907) and Petri (1913) made further studies of the relationship of the fungus to the leaf tissue. Tobler and Rossi-Ferrini (1906–07) investigated its control with Bordeaux mixture.

Leaf spot has been known in California since 1899 (Bioletti and Colby), but according to all early reports it seldom caused much damage. However, outbreaks between 1941 and 1949 reduced the productivity of some trees as much as 20 percent. Yearly losses of 9 to 15 percent of the leaves followed by the death of 10 to 20 percent of the fruiting twigs occurred in some orchards. Also, defoliation caused by the leaf-spot pathogen often leads to excessive infection by the olive-knot bacterium.

Symptoms. Although the disease lesions most often occur on the leaf blade, they are sometimes found on the leaf petiole, the fruit, and the fruit stem. They are at first inconspicuous, superficial, sooty blotches, but later develop into muddy green to almost black circular spots 2 to 10 mm in diameter. A faint yellow halo sometimes occurs in the leaf tissue around the spot. When the lesions are numerous, the leaf becomes yellow and soon falls. The dark green to black spots on the yellow background of the infected leaf blades (**fig. 2**) are said to resemble the spots on a peacock's tail, hence the name "peacock spot" (Bioletti and Colby 1899).

Lesions are notably more abundant on leaves in the lower part of the tree; many of the twigs in these parts are completely defoliated. The defoliated twigs die in large numbers during the summer.

Causal organism. The causal fungus of olive leaf spot was originally named *Cycloconium oleaginum* by Castagne (1845), who found it in material collected near Marseilles, France. Boyer (1891) described the fungus much more completely in 1891. The mycelium, composed of hyaline, septate, much-branched, radiating hyphae, is immersed in the outermost layer of the two-cell-thick leaf epidermis. This layer of epidermal cells is embedded in cutin and other fatty materials (Miller 1949). Slender hyphal branches from the mycelium penetrate upward through the leaf cuticle and end in globose to flask-shaped light brown conidiophores from which arise, apically and singly, the obclavate to pyriform, two-celled, light brown conidia (14 to 27 μm long and 9 to 15 μm broad). Several apical protrusions or necks may occur on the conidiophores that have borne a succession of conidia.

In 1953, Hughes suggested that this fungus properly belonged in the genus *Spilocaea*, which Fries erected in 1819 on the type species *S. pomi*. His view has generally been accepted, and the binomial of the fungus thus becomes *S. oleaginea* (Cast.) Hughes.

In the living leaf the fungus seldom penetrates the lower layer of epidermal cells. But after the leaf falls from the tree and dies, the fungus grows into pallisade and mesophyll cells and at times forms dense masses of stromatic tissue. The fungus produces similar masses of stromatic tissue in culture (Miller 1949; Wilson and Miller 1949). Though Petri (1913) suggested that such structures are sporocarps arrested in a primary stage of development, none has been found to develop to maturity.

Hosts. The edible olive, *Olea europea*, is the principal host of the fungus. Species of *Phillyrea* in Mediterranean countries are attacked by a fungus that has been designated *Cycloconium oleaginum* var. *phillyrea* (Desm.) Nic. and Agg.

Olive cultivars differ widely in susceptibility to infection. In California, Mission is severely infected and Manzanillo occasionally, while other important cultivars (Sevillano, Barouni, Ascolano, and Nevadillo Blanco) are seldom seriously affected. In Egypt, Manzanillo is considered to be highly resistant (Zayed et al. 1980).

Disease development. The fungus survives unfavorable periods, such as the hot, dry summers of interior California, in the affected leaves on the tree. Although newly formed lesions sporulate freely during spring, few spores are produced during the summer. In October or November the margins of the lesions extend laterally into adjacent leaf tissue, and a new crop of conidia develops there. These spores are the principal if not the only inoculum for infection. They are spread about the tree, mostly in a downward direction, by rain; lateral spread is notably limited. Apparently, they are not readily detached from the conidiophores by air currents, and thus are not easily airborne. What lateral dissemination does occur probably results from the wind carrying spore-laden droplets of rainwater.

Under California conditions a few new lesions may develop in autumn, but the greatest disease development occurs in spring. The occurrence of infection in fall, winter, and spring, however, is associated with periods of rainfall; the disease does not develop during rainless periods. Temperature is an important but not a limiting factor in the development of the fungus. High temperatures appear to restrict spore germination and mycelial growth more than do low temperatures. Though both processes occur quite readily at 8.9°C, the optimum temperature is near 21°C. There is evidence that infection occurs during the coolest part of the California winter, but the lesions do not develop to a visible stage until spring. Lesions resulting from infection in the spring become visible within a few weeks. The major effect of temperature, therefore, is in the rate at which infection occurs and symptoms develop.

Recent studies in China (Chen and Zhang 1983; Chen, Zhang, and Zhang 1981) and Lebanon (Saad and Masri 1978) have provided new information on the biology of the pathogen and on the epidemiology of the disease. The conidia germinate at 6° to 28°C, with an optimum temperature for both germination and appressorium formation of approximately 20°C. Germination is 62 percent at a relative humidity of 100 percent, but only 5 percent at 90 percent relative humidity. The conidia germinate better in a 1 percent glucose solution than in water. The minimum duration of leaf wetness for infection is 48 hours at 16°, 24 hours at 20°, and 36 hours at 24°C. The incubation period at 10° to 20°C is 12 days, and at 21° to 25°C, 16 to 19 days. In Lebanon inoculum density is highest in spring and fall; in China peak conidial production occurs in early June to early July and again in early October to early November. In China the optimum conditions for disease development and spread are temperatures of 16° to 22°C and a relative humidity of 80 to 85 percent.

Control. The elimination of the Mission cultivar from an orchard would largely solve the leaf-spot problem in California, but this could hardly be done, since Mission is very important in oil production and pickling.

Control by fungicides has been demonstrated (Papo and Peleg 1952; Tobler and Rossi-Ferrini 1906–07; Wilson and Miller 1949). Bordeaux mixture 10–10–100 or lime sulfur 3 percent applied in early autumn before the winter rains begin effectively prevents infection. To avoid copper injury, a 5–10–100 Bordeaux mixture or the equivalent of fixed copper can be used. Lime sulfur, in addition to its protective value, exhibits distinct eradicatory action against the fungus in the leaf (Wilson 1950). Other reportedly effective fungicides in the Mediterranean region are captafol and cuprozineb (Cu 36 percent, zineb 15 percent) (D'Armini 1970; Jaidi 1968).

Current control recommendations in California call for a postharvest application and an early spring application of either Bordeaux mixture or a fixed copper compound. These sprays are also effective against olive knot.

REFERENCES

Bioletti, F. T., and G. E. Colby. 1899. Olives. *Calif. Agric. Exp. Stn. Bull.* 123, 34 pp.

Boyer, G. 1891. Recherches sur les maladies de l'olivier: Le *Cycloconium oleaginum J. Bot.* 5:434–40.

Brizi, U. 1899. [Bird's eye of olives.] *Staz. Sper. Agric. Italy* 32:329–98.

Castagne, L. 1845. *Catalogue des plants qui croissent naturellement aux environs des Marseilles.* p. 200. Reviewed by Boyer (1891).

Chen, S., and J. Zhang. 1983. [Studies on olive peacock's eye disease. II. Infection cycle and epidemiology.] *Acta Phytopathol. Sinica* 13(1):31–40. (*Rev. Plant Pathol.* 62:3921. 1983.)

Chen, S., J. Zhang, and L. Zhang. 1981. [Studies on olive peacock's eye disease. I. Biological characteristics of the pathogen.] *Acta Phytopathol. Sinica* 11(3):37–42. (*Rev. Plant Pathol.* 61:3564. 1982.)

D'Armini, M. 1970. La difesa dell'olivo dallocchio di pavone. *Inf. Agrar. Verona* 26:2271–72. (*Rev. Plant Pathol.* 51:534. 1972.)

Ducomet, V. 1907. Recherches sur le développement de quelques champignons parasites a thalle subcuticulaire. *Ann. Ecol. Nat. Agr. Rennes* 1907:380–91.

Hughes, S. J. 1953. Some foliicolus Hyphomycetes. *Can. J. Bot.* 31(5):560–76.

Jaidi, A. 1968. Quelques observations biologiques et essais de traitement sur l'oeil de paon de l'olivier. *Al-Awamia* 27:41–50. (*Rev. Plant Pathol.* 49:3413. 1970.)

Miller, H. N. 1949. Development of the leaf spot fungus in the olive leaf. *Phytopathology* 39:403–10.

Papo, S., and J. Peleg. 1952. Trials in control of the olive leaf spot, caused by the fungus *Cycloconium oleaginum.* Israel Div. Plant Protect. Bull. 34, 16 pp.

Petri, L. 1913. *Studie sulle malattie dell olivo, III. Alcune richerche sulle biologia del* Cyclconium oleaginum Cast. R. Staz. Patol. Veg. Roma Mem., 136 pp.

Saad, A. T., and S. Masri. 1978. Epidemiological studies on olive leaf spot incited by *Spilocaea oleaginea* (Cast.) Hughes. *Phytopathol. Medit.* 17:170–73. (*Rev. Plant Pathol.* 60:3268. 1981.)

Tobler, O., and U. Rossi-Ferrini. 1906–07. [The use of Bordeaux mixture for the control of *Cycloconium* on the olive.] *R. Acad. Econ. Agr. Georg. Frienze Atti* 5:326–37.

Wilson, E. E. 1950. The protective and eradicative actions of lime-sulfur and Puratized in controlling a fungus leaf-spot disease of olive (abs.). *Phytopathology* 40:32.

Wilson, E. E., and H. N. Miller. 1949. Olive leaf spot and its control with fungicides. *Hilgardia* 19:1–24.

Zayed, M. A., H. M. El-Saied, A. S. Ali, and K. S. Saied. 1980. Reaction of olive cultivars to *Cycloconium oleaginum* Cast., and chemical control of live leaf spot disease in Egypt. *Egypt. J. Phytopathol.* 12:49–56. (*Rev. Plant Pathol.* 62:1133. 1983.)

Shoestring (Armillaria) Root Rot of Olive

Armillaria mellea (Vahl.:Fr.) Kummer

Armillaria mellea occasionally attacks the olive tree and, in time, kills it. The symptoms of the disease are similar to those on stone-fruit trees. White to yellowish fan-shaped mycelial mats are present between the bark and wood of the affected roots, and dark brown to black rhizomorphs are present on the surface. Early recognition of the disease and removal of infected roots before the fungus girdles the crown may prolong the life of the tree (see Chapter 4 for further discussion).

Verticillium Wilt of Olive

Verticillium dahliae Kleb.

This disease (a general discussion of which appears in Chapter 4) has become of economic importance on olive, particularly in the San Joaquin Valley, with the introduction of a more pathogenic strain of the fungus (Schnathorst and Sibbett 1971).

The external symptoms (**fig. 3**) are similar to those in stone-fruit trees. The leaves on one or more branches suddenly wilt during the summer. Discoloration of the sapwood, which occurs extensively in apricot and almond, does not occur in olive. The death of mature trees with *Verticilium* infection is probably more common in olive than in stone fruit.

The fungus dies in the aerial parts of the tree during the summer, but remains viable in the roots from one season to the next (W. C. Schnathorst, unpublished). The recurrence of the disease in infected trees is due, therefore, to either a reactivation of the fungus in the roots or new root infection by pathogen propagules in the soil.

Attempts have been made during the past 20 years to develop olive rootstocks resistant to Verticillium wilt (Hartmann, Schnathorst, and Whistler 1971; Wilhelm and Taylor 1965). The best is a clonally propagated seedling called Oblonga. But recent studies (W. C. Schnathorst, unpublished) have shown that the virulent T-1 strain of the pathogen will pass through the rootstock and cause wilt symptoms in the scion. Tests are in progress to obtain a seedling of Oblonga with greater resistance than the original selection.

Another approach to control has been through the use of sod culture instead of clean cultivation (Ashworth and Wilhelm 1971). But recent studies have failed to substantiate the initial claim of reduced wilt in orchards with a sod cover crop (Ashworth et al. 1982).

REFERENCES

Ashworth, L. J., Jr., S. A. Gaona, G. Weinberger, G. S. Sibett, and R. M. Beede. 1982. Research on Verticillium wilt disease: An update on the use of polyethylene tarps for weed control and a summary of information on weeds, weed control, and wilt. *Proc. Calif. Pistachio Symp.* Fresno 34–43.

Ashworth, L. J., Jr., and S. A. Wilhelm. 1971. Verticillium wilt of olives: Sod culture offers a promise of control. *Calif. Olive Ind. News* 25(1).

Hartmann, H., W. C. Schnathorst, and J. Whistler. 1971. Oblonga, a clonal olive rootstock resistant to Verticillium wilt. *Calif. Agric.* 25(6):13–15.

Schnathorst, W. C., and G. S. Sibbett. 1971. The relation of strains of *Verticillium albo-atrum* to severity of Verticillium wilt in *Gossypium hirsutum* and *Olea europaea* in California. *Plant Dis. Rep.* 55:780–82.

Wilhelm, S., and J. B. Taylor. 1965. Control of Verticillium wilt of olive through natural recovery and resistance *Phytopathology* 55:310–16.

Miscellaneous Fungal Diseases of Olive

Diseases of minor importance reported on olive in California are anthracnose (*Glomerella cingulata* [Ston.] Spauld. & Schrenk), root rot (*Phytophthora* sp.), wood rot (*Abortiporus [Polyporus] biennis* [Bull.:Fr.] Sing.), and wood rot and dieback (*Omphalotus olearius* [DC.:Fr.] Sing.).

Diseases of olive not known in California but present in other parts of the United States are black leaf spot (*Zukalia purpurea* [Ellis & G. Martin] Theiss.), Florida; Phyllosticta leaf spot (*Phyllosticta oleae* Ellis & G. Martin), Florida; Phymatotrichum root rot (*Phymatotrichopsis [Phymatotrichum] omnivora* [Duggar] Henebert), Texas; Pythium crown and root rot (*Pythium* sp.), Florida; Septoria leaf spot (*Septoria serpentaria* Ellis & G. Martin), Florida; and wood rot (*Ganoderma lucidum* [Curtis:Fr.] P. Karst.), Arizona.

REFERENCES

Farr, D. F., G. F. Bills, G. P. Chamuris, and A. Y. Rossman. 1989. *Fungi on plants and plant products in the United States.* APS Press, American Phytopath. Soc., St. Paul, MN. 1252 pp.

French, A. M. 1987. *California plant disease host index. Part 1: Fruits and nuts.* Calif. Dept., Food & Agric., Sacramento, CA 39 pp.

VIRUS AND VIRUSLIKE DISEASES OF OLIVE

Sickle Leaf

Graft-Transmissible Pathogen

The name "sickle leaf" is descriptive of this disease because of the laterally curved shape of affected leaves. The disease is rather common in commercial plantings, and there is evidence that the same disease occurs in Chile, Italy, and Portugal (Thomas 1958; Waterworth and Monroe 1975). Although Thomas (1958) reported some branch stunting and reduction in fruiting, the current consensus among knowledgeable California agriculturists is that the disease causes little if any tree damage.

Symptoms. In addition to the sickle shape and reduced size of the leaves, the leaf blade is often chlorotic, mostly on the inner side of the curve (**fig. 4**). Affected branches may be stunted and the amount of fruit reduced (Thomas 1958). The disease appears to be systemic, but the symptoms are not evident on all branches. There usually are many more symptomless than symptomatic leaves on an affected tree. No fruit symptoms have been observed.

In California, sickle leaf appears to be present only on the Mission cultivar. However, somewhat similar symptoms have been seen on Sevillano.

Cause. Sickle leaf is caused by an unidentified graft-transmissible pathogen. Although the symptoms induced by this agent and its infectious nature suggest that it may be a virus (Martelli 1981; Thomas 1958), there is no experimental evidence to support such a hypothesis. Waterworth and Monroe (1975) were unable to detect either virus particles or mycoplasmalike bodies in leaf-dip preparations from affected olives. The incubation period varies from seven months to three years, depending on the cultivar inoculated (Thomas 1958; Waterworth and Monroe 1975). By inarch grafting, Thomas (1958) transmitted the disease from Mission to the cultivars Sevillano C and Mission Nobs. No transmission occurred when cultivars Manzanillo Sharpe and Oblonga, oleander (*Nerium oleander*), and privet (*Ligustrum* sp.) were similarly inoculated. Mechanical transmission to 16 herbaceous plants yielded negative results (Waterworth and Monroe 1975). The pathogen appears to move very slowly in infected trees. It was inactivated in young olive trees subjected to 37°C for three or more weeks (G. Nyland, unpublished).

Control. Propagation material should be obtained from trees free of sickle-leaf symptoms. Thermotherapy could be used to eliminate the pathogen from newly developed or introduced cultivars.

REFERENCES

Martelli, G. P. 1981. Le virosi dell'olivo: Esistono? *Inform. Fitopatol.* 31:97–100.

Thomas, H. E. 1958. Sickle leaf of olive. *Plant Dis. Rep.* 42:1154.

Waterworth, H. E., and R. L. Monroe. 1975. Graft transmission of olive sickle leaf disorder. *Plant Dis. Rep.* 59:366–67.

Abiotic Diseases of Olive

Shotberry of Olive

Small, almost spherical fruit develop in the fruit clusters (**fig. 5**). Sevillano and Manzanillo exhibit the disorder most commonly. Affected fruit do not develop seed in the pit, so it seems likely that the disorder is a delayed form of pistil abortion or possibly a form of parthenocarpy. Presumably, the pistil develops enough to stimulate the fruit to considerable growth and premature ripening before it aborts. Many of the undersized fruit drop early, and no way to bring such fruit to normal size is known.

Soft Nose of Olive

Soft nose or blue nose is a disorder of minor importance affecting fruit of the Sevillano cultivar (**fig. 6**). As the fruit begin to ripen, a large area at the stylar end becomes pinkish red and then blue. Later the flesh in this area turns dark and shrivels, and the fruit become unfit for pickling. The cause of the disorder is unknown but the problem is more prevalent in heavy crop years. Apparently it is nonparasitic and may be associated with water stress or heavy nitrogen fertilization. It is more severe on young trees than on older ones, and symptoms first appear during the harvesting period.

Split-Pit of Olive

Split-pit is largely a problem on Sevillano and is serious only in some years. Fruit are bluntly flattened, and the pits are split along the suture. The cause is not known, but is suspected to be environmental; some workers believe that heavy irrigation of dry soil at the time of pit hardening may induce it.

REFERENCES

Hartman, H. T. 1953. Olive production in California. *Calif. Agric. Exp. Stn. Man.* 7, 59 pp.

Hartman, H. T., K. W. Opitz, and J. A. Beutel. 1980. Olive production in California. *Univ. Calif. Div. Agric. Sci. Leaf.* 2474. 64 pp.

Chapter 8

The Fig
and Its
Diseases

Page numbers in bold indicate photographs.

INTRODUCTION	354	
REFERENCES	355	
BACTERIAL DISEASES OF FIG	355	
Bacterial Canker of Fig	355	
Xanthomonas campestris pv. *fici* (Cavara) Dye		
REFERENCES	355	
FUNGAL DISEASES OF FIG	356	
Botrytis and Sclerotinia Limb Blight or Dieback of Fig	356	**436**
Botrytis cinerea Pers.:Fr., *Sclerotinia sclerotiorum* (Lib.) dBy.		
REFERENCES	356	
Branch Wilt of Fig	356	
Hendersonula toruloidea Nattrass		
REFERENCES	356	
Endosepsis of Fig	356	**436**
Fusarium moniliforme Sheldon		
REFERENCES	358	
Phomopsis Canker of Fig	358	**436**
Phomopsis cinerascens Trav.		
REFERENCES	360	
Smut and Mold of Fig	360	**437**
Aspergillus niger van Tiegh. and Other Fungus Species		
REFERENCES	361	
Souring of Fig	361	**437**
Yeast Fungi		
REFERENCES	362	
Surface Mold and Alternaria Rot of Fig	362	**437**
Cladosporium herbarum (Pers.) Lindt., *Alternaria alternata* (Fr.) Keissl.		
REFERENCES	363	
VIRUS AND VIRUSLIKE DISEASES OF FIG	363	
Fig Mosaic	363	**437**
Graft- and Mite-Transmissible Pathogen		
REFERENCES	364	
MISCELLANEOUS DISEASES OF FIG	365	
Crown gall, shoestring root rot, Cytospora canker, twig blight, thread blight, leaf blight, rust, anthracnose, leaf spot, limb blight, Tubercularia canker.		
REFERENCES	366	

Introduction

The fig (*Ficus carica* L.) is a native of southwestern Asia, and probably was first cultivated in Saudi Arabia and neighboring countries. It is now grown in all Mediterranean countries, Iran, the USSR, Afghanistan, India, China, Japan, Australia, and South Africa as well as the Americas.

Fig trees were introduced into the Western Hemisphere about 1520 when they were brought to the West Indies from Europe. Today they are grown commercially in Chile, Mexico, Argentina, and the United States. The principal fig-producing states are California, Texas, Alabama, Georgia, Louisiana, Mississippi, and North Carolina.

Although the fig was brought into California by Franciscan missionaries around 1769, commercial production of some of the cultivars did not begin until the middle of the nineteenth century. Later, new cultivars from the eastern United States and Spain were planted, and the practice of introducing the better European cultivars continued till the end of the nineteenth century.

Today, California has about 18,200 bearing acres of figs, 98 percent of the acreage being located in four San Joaquin Valley counties. The most important cultivars are Calimyrna (California Smyrna, or Lob Injir), Adriatic (White Adriatic, Verdone, Nebian, or Grosse Verti), Kadota (Dottato, White Pacific, or White Endrich), and Mission (Black Mission).

Figs are propagated by means of rooted cuttings. Sections of branches 9 to 12 inches long and ½ to ¾ inch in diameter placed in moist soil will produce roots readily. The plants are grown for one year in the nursery before being field-planted.

An understanding of diseases of fig fruit requires knowledge of the structural peculiarities of the fruit and the role played by the fig wasp, *Blastophaga psenes* L., in the development of the fruit of certain cultivars. The structure popularly regarded as a fruit is known to horticulturists as a syconium. This is a hollow receptacle with an apical orifice or "eye." In an immature syconium, the eye is closed by overlapping bracts, but as the fruit mature an opening 2 to 10 mm in diameter is formed by the loosening of the bracts. The true fruit are borne on the inner wall of the syconium and, when mature, each consists of a stalk, the remains of the flower perianth, some parenchymatous tissue of the outer ovary wall, and one seed.

Ficus carica consists of two botanically different groups, the caprifig and the edible fig. For the most part, syconia of the caprifig are inedible. The two groups also differ with respect to the kind of flowers present in the syconium. The caprifig syconium produces short-stiled pistillate flowers over a greater part of the inner wall and staminate flowers around the eye. Trees of this group are necessary to the production of the edible fruit of certain fig cultivars because their syconia are the habitat for the wasp *B. psene* L., whose activity results in the pollination (caprification) of the female flowers of the edible fig. Edible figs are of three types: Smyrna, San Pedro, and the common type (self-pollinated). Syconia of the Smyrna type produce only long-stiled pistillate flowers, which require caprification before the syconia will develop to maturity. Syconia of the San Pedro type are of two kinds: those of the first crop, which contain only long-stiled pistillate flowers; and those of the second crop, which are borne on wood of the current season. These contain long-stiled pistillate flowers, but require pollination before the syconia will develop to maturity. Fruit of all cultivars of the common type (self-pollinated) develop to maturity without pollination, and caprification by the fig wasp is not necessary to production of a crop.

Development of two distinct crops of fruit—the first on the wood of the previous season and the second on wood of the current season—is common among certain cultivars of all types, including the caprifig. In California, the first crop of edible figs (known as the "breba" crop) matures from May to June; the second crop matures from August to October. Certain cultivars of edible figs sometimes produce a third crop, which matures some weeks after the second crop; this third crop is borne on wood of the current season and thus is nothing more than an extension of the second crop. The caprifig commonly develops three more or less distinct crops in the interior valleys of California. Because all three crops are associated with development of the wasp, they have been given corresponding names: the spring crop, which develops in March or

April and remains on the tree until June, is called the "profichi"; the summer crop, which develops in July and remains on the tree until November, is called the "mammoni"; the winter crop, which develops in November and remains on the tree throughout the winter, is called the "mamme."

Blastophaga psenes passes the winter as larvae in parasitized pistillate flowers of the mamme syconia. It goes into the pupal stage in early spring and emerges as a winged adult at the same time the profichi crop is developing. The female, previously fertilized by the male while still in the "inhabited" flowers of the mamme syconium, enters a profichi syconium through the orifice and deposits her eggs in the pistillate flowers. These eggs hatch about June, and the fertilized adult female emerges sometime later and lays her eggs in a syconium of the mammoni crop. The new population of fertilized females emerges from the mammoni syconia in later summer and enters the mamme syconia, where they lay their eggs, thereby completing the yearly cycle.

Caprifigs formerly were planted in rows or in groups among trees of the edible Smyrna type, but after it was found that the insect transmitted spores of the endosepsis fungus *Fusarium moniliforme* from the caprifig to the edible fig, many orchardists removed the caprifig trees from the orchards and grew them in separate plantings. The practice of caprification now followed is to pick the fruit of the profichi crop (the only crop with viable pollen) just before the male wasps begin to emerge and as the staminate flowers begin to shed pollen. The caprifigs are placed in wire baskets and the baskets are hung in the trees of the edible cultivars. As the female wasp emerged from the profichi fruit her body becomes dusted with pollen. She immediately enters a syconium of the edible fig and attempts to lay her eggs in the flower ovaries by inserting her ovipositor down the stilar canal. She is unable to do so, owing to the long stiles possessed by flowers of the edible sorts. In her vain attempt at oviposition, however, pollen from her body is deposited on the stigmas of the florets, thereby completing the process of caprification.

REFERENCES

Condit, Ira J. 1947. *The fig.* Chronica Botanica, Waltham, Mass., 206 pp.

Eisen, Gustav. 1896. Biological studies on figs, caprifigs, and caprification. *Proc. Calif. Acad. Sci.* 5:897–1003.

Obenauf, G., M. Gerdts, G. Leavitt, and J. Crane. 1978. Commercial dried fig production. *Univ. Calif. Div. Agric. Sci. Leaf.* 21051. 30 pp.

BACTERIAL DISEASES OF FIG

Bacterial Canker of Fig

Xanthomonas campestris pv. *fici* (Cavara) Dye

In 1948, a bacterial canker of the fig originally described in Italy was reported by Hansen as occurring in California on the white Adriatic fig. In Italy, the disease is said to cause dark lesions on leaves and elongated lesions on new shoots, followed by wilting of these parts. In California, the disease is characterized by the development of roughly circular necrotic areas in the inner bark of older limbs. A slight cracking of the outer bark is the only external evidence of the cankers (Hansen 1948; Petri 1906).

It is said (Krieg 1984) that the causal organism was named *Bacterium fici* by Cavara in Italy, where it causes a leaf spot and shoot wilt of *Ficus carica*. Only a meager description of this organism exists. It was later transferred to the genus *Phytomonas*, but with the elimination of this genus in 1939, the organism was placed in the new genus *Xanthomonas* and now is considered a pathovar of *X. campestris* (Krieg 1984). In recent years this disease has attracted little attention in California or elsewhere.

REFERENCES

Hansen, H. N. 1948. A canker disease of figs (abs.). *Phytopathology* 38:914–15.

Krieg, N. R., ed. 1984. *Bergey's manual of systematic bacteriology*, vol. 1. Williams & Wilkins, Baltimore, 964 pp.

Petri, L. 1906. Investigations on the bacteriosis of figs. *Atti R. Acad. Lincei, Rend. Cl. Sci. Fis., Mat. Nat.* 5 ser. 15, II, 10:644–51.

Pilgrim, A. J. 1950. Bacterial canker of Adriatics. *Proc. Fourth Ann. Res. Conf. Calif. Fig Inst.*: 32–33.

Fungal Diseases of Fig

Botrytis and Sclerotinia Limb Blight or Dieback of Fig

Botrytis cinerea Pers.:Fr.,
Sclerotinia sclerotiorum (Lib.) dBy.

In 1919, Condit and Stevens reported that a dieback of the new shoots of fig trees was common in California. It was thought that the shoots were first weakened by *Botrytis* and thus rendered more susceptible to sunburning and attack by other fungi. Shoot dieback after winter injury to fruit or branches can be caused by *Botrytis* (English 1962; Smith 1941) (**fig. 1**).

In 1931, Taubenhaus and Ezekiel showed by inoculations that limb blight of the Magnolia cultivar of fig in Texas was caused by the fungus known as *Sclerotinia sclerotiorum* (Lib.) dBy. or *Whetzelinia sclerotiorum* (Lib.) Korf and Dumont. The disease was characterized by a sudden wilting of the foliage of new shoots in the spring. The bark of affected branches was water-soaked and overrun by a thick white growth of the fungus. In later stages of the disease, numerous sclerotia developed in and on affected portions of the limbs.

Stems and "stick tight" figs affected with *Botrytis* or *Sclerotinia* should be removed (Smith 1941).

REFERENCES

Condit, I. J., and H. L. Stevens. 1919. "Dieback" of the fig in California. *Fig and Olive J.* 4:11–12.

English, H. 1962. Canker and dieback disorders of fig trees. *Proc. 16th Ann. Res. Conf. Calif. Fig Inst.*: 13–15.

Smith, R. E. 1941. *Diseases of fruit and nuts.* Calif. Agric. Ext. Serv. Circ. 120, 168 pp.

Taubenhaus, J. J., and W. N. Ezekiel. 1931. A Sclerotinia limb blight of figs. *Phytopathology* 21:1191–97.

Branch Wilt of Fig

Hendersonula toruloidea Nattrass

Branch wilt, produced by the fungus *Hendersonula toruloidea* Nattrass, has been found on Kadota fig trees in the San Joaquin Valley (Paxton, Wilson, and Davis 1964; Warner 1952; Wilson 1952). It has not, however, become of great significance to the fig industry. See the description of the disease and the fungus under "Branch Wilt of English Walnut" in Chapter 5.

REFERENCES

Paxton, J. D., E. E. Wilson, and J. R. Davis. 1964. Branch wilt disease of fig caused by *Hendersonula toruloidea*. *Plant Dis. Rep.* 48:142.

Warner, R. M. 1952. Some observations on branch wilt in figs. *Proc. Sixth Ann. Res. Conf. Calif. Fig Inst.*: 24–25.

Wilson, E. E. 1952. Factors affecting the branch wilt disease in walnuts. *Proc. Sixth Ann. Res. Conf. Calif. Fig Inst.*: 21–23.

Endosepsis of Fig

Fusarium moniliforme Sheldon

Endosepsis (internal rot)—also called pink rot, brown rot, soft rot, and eye-end rot—was described and named by Caldis in 1925 and 1927. It occurs wherever figs are grown in California but has not been reported in other parts of the United States or from other countries (Caldis 1927; Smith and Hansen 1931). Before control measures using mercury fungicides were developed, endosepsis losses accounted for 30 to 50 percent of the crop in many California orchards (Hansen 1927, 1928; Smith and Hansen 1931). In early 1972, Gerdts (1972) developed a substitute treatment using benomyl, but evaluations in 1977 showed mold counts from 17 to 54 percent. A year later, Obenauf (1978) showed that the benomyl treatment reduced the incidence of *Fusarium* but not of other molds. In 1982, Obenauf et al. published results showing the benefits of fungicide mixtures in reducing the decay of mamme caprifigs caused by a complex of molds. Since that time, the mamme figs have been treated with a combination of fungicides, yet endosepsis and the rot caused by a complex of molds were not adequately controlled. R. Klamm, Managing Director of the California Fig Institute, provided data (personal communication) showing caprifig endosepsis ranges of 3 to 10 percent from 1963 to 1977, 10 to 28 percent from 1978 to 1985, and almost 60 percent in 1986. He also reported an average 25 percent "cull-out" in Calimyrna figs, a significant portion of which was due to endopsis, with an exceptionally high loss of 45 percent in 1986.

Symptoms. The first observable symptoms of endosepsis on the edible female Calimyrna fig are found in the fruit (syconium) as it begins to ripen. Brown streaks develop along the normal white stalks of the female florets. Later, as the number of infected florets increases, the diseased areas are clearly visible as yellowish brown spots. At that time no external symptom is visible, but as the fruit ripens indefinite water-soaked spots appear on the skin. The spots, which are usually most numerous around the eye (orifice) or along the neck of the fruit, gradually develop a bright pink, brown, or purplish color. At this point the fruit may split across the eye end in which case a drop of amber gumlike substance appears in the eye. Infected fruit are insipid and lack the characteristic sweetness and flavor of normal figs, and are off-color internally (**fig. 2A**). The endosepsis symptoms may not be easily detected because of other mold organisms and yeasts that develop on affected fruit.

Also affected with *Fusarium* are the male caprifigs (**fig. 2B**), the mammoni or summer crop, the mamme or winter crop, and the profichi or spring crop. Upon splitting these fresh figs, one can macroscopically observe fungal growth (*Fusarium*) within the fruit, and upon plating the tissue one can identify cultures of other molds such as *Cladosporium herbarum*, *Penicillium* spp., *Alternaria alternata*, *Botrytis cinerea*, *Aspergillus niger*, and yeasts. Dry mummified figs of the previous crop remaining on the tree also show the presence of these molds. In 1987, *F. moniliforme* was isolated from 100 percent of the mammoni crop left on the tree in one orchard; Stanford and Roeding caprifigs (mamme crop) averaged about 40 percent infection, the same as that found on wild caprifigs (T. Michailides, personal communication).

Causal organism. Although Caldis (1927) identified the causal fungus as *Fusarium moniliforme* Sheldon var. *fici* n. var., any clone of *F. moniliforme* apparently can cause endosepsis. The clone described by Caldis produced macroconidia that are sickle-shaped, attenuate, subpedicellate, 3- to 5-septate, 20 to 52 μm long, and 2 to 5 μm broad. Microconidia are produced in false heads or in chains on white to dark maroon aerial hyphae and are ovoid-fusoid, 5 to 11 μm long, and 2 to 3 μm broad.

Disease development. The fungus survives during winter in the mamme crop on the caprifig tree. Fungal conidia are introduced into these fruit in the fall by the female fig wasps (*Blastophaga psenes*) (**fig. 2C**) when they enter to lay eggs (Caldis 1925; Simmons and Fisher 1947; Gerdts and Clark 1979). The fungus is in turn transferred to successive caprifig crops (profichi, mammoni) during the spring and summer. Its introduction into edible cultivars occurs when the wasps emerge from the infected profichi crop and enter the fruit of cultivars that require pollination. Caldis (1927) showed that female wasps emerging from diseased caprifigs bear conidia of *F. moniliforme* on their bodies, and proved that fruit that had not been entered by the wasp were free of the fungus. He also showed that endosepsis in parthenocarpic cultivars (common types, such as Mission) occurred only after wasps contaminated with spores of *F. moniliforme* entered the fruit.

The female wasp enters the fruit of the edible fig when these structures are still green (May to June), and after trying in vain to lay her eggs in the florets, she dies. Although the fungus develops saprophytically on the dead body of the wasp, it cannot invade the fig tissue until the fruit begins to ripen. Subsequent development of the disease is usually rapid, depending on the weather conditions. If, however, the weather hastens ripening (thus producing a rapid increase in sugar concentration) fungal development may be checked and the fruit may dry without appreciable loss in quality.

Control. Once it became known that the caprifig harbored the fungus from one year to the next, efforts were directed toward eliminating this source of inoculum. Caprifig trees were removed from the orchards of edible figs and separate plantings were established. The caprifig trees are occasionally sprayed with whitewash to destroy lichens that might harbor possible insect vectors of molds. Moldy fruit (mamme crop) or fruit injured by frost are removed from the trees.

Around the end of March or the beginning of April, before the male wasps begin to emerge from the caprifig (mamme), the figs are picked and treated with a chemical to kill or suppress the fungus inside them (Hansen 1927, 1928). The treatment involves cutting the fresh mamme caprifigs in half, with the incision starting at the neck of the fig and continuing beside the eye to the neck on the opposite side. The fruit are then split to reduce the likelihood of injury to the wasps inside. Next, the halves are dipped in a chemical suspension, such as a mixture of benomyl, chlorothalonil, and dicloran, and hung back on the caprifig trees so that the issuing wasps are free of *F. moniliforme* and other molds as they enter the profichi crop. The mamme figs are introduced twice into the caprifig trees to ensure proper pollination of the profichi crop. This process is used to enable the introduction of mold-free profichi figs into Calimyrna orchards during the month of June when the second-crop Calimyrna fruit are receptive to pollen (Warner 1950).

Research to assess current fungicidal treatments for the reduction of molds in the caprifig and endosepsis in the Calimyrna fig was reopened in 1987. Evidence has shown that benomyl dip treatments of the mamme crop reduced the spore populations of *Fusarium* species (*F. moniliforme, F. solani,* and *F. episphaeria*) but not those of *Alternaria* or *Cladosporium*. These treatments reduced the incidence of *Fusarium* on emerged fig wasps, but did not significantly reduce endosepsis on the Calimyrna crop. Careful selection of apparently disease-free (without internal brown symptoms) mamme crop figs was important in obtaining *Fusarium*-free fig wasps to pollinate the profichi crop. Yet the presence of *Fusarium* spores on surfaces of both caprifig and Calimyrna trees, which could account for recontamination of the fig wasps (Michailides and Ogawa 1988), was disturbing.

To effectively reduce the presence of *Fusarium*-contaminated fig wasps during pollination of the profichi crop, the unharvested mamme crop in the orchard must be eliminated before disease-free mammes are reintroduced in the orchard. In addition, elimination of mold organisms on the caprifig trees (1988 studies showed that chlorine solutions were not effective for extended periods) is required to prevent the recontamination of emerging fig wasps before they enter the profichi crop.

REFERENCES

Caldis, P. D. 1925. A rot of the Smyrna fig in California. *Science* 62:161–62.

———. 1927. Etiology and transmission of endosepsis in the fruit of the fig. *Hilgardia* 2:287–328.

Gerdts, M. H. 1972. Caprifig fungicide experiments. *Proc. Calif. Fig Res. Inst.*:14–15.

Gerdts, M., and J. K. Clark. 1979. Caprification: A unique relationship betwen plant and insect. *Calif. Agric.* 33(11/12):12–14.

Hansen, H. N. 1927. Control of internal rot of caprified figs. *Phytopathology* 17:199–200.

———. 1928. Endosepsis and its control in caprifigs. *Phytopathology* 18:931–38.

Michailides, T. J., and J. M. Ogawa. 1988. Investigations on the correlation of fig endosepsis on Calimyrna fig with caprifig infestations by *Fusarium moniliforme*. California Fig Institute, Fresno *Fig Research Report: Crop year 1988.* 31 pp.

Obenauf, G. L. 1978. Review of fig spoilage research. *Proc. Calif. Fig Res. Inst.*:48–51.

Obenauf, G. L., J. M. Ogawa, K. Lee, and C. A. Frate. 1982. Fungicide control of molds that attack caprifigs. *Plant Dis.* 66:566–67.

Simmons, P., and C. K. Fisher. 1947. Caprification of Calimyrna figs. *Calif. Dept. Agric. Bull.* 36(3):115–21.

Smith, R. E., and H. N. Hansen. 1931. Fruit spoilage diseases of the fig. *Univ. Calif. Agric. Exp. Stn. Bull.* 506, 84 pp.

Warner, R. M. 1950. Caprifig care and culture: Seasonal tips on clean-up. *Calif. Fruit Grape Grower*, March 1950.

Phomopsis Canker of Fig

Phomopsis cinerascens Trav.

Phomopsis canker (also called "fig canker") affects all commercial fig cultivars in California. Kadota is the most seriously affected because of certain pruning practices. Calimyrna is damaged to some extent, but Mission and Adriatic are rarely attacked.

The first report of this disease came from Italy in 1876. It was observed in southern France in 1912, a few years later in England, and in more recent years in North and South Africa, Denmark, the southern coast of the Crimea, and Brazil. The first published report from the United States was by Smith in California in 1941. Hansen (1949a) indicated, however, that the disease was observed in California on trees of the Adriatic cultivar as early as 1918, and that it was epidemic in a Kadota orchard just south of Stockton in 1936 (Hansen 1949c). In the 1940s, the disease became prevalent in the extensive Kadota plantings in Merced County (English 1951).

Kadota trees, which are subject to heavy annual pruning (**fig. 3B**), may have as many as 100 cankers, some of which eventually girdle branches (English 1953). The most important cause of crop loss is the reduction of fruiting wood.

In many orchards with a history of heavy disease occurrence, the problem is nonexistent today. The natural decline in incidence of this disease warrants further investigation which might well provide a new approach to the control of other canker diseases.

Symptoms. Localized dead areas in bark and wood develop at pruning wounds (**fig. 3A**) or at injuries caused by some other agency such as frost (English 1951). Cankers are extremely difficult to see during the year of infection because there is very little discoloration of the outer bark. In succeeding years, however, the bark in the older portions of the cankers becomes bleached, cracked, and somewhat sunken, and upon close examination the minute black fruiting bodies of the causal fungus can be seen in the outer bark layers. Yearly elongation of cankers is evidenced by distinct zonations, with the outermost zone sporulating most profusely. The cankers become more or less elliptical in shape because they extend more rapidly up-and-down than side-to-side. Many of the branches are eventually girdled, and the withered brown foliage remains hanging on these dead branches during the summer. New shoots and foliage that develop on partially girdled branches are markedly dwarfed. Cankers occasionally are found around buds, suggesting infection through leaf scars (English 1951).

Causal organism. The causal fungus is *Phomopsis cinerascens* Trav. (Grove 1935). Synonyms are *Phoma cinerascens* Sacc. (Saccardo 1895) and *Phoma ficus* Cast. (Smith 1941). The perfect stage of this fungus has been reported as *Diaporthe cinerascens* Sacc. (Grove 1935).

The pycnidia are gregarious, immersed in the bark, globose-depressed, 250 to 100 μm in diameter, blackish, and at length emerging by the ostiole. The α-spores are elliptic-fusoid, 6 to 6 by 2 to 2.5 μm, somewhat obtuse at one or both ends, often biguttulate, and sometimes exuding as a pallid globule. The condiophores are crowded, filiform-subulate, 15 to 20 by 1.5 to 2 μm, nearly straight, and faintly colored at their base. The β-spores are filiform, 20 to 25 by 1 μm, mostly hooked, on short pedicels, with occasional α-spores intermixed (Grove 1935).

Disease development. The fungus survives from one year to the next in cankers on the trees, or on infected branches cut from the tree and left in the orchard (Hansen 1949a). Pycnidiospores are first produced on the cankers during wet spells the winter after infection (English 1951; Livshits and Pupysheva 1949). Splashing rains are probably the chief means of pycnidiospore dispersal, but pruning tools undoubtedly play a part in disseminating the spores from one cut to another and from one tree to another (Hansen 1949a). Insects (bark beetles) and birds may also help spread the spores. The spores are produced in a sticky mass that becomes firm and adherent when dry, so the wind probably does not play an appreciable part in their dissemination (English 1951; Livshits and Pupysheva 1949).

New pruning cuts, given proper moisture and temperature, are susceptible to infection. Inoculation tests have shown Kadota trees to be practically immune to infection from April 1 through the growing season, but from around November 1 until at least February they are highly susceptible (English 1951, 1952a). The fungus advances most rapidly on the upper side of the branch.

Wounds and bark killed by frost and sunburn are the chief areas of infection (English 1951; Hansen 1949b; Livshits and Pupysheva 1949). Leaf-scar infections are of minor importance (English 1951).

External symptoms on pruned stubs appear a year after infection, although internal symptoms are seen much earlier. Artificial inoculations of the bark during the dormant season produce visible cankers within a few months.

In culture, growth of the fungus is most rapid at 25°C. Growth is relatively slow at 4.4° and at or above 30°C.

The failure of rapid canker extension between April and October apparently results from the active growth of the host.

Control. Pruning late in the dormant season (late March to early April) has usually resulted in a marked reduction in disease, but is often undesirable because it delays fruit maturity (English 1953, 1958). Removing cankered branches and chiseling away smaller cankers late in the dormant season is beneficial. Trees with many large cankers on the scaffold branches should be cut off just above the ground and new trees developed from the stumps (English 1952a; Hansen 1949b). Pruning tools used in these operations should be sterilized with 5.25 percent sodium hypochlorite (chlorine bleach) diluted 1:9 with water (Pilgrim 1950).

In estabishing a new orchard, obtain planting stock from an area, such as Tulare County, where the disease is extremely rare. Trees should be planted on good soils and kept in a vigorous condition because such trees appear to be the least susceptible to the disease (English 1960). Prunings from diseased trees should be removed from the orchard and burned (Hansen 1949a).

Studies by English (1951, 1952b, 1953, 1960) indicated that fungicides at economically feasible concentrations do not provide satisfactory control. A single organic mercury spray applied within 24 to 48 hours after pruning Kadota trees provided better control than did any other spray material (English 1953; Gilmer 1957). Plastic caps gave effective control, but an elastic film applied as a liquid did not afford protection (English 1960). Mercury fungicide prevented sporulation in some years (English 1952b).

No control is effected by wound dressings, sodium pentachlorophenoxide, liquid lime sulfur, dinitrocresol, dodine, or metiram. Bordeaux mixture spray, recommended by Hansen (1949a), has also proved to be ineffective (English 1953, 1958, 1960). Because of the current low incidence of the disease, control measures are not required at this time.

REFERENCES

English, H. 1951. Phomopsis canker: A progress report. *Proc. Fifth Ann. Res. Conf. Calif. Fig Inst.*:45–48.

———. 1952a. Phomopsis canker of figs (abs.). *Phytopathology* 42:513.

———. 1952b. Pruning and spraying experiments for Phomopsis canker. *Proc. Sixth Ann. Res. Conf. Calif. Fig Inst.*:16–19.

———. 1953. Further work on the control of Phomopsis canker. *Proc. Seventh Ann. Res. Conf. Calif. Fig Inst.*:12–15.

———. 1958. Physical and chemical methods of reducing Phomopsis canker infection in Kadota fig trees (abs.). *Phytopathology* 48:392.

———. 1960. Experiments on the control of Phomopsis canker in Kadota fig trees. *Proc. 14th Ann. Res. Conf. Calif. Fig Inst.*:13–15.

Gilmer, R. 1957. Phomopsis canker on Kadota figs. *Proc. 11th Ann. Res. Conf. Calif. Fig Inst.*: 26–28.

Grove, W. B. 1935. In *British stem-and-leaf fungi*, vol. 1:186–87. Cambridge University Press.

Hansen, H. N. 1949a. Phomopsis canker of fig. *Calif. Agric.* 3(11):13–14.

———. 1949b. Control and cure of Phomopsis canker of fig. *Calif. Fruit Grape Grower* 3(9):7.

———. 1949c. Canker diseases in figs. *Proc. Third Ann. Res. Conf. Calif. Fig Inst.*:18–19.

Livshits, I. Z., and L. I. Pupysheva. 1949. Fig canker and measures of its control. *Sad i Ogorod* (3):25–27.

Pilgrim, A. J. 1950. Phomopsis canker of Kadotas. *Proc. Fourth Ann. Res. Conf. Calif. Fig Inst.*: 30–32.

Saccardo, P. A. 1895. *Sylloge Fungorum* 11:486.

Smith, R. E. 1941. *Diseases of fruits and nuts.* Calif. Agric. Ext. Serv. Circ. 120, 168 pp.

Warner, R. M. 1952. Injection of fig trees for Phomopsis canker. *Proc. Sixth Ann. Res. Conf. Calif. Fig Inst.*:20–21.

Smut and Mold of Fig

Aspergillus niger van Tiegh. and
Other Fungus Species

"Fig smut," caused chiefly by *Aspergillus niger* van Tiegh., is an unfortunate choice of name because the name "smut" has been used historically for a group of cereal diseases caused by fungi unrelated to *Aspergillus*. The name fig "mold" refers to a number of internal rots other than endosepsis caused by such common facultative parasites as *Aspergillus, Botrytis, Cladosporium (Hormodendrum), Penicillium,* and *Rhizopus* (Buchanan, Sommer, and Fortlage 1975; (Hansen and Davey 1932; Phillips, Smith, and Smith 1925; Smith 1941). Both smut and mold are troublesome in the principal fig-growing districts of California. These fungi occur both in the edible fig and the caprifig, and probably occur worldwide, though foreign literature is vague on this point. Edgerton (1911) reported a soft rot caused by *R. stolonifer* to be common in Louisiana; the same fungus causes a storage rot of fig in India (Wani and Thirumalachar 1973). In Antibes, decay losses are caused by *Botrytis cinerea, Alternaria alternata,* and species of *Penicillium, Mucor,* and *Rhizopus* (Ricci 1972) **(fig. 4B)**.

Symptoms. Smut is distinguishable by the presence of black powdery masses of spores of *A. niger* in the pulp of the fruit cavity (**fig. 4A**). With mold, the pulp is discolored dirty gray, greenish, or yellowish, depending on the species of fungus involved.

Disease development. The causal fungi are present in the soil and decaying organic matter. They are also present in affected caprifigs. Phillips, Smith, and Smith (1925) and Smith and Hansen (1931) blamed *Drosophila ampelophila* and the dried-fruit beetle, *Carpophilus hemipterus*, for carrying the spores into the cavities of the edible fig. According to Hansen (1929) and Hansen and Davey (1932), thrips and predaceous mites carry spores of organisms capable of producing spoilage. These small arthropods enter the green fig long before the scales around the orifice of the fruit loosen sufficiently to permit entry of larger types. The time at which the interior of the fig fruit becomes contaminated by fungus spores corresponds closely with the time of infestation by thrips and mites.

Aspergillus niger and *Botrytis cinerea* may attack fig fruit through external injuries. Subsequent spread of the mold, especially with *Botrytis*, may occur when fruit are in contact.

Control. Reduction of the insect and mite population in fig orchards is helpful in preventing smut and mold. It is advisable to practice clean cultivation to destroy the weed hosts of the thrips and mites, and to remove fallen figs from the ground to eliminate the source of fungi, the fruit fly, and the dried-fruit beetle. A recent study (Obenauf et al. 1982) has shown that molds in caprifigs can be reduced but not effectively controlled by the combined use of certain fungicides.

REFERENCES

Buchanan, J. R., N. E. Sommer, and R. J. Fortlage. 1975. *Aspergillus flavus* infection and aflatoxin production in fig fruits. *Appl. Microbiol.* 30:238–41.

Edgerton, C. W. 1911. Diseases of the fig tree and fruit. *Louisiana Agric. Exp. Stn. Bull.* 126, 20 pp.

Hansen, H. N. 1929. Thrips as carriers of fig-decaying organisms. *Science* 69:356–57.

Hansen, H. N., and A. E. Davey. 1932. Transmission of smut and molds of fig. *Phytopathology* 22:247–52.

Obenauf, G. L., J. M. Ogawa, K. Lee, and C. A. Frate. 1982. Fungicide control of molds that attack caprifigs. *Plant Dis.* 66:566–67.

Phillips, E. H., R. E. Smith, and E. H. Smith. 1925. Fig smut. *Univ. Calif. Agric. Exp. Stn. Bull.* 387, 38 pp.

Ricci, P. 1972. Observations sur la pourriture des figues fraîches après la récolte. *Ann. Phytopathol.* 4:109–117. (*Rev. Plant Pathol.* 51:4221. 1972).

Smith, R. E. 1941. *Diseases of fruit and nuts.* Calif. Agric. Ext. Circ. 120, 168 pp.

Smith, R. E., and H. N. Hansen. 1931. Fruit Spoilage diseases of figs. *Univ. Calif. Agric. Exp. Stn. Bull.* 506, 84 pp.

Wani, D. D., and M. H. Thirumalachar. 1973. Control of anthracnose disease of figs by fungicides and antifungal antibiotic aureofungin. *Hindustan Antibiotics Bull.* 15(3):79–80. (*Rev. Plant Pathol.* 53:1030. 1974).

Souring of Fig

Yeast Fungi

Souring or fermentation of the flesh of the fig fruit, a widespread disease of long standing (Caldis 1930; Haring 1922; Howard 1901), occurs in all three types of fig. It is caused by various yeasts.

Symptoms. First symptoms are noticeable when the fruit begin to ripen and the "eye" is open. The inner flesh (pulp) of the fruit first develops a pink color, and later becomes water-soaked. A pink liquid exudes from the eye. Gas bubbles detectable in the pulp give off a strong alcohol odor. In late stages of the disease a scum of the yeast organisms collects in the watery pulp. Affected fruit lose their firmness and later dry up on the twig (**fig. 5**).

Disease development. The dried-fruit beetle (*Carpophilus hemipterus*) and fruit flies (*Drosophila* spp.) are responsible for disseminating the yeast cells from fruit to fruit. These insects are attracted to fermenting fruit, and since they are small enough to enter the eye of the fruit after the eye scales are shed, they can carry the yeast into the fruit.

Control. Little can be done to control the disease, but recent research (Obenauf et al. 1982) suggests that its incidence in Smyrna-type figs could be reduced but not effectively controlled through the use of certain fungicides.

REFERENCES

Caldis, P. .D. 1930. Souring of figs by yeasts and the transmission of the disease by insects. *J. Agric. Res.* 40:1031–51.

Haring, C. M. 1922. Splitting and souring of Smyrna figs. *Calif. Univ. Agric. Exp. Stn. Ann. Rept.* 1920–21:80–81.

Howard, L. O. 1901. Smyrna fig culture in the United States. *U. S. Dept. Agric. Yearbook* 1900:79–106.

Obenauf, G. L., J. M. Ogawa, K. Lee, and C. A. Frate. 1982. Fungicide control of molds that attack caprifigs. *Plant Dis.* 66:566–67.

Surface Mold and Alternaria Rot of Fig

Cladosporium herbarum (Pers.) Lindt.,
Alternaria alternata (Fr.) Keissl.

"Surface mold" and "Alternaria rot" are good descriptive terms for fruit infections by *Cladosporium* and *Alternaria*, respectively. Other terms used are spotting, smudge, mildew, and black spot (Brooks and McColloch 1938; English 1954; Smith and Hansen 1931). In the handbook published in 1971 (Smooth, Houck, and Johnson 1971), the use of the name "Cladosporium spot" is confusing because the term "spot" has been used to describe fruit infections caused by *Alternaria*.

All fig cultivars are susceptible to both pathogens, but the disease is most severe on Kadota. In California, the moldy condition on Kadota fruit was first brought to the attention of researchers in the fall of 1951 with hopes that they would develop control measures (English 1953). The loss of fruit from Alternaria rot relates directly to the weather during fruit harvest in August and September—rains at this time cause moldy fruit and severe splitting of fruit.

Symptoms and pathogens. The surface mold, caused by *Cladosporium herbarum* (Pers.) Lindt. (Brooks and McColloch 1938), develops on both immature and mature fruit and is largely restricted to the upper and outer surface of the fruit (English 1954). The disease first appears as dark olive-green specks, and as the lesions enlarge they become slightly depressed and turn into a yellowish olive smudge. Conidiophores arise in tufts, erect or nearly so, from nodular masses of cells embedded in the epidermis; they are septate and brownish olive in color and are 3 to 4 μm in diameter. The catenulate conidia range from cylindrical to subfusiform, elliptical, ovoid, or subglobose, and may become uni- or multi-septate. The conidia measure 3 to 7.2 μm long by 2 to 5.6 μm wide. They are at first hyaline to pale greenish yellow, and eventually olivaceous.

Alternaria rot is first evident as water-soaked areas that soon become slightly depressed and dark as the olivaceous spores are produced (**fig. 4B,6**). The organism penetrates the skin and the mycelium is found in abundance several cells deep, with scattered hyphae much deeper in the tissue. The conidiophores are 25 to 250 μm long and 3 to 5 μm in diameter, and sometimes moderately echinulate. The conidia are produced in long, sometimes branched chains. They are obclavate to pyriform and muriform, echinulate, at first pale to amber-yellow but later olivaceous to fuscous, usually with short beaks, longitudinally and transversely septate, and measure 20 to 75 μm long by 8 to 17 μm wide. Mature conidia finally become distorted and swollen, and septa are formed at various angles. The fungus was originally identified as *Alternaria tenuis* Nees (Brooks and McColloch 1938), a binomial that has since been changed to *A. alternata* (Fr.) Keissl.

Disease development. The causal fungi are commonly found in nature affecting other crops and on dying or dead plant tissue of various kinds. On figs, *Cladosporium* and *Alternaria* can be found independently or together on the fruit surface, with *Cladosporium* usually predominating; the only active decay or rot is produced by *Alternaria* (English 1953). *Cladosporium* is found almost entirely on low-vigor trees producing subnormal vegetative growth (English 1954); *Alternaria* is found more often in trees having dense foliage (Smith and Hansen 1931). *Cladosporium* occurs on fruit at all stages of maturity, while *Alternaria* is serious only on ripe fruit (English 1954). Because spores of both fungi require high humidity or free moisture for germination and infection, the diseases in California usually do not develop until late August, becoming progressively worse in September. Water droplets are detrimental to production of appresoria by *Alternaria* germ tubes, while relative humidities above 90 percent favor both germination and infection of tomato fruit (Koepsell 1968).

Control. Field infection by *Cladosporium* is reduced by spray applications of zinc coposil and zineb in the first week of September. Alternaria is controlled with two or three applications of zineb during the same period (English 1954). Clorothalonil sprays are also effective in controlling the rot caused by this fungus (Bewaji, English, and Schick 1977). Alternaria rot is also reduced by picking the fruit before it becomes overripe. On figs picked for immediate marketing, maneb and captan field sprays effectively reduce decay for at least two days at room temperature following storage at 2°C (Harvey 1957). Sanitation of the picking buckets and field boxes and fungicidal dips also reduce decay during storage. A modified atmosphere containing 23 percent or more carbon dioxide during storage and transit is as effective in controlling Alternaria rot as immediate storage at 0°C (Brooks and McColloch 1938).

REFERENCES

Bewaji, O., H. English, and F. J. Schick. 1977. Control of Alternaria surface rot of Kadota figs. *Plant Dis. Rep.* 61:351–55.

Brooks, C., and L. P. McColloch. 1938. Spotting of figs on the market. *J. Agric. Res.* 56:473–88.

English, W. H. 1953. Sprays for reducing surface mold and rot of Kadota figs. *Proc. Seventh Ann. Res. Conf. Calif. Fig Inst.*:17–20.

———. 1954. Further experiments on the control of surface mold and rot of Kadota figs. *Proc. Eighth Ann. Res. Conf. Calif. Fig Inst.*:16–20.

Harvey, J. M. 1957. The effect of handling practices and fungicide treatments on market decay losses in fresh figs. *Proc. 11th Ann. Res. Conf. Calif. Fig Inst.*:23–26.

Koepsell, P. A. 1968. The etiology of *Geotrichum* and *Alternaria* molds and their effects on mechanical harvesting of tomato. Ph.D. dissertation, University of California, Davis.

Smith, R. E. 1941. *Diseases of fruits and nuts.* Calif. Agric. Ext. Serv. Circ. 120, 168 pp.

Smith, R. E., and H. N. Hansen. 1931. Fruit spoilage diseases of figs. *Univ. Calif. Agric. Exp. Stn. Bull.* 506:49–51.

Smooth, J. J., L. G. Houck, and H. B. Johnson. 1971. Market diseases of citrus and other subtropical fruits. *USDA Agric. Handbook* 398:63–66.

Virus and Viruslike Diseases of Fig

Fig Mosaic

Graft- and Mite-Transmissible Pathogen

Mosaic disease is common on caprifig and edible fig trees in California and in most countries where figs are cultivated, including England, China, India, Hungary, and Australia, and several countries in the Mediterranean region (Blodgett and Gomec 1967; Harvey 1963). On long-established trees in the eastern United States, mosaic symptoms were not observed, but they did occur on fig plants imported from California (Condit and Horne 1941).

The host range of the pathogen embraces many species of *Ficus* including *F. carica* L. (the edible fig), *F. stipulata* Thunb., *F. altissima* Blume, and *F. krisna* DC. (Burnett 1962; Cooper 1979). Other moraceous hosts include mulberry (*Morus indica*) and *Cudrania tricuspidata*. Condit and Horne (1933, 1943) reported that of the edible fig cultivars Mission was highly susceptible and Turkey moderately so; Kadota, Calimyrna, and White San Pedro could become infected but, apparently, with little adverse effect. Among caprifig cultivars, Sampson was considered highly susceptible and Roeding No. 1 moderately resistant. An entire-leaf form of caprifig (*F. palmata*) was reported to be immune (Smith 1972).

Economic losses resulting from fig mosaic have been difficult to assess. Even so, the leaves on certain cultivars may be severely malformed and may drop prematurely; there may also be a spotting and shedding of immature fruit. This suggests that fruit yield is likely to be reduced (Condit 1941; Condit and Horne 1933, 1943).

Symptoms. The disease is manifested on the leaves and fruit. On the leaves, the mosaic spots are characterized by a pale green to light yellow color (**fig. 7**), in contrast to the normal green of the healthy foliage. The borders of the mosaic spots usually are not sharply delineated, and the light yellow color blends gradually into the green tissue. These spots may be small, of more or less uniform size, sometimes vein-delimited, and densely scattered over the leaf surface. In other cases, they are of various sizes with an indefinite outline, or may occur as irregular patches of light green scattered widely in the leaf with no relation to the veins. Often these spots are bordered by dead epidermal or subepidermal cells, which give them a rust-colored margin. A malformation of the foliage is frequent, with an infinite variety of leaf shapes and sizes. In some instances only certain twigs show the disease while others appear healthy (Condit and Horne 1933, 1943). Mosaic on the fruit is very similar to that on the leaves, and premature dropping of the fruit is attributed to the presence of mosaic in both the leaves and the fruit. Fruit of the profichi crop of the Sampson caprifig are often markedly distorted and tend to drop prematurely (Condit and Horne 1943).

Cause. The causal agent of the common miteborne fig mosaic has not been identified, but the symptoms suggest it may be a virus. The pathogen can be vectored by both the larval and adult stages of the mite *Aceria ficus* (Cotte) (Flock and Wallace 1955). The pathogen is also transmitted by grafting, but apparently is neither seedborne nor mechanically transmitted (Smith 1972). In contrast, a fig mosaic disorder in Italy is associated with a totally different agent, sowbane mosaic virus (SMV) (Quacquarelli 1971). Easily transmitted by sap inoculations, SMV is seedborne and is vectored by a variety of insects (Kado 1971) but not by mites. A recent report by Grbelja and Eric (1983) indicates the association of a potyvirus with a type of fig mosaic occurring in Yugoslavia. This filamentous (750 to 800 nm) virus was transmitted in a nonpersistent manner by aphids (*Myzus persicae*); it also could be mechanically transmitted to a few *Nicotiana* species.

Epidemiology. In California, the disease appears to be widespread, which is not surprising since it is vectored by a widely distributed eriophyid mite. This mite infests bud scales and young leaves and at times causes the abscission of leaves and stunting of twigs; on the Adriatic cultivar it can cause early defoliation of the terminal twigs. The pale yellow mites range from 0.005 to 0.008 inch long and have two pairs of legs. The feeding of mosaic-free mites on disease-free seedling plants induces leaf distortion and slight chlorosis and russeting, but no mosaic symptoms (Flock and Wallace 1955). These symptoms evidently represent a toxicogenic reaction of the plant.

In transmission tests, increasing the number of pathogen-containing mites per plant from 1 to 200 resulted in increases in mosaic symptom expression from 20 to 100 percent within 10 days and 70 to 100 percent within 90 days (Flock and Wallace 1955). All cuttings made by Flock and Wallace (1955) from diseased trees expressed mosaic symtoms. In previous studies by Condit and Horne (1933), 16 of 100 cuttings from four diseased trees showed no positive indications of mosaic and 10 had insufficient leaf growth for disease assessment. It appears that the pathogen is not always uniformly distributed in infected trees.

Extensive studies on pathogen-vector relationships were made by Proeseler (1969, 1972) in Hungary. He reported that the minimum feeding time for acquisition of the fig-mosaic pathogen was 5 minutes. Acquisition was most likely from terminal buds, followed by the lower surface of leaf sectors with symptoms, and then leaf petioles. The probability of pathogen uptake was twice as great at 20° or 30°C as at 5° or 10°C. The pathogen had a short circulation period of a few hours in the vector. It persisted in *A. ficus* for 6 to 10 days at room temperature and for 20 days at 5°C; pathogen multiplication in the vector was not proved. Transmission could be obtained with a single ineffective mite, and the minimum feeding time for transmission was less than 15 minutes. All developmental stages of the mite were transmissive, and the pathogen was retained during moulting. According to Oldfield (1970), the pathogen is not transmitted through the eggs.

Control. No control measures are known. However, the use of certified nursery stock is indicated, perhaps in conjunction with an eriophyid mite-control program. The pesticide aldicarb has shown some promise in preventing disease transmission (Proeseler 1972).

REFERENCES

Ainsworth, G. C. 1935. Fig mosaic. *J. R. Hort. Soc.* 60:522–23.

Blodgett, E. C., and B. Gomec. 1967. Fig mosaic. *Plant Dis. Rep.* 51:893–96.

Burnett, H. C. 1960. Species of *Ficus* susceptible to the fig mosaic virus. *Proc. Florida State Hort. Soc.* 73:316–20.

———. 1962. Additional hosts of the fig mosaic virus. *Plant Dis. Rep.* 46:693.

Condit, I. J. 1941. Fig culture in California. *Calif. Agric. Ext. Serv. Circ.* 77, 67 pp.

Condit, I. J., and W. T. Horne. 1933. A mosaic of the fig in California. *Phytopathology* 23:887–96.

———. 1941. Further notes on fig mosaic. *Phytopathology* 31:561–63.

———. 1943. Mosaic spots of fig fruits. *Phytopathology* 33:719–23.

Cooper, J. I. 1979. *Virus diseases of trees and shrubs.* Institute of Terrestrial Ecology, Unit of Invertebrate Virology, Oxford, 74 pp.

Flock, R. A., and J. M. Wallace. 1955. Transmission of fig mosaic by the eriophyid mite *Aceria ficus. Phytopathology* 45:52–54.

Grbelja, J., and Z. Eric. 1983. Isolation of a potyvirus from *Ficus carica* L. *Acta Bot. Croatia* 42:11–14. (*Rev. Plant Pathol.* 63:3476. 1984).

Harvey, H. L. 1963. Fig mosaic. *J. Agric. W. Austral.* 4:728–29.

Kado, C. I. 1971. Sowbane mosaic virus. *Commonw. Mycol. Inst./Assoc. Appl. Bact. Descrip. Plant Viruses* 64, 4 pp.

Oldfield, G. N. 1970. Mite transmission of plant viruses. *Ann. Rev. Entomol.* 15:343–80.

Proeseler, G. 1969. Zur Ubertragung des Feigenmosaikvirus durch die Gallmilbe *Aceria ficus* Cotte. *Zentbl. Bakt. Parasitkde Abt.* 2, 123(3):288–92. (*Rev. Plant Pathol.* 50:193. 1971).

———. 1972. Beziehungen zwischen Virus, Vektor und Wirtspflanze am Beispiel des Feigenmosaik-Virus *Aceria ficus* Cotte (Eriophyoidea). *Acta Phytopathol. Acad. Sci. Hung.* 7 (1/3):179–86. (*Rev. Plant Pathol.* 52:2698. 1973).

Quacquarelli, A. 1971. Il mosaico del fico el il virus latente del Chenopodium. [Fig mosaic and sowbane mosaic virus.] *Phytopathol. Medit.* 10:283–86.

Smith, K. M. 1972. *A textbook of plant virus diseases.* Academic Press, New York, 684 pp.

Smith, R. E. 1941. *Diseases of fruits and nuts.* Calif. Agric. Ext. Serv. Circ. 120, 168 pp.

MISCELLANEOUS DISEASES OF FIG

Diseases of minor importance on fig in California which are discussed more fully on other crops in this book are crown gall (*Agrobacterium tumefaciens* [E. F. Smith & Town.] Conn), shoestring root rot (*Armillaria mellea* [Vahl:Fr.] Kummer), and Cytospora canker (*Cytospora* sp.). Twig blight, caused by *Gibberella baccata* (Wallr.) Sacc. (anam. = *Fusarium lateritium* Nees) has also been reported.

Diseases not present in California but important in other states are thread blight, leaf blight, and rust, while those of minor importance are anthracnose, leaf spot, limb blight, and Tubercularia canker.

Thread blight, caused by *Pellicularia koleroga* Cooke, is a fungus disease occurring on many plant species throughout the wet tropical and subtropical regions of the world. It was first described in 1878 on coffee in India. Weber (1927) was first to report the disease on fig in Florida. It is a destructive disease of fig in states bordering the Gulf of Mexico, where diseased leaves abscise but are held to the twig by wefts of loosely interwoven "rhizomorphic" strands. These strands extend upward from sclerotia on the twigs and overrun the fruit, which wither and dry on the tree. Control measures involve the destruction or inactivation of the sclerotia (Tims and Mills 1943).

Leaf blight was first described by Matz (1917) in Florida, and later was found in Louisiana and other southeastern states. The disease symptoms are similar to those of thread blight, but the brown sclerotia are smaller and develop over the necrotic areas of infected fruit and twigs. The causal organism is *Corticium microsclerotium* G. F. Weber. The fungus is spread by windborne sclerotia and by windborne fragments of diseased leaves, with basidiospores playing only a minor role. Control measures include fungicide sprays to destroy the sclerotia and mycelium of the fungus on the tree to protect the healthy leaves from infection (Tims and Mills 1943).

Fig rust is common on various *Ficus* species in tropical and subtropical regions of the world. It is said to be troublesome on *F. carica* in the states along the Gulf of Mexico, often causing severe defoliation of the trees (Edgerton 1911; Lanham, Wyche, and Stansel 1927). It occurs rarely if at all in California. Symptoms of the disease include dying of the leaf blade from the margin inward and, eventually, defoliation. The rust pustules are numerous, small, raised, and light salmon-colored, and develop on the undersurface of the leaves. The causal organism is *Cerotelium* (*Physopella*) *fici* (E. J. Butler) Arth. The rust fungus produces only uredospores, which are ellipsoid or obovate-globoid, 14 to 23 by 18 to 32 µm, and pale yellow, and teliospores, which are broadly ellipsoid or oblong, 10 to 13 by 15 to 22 µm, colorless and produced in chains of 2 to 7.

Anthracnose is caused by *Glomerella cingulata* (Ston.) Spaulding & Schrenk and occurs in the fig-growing states that border the Gulf of Mexico. This fungus also causes bitter rot in apple and peach fruit. The disease occurs on leaves and fruit, forming grayish spots with dark brown margins that, in humid weather, are covered with pink fruiting structures (acervuli) of the fungus (Edgerton 1911; Matz 1918).

Leaf spots reported are Cephalosporium leaf spot, caused by *Acremonium zonatum* (Sawada) W. Gams (=*Cephalosporium fici* Tims & Olive), and Ormathodium leaf spot, caused by *Gonatophragmium mori* (Sawada) Deighton (=*Ormathodium fici* Tims & Olive).

Limb blight is caused by *Erythricium salmonicolor* (Berk. & Broome) Burdsall, and differs from thread blight and leaf blight in that the fungus invades and kills branches (Matz 1918).

Tubercularia canker is caused by *Tubercularia fici* Edgerton. Infection occurs in fruit scars, and the surrounding healthy tissue responds by forming a club-shaped enlargement which eventually sloughs off (Edgerton 1911).

REFERENCES

Edgerton, C. W. 1911. Diseases of the fig tree and fruit. *Louis. Agric. Exp. Stn. Bull.* 126. 20 pp.

Farr, D. F., G. F. Bills, G. P. Chamuris, and A. Y. Rossman. 1989. *Fungi on plants and plant products in the United States.* APS Press, American Phytopathological Society, St. Paul, MN. 1252 pp.

French, A. M. 1987. *California plant disease index. Part 1: Fruits and nuts.* California Department of Food and Agriculture, Sacramento, CA 95814. 39 pp.

Lanham, W. B., R. H. Wyche, and R. H. Stansel. 1927. Spraying for the control of leaf rust. *Tex. Agric. Exp. Stn. Circ.* 47. 8 pp.

Matz, J. 1917. A Rhizoctonia of the fig. *Phytopathology* 7:110–18.

Tims, E. C., and P. J. Mills. 1943. Corticium leaf blights of fig and their control. *Louis. Agric. Exp. Stn. Bull.* 367. 19 pp.

Weber, G. F. 1927. Thread blight, a fungus disease of plants caused by *Corticium stevensii* Burt. *Flor. Univ. Agric. Exp. Stn. (Gainesville) Bull.* 186:141–62.

Chapter 9

Minor Crops and Their Diseases

9

Page numbers in bold indicate photographs.

CHESTNUT AND ITS DISEASES	370	**438**
FEIJOA AND ITS DISEASES	370	**438**
FILBERT AND ITS DISEASES	371	
PECAN AND ITS DISEASES	371	
PERSIMMON AND ITS DISEASES	372	**438**
POMEGRANATE AND ITS DISEASES	372	**438**
REFERENCES	373	

Chestnut and Its Diseases

Chestnuts (*Castanea* spp.) belong in the family Fagaceae along with beech and oak. Approximately 135 acres are grown in California (Anonymous 1988). The sweet, rich flavor of the nuts makes them highly prized for roasting in the shell and for cooking.

Pathogens isolated from samples submitted for identification to the State Department of Food and Agriculture include *Armillaria mellea, Botryosphaeria dothidea, Dematophora necatrix, Coryneum castanicola, C. kunzei, C. pustulatum, Cryphonectria parasitica, Diaporthe eres, Hendersonula toruloidea, Laetiporus sulphureus, Marssonina ochroleuca, Agrobacterium tumefaciens,* and *Phoradendron flavescens* var. *macrophyllum* (French 1987). Also, a kernel mold condition involving *Alternaria* sp. and several unidentified fungi has recently been encountered (**fig. 1**).

Chestnut blight, caused by *Cryphonectria (Endothia) parasitica* (Murr.) Barr., is most destructive in the eastern United States and has reduced the U.S. chestnut culture to a position of minor importance. The disease was first discovered in New York in 1904, and within 30 years every stand of chestnut trees had been infected or killed. Genetic resistance and biocontrol measures offer some promise for reducing losses from this disease (Kuhlman 1983). Although the American cultivars are highly susceptible, the edible Asiatic chestnuts (*C. japonica* and *C. mollissima*) are resistant. Details on this disease, which has been reported from California (U.S. Department of Agriculture 1960), are found in the publication by Hepting (1971).

Feijoa and Its Diseases

The feijoa (*Feijoa sellowiana* Berg), often called pineapple guava, is a member of the myrtle family (Myrtaceae). It is a native of southern Brazil, western Paraguay, Uruguay, and parts of Argentina. The feijoa was introduced into Europe in the late 1800s and into New Zealand and the United States (California) in 1900. In California it has been grown largely as an ornamental. The feijoa has recently become more popular as a fruit crop in New Zealand, Japan, the southern United States, and California. It is also grown to some extent for this purpose in Greece, Spain, Israel, and the USSR. In New Zealand, research on the development of new, commercially acceptable fruit types was conducted by Hayward Wright for over 50 years and resulted in the release of the improved cultivars Triumph and Mammoth. These two cultivars are self-sterile, however, and cross-pollination is essential. Two new cultivars, Apollo (self-fertile) and Gemini (partially self-fertile), have recently been released in New Zealand. Orchard plantings of feijoa in California now total approximately 900 acres, with 500 acres in production.

The oblong fruit, about 5 cm long and dull green with a crimson blush, has a translucent, melting pulp and is harvested in October, November, and December. The hardy evergreen tree is currently grown under irrigation in the Sacramento and San Joaquin valleys. It produces its first commercial crop three years after planting, and reaches a full production of about 10 tons per acre in eight years. The tree requires a dry summer climate and is tolerant of cold temperatures ($-10°C$), but autumn frost could be hazardous to the fruit. The feijoa is usually propagated from seed, which comes fairly true to type; trees propagated by cuttings produce less root suckering. The trees are planted with a spacing of 12 to 15 by 15 to 20 feet, for a total of about 150 to 250 plants per acre (Anonymous 1985).

The feijoa is considered resistant to nematodes and partially resistant to Phytophthora root and crown rot. Fruit rots could be a problem as mature fruit have a tendency to fall from the tree and become injured.

The highly aromatic fruit could attract insects contaminated with fruit-rotting organisms, and some fruit with stylar end infections have been encountered (fig. 2). Fruit must be kept in a cool place until sufficiently soft for eating. The only recorded diseases of feijoa in California are shoestring root rot (*Armillaria mellea*), Phytophthora root and crown rot (*Phytophthora* sp.), Pythium root rot (*Pythium* sp.), and fruit rots caused by *Botrytis cinerea* Pers.:Fr., *Glomerella cingulata* (Ston.) Spauld. & Schrenk, and *Penicillium expansum* Lk. (French 1987).

Filbert and Its Diseases

Filbert is a type of hazelnut (*Corylus* spp.) belonging to the birch family, Betulaceae. There are only about 5 acres of filbert orchard in California. The major production areas in the United States are Oregon and Washington, where the most important cultivar is Barcellona. The first culture of *Corylus* species seems to have begun in Italy as early as 1671.

Fungus pathogens reported in California on filbert are *Armillaria mellea*, *Dothidea corylina*, *Gnomonia setacea*, *Phellinus ferreus*, *Physalospora obtusa*, and *Septoria corylina*. Powdery mildew (*Phyllactinia guttata*) also has been identified (French 1987). The dieback canker disease caused by *Apioporthe anomala* (Pk.) Hoehn (*Anisogramma anomala*) is a serious disease in the eastern United States (Barss 1930; U.S. Department of Agriculture 1960) and a few years ago was introduced into Washington (Davison and Davidson, Jr. 1973) and Oregon (Cameron 1976). The disease has caused heavy damage in Washington and continues to spread in Oregon with no effective control (Gottwald and Cameron 1980).

Ascopores, the only type of inoculum, are discharged for infection of galled buds infested by an eriophyid mite (*Phytocoptella avellanae*) or other injured tissue during the rainy period from the end of November to the beginning of April, with natural field infection at its highest incidence from February through May (Gottwald and Cameron 1980b). The long incubation period (12 to 16 months) for symptom expression as well as the perennial nature of the cankers make studies on epidemiology and control difficult (Gottwald and Cameron 1980a). Barcellona, the main commercial cultivar, is moderately susceptible while the pollenizer cultivars Daviana and DuChilly are highly susceptible because of greater mite infestation and bud galling (Gottwald and Cameron 1980). The disease has not been reported in California (French 1987).

Pecan and Its Diseases

The pecan, *Carya illinoensis* (Wang.) K. Koch, is native to the United States, with most production in the southern states from Oklahoma to Florida. California has about 2,550 acres of pecans, mostly in young trees, located principally in the southern San Joaquin Valley (Sibbett, Thompson, and Troiani 1987).

Crown gall, *Agrobacterium tumefaciens*, was reported by Smith (1941) to occur on pecan in California, but the trees are not highly susceptible to the disease. Smith also noted that pecan roots appear to be very resistant to *Armillaria mellea*. *Verticillium dahliae* and *Alternaria* sp. are other fungal pathogens reported in California (French 1987). The most important disease problem in California is little-leaf or rosette, caused by zinc deficiency (Chandler 1947). Scab, caused by *Cladosporium caryigenum* (Ell. et Lang.) Gottwald, attacks leaves, shoots, and nuts, and is the most important disease of pecan in the southern states. *Cladosporium caryigenum* has not been reported in California, but a possibly identical fungus under the name *Fusicladium* sp. recently was recorded (French 1987).

Persimmon and Its Diseases

The persimmon (*Diospyros* spp.) belongs in the family Ebenaceae. The most important species grown for its fruit is the Oriental persimmon or Kaki (*D. kaki* L.). The common American species, *D. virginiana* L., was brought to California by the first American settlers (trees of this species were planted by General Bidwell on Rancho Chico, and by 1889 were 30 to 40 feet tall). Until the Oriental or Japanese persimmon was introduced in 1870, however, there was little incentive for extensive planting. In the late 1870s the University of California Agricultural Experiment Station became interested in the oriental species. The first large order of 3,000 trees, comprising many cultivars, was received in San Francisco in 1880.

The first Oriental cultivars grown in California were Hachiya, Tanenashi (seedless), and Yemon. Later came the Fuyu, which is nonastringent. Private importation ceased in 1919 because of quarantine laws. In 1910 there were 3,274 bearing trees. By 1930 there were more than 98,000 trees in bearing and more than 96,000 nonbearing in the southern California counties of San Bernardino, Los Angeles, Orange, and San Diego and the northern counties of Placer and Butte (Butterfield 1937–38). In 1982, there were 770 acres of persimmon trees in the state. The cultivars currently grown in California are discussed by Ryugo et al. (1988).

The fruit is harvested for the fresh market or for drying from late September to early December. The most important cultivars, such as Hachiya, Tanenashi, and Fuyu, bear only pistillate flowers and have a strong tendency to set fruit parthenocarpically. To reduce astringency, Hachiya fruit are treated with ethylene or alcohol. In Japan, Hachiya are kept in cold rooms in an atmosphere high in carbon dioxide; fruit so stored will keep for several months and become nonastringent (Chandler 1947).

The pathogens reported in California on persimmon are *Agrobacterium tumefaciens*, *Phytophthora* sp., *Armillaria mellea*, *Botrytis cinerea* (**fig. 3**), *Cladosporium* sp., *Phoma diospyri*, and *Viscum album* (French 1987; Smith 1913; U.S. Department of Agriculture 1960). In 1933, the fungus pathogen *Cephalosporium diospyri* Crandall was discovered on American persimmon in Tennessee. Since then, the wilt disease it causes has spread in the southern states and, as a result, introduction of the American persimmon into California is prohibited (P. W. Hiatt and S. M. Gotan, personal communication). The Oriental persimmon, however, appears to be resistant to this fungus. An abiotic disorder of considerable concern is sunburn of the fruit (**fig. 4**).

Pomegranate and Its Diseases

The pomegranate, *Punica granatum* L., native to Persia, is one of the oldest edible fruits known. It was introduced into California by the first Spanish settlers when Mission San Diego was founded in 1769. In 1859 there were 3,149 trees in California. According to the U.S. census for 1910, there were 1,771 bearing and 2,745 nonbearing trees in the state, and by 1930 the number of bearing trees had increased to 110,518. The bulk of the acreage was and is in Tulare and Fresno counties, where in 1982 there were 4,160 acres. Fruit was shipped from the Sackett orchard on Putah Creek, Solano County, before 1889. Of the cultivars, Spanish Ruby was the early favorite, and Sweet Fruited was later tested. Wonderful has proved to be the cultivar best suited for California (Butterfield 1937–38).

The crop requires plenty of summer heat, yet can withstand low temperatures of $-9.4°$ to $-12.2°C$. The fruit usually is harvested from mid-September to early November. Botanically a berry, the fruit has a thick leathery rind and therefore is inconvenient to eat and considered by some to taste insipid. When ripe, the fruit tend to crack, producing an avenue for infection by rot organisms.

Smut disease of the fruit (**fig. 5**), caused by *Aspergillus niger* van Tiegh., has been reported in California (Smith 1941). It was the only disease of the pomegranate fruit seen by an American visitor in southern Iran in 1974. Also reported on cracked fruit are *Alternaria* sp., *Botrytis cinerea*, *Nematospora coryli*, and *Penicillium expansum* (U.S. Department of Agriculture 1960) (**fig. 5**). The best way to prevent infection by these fungi is to harvest the fruit as soon as it is ripe. Other pathogens reported on pomegranate in California are *Agrobacterium tumefaciens*, *Armillaria mellea*, *Fusarium solani*, and *Rhizoctonia solani* (French 1987). Also, the senior author has recently identified the powdery mildew fungus *Sphaerotheca pannosa* on this host (**fig. 6**).

REFERENCES

Anonymous. 1985. Feijoa. *Calif. Fruit Grower* 61:24.

———. 1988. Chestnuts as an alternative crop. *Fruit Findings*, Jan. 1988, 6 pp.

Barss, H. P. 1930. Eastern filbert blight. *Calif. Agric. Dep. Bull.* 19:489–90.

Butterfield, H. M. 1937–38. History of deciduous fruits in California. *Blue Anchor*. vols. 14 and 15, 38 pp.

Cameron, H. R. 1976. Eastern filbert blight established in the pacific northwest. *Plant Dis. Rep.* 60:737–40.

Chandler, W. H. 1947. *Deciduous orchards*. Lee & Feibiger, Philadelphia, 437 pp.

Davison, A. D., and R. M. Davidson, Jr. 1973. Apioporthe and Monochaetia cankers reported in western Washington. *Plant Dis. Rep.* 57:522–23.

French, A. M. 1987. California plant disease host index. Part 1: Fruits and nuts. *Calif. Dept. Food Agric. Div. Plant Ind.*, 39 pp.

Gottwald, T. R., and H. R. Cameron. 1980a. Infection site, infection period, and latent period of canker caused by *Anisogramma anomala* in European filbert. *Phytopathology* 70:1083–87.

———. 1980b. Disease increase and the dynamics of spread of canker caused by *Anisogramma anomala* in European filbert in the pacific northwest. *Phytopathology* 70:1087–92.

Griffin, G. J., F. V. Hebard, P. W. Wendt, and J. R. Elkins. 1983. Survival of American chestnut trees: Evaluation of blight resistance and virulence in *Endothia parasitica*. *Phytopathology* 73:1084–92.

Hepting, G. H. 1971. *Diseases of forest and shade trees of the United States*. (USDA Handbook 386) U.S. Government Printing Office, Washington, D.C., 658 pp.

Kuhlman, E. G. 1983. Effects of hypovirulence in *Cryphonectria parasitica* and of secondary blight infections on dieback of American chestnut trees. *Phytopathology* 73:1030–34.

LaRue, J. H. 1980. Growing pomegranates in California. *Univ. Calif. Div. Agric. Sci. Leaf.* 2459. 8 pp.

Maranto, J. 1987. Feijoa—A new fruit in Kern County. *Calif.-Ariz. Farm Press* 9(24):15, 23.

Ryugo, K., C. A. Schroeder, A. Sugiura, and K. Yonemori. 1988. Persimmons for California. *Calif. Agric.* 42(4):7–9.

Sibbett, G. S., T. E. Thompson, and N. Troiani. 1987. New pecans for California. *Calif. Agric.* 41(11/12):27–30.

Smith, C. O. 1913. Some successful inoculations with the peach crown gall organism and certain observations upon retarded gall formation. *Phytopathology* 3:59–60.

Smith, R. E. 1941. *Diseases of fruits and nuts*. Calif. Agric. Ext. Serv. Circ. 120, 168 pp.

U.S. Department of Agriculture. 1960. *Index of plant diseases in the United States*. (USDA Agric. Handbook 165) U.S. Government Printing Office, Washington, D.C., 531 pp.

Chapter 10

Pesticides

| How Fungicides Protect Plants | 378 |

| Pesticide Groups Used in Disease Control | 379 |

Inorganic Fungicides and Bactericides	379
Organic Fungicides	379
Fungicides for Postharvest Diseases	380
Other Fungicides	380
Antibiotics	380

Resistance of Plant Pathogens to Pesticides	381
REFERENCES	382

Natural and synthetic pesticides are extremely important in controlling diseases of fruit and nut crops. When properly applied, chemical treatments are effective, economical, and safe. They play an integral part in disease management programs, from crop production to postharvest marketing. The treatment strategies for each formulation are based on knowledge of the host-parasite relationships occurring within the existing environmental parameters, as indicated in this test for specific crop-disease combinations. At present, pesticides are essential in controlling most of the major diseases of fruit and nut crops in California, and in many instances they provide the only means of preventing crop losses.

Fungicides—chemicals that show activity against fungi—are useful in the control of many plant diseases. Other pesticides such as bactericides and nematicides are used to manage diseases caused, respectively, by bacteria and nematodes. Also, a few antibiotic compounds are now available for controlling diseases caused by bacteria, fungi, and mycoplasmalike organisms. There are no known "viricides" that can be used directly to control virus diseases, but the vectors of these diseases are often controlled through the use of nematicides or insecticides. Since fungicides constitute by far the most important group of pesticides used for disease control in fruit and nut crops, this discussion will be largely restricted to that group of chemicals.

How Fungicides Protect Plants

Most fungicides currently used to control plant diseases are considered "fungistats" because they do not actually kill the fungus. Examples are ziram and several copper compounds that are used to control the shothole disease (*Wilsonomyces*) of stone fruit. Here the presence of the chemical on the plant surface prevents spore germination or the further growth of germ tubes, thereby protecting the host from infection. Some plant pathologists call fungicides of this type "contact fungicides" and separate them from "systemic fungicides," which penetrate and move more or less systemically in the plant. These terms appear to have been borrowed from the entomologists, whose "contact insecticides" kill or suppress insects upon contact and whose "systemic insecticides" move within the plant and have an effect on insects that feed upon its tissue. For plant pathogens, there appears to be no clear separation between the two terms, except that in one case the chemical moves within the plant tissue where it comes in contact with the pathogen. With both contact and systemic materials, the plant is being protected from infection or colonization. Both types can therefore be called protectant fungicides.

Fungicides with systemicity in plant tissue are few in number but play an important role in disease control. For postharvest decay control, DCNA and iprodione penetrate into fruit tissue and suppress established infections of both *Rhizopus* and *Monilinia*. Control of brown-rot blossom blight requires multiple applications of nonsystemic fungicides (coppers, dithiocarbamates, captan), but the benzimidazole compounds have local systemic action, and an application when the blossoms are still closed protects the anthers and stigmas of open blossoms. An experimental fungicide, Stauffer Chemical Company's SC-0858, has been shown to have more systemicity in blossoms and fruit than the benzimidazoles. Other examples are oxycarboxin and oxytetracycline used as tree injections to control, respectively, silver leaf (*Chondrostereum*) and X-disease (caused by a mycoplasmalike organism) on peach trees.

Some fungicides (e.g., sodium hypochlorite) that have the ability to kill the pathogen can be used to disinfest the plant surface. A few fungicides are capable of killing the pathogen within the host plant (disinfection or eradication); examples are monocalcium arsenite and sodium pentachlorophenoxide, which have been used in the control of stone-fruit brown rot. However, because of health hazards, neither of these materials is currently available for use on these crops. The fungicides triforine and dodine can be applied to an apple tree within a few days after infection by the apple scab fungus and still control the scab disease. These materials show eradicative or suppressive action, sometimes called "kickback" action. An eradicant kills the pathogen, and a suppressant (fungistat) prevents pathogen growth as long as the toxicant is present in sufficient concentration.

In some cases, fungicides prevent fungal sporulation (genestatic effects), such as the action of benzimidazole fungicides on *Monilinia*, *Botrytis*, and *Penicillium*. Some of these materials are also known for their localized systemic activity and can reduce the rate of growth of the fungus within the plant (DCNA and benomyl on fresh stone fruit infected with *Monilinia*). More fungicides with both genestatic and systemic activity are needed in order to enable growers to achieve disease protection with fewer pesticide applications. Fungicides that kill the pathogen (eradicants) would be most useful for reducing the primary inoculum of the pathogen, and thereby reducing the chance of epidemics, especially with diseases for which secondary infection is important.

At times, two or more chemicals are combined in order to increase the toxic action against the pathogen, but currently combination treatments are used primarily to control more than one pathogen, as in the postharvest decay control of *Rhizopus*, *Monilinia*, and *Botrytis*; here DCNA is combined with triforine or iprodione. To control both *Monilinia* and *Wilsonomyces* in stone fruit, benomyl and ziram or captan are combined in sprays applied during the blossoming period. When using fungicide combinations or any other mixtures, growers or advisors should consult compatibility charts.

PESTICIDE GROUPS USED IN DISEASE CONTROL

Inorganic Fungicides and Bactericides

Sulfurs (pre-1882) and coppers (since 1882), although introduced more than a century ago, still play a major role in disease management. Sulfurs continue to be among the least expensive materials, and their formulations as dusts, wettable powders, and liquids (sodium and calcium polysulfides) are most effective in the control of powdery mildews and rusts on stone fruit. Also, the eradicative action of liquid lime sulfur has been shown on *Monilinia* infections on peach fruit and *Venturia* infections on pear twigs. Copper compounds are used to control bacterial diseases such as fireblight of pome fruit, bacterial canker of stone fruit, and walnut blight, and fungal diseases of stone fruit such as brown rot, shothole, and peach leaf curl. Copper formulations have long residual activity and are not hazardous to humans or the natural fauna and flora. The copper formulations have changed from the early developed Bordeaux mixture (copper sulfate, lime, and water) to the fixed coppers, which can be used without premixing. Some of the formulations on the market are basic copper sulfate, copper oxychloride sulfate, and copper hydroxide. Sulfur and copper formulations are generally regarded as safe for humans and the environment, except for the bisulfites and sulfur dioxide which have been used to control certain fruit diseases. Another inorganic is sodium hypochlorite; it has been used primarily to control postharvest diseases.

Organic Fungicides

The organic fungicides were introduced in the early 1930s, and currently number over 100 chemical compounds. The first groups to be introduced were the dithiocarbamates (e.g., ziram) and the ethylenebisdithiocarbamates (maneb, mancozeb, and nabam plus coordinated salts). The basic dithiocarbamates are active but unstable, but in coordination with metal salts they are highly active and have excellent residual qualities. Ziram is used extensively for the control of shothole on stone fruit, peach leaf curl, and scab of almond, while ethylenebisdithiocarbamate formulations are useful for control of rust diseases.

The quinone dichlone was introduced in 1946 and provided kickback action in the control of brown-rot blossom blight of stone fruit. A few years later, a mixture of dichlone and sulfur was introduced for the control of both *Monilinia* blossom blight and fruit rot of stone fruit. Dichlone has had only limited usage because it irritates the exposed skin of applicators and is occasionally phytotoxic.

The imidazolines—captan and its analog captafol—introduced in 1949, highlighted what was almost a new era in the control of plant diseases, and their use ranked next to that of the dithiocarbamates. They controlled a wide spectrum of stone-fruit diseases, including brown rot, shothole, leaf curl, scab, and gray mold (*Botrytis*). These compounds have limited residual half-lives when exposed to water or ultraviolet light and have low mammalian toxicities.

Introduced in the 1960s, the benzimidazoles (benomyl and thiophanate-methyl) appeared to be model toxicants that provided outstanding fungicidal action against a wide spectrum of pathogens and controlled such fruit diseases as brown rot, the powdery mildews, apple scab, and Botrytis blossom blight.

The required dosage was small (1 pound active ingredient per acre) compared to that of other fungicides (4 pounds of captan to 40 pounds of sulfur or copper sulfate in combination with lime); the chemicals were locally systemic and reduced fungal sporulation. The downfall of the benzimidazoles came with the emergence of resistant strains of pathogens under extensive and intensive usage of these compounds, and today their use is severely curtailed. Mixing the benzimidazoles with other fungicides having different modes of action is recommended by the manufacturers, but such mixtures have not prevented or delayed the development of resistant fungal strains of *Monilinia*.

More recently, the sterol biosynthesis inhibitors (SBIs) and dicarboximides have replaced the benzimidazoles for many uses. The SBI compounds—triazoles (bitertanol, folicur, triadimefon, propiconazole, etaconazole, penconazole, flusilazole, myclobutanil, and diniconazole), pyrimidines (fenarimol and nuarimol), piperazines (triforine), and imidazoles (imazalil, prochloraz, and triflumizole)—have similar modes of action, but present different activities toward plant pathogens. Imazalil and folicur are active against *Monilinia* and *Botrytis*, whereas triforine is not active against *Botrytis*. Of the SBI compounds, only triforine is currently (early 1988) registered for use on stone-fruit crops. The dicarboximides (iprodione and vinclozolin) appear to have broader activity than the SBIs and provide excellent control of such pathogens as *Monilinia* and *Botrytis*. Also, iprodione is active against *Rhizopus*, *Wilsonomyces*, *Alternaria*, and *Botryosphaeria*. Iprodione-resistant *Monilinia* has been detected in orchards in Australia but not in California as of early 1988.

Fungicides for Postharvest Diseases

Among the postharvest control chemicals, hypochlorous acid is a highly reactive and nonspecific oxidizing agent with no suspected residues; it is applied to fruit in dips, sprays, or drenches. In order to maintain essential levels of active hypochlorous acid, the acidity of the solutions is held near or slightly above pH 7. An acidic solution releases chlorine gas, and at a basic pH only a minimal amount of the active hypochlorous acid is present.

Sodium orthophenylphenate (SOPP) has been used extensively in dips, sprays, and fruit waxes. Low dosages of SOPP are ineffective, and high dosages can cause phytotoxicity.

The introduction of DCNA (dicloran, a dichloronitroanaline compound) made possible the control of Rhizopus rot caused by *Rhizopus stolonifer* and, in addition, some control of brown rot. At first DCNA was used in combination with captan, later with benomyl until benomyl-resistant strains were detected, and currently, to ensure effective control of *R. stolonifer* and benomyl-resistant *Monilinia* spp., DCNA is combined with triforine or iprodione. A recent study of fruit volatiles by Wilson, Franklin, and Otto (1987) showed that several of these naturally occurring compounds are inhibitory to spore germination or growth of *B. cinerea* and *M. fruticola*. Research on the possible application of these findings to postharvest disease control would appear to be justified.

Other Fungicides

Two materials that are being tested extensively for the control of soilborne diseases are metalaxyl and aluminum ethylphosphite. As soil fumigants, methyl bromide and carbon disulfide are used to control shoestring (Armillaria) root rot. For foliage diseases, dodine (a guanidine compound) serves to effectively control apple and pear scab.

Antibiotics

Antibiotics are compounds produced by microorganisms, mostly bacteria, that kill or inhibit other microorganisms. They have been useful in the control of several bacterial diseases, two diseases caused by mycoplasmalike organisms, and a few fungal diseases. One antibiotic, streptomycin, is a material excreted from a species of *Streptomyces*; it has been used to control such bacterial diseases as walnut blight and fireblight of pome fruit. Streptomycin-resistant strains of the fireblight pathogen have curtailed its use; currently, a mixture of streptomycin and oxytetracycline (another antibiotic) is being used in an effort to overcome this difficulty.

Oxytetracycline is active against both bacteria and mycoplasmalike organisms (MLOs) and, in addition to fireblight, has been used in the control of pear decline and X-disease of peach and sweet cherry. A solution of the compound is injected into the xylem either by gravity or with a high-pressure injector; it moves in this vascular tissue and comes into contact with the MLOs harbored in the phloem. No resistant MLO strains have been detected. An antifungal antibiotic, cycloheximide, is effective in control of cherry leaf spot (*Blumeriella*) and some of the powdery mildews. Another antibiotic is the recently identified antifungal compound pyrrolnitrin produced by *Pseudomonas cepacia* (Janisiewicz and Roitman 1988). This compound has been highly effective in preventing infection of apple and pear fruit by *Botrytis cinerea* and *Penicillium expansum*.

Resistance of Plant Pathogens to Pesticides

Pathogen resistance has become a major problem in the management of diseases of fruit and nut crops, with failures in benzimidazole control of brown rot in stone fruit, and in streptomycin control of fireblight in pome fruit. There also is evidence that pathogens are developing resistance to other fungicides that are highly specific in their mode of action, such as the dicarboximides and, to a lesser degree, the sterol-biosynthesis inhibitors (Köller and Scheinpflug 1987). Although many of the chemicals in the newer groups of fungicides vary in chemical structure, they often have similar modes of action. Because of this, selective strains of fungi are resistant to all benzimidazole formulations (benomyl, thiophanate-methyl, and carbendazim) or to the dicarboximides iprodione and vinclozolin, or DCNA. Multiple resistance between benzimidazoles and dicarboximides has been reported for *Botrytis*.

Critical studies are needed to develop strategies that delay the development of pathogen resistance resulting from repeated applications of closely related fungicides. In managing resistant strains, the baseline sensitivity of the pathogen to a new chemical should be established before the chemical is introduced for use on a commercial scale. Limiting the number of applications of a given chemical would prolong its useful life. For example, benomyl was used in multiple applications to control brown rot on peaches and resistant strains appeared four years after its first use; yet on prune, only a single yearly application is made, and resistant strains of *M. laxa* still have not developed after 15 years. Multiple applications of a single fungicide often result in the rapid selection of resistant strains. If combinations of fungicides with different modes of action are considered, all chemicals must be highly and equally effective in order to delay the development of resistance. If two highly effective chemicals are used, it might be less costly to alternate applications instead of using mixtures.

When postharvest decay control is of top priority, it may be more judicious to use a disease-control chemical only in a postharvest treatment. Such has been the case in controlling postharvest Penicillium decay of apples with benzimidazoles in orchards where scab has not been a major problem.

Unless more extensive efforts are undertaken to resolve the problem of pathogen resistance to pesticides, few or no effective chemical treatments will be available. Recently, a new fungicide was developed that is most effective against strains of *Botrytis* resistant to the benzimidazoles (Kato et al. 1983). Also, a chemical used as a wetting agent was found to effectively control benomyl-resistant *Penicillium*, but only those strains with high levels of resistance. These examples should provide encouragement for further investigations on delaying the selection of pesticide-resistant pathogens or reducing the existing levels of resistant strains.

Condensed information on representative groups of fungicides and bactericides has been presented by Ogawa, Cañez, and Walls (1987). In an extensive table, they give the origin of these chemicals, their common, trade, and chemical names, their mechanism of action, and their mammalian activity.

REFERENCES

Gruzdyev, G. S., V. A. Zinchencko, V. A. Kalinin, and R. I. Slovtsov. 1983. *The chemical protection of plants*. Mir Publishers, Moscow, 471 pp.

Craigmill, A. L. 1982. Toxicology, The science of poisons. *Univ. Calif. Div. Agric. Sci. Leaf.* 21221. 12 pages.

Ferguson, M. P., and H. G. Alford, eds. 1986. Microbial/biorational pesticide registration. *Univ. Calif. Div. Agric. Nat. Resour. Spec. Publ.* 3318. 53 pp.

Janisiewicz, W. J., and J. Roitman. 1988. Biological control of blue mold and gray mold on apple and pear with *Pseudomonas cepacia*. *Phytopathology* 78:1697–1700.

Kato, T., K. Suzuki, J. Takahashi, and K. Kamoshita. 1983. Negative cross resistance between benzimidazoles and methyl N-(3-5 dichlorophenyl)-carbamate (abs.). *Proc. Fourth Internat. Cong. Plant Pathol.*:17.

Köller, W., and H. Scheinpflug. 1987. Fungal resistance to sterol biosynthesis inhibitors: A new challenge. *Plant Dis.* 71:1066–74.

Marer, P. J., M. L. Flint, and M. W. Stimman. 1988. *The safe and effective use of pesticides*. Univ. Calif. Div. Agric. Nat. Resour. Publ. 3324. 387 pp.

Ogawa, J. M., V. M. Cañez, Jr., and K. M. Walls. 1987. Fungicides and bactericides used for management of plant diseases. In *Fate of pesticides in the environment*, 25–45, 149–55. Univ. Calif. Div. Agric. Nat. Res. Pub. 3320.

Ogawa, J. M., J. D. Gilpatrick, and L. Chiarappa. 1977. Review of plant pathogens resistant to fungicides and bactericides. *FAO Plant Protect. Bull.* 25:97–111.

Ogawa, J. M., W. D. Gubler, and B. T. Manji. 1988. Effect of sterol biosynthesis inhibitors on diseases of stone fruits and grapes in California. In *Sterol biosynthesis inhibitors*, ed. D. Berg and M. Plempel. Ellis Horwood Limited, Chichester, England. Chap. 9:262–87.

Ogawa, J. M., and B. T. Manji. 1984. Control of postharvest diseases by chemical and physical means. In *Postharvest pathology of fruits and vegetables: Postharvest losses in perishable crops*, 55–66. Univ. Calif. Agric. Exp. Stn. Bull. 1914.

Ogawa, J. M., B. T. Manji, J. E. Adaskaveg, and T. J. Michailides. 1988. Population dynamics of benzimidazole-resistant *Monilinia* species on stone fruit trees in California. In *Fungicide resistance in North America*, ed. by C. J. Delp. Chap. 13:36–39. The American Phytopathological Society, APS Press. 133 pp.

Wilson, C. L., J. D. Franklin, and B. E. Otto. 1987. Fruit volatiles inhibitory to *Monilinia fructicola* and *Botrytis cinerea*. *Plant Dis.* 71:316–19.

Chapter 2. Pome Fruit and Their Diseases

Fig. 1. Bacterial blast and canker of pear.
page 13

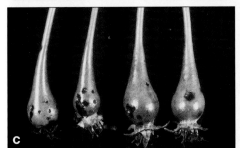
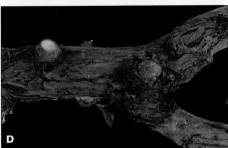

A. Blasted and healthy blossom clusters.

B. Affected twig. Note the papery bark condition.

C. Necrotic spotting on immature fruit.

D. Canker on branch.

Fig. 2. Fireblight of pear.
page 16

A. Young tree with blighted branches.

B. Blighted blossom clusters and healthy "rat-tail" bloom.

C. Blighted fruit showing bacterial ooze.

Fig. 3. Hairy root of apple.

page 25

Note the excessive development of fibrous roots.

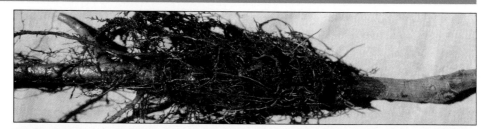

Fig. 4. Pear decline.

page 27

A. Reduced shoot growth of affected tree (center); adjacent trees are healthy.

B. Tree in late summer with advanced decline. Note the premature reddening of foliage. (Courtesy of G. Nyland.)

C. Bark removed to show brown line at the graft union of a domestic pear on Oriental pear rootstock. (Courtesy of W. J. Moller.)

D. Adult pear psylla, the vector of the pear decline pathogen. (Courtesy of Jack Kelly Clark and A. H. Purcell.)

E. Apple decline as observed in Washington state. (Courtesy of C. L. Parish.)

Fig. 5. Anthracnose and perennial canker of apple.

page 32

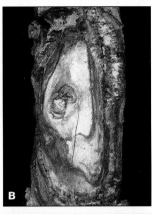

A. Branches with anthracnose cankers. (Courtesy of Mid-Columbia Experiment Station, Hood River, Oregon.)

B. Branch with perennial canker centered at pruning wound. (Courtesy of G. G. Grove.)

Fig. 6. Dematophora root rot of apple.

page 39

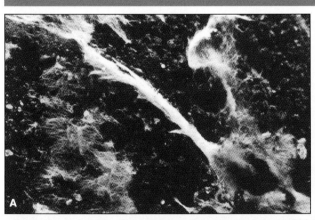

A. Typical white, wefty mycelium of the fungus in soil around rotted roots. (Courtesy of R. D. Raabe.)

B. White mycelial strands of the pathogen on decorticated wood of a rotted root. (Courtesy of J. E. Adaskaveg.)

Fig. 7. European (Nectria) canker of apple.

page 41

A. Shoot with small cankers centered at leaf scars.

B. Branch with canker centered at pruning wound.

C. Calyx-end rot of Red Delicious apple.

– continued

Fig. 7. European (Nectria) canker of apple — *continued*.

page 41

D. Golden Delicious apple with decay lesions at lenticels.

E. Sporodochia of the pathogen on killed bark.

F. Small red perithecia of Nectria on a canker.

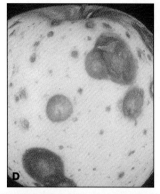

Fig. 8. Fabraea leaf spot on pear leaves.

page 46

Lesions on leaves (left) and fruit (right).

Fig. 9. Phytophthora root and crown rot of apple and pear.

page 48

A. Decline in young apple trees associated with poor soil drainage and Phytophthora infection.

B. Apple seedling roots (right) destroyed by Phytophthora.

C. Typical crown rot symptoms (with outer bark removed) at base of pear trunk.

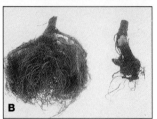

Fig. 10. Powdery mildew of apple and pear.

page 50

A. Terminal leaves of apple with mildew (left) and healthy terminal with fruit (right).

B. Mildew mycelium and sporulation on lower surface of apple leaf.

C. Typical fruit russeting on apple.

D. Young pear shoot and leaves with mildew.

Fig. 11. Pear trellis rust.

page 54

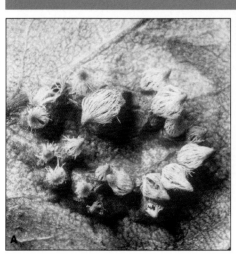
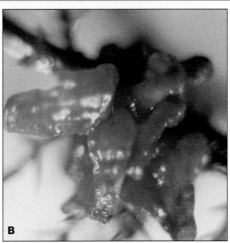

A. Aecial stage of rust fungus on lower leaf surface. (Courtesy of California Department of Food and Agriculture.)

B. Gelatinous telial horns on juniper. (Courtesy of Z. Punja.)

Fig. 12. Sappy bark of apple.

page 56

A. Dieback of scaffold branches and new scions, resulting infection of large cuts made during topworking operation.

B. Dieback of trunk from infection of large pruning wound. Note the papery bark condition.

C. Sporophores of *Trametes versicolor* on killed scaffold branch.

Fig. 13. Scab of apple.

page 58

A. Heavily infected young fruit.

B. Scab lesions on fruit and leaves.

C. Well-developed lesions on upper leaf surface.

D. Erumpent grayish black pseudothecia of the pathogen on a dead overwintered apple leaf. (Courtesy of M. Szkolnik.)

E. Green-tip stage of bud development—a critical time for the first fungicide application.

Pome Fruit and Their Diseases • 389

Fig. 14. Pear scab.
page 67

Fruit and twig lesions of pear scab.

Fig. 15. Southern blight of apple.
page 70

Note the white mycelium and small brown sclerotia on the necrotic base of a young tree. (Courtesy of W. Asai.)

Fig. 16. Apple flat limb.
page 75

Severe distortion of scaffold branches of an affected apple tree. (Courtesy of C. L. Parish.)

Fig. 17. Apple graft-incompatibility.
page 75

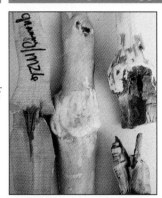

Symptoms on three Granny Smith/M26 trees from which the bark has been removed. Note the overgrowth at the union and the inner xylem necrosis. (Courtesy of J. K. Uyemoto.)

Fig. 18. Apple green crinkle.
page 76

Symptoms on two apple cultivars. (Photograph at far left courtesy of T. J. Michailides.)

Fig. 19. Apple mosaic.

page 76

Leaf symptoms induced by different strains of the virus.

Fig. 20. Apple star crack.

page 79

Symptoms on half-grown fruit.

Fig. 21. Apple stem pitting.

page 79

Bark removed to show the pitting and grooving of the wood of a susceptible scion cultivar (right) and of a commonly used indicator, Virginia crab (far right). (Courtesy of C. L. Parish.)

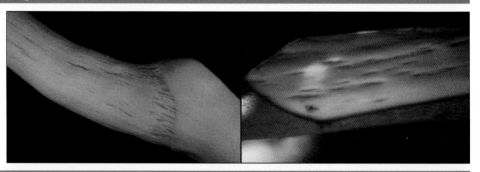

Fig. 22. Apple union necrosis and decline.

page 80

(Photographs courtesy of C. L. Parish.)

A. Diseased (left) and healthy five-year-old Red Delicious apple trees on MM 106 rootstock.

B. Xylem necrosis and breakage at the union of a Red Delicious tree on MM 106.

Fig. 23. Pear blister canker.

page 82

Typical blistering and cracking of the bark on the scaffold branch of a Bartlett tree. (Courtesy of C. L. Parish.)

Fig. 24. Pear stony pit.

page 82

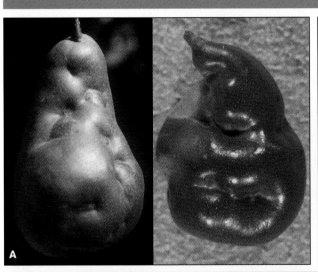

A. Typical fruit symptoms on cultivars Bosc (far left) and Rogue Red. (Rogue Red courtesy of D. Sugar.)

B. Markedly roughened bark ("oak bark") of a diseased scaffold branch.

Fig. 25. Pear vein yellows.

page 83

Leaf symptoms on the indicator pear Nouveau Poiteau. Note the yellow flecks along some of the smaller leaf veins. (Courtesy of J. K. Uyemoto.)

Fig. 26. Viruslike diseases of apple in the Pacific Northwest.

page 74

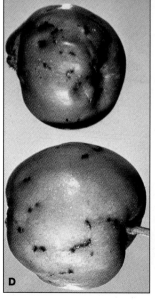

These diseases have not yet been reported in California. (Photographs courtesy of C. L. Parish.)

A. Symptoms of blister bark on Red Delicious.

B. False sting on Red Delicious.

C. Flat apple disease. Note the similarity of symptoms of fruit from trees inoculated with either the flat apple pathogen or cherry rasp leaf virus.

D. Rough skin disease on Golden Delicious.

E. Symptoms of russet ring on cv. Pacific Pride.

Pome Fruit and Their Diseases • 393

Fig. 27. Bitter pit of apple and pear.

pages 87, 119

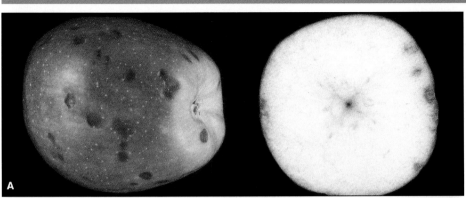

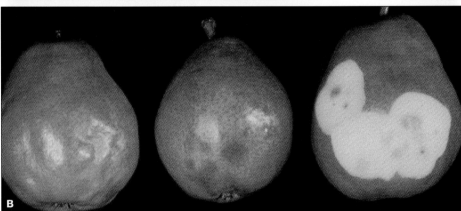

A. External and internal symptoms of bitter pit on Red Delicious apple.

B. Bitter pit of Anjou pear. (Courtesy of S. W. Porritt, M. Meheriuk, and P. D. Lidster.)

Fig. 28. Black-end of Bartlett pear.

page 91

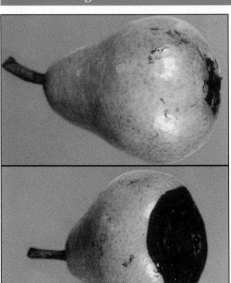

Typical stylar-end necrosis. (Courtesy of K. Ryugo.)

Fig. 29. Internal bark necrosis of apple.

page 92

Note roughened bark surface and small necrotic lesions in the phloem.

Fig. 30. Water core of apple.

page 95

A. Fruit with water-soaked areas in the exposed flesh.

B. Core and vascular water core breakdown in Delicious. (Courtesy of S. W. Porritt, M. Meheriuk, and P. D. Lidster.)

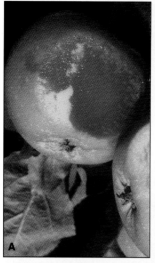

Fig. 31. Alternaria rot.

page 97

A. Rot centered at a skin puncture in a Fuji apple.

B. Rot developing in the core area of an apple with an open calyx sinus.

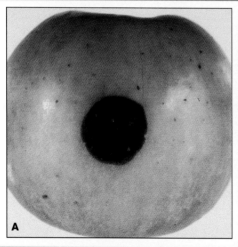
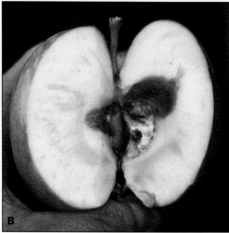

Fig. 32. Blue mold rot.

page 99

Rot developing at the stem end of a pear.

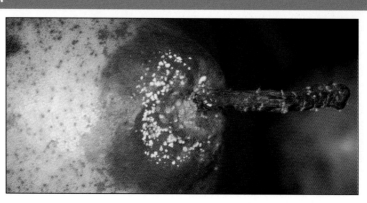

Pome Fruit and Their Diseases • 395

Fig. 33. Bull's-eye rot.
page 103

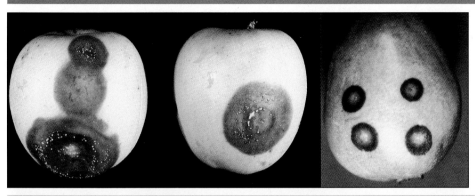

Symptoms on Golden Delicious apple (center and far left) and Anjou pear. (Courtesy of G. G. Grove and R. A. Spotts, respectively.)

Fig. 34. Gray mold rot of apple and pear.
page 104

A. Nest rot with sporulation on Golden Delicious apples.

B. Nest rot in packed Anjou pears. (Courtesy of R. A. Spotts.)

Fig. 35. Stem-end decay.
page 107

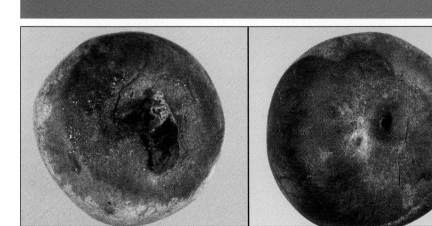

Symptoms in apple, caused by *Phomopsis mali*. (Courtesy of N. F. Sommer.)

Fig. 36. Miscellaneous fungal rots of pome fruit.

page 108

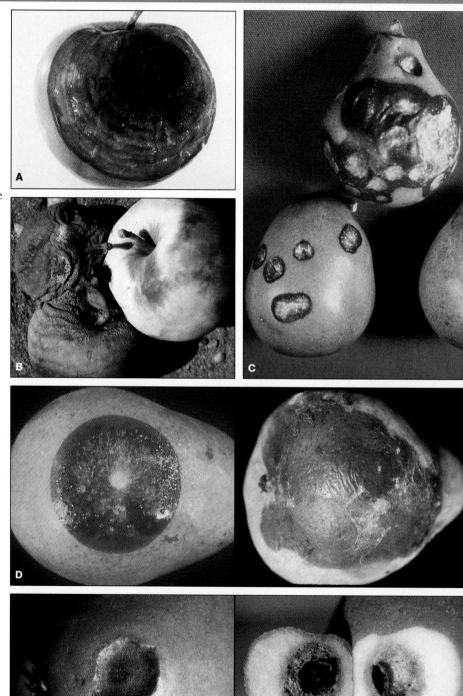

A. Black rot of apple. Note pycnidia of the pathogen forming in the older portion of the lesion. (Courtesy of T. J. Michailides.)

B. Apple fruit with brown rot decay. Note the buff-colored sporulation of *Monilinia fructicola*.

C. Coprinus rot on Anjou pears. Note the circular, sunken lesions with brown margins and lighter centers. (Courtesy of R. A. Spotts.)

D. Mucor (right) (courtesy of T. J. Michailides) and Rhizopus (far right) rots of pears.

E. Side rot of pear caused by *Phialophora malorum*.

Pome Fruit and Their Diseases • 397

Fig. 37. Carbon dioxide injury in stored apples and pears.

pages 112, 119

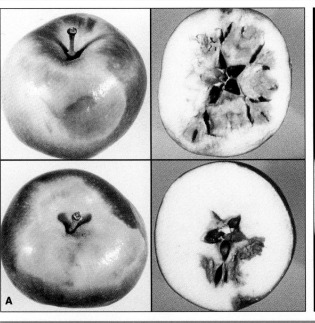
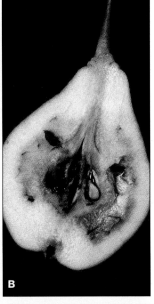

(Photographs courtesy of S. W. Porritt, M. Meheriuk, and P. D. Lidster.)

A. External and internal symptoms in McIntosh apples.

B. Internal symptoms in Clapp Favorite pears.

Fig. 38. Chemical injury to stored apples and pears.

page 112

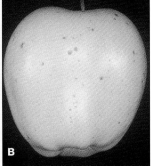
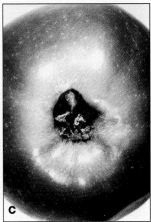

(Photographs courtesy of S. W. Porritt, M. Meheriuk, and P. D. Lidster.)

A. Ammonia injury on Golden Delicious apple and Bartlett pear.

B. Calcium chloride injury on Golden Delicious.

C. Calcium chloride injury on Spartan apple.

– continued

Fig. 38. Chemical injury to stored apples and pears – *continued.*

page 112

D. Diphenylamine injury on Delicious apple.

E. Ethoxyquin injury on Winesap apple.

F. Ethoxyquin injury at contact points between Anjou pears.

G. Sodium carbonate injury to lenticels of Bartlett pear.

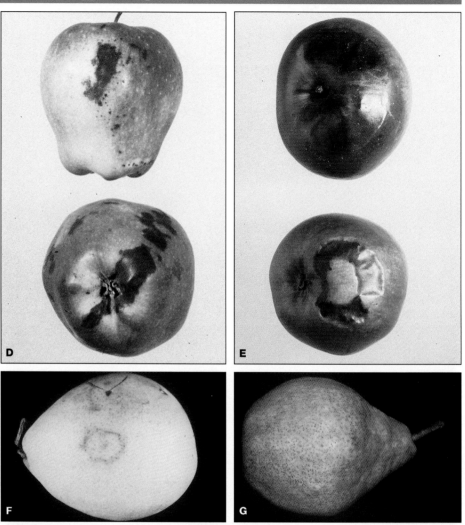

Fig. 39. Friction marking.

pages 113, 118

Symptoms on Granny Smith apple (right) and Bartlett pear (far right). (Courtesy of P. A. Mauk and of S. W. Porritt, M. Meheriuk, and P. D. Lidster, respectively.)

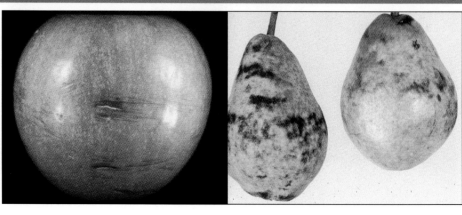

Fig. 40. Heat (hot water) injury.

page 113

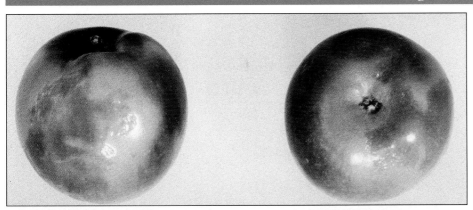

Symptoms on stored McIntosh apples. (Courtesy of S. W. Porritt, M. Meheriuk, and P. D. Lidster.)

Fig. 41. Jonathan spot of apple.

page 113

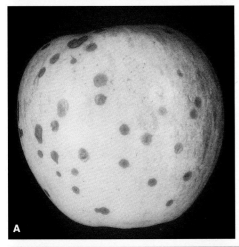
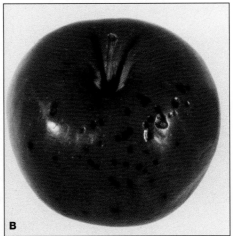

A. On cv. Jonathan.

B. On cv. Winesap. (Courtesy of S. W. Porritt, M. Meheriuk, and P. D. Lidster.)

Fig. 42. Low oxygen (alcohol) injury.

page 113

Symptoms on McIntosh apple. (Courtesy of S. W. Porritt, M. Meheriuk, and P. D. Lidster).

Fig. 43. Scald (superficial scald).

page 114

Rome Beauty (top right) and Granny Smith (bottom right) apples and Anjou pear (far right). (Photographs at bottom right and far right courtesy of S. W. Porritt, M. Meheriuk, and P. D. Lidster.)

Fig. 44. Soft scald.

page 114

Symptoms in Delicious apple. (Courtesy of S. W. Porritt, M. Meheriuk, and P. D. Lidster.)

Fig. 45. Sunscald.

page 114

Symptoms in Granny Smith apple.

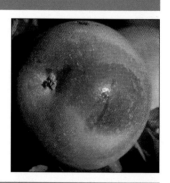

Fig. 46. Core browning.

page 115

Symptoms in Yellow Newtown (Pippin) apple. (Courtesy of S. W. Porritt, M. Meheriuk, and P. D. Lidster.)

Fig. 47. Internal browning (flesh browning).

page 115

Symptoms in Delicious apple. (Courtesy of S. W. Porritt, M. Meheriuk, and P. D. Lidster.)

Fig. 48. Internal breakdown in apple and pear.

pages 116, 119

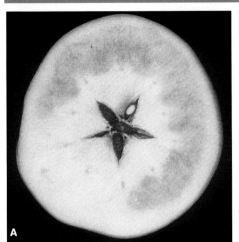
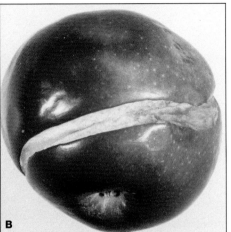

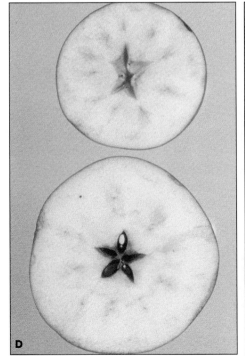

(Photographs courtesy of S. W. Porritt, M. Meheriuk, and P. D. Lidster.)

A. Low-temperature breakdown in Yellow Newtown apple.

B. Mealy breakdown in McIntosh apple.

C. Senescent breakdown in Spartan apple.

D. Vascular breakdown in McIntosh apple.

E. (*page 119*) Core breakdown in Bartlett pear.

Fig. 49. Freezing injury in apple and pear.

page 117

A. Apple with band of russeted tissue resulting from spring frost.

B. Freezing injury in stored McIntosh apples (left) and Bartlett pears. (Courtesy of S. W. Porritt, M. Meheriuk, and P. D. Lidster.)

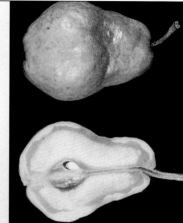

Fig. 50. Senescent scald.

page 118

Symptoms in Bartlett pears.

Fig. 51. Premature ripening (pink end).

page 119

Symptoms in Bartlett pears.

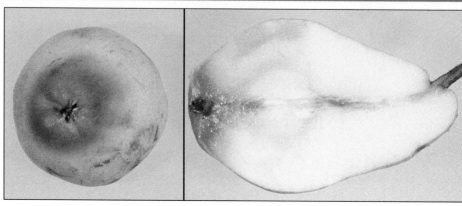

Chapter 3. Stone Fruit and Their Diseases

Fig. 1. Foamy canker of almond.
page 132

Reddish gummy exudate draining from cankered area of branch.

Fig. 2. Leaf scorch of almond.
page 133

A. Almond shoot with typical leaf symptoms.

B. Upper (far left) and lower (left) portions of a healthy and a diseased tree (diseased tree is at right in both photographs). Note the pronounced stunting of the affected tree.

C. Close-up view of affected leaves. Note scalded appearance.

D. Diseased wood, detected by immersion in acidified methanol, shows distinct reddish streaks; segment at right is healthy. (Courtesy of S. M. Mircetich.)

E. Two important vectors of the pathogen are the green sharpshooter (top) and the meadow spittlebug (bottom). (Courtesy of A. H. Purcell and Jack Kelly Clark.)

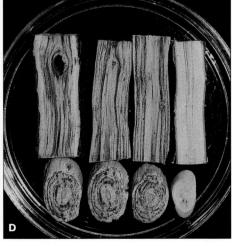

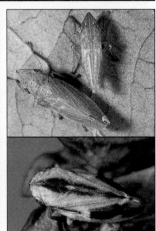

Fig. 3. X-disease of cherry and peach.

page 136

A. Death of branches of sweet cherry on mahaleb rootstock.

B. Small, off-color Bing cherries (left) affected with the disease; fruit at right are healthy.

C. Branch of tree at right shows typical symptoms caused by the common strain of the X-disease MLO; healthy tree is at left.

D. Peach tree at left shows early symptoms caused by the PYLR strain of the pathogen; healthy tree is at right.

E. Close-up view of peach leaves affected with the PYLR strain.

F. Adult and nymph of the leafhopper *Fieberiella florii*, probably the most important vector of the pathogen. (Courtesy of A. H. Purcell and R. Coville.)

Fig. 4. Brown rot of stone fruit.

page 142

A. Blossom and twig blight of almond caused by *Monilinia laxa*.

B. Sporodochia of *M. laxa* on apricot twig.

C. Sporulation of *M. laxa* on killed apricot blossom and incipient infections of adjacent green fruit. (Courtesy of G. Tate.)

D. Quiescent infections in French prune caused by *M. fructicola*.

E. Peach fruit rot caused by *M. fructicola*.

F. Rotted peach fruit and dead hangers caused by *M. fructicola*.

G. Blossoming peach shoot with an adjacent brown rot mummy, and a piece of surface soil with apothecia of *M. fructicola*.

H. Critical blossom stages for the first spring fungicide application for brown rot control: red bud of apricot (left), and pink bud of almond (center) and peach (right). Note the highly susceptible protruding anthers of the peach buds.

Fig. 5. Ceratocystis canker of almond and prune.

page 153

A. Scaffold of Mission almond tree dying from infection originating in shaker injury.

B. Wine-colored internal bark of a canker in French prune.

C. An improved shaker that causes minimum injury to the bark, thereby reducing canker incidence.

Fig. 6. Cytospora canker of stone fruit.

page 156

A. Branch of President plum killed by *Cytospora leucostoma*.

B. Periderm of infected branch removed to reveal gray pycnidia of the pathogen.

C. White-tipped pycnidia protruding from the outer bark of an infected branch.

D. Extruded spore tendrils on a diseased peach scaffold.

Fig. 7. Dothiorella canker of almond.

page 161

A. Killed scaffold branches of cv. Nonpareil.

B. Typical gummy canker on the trunk of a Nonpareil tree.

C. Outer bark of canker removed to show necrotic inner bark.

Fig. 8. Eutypa dieback of apricot.

page 164

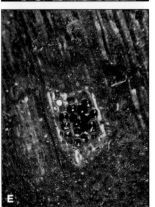

A. Extensive killing of scaffold branches. (Courtesy of W. J. Moller.)

B. Recently killed apricot branch.

C. Gummy canker spreading from infected pruning wound. (Courtesy of W. J. Moller.)

D. Outer bark of canker removed to show zonate, necrotic inner bark.

E. Black stroma of the pathogen on a dead, decorticated branch. A small area of the surface has been removed to reveal the perithecial locules. (Courtesy of W. J. Moller.)

Fig. 9. Green fruit rot of stone fruit.

page 169

A. Young apricot fruit (right) invaded by *Botrytis cinerea* from infected blossom.

B. Infection of apricot fruit from Botrytis-infected blossom parts.

C. Young sweet cherry fruit rotted by Botrytis and Monilinia.

D. Decay of green apricots following invasion by *Sclerotinia sclerotiorum* from infected blossoms.

E. Small, cream-colored apothecium of *S. sclerotiorum* (below peach pit) and two large, brown apothecia of *Monilinia fructicola*.

Fig. 10. Hull rot of almond.

page 171

A. Rhizopus-infected hull (sporulating beneath the ruptured hull) with twig dieback.

B. Harvested nuts with Rhizopus mold on the shell.

C. Hull rot caused by *Monilinia fructicola*. Note the buff-colored sporulation.

D. Splitting of the hull opens the way for infection by hull-rot fungi.

Fig. 11. Leaf blight of almond.

page 172

A. Sudden death of leaves caused by infection at base of leaf petiole.
B. Pathogen in blighted petioles attached to twig kills buds and forms small, dark, cortical lesions.

Fig. 12. Leaf curl (*Taphrina cerasi*) of cherry.

page 173

Diseased leaves at left, healthy leaf at right. (Courtesy of H. R. Cameron.)

Fig. 13. Leaf curl (*Taphrina deformans*) on peach and apricot.

page 174

A. Heavy infection of the highly susceptible Fay Elberta peach.

B. Extensive development of the disease from systemic invasion of the young green shoot. The whitish appearance of some leaves is due to the presence of a layer of asci.

C. Fruit symptoms on Red Haven peach; healthy fruit is at right.

D. Winter symptoms of a peach shoot and leaves killed by the disease the preceding spring. The blackish appearance of the leaves indicates invasion by a secondary fungus.

E. Whitish, distorted leaves of apricot. (Photographed in Australia.)

Fig. 14. Leaf spot of cherry.
page 177

A. Characteristic spotting and yellowing of infected sour cherry leaves.

B. Cherry fruit with heavy pedicel infection and a few fruit lesions.

C. Infected sour cherry leaves showing typical small, white conidial pustules on the lower side (far left) and purplish spots on the upper surface (left).

Fig. 15. Phytophthora canker.
page 178

Profuse gumming associated with canker at almond pruning wound. (Courtesy of R. M. Bostock.)

Fig. 16. Plum pockets.
page 180

Healthy plum fruit is seen above infected plum fruit.

Fig. 17. Powdery mildew of stone fruit.

page 181

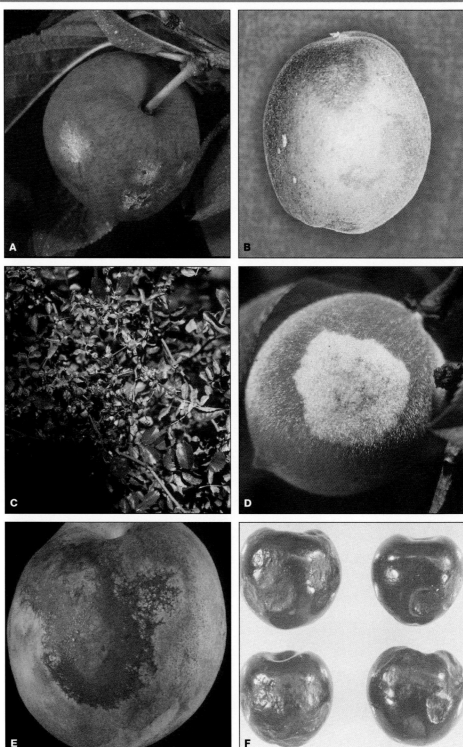

A. Plum fruit with infections of *Sphaerotheca pannosa*. Note the absence of mildew on the foliage.

B. Apricot fruit with mildew lesions caused by *S. pannosa*.

C. A rose bush infected with *S. pannosa*; a common source of primary inoculum for plum and apricot.

D. Young peach fruit with a typical *S. pannosa* infection.

E. Peach fruit with a typical brown, scabby lesion caused by *Podosphaera leucotricha*.

F. Symptomatic sweet cherries with infections of *P. tridactyla*.

Fig. 18. Russet scab of prune.

page 187

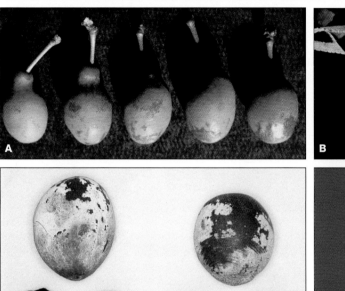

A. Typical symptoms on immature fruit; healthy fruit is at left.

B. Mechanical damage (not russet scab) resulting from friction with an adjacent leaf.

C. Symptoms on mature and dried fruit.

D. Full bloom in French prune, the critical stage for spraying to control russet scab.

Fig. 19. Rust of stone fruit.

page 188

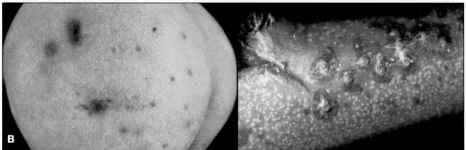

A. Typical symptoms on the upper (far left) and lower leaf surface of prune. Note the uredial and telial sori on the underside.

B. Lesions on a peach fruit and a current-season shoot.

Fig. 20. Scab of almond and peach.

page 191

A. Typical dark lesions on almond fruit.

B. Diffuse, olivaceous lesions on the undersurface of an almond leaf.

C. Discrete lesions on an almond shoot. The dark margin of the lesions results from the abundant production of conidia.

D. Peach fruit with typical scab lesions.

Fig. 21. Shothole of stone fruit.

page 193

A. Almond leaves with typical shothole lesions.

B. Close-up of an almond leaf lesion with black sporulation of the pathogen near the center.

C. Overwintered peach shoot with a characteristic circular lesion.

– continued

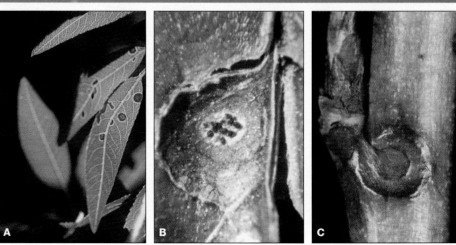

Fig. 21. Shothole of stone fruit — *continued*.

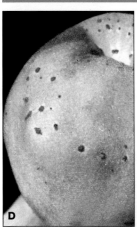  

page 193

D. Peach fruit with numerous infections.

E. Apricot with dormant bud blight and heavy infection of adjacent fruit.

F. Sweet cherry fruit with black, depressed shothole lesion surrounded by a highly pigmented area.

Fig. 22. Almond union disorders.

 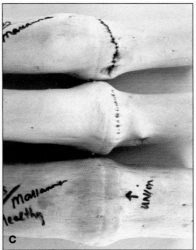
  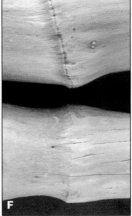

page 198

(Photographs courtesy of J. K. Uyemoto.)

A. Young tree dying from almond brownline and decline (ABLD) disease.

B. Cultivar Carmel/Marianna 2624 affected with ABLD. Note the necrosis of phloem at the graft union.

C. With bark removed, young Peerless/Marianna 2624 trees show brownline symptoms of ABLD; bottom tree is healthy.

D. Symptoms of a second union disorder on a young tree (foreground) of Carmel/Marianna 2624.

E. Leaf symptoms of the second union disorder on a Carmel/Marianna 2624 tree.

F. Mild etching of the xylem at the union (top) of a Carmel/Marianna 2624 tree affected with the second union disorder; bottom tree is healthy.

Fig. 23. Cherry rasp leaf.

page 200

The underside of a sweet cherry leaf shows prominent enations and leaf distortion characteristic of the disease. (Courtesy of J. K. Uyemoto.)

Fig. 24. Three graft-transmissible diseases of cherry with mottle-leaf symptoms.

pages 199, 200

Cherry mottle leaf (right), cherry necrotic rusty mottle (center), and cherry rusty mottle (far right). (Left and center courtesy of J. K. Uyemoto; right courtesy of C. L. Parish.)

Fig. 25. Cherry stem pitting.

page 201

A. Foreground tree shows sparse canopy associated with the disease; trees at left are healthy.

B. Bark near base of sweet cherry trunk removed to show pitting and grooving of the woody cylinder.

Fig. 26. Peach mosaic.

page 202

(Photographs courtesy of H. Larsen.)

A. Color breaking in the petals of a peach blossom.

B. Healthy (top) and diseased peach leaves.

Fig. 27. Prune brownline.

page 204

Declining French prune on Myrobalan rootstock. (Courtesy of S. M. Mircetich.)

Fig. 28. Peach stunt (prune dwarf).

page 206

Tree at left-center is affected with peach stunt; tree at right is healthy. (Courtesy of C. Kuske.)

Fig. 29. Prunus necrotic ringspot.

page 209

A. Spotted and shotholed foliage of sweet cherry (left) and almond, typical shock symptoms.

B. Distorted, chlorotic sweet cherry leaves due to the cherry rugose mosaic strain. (Courtesy of J. K. Uyemoto.)

C. Almond leaves with symptoms caused by the almond calico strain.

Fig. 30 Prunus stem pitting.

page 212

A. Peach tree with advanced symptoms; healthy tree is in the background.

B. Bark removed from the trunk of a peach tree to show typical xylem pitting. (Courtesy of B. Jaffee.)

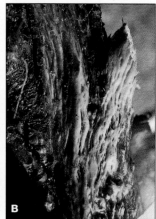

Fig. 31. Yellow bud mosaic.

page 214

A. Healthy and diseased peach trees. Insert: yellow bud symptoms of opening leaf buds of peach.

B. Declining sweet cherry tree affected with the disease.

Fig. 32. Chilling canker of peach.

page 218

Cankering and killing of Lovell seedlings subjected to chilling.

Fig. 33. Corky growth.

page 218

Symptoms on Jordanolo almond kernel and inner surface of endocarp.

Fig. 34. Crinkle leaf and deep suture of sweet cherry.

page 219

A. Characteristic deformity of leaves affected with crinkle leaf.

B. Fruit affected with crinkle leaf disorder; healthy fruit is at upper left.

C. Deep suture: affected (left) and healthy (right) fruit and leaf.

Fig. 35. Nectarine pox.

page 220

Two types of disease symptom.

Fig. 36. Noninfectious bud failure of almond.

page 221

A. Dead and living shoots on a severely affected branch.

B. Rough-bark symptom on cv. Jordanolo.

Fig. 37. Noninfectious plum shothole.

page 225

Leaf symptoms on cv. Eldorado.

Fig. 38. Nonproductive syndrome of almond.

page 225

A. Mission tree with nonproductive syndrome (left); tree at right is healthy.

B. Fruit from a healthy Mission tree (left) and a symptomatic tree (right).

C. Nuts from a healthy Mission tree (top) and a diseased tree (bottom).

Fig. 39. Rusty blotch.
page 226

Disease symptoms on the leaves of Santa Rosa plum. (Courtesy of T. S. Pine.)

Fig. 40. Postharvest diseases.
pages 227–236

A. Molds on fresh cherries are *Rhizopus stolonifer* (R), *Monilinia fructicola* (M), *Botrytis cinerea* (B), *Alternaria alternata* (A), and *Penicillium* sp. (P).

B. Peach fruit rots caused by *Gilbertella persicaria* (left) and *Rhizopus stolonifer* (right).

C. *Mucor piriformis* sporulating on an infected peach.

– continued

Fig. 40. Postharvest diseases — *continued.*

pages 227–236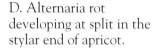

D. Alternaria rot developing at split in the stylar end of apricot.

E. Nectarines with contact spread of brown rot caused by *Monilinia fructicola*.

F. Nectarine fruit decay caused by *Penicillium* sp.

G. Fungal decay in French prunes held too long before dehydration.

H. Slip-skin maceration of a dried prune resulting from fungal decay.

I. Enzymatic softening of canned apricots (left) when one fruit decayed by *Rhizopus arrhizus* was added to the can; healthy canned product is at right.

Chapter 4. Diseases Attacking Several Genera of Fruit and Nut Trees

Fig. 1. Bacterial canker and blast.

page 246

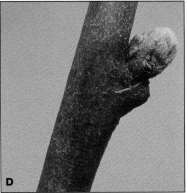

A. Blackened prune blossoms at left killed by *Pseudomonas*, blossom at right blighted by *Monilinia*. Note the abundant production of conidia.

B. Almond leaves blasted and spotted by *Pseudomonas*. Note the light-green halo around the necrotic leaf spots.

C. Infected fruit (top) and healthy fruit and pedicels of sweet cherry.

D. Early symptoms of leafscar infection and bud killing on peach shoot. (Courtesy of L. Gardan and A. Vigouroux.)

E. Young almond trees in the San Joaquin Valley killed by the disease.

– continued

Fig. 1. Bacterial canker and blast — *continued*.

page 246

F. Young almond tree killed by the "sour sap" phase of the disease. Note the stained trunk that results from bleeding.

G. Gummy canker on an almond branch. The outer bark has been removed to show the cortical symptoms.

H. The outer bark of a cankered apricot branch has been removed to show the necrotic flecks and streaks in the cortex.

I. Sweet cherry tree severely cankered in the crotch area. Note that one scaffold limb has been killed and removed.

J. A dying young French prune tree infected initially by *Pseudomonas* and secondarily by *Cytospora*. Note the characteristic sprouting from the base of the trunk.

Diseases Attacking Several Genera of Fruit and Nut Trees • 425

Fig. 2. Crown gall.

page 257

A. Young almond tree with galls on the roots.

B. Large gall at the base of a mature almond tree.

Fig. 3. Phytophthora root and crown rot.

page 263

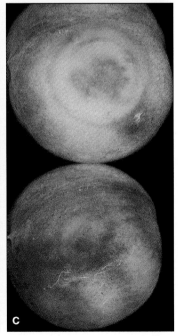

A. Peach (top) and sweet cherry trees killed by the disease.

B. Outer bark has been removed to show symptoms of crown rot on a young peach tree. (Courtesy of S. M. Mircetich.)

C. Phytophthora fruit rot of peach resulting from the use of contaminated sprinkler-irrigation water.

Fig. 4. Replant diseases.

page 276

A. Replanted apple trees in the foreground are third-leaf Gala/7a in central Washington. The tree at left is in soil that was fumigated preplant with methyl bromide; the tree at right is planted in untreated soil. (Courtesy of T. J. Smith.)

B. Peach replant disorder in a site near Davis, California: top, second-leaf trees in virgin soil (trees were removed and the enlarged site was replanted at the end of the season); center, replanted trees in autumn of first year's growth (note improved growth of trees in virgin soil, foreground and extreme right); bottom, trees in second leaf continue to show the growth inhibition associated with the replant problem. (Courtesy of A. E. Gilmore.)

Fig. 5. Shoestring (Armillaria) root rot.

page 280

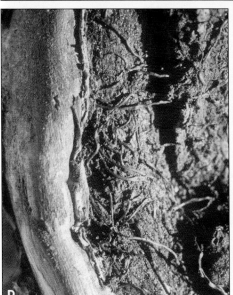

A. Young almond tree killed by Armillaria. Note the cluster of sporophores at the base of the trunk.

B. Juvenile sporophores ("buttons") developing on the trunk base of a dead almond tree.

C. Mature sporophores showing annulus and typical honey coloration.

D. Rhizomorphs on surface of dead root.

E. Typical creamy-white mycelial plaques found under killed bark.

Fig. 6. Silver-leaf disease.

page 285

A. Nectarine tree (right-center) with typical early symptoms; tree at extreme right is healthy. Inoculum had spread into orchard from infected willow stump in background. (Photographed in Chile.)

B. Diseased (top) and healthy peach leaves, and a piece of killed bark bearing sporophores of the pathogen.

C. Mycelium of the pathogen growing from the discolored wood of a peach scaffold branch. (Courtesy of W. J. Moller.)

 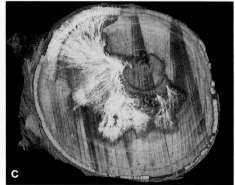

Fig. 7. Verticillium wilt.

page 287

A. One-sided wilting of an almond tree.

B. Characteristic darkening of xylem in an infected almond branch.

Fig. 8. Wood decay fungi in orchard trees.

page 292

A. Basidiocarps of *Oxyporus latemarginatus* at the base of an infected almond tree.

B. Wind-thrown almond tree with butt and root rot caused by *O. latemarginatus*, and a close-up view of the white wood rot in the base of the trunk.

C. Basidiocarp of *Perennoporia medulla-panis* on an almond trunk.

D. Broken scaffold of prune affected with wood decay resulting from infection by *P. medulla-panis* at an injury caused by the support band.

E. Dieback in prune associated with wood decay caused by *P. medulla-panis*.

F. Almond tree with a basidiocarp of *Laetiporus sulphureus* (top) and wind-twisting injury of a scaffold branch with brown heart rot caused by this fungus (bottom).

Fig. 9. Basidiocarps of three wood-decay fungi in peach.

page 294

(Photographs courtesy of J. E. Adaskaveg.)

A. *Ganoderma brownii* on a tree trunk.

B. *Trametes hirsuta* on a rotted stump.

C. *Schizophyllum commune* on a slab of bark.

Fig. 10. Basidiocarps of wood-decay fungi on four orchard crops.

page 294

(Photographs courtesy of J. E. Adaskaveg.)

A. *Phellinus robustus* on almond.

B. *Fomitopsis cajanderi* on prune.

C. *Pleurotus ostreatus* on walnut.

D. *Inonotus cuticularis* on fig.

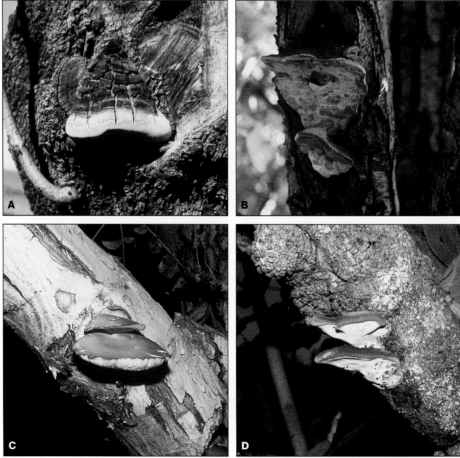

Chapter 5. The English Walnut and Its Diseases

Fig. 1. Walnut blight.
page 305

A. Fruit with incipient infection (left) and large stylar-end lesion (right).

B. Blighted walnut catkin.

C. Pistillate flowers at the critical stage for the first application of copper for disease control.

Fig. 2. Deep-bark canker of English walnut.
page 309

A. Severely affected Hartley tree.

B. Typical bleeding canker (top) and outer bark of a canker removed to show extensive phloem necrosis (bottom).

C. Characteristic cracking of the bark and sprouting of the rootstock below the affected area.

Fig. 3. Shallow-bark canker of English walnut.

page 312

A. Black exudate from small bark cracks associated with this disease.

B. Periderm removed to show necrosis of outer bark; infection apparently initiated at injuries caused by the pecking of sapsuckers.

Fig. 4. Branch wilt of English walnut.

page 314

A. Death of a branch in summer. Note leaf retention.

B. Ruptured periderm of dead branch reveals masses of black arthrospores.

C. Cross-section of severely affected branch showing only a small amount of living tissue.

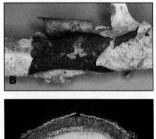
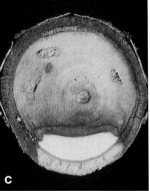

Fig. 5. Downy leaf spot of English walnut.

page 315

Typical lesions on the underside of leaves.

Fig. 6. Kernel mold of English walnut.

page 316

Typical symptoms of kernel mold.

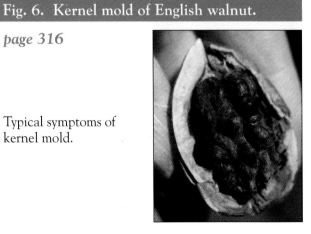

Fig. 7. Blackline of English walnut.

page 319

A. Decline in vigor of affected tree and suckering of the black walnut rootstock.

B. Blackline at graft union of English walnut and Northern California black walnut rootstock. (Courtesy of S. M. Mircetich.)

C. English walnut on Paradox rootstock showing necrosis at the union and in the cortex of the rootstock. (Courtesy of S. M. Mircetich.)

Fig. 8. Typical symptoms of sunburn injury.

page 322

Injury to the husks (far left) and kernels of the English walnut. (Courtesy of Jack Kelly Clark.)

Chapter 6. The Pistachio and Its Diseases

Fig. 1. Botryosphaeria panicle and shoot blight of pistachio.

page 329

A. Typical brown lesions on the fruit and leaves.

B. Lower cluster is dying from girdling of the rachis.

C. Blighted shoot with numerous black pycnidia that have extruded white masses of conidia.

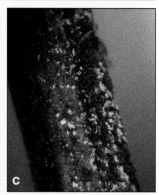

Fig. 2. Botrytis blight of pistachio.

page 331

Diseased male inflorescence and shoot of Peters cultivar.

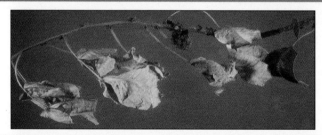

Fig. 3. Verticillium wilt of pistachio.

page 332

A. Early symptoms of wilt.

B. Trunk of diseased tree with bark and outer layer of wood removed to show the brownish black xylem.

Chapter 7. The Olive and Its Diseases

Fig. 1. Olive knot.
page 341

Nodal galls resulting from leaf scar infection.

Fig. 2. Olive leaf spot disease.
page 344

Typical foliage symptoms.

Fig. 3. Verticillium wilt.
page 347

Olive trees affected by Verticillium wilt.

Fig. 4. Sickle leaf of olive.
page 348

Disease symptoms (right) in contrast to healthy foliage.

Fig. 5. Shotberry.
page 349

Symptoms on Sevillano olive. Green fruit just below center is healthy.

Fig. 6. Soft nose.
page 349

Symptoms in the Sevillano cultivar.

Chapter 8. The Fig and Its Diseases

Fig. 1. Botrytis limb blight of fig.

page 356

Branch is being killed after infection of attached fruit. Note grayish fungal sporulation on necrotic bark.

Fig. 2. Endosepsis of fig.

page 356

A. Healthy (left) and diseased Calimyrna figs.

B. External and internal symptoms of caprifigs infected with *Fusarium moniliforme*.

C. The tiny fig wasp, the vector of the endosepsis pathogen.

Fig. 3. Phomopsis canker of Kadota fig.

page 358

A. Cankers developing from pruning-wound infections. Note the creamy-white conidial masses extruded from the pycnidia.

B. Winter view of a typical, heavily pruned tree.

Fig. 4. Smut and mold of fig.

page 360

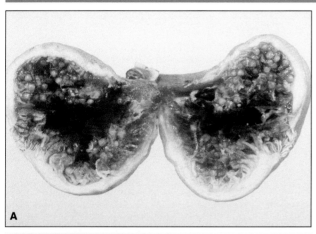

A. Fruit has been split to show the powdery black mass of spores of the smut fungus *Aspergillus niger*.

B. Fruit affected with three molds: *Rhizopus stolonifer* (upper fruit); *Botrytis cinerea* (upper part of lower fruit); and *Alternaria alternata* (black lesion).

Fig. 5. Souring of fig.

page 361

Fruit at left is collapsing from souring disease. Note the fermented juice that has dripped onto the leaf below the affected fruit.

Fig. 6. Alternaria rot.

page 362

Kadota fig with many Alternaria rot lesions.

Fig. 7. Fig mosaic leaf patterns.

page 363

Patterns seen with reflected light (left) and transmitted light (right).

Chapter 9. Minor Crops and Their Diseases

Fig. 1. Chestnut kernel mold.

page 370

Nut and kernel specimens at right are healthy; those at far right are infected with *Alternaria* sp.

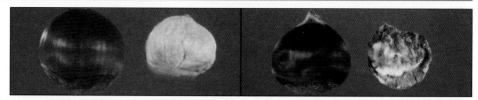

Fig. 2. Stylar-end rot of feijoa fruit.

page 370

This rot is caused by an unidentified organism; healthy fruit is at far right.

Fig. 3. Botrytis rot of persimmon.

page 372

This instance of Botrytis rot is centered at a mechanical injury.

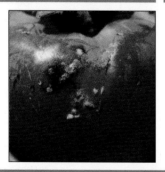

Fig. 4. Sunburn of persimmon.

page 372

Severe sunburn on Hachiya fruit.

Fig. 5. Fungal diseases of pomegranate.

page 372

Cracked pomegranate fruit infected with *Aspergillus niger* (left) and *Botrytis* (grayish white sporulation) and *Penicillium* (greenish sporulation) (right).

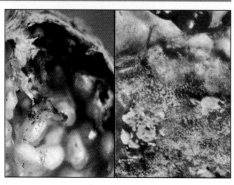

Fig. 6. Powdery mildew.

page 372

Pomegranate fruitlet and shoot apex affected with powdery mildew.

Index

Numbers in **bold** indicate photographs.

A

Abiotic orchard disorders
　See also specific disorders; specific fruit
　of olive, 349, **435**
　of pome fruit, 87–96
　of stone fruit, 218–226
　of walnut, 322
Abiotic storage disorders
　See also specific disorders
　of apple, 111–118
　bitter pit
　　of apple, 87–89, **393**
　　of pear (Anjou pit, cork spot), 119, **393**
　carbon dioxide injury
　　to apple, 112, **397**
　　to pear, 119, **397**
　chemical injury
　　to apple, 112–113, **397–398**
　　to pear, 118, **397**, **398**
　core breakdown of pear, 119
　core browning, 115, **400**
　freezing injury
　　to apple, 117–118, **402**
　　to pear, 119, **402**
　friction marking
　　of apple, 113, **398**
　　of pear, 118, **398**
　heat injury, 113
　internal breakdown, 116–117, **401**
　internal browning, 115–116, **400**
　internal carbon dioxide injury, 116, **397**
　Jonathan spot, 113, **399**
　low oxygen (alcohol) injury, 113, **399**
　of pear, 118–119. *See also specific disorders*
　premature ripening (pink end), 119, **402**
　scald, 114, **400**
　senescent scald, 118, **402**
　soft scald, 114, **400**
　sunscald, 114–115, **400**
　superficial scald, 118–119, **400**
　water core, 117, **394**
ABLD (almond brownline and decline), 198
Aceria ficus, fig mosaic and, 364
Acer spp. (Maple), Verticillium wilt and, 289
Acremonium breve, for gray mold control, 105
Acremonium zonatum, fig leaf spot and, 366
Actidione, for powdery mildew, 185
Actinomycetes, replant problems and, 278
Acylalanines, for Phytophthora root and crown rot, 270–271
Aecidium blasdaleanum, 55
Aflatoxins, 274, 275
Agaricus melleus, 281
Agrobacterium spp., 25
　A. radiobacter, 258

　　for crown gall control, 260–261, 309
　　for hairy root, 26
　A. rhizogenes
　　A. tumefaciens vs., 258
　　as crown gall protection, 132, 309
　　hairy root caused by, 25–26
　A. tumefaciens, crown gall and, 15–16, 132, 257–261, 309
Agrocin 84
　for crown gall, 260, 261
　for hairy root, 26
Alcohol injury. *See* Low oxygen injury
Aldicarb, fig mosaic and, 364
Alfalfa, diseases of
　Coprinus rot of pome fruit and, 108
　leaf scorch of almond and, 133, 134
Allyl bromide
　for Dematophora root rot, 41
　for Verticillium wilt, 290
Almond (*Prunus dulcis*), 128
　See also Stone fruit
　anthracnose of, 141
　bacterial canker and blast of, 246, 247, 250, 251, 254, **423–424**
　brownline and decline of, 198
　brown rot of, 142–145, **405**
　bud failure of
　　infectious, 222
　　noninfectious, 221–224, **419**
　Ceratocystis (mallet wound) canker of, 153–155, **406**
　Cercospora leaf spot of, 156
　corky kernel growth on, 218–219
　crazy top of, 221, 222
　crown gall of, 259, 260, **425**
　Dematophora root rot of, 39–41
　Dothiorella canker of, 161–163, 242, **407**
　dry canker of, 153–154
　Eutypa dieback of apricot and, 166
　foamy canker of, 132–133, **403**
　green-fruit rot of, 169–170
　hull rot of, 171, **409**
　Jordanolo, corky growth and, 219, **418**
　leaf blight of, 172–173, **409**
　leaf curl of, 174
　leaf scorch of, 133–135, **403**
　mistletoe and, 241
　mycotoxins and, 274
　nonproductive syndrome of, 225–226, **420**
　peach mosaic of, 202, 203
　Phytophthora pruning wound canker of, 178–180, 267, 271, **411**
　Phytophthora root and crown rot and, 180, 264, 265, 267, 269
　powdery mildew of, 181, 185
　prunus necrotic ringspot of, 209–211, **417**
　　bud failure and, 222
　Pythium root rot of, 241
　replant problems and, 276

　rootstocks, 128
　　Phytophthora rot and, 269
　rust of, 188–190
　scab of, 191–192, **414**
　shoestring (Armillaria) root rot of, 193, 282, 283, **427**
　shothole of, 193–196, **414**
　tomato ringspot virus and, 205
　union disorders of, 198, **415**
　Verticillium wilt of, 287–288, **428**
　yellow bud mosaic virus of, 205, 214–216
Almond brownline and decline (ABLD), 198
Almond calico strain, 209, 210–211
　bud failure and, 222
"Almond decline." *See* Almond, leaf scorch of
Almond mosaic. *See* Prunus necrotic ringspot
Alphelinas mali, for woolly apple aphid control, 37
Alternaria leaf and fruit spot, of pistachio, 334
Alternaria spp.
　A. alternata
　　apple or pear rot and, 97–98, **394**
　　apricot rot and, **422**
　　cherry decay and, 228, 229–230, **421**
　　endosepsis of fig and, 357, 358
　　fig decay and, 360
　　fig surface mold and rot and, 362–363, **437**
　　pistachio leaf and fruit spot and, 334
　　prune decay and, 235
　A. tenuis, 97
　chestnut kernel mold and, **438**
　mycotoxins and, 274
　nectarine and peach decay and, 233
　plum decay and, 234
　sweet cherry decay and, 228, 229–230
　walnut kernel mold and, 316
Aluminum ethylphosphite, 380
Aluminum tris-(o-ethyl phosphonate). *See* Fosetyl-Al
Amelanchier spp. (Serviceberry)
　Fabraea leaf spot and, 46
　Kern's pear rust and, 55
　Pacific Coast pear rust and, 54
　as *Pezicula malicorticis* host, 35
　Rocky Mountain pear rust and, 55
American brown rot, of stone fruit, 142, 145–149, **405**
American plum line pattern virus, apple mosaic virus and, 77
2-aminobutane
　for blue-mold or soft rot, 101
　postharvest dip, for European (Nectria) canker, 45
Ammonia, chemical injury and, 112, 113, **397**
Ammoniation, mycotoxins and, 275

Ammonium sulfate, internal bark necrosis (measles) and, 93
AMV. See Apple mosaic virus
Amygdalus communis. See Almond
Anemone spp., rust and, 189
Anisogramma anomala, 371
Anjou pit. See Bitter pit, of pear
Anthracnose
　of almond, 141
　false, 32
　of fig, 366
　olive and, 347
　of pome fruit, 32–38, **385**
　of stone fruit, 241
　of walnut, 313
Antibiotics, 380–381
　See also specific kinds
　for blight of walnut, 307
　for blister spot of pome fruit, 122
　deep-bark canker of walnut and, 311
　for nectarine and peach decay prevention, 233
Antisporulants
　for European (Nectria) canker, 41–45
　for perennial canker, 37
Aphids, woolly apple, 35, 36, 37
Apioporthe anomala, filbert dieback canker and, 371
Apiosporina morbosa, black knot and, of plum and cherry, 241–242
Apis pomi, apple mosaic virus and, 78
Apple (*Malus domestica*), 10-11
　See also Pome fruit
　abiotic disorders of, 87–96, **393, 394**
　　in storage, 111–118
　Alternaria rot of, 97–98, **394**
　anthracnose and perennial canker of, 32–38, **385**
　bacterial canker and blast of, 247
　bacterial diseases of, 13–26. See also specific diseases
　bitter pit of, 87–89, **393**
　bitter rot of, 121
　black pox of, 122
　black root rot of, 122
　black rot of, 108, 121, **396**
　black spot of. See Apple, scab of
　blister bark of, **392**
　blister spot of, 13, 122
　blossom blast and canker of, 13–14
　blossom-end rot of, 122
　blotch of, 122
　blue-mold or soft rot of, 99–101
　Brook's spot (Phoma fruit spot) of, 122
　brown rot of, 108, 145, **396**
　budding, stem pitting and, 79
　bull's-eye rot of, 37–38, 103–104, **395**
　carbon dioxide injury to, 112, 116, **397**
　cedar apple rust of, 122
　chemical injury to, 112–113, **397–398**
　chip budding, stem pitting and, 79
　Cladosporium-Hormodendrum rots of, 108
　Clitocybe root rot of, 122
　collar rot of, Phytophthora root and crown rot, 48, 263, 268, 270
　Coprinus rot of, 108–109
　core browning of, 115, **400**
　crown gall of, 15, 260
　cultivation history, 10–11
　Dematophora root rot of, 39–41, **385**
　Dothiorella canker and, 162, 163
　European (Nectria) canker of, 41–45, **385–386**
　Fabraea leaf spot of, 46
　false sting of, 76, **392**
　fireblight of, 16–21
　flat apple disease of, **392**
　flat limb of, 75, **389**
　fly speck of, 121
　freezing injury to, 117–118, **402**
　friction marking on, 113, **398**
　fruit crinkle, 76
　fruit rot on, 37–38. See also Apple, postharvest disorders of
　fungal diseases of, 32–72. See also specific diseases
　graft-incompatibility in, 75–76, **389**
　gray mold rot of, 104–105, **395**
　green crinkle of, 76, **389**
　hairy root of, 25–26, **384**
　heat injury to, 113, **399**
　insects on, anthracnose and perennial canker and, 35
　internal bark necrosis (measles) of, 92–94, **393**
　internal breakdown of, 116–117, **401**
　internal browning of, 115–116, **400**
　internal carbon dioxide injury (brown heart) to, 116, **397**
　Jonathan spot of, 113, **399**
　low oxygen (alcohol) injury to, 113, **399**
　mosaic virus of, 76–78, **390**
　Mucor rot of, 109–110
　mycotoxins and, 274
　Nectria blight of, 122
　Nectria (European) canker of, 41–45, **385–386**
　pear decline and, 27, **384**
　Phomopsis rot of, 107, **395**
　Phytophthora root and crown rot of, 48–49, 263, 264, 267, 268, 270, 271, **386**
　postharvest disorders of, 111–118
　　abiotic, 111–118
　　Alternaria rot, 97–98, **394**
　　bitter pit, 87–89, **393**
　　black rot, 108, **396**
　　blue-mold or soft rot, 99–101
　　brown rot, 108, 145, **396**
　　bull's-eye rot, 37–38, 103–104, **395**
　　carbon dioxide injury, 112, **397**
　　chemical injury, 112–113, **397–398**
　　Cladosporium-Hormodendrum rots, 108
　　Coprinus rot, 108–109
　　core browning, 115, **400**
　　freezing injury, 117–118, **402**
　　friction marking, 113, **398**
　　gray mold rot, 104–105, **395**
　　heat injury, 113
　　internal breakdown, 116–117, **401**
　　internal browning, 115–116, **400**
　　internal carbon dioxide injury, 116, **397**
　　Jonathan spot, 113, **399**
　　low oxygen (alcohol) injury, 113, **399**
　　Mucor rot, 109–110
　　mycotoxins, 274
　　Phomopsis rot, 107, **395**
　　scald, 114, **400**
　　side rot, 110
　　soft scald, 114, **400**
　　spot rot, 104
　　Stemphylium and Pleospora rots, 110
　　sunscald, 114–115, **400**
　　water core, 95–96, **394**
　powdery mildew of, 50–52, **387**
　　rusty spot of peach and, 184–185
　replant diseases of, 276–279, **426**
　rootstocks, 11
　　Dematophora root rot and, 40
　　graft incompatibility, 75–76, **389**
　　Phytophthora rot and, 49, 268–269, 270
　　stem pitting and, 79
　　union necrosis and decline and, 80, 81, **390**
　rough skin disease of, **392**
　rubbery wood of, 78–79
　russet ring of, **392**
　sappy bark of, 56–57, **388**
　scab of, 58–63, **388**
　scab-resistant cultivars of, anthracnose and perennial canker and, 35–36
　scald of, 112, **400**
　scurfy-bark canker of, 13
　shoestring (Armillaria) root rot of, 70, 282
　side rot of, 110
　silver-leaf of, 285, 286
　soft scald of, 114, **400**
　　Cladosporium-Hormodendrum rots and, 108
　sooty blotch of, 122
　southern blight of, 70–72, **389**
　star crack of, 79, **390**
　Stemphylium and Pleospora rots of, 110
　stem pitting of, 79–80, **390**
　　pear vein yellows and red mottle and, 83
　　quince sooty ring spot and, 84
　sunscald of, 114–115, **400**
　thread blight of, 122
　twig blight of, 16
　union necrosis and decline of, 80–81, **390**
　virus or viruslike diseases of, 74, **392**. See also specific diseases
　water core of, 95–96, **394**
　white root rot of, 122–123
　white rot of, 121
　witches'-broom of, replant problems and, 276
　wood-rot disorders of, 294, 295
　X-spot of, 123
Apple canker. See European (Nectria) canker
Apple mosaic virus (AMV), 76–78, 209, **390**
　Tulare, 77
Apple star crack, 79, **390**
Apple stem pitting, 79–80, **390**

pear vein yellows and red mottle and, 83
Apple-tree ("Northwestern") anthracnnose, 32
Apricot (*Prunus armeniaca*), 128–129
  See also Stone fruit
  Alternaria rot of, **422**
  bacterial canker and blast of, 246, 247, 251, 253, **424**
  bacterial canker of, Eutypa dieback vs., 164
  bacterial spot of, 241
  brown rot of, 142–145, **405**
  Cercospora leaf spot of, 156
  cherry rusty mottle virus and, 201
  crown gall and, 13
  Dematophora root rot of, 39–41
  Dothiorella canker and, 162
  Eutypa dieback of, 164–167, **407**
  fruit rot of, 241
  fungal diseases of, 141–197, 241. See also specific diseases
  green-fruit rot of, 169–170, **408**
  gummosis of, prune dwarf virus and, 206
  harvesting techniques, 231
  leaf curl of, **410**
  peach mosaic of, 202, 203
  as *Pezicula malicorticis* host, 35
  Phloeosporella shothole of, 241
  Phyllosticta leaf spot of, 241
  Phytophthora pruning wound canker and, 178
  Phytophthora root and crown rot of, 265, 267, 269, 270
  postharvest decay of, 231
  powdery mildew of, 181–186, **412**
  prune dwarf virus of, 206–208
  prunus stem pitting of, 212–214
  replant problems and, 276
  ring pox of, 241
  rootstocks, Phytophthora root and crown rot and, 269, 270
  rust of, 188–190
  scab of, 191–192
  shoestring (Armillaria) root rot of, 193, 282, 283
  shothole of, 193–196, **415**
  silver-leaf of, 197
  stem pitting virus of, 205
  tomato ringspot virus and, 205
  trunk and limb gall of, 241
  trunk rot of, 241
  Verticillium wilt of, 287–288, 289, 290
  virus or viruslike diseases of, 202, 203, 205, 241
  yellow bud mosaic virus and, 214
Apricot brown rot. See Brown rot, European
Apricot ring pox, 241
Armillaria root rot. See Shoestring (Armillaria) root rot
Armillaria spp.
  A. mellea
    Dematophora root rot vs., 40
    shoestring root rot and, 70, 193, 280–283, 318, 347

  A. tabescens, Clitocybe root rot and, 122, 242
  crown gall and, 257
  wood decay and, 292, 293, 295, 296
Aronia spp., Fabraea leaf spot and, 46
Ascospora beijerinckii, shothole and, 194
Ash. See *Fraxinus* spp.
Ash, mountain. See *Sorbus* spp.
Asian pear
  See also Pear black-end and, 91
  cultivation history, 11
  fireblight and, 19
  scab and, 67–68
  shoestring (Armillaria) root rot and, 70
  water core and, 95
Aspergillus spp.
  A. chevalieri, mycotoxins and, 235
  A. niger
    endosepsis of fig and, 357
    fig smut and mold and, 360–361, **437**
    pomegranate smut disease and, 373, **438**
  apricot decay and, 231
  mycotoxins and, 274, 275
  nectarine and peach decay and, 233
  plum decay and, 234
  prune decay and, 235, 236
  walnut kernel mold and, 316
Aspidiotus perniciosus, Cytospora canker and, 158
Athelia spp.
  A. bombacina, for scab, of apple, 63
  A. rolfsii, southern blight and, 71
Aureobasidium pullulans, for scab, of apple, 63
Avocado
  Dematophora root rot of, 39, 41
  Dothiorella canker and, 163
  Verticillium wilt of, 287

B

Bacillus spp.
  B. amylovorus, 17. See also *Erwinia amylovora*
  B. subtilis
    for European (Nectria) canker, 45
    for stone fruit decay, 149, 234
  B. vulgatus, for Verticillium wilt, 290
Bacterial diseases, 246–261
  bacterial canker and blast, 246–254, **423–424**
    of almond, 246, **423–424**
    Cytospora canker and, 158, 246, **424**
    of pear, 246–247
    of pome fruit, 13–14
    of stone fruit, 132, 158, 246, 247–254, **423, 424**
  bacterial canker of fig, 355
  bacterial spot of stone fruit, 241
  blossom blast and canker of pome fruit, 13–14, **383**
  crown gall, 257–261
    of almond, 259, 260, **425**
    of English walnut, 309
    of fig, 365

  of pome fruit, 15–16
  of stone fruit, 132
  deep-bark canker of English walnut, 309–311, **431**
  of fig, 355, 365
  fireblight of pome fruit, 16–21, **383**
  foamy canker of almond, 132–133, **403**
  hairy root of apple and quince, 25–26, **384**
  leaf scorch of almond, 133–135, **403**
  olive knot, 341–343, **435**
  phony peach, 241
  of pome fruit, 13–26
    bacterial blast and canker, 13–14, 246–254, **383**
    crown gall, 15–16, 257–261
    fireblight, 16–21, **383**
    hairy root, 25–26, **384**
    replant problems and, 278
    shallow-bark canker of English walnut, 312–323, **432**
  of stone fruit, 132–135, 241
    bacterial canker and blast, 132, 158, 246, 247–254, **423, 424**
    bacterial spot, 241
    chilling injury and, 218
    crown gall, 132, 257–261, **425**
    foamy canker of almond, 132–133, **403**
    leaf scorch of almond, 133–135, **403**
    phony peach, 241
  of walnut, 305–313
    blight, 305–308, **431**
    crown gall, 309
    deep-bark canker, 309–311, **431**
    shallow-bark canker, 312–313, **432**
Bactericides, 379
Bacterium fici. See *Xanthomonas campestris*
Bacterium savastanoi, 341
Bacticin, for crown gall, 260
Baldwin spot of apple. See Bitter pit
Balsamorhiza sagittata (Balsam root), cherry rasp leaf virus and, 200
Band canker. See Dothiorella canker
Barberry, Verticillium wilt and, 289
Bark, papery or sappy, of apple, 56–57, **388**
Bark measles, of pear, 81–82
Basidiomycetes spp., wood decay and, 291–298
Beans, tomato ringspot virus and, 205, 215
Bees, fireblight and, 18, 19
Beets, sugar, southern blight and, 72
Benomyl, 379
  See also Benomyl drench; Benzimidazole fungicides
  for anthracnose, of English walnut, 313
  for Botryosphaeria panicle and shoot blight of pistachio, 330–331
  for Botrytis blossom and shoot blight of pistachio, 332
  for brown rot, of stone fruit, 145, 148
  for bull's-eye rot prevention, 103
  for Cytospora canker, 159
  for downy leaf spot of walnut, 316
  for European (Nectria) canker, 44, 45
  for Eutypa dieback, 167

for Fabraea leaf spot, 47
for green-fruit rot, 170
nectarine pox and, 221
for perennial canker control, 37
for Phomopsis rot prevention, 107
for postharvest decay prevention
  in apricot, 231
  in nectarine and peach, 233, 234
  in sweet cherry, 230
for powdery mildew, 52, 185
resistance to, 381
russet scab of prune and, 188
for scab
  of apple, 62, 63
  of loquat, 67
  of pear, 69
  of stone fruit, 192
side rot and, 110
for stone fruit, brown rot of, 145
for Verticillium wilt, 291
Benomyl drench
for Alternaria rot, 98
for blue-mold or soft rot, 101
chemical injury and, 113
for endosepsis of fig, 356–357
for gray mold rot, 105
Benzalkonium chloride, for fireblight, 21
Benzimidazole fungicides, 378, 379, 380
  See also Benomyl
  Alternaria rot and, 98
  blue-mold or soft rot and, 101
  for brown rot, of stone fruit, 148
  for bull's-eye rot prevention, on apple, 38
  for Eutypa dieback, 167
  for gray mold rot, 105
  for perennial canker, on pome fruit, 37
  postharvest decay and, in apricot, 231
  resistance to, 227, 381
  russet scab of prune and, 188
  for scab
    of pear, 69
    of stone fruit, 192
Benzoates, and powdery mildew, 52
Berberis spp. (Barberry), Verticillium wilt and, 289
Betula spp. (Birch)
  cherry leaf roll virus and, 320
  European (Nectria) canker and, 44
  silver leaf and, 285
Binapacryl, for powdery mildew, 52
Biological controls
  for Armillaria root rot, 283
  for blue-mold or soft rot, 101
  for brown rot, of stone fruit, 145, 149
  for chestnut blight, 370
  for crown gall, 260–261
    of stone fruit, 132
  for Cytospora canker, of stone fruit, 159
  for Dematophora root rot, 41
  for European (Nectria) canker, 45
  for Eutypa dieback, 167
  for fireblight, 21
  for gray mold rot, 101, 105
  of Monilinia decay, in nectarine and peach, 234

mycotoxins and, 275
for Phytophthora root and crown rot, 49, 271
for replant problems, 279
for sappy bark, 57
for scab, of apple, 63
for silver-leaf, 286
for southern blight, 72
for Verticillium wilt, 290
Birch. See *Betula* spp.
Birds
  fireblight and, 19
  olive knot and, 342
  shallow-bark canker of walnut and, 312–313
Bird's-eye spot. See Leaf spot, of olive
Bisulfites, 379
Bitertanol, 380
  for powdery mildew, 52
  for scab, of pear, 69
Bitter pit
  of apple, 87–89, **393**
  of pear, 87–89, 119, **393**
Bitter rot
  of peach and nectarine, 241
  of pome fruit, 121
Blackberry
  cherry leafroll virus and, 320
  Dematophora root rot and, 40–41
  Pierce's disease and, 134
Black-end, of pear, 91, **393**
Black heart. See Verticillium wilt
Black knot, of plum and cherry, 241–242
Blackline, of English walnut, 319–321, **433**
Black pox, of pome fruit, 122
Black root rot, of pome fruit, 122
Black rot
  of apple, 108, 121, **396**
  of pome fruit, 108, 121, **396**
Black spot, of apple. See Scab
Bladder plum. See Plum pockets
Blast, bacterial canker and, 246–254
  Cytospora canker and, 158, 246
  of pome fruit, 13–14, 246–247
  of stone fruit, 132, 158, 246, 247–254
Blast, blossom, of pome fruit, 13–14
Blastomyces spp., for Verticillium wilt, 290
Blastophaga psenes (Fig wasp), 354, 355, 357, 358
Blight, walnut, 305–308, **431**
"Blight cutters," 20
Blister bark of apple, **392**
Blister canker of pear, 81, 82, **391**
Blister spot of pome fruit, 13, 122
Bloom removal, for fireblight control, 21
Blossom blast and canker
  of pear, **383**
  of pome fruit, 13–14
Blossom blight
  of pome fruit. See Fireblight
  of stone fruit, postharvest gray mold and, 228
Blossom blight, brown rot or Monilinia. See Brown rot, European
Blossom-end rot, of pome fruit, 122

Blossom rot, of stone fruit. See Green-fruit rot
Blotch, of pome fruit, 122
Blueberry, Dothiorella canker and, 163
Blumeriella jaapii, leaf spot of cherry and, 177–178
Boarder tree, prune dwarf virus and, 206
Bordeaux mixture, 379
  for anthracnose
    of almond, 141
    of English walnut, 313
    of pome fruit, 37
  for bacterial canker and blast, 253, 254
    of pome fruit, 14
  for blight of walnut, 307, 388
  for branch wilt, of walnut, 315
  for brown rot, of stone fruit, 144
  downy leaf blight of walnut and, 316
  for Fabraea leaf spot, of pome fruit, 47
  for fireblight, of pome fruit, 20, 21
  for leaf spot of cherry, 178
  for leaf spot of olive, 344, 346
  for melaxuma twig blight, 317
  for olive knot, 343
  for Pacific Coast pear rust, 55
  Phomopsis canker of fig and, 360
  for Phytophthora root and crown rot, of pome fruit, 49
  for plum pockets, 181
  for shothole, 195
  for stone fruit, brown rot of, 144
Borers, Cytospora or Rhodosticta canker and, 158
Boron deficiency
  bitter pit of apple or pear and, 88
  internal bark necrosis and, 92, 93
  pear blossom blast and, 14
  symptoms of, 93
Boron excess, core browning and, 115
Botran. See DCNA
Botryosphaeria spp.
  B. *dothidea,*
    pistachio and, 329–331, **434**
    Dothiorella canker and, 161–163, 242
    white rot and, 121
  B. *obtusa,* black rot and, 108, 121
  gummosis of peach and, 242
Botryotinia fuckeliana, 104, 169
Botrytis blossom and shoot blight of pistachio, 331–332, **434**
Botrytis limb blight or dieback of fig, 356, **436**
Botrytis rot of persimmon, **438**
Botrytis spp.
  apple decay and, 104–105
  apricot decay and, 231
  B. *cinerea,*
    blossom and shoot blight of pistachio and, 331–332
    endosepsis of fig and, 357
    gray mold and, 104–105
    green-fruit rot and, 169–170, **408**
    limb blight or dieback of fig and, 356, **436**
  B. *mali,* 104

fig mold and, 360, 361, **437**
nectarine and peach decay and, 232, 233, 234
in pomegranate, **438**
sweet cherry decay and, 228–229, 230, **408, 421**
Box rot, of prunes, 236
Boysenberry
Dematophora root rot and, 39, 41
Brambles
cherry leafroll virus and, 320
crown gall of, 258
Branch wilt
of English walnut, 314–315, **432**
Dothiorella canker of almond and, 162
of fig, 356
Breakdown, internal, 116–117
Brook's spot (Phoma fruit spot), of apple, 122
Brown core. *See* Core browning
Brown heart. *See* Carbon dioxide injury
Brownline, prune, 204–206, **417**
Brownline and decline, almond (ABLD), 198
Brown rot
of fig. *See* Endosepsis
of pome fruit, 108, **396**
of stone fruit, 142–149, **405**
American, 145–149
American vs. European, 142, 143–144, 146
cherry, 229
European (Brown rot blossom blight), 142–145
fungicides for, 378
green-fruit rot and, 169, **408**
hull rot of almond and, 171
postharvest, 228, 229
Buckskin. *See* X-disease
Bud failure of almond
infectious, 222
noninfectious, 221–224, **419**
Bull mission, of almond, 225–226
Bull's-eye rot, of pome fruit, 34, 36, 37–38, 103–104, **395**
Butternut, Dematophora root rot and, 40

C

Calcium, nectarine and peach decay and, 234
Calcium chloride dip
for bitter pit, 89
for blue-mold or soft rot, 101
chemical injury and, 112, **397**
for gray mold rot, 105
Calcium cyanimide, for brown rot of stone fruit, 148
Calcium deficiency
bacterial canker and blast and, 250
bitter pit of apple or pear and, 89
Calcium hydroxide, for internal bark necrosis (measles), 93
Calcium hypochlorite, brown rot of stone fruit and, 149
Calcium levels, water core of apple and, 95
Calcium nitrate

for bitter pit of apple or pear, 89
chemical injury and, 112, 113
Calcium polysulfides, 379
for leaf curl of peach, 176
Calcium salts
chemical injury and, 112, 113, **397**
internal breakdown and, 117
soggy breakdown and, 117
California peach blight. *See* Shothole
Calyx rot. *See* Green-fruit rot
Camarosporium pistaciae, pistachio blight and, 335
Candida spp.
C. albicans, peach fruit rot and, 241
C. chalmersi, prune decay and, 235
C. krusei, prune decay and, 235
C. parapsilosis, prune decay and, 235
Canker disease, bacterial, 246–254
See also Bacterial diseases, bacterial blast and canker of pome fruit
perennial, 32–38
Pseudomonas, 13–14
of stone fruit, 132
Capopsylla pyricola, 28. *See also* Psylla
Captafol, 379–380
for anthracnose, of almond, 141
for black rot prevention, 108
for Botryosphaeria panicle and shoot blight of pistachio, 330–331
for Botrytis blossom and shoot blight of pistachio, 332
for Cytospora canker, 159
for European (Nectria) canker, 44, 45
for Fabraea leaf spot, 47
for leaf curl of nectarine and peach, 176
for leaf spot of olive, 346
for Phytophthora root and crown rot, 270–271
for silver-leaf, 286
Captan, 379–380
See also Captan drench
for anthracnose, 37
of almond, 141
for Botryosphaeria panicle and shoot blight of pistachio, 330–331
for brown rot, of stone fruit, 145, 148, 149
for bull's-eye rot, on apple, 37
in combined treatments, 379
for European (Nectria) canker, 45
for Fabraea leaf spot, 47
for fig surface mold and Alternaria rot, 363
for leaf blight of almond, 173
for leaf curl, of nectarine and peach, 176
for leaf spot of cherry, 178
nectarine pox and, 221
for postharvest decay prevention
in apricot, 23
in nectarine and peach, 233
in sweet cherry, 230
for russet scab of prune, 188
for scab
of apple, 62
of stone fruit, 192
for shothole, 195, 196
Captan drench or dip

for blue-mold or soft rot, 101
for gray mold rot, 105
prune decay and, 236
Captan-triforine mixture, for scab, of pear, 69
Carbamates
for postharvest decay prevention, in sweet cherry, 230
Carbendazim
for Dematophora root rot, 41
for European (Nectria) canker, 44, 45
for scab, of pear, 69
Carbon dioxide
for European (Nectria) canker (fruit rot) control, 45
Jonathan spot and, 113
for soft scald prevention, 114
Carbon dioxide injury
to apple, 112, 116, **397**
to pear, 119, **397**
Carbon disulfide, 380
for Armillaria root rot, 282–283
for Dematophora root rot, 41
Carpophilus spp.
C. freemani, Ceratocystis canker and, 154–155
C. hemipterus
Ceratocystis canker and, 154–155
fig smut and mold and, 361
souring of fig and, 361
Carrots, southern blight and, 71, 72
Carya illinoensis. *See* Pecan
Castanea spp. *See* Chestnut
Catastrophic diseases, pear decline, 27–30, **384**
Ceanothus spp., Eutypa dieback and, 166
Cedar apple rust, 122
Cephalosporium fici, fig leaf spot and, 366
Ceratobasidium stevensii, thread blight and, 122
Ceratocystis fimbriata, 153–155
Ceratocystis (mallet wound) canker
of almond and prune, 153–155, **406**
Phytophthora canker vs., 179
Cercospora spp.
C. circumscissa
anthracnose of almond and, 141
Cercospora leaf spot and, 156
C. cladosporioides, Cercospora leaf and fruit spot and, 344
Cerotelium fici, fig rust and, 365
CGA-64251 drench, for blue-mold or soft rot, 101
Chaenomeles spp.
apple mosaic virus and, 78
C. japonica
Pacific Coast pear rust and, 54
quince sooty ring spot and, 84
Chaetomella, prune decay and, 235
Chaetomium globosum, for scab, of apple, 63
Chalaropsis spp., mallet wound and, 154
Chemical injury
to apple, 112–113, **397–398**
to pear, 118, **397, 398**
wood-decay fungi and, 294
Chenopodium spp.

C. *amaranticolor*,
  prunus necrotic ringspot and, 211
  tomato ringspot virus and, 80, 205
C. *quinoa*,
  cherry rasp leaf virus and, 200
  tomato ringspot virus and, 80, 205, 213
Cherry (*Prunus* spp.), 129–130
  See also Cherry, sour; Cherry, sweet; Stone fruit
  bacterial canker and blast of, 251, 253, **423, 424**
  bark blister of, 199
  black knot of, 241–242
  brown rot of, 142–149
  Cercospora leaf spot of, 156
  cherry mottle leaf (CML) virus of, 198–199, **416**
  cherry stem pitting (CSP) virus of, 201, **416**
  crown gall of, 259, 260
  damping-off and root rot of, 241
  Dematophora root rot of, 39–41
  Duke, 129
  Eola rasp leaf of, 214, 216
  flat limb virus and, 75
  fungal diseases of, 141–197, 241. *See also specific diseases*
  green-fruit rot of, 169–170, **408**
  leaf curl of, 173–174, **409**
  leaf spot of, 177–178, **411**
  mycoplasmalike disease of, X-disease, 136–139, **404**
  necrotic rusty mottle (NRM) of, 199, **416**
  as *Pezicula malicorticis* host, 35
  Phytophthora root and crown rot of, 181–186, 264, 265, 267, 268, 269, 270
  powdery mildew of, 181–186, **412**
  prunus stem pitting of, 212–214
  Pythium root rot of, 241
  rasp leaf of, 200, **416**
  replant diseases of, 276–279
  rootstocks, 129
    bacterial canker and blast and, 251
    Dematophora root rot and, 40
    Phytophthora root and crown rot and, 269, 270
  rust of, 188–190
  rusty mottle of, 200–201, **416**
  scab of, 191
  shothole of, 193–194, **415**
  Verticillium wilt of, 287, 288
  virus or viruslike diseases of, 198–216, 241. *See also specific diseases*
  walnut blackline and, 320
  X-disease of, 136–139, **404**
Cherry, sour (*Prunus cerasus*), 129–130
  See also Cherry; Stone fruit
  Dematophora root rot of, 39–41
  prune dwarf virus on, 206–208
  prunus stem pitting of, 212–214
  silver-leaf of, 285
  yellows of, 206, 207
Cherry, sweet (*Prunus avium*), 129
  See also Cherry; Stone fruit
  Alternaria decay of, 229–230, **421**

bacterial canker and blast of, 251, 252, 253, **423, 424**
black root rot of, 122
Botrytis decay of, 228–229, **421**
cherry mottle leaf (CML) of, 198–199, 205, **416**
crinkle leaf and deep suture of, 219–220, **419**
Dematophora root rot of, 39–41
Dothiorella canker and, 163
gray-mold rot of, 228–229
leaf curl of, 173–174
Monilinia decay (brown rot) of, 228, 229, **421**
mycoplasmalike disease of, 136–139, **404**
necrotic rusty mottle (NRM) of, 199, **416**
Penicillium decay of, 228, 229
Phytophthora root and crown rot of, 264, 265, 269, 270
postharvest diseases of, 227–231, **421**
prune dwarf virus on, 206–208
prunus necrotic ringspot of, 209–211, **417**
prunus stem pitting of, 212–214
rasp leaf of, 200, **416**
Rhizopus decay of, 228, 229, **421**
rootstocks, 129
  Phytophthora rot and, 269, 270
shoestring (Armillaria) root rot of, 193, 282, 283
tatter leaf of, prune dwarf virus and, 206
Verticillium wilt of, 287, 288
walnut blackline and, 320
X-disease of, 136–139, **404**
yellow bud mosaic virus of, 214–216, **418**
Cherry leafroll virus (CLRV), blackline and, 319–321
Cherry mottle leaf virus, tomato ringspot virus and, 205
Cherry rasp leaf virus (CRLV), 200, **416**
Cherry recurrent ringspot virus, 209
Cherry ringspot. *See* Prunus necrotic ringspot
Cherry rugose mosaic strain (CRMS), 209, 210
Cherry rusty mottle (CRM), 200–201, **416**
Cherry stem pitting (CSP), 201, **416**
Chestnut (*Castanea* spp.), diseases of, 285, 370, **438**
Chickweed. *See Stellaria media*
Chilling canker of peach, 218, **418**
Chip budding, apple stem pitting and, 79
Chlorine solution washes
  Alternaria rot and, 98
  for blue-mold or soft rot, 101
  endosepsis of fig and, 358
  for gray mold rot, 105
  for Mucor rot, 109
  nectarine and peach decay and, 233, 234
  prune decay and, 236
  sweet cherry decay and, 230
Chlorobromopropene, for Verticillium wilt, 290
Chloropicrin, soil fumigation with,
  for Armillaria root rot, 283
  for bacterial canker and blast, 253
  for crown gall, 260

for Dematophora root rot, 41
for hairy root, 26
for replant problems, 278, 279
for southern blight, 72
for Verticillium wilt, 290, 333
Chlorothalonil
  for anthracnose, of black walnut, 313
  for Botryosphaeria panicle and shoot blight of pistachio, 330–331
  for Fabraea leaf spot, 47
  for fig surface mold and Alternaria rot, 363
  for leaf curl of peach, 176
  for russet scab of prune, 188
  for scab, of stone fruit, 192
  for shothole, 195, 196
Chlorotic mottle, prune dwarf virus and, 206
Chlorotic ringspot, prune dwarf virus and, 206
Chokecherry
  cherry rusty mottle virus and, 201
  Eutypa dieback of apricot and, 166
  X-disease of cherry and peach and, 136, 138–139
Chondrostereum purpureum
  silver-leaf and, 197, 285–286
  systemic pesticide for, 378
  wood decay and, 292, 294, 296
Chrysanthemum, Verticillium wilt and, 289
Chrysosporium, prune decay and, 235
Chymomyza procnemoides, Ceratocystis canker and, 154–155
Citrus
  Dematophora root rot and, 39, 41
  Dothiorella canker and, 163
  fungal diseases of
    Dematophora root rot, 39, 41
    Verticillium wilt, 287
  replant diseases of, 276–279
Cladosporium-Hormodendrum rots, of pome fruit, 108
Cladosporium spp.
  apricot decay and, 231
  C. carpophilum, scab and, 191–192
  C. caryigenum, pecan scab and, 371
  C. herbarum, 108
    endosepsis of fig and, 357, 358
    fig surface mold and rot and, 362–363
    mycotoxins and, 274
  fig mold and, 360
  prune decay and, 235
Clasterosporium carpophilum, shothole and, 194
Clathridium corticola, 172
Clay soil, European (Nectria) canker and, 44
Clitocybe root rot
  of pome fruit, 122
  of stone fruit, 242
Clitocybe tabescens, 122, 242
  peach replant disease and, 277
Clover
  apple union necrosis and decline and, 81
  Botrytis infection and, 104
CLRV. *See* Cherry leafroll virus
Cobalt irradiation, for apple rubbery wood, 79

Coccomyces hiemalis, leaf spot of cherry and, 177–178
Colladonus spp., X-disease of cherry and peach and, 138
Collar rot of apple, 48. *See also* Phytophthora root and crown rot
Colletotrichum gloeosporioides, anthracnose of almond and, 141
Copper compounds, 379
  for European (Nectria) canker, 44, 45
Copper crystals, for Phytophthora root and crown rot, 270–271
Copper, fixed, 379
  for anthracnose, of pome fruit, 37
  for bacterial blast and canker, of pome fruit, 14
  for bacterial canker and blast, 253
  for blight of walnut, 308
  for Botryosphaeria panicle and shoot blight of pistachio, 330–331
  for brown rot, of stone fruit, 145
  for leaf spot of olive, 346
  for Phytophthora root and crown rot, 49
  for russet scab of prune, 188
  for shothole, 195
Copper hydroxide, 379
  for Botrytis blossom and shoot blight of pistachio, 332
  for fireblight of pome fruit, 21
Copper-impregnated paper wrappers, for gray mold prevention, 105
Copper-lime dust
  for fireblight of pome fruit, 20
  for walnut blight, 307
Copper oxychloride, for bacterial blast and canker, 14
Copper oxychloride sulfate, 379
  for fireblight, of pome fruit, 21
Copper pastes, for European (Nectria) canker, 45
Copper sprays, 379
  for bacterial canker and blast, 253
  for blight of walnut, 307
  for brown rot, of stone fruit, 148
  for Cercospora leaf spot, of stone fruit, 156
  deep-bark canker of walnut and, 311
  for Fabraea leaf spot, 47
  for fireblight, 20, 21
  for leaf curl
    of cherry, 174
    of nectarine and peach, 176
  nectarine pox and, 221
  for Phytophthora root and crown rot, 270–271
  for postharvest decay prevention, in sweet cherry, 230
  for rust protection, 54
  for shothole, 195
Copper sulfate, 379. *See also* Bordeaux mixture
Coprinus rot, of pome fruit, 108–109, **396**
Coprinus, 108
  wood decay and, 296
Core breakdown of pear, 119
Core browning of apple, 115, **400**

Core flush. *See* Core browning
Coremium spp., 100
Coriolus versicolor, 56
Cork rot of apple or pear. *See* Alternaria rot
Cork spot of pear. *See* Bitter pit
Corky growth of almond kernel, 218–219, **418**
Cornus (Dogwood)
  cherry leafroll virus and, 320
  leaf blight of almond and, 172
  melaxuma twig blight and, 317
Corticium spp.
  C. galactinum, white root rot and, 122
  C. microsclerotium, fig leaf blight and, 365
  C. rolfsii, 71
  C. stevensii, thread blight and, 122
Corylus spp. *See* Filbert
Coryneum beyerinckii, shothole and, 194
Corynosis. *See* Shothole
Cotoneaster spp.
  Fabraea leaf spot and, 46
  fireblight and, 18
Cotton, Verticillium wilt and, 197, 288, 289, 290, 332, 333
Cover crops
  Armillaria root rot and, 282
  Botrytis infection and, 104
  Verticillium wilt of olive and, 347
Cowpeas
  apple mosaic virus and, 77
  tomato ringspot virus and, 205, 215
Crab apple (*Malus* spp.)
  anthracnose and perennial canker and, 35
  pear rusts and, 54, 55
  as *Pezicula malicorticis* host, 35
  Phytophthora root and crown rot and, 265
  stem pitting and, 79–80
Crataegus spp. (Hawthorn)
  apple mosaic virus and, 78
  apple stem pitting and, 79
  C. douglasii, Pacific Coast pear rust and, 54
  C. oxyacantha (English hawthorn), fireblight and, 18
  European (Nectria) canker and, 44
  Fabraea leaf spot and, 46
  Pezicula malicorticis and, 35
  Rocky Mountain pear rust and, 55
Crazy top of almond, 221, 222
m-cresol, in olive knot treatment, 343
Criconemella spp.
  C. xenoplax,
    bacterial canker and blast and, 250, 253
    Cytospora canker and, 159
    replant diseases and, 277
Crinkle leaf and deep suture of sweet cherry, 219–220, **419**
CRLV. *See* Cherry rasp leaf virus
CRM. *See* Cherry rusty mottle
CRMS. *See* Cherry rugose mosaic strain
Crop rotation, southern blight and, 72
Crown gall, 257–261
  of almond, 259, 260, **425**
  of English walnut, 309
  of fig, 365
  hairy root vs., 25

  of pome fruit, 15–16
  of stone fruit, 132, 257–261, **425**
Crown rot. *See* Phytophthora root and crown rot
Cryptococcus laurentii, for blue-mold control, 101
Cryptosporiopsis spp.
  C. corticola, European (Nectria) canker and, 44
  C. curvispora, 33, 34, 35
CSP. *See* Cherry stem pitting
Cucumber *Cucumis sativus*
  apple mosaic virus and, 77
  cherry rasp leaf virus and, 200
  prunus necrotic ringspot and, 211
  prunus stem pitting and, 213
  tomato ringspot virus and, 215, 216
  Verticillium wilt and, 288
Cultural practices
  Armillaria root rot and, 282
  bacterial canker and blast and, 252
  crown gall and, 259–260
  grapes, leaf scorch (Pierce's disease) and, 134
  Phytophthora root and crown rot and, 270, 271
  pome fruit
    apple union necrosis and decline and, 81
    European (Nectria) canker and, 44
    Fabraea leaf spot and, 47
    fireblight and, 19
    sappy bark and, 57
    southern blight and, 72
  stone fruit
    bacterial blast and canker and, 132
    brown rot and, 148
    Eutypa dieback and, 167
    green-fruit rot and, 169
    leaf scorch of almond and, 134
    noninfectious bud failure of almond and, 223–224
    X-disease of cherry and peach and, 139
  Verticillium wilt of olive and, 347
  walnut blight and, 307
  walnut branch wilt and, 315
  walnut kernel mold and, 316
  wood-rot fungi and, 294–295
Cupric hydroxide
  for anthracnose of English walnut, 313
  Phytophthora pruning wound canker and, 179–180
Cuprozineb, for leaf spot of olive, 346
Curly leaf. *See* Crinkle leaf
Currant, silver-leaf and, 285
Cycloconium oleaginum. *See Spilocaea oleaginea*
Cycloheximide, 381
  for powdery mildew, 185
Cydonia oblonga. *See* Quince
Cylindrocarpon mali, European (Nectria) canker and, 42
Cylindrosporium padi, 177, 178
Cytospora canker
  of English walnut, 318
  of fig, 365
  of stone fruit, 156–160, **406**

Cytospora spp.
See also Cytospora canker
C. *cincta*, 157, 158, 159
C. *leucostoma*, 156–160
  bacterial canker and blast and, 246
C. *rubescens*, 157
  prune infected by, **424**
Cytosporina sp., Eutypa dieback and, 164

D

2,4-D, for Verticillium wilt, 290
Dacus olea (Olive fly), olive knot and, 342
Dagger nematodes
  cherry rasp leaf virus and, 200
  prune brownline virus and, 205
  tomato ringspot virus and, 205
Dandelion. See *Taraxacum officinale*
DCNA, 378, 379, 380
  for apricot decay, 231
  for brown rot of stone fruit, 149
  for green-fruit rot, 170
  for nectarine and peach decay, 233, 234
  for plum decay, 234–235
  for sweet cherry decay, 230
DD fumigant, for yellow bud mosaic virus, 216
Decenylsuccinic acid, chilling canker of peach and, 218
Decline
  in pear, 27–30, **384**
  wood decay and, 292
Deep-bark canker of English walnut, 309–311, **431**
Deep suture, of sweet cherry, 219–220
Dehydroacetic acid, for postharvest decay prevention, in sweet cherry, 230
Dematophora root rot
  Armillaria root rot vs., 40
  of English walnut, 318
  of pome fruit, 39–41
Detergents, chemical injury and, 112, 113
Diaporthe spp.
  D. *cinerascens*, 359
  D. *perniciosa*, 107
Dibotryon morbosum, black knot and, of plum and cherry, 241–242
Dibromochloropropane, bacterial canker and blast and, 253
Dicarboximides, 380
  for brown rot, of stone fruit, 145, 148–149
  resistance to, 381
Dichlone, 379
  for anthracnose, of pome fruit, 37
  for brown rot, of stone fruit, 145, 148
  for leaf blight of almond, 173
  for leaf curl of nectarine and peach, 176
  for leaf spot of cherry, 178
  for scab, of stone fruit, 192
  for shothole, 195
1, 3-dichloropropene
  for bacterial canker and blast, 253
  for replant problems, 278

Dichloropropane-dichloropopene mixture, soil fumigation with, for hairy root, 26
Dicloran. See DCNA
Dieback
  of fig, 356
  wood decay and, 292
Dieback canker of filbert, 371
Dimethyl acrylates, for powdery mildew, 52
Dimethyldithiocarbamate
  ferric. See Ferbam
  for leaf curl, of nectarine and peach, 176
n-(2,6,-dimethyl-phenyl)-n-(methoxyacetyl) alanine methyl ester. See Metalaxyl
Diniconazole, 380
Dinitro-alkyl-phenyl-acrylate. See Binapacryl
Dinitro-o-cresol compounds (DNOC)
  for crown gall, 260
  for powdery mildew, 52
  for scab, 63
Dinitro sprays, for leaf spot of cherry, 178
Dinocap, for powdery mildew, 51, 52, 185, 186
Diospyros spp. See Persimmon
Diphenylamine (DPA)
  apple injury by, **398**
  dip treatment with
    for blue-mold or soft rot, 101
    for bull's-eye rot, 38
    chemical injury and, 112
    for gray mold rot, 105
    scald and, 114
  wrap with, scald and, 114
Diplocarpon sp., Fabraea leaf spot and, 46–47
Diplodia canker and dieback of English walnut, 318
Dip treatments
  See also Washes; specific chemicals
  for Alternaria rot, 98
  for bitter pit, 89
  for blue-mold or soft rot, 101
  for bull's-eye rot, 38, 103
  chemical injury and, 112–113
  for European (Nectria) canker (fruit rot), 45
  for fig surface mold and Alternaria rot, 363
  for gray mold rot, 105
  for nectarine and peach, 233
  for postharvest decay prevention, in nectarine and peach, 233
Disinfestants, for fireblight control, 20
Dithane Z78, for Verticillium wilt, 290
Dithianon, for European (Nectria) canker, 44
Dithiocarbamates, 379
  for Coprinus rot, 109
Dithiocarbamate sprays, for bull's-eye rot prevention, 103
DNOC. See Dinitro-o-cresol compounds
Dodine, 378, 380
  for anthracnose of English walnut, 313
  for Fabraea leaf spot, 47
  for leaf curl of peach, 176
  for leaf spot of cherry, 178
  for scab
    of apple, 62
    of loquat, 67

  of pear, 69
Dormant sprays
  brown rot of stone fruit and, 144
  leaf curl of peach and, 176
  powdery mildew of pome fruit and, 52
  for shothole, 195
Dothiorella canker of almond, 161–163, 242, **407**
Dothiorella gregaria, melaxuma twig blight and, 316–317
Downy leaf spot of English walnut, 315–316, **432**
DPA. See Diphenylamine
Draeculacephala minerva, leaf scorch bacteria and, 134
Dried fruit
  decay and, 235
  mycotoxins and, 274–275
Dried-fruit beetle. See *Carpophilus hemipterus*
Drip irrigation, scab of almond and, 191
Drosophila spp.
  D. *ampelophila*, fig smut and mold and, 361
  D. *melanogaster*, Ceratocystis canker and, 154–155
  souring of fig and, 361

E

EBI fungicides. See Ergosterol biosynthesis-inhibiting fungicides
Elder, American, Pierce's disease and, 134
Elderberry, cherry leafroll virus and, 320
Elgetol Extra. See Dinitro-o-cresol compounds
Elgetol-methanol mixture, olive knot and, 343
Endoconidiophora spp., 154
Endosepsis of fig, 356–358, **436**
Enterobacter aerogenes
  for Phytophthora rot, 49
  for replant problems, 279
Entomosporium mespili, 46
Eola rasp leaf, of cherry, 214, 216
Epinasty, quince sooty ring spot and, 83
Eradicants, 378, 379
Ergosterol biosynthesis-inhibiting (EBI) fungicides
  See also Sterol biosynthesis inhibiting fungicides
  scab of apple and, 62, 63
  scab of pear and, 69
Eriobotrya japonica. See Loquat
Eriophyes spp. (Eriophyid mites)
  cherry mottle leaf and, 199
  fig mosaic and, 364
  filbert dieback canker and, 371
  peach mosaic and, 203
Eriosoma lanigerum. See Aphids, woolly apple
Erwinia spp.
  E. *amylovora*, fireblight of pome fruit and, 16–21
  E. *nigrifluens*, shallow-bark walnut canker and, 309, 312–313
  E. *rhapontici*, 310

E. rubrifaciens, deep-bark walnut canker and, 309–311
Erythricium salmonicolor, fig limb blight and, 366
Etaconazole, 380
 for powdery mildew, 52
 prune decay and, 236
Ethephon, walnut kernel mold and, 316
Ethoxyquin
 dip treatment with
  for bull's-eye rot, 38
  chemical injury and, 112, 113, **398**
  for gray mold rot, 105
  scald and, 114
 wrap impregnated with, scald and, 114
Ethylenebisdithiocarbamates, 379
Ethylene bromide, for Dematophora root rot, 41
Ethylene oxide, for prune mold prevention, 236
Ethyl phosphites, for Phytophthora root and crown rot, 270–271
European brown rot. *See* Brown rot
European (Nectria) canker, of pome fruit, 41–45, **385–386**
European plum line pattern virus, apple mosaic virus and, 77
Eutypa dieback, of apricot, 164–167, **407**
Eutypa spp., 164–167
Euzophera semifuneralis, Ceratocystis canker and, 154–155
Exoascus deformans, 174
Exosporina fawcetti, walnut branch wilt and, 314–315
Eye-end rot of fig. *See* Endosepsis
"Eye-rot," European (Nectria) canker and, 42

F

Fabraea leaf spot, 46–47, **386**
Fabraea maculata, 46
 peach leaf blight and, 241
Fallowing, for yellow bud mosaic virus, 216
False anthracnose, 32
False sting of apple, 76, **392**
Feijoa (*Feijoa sellowiana*), 370
 diseases of, 39–41, 370–371, **438**
 mycotoxins and, 274
Fenaminosulf, for Phytophthora root and crown rot, 270–271
Fenamiphos fumigation, bacterial canker and blast and, 253
Fenarimol, 380
 for powdery mildew, 52
 for scab, of pear, 69
Ferbam
 for brown rot of stone fruit, 148
 for Fabraea leaf spot, 47
 for fireblight, 21
 for green-fruit rot, 170
 for leaf spot of cherry, 178
 for scab, of pear, 69
 for shothole, 195
Fermate, for Verticillium wilt, 290

Ferric dimethyl-dithiocarbamate. *See* Ferbam
Fertilizing
 for Armillaria root rot, 282
 bacterial canker and blast and, 252
 European (Nectria) canker and, 44
 internal bark necrosis (measles) and, 93
 for olive knot and, 343
 replant disease and, 277, 279
 southern blight and, 72
 Verticillium wilt and, 290
Fieberiella florii (Leafhoppers), X-disease of cherry and peach and, 138, 139, **404**
Fig (*Ficus carica*), 354–355
 Alternaria rot of, 362–363, **437**
 anthracnose of, 366
 bacterial canker of, 355
 Botrytis and Sclerotinia limb blight or dieback and, 356, **436**
 crown gall of, 365
 Cytospora canker of, 365
 Dematophora root rot of, 39–41
 dried, mycotoxins and, 274
 endosepsis of, 356–358, **436**
 fungal diseases of, 356–363. *See also specific diseases*
 leaf blight of, 365
 leaf spots of, 366
 limb blight of, 366, **436**
 mosaic disease of, 363–364, **437**
 Phomopsis canker of, 358–360, **436**
 rust disease of, 365
 shoestring (Armillaria) root rot of, 282, 283, 365
 smut and mold of, 360–361, **437**
 souring of, 361, **437**
 thread blight of, 365
 Tubercularia canker of, 366
 twig blight of, 365
 virus or viruslike diseases of, 363–364
 wood decay of, **430**
Fig wasps, 354, 355, 357, 358, **436**
Filbert (*Corylus* spp.)
 diseases of, 371
 mycotoxins and, 274
Fireblight
 loquat, 13
 pear, **383**
  rootstocks and, 11, 12
 of pome fruit, 16–21
Firethorn. *See Pyracantha* spp.
Flat apple disease, **392**
 cherry rasp leaf and, 200
Flat limb, apple, 75, **389**
Floccaria glauca, 100. *See also Penicillium expansum*
Flood irrigation, Phytophthora root and crown rot and, 49, 267
Flusilazole, 380
 for leaf curl of peach, 176
 for powdery mildew, 185
 for scab, of apple, 62
Fly speck, of pome fruit, 121
Foamy canker of almond, 132–133, **403**
Folcid, for russet scab of prune, 188
Folicur, 380

Folpet
 for anthracnose of almond, 141
 for powdery mildew, 185
Fomitopsis cajanderi, wood decay and, 294, 296, **430**
Fools. *See* Plum pockets
Formaldehyde
 for Dematophora root rot, 41
 for replant problems, 278, 279
Fosetyl-A1
 Phytophthora pruning wound canker and, 179–180, 271
 for Phytophthora root and crown rot, 49, 271
Fragaria virginiana (Wild strawberry), apple union necrosis and decline and, 81. *See also* Strawberry
Fraxinus spp. (Ash), olive knot and, 342
Freezing injury
 anthracnose and perennial canker and, 36
 to apple, 117–118, **402**
 bacterial blast and canker and, 14
 chilling canker of peach, 218, **418**
 European (Nectria) canker and, 43
 leaf spot of cherry and, 177
 olive knot and, 342
 to pear, 117, 119, **402**
Friction marking
 of apple, 113, **398**
 of pear, 118, **398**
Fruit blight. *See* Fireblight
Fruit brown rot. *See* Brown rot, American
Fruit dips. *See* Dip treatments
Fruit flies. *See Drosophila* spp.
Fruit rot
 dip treatments for. *See* Dip treatments
 European (Nectria) canker and, 42, 44, 45
 of feijoa, 370–371
 of pome fruit
  Alternaria rot, 97–98, **394**
  anthracnose and perennial canker, 32–38
  bull's-eye rot, 34, 36, 37–38, 103, **395**
 of stone fruit, 241, **421–422**
 green-fruit rot, 169–170, **408**
Fruit spot
 of apple. *See* Bitter pit
 of olive, 344
Fruit-stem scars
 anthracnose and perennial canker and, 36
 European (Nectria) canker and, 43, 44
Fruit wash. *See* Postharvest wash
Fungal diseases, 263–298
 See also specific fungi
 of almond, 161–163
  shoestring (Armillaria) root rot, 193, 282, 283, **427**
  Verticillium wilt, 287–288, **428**
 Alternaria leaf and fruit spot of pistachio, 334
 Alternaria rot
  See also Alternaria spp.
  of apple and pear, 97–98, **394**
  of apricot, **422**
  of fig, 362–363, **437**

anthracnose
  of almond, 141
  false, 32
  of fig, 366
  olive and, 347
  of pome fruit, 32–38, **385**
  of stone fruit, 241
  of walnut, 313
bitter rot of pome fruit, 121
black knot of plum and cherry, 241–242
black pox, of pome fruit, 122
black rot, of pome fruit, 108, 121, **396**
Botryosphaeria panicle and shoot blight, of pistachio, 329–331, **434**
Botrytis blossom and shoot blight of pistachio, 331–332, **434**
Botrytis limb blight or dieback of fig, 356, **436**
Botrytis rot of persimmon, **438**
branch wilt
  of fig, 356
  of walnut, 314–315, **432**
brown rot of pome fruit, 108, **396**
Ceratocystis (mallet wound) canker of almond and prune, 153–155, **406**
  Phytophthora canker vs., 179
Cercospora leaf spot of stone fruit, 156
of chestnut, 370
Clitocybe root rot
  of pome fruit, 122
  of stone fruit, 242
Cytospora canker
  of English walnut, 318
  of fig, 365
  of stone fruit, 156–160, **406**
Dematophora root rot
  Armillaria root rot vs., 40
  of English walnut, 318
  of pome fruit, 39–41, **385**
Diplodia canker and dieback of walnut, 318
Dothiorella canker of almond, 161–163, 242, **407**
downy leaf spot of English walnut, 315–316, **432**
endosepsis of fig, 356–358, **436**
  of English walnut, 313–318
European (Nectria) canker of pome fruit, 41–45, **385–386**
Eutypa dieback
  of pistachio, 335
  of stone fruit, 164–167, **407**
Fabraea leaf spot of pome fruit, 46–47, **386**
  of feijoa, 370, 371
  of fig, 356–363
  of filbert, 371
fly speck of pome fruit, 121
fruit rots. *See* Fruit rot
fruit spot of apple. *See* Bitter pit
fruit spot of olive, 344
Fusarium root rot
  of English walnut, 318
  of peach, 241
Fusicoccum canker of peach, 242
Gibberella bud rot and shoot blight, 241

green-fruit rot of stone fruit, 169–170, **408**
gummosis of peach, 242
hull rot of almond, 171, **409**
kernel mold
  chestnut, **438**
  of English walnut, 316, **432**
kernel rot of almond. *See* Anthracnose
leaf blight
  of almond, 172–173, **409**
  of fig, 365
  of peach, 241
leaf curl
  of apricot, **410**
  of cherry, 173–174, **409**
  of nectarine and peach, 174–176, **410**
leaf and fruit spot of olive, 344
leaf spot
  of cherry, 177–178, **411**
  downy, of walnut, 315–316, **432**
  of fig, 366
  of olive, 342, 344–346, **435**
limb blight of fig, 366, **436**
melaxuma canker and twig blight of walnut, 316–317
of olive, 344–348, **435**
of pecan, 371
of persimmon, 372, **438**
pesticide resistance and, 381
Phloeosporella shothole, of apricot, 241
Phomopsis canker of fig, 358–360, **436**
Phomopsis rot of apple, 107, **395**
Phyllosticta leaf spot of apricot, 241
Phytophthora pruning wound canker of almond, 178–180, 267, **411**
Phytophthora root and crown rot, 263–271
  of English walnut, 317
  feijoa and, 370, 371
  olive and, 347
  of peach, 180, 264, 265, 267, 268, 269, 270, **425**
  of pistachio, 335
  of pome fruit, 48–49, **386**
  of stone fruit, 180, **425**
  wood-decay fungi and, 294
of pistachio, 329–335, **434**
plum pockets, 180–181, **411**
of pome fruit, 32–72, 121, 122, 263–298
  postharvest, 97–110
of pomegranate, 373, **438**
postharvest
  of apricot, 231, **422**
  of cherry, 228–231, **421**
  mycotoxins, 274–275
  of nectarine and peach, 231–234, **421, 422**
  of plum, 234–235
  of pome fruit, 97–110
  of prune, 235–236, **422**
  of stone fruit, 227–236, **421–422**
powdery mildew
  of pome fruit, 50–52, **387**
  of pomegranate, **438**
  rusty spot of peach and, 184–185
  of stone fruit, 181–186, **412**

Pythium root rot of almond and cherry, 241
replant diseases and, 277
Rhodosticta canker of stone fruit, 156–160
russet scab of prune, 187–188, **413**
rusts
  cedar-apple, 122
  of fig, 365
  of pome fruit, 54–55, **387**
  of stone fruit, 188–190, **413**
sappy bark of pome fruit, 56–57, **388**
scab
  of almond and stone fruit, 191–192, **414**
  of apple, 58–63, **388**
  of loquat, 66–67
  of pear, 67–69, **389**
  of pecan, 371
  russet, of prune, 187–188
Sclerotinia limb blight or dieback of fig, 356
shoestring (Armillaria) root rot, 280–283
  of almond, 193, 282, 283, **427**
  Dematophora root rot vs., 40
  of English walnut, 318
  of fig, 365
  of olive, 347
  of pome fruit, 70
  of stone fruit, 193
shothole of stone fruit, 193–196, **414–415**.
  See also Leaf spot of cherry
silver-leaf, 285–286, **428**
  of stone fruit, 197
  systemic pesticide for, 378
  wood decay and, 294
smut of pomegranate, 373
smut and mold of fig, 360–361, **437**
souring of fig, 361, **437**
southern blight of apple, 70–72, **389**
of stone fruit, 141–197, 241–242, 263–298
surface mold and Alternaria rot, of fig, 362–363
thread blight
  of fig, 365
  of pome fruit, 122
trunk and limb gall of apricot, 241
trunk rot of apricot, 241
twig blight of fig, 365
Verticillium wilt, 287–291
  of almond, 287–288, **428**
  of olive, 347, **435**
  of pistachio, 332–333, **434**
  of stone fruit, 197
of walnut, 313–318, **432**
white rot of pome fruit, 121
wood decay, 291–298, **429–430**
wood rot
  of English walnut, 318
  olive and, 347
Fungicidal sprays, 378–379. *See also specific diseases*
Fungicide dips. *See* Dip treatments
Fungicides, 378–381
Fungicide-wax applications, postharvest
  for apricot, 231
  for nectarine and peach, 233, 234

for plum, 235
Fungistats, 378
Fusarium root rot
　of English walnut, 318
　of peach, 241
Fusarium spp.
　F. lateritium, for Eutypa dieback, 167
　F. moniliforme,
　　fig endosepsis and, 356–358, **436**
　　fig wasps and, 355
　　mycotoxins and, 274
　F. solani, mycotoxins and, 274
　　peach fruit rot and, 241
　　peach root rot and, 241
　　replant diseases and, 277
Fusicladium spp., scab and
　of almond and stone fruit, 192
　of apple, 58
　of loquat, 66
　pecan and, 371
　of pear, 68
Fusicoccum amygdali, 242
Fusicoccum canker of peach, 242

G

Gallex, for olive knot, 343
Galltrol-A, for crown gall, 309
Gamma radiation. *See* Irradiation
Ganoderma spp., wood decay and, 293, 294, 296, **430**
Genestatic pesticides, 379
Gibberella baccata, 241
Gibberella bud rot and shoot blight, 241
Gibberellic acid
　for bacterial blast and canker prevention, in pome fruit, 14
　for chilling injury in peach, 218
　soggy breakdown and, 117
Gilbertella persicaria
　apricot decay and, 231
　nectarine and peach decay and, 232, 233, **421**
Girdle of walnut. *See* Blackline
"Glassiness." *See* Water core
Gliocladium spp., for Verticillium wilt, 290
Gloeodes pomigena, sooty blotch and, 122
Gloeosporium spp., anthracnose and
　of almond, 141
　of pome fruit, 32, 33, 34, 35
Glomerella cingulata
　bitter rot and
　　of pome fruit, 121
　　of stone fruit, 241
　fig anthracnose and, 366
Gnomonia leptostyla, anthracnose of walnut and, 313
"Golden death." *See* Almond, leaf scorch of
Gomphrena globosa
　prunus necrotic ringspot and, 211
　tomato ringspot virus and, 80
Gonatophragmium mori, fig leaf spot and, 366
Gooseberry, silver leaf and, 285
Grape

　crown gall of, 258, 260
　Dematophora root rot of, 39–41
　Eutypa dieback and, 164, 166, 167
　leaf scorch of, leaf scorch of almond and, 133, 134
　raisin dehydration process, mycotoxins and, 275
　shoestring (Armillaria) root rot of, 280
　triadimefon and, 185
Grapefruit, Dematophora root rot and, 41
Graphium spp., mallet wound and, 154
Gray mold rot
　of apple, 104–105, **395**
　of pear, **395**
　of sweet cherry, 228–229
Green crinkle, apple, 76, **389**
Green-fruit rot of stone fruit, 169–170, **408**
Green ring mottle virus (GRMV), prune dwarf virus and, 207
Green sharpshooter, leaf scorch bacteria and, 134, **403**
Green Valley strain. *See* X-disease of cherry and peach
GRMV. *See* Green ring mottle virus
Growth regulators
　for Verticillium wilt, 290
　walnut kernel mold and, 316
Grubs, white, hairy root and, 26
Guava, pineapple. *See* Feijoa
Gummosis
　See also Bacterial diseases, bacterial canker and blast
　of almond. *See* Anthracnose
　of apricot. *See* Prune dwarf virus
　of peach, 242
Gymnosporangium spp., rust and, 54–55
　cedar apple, 122
Gynotelium blasdaleanum, 55

H

Hadromycosis, Verticillium. *See* Verticillium wilt
Hail, fireblight and, 21
Hairy root, of apple and quince, 25–26
Hard-end. *See* Black-end
Harvesting techniques
　for almonds, 128
　for pome fruit
　　blue-mold or soft rot and, 100
　　friction marking and, 113
　　Mucor rot and, 109
　for stone fruit
　　almonds, 128, 153, 155
　　brown rot and, 149
　　Ceratocystis (mallet wound) canker and, 153, **406**
　　hull rot of almond and, 171
　　plums, 234
　　postharvest decay and, 227, 230–231
　　prunes, 131, 153, 155, 235, 236
　　rust and, 188
　　sweet cherry, postharvest decay and, 227, 230–231

　for walnut
　　deep-bark canker of walnut and, 311
　　walnut kernel mold and, 316
Hawthorn. *See Crataegus* spp.
Heat injury, to apple, 113
Heat therapy
　See also Thermotherapy
　for apple rubbery wood, 79
　for crown gall control, 261
　for replant problems, 279
Helminthosporium papulosum, black pox and, 122
Henbit. *See Laminum amplexicaule*
Hendersonia rubi, leaf blight of almond and, 172
Hendersonula toruloidea
　Dothiorella canker of almond and, 162
　fig branch wilt and, 356
　walnut branch wilt and, 314
Hepatica spp., rust and, 189
Herbicides, for southern blight, 72
Heterobasidion annosum, apricot trunk rot and, 241
Heteromeles arbutifolia (Toyon)
　fireblight and, 18
　scab of loquat and, 66
Hexaconazole
　for leaf curl of peach, 176
　for powdery mildew, 185
　for shothole, 195
Historical background. *See specific fruits*
Honeybees
　fireblight and, 18, 19
　prunus necrotic ringspot and, 211
Hop mosaic virus, 209
Hormodendrum rot of pome fruit, 108
Hormodendrum spp.
　fig mold and, 360
　H. cladosporioides, 108
Hot-water dips. *See* Washes
Hot-water injury, 113, **399**
Hull rot of almond, 171, **409**
Hydrochloric acid, in fireblight treatment, 20
8-hydroxyquinoline, for perennial canker, 37
Hypochlorous acid, 380

I

Imazalil, 380
　for Alternaria rot, 98
　for gray mold rot, 105
Imidazole, 380
　for brown rot of stone fruit, 145
Imidazolines, 379–380
Impact bruising, of apple, 113
India-hawthorn (*Raphiolepis indica*), Fabraea leaf spot and, 46
Inonotus cuticularis, fig wood decay and, **430**
Inorganic fungicides, 379
Insecticides. *See* Pesticides
Insects
　See also specific kinds
　Agrobacterium rhizogenes and, 26
　anthracnose and perennial canker and, 36

apple mosaic virus and, 77–78
black rot and, 108
Ceratocystis canker and, 154–155
cherry mottle leaf and, 199
Cladosporium-Hormodendrum rots and, 108
Cytospora or Rhodosticta canker and, 158, 159
European (Nectria) canker and, 43
feijoa and, 371
fig disease and, 354, 357, 358, 361, 364
fireblight and, 18, 21
hairy root and, 26
hull rot of almond and, 171
leaf scorch bacteria and, 134, **403**
mycotoxins and, 274, 275
nectarine and peach decay and, 232
olive knot and, 342
peach mosaic and, 203
pear decline and, 27–28, **384**
predaceous. See Predaceous insects
prune brownline virus and, 205
prunus necrotic ringspot and, 211
prunus stem pitting and, 213
russet scab of prune and, 187
tomato ringspot virus and, 205
walnut blight and, 307
X-disease of cherry and peach and, 138
Internal bark necrosis (measles) of apple, 92–94, **393**
Internal breakdown of apple, 116–117, **401**
Internal browning of apple, 115–116, **400**
Internal carbon dioxide injury (brown heart) of apple, 116, **397**
Internal rot of fig. See Endosepsis
Iprodione, 378, 379, 380
 for Alternaria rot
  in pome fruit, 98
  in sweet cherry, 230
 for blue-mold or soft rot, 101
 for Botryosphaeria panicle and shoot blight of pistachio, 330–331
 for brown rot of stone fruit, 145, 148–149
 for gray mold rot, 105
 for shothole, 195
 for stone fruit decay prevention, 230
  in apricot, 231
  in nectarine and peach, 234
  in plum, 235
  in sweet cherry, 230
Iron excess, internal bark necrosis and, 92, 93
Iron sulfate, for silver-leaf, 286
Irradiation
 for blue-mold or soft rot, 101
 for crown gall resistance, 260
 nectarine and peach decay and, 234
 for replant problems, 279
 sweet cherry decay and, 230
Irrigation
 See also Drip irrigation; Flood irrigation; Sprinkler irrigation
 Ceratocystis (mallet wound) canker and, 155
 for frost protection. See Frost protection

hull rot of almond and, 171
olive knot and, 343
pear decline and, 30
Phytophthora root and crown rot and, 49, 267, **425**
split-pit of olive and, 349
water core of apple and, 95
wood-decay fungi and, 294, 295
yellow bud mosaic and, 216

J

Jacket rot. See Green-fruit rot
Jonathan breakdown, 116
Jonathan spot, of apple, 113, **399**
Juglans spp. See Walnut
 J. cinerea. See Butternut
Juniperus spp. (Juniper), pear rusts and, 54, 55

K

K84 strain of *Agrobacterium radiobacter*
 for crown gall, 260, 261
 for hairy root, 26
K1026 strain of *Agrobacterium radiobacter*, for crown gall, 260–261
Kageneckia oblonga, scab of loquat and, 66
Kanamycin, for bacterial canker and blast, 253
Kernel mold
 chestnut, 438
 of English walnut, 316, **432**
Kernel rot of almond *See* Anthracnose
Kern's pear rust, 55
Kerosene, for olive knot, 343
"Kickback" action, 378
K sorbate. See Potassium sorbate

L

Lace leaf. See Prunus necrotic ringspot
Lacy scab of prune. See Russet scab of prune
Laetiporus sulphureus, wood decay and, 294, 297, **429**
Lambertella spp.
 L. corni-maris, stone fruit brown rot and, 145
 L. pruni, apricot and plum rot and, 241
Lambert mottle (LM) of cherry, 199
Laminum amplexicaule (Henbit), apple union necrosis and decline and, 81
Leaf blight
 of almond, 172–173, **409**
 of fig, 365
 of peach, 241
Leaf casting yellows. See X-disease of cherry and peach
Leaf curl
 of apricot, **410**
 of cherry, 173–174, **409**
 of nectarine and peach, 174–176, **410**

Leaf and fruit spot of olive, 344
Leafhoppers
 leaf scorch bacteria and, 134
 X-disease of cherry and peach and, 138, **404**
Leaf roll, yellow. See X-disease of cherry and peach
Leaf rust of pistachio, 335
Leaf spot
 of cherry, 177–178, **411**
 downy, of English walnut, 315–316, **432**
 of fig, 366
 of olive, 344–346, **435**
 olive knot and, 342
Lecithin, chemical injury and, 113
Legumes, Dematophora root rot and, 41
Lemon
 rough, Dematophora root rot and, 41
 shoestring (Armillaria) root rot of, 280
Leucostoma persoonii, 157, 158
Libertella blepharis, 164
Libocedrus decurrens, pear rust and, 55
Lichens, endosepsis of fig and, 358
Ligustrum spp. (Privet)
 cherry leafroll virus and, 320
 sickle leaf of olive and, 348
Lilac, silver leaf and, 285
Limb blight of fig, 366, **436**
Lime sulfur, 379
 for anthracnose of almond, 141
 for brown rot of stone fruit, 148, 149
 for leaf spot
  of cherry, 178
  of olive, 346
 for powdery mildew, 185
 for scab
  of apple, 62
  of pear, 69
Litargus balteatus, Ceratocystis canker and, 154–155
Little-leaf of pecan, 371
LM (Lambert mottle) of cherry, 199
Loganberry, Dematophora root rot and, 39, 41
Longidorus macrosoma
 pear stony pit and, 83
 prunus necrotic ringspot and, 211
Long leaf. See Deep suture
Loquat, (*Eriobotrya* spp.), 12–13
 bacterial canker and blast of, 247
 blossom blast and canker of, 13–14
 Dematophora root rot of, 39–41
 Fabraea leaf spot and, 46
 fireblight and, 13, 18
 Phytophthora root and crown rot of, 48, 49
 scab of, 13, 66–67
 twig blight of, 16
Low oxygen (alcohol) injury, to apple, 113, **399**
Low-temperature (soggy) breakdown of apple, 116, 117

M

Macrophoma curvispora, 34
Macrosiphon eriosoma, apple mosaic virus and, 78
Magnesium, soggy breakdown and, 117
Mallet wound canker. See Ceratocystis canker
Malus spp. See Apple; Crab apple
Malva parviflora, yellow bud mosaic and, 216
Mancozeb, 379
   for anthracnose, 37
   for fireblight, 21
   for gray mold rot, 105
   for rust of stone fruit, 190
Maneb, 379
   for bull's-eye rot on apple, 37
   for fig surface mold and Alternaria rot, 363
   for fireblight, 21
   for Phytophthora root and crown rot, 270–271
   for scab of apple, 62
Manganese excess, internal bark necrosis and, 92, 93
Mango
   Dematophora root rot and, 39–41
   Verticillium wilt and, 287
Maple. See *Acer* spp.
Maple leaf. See Crinkle leaf
Marssonina juglandis, anthracnose of walnut and, 313
MARYBLYT model, in blight prediction, 21
Mealy (senescent) breakdown of apple, 116, 117, **401**
Measles of apple. See Internal bark necrosis
Mechanical injury
   olive knot and, 342
   to pome fruit
      black rot and, 108
      blue-mold or soft rot and, 100
      Cladosporium-Hormodendrum rots and, 108
      European (Nectria) canker and, 43
      Mucor rot and, 109
   to stone fruit
      brown rot and, 147
      Ceratocystis (mallet wound) canker and, 153, 155
      Cytospora or Rhodosticta canker and, 158
      prune decay and, 235
      wood-decay fungi and, 294
Medlar (*Mespilus* spp.), Fabraea leaf spot and, 46
Melaxuma canker and twig blight of walnut, 316–317
Meloidogyne spp., replant diseases and, 277
2-mercaptobenzothiazole, friction marking and, 118
Mercuric chloride/mercuric cyanide, for fireblight, 20
Mercury fungicides
   endosepsis of fig and, 356
   Phomopsis canker of fig and, 360
   for scab of apple, 62
Mespilus spp. See Medlar
Metalaxyl, 380
   for Phytophthora root and crown rot, 49, 271
Metal salts, Nabam and. See Nabam plus salts
Metham-sodium soil fumigation, for southern blight, 72
Methyl bromide soil fumigation, 380
   for bacterial canker and blast, 253
   for crown gall, 260
   for Dematophora root rot, 41
   as pear crown gall preventive, 15
   for Phytophthora root and crown rot, 49, 270
   for replant problems, 278, **426**
   shoestring (Armillaria) root rot and, 193, 282, 283
   for southern blight, 72
   for Verticillium wilt, 290, 333
   for yellow bud mosaic virus, 216
Micrococcus amylovorus, 17. See also *Erwinia amylovora*
Microstroma juglandis, downy leaf spot of walnut and, 315–316
Mild rusty mottle (MRM) of cherry, 200–201
Mill's Chart, for scab infection, 60
Mistletoe, leafy, stone fruit and, 241
Mites
   fig mosaic and, 364
   peach mosaic and, 203
   pear decline and, 30
   predaceous, fig smut and mold and, 361
   prunus necrotic ringspot and, 211
MLO. See Mycoplasmalike organisms
Mock plums. See Plum pockets
Monilia fructigena, 146
Monilinia spp.
   apricot decay and, 231
   brown rot and
      of pome fruit and, 108, **396**
      of stone fruit and, 142–149, 228, 229, 230, **405, 421, 422**
   cherry decay and, 228, 229, 230, **408, 421**
   green-fruit rot and, 169–170, **408**
   hull rot of almond and, 171, **409**
   nectarine or peach decay and, 232, 233, 234, **405, 422**
   plum decay and, 234
   prune decay and, 235, 236, **405, 423**
Monocalcium arsenite or meta-aresenite, 378
   for brown rot of stone fruit, 145
   for scab of apple, 63
Monochaetia rosenwaldia, apricot trunk and limb gall and, 241
Monosodium phosphate plus urea, for scab, 192
Mosaic virus
   apple (AMV), 76–78, 209, **390**
   fig, 363–364, **437**
   peach, 202–203, **416**
   prune. See Prune dwarf virus
   prunus necrotic ringspot virus and, 209
   yellow bud, 205, 214–216, **418**
Mountain ash. See *Sorbus* spp.
MRM. See Mild rusty mottle
Mucor rot
   of pome fruit, 109–110, **396**
   of stone fruit, 231, 233, 234, 235
   nectarine and peach, 233, **421**
Mucor spp.; *M. piriformis*
   in apricot, 231
   fig decay and, 360
   in nectarine and peach, 233, **421**
   in plum, 234
   in pome fruit, 109–110
   in prune, 235
Muir peach dwarf virus. See Prune dwarf virus
Mulberry, Dematophora root rot in, 41
Mushroom root rot. See Shoestring (Armillaria) root rot
Muskmelon, Verticillium wilt and, 289
Myclobutanil, 380
   for leaf curl of peach, 176
   for powdery mildew, 185
Mycocentrospora cladosporioides, olive leaf and fruit spot and, 344
Mycoplasmalike organisms (MLO)
   apple flat limb and, 75, **389**
   apple rubbery wood and, 78
   non-California stone fruit diseases and, 241
   peach rosette and, 241
   pear decline and, 27–30, **384**
   X-disease of cherry and peach and, 136–139, **404**
Mycosphaerella pistacearum, 334–335
Mycosphaerella pomi, Brook's spot and, 122
Mycostatin, for nectarine and peach decay, 233
Mycotoxins, 274–275
   pesticide residues vs., 227, 275
Myzus persicae, apple mosaic virus and, 78

N

Nabam plus salts, 379
   for brown rot of stone fruit, 145, 148
   for leaf curl of nectarine and peach, 176
   for shothole, 196
Napa Valley strain. See X-disease of cherry and peach
Navel orange worm, hull rot of almond and, 171
Necrotic ringspot strains (NRSS), 210
Necrotic rusty mottle (NRM) of cherry, 199, **416**
Nectarine
   See also Stone fruit
   (*Prunus persica* var. *nectarina*), 130
   abiotic disorders of, nectarine pox, 220–221, **419**
   anthracnose and, 141, 241
   bacterial canker and blast of, 247
   brown rot of, 142–149
   Cercospora leaf spot of, 156
   leaf curl of, 174–176
   mosaic virus of, 202–203
   nectarine pox of, 220–221, **419**
   Phytophthora root and crown rot of, 264
   postharvest diseases of, 231–234, **422**

powdery mildew of, 181–186
prunus stem pitting of, 212–214
rust of, 188–190
scab of, 191–192
shoestring (Armillaria) root rot of, 193
shothole of, 193–196
silver-leaf of, **428**
Verticillium wilt of, 287
X-disease of, 136–139
yellow bud mosaic virus of, 214–216
Nectria blight of pome fruit, 122
Nectria (European) canker, of pome fruit, 41–45, **385–386**
Nectria spp.
  European (Nectria) canker and, 41–45, **386**
  nectria blight and, 122
Nematicides, for southern blight, 72
Nematodes
  apple union necrosis and decline and, 81
  bacterial canker and blast and, 250, 252, 253
  cherry rasp leaf virus and, 200
  Cytospora canker and, 159
  pear decline and, 28
  pear stony pit and, 83
  prune brownline virus and, 205
  prunus necrotic ringspot and, 211
  prunus stem pitting and, 213
  replant diseases and, 276, 277
  shoestring (Armillaria) root rot and, 193
  tomato ringspot virus and, 205, 213, 216
  walnut blackline and, 320–321
Neofabraea spp., 32, 34, 35
Nerium oleander (Oleander)
  olive knot and, 341–342
  sickle leaf of olive and, 348
Nicotiana spp., tomato ringspot virus and, 80, 215
Nigrospora oryzae, X-spot and, 123
Nitidulid beetles
  Ceratocystis canker and, 154–155
  hull rot of almond and, 171
  nectarine and peach decay and, 232
Nitrogen excess, water core of apple and, 95
Nitrogen fertilization
  for Armillaria root rot, 282
  bacterial canker and blast and, 250
  pear decline and, 30
  southern blight and, 72
Noninfectious bud failure of almond, 221–224, **419**
Noninfectious plum shothole, 225, **420**
Nonproductive syndrome of almond, 225–226, **420**
Norbac, for crown gall, 309
NRM. *See* Necrotic rusty mottle
NRSS. *See* Necrotic ringspot strains
Nuarimol, 380
Nustar, for brown rot of stone fruit, 149
Nuts. *See specific kinds*

O

Oak. *See Quercus* spp.
Oak root fungus disease. *See* Shoestring (Armillaria) root rot
Oiled wrappers
  internal browning and, 115
  scald prevention by, 114
Oleander. *See Nerium oleander*
Olive (*Olea europaea*), 340
  abiotic diseases of, 349, **435**
  cherry leafroll virus and, 320
  Dematophora root rot of, 39–41
  leaf and fruit spot of, 344
  leaf spot of, 344–346, **435**
    olive knot and, 342, **435**
  olive knot of, 341–343, **435**
  shoestring (Armillaria) root rot of, 282, 347
  shotberry of, 349, **435**
  sickle leaf of, 348, **435**
  soft nose of, 349, **435**
  split-pit of, 349
  Verticillium wilt of, 287, 289, 290, 347, **435**
  virus and viruslike diseases of, sickle leaf, 348, **435**
Olive fly, olive knot and, 342
Ophiostoma, 154
Orange, Dematophora root rot and, 41
Organic fungicides, 379–380
Organic produce, postharvest decay and, 227
Oriental fruit moth, nectarine and peach decay and, 232
Ormathodium leaf spot, of fig, 366
Overhead watering. *See* Sprinkler irrigation
Oxalis corniculata (Creeping woodsorrel), apple union necrosis and decline and, 81
Oxycarboxin, 378
Oxyporus spp.
  Armillaria root rot vs., 281
  wood decay and, 294, 297, **429**
Oxytetracycline (Terramycin), 378, 381
  for bacterial canker and blast, 253
  for fireblight, 20
  for pear decline, 30
  for X-disease of cherry and peach, 139
Ozone treatments, nectarine and peach decay and, 234

P

Pacific Coast pear rust, 54–55
Packing techniques
  See also Harvesting techniques for pome fruit
  blue-mold or soft rot and, 100
  friction marking and, 113, 118
  internal browning and, 115
  Mucor rot and, 109
  scald and, 114
  stone fruit decay and
    apricot, 231
    nectarine and peach, 232, 234
    sweet cherry, 230
Painting, for wood-decay protection, 295
Papery bark. *See* Sappy bark
Paraffin-oil-water emulsion, for olive knot, 343
Parasitic insects, for woolly apple aphid control, 37
Parasitic seed plants of stone fruit, 241
Paratylenchus neoamblycephalus
  bacterial canker and blast and, 250
  Cytospora canker and, 159
PDV. *See* Prune dwarf virus
Peach (*Prunus persica*), 130
  See also Stone fruit
  abiotic disorders of
    chilling canker, 218, **418**
    nectarine pox and, 221
  anthracnose or bitter rot of, 141, 241
  bacterial canker and blast of, 246, 247, 250, 251, 252, 253, **423**
  chilling injury and, 218
  bacterial spot of, 241
  brown rot of, 232, 233, 234, **405**
  Ceratocystis (mallet wound) canker of, 155
  Cercospora leaf spot of, 156
  cherry rusty mottle virus and, 201
  chilling canker of, 218, **418**
  Clitocybe root rot and, 242
  crown gall of, 259, 260
  Cytospora canker of, 156–160
  damping-off and root rot of, 241
  decline of, cherry rasp leaf virus and, 200
  Dematophora root rot of, 39–41
  Dothiorella canker and, 162, 163
  Eutypa dieback of apricot and, 166
  fruit rot of, 241, **421**
  fungal diseases of, 141–197, 241, 242
  Fusarium root rot of, 241
  Fusicoccum canker of, 242
  Gibberella bud rot and shoot blight of, 241
  gummosis of, 242
  leaf blight of, 241
  leaf curl of, 174–176, **410**
  mosaic virus of, 202–203, **416**
  Muir dwarf of, prune dwarf virus and, 206
  nectarine pox and, 221
  peach blotch of, 241
  peach rosette of, 241
  peach wart of, 241
  as *Pezicula malicorticis* host, 35
  phony, 241
    leaf scorch of almond and, 133
  Phytophthora pruning wound canker and, 178
  Phytophthora root and crown rot of, 180, 264, 265, 267, 268, 269, 270, **425**
  powdery mildew of, 181–186, **412**
  prune dwarf virus on, 206–208
  prunus necrotic ringspot of, 209–211
  prunus stem pitting of, 212–214, **418**
  rasp leaf virus on, 200
  replant diseases of, 276–279, **426**
  rootstocks, 130
  Phytophthora root and crown rot and, 269, 270

rust of, 188–190, **413**
rusty spot of, powdery mildew of apple and, 184–185
scab of, 191–192, **414**
shoestring (Armillaria) root rot of, 193, 281, 282, 283
short-life of, 250, 252
shothole of, 193–196, **414–415**
silver-leaf of, 197, 285–286, **428**
stunt virus of, 209–210, **417**
  prune dwarf virus and, 206, **417**
  tomato ringspot virus and, 205
Verticillium wilt of, 287, 289
virus or viruslike diseases of, 198–216, 241
wood-rot disorders of, 294, **430**
X-disease of, 136–139, 378, **404**
yellow bud mosaic of, 205, 214–216, **418**
yellow leaf roll disease of, pear decline and, 29
Peach blight. *See* Shothole
Peach blotch, 241
Peach brown rot. *See* Brown rot, American
Peach mosaic, 202–203, **416**
Peach mule's ear. *See* Prunus necrotic ringspot
Peach necrotic leaf spot, 209. *See also* Prunus necrotic ringspot
Peach rosette, 241
Peach stem pitting. *See* Prunus stem pitting
Peach stunt disease. *See* Prune dwarf virus
Peach twig borer, postharvest decay and, 232
Peach wart, 241
Peach willow twig. *See* Prunus necrotic ringspot
Peach yellow leaf roll (PYLR) strain of X-disease, 136, **404**
Peacock spot. *See* Leaf spot, of olive
Pear, 11–12
  See also Asian pear; Pome fruit
  abiotic disorders of, 87–94, 118–119, **393**
  Alternaria rot of, 97–98, **394**
  anthracnose and perennial canker of, 33, 35
  apple mosaic virus and, 78
  apple rubbery wood and, 78
  apple stem pitting and, 79
  bacterial blast and canker of, 13–14, 246–247, **383**
  bark measles of, 81–82
  bitter pit of, 87–89, 119, **393**
  black-end of, 91, **393**
  black root rot of, 122
  black rot of, 108
  blister canker of, 82, **391**
    bark measles and, 81, 82
  blossom blast and canker of, 13, 14, 246–247, **383**
  blossom blight of, apricot brown rot and, 144
  blue-mold or soft rot of, 99–101, **394**
  brown rot of, 108, **396**
  carbon dioxide injury to, 119, **397**
  chemical injury to, 118, **397, 398**
  Cladosporium-Hormodendrum rots of, 108
  Coprinus rot of, 108–109, **396**
  core breakdown of, 119
  crown gall of, 15–16, 260
  cultivation history, 11–12
  decline, 27–30, 264, **384**
  Dematophora root rot of, 39–41
  European (Nectria) canker of, 41–45
  Fabraea leaf spot of, 46–47, **386**
  fireblight of, 16–21, **383**
    rootstocks and, 11, 12
  flat limb of, 75
  freezing injury to, 117, 119, **402**
  friction marking on, 118, **398**
  fruit rot disorders of, perennial canker and, 33
  fungal diseases of, 32–72, 121, 122, 263–298
  gray mold of, 104–105, **395**
  insects on, pear decline and, 27, 28, 29, 30, **384**
  Mucor rot of, 109–110, **396**
  mycotoxins and, 274
  as *Pezicula malicorticis* host, 35
  Phytophthora root and crown rot of, 48, 264, 265, 267, 269, **386**
  postharvest disorders of, 111–118
    abiotic, 118–119
    Alternaria rot of, 97–98, **394**
    bitter pit (Anjou pit; cork spot), 87–89, 119, **393**
    black rot, 108
    blue-mold or soft rot, 99–101, **394**
    brown rot, 108, **396**
    carbon dioxide injury, 119, **397**
    chemical injury, 118, **397, 398**
    Cladosporium-Hormodendrum rots, 108
    Coprinus rot, 108–109, **396**
    core breakdown, 119
    freezing injury, 119, **402**
    friction marking, 118, **398**
    gray mold rot, 104–105, **395**
    Mucor rot, 109–110, **396**
    mycotoxins, 274
    premature ripening (pink end), 119, **402**
    senescent scald, 118, **402**
    side rot, 110, **396**
    Stemphylium and Pleospora rots, 110
    superficial scald, 118–119, **400**
  powdery mildew of, 50, **387**
  premature ripening (pink end) of, 119, **402**
  Pseudomonas syringae infections, 13, 14, 132, 246–254
  replant problems and, 276, 277
  Rhizopus rot of, **396**
  rootstocks, 11, 12
    black-end and, 91
    Dematophora root rot and, 40
    fireblight and, 19–20
    pear decline and, 28, 29, 30, **384**
    Phytophthora root and crown rot and, 269
  rusts of, 54–55
    pear trellis rust, 54, **387**
  scab of, 67–69, **389**
  senescent scald of, 118, **402**
  shoestring (Armillaria) root rot of, 70, 282, 283
  side rot of, 110, **396**
  Stemphylium and Pleospora rots of, 110
  stony pit of, 82–83, **391**
  twig blight of, 16
  vein yellows and red mottle of, 83, **391**
  vigor rating of, 30
  virus or viruslike diseases of, 74–84
"Pear blight." *See* Fireblight
Pear decline, 27–30, **384**
Pear trellis rust, 54, **387**
Pecan (*Carya illinoensis*)
  diseases of, 371
  mycotoxins and, 274
Pellicularia koleroga, thread blight and, 365
Penconazole, 380
  for leaf curl of peach, 176
  for powdery mildew, 185
Penicillin sprays, for walnut blight, 307
Penicillium spp.
  endosepsis of fig and, 357
  fig mold and, 360
  mycotoxins and, 274
  nectarine and peach decay and, 232, **422**
  P. expansum,
    apricot decay and, 231
    blue-mold or soft-rot and, 99–101
    cherry decay and, 228, 229, **421**
    mycotoxins and, 274
    replant diseases and, 277
  P. frequentans, brown rot of stone fruit and, 145
  P. glaucum, 100
    plum decay and, 234
    in pomegranate, **438**
    prune decay and, 235, 236
    walnut kernel mold and, 316
Peppervine, Pierce's disease and, 134
Peptographium spp., wounded tissues and, 154
Perennial canker of pome fruit, 32–38
Perennoporia medulla-panis, wood decay and, **429**
Persimmon (*Diospyros* spp.), 372
  diseases of, 372
    Botrytis rot, **438**
    shoestring (Armillaria) root rot, 283
    Verticillium wilt, 287
  sunburn and, 372, **438**
Pesticides, 378–381
  nectarine pox and, 221
  pathogen resistance to, 380, 381
  residues of, mycotoxins vs., 227, 275
  restrictions on, pear decline and, 30
  for southern blight, 72
  for woolly apple aphid control, 37
  for X-disease of cherry and peach, 138, 139
Petroleum oil, miscible tar and,
  for European (Nectria) canker, 45
  for perennial canker, 37
Pezicula malicorticis, 33, 34, 35, 38
Phage, for fireblight control, 21
Phellinus spp., wood decay and, 294, 297, **430**
Phenols
  for olive knot, 343
  for Verticillium wilt, 290

Phenylmercuric chloride, for scab
   of apple, 63
   of pear, 69
Phialophora malorum, side rot and, 110, **396**
Phillyrea, olive leaf spot fungus and, 343
Phloem canker. *See* Deep-bark canker
Phloeosporella padi
   leaf spot of cherry and, 178
   shothole of apricot and, 241
Phoma fruit spot (Brook's spot), of apple, 122
Phoma spp. *See Phomopsis* spp.
Phomopsis canker of fig, 358–360, **436**
Phomopsis rot of apple, 107, **395**
Phomopsis spp.
   P. ambigua, 107
   P. cinerascens, in fig, 358–360
   P. mali, in apple, 107, 395
Phony peach, 241
   leaf scorch of almond and, 133
Phoradendron flavescens (Leafy mistletoe), stone fruit, 241
Phosphorus deficiency, sappy bark and, 57
Photinia spp., Fabraea leaf spot and, 46
Phragmidium libocedri, 55
Phyllosticta circumcissa, leaf spot of apricot and, 241
Physalospora spp.
   P. gregaria, 317
   P. obtusa, 108
Physiological leaf drop, prune dwarf virus and, 206
Physopella fici, fig rust and, 365
Phytocoptella avellanae, filbert dieback canker and, 371
Phytomonas spp.
   P. rhizogenes, 25. *See also Agrobacterium rhizogenes*
   P. savastanoi, 341
Phytophthora pruning wound canker of almond, 178–180, 267, **411**
Phytophthora root and crown rot, 263–271
   of English walnut, 180
   feijoa and, 370, 371
   olive and, 347
   of peach, 180, 264, 265, 267, 268, 269, 270, **425**
   pistachio and, 335
   of pome fruit, 48–49, **386**
   of stone fruit, 180, **425**
   wood-decay fungi and, 294
Phytophthora spp., 48, 49, 265, 268, 269
   See also Phytophthora root and crown rot
   Armillaria root rot vs., 281
   P. cactorum, 48, 265, 268, 269
      pear decline and, 28, 264
   P. syringae, 265
      pruning wound canker of almond and, 178–180
      root and crown rot and, 263–271
         of English walnut, 317
         of pome fruit, 48–49
         of stone fruit, 180
Pichia spp., prune decay and, 235
Pierce's disease. *See* Grape, leaf scorch of
Pileolaria terebinthi, leaf rust and, 335

Pimple canker. *See* Internal bark necrosis (measles)
Pine, Armillaria root rot and, 283, 284
Pineapple guava. *See* Feijoa
Pink end of pear. *See* Premature ripening
Pink rot of fig. *See* Endosepsis
Pin nematode, Cytospora canker and, 159
Piperazine, 380
   for brown rot of stone fruit, 145
Pistachio (*Pistacia vera*), 328
   Alternaria leaf and fruit spot of, 334
   Botryosphaeria panicle and shoot blight of, 329–331, **434**
   Botrytis blossom and shoot blight of, 331–332, **434**
   Dematophora root rot of, 39–41
   Dothiorella canker and, 163
   Eutypa dieback of, 335
   leaf rust of, 335
   mycotoxins and, 274
   Phytophthora root and crown rot of, 335
   Septoria leaf spot of, 334–335
   Verticillium wilt of, 287, 289, 290, 332–333, **434**
Plantago major (Plantain)
   apple union necrosis and decline and, 81
   cherry rasp leaf virus and, 200
Plant cancer. *See* Crown gall
Planting sites
   bacterial canker and blast and, 252
   European (Nectria) canker and, 44
   Phytophthora rot and, 270
   sappy bark and, 57
   southern blight and, 72
   tomato ringspot virus and, 81
Pleospora fructicola, 110
Pleospora rot of pome fruit, 110
Pleurotus ostreatus, walnut wood decay and, **430**
Plowing, southern blight and, 72
Plum; Prune (*Prunus salicina; P. domestica*), 130–131
   See also Stone fruit
   abiotic disorders of
      noninfectious plum shothole, 225, **420**
      plum rusty blotch, 226, **421**
   apple mosaic virus and, 77, 78
   bacterial canker and blast of, 251, 252, 253, **423, 424**
   bacterial spot of, 241
   black knot of, 241–242
   box rot of, 236
   brown rot of, 142–149, **405**
   Ceratocystis (mallet wound) canker of prune, 153–155, **406**
   cherry rusty mottle virus and, 201
   crown gall and, 259
   Cytospora or Rhodosticta canker of, 156–160, **406**
   damping-off and root rot of, 241
   Dematophora root rot of, 39–41
   Dothiorella canker and, 162, 163
   dried, mycotoxins and, 274
   fruit rot of, 241, **422**
   fungal diseases of, 141–197, 263–298

   green-fruit rot of, 169–170
   leaf scald of, leaf scorch of almond and, 133
   leafy mistletoe and, 241
   mycotoxins and, 274
   noninfectious plum shothole of, 225, **420**
   peach mosaic and, 202, 203
   as *Pezicula malicorticis* host, 35
   Phytophthora pruning wound canker and, 178
   Phytophthora root and crown rot and, 267, 269, 270
   plum pockets of, 180–181, **411**
   postharvest disorders of, 234–236, **422**
      mycotoxins and, 235, 274
   powdery mildew of, 181–186, **412**
   prune brownline of, 204–206, **417**
      almond brownline vs., 198
   prune cultivation, 131
   prune dehydration process, decay and, 235
   prune diamond canker and, 241
   prune dwarf virus (PDV) of, 206–208
   prunus stem pitting of, 212–214
   replant problems and, 276, 278
   russet scab of prune, 187–188, **413**
   rust of, 188–190, **413**
   rusty blotch of, 226, **421**
   scab of, 191–192
   shoestring (Armillaria) root rot of, 193, 281, 282, 283
   shothole of, 194
   Verticillium wilt of, 287
   virus or viruslike diseases of, 198–216, 241
      apple mosaic virus and, 77, 78
   wood-rot disorders of, 294, **429, 430**
   X-disease of cherry and peach and, 136
   yellow bud mosaic virus and, 216
Plum line pattern virus, 77, 209
PMV. *See* Peach mosaic
PNRSV. *See* Prunus necrotic ringspot virus
Podosphaera species
   powdery mildew and, **412**
      of pome fruit, 50–52
      of stone fruit, 181–186
Polyethylene mulch, for Verticillium wilt, 290–291, 333
Polyethylene wrappers
   for apricot, 231
   nectarine and peach decay and, 234
Polyporus spp.
   P. versicolor, 56
   wood decay and, 297
Polystictus versicolor, 56
Polyvinyl acetate, for wood-decay prevention, 295
Pome fruit, 10–13
   See also Apple; Loquat; Pear; Quince
   abiotic disorders of, 87–96
      bitter pit, 87–89, 393
      black-end of pear, 91, 393
      internal bark necrosis (measles) of apple, 92–94, 393
      in storage, 111–119. *See also specific disorders*
      water core of apple, 95–96, 394

bacterial diseases of
  bacterial blast and canker, 13–14, 246–254, **383**
  crown gall, 15–16, 257–261
  fireblight, 16–21, **383**
  hairy root, 25–26
fungal diseases of, 32–72, 121, 122, 263–298
  Alternaria rot, 97–98, **394**
  anthracnose and perennial canker, 32–38, **385**
  bitter rot, 121
  black pox, 121, 122
  black root rot, 121, 122
  black rot, 108, 121, **396**
  blister spot, 121, 122
  blossom-end rot, 121, 122
  blotch, 121, 122
  Brook's spot (Phoma fruit spot), 121, 122
  brown rot, 108, **396**
  cedar apple rust, 121, 122
  Cladosporium-Hormodendrum rots, 108
  Clitocybe root rot, 121, 122
  Coprinus rot, 108–109, **396**
  Dematophora root rot, 39–41, **385**
  European (Nectria) canker, 41–45, **385–386**
  Fabraea leaf spot, 46–47, **386**
  fly speck, 121
  Mucor rot, 109–110, **396**
  Nectria blight, 122
  Phomopsis rot, 107, **395**
  Phytophthora root and crown rot, 48–49, 263–271, **386**
  postharvest, 97–110
  powdery mildew, 50–52
  rusts, 54–55
  sappy bark of apple, 56–57, **388**
  scab of apple, 58–63, **388**
  scab of loquat, 66–67
  scab of pear, 67–69, **389**
  shoestring (Armillaria) root rot, 70
  side rot, 110, **396**
  silver-leaf, 285, 286
  sooty blotch, 122
  southern blight, 70–72
  Stemphylium and Pleospora rots, 110
  thread blight, 122
  Verticillium wilt, 287
  white root rot, 122–123
  white rot, 121
  X-spot, 123
mycoplasmalike disease of, pear decline, 27–30, **384**
postharvest disorders of, 97–110, 111–119
  Alternaria rot, 97–98, **394**
  black rot, 108, 121, **396**
  blue-mold or soft rot, 99–101, **394**
  brown rot, 108, **396**
  bull's-eye rot, 103–104, **395**
  Cladosporium-Hormodendrum rots, 108
  Coprinus rot, 108–109, **396**
  gray mold rot, 104–105, **395**
  Mucor rot, 109–110, **396**
  mycotoxins, 274–275

Phomopsis rot, 107, **395**
side rot, 110, **396**
Stemphylium and Pleospora rots, 110
replant diseases of, 276–279, **426**
virus or virus like diseases of, 74–84
  apple flat limb, 75, **389**
  apple graft-incompatibility, 75–76, **389**
  apple green crinkle, 76, **389**
  apple mosaic virus (AMV), 76–78, 209, **390**
  apple stem pitting, 79–80, **390**
  apple union necrosis and decline, 80–81, **390**
  blister bark of apple, **392**
  false sting of apple, 76, **392**
  flat apple disease, 200, **392**
  pear bark measles, 81–82
  pear blister canker, 82, **391**
  pear stony pit, 82–83, **391**
  pear vein yellows and red mottle, 83, **391**
  quince sooty ring spot, 83–84
  rough skin disease of apple, **392**
  russet ring of apple, **392**
Pomegranate (*Punica granatum*), 372–373
  diseases of, 373, **438**
  mycotoxins and, 274
Populus spp. (Poplar)
  European (Nectria) canker and, 44
  melaxuma twig blight and, 317
  silver leaf and, 285, 286
Poria ambigua, 294
Postassium deficiency, Cytospora canker and, 159
Postassium sorbate, for prune mold prevention, 236
Postharvest disorders
  abiotic. *See* Abiotic storage disorders
  of fig, endosepsis, 356–358, **436**
  fungicides for, 378, 380
  mycotoxins, 274–275
    pesticide residues vs., 227
    prune processing and, 235
  pesticide resistance and, 381
  of pome fruit, 97–110, 111–119
    Alternaria rot, 97–98, **394**
    black rot, 106, 108, 121, **396**
    blue-mold or soft rot, 99–101, **394**
    brown rot, 108, **396**
    bull's-eye rot, 103–104, **395**
    Cladosporium-Hormodendrum rots, 108
    Coprinus rot, 108–109, **396**
    gray mold rot, 104–105, **395**
    Mucor rot, 109–110, **396**
    mycotoxins, 274–275
    Phomopsis rot, 107, **395**
    side rot, 110, **396**
    Stemphylium and Pleospora rots, 110
    water core of apple, 95, **394**
  of stone fruit, 227–236, **421–422**
    apricot, 231, **405**, **422**
    cherry, 227–231, **421**
    mycotoxins, 274–275
    nectarine or peach, 231–234, **405**, **421**
    plum, 234–235

prune, 235–236, **405**, **422**
Postharvest drench. *See* Dip treatments
Postharvest wash. *See* Washes
Potassium
  bacterial canker and blast and, 250
  soggy breakdown and, 117
  Verticillium wilt and, 289, 290
  in pistachio, 333
Potassium azide, for southern blight, 72
Potassium sorbate treatment, prune decay and, 236
Potato, Verticillium wilt and, 287, 288, 289, 290
Powdery mildew
  of pome fruit, 50–52, 62, **387**
  of pomegranate, **438**
  rusty spot of peach and, 184–185
  of stone fruit, 181–186, **412**
Pratylenchus spp., replant diseases and, 277
Predaceous insects
  fig smut and mold and, 361
  for woolly apple aphid control, 37
Premature ripening of pear, 119
Prionus beetle, shoestring (Armillaria) root rot of pear and, 70
Privet. *See* Ligustrum spp.
Prochloraz, 380
  for Alternaria rot, 98
  for blue-mold or soft rot, 101
  for brown rot of stone fruit, 149
  for scab, of pear, 69
Propiconazole, 380
Propylene oxide, for prune mold prevention, 236
Protectant pesticides, 378
Prune brownline, 204–206, **417**
  almond brownline vs., 198
Prune dwarf virus (PDV), 206–208
Prune. *See* Plum; Prune
Pruning
  anthracnose and perennial canker and, 36, 37
  apple mosaic virus and, 77–78
  bacterial canker and blast and, 250–251, 253–254
  bitter pit of apple or pear and, 88
  "blight cutters," 20
  crinkle leaf of cherry and, 220
  Cytospora canker and, 158, 159
  Dothiorella canker and, 162
  European (Nectria) canker and, 43, 44, 45
  Eutypa dieback and, 164, 166, 167, **407**
  olive knot and, 342, 343
  pear decline and, 30
  for perennial canker control, 37
  Phomopsis canker of fig and, 359, 360, **436**
  Phytophthora canker of almond and, 178–180, 267, 271, **411**
  pome fruit, sappy bark and, 56
  powdery mildew and, 52
  sappy bark and, 57
  silver-leaf and, 285, 286
  wood-decay fungi and, 294, 295

Prunus necrotic ringspot virus (PNRSV), 209–211, **417**
   prune dwarf virus and, 206, 207
Prunus spp.
   bacterial canker and blast and, 246
   cherry leaf roll virus and, 320
   P. armeniaca. See Apricot
   P. avium. See Cherry, sweet
   P. cerasus. See Cherry, sour
   P. dulcis. See Almond
   P. persica. See Peach
   P. persica var. *nectarina.* See Nectarine
   P. salicina; P. domestica. See Plum; Prune
Prunus stem pitting (PSP), 212–214, **418**
   cherry stem pitting and, 201, **418**
Pseudomonas spp.
   bacterial canker and blast and, 247–248, **423–424**
   P. cepacia, for blue-mold or gray mold, 101
   P. syringae
      bacterial canker and blast and, 13–14, 132, 246–254, **423**
      Cytospora leucostoma and, 158
      Fusicoccum canker and, 242
      olive knot and, 341–343
      in pome fruit, 13–14, 246–254
      in stone fruit, 132, 158, 246, 247–254, **423**
   P. syringae pv. *papulans,* blister spot and, 122
Psylla
   monitoring and control of, 30
   pear decline and, 27, 28, 29, 30, **384**
Puccinia pruni-spinosae, rust and, 188
Punica granatum. See Pomegranate
Purrolnitrin, 381
PYLR strain. See X-disease of cherry and peach
Pyracantha spp. (Firethorn)
   Fabraea leaf spot and, 46
   fireblight and, 18
   flat limb and, 75
   scab of loquat and, 66
Pyridinitril, for European (Nectria) canker, 44
Pyrifenox, for scab, of apple, 62
Pyrimidine, 380
   for brown rot of stone fruit, 145
Pyronia veitchii
   pear vein yellows and red mottle and, 83
   quince sooty ring spot and, 84
Pyrus spp. See Pear
Pythiacystis canker. See Phytophthora root and crown rot
Pythium spp.
   almond and cherry root rot and, 241
   replant diseases and, 277

Q

Quercus spp. (Oak), silver-leaf and, 285
Quince (*Cydonia oblonga*), 12
   See also Pome fruit

   anthracnose and perennial canker of, 33, 35
   apple mosaic virus and, 78
   blue-mold or soft rot of, 99–101
   Brook's spot (phoma fruit spot) of, 122
   brown rot of, 108
   crown gall of, 258
   Dematophora root rot of, 39–41
   fireblight and, 16, 18
   flat limb of, 75
   flowering
      European brown rot and, 142, 144
      as *Pezicula malicorticis* host, 35
   fruit rot disorders of, perennial canker, 33
   hairy root of, 25–26
   as pear rootstock, 12
   pear stony pit and, 82
   as *Pezicula malicorticis* host, 35
   postharvest disorders of
      blue-mold or soft rot, 99–101
      brown rot, 108
   powdery mildew of, 50
   sooty ring spot of, 83–84
   twig blight of, 16
   virus or viruslike diseases of, 74, 75, 78, 82, 83–84
Quinones, for Verticillium wilt, 290

R

Radiation. See Irradiation
Raphiolepis indica. See India-hawthorn
Raspberry, cherry leafroll virus and, 320
Rasp leaf, of cherry, 200, **416**
Red bud. See Crinkle leaf
Refrigerants, chemical injury and, 112, 113, **397**
Replant diseases, of fruit trees, 276–279, **426**
Rhizoctonia spp., replant diseases and, 277
Rhizomorpha necatrix, 40
Rhizomorpha spp., shoestring (Armillaria) root rot and, 281
Rhizopus spp.
   apricot decay and, 231, **422**
   cherry decay and, 228, 229, 230, **421**
   fig mold and, 360
   hull rot of almond and, 171, **409**
   mycotoxins and, 275
   nectarine and peach decay and, 232, 233, 234, **421**
   pear decay and, **396**
   plum decay and, 234
   prune decay and, 235, 236
   R. arrhizus,
      apricot decay and, 231, **422**
      peach and nectarine decay and, 232–233
   R. oryzae, peach and nectarine decay and, 232
   R. stolonifer, 229, 232, 233
      apricot decay and, 231
      cherry decay and, 228, 229, 230, **421**
      Mucor rot and, 109, **396**

      peach and nectarine decay and, 232, **421**
      plum decay and, 234
      prune decay and, 236
      walnut kernel mold and, 316
Rhizosphere organisms, replant problems and, 278
Rhododendron, silver leaf and, 285
Rhodosticta canker, of stone fruit, 156–160
Rhodosticta quercina, 157, 159
Rhubarb, cherry leafroll virus and, 320
Rickettsia-like organisms, apple flat limb and, 75
Ring nematode
   bacterial canker and blast and, 252, 253
   Cytospora canker and, 159
   Fusicoccum canker and, 242
Ringspot virus
   prunus necrotic (PNRSV), 209–211
   tomato. See Tomato ringspot virus
   "yellows" strain of. See Prune dwarf virus
Ripening, premature, of pear (pink end), 119, **402**
Rocky Mountain pear rust, 55
Rodent injury, Armillaria root rot vs., 281
Root rot. See specific types
Rootstocks
   bacterial canker and blast and, 250–251, 252
   crown gall and, 259
   Phytophthora rot and, 268–270
   shoestring (Armillaria) root rot and, 283
   Verticillium wilt and, 287, 290
Root tumors. See Crown gall
Rosa spp. (Rose)
   crown gall and, 260
   hairy root and, 25–26
   leaf blight of almond and, 172
   mosaic virus of, 209
      apple mosaic virus and, 77, 78
   powdery mildew and, 182, 184, 185, 186
   Verticillium wilt and, 289
Rosellinia necatrix, Dematophora root rot and, on pome fruit, 39–41
Rosette
   of peach, 241
   of pecan, 371
Rosin, sardine oil and, for woolly apple aphids, 37
Rough leaf. See Deep suture
Rough skin disease of apple, **392**
Rubus spp., leaf blight of almond and, 172
Rumex acetosella (sorrel), apple union necrosis and decline and, 81
Russet ring of apple, **392**
Russet scab of prune, 187–188, **413**
Rusts, 54–55
   cedar-apple, 122
   of fig, 365
   of pome fruit, 54–55, **387**
   of stone fruit, 188–190, **413**
Rusty mottle, cherry, 200–201, **416**
Rusty spot of peach, powdery mildew and, 184–185

S

Saccharomyces spp., prune decay and, 235
Saline soils, internal bark necrosis and, 92
Salix spp. (Willow)
    leaf blight of almond and, 172
    melaxuma twig blight and, 317
    silver-leaf and, 285, 286
San Jose scale, Cytospora canker and, 158
Sappy bark of apple, 56–57, **388**
Sardine oil, rosin and, for woolly apple aphids, 37
SBIs. *See* Sterol biosynthesis inhibitors (SBIs)
SC-0858, 234, 378
    for Botryosphaeria panicle and shoot blight of pistachio, 330–331
    for brown rot of stone fruit, 149
Scab
    of almond and stone fruit, 191–192, **414**
    of apple, 68–63, **388**
    of loquat, 13, 66–67
    of pear, 67–69, **389**
    of pecan, 371
    russet, of prune, 187–188
Scald
    See also specific scalds
    of almond, leaf scorch and, 133
    of apple, 112, 114–115, 117, **400**
    control of, chemical injury and, 112
    of pome fruit, 112, 114–115, 117, 118–119, **400, 402**
Scald, senescent, of pear, 118, **402**
Scald, soft, of apple, 114, 117, **400**
Scald, superficial, of pear, 118–119, **400**
Scale insects, Cytospora canker and, 158
Scale mites, cherry mottle leaf and, 199
Schizophyllum commune, wood decay and, 294, 297, **430**
Schizothyrium pomi, fly speck and, 121
Sclerotinia limb blight or dieback of fig, 356
Sclerotinia spp.
    brown rot of stone fruit and, 142, 143, 146
    S. fuckeliana, 104
    S. sclerotiorum
        blossom-end rot and, 122
        green-fruit rot and, 169–170, **408**
        limb blight or dieback of fig and, 356
Sclerotium rolfsii, southern blight of apple and, 70–72
Scolytus rugulosus
    Ceratocystis canker and, 154–155
    Cytospora canker and, 158
Scytinostroma galactinum, white root rot and, 122
Seimatosporium lichenicola, leaf blight of almond and, 172–173
Semicarbazone, for powdery mildew, 185
Senescent breakdown, 116, 117, **401**
Senescent scald of pear, 118, **402**
Septoria leaf spot of pistachio, 334–335
Septoria spp.
    S. cerasina, 177
    S. pistaciarum, 334–335
Serviceberry. *See Amelanchier* spp.
Shallow-bark canker of walnut, 312–313, **432**

Sharpshooter, green, leaf scorch bacteria and, 134, **403**
Shell perforation of walnut, 322
Shoestring (Armillaria) root rot, 280–283
    of almond, 193, 282, 283, **427**
    Dematophora root rot vs., 40
    of English walnut, 318
    of fig, 365
    of olive, 347
    of pome fruit, 70
    of stone fruit, 193
Short-life
    of peach, 250, 252
    wood decay and, 293
Shotberry of olive, 349, **435**
Shothole; Shothole disease
    of apricot, 241
    of stone fruit, 193–196, **414–415**. *See also* Leaf spot of cherry
Shothole borer, Cytospora canker and, 158
Sickle leaf of olive, 348, **435**
Side rot of pome fruit, 110, **396**
Silver-leaf, 285–286, **428**
    of stone fruit, 197
    systemic pesticide for, 378
    wood decay and, 294
Smut disease of pomegranate, 373
Smut and mold of fig, 360–361, **437**
Sodium carbonate injury, **398**
Sodium chlororthophenphenate, fruit wash with, for blue-mold or soft rot, 101. *See also* Sodium orthophenylphenate
Sodium dinitro-cresol-methanol. *See* Dinitro-o-cresol compounds
Sodium hypochlorite, 378, 379
    for blue-mold or soft rot, 100–101
    for crown gall prevention, 15, 259
    deep-bark canker of walnut and, 311
Sodium N-methyldithiocarbamate, for Armillaria root rot, 283
Sodium orthophenylphenate (SOPP), 380
    chemical injury and, 112
    fruit wash with
        for gray mold rot, 105
        for nectarine and peach, 233
Sodium pentachlorophenoxide, 378
    for brown rot of stone fruit, 145, 148
    for leaf blight of almond, 173
    shothole and, 196
Sodium polysulfides, 379
Sodium salts, chemical injury and
    to apple, 112
    to pear, 118, **398**
Soft nose of olive, 349, **435**
Soft rot
    Eutypa dieback and, 167
    of fig. *See* Endosepsis
Soft scald of apple, 114, 117, **400**
Soggy (low-temperature) breakdown of apple, 116, 117
Soil-borne diseases
    See also specific diseases
    cherry stem pitting (CSP), 201, **416**
    Dematophora root rot, 40
    hairy root, 26

Mucor rot, 109, **396**
    mycotoxins and, 275
    pesticides for, 380
    Phytophthora pruning wound canker, 178–180
    Phytophthora root and crown rot, 49, 266–268
    prune brownline, 205, 206, **417**
    replant problems and, 276–279, **426**
    shoestring (Armillaria) root rot, 193
    southern blight, 72
    tomato ringspot virus, 205, 216
    Verticillium wilt, 287–291
    yellow bud mosaic virus, 216, **418**
Soil drainage, Phytophthora rot and, 49, 270
Soil fertility. *See* Soil nutrition
Soil fumigation
    for bacterial canker and blast, 132, 253
    for cherry rasp leaf virus, 200
    for crown gall, 132, 260
    for Dematophora root rot, 41
    fungicides used for, 380. *See also specific chemicals*
    for hairy root, 26
    methods, 282–283
    for Phytophthora root and crown rot, 49, 270
    for prune brownline, 206
    for prunus stem pitting, 213
    for replant problems, 278, 279, **426**
    shoestring (Armillaria) root rot and, 193, 282–283
    for southern blight, 72
    for Verticillium wilt, 290–291, 333
    for yellow bud mosaic virus, 216
Soil nutrition
    See also specific nutrients
    bacterial canker and blast and, 250
    olive knot and, 343
    replant diseases and, 278
Soil solarization
    for Dematophora root rot, 41
    for southern blight, 72
    for Verticillium wilt, 290–291, 333
Soil temperature. *See* Temperature
Soil types
    bacterial canker and blast and, 252
    European (Nectria) canker and, 44
    Verticillium wilt and, 289
Solarization of soil. *See* Soil solarization
Sooty blotch of pome fruit, 122
Sooty ring spot of quince, 83–84
SOPP. *See* Sodium orthophenylphenate
Sorbic acid, for prune mold prevention, 236
Sorbus spp. (Mountain ash)
    apple mosaic virus and, 78
    apple stem pitting and, 79
    European (Nectria) canker and, 44
    Fabraea leaf spot and, 46
    flat limb and, 75
    Pacific Coast pear rust and, 54
    as *Pezicula malicorticis* host, 35
    Rocky Mountain pear rust and, 55
Sorrel. *See Rumex acetosella*

Sour cherry yellows, prune dwarf virus and, 206, 207
Souring of fig, 361, **437**
Sour sap. *See* Bacterial canker and blast; Phytophthora root and crown rot
Southern blight of apple, 70–72, **389**
Sowbane mosaic virus (SMV), fig mosaic and, 364
Sphaeroblasts, for apple rubbery wood, 79
Sphaeropsis malorum, 108
Sphaerotheca spp.
    S. leucotricha, 51. *See also Podosphaera* spp.
    S. pannosa, powdery mildew and, 181–186
Spilocaea spp.
    S. eriobotryae, 66
    S. oleaginea, 342, 344–346
    S. pomi, 58
    S. pyracanthae, loquat scab and, 66–67
Spiroplasma citri, pear decline and, 29
Spiroplasmas, pear decline and, 29
Spittlebugs, leaf scorch bacteria and, 134, **403**
Split-pit of olive, 349
Sporobolomyces roseus, prune decay and, 236
Sporotrichum rot. *See* Side rot
Sporotrichum spp., 110
Sprinkler irrigation
    See also Irrigation
    Botryosphaeria panicle and shoot blight of pistachio and, 329, 330
    Eutypa dieback and, 166
    Fabraea leaf spot and, 47
    Phytophthora root and crown rot and, 49, 267, **425**
    russet scab of prune and, 187
    scab of apple and, 58, 62
    scab of stone fruit and, 68, 191, 192
    shothole and, 195, 196
    stone fruit diseases and, 241
Spur blight. *See* Fireblight
Stanley constriction disease. *See* Prune brownline
Star crack, apple, 79, **390**
Steam-sterilized fruit boxes, blue-mold or soft rot and, 100
Stecklenberger disease. *See* Prunus necrotic ringspot
Stellaria media (Chickweed)
    apple union necrosis and decline and, 81
    yellow bud mosaic and, 216
Stemphylium congestum, 110
Stemphylium rot of pome fruit, 110
Stem pitting
    of apple, 79–80
        pear vein yellows and red mottle and, 83
        quince sooty ring spot and, 84
    of cherry (CSP), 201, **416**
    of prunus (PSP), 212–214, **418**
Sterol biosynthesis inhibitors (SBIs), 380
    for brown rot of stone fruit, 145, 148–149
    for Coprinus rot, 109
    for leaf curl of peach, 176
    for powdery mildew, 52, 185
    resistance to, 381
    for scab of apple, 62, 63

Stigmina spp., shothole and, 194
    S. carpophila, Cercospora leaf spot and, 156
"Stippen." *See* Bitter pit
Stone fruit, 128–131
    abiotic disorders of, 218–226
        bud failure of almond, 221–224
        chilling canker of peach, 218, **418**
        corky growth on almond, 218–219, **418**
        crinkle leaf and deep suture of sweet cherry, 219–220, **419**
        nectarine pox, 220–221, **419**
        noninfectious plum shothole, 225, **420**
        nonproductive syndrome of almond, 225–228, **420**
        plum rusty blotch, 226, **421**
    bacterial diseases of, 132–139
        bacterial canker and blast, 132, 246, 247–254, **423–424**
        crown gall, 132, 257–261, **425**
        leaf scorch of almond, 133–135, **403**
    fungal diseases of, 141–197, 263–298
        anthracnose of almond, 141
        anthracnose or bitter rot, 241
        black knot, 241–242
        brown rot, 142–149
        Ceratocystis (mallet wound) canker, 153–155, **406**
        Cercospora leaf spot, 156
        Clitocybe root rot, 242
        Cytospora canker, 156–160, **406**
        damping-off and root rot, 241
        Dothiorella canker, 161–163, **407**
        Eutypa dieback, 164–167, **407**
        fruit rots, 241
        Fusarium root rot, 241
        Gibberella bud rot and shoot blight, 241
        green-fruit rot, 169–170, **408**
        hull rot of almond, 171, **409**
        leaf blight of almond, 172–173, **409**
        leaf blight of peach, 241
        leaf curl of apricot, **410**
        leaf curl of cherry, 173–174, **409**
        leaf curl of nectarine and peach, 174–176, **410**
        leaf spot of cherry, 177–178, **411**
        Phloeosporella shothole, 241
        Phyllosticta leaf spot, 241
        Phytophthora pruning wound canker, 178–180, **411**
        Phytophthora root and crown rot, 180, 263–271, **425**
        plum pockets, 180–181, **411**
        powdery mildew, 181–186
        Pythium root rot, 241
        Rhodosticta canker, 156–160
        russet scab of prune, 187–188, **413**
        rust, 188–190, **413**
        scab, 191–192, **414**
        shoestring (Armillaria) root rot, 193
        shothole, 193–196, **414–415**
        silver-leaf, 197, 285–286, **428**
        trunk and limb gall, 241
        trunk rot of apricot, 241
        Verticillium wilt, 197, 287, **428**

    mycoplasmalike disease of, X-disease of cherry and peach, 136–139, **404**
    Phytophthora root and crown rot of, 180, 263–271, **425**
    postharvest disorders of, 227–236, **421–422**
    mycotoxins, 274–275
    replant diseases of, 276–279, **426**
    virus or viruslike diseases of, 198–216, 241
        almond union disorders, 198, **415**
        cherry mottle leaf, 198–199, **416**
        cherry necrotic rusty mottle, 199, **416**
        cherry rasp leaf, 200, **416**
        cherry rusty mottle, 200–201, **416**
        cherry stem pitting, 201, **416**
        peach mosaic, 202–203, **416**
        prune brownline, 204–206, **417**
        prune dwarf, 206–208
        prunus necrotic ringspot, 209–211, **417**
        prunus stem pitting (PSP), 212–214, **418**
        yellow bud mosaic, 214–216, **418**
Stony pit, pear, 82–83, **391**
"Stop-drop" sprays, water core of apple and, 95
Storage disorders. *See* Abiotic storage disorders; Postharvest disorders
Storage scald. *See* Scald
Strawberry
    apple mosaic virus and, 78
    Dematophora root rot of, 39–41
    Verticillium wilt and, 197
    wild. *See Fragaria* spp.
Streptomycin, 380–381
    for bacterial canker and blast, 14, 253, 254
    for crown gall, 260
    for fireblight, 20, 21
    resistance to, 381
    for walnut blight, 307
Stylar-end rot of feijoa, **438**
Sulfur dioxide, 379
    for prune decay prevention, 236
Sulfur pesticides, 379
    See also Lime sulfur; Sulfur, wettable
    for brown rot of stone fruit, 148
    for leaf curl of cherry, 174
    for nectarine and peach decay prevention, 233
    nectarine pox and, 221
    for postharvest decay prevention
        in apricot, 231
        in nectarine and peach, 233
        in sweet cherry, 230
    for powdery mildew, 51, 52, 185, 186
    for rust of stone fruit, 190
    for scab
        of pear, 69
        of stone fruit, 192
    Verticillium wilt and, 290
Summer wilt, 28. *See also* Pear decline
Sunburn injury
    anthracnose and perennial canker and, 36
    Cytospora canker and, 159
    to English walnut, 322, **433**
    persimmon and, 372, **438**
    wood decay and, 294

Sunscald of apple, 114–115, **400**
Superficial scald of pear, 118–119, **400**. *See also* Scald
Suppressant pesticides, 378
Surgery, for anthracnose and perennial canker, 37
Sycamore, silver leaf and, 285
Syringomycin, bacterial canker and blast and, 248, 251
Syringotoxin, bacterial canker and blast and, 251
Systemic pesticides, 378, 379

T

Talaromyces flavus, for Verticillium wilt, 290
Taphrina spp.
   leaf curl and
      on apricot, **410**
      on cherry, 173–174, **409**
      on nectarine and peach, 174–176, **410**
   plum pockets and, 180–181
Taraxacum officinale (Dandelion)
   apple union necrosis and decline and, 81
   cherry rasp leaf virus and, 200
   prunus stem pitting and, 213
Tar, miscible, petroleum oil and,
   for European (Nectria) canker, 45
   for perennial canker, 37
Tarsonemus spp., Ceratocystis canker and, 154–155
Tatter leaf, prune dwarf virus and, 206. *See also* Prunus necrotic ringspot
TBZ, for scab of loquat, 67
Telone fumigation, for yellow bud mosaic virus, 216
Terramycin. *See* Oxytetracycline hydrochloride
2,3,4,5-tetrachlorophenol, for European (Nectria) canker, 45
Tetracycline
   for crown gall, 260
   for leaf scorch of almond, 135
   for X-disease of cherry and peach, 139
Texas mosaic. *See* Peach mosaic
Thalictrum spp., rust and, 189
Thermotherapy
   See also Heat therapy
   for apple star crack, 79
   pear vein yellows and red mottle and, 83
   for sickle leaf of olive, 348
Thiabendazole
   for gray mold rot, 105
   for scab of pear, 69
Thielaviopsis basicola, replant diseases and, 277
Thiophanate-methyl, 380
   See also Benzimidazole fungicides
   for brown rot of stone fruit, 148
   for European (Nectria) canker, 45
   for scab
      of apple, 63
      of loquat, 67
      of stone fruit, 192
   for sweet cherry decay, 230–231
Thiram, for Fabraea leaf spot, 47
Thread blight
   of fig, 365
   of pome fruit, 122
Thrips
   fig smut and mold and, 361
   russet scab of prune and, 187
TmRSV. *See* Tomato ringspot virus
Tobacco, tomato ringspot virus and, 205, 215–216
Tomato
   Phytophthora root and crown rot and, 267
   Verticillium wilt and, 197, 287, 288, 289, 290
Tomato ringspot virus (TmRSV)
   apple union necrosis and decline and, 80–81
   prune brownline and, 204–206
   prunus stem pitting and, 212–214
   replant disease and, 278
   yellow bud mosaic (YBMV) strain, 214–216
Torulopsis spp., prune decay and, 235
Toxic chemicals, anthracnose and perennial canker and, 36
Toxins
   See also Mycotoxins
   bacterial canker and blast and, 251
   hull rot of almond and, 171, **409**
   replant diseases and, 277–278
Toyon. *See* Heteromeles arbutifolia
Trametes spp.
   sappy bark and, 56–57
   wood decay and, 294, 295, 298, **430**
Tranzschelia spp., rust and, 188–189
Triadimefon, 380
   for powdery mildew, 52, 185, 186
   for shothole, 195
Triazole, 380
   for brown rot of stone fruit, 145
Tricarbamix, for bull's-eye rot of apple, 37
Trichoderma spp.
   mycotoxins and, 274
   T. harzianum
      Dematophora root rot and, 41
      gray mold and, 105
      for southern blight, 72
   T. viride
      Armillaria and, 283
      sappy bark and, 57
      for scab of apple, 63
      for silver-leaf, 286
Trichosporon behrendii, prune decay and, 235
Trichothecium roseum, peach fruit rot and, 241
Triflumizole, 380
Trifolium pratense (Red clover), apple union necrosis and decline and, 81
Triforine, 378, 379, 380
   for brown rot of stone fruit, 145, 148–149
   for fireblight, 21
   for leaf curl of peach, 176
   nectarine pox and, 221
   for postharvest decay prevention
      in nectarine and peach, 234
      in plum, 235
      in sweet cherry, 230
   for powdery mildew, 185, 186
   for scab
      of apple, 62
      of pear, 69
   for shothole, 195–196
Tubercularia spp.
   T. fici, Tubercularia canker and, 366
   T. vulgaris, Nectria blight and, 122
Tulare apple mosaic, 77
Twig blight
   See also Fireblight
   of fig, 365
   of pome fruit, 16
Twist. *See* Flat limb

U

Union disorders of almond, 198, **415**
Union necrosis and decline of apple, 80–81, **390**
Urea foliar sprays, for scab
   of apple, 63
   of pear, 69

V

Valsa persoonii, 157
Vasates jockeui, prunus necrotic ringspot and, 211
Vascular breakdown, 116
Vascular wilt. *See* Verticillium wilt
Venturia spp.
   European (Nectria) canker and, 43–44
   scab and, 192
      of apple, 59–63
      of pear, 67–69
Verticillium spp.
   V. albo-atrum, 197, 288–289
   V. dahliae, 197, 287–291
Verticillium wilt, 287–291
   of almond, 287–288, **428**
   of olive, 347
   of pistachio, 332–333, **434**
   of stone fruit, 197
Vetch, *Botrytis* infection and, 104
Vinca rosea, tomato ringspot virus and, 80
Vinclozolin, 380
   for blue-mold or soft rot, 101
   for brown rot of stone fruit, 145, 148–149
Virginia creeper, Pierce's disease and, 134
Virus or viruslike diseases
   See also specific viruses
   almond union disorders, 198
   apple flat limb, 75, **389**
   apple graft-incompatibility, 75–76, **389**
   apple mosaic virus (AMV), 76–78, 209, **390**
   apple rubbery wood, 78–79
   apple star crack, 79, **390**
   apple stem pitting, 79–80, **390**
   apple union necrosis and decline, 80–81, **390**

apricot ring pox, 241
blackline of walnut, 319–321, **433**
blister bark of apple, **392**
cherry mottle leaf, 198–199, **416**
cherry necrotic rusty mottle, 199, **416**
cherry rasp leaf, 200, **416**
cherry rusty mottle, 200–201, **416**
cherry stem pitting, 201, **416**
false sting of apple, 76, **392**
fig mosaic, 363–364, **437**
flat apple disease, 200, **392**
of olive, sickle leaf, 348
peach blotch, 241
peach mosaic, 202–203, **416**
peach wart, 241
pear bark measles, 81–82
pear blister canker, 82, **391**
pear decline and, 28
pear stony pit, 82–83, **391**
pear vein yellows and red mottle, 83, **391**
of pome fruit, 74–84, **389–392**
prune brownline, 204–206, **417**
prune diamond canker, 241
prune dwarf, 206–208
prunus necrotic ringspot, 209–211, **417**
prunus stem pitting (PSP), 212–214, **418**
quince sooty ring spot, 83–84
replant disease and, 278
rough skin disease of apple, **392**
russet ring of apple, **392**
sickle leaf, of olive, 348
of stone fruit, 198–216, 241, **416–418**
of walnut
  blackline, 319–321, **433**
  flat limb, 75
yellow bud mosaic, 214–216, **418**
Viscum album var. *album* (European mistletoe), on almond, 241
Vorlex, for shoestring (Armillaria) root rot, 283

W
Walnut
  See also Walnut, black; Walnut, English or Persian
  rootstocks, 304
    blackline and, 319, **433**
    crown gall and, 309
    Phytophthora root and crown rot and, 267, 269–270
    shoestring (Armillaria) root rot and, 282, 283, 304, 318
  wood decay of, **430**
Walnut, black (*Juglans nigra; J. hindsii; J. californica*)
  See also Walnut
  Dematophora root rot and, 40
  shoestring (Armillaria) root rot and, 282, 283
  southern blight and, 71
Walnut, English or Persian (*Juglans regia*), 304
  See also Walnut
  abiotic diseases of, 322

anthracnose of, 313
bacterial diseases of, 305–313, **431, 432**
blackline of, 319–321, **433**
blight of, 305–308, **431**
branch wilt of, 314–315, **432**
  Dothiorella canker of almond and, 162
crown gall of, 309
Cytospora canker of, 318
deep-bark canker of, 309–311, **431**
Dematophora root rot and, 40
downy leaf spot of, 315–316, **432**
fungal diseases of, 313–318, **432**
kernel mold of, 316, **432**
melaxuma canker and twig blight of, 316–317
mycotoxins and, 274
Phytophthora root and crown rot of, 316
shallow-bark canker of, 312–313, **432**
shell perforation of, 322
shoestring (Armillaria) root rot and, 282, 283, 318
sunburn of, 322, **433**
Verticillium wilt of, 287
virus diseases of, 319–321
walnut blight of, 305–308, **431**
Washes
  See also Dip treatments
  for blue-mold or soft rot, 100, 101
  gray mold and, 104
  hot water
    injury by, 113, **399**
    for nectarine and peach decay, 234
  prune decay and, 236
  for stone fruit, 230
Water core of apple, 95–96, **394**
Watering. *See* Irrigation; *specific methods*
Water table, European (Nectria) canker and, 44
Waxes
  for apricot decay prevention, 231
  chemical injury and, 112, 113
  in olive knot treatment, 343
Wax-fungicide applications
  for apricot, 231
  for nectarine and peach, 233, 234
  for plum, 235
Weeds
  apple union necrosis and decline and, 81
  cherry rasp leaf virus and, 200
  green-fruit rot and, 169
  leaf scorch bacteria and, 134
  prune brownline and, 205
  prunus stem pitting and, 213
  tomato ringspot virus and, 205, 213
  yellow bud mosaic and, 216
Western X; Western X little cherry. *See* X-disease of cherry and peach
Wet feet. *See* Phytophthora root and crown rot
White root rot of pome fruit, 122–123
White rot, of pome fruit, 121
Whitewashing
  bacterial canker and blast and, 253
  Cytospora canker and, 159
  endosepsis of fig and, 358

Willow. *See Salix* spp.
Willows. *See* Prune dwarf virus
Wilsonomyces carpophilus
  anthracnose of almond and, 141
  Cercospora leaf spot and, 156
  noninfectious plum shothole vs., 225
  shothole of stone fruit and, 193–196
Wilt and decline. *See* X-disease of cherry and peach
Wind
  brown rot of stone fruit and, 144, 147
  Eutypa dieback and, 166
  fireblight and, 18–19
  green-fruit rot and, 169
  olive knot and, 342
  powdery mildew and, 184
  silver-leaf and, 285
  walnut blight and, 307
  walnut deep-bark canker and, 310
Winter chilling deficiency, pear blossom blast and, 14
Winter injury
  See also Freezing injury
  apple rubbery wood and, 78
  Cytospora or Rhodosticta canker and, 158
  leaf spot of cherry and, 177
Winters disease. *See* Yellow bud mosaic
Winters peach mosaic. *See* Yellow bud mosaic
Witches'-broom
  apple replant problems and, 276
  Kern's pear rust and, 55
  leaf curl of cherry, 173–174
  Pacific Coast pear rust and, 55
  plum pockets and, 181
Wood-rot disorders, 291–298, **429–430**
  olive and, 347
Woodsorrel, creeping. *See Oxalis corniculata*
Woolly apple aphids, anthracnose and perennial canker and, 35, 36, 37
Wounds. *See* Mechanical injury; Pruning

X
Xanthan gum, chemical injury and, 113
Xanthomonas campestris
  bacterial canker of fig and, 355
  bacterial spot and, 241
  blight of English walnut and, 305–308
X-disease of cherry and peach, 136–139, 378, **404**
Xiphinema spp. (Nematodes)
  prune brownline virus and, 205
  replant diseases and, 277
  tomato ringspot virus and, 205, 213, 216
  walnut blackline and, 320–321
  X. americanum
    cherry rasp leaf virus and, 200
    prunus stem pitting and, 213
  X. californicum
    apple union necrosis and decline and, 81
    yellow bud mosaic and, 216
X-spot of pome fruit, 123
Xylaria mali, black root rot and, 122

Xylella fastidiosa, leaf scorch of almond and, 133–135
1,2,3,4-xylenol, in olive knot treatment, 343

Y

Yeast infections
  of fig, 361
    endosepsis and, 357
  prune decay and, 235
  sweet cherry decay and, 230
Yellow bud mosaic virus, 205, 214–216, **418**
Yellow leaf
  of cherry. *See* Leaf spot of cherry
  prune dwarf virus and, 206
Yellow leaf roll. *See* X-disease of cherry and peach
Yellows, leaf casting. *See* X-disease of cherry and peach
Yellows, pear vein, red mottle and, 83, **391**

Z

Zinc chloride, for fireblight, 20
Zinc coposil and zineb, for fig surface mold and Alternaria rot, 362
Zinc deficiency, pecan and, 371
Zinc sulfate, for bacterial canker and blast, 253
Zineb
  for Fabraea leaf spot, 47
  for fig surface mold and Alternaria rot, 362–363
  for shothole, 196
Ziram, 378, 379
  for anthracnose, 37
  for bull's-eye rot prevention, 37
  chilling canker of peach and, 218
  for Coprinus rot prevention, 109
  for leaf blight of almond, 173
  nectarine pox and, 221
  for scab
    of loquat, 67
    of pear, 69
    of stone fruit, 192
  for shothole, 195, 196
Zygophiala jamaicensis, fly speck and, 121
Zymononas sp., foamy canker of almond and, 133